Impact of Littoral Environmental Variability on Acoustic Predictions
and Sonar Performance

Impact of Littoral Environmental Variability of Acoustic Predictions and Sonar Performance

Edited by

Nicholas G. Pace

and

Finn B. Jensen
SACLANT Undersea Research Centre,
La Spezia, Italy

SPRINGER SCIENCE+BUSINESS MEDIA, B.V

A C.I.P. Catalogue record for this book is available from the Library of Congress.

Additional material to this book can be downloaded from http://extras.springer.com.

ISBN 978-94-010-3933-8   ISBN 978-94-010-0626-2 (eBook)
DOI 10.1007/978-94-010-0626-2

Printed on acid-free paper

TABLE OF CONTENTS

Including a CD-Rom with a.o. color images

PREFACE

The limiting influence of the environment on sonar has long been recognised as a major challenge to science and technology. As the area of interest shifts towards the littoral, environmental influences become dominant both in time and space. The manyfold challenges encompass prediction, measurement, assessment and adaptive responses to maximize the effectiveness of systems. Although MCM and ASW activities are dominated in different ways and scales by the environment, both warfare areas have had to consider the significantly changing requirements posed by operations in the littoral. The fundamental scientific issues involved in developing models relating acoustics to the environment are matched in difficulty by the need for data for their validation and eventual practical use for prediction. In many instances the need is for on-line adaptation of systems to changing circumstances whilst other needs are for the longer term planning activities.

This book and the attached full-color CD are the proceedings of a conference organised by the SACLANT Undersea Research Centre, held at Villa Marigola, Lerici, Italy, on 16–20 September 2002. The fundamental problems associated with environmental variability and sonar were explored at a previous SACLANTCEN conference in 1990.[1] These problems have not gone away but, on the one hand are exaggerated by the move to the littoral and on the other hand, are open to treatment in new ways that advances in technology and computer power allow.

Oceanographers, acousticians, sonar systems engineers and operators need closer understanding of each others problems; we hope that this conference provides some momentum in this direction. One of us (FBJ) is going to chair a special session at the ASA meeting in December 2002 in Cancun, Mexico, which in many ways can be seen as a continuation of this conference both in topic and attendees.

The support of ONR through the good offices of Jeff Simmen is much appreciated.

N.G. Pace and F.B. Jensen
La Spezia, May 2002

---

[1] *Ocean Variability & Acoustic Propagation*, edited by John Potter and Alex Warn-Varnas (Kluwer Academic Publishers, The Netherlands, 1990) 608 pp.

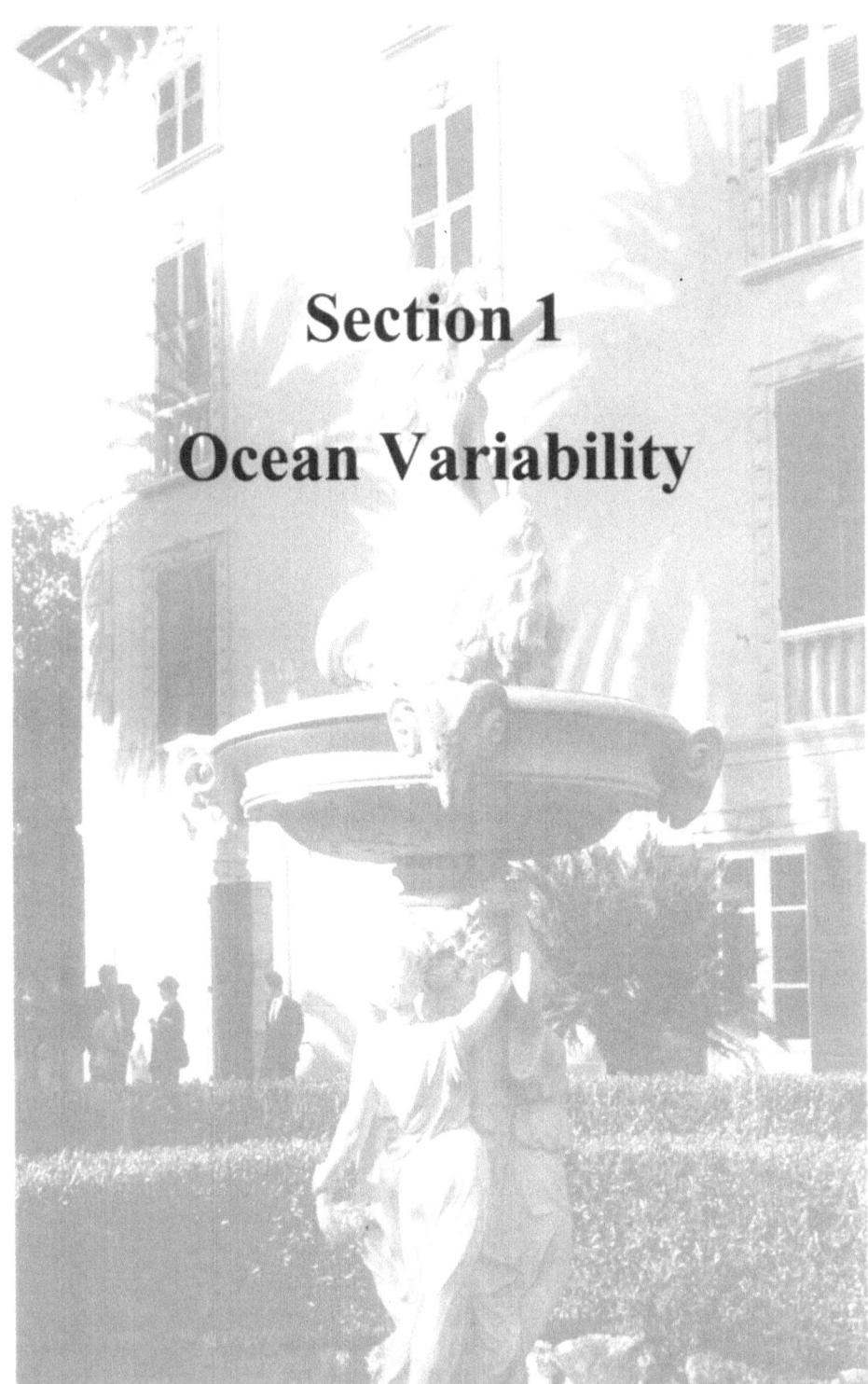

Section 1

Ocean Variability

ACOUSTIC EFFECTS OF ENVIRONMENTAL VARIABILITY IN THE SWARM, PRIMER AND ASIAEX EXPERIMENTS

J. LYNCH, A. FREDRICKS, J. COLOSI, G. GAWARKIEWICZ AND A. NEWHALL

Woods Hole Oceanographic Institution (WHOI), Woods Hole, MA 02543
E-mail: jlynch@whoi.edu

C.S. CHIU

Naval Postgraduate School (NPS), Monterey, CA 93943
E-mail: chiu@nps.navy.mil

M. ORR

Naval Research Laboratory (NRL), Washington, D.C. 20375
E-mail: orr@wave.nrl.navy.mil

We present an overview of how the coastal oceanographic environment affected acoustic propagation and scattering in three recent major field experiments: SWARM (1995), PRIMER (1996–97), and ASIAEX (2000–01). In all three of these experiments, low frequency sound (50–600 Hz) was transmitted through strong coastal oceanographic features to array receivers. The differences and similarities of these experiments will be emphasized. In particular, we will focus on the effects of coastal oceanography, i.e. fronts, eddies, and internal waves. We begin with the SWARM experiment, which examined the effects of internal waves, particularly non-linear internal waves, on acoustic propagation and scattering. This experiment had excellent environmental support along a single across-shelf propagation path, and we were able to make good progress understanding the sound scattering within the context of this limited geometry. In the PRIMER experiment, we looked at oceanographic effects in a fully three-dimensional configuration, being supported by numerous environmental moorings and Sea Soar (undulating CTD) high-resolution hydrography. This experiment observed the effects of both the shelfbreak front, eddies and filaments, and nonlinear internal waves on the acoustic field. Finally, we examine the recent ASIAEX experiment, which again dealt with fully three-dimensional oceanography and along- and across-shelf acoustic propagation. ASIAEX had perhaps the best environmental support of the three experiments, including thirty environmental moorings, Sea Soar hydrography, satellite remote sensing, and acoustic flow visualization surveys. The acoustic monitoring included both a vertical and horizontal array, and moored and towed sources. Oceanographically, this experimental site featured some of the strongest non-linear internal waves we have seen to date.

1 Introduction

The Yellow Sea experiments of Zhou *et al.* in the 1980's, published in the Journal of the Acoustical Society of America in 1991 [1], created a major stir in the field of low frequency (50–1000 Hz) shallow water acoustics. Field data showed anomalous propagation loss

N.G. Pace and F.B. Jensen (eds.), Impact of Littoral Environmental Variability on Acoustic Predictions and Sonar Performance, 3-10.

effects of up to 30 dB (depending on frequency) along with strong azimuthal dependence of the propagation loss, with the likely cause being groups of strong nonlinear internal waves often called "soliton trains." Using a finite-lattice, Bragg resonance scattering theory, Zhou et al. [1] were able to explain these data rather well, and to first order, this looked like an effect that was found and quickly understood. However, given the magnitude and potential importance of the effect, and the fact that not all aspects of the scattering were measured by Zhou et al., further examination was suggested. This motivated a number of US experiments, including the SWARM (Shallow Water Acoustic Random Medium) experiment.

2   The SWARM experiment

The SWARM experiment [2–5] studied across-shelf propagation and scattering through soliton trains, and had two major objectives. The first objective was to look at acoustic intensity fluctuations as a function of time, frequency, and distance from the source, and was spearheaded by the NRL acoustics group. This work was motivated in part by the theoretical work of Creamer [6], who's extension of Dozier and Tappert's deep-water fluctuation work [7] to shallow water predicted that the scintillation index (the normalized variance of the intensity) should increase exponentially with range. This intensity variability analysis of the SWARM data will be reported in another paper from this conference [8], and so will not be pursued further here.

The WHOI group examined the time spreading of mode filtered acoustic pulses in SWARM. It was observed that the pulse spreading measured oscillated very strongly at the M2 (semidiurnal) tidal period. Moreover, it was seen, again in the context of mode filtered data, that the spreading was the largest when the scatterer (the soliton train) was in the vicinity of the acoustic receiver array. This "near receiver dominance" was explained by Headrick et al., and we refer the reader to those papers [3,4].

The NPS group concentrated on the physical oceanography, and in some sense, this topic contained the most surprising results [2]. To begin, the internal-wave field one observed at the SWARM site, while being dominated by non-linear solitons, was very different from the Yellow Sea internal-wave field. Rather than the neat, evenly spaced "ducks-in-a-row" seen in the Yellow Sea, the SWARM wave field was temporally and spatially complex due to multiple sources. Though the waves generated along the shelfbreak were dominant, waves generated by local canyons also contributed substantially to the total field. Secondly, the wave-field varied significantly in time. Thus, the internal-wave field in this region was unpredictable without further information on the generation processes occurring at remote sites. These space-time variations in the internal-wave field indicated to researchers that the oceanography and the acoustic scattering theory that applied to Zhou's Yellow Sea experiment was not universally applicable, and that more investigations were warranted into the coastal internal-wave field and the acoustic scattering theory needed to describe its effects.

3   The New England shelfbreak PRIMER experiment

Internal waves are but one variety of oceanographic variability that can have a strong affect on acoustics. Coastal fronts and eddies are also of importance to acoustics, and are common features as well. The WHOI and NPS groups had previously examined the acoustic effects of shallow water fronts in the 1992 Barents Sea Polar Front experiment [9–11], but this was done for a rather weak frontal system, and without large-scale oceanographic support. A stronger frontal system, with 50 cm/s flows and frontal eddies, was conveniently available for study in the New York Bight, and so it was decided to examine that region.

In the summer of 1996 and the winter of 1997, two large acoustics-plus-oceanography field experiments were undertaken in the vicinity of the shelfbreak front south of Martha's Vineyard, MA. In these efforts, three moored sources seaward of the front transmitted signals for two weeks to two vertical line array (VLA) receivers shoreward of the front, while the oceanography was measured concurrently by a wide variety of means, including the Sea Soar towed CTD instrument, which provided three-dimensional maps of the sound-speed field on a daily basis. We will briefly examine some of the more interesting results from the summer 1996 experiment here, and refer the reader wanting further detail to the literature [12–15].

An interesting study we are undertaking with the PRIMER data is that of the amplitude fluctuations observed. One part of the work we are engaged in is trying to explain the amplitude fluctuations observed at the receiver in terms of an "energy budget" (i.e. energy conservation) made at that location. Looking at either of the two PRIMER receiver sites, we note trivially that acoustic energy passing by an array has to either hit the receiver, or go above it or below it. The total energy impinging upon the receiver is calculated by integrating both over the temporal arrival pattern (as we sent pulses) and over the spatial extent of the array. The energy going either above the vertical array receiver (which goes from the bottom to 40 m water depth) or below the array (through the bottom) is not measured, but can be estimated by computer calculations, which we tune to match the environment of the site and the measured array data. By combining data and model estimates, we can attempt to understand the distribution of the energy in the water column and in time, and thus the acoustic intensity fluctuations. Moreover, the models provide insight as to the physical causes of the fluctuations.

We now show an example of how energy conservation may be related to the amplitude fluctuations. In Fig. 1 we display the temporally and spatially integrated arrivals seen at the northeastern vertical line array in the summer 1996 PRIMER experiment. Immediately evident are the fluctuations in the energy with a large M2 tidal component and also with higher and lower frequency components. The energy changes seen can be due to energy going above and below the receiver, as mentioned, and also due to time dependent changes in bottom loss along the acoustic track (we disregard 3-D effects). In order to explain the origin of these variations seen in the data, we have simulated the acoustic propagation through a temporally and spatially evolving internal wave field over a full M2 semidiurnal tidal cycle (12.42 hours), and then examined the distribution of energy seen at the array.

Our initial results indicate that energy going below the array in the bottom is minimal, not surprising since the most energetic acoustic modes have turning points above the bottom. Somewhat more surprising is the result that the amount of bottom loss along the track is not varying much with time, but rather that the energy is being redistributed above the receiver in a time varying fashion. This seems to indicate that solitons coming near the receiver at the semidiurnal tidal frequency are sending energy via mode coupling to higher order modes, which puts energy higher up in the water column. This tentative result is still being verified at the time of writing. Whatever the final outcome, it is evident that such a computer analysis, especially when combined with data, can be very useful in understanding the nature of the energy/intensity fluctuations seen.

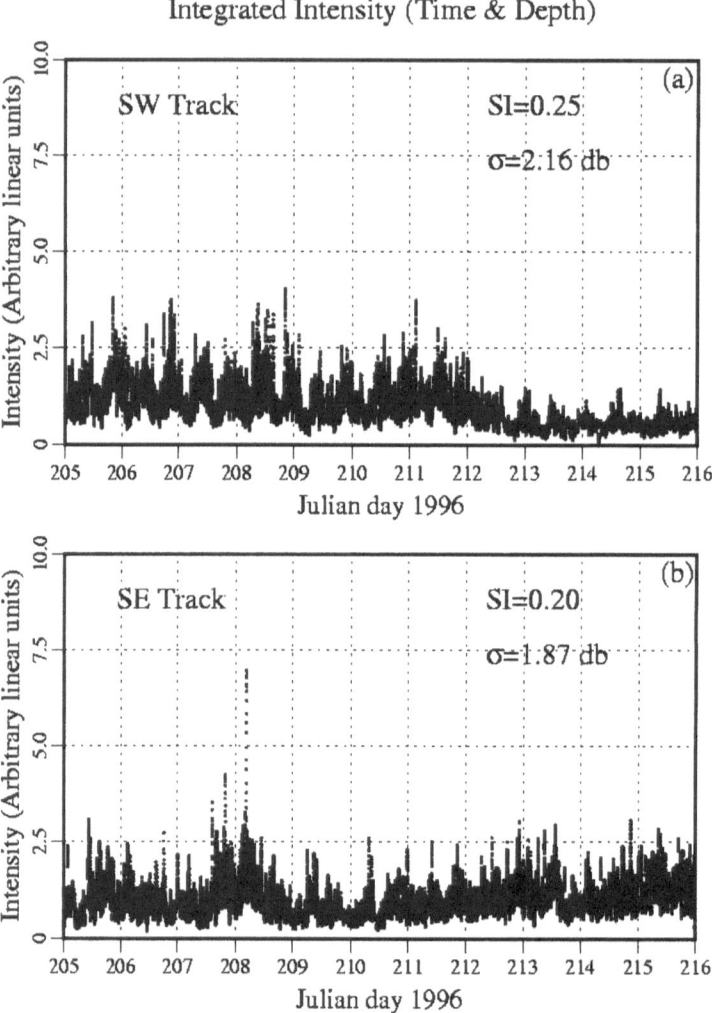

Figure 1. 1996 PRIMER experiment time series of integrated intensity at a point receiver from two 400 Hz sources, one in the southwest corner of the experimental area (a) and one in the southeast corner (b). SI is the scintillation index, and σ is the intensity fluctuation in dB.

Another interesting intensity fluctuation study made with the PRIMER data is relating the amount of time spreading seen in the signals to the peak acoustic intensity. As a very simple first guess, we have first considered an energy conserving "accordion model" that says that the area under the received pulse at a given hydrophone is conserved in time, so that if the pulse width is expanded, its peak is correspondingly decreased. (This is a phase modulation effect.) In such a model, the time spreading and the acoustic intensity at a given hydrophone are anti-correlated. We can look at this anti-correlation directly with our data and it seems to work fairly well, as shown in Fig. 2. This would indicate that a first order effect in changing the amplitude for a point receiver is the time spreading of the pulse. The fact that this correlation is not perfect suggests that there are other mechanisms (amplitude modulation) also coming into play. We are hypothesizing, based on our first example, that the vertical redistribution of the energy in the water column is the other principal effect. This remains to be seen, and is one of the foci of our ongoing research.

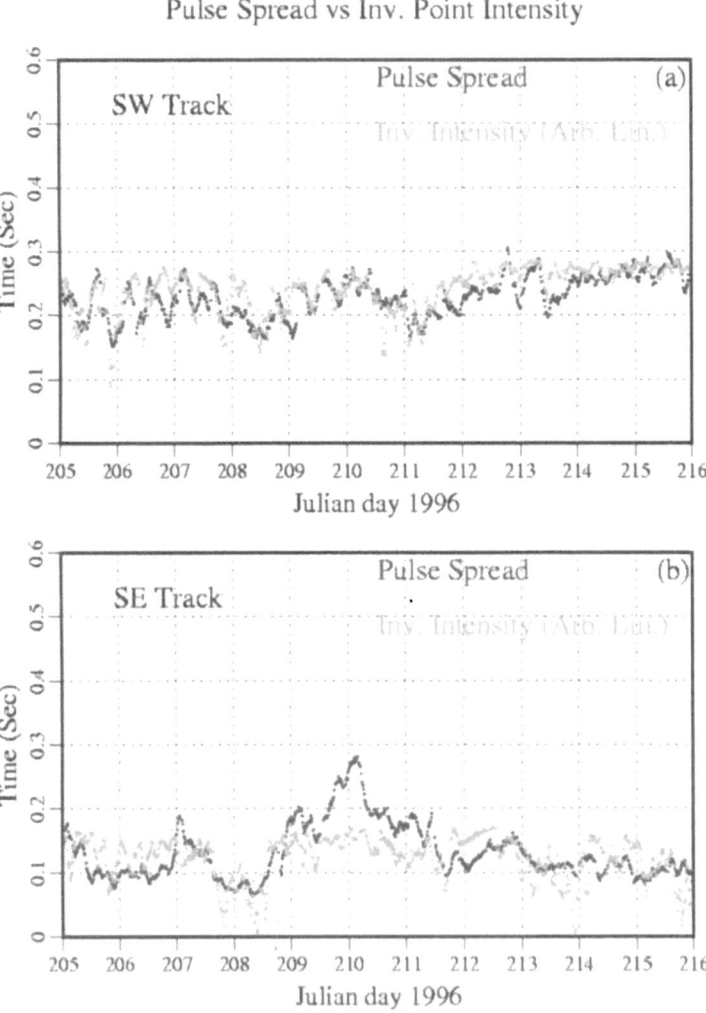

Figure 2. Time series of pulse spreading (sec) versus inverse measured intensity (arbitrary linear units) for the two sources discussed in Fig. 1.

The other topic of interest from PRIMER is the physical oceanography and its variability, which we are endeavoring to correlate to the acoustic field and its time and space variability. The PRIMER experimental site featured extremely variable and complicated coastal oceanography, with the prime features being the shelfbreak front, Gulf Stream eddy forcing from the continental slope, and the nonlinear internal tide (i.e. the soliton trains and tidal bore). The summer Shelfbreak front oceanography has been examined by Gawarkiewicz et al. [15], and we will just mention some of his more interesting results which are of importance to the low-frequency acoustic field. To begin with, the sound-speed field was found to be very inhomogeneous, both in the across-front and the along-front directions, somewhat at variance with popular wisdom. The correlation lengths and times of the oceanography near the front were estimated to be of order 7–15 km and 1–2 days, as measured by the Sea Soar hydrography and moorings. The warm bottom layer beneath the front was seen to be a consistent and acoustically important oceanographic entity. The eddy field spawned by the Gulf Stream was seen to interact with the front in a very complicated fashion, modulating its shape and thus its acoustic effects, and this is the subject of intense further study. Regarding the internal-wave field, Colosi et al. [16] have studied it from moored records and have again found it to be extremely variable, both in time and space. Significant interaction between the internal tides and the larger scale oceanography is seen, leading to a picture of a field that most likely has to be described by a combination of deterministic and random contributions. There is much more that can be said regarding the PRIMER oceanography, but due to the limits of space, we will just refer the reader to some of the literature for now [15,16].

4   The ASIAEX experiment

The final experiment we will touch on here is the recent ASIAEX experiment [17], which was conducted in the South China Sea. Since this experiment is rather recent, we have fewer detailed results to show, but we can at least talk about what the data set includes and the analyses it will support.

The biggest improvements in ASIAEX over the previous shallow water experiments, as regards acoustics, were: 1) the inclusion of a 400 m length, 32 element horizontal array, in addition to a 79 m long, 16 element vertical array — previous experiments only had 16 element vertical arrays, 2) more acoustic frequencies, filling the 50–600 Hz band — previous experiments had only 224 Hz and 400 Hz transmissions, 3) a longer overall time series (three weeks), so that we could examine a full spring-neap tidal cycle, 4) longer duration moored transmissions, so that we could unambiguously look at temporal decorrelation times for the acoustic field, and 5) a variety of towed source tracks, allowing us to look at the range variability of the acoustic field, which cannot be done with purely moored transmissions. Turning to the oceanography, we had far more oceanographic environmental data than any of the previous efforts. In addition to having thirty oceanographic moorings deployed in the experimental region, along with intense satellite imagery support, we also had three ships performing measurements from onboard, including Sea Soar, ADCP current measurements, and high frequency acoustic flow visualization. Geologically, measurements included a high-resolution bathymetry survey, chirp sonar imagery along the fixed acoustic paths, and numerous cores. Thus, our ASIAEX data set is the most complete coastal acoustics-plus-environment measurement set in our possession to date.

Given the new dimension of the horizontal array, one of the first things we are looking at in ASIAEX is the coherence of the acoustic signal across the array. In Fig. 3, we show a typical 224 Hz pulse arrival structure across the horizontal array. Due to the array having a somewhat irregular shape, and to the sources being off-broadside, the acoustic pulses arrive skewed in time. This is being corrected by time delay beamforming, or more prosaically, lining up the initial peaks in this figure. Looking at the arrivals in Fig. 3, one is struck at their general regularity, with some difference in the later arrivals. This general regularity indicates a rather high degree of coherence of the signal across the array (with exact numerical estimates of that number forthcoming), a result which is somewhat surprising, given that the ASIAEX was located in a very active oceanographic area. It is our conjecture that the near-bottom position of both source and receiver is mitigating against seeing strong scattering effects, and we propose to test this by looking at the variability in the arrivals for receiver elements higher in the water column in the vertical array.

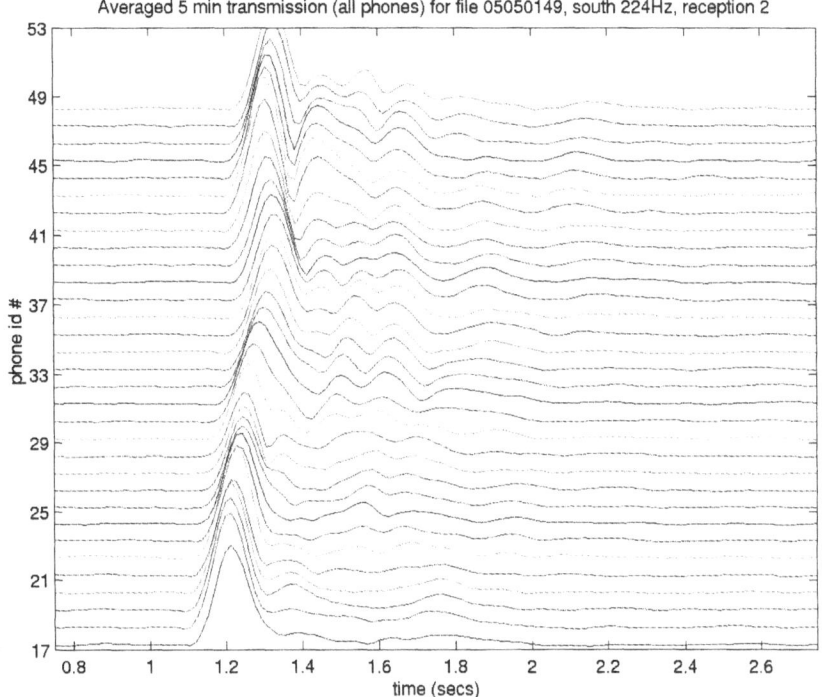

Figure 3. Typical arrival structure seen across the bottom-lying horizontal array in ASIAEX.

There are numerous other topics in ASIAEX that we will be looking at, but rather than enumerate them, we will just mention that the intercomparison of various acoustic effects between the SWARM, PRIMER, and ASIAEX sites is one topic that has been enabled by finally having high quality data from multiple sites. One of the key issues in shallow water is the generality and transportability of results from site to site, and we hope that these three data sets will allow us to begin answering that question.

Acknowledgements

We would like to thank our numerous colleagues and shipmates from the SWARM, PRIMER and ASIAEX projects for their hard work and collegiality. The three projects described in this report (SWARM, PRIMER, and ASIAEX) all were supported by the Office of Naval Research.

References

1. Zhou, J.X., Zhang, X.S. and Rogers, P., Resonant interaction of sound waves with internal solitons in the coastal zone, *J. Acoust. Soc. Am.* **90**(4), 2042–2054 (1991).

2. Apel, J., Badiey, M., Chiu, C., Finette, S., Headrick, R., Kemp, J., Lynch, J., Newhall, A., Orr, M., Pasewark, B., Tielbuerger, D., Turgut, A., von der Heydt, K. and Wolf, S., An overview of the 1995 SWARM shallow water internal wave acoustic scattering experiment, *IEEE J. Oceanic Eng.* **22**(3), 465–500 (1997).

3. Headrick, R., Lynch, J. and the SWARM group, Acoustic normal mode fluctuation statistics in the 1995 SWARM internal wave scattering experiment, *J. Acoust. Soc. Am.* **107**(1), 201–220 (2000).

4. Headrick, R., Lynch, J. and the SWARM group, Modeling mode arrivals in the 1995 SWARM experiment acoustic transmissions, *J. Acoust. Soc. Am.* **107**(1), 221–236 (2000).

5. Tielbuerger, D., Finette, S. and Wolf, S., Acoustic propagation through an internal wavefield in a shallow water waveguide, *J. Acoust. Soc. Am.* **101**(2), 789–808 (1997).

6. Creamer, D., Scintillating shallow water waveguides, *JASA* **99**(5), 2825–2838 (1996).

7. Dozier, L. and Tappert, F., Statistics of normal mode amplitudes in a random ocean, *J. Acoust. Soc. Am.* **63**(2), 353–365 (1978).

8. Pasewark, B., Wolf, S., Orr, M. and Lynch, J., Acoustic intensity variability in a shallow water environment. In *Impact of Littoral Environmental Variability on Acoustic Predictions and Sonar Performance*, edited by N.G. Pace and F.B. Jensen (Kluwer, The Netherlands, 2002) pp. 11–18.

9. Lynch, J., Jin, G., Pawlowicz, R., Ray, D., Plueddemann, A., Chiu, C.S., Miller, J., Bourke, R., Parsons, A.R. and Muench, R., Acoustic travel time perturbations due to shallow water internal waves and internal tides in the Barents Sea Polar Front: Theory and experiment, *J. Acoust. Soc. Am.* **99**(2), 803–821 (1996).

10. Parsons, A.R., Bourke, R., Muench, R., Chiu, C.-S., Lynch, J.F., Miller, J., Plueddemann, A. and Pawlowicz, R., The Barents Sea Polar Front in summer, J. Geophys. Res. **101**(C6), 14201–21 (1996).

11. Gawarkiewicz, G. and Plueddemann, A.J., Topographic control of thermohaline frontal structure in the Barents Sea Polar Front on the south flank of Spitsbergen Bank, *J. Geophys. Res.* **100**(C3), 4509–4524 (1995).

12. Sperry, B., Analysis of acoustic propagation in the region of the New England continental shelfbreak. MIT/WHOI Joint Program Ph.D. Thesis, 1999.

13. Newhall, A., von der Heydt, K., Sperry, B., Gawarkiewicz, G. and Lynch, J., Preliminary acoustic and oceanographic observations from the winter PRIMER experiment. WHOI Technical Report WHOI-98-19, 1998.

14. Lynch, J., Spatial and temporal variations in acoustic propagation characteristics at the New England shelfbreak front, *IEEE J. Oceanic Eng.* (submitted 2001).

15. Gawarkiewicz, G., Bahr, F., Brink, K., Beardsley, R., Caruso, M., Lynch, J. and Chiu, C.S., A large amplitude meander of the shelfbreak front during summer south of New England: Observations from the shelfbreak PRIMER experiment, *J. Geophys. Res.* (submitted 2002).

16. Colosi, J., Beardsley, R., Lynch, J., Gawarkiewicz, G., Chiu, C.-S. and Scotti, A., Observations of nonlinear internal waves on the outer New England continental shelf during the summer shelfbreak PRIMER study, *J. Geophys. Res.* **106**(C5), 9587–9601 (2001).

17. Newhall, A., Costello, L., Duda, T., Gawarkiewicz, G., Irish, J., Kemp, J., McPhee, N., Liberatore, S., Lynch, J., Ostrom, W., Trask, R. and von der Heydt, K., Preliminary acoustic and oceanographic observations from the ASIAEX 2001 South China Sea experiment. WHOI Technical Report WHOI-2001-12, 2001.

ACOUSTIC INTENSITY VARIABILITY
IN A SHALLOW WATER ENVIRONMENT

BRUCE H. PASEWARK, STEPHEN N. WOLF AND MARSHALL H. ORR
Naval Research Laboratory, Acoustics Division, Code 7120, Washington D.C. 20375, USA
E-mail: pasewark@wave.nrl.navy.mil

JAMES F. LYNCH
Woods Hole Oceanographic Institution, Department of Applied Ocean Physics and Engineering, Woods Hole, MA 02543, USA

Acoustic signals with center frequencies 224 and 400 Hz were recorded for 63-hours during an experiment on the New Jersey Shelf, USA (SWARM95). Acoustic energy statistics have been extracted for both narrowband and broadband signals at a fixed range of 42 km. The statistics have been found to be non-stationary and depth dependent. There is frequency and bandwidth dependence to the signal properties and no unique probability distribution representation.

1 Introduction

Narrowband and broadband acoustic signal scintillation index and intensity probability distributions extracted from a 63-hour section of the Shallow Water Acoustic Random Medium 1995 (SWARM95) experimental dataset are presented.

The (SWARM95) experiment was performed from the later part of July through early August of 1995. The experiment was located on the New Jersey Shelf off the east coast of the United States [1]. The experiment investigated the impact of water column sound speed variability induced by a variety of fluid processes on the spatial and temporal variability of acoustic signals. During the measurement period the sound speed field was perturbed by linear and nonlinear internal waves and internal tides.

The SWARM95 experiment used several acoustic sources and receivers. The acoustic signals presented in this paper were generated by two broadband acoustic sources and received by a 32-element acoustic vertical line array (AVLA). The acoustic sources projected pseudo random number (PRN) signals centered at 224 Hz (16 Hz bandwidth) and 400 Hz (100 Hz bandwidth) and were moored in ~54.5 m of water. The source depths were 48 m and 29 m, respectively. Source levels for both projectors were approximately 181 dB re 1μPa @ 1m. The AVLA receiver was moored 42 km seaward of the acoustic sources in 89 m of water. The AVLA receiver spanned the water column from 23 to 85 m with elements equally spaced every 2 m.

The narrowband acoustic energies were extracted from a single frequency bin of a Fast Fourier Transform (FFT) of the calibrated hydrophone time-series. The FFT length was equal to the duration of the PRN sequence (3.9375 s for the 224 Hz data and 5.110 s for the 400 Hz data). The single frequency signal was squared and calibrated to provide

11

N.G. Pace and F.B. Jensen (eds.), Impact of Littoral Environmental Variability on Acoustic Predictions and Sonar Performance, 11-18.

the narrowband energy. The broadband acoustic energy was calculated by cross correlating the received acoustic signal with the transmitted PRN sequence (i.e. replica correlation). The resulting cross correlation time series was integrated over the PRN sequence duration and calibrated to provide total broadband energy.

2   Acoustic energy probability distribution function histograms

The narrowband (Figs. 2 and 3) and broadband (Figs. 4 and 5) acoustic energy probability distribution functions are presented as histograms of the acoustic energy versus the number of received PRN sequences with that energy. The narrowband energy histograms exhibit the exponential energy (or Rayleigh pressure) distribution expected for an ensemble of vectors obtained by coherently adding randomly phased sinusoids of the same frequency [2]. For this narrowband case, each sample has two degrees of freedom. The exponential-like distribution is found at both signal frequencies and for all receiver depths.

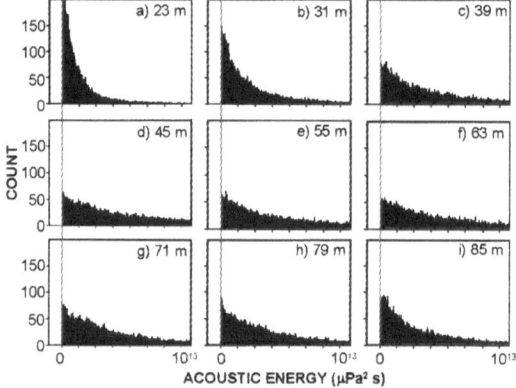

Figure 1. Histograms of the 224 Hz narrowband energies measured over 63 hours (21,593 data points) at nine water depths. Histogram bin size is 10^{10} $\mu Pa^2 s$.

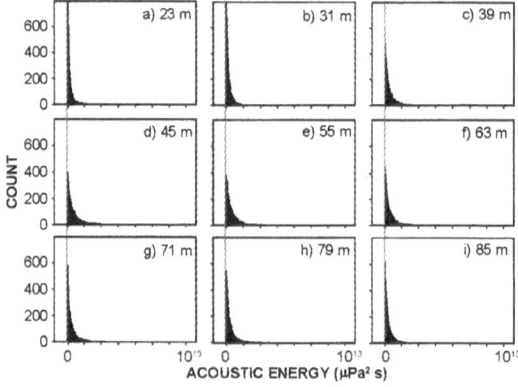

Figure 2. Histograms of the 400 Hz narrowband energies measured over 63 hours (13,245 data points) at nine water depths. Histogram bin size is 10^{10} $\mu Pa^2 s$.

In the case of the broadband replica correlation processing, each energy sample is formed by adding the narrowband components over the entire signal bandwidth. Not all of these signal components are statistically independent. The number of statistically independent components realized over the bandwidth will be equal to the number of temporally resolved arrivals observed in the time correlated matched filter output. Each statistically independent component in the energy sum increases its number of degrees of freedom by two and the energy distribution is described by the Chi-squared distribution with N degrees of freedom [2]. As the number of degrees of freedom becomes very large, the distribution approaches the lognormal distribution. In the matched filter time domain data (not shown here), the 224 Hz signals typically showed two or three temporally resolved pulse arrivals, with the number usually smaller near the boundaries and larger near the center of the water column. At 400 Hz, a larger number (4–10) of pulse components was typically seen. The increase in number is attributed to the larger number of acoustic modes of propagation, the larger signal bandwidth, and the variable presence of signal components scattered by internal waves.

The significant difference in the number of degrees of freedom for the 224 Hz and 400 Hz signals produces a significant difference in their energy distribution functions. Near the boundaries at 224 Hz, we find an exponential-like distribution associated with a single degree of freedom. At mid depth, the peak in the distribution moves to a small, but nonzero value. The mode (most probable value) continues to increase as we consider the 400 Hz data. The complexity of the pulse structure in the 400 Hz data was observed to be greatest near mid-depth and to vary considerably over periods of time approaching tidal cycles, consequently the number of degrees of freedom in the signal should be expected to be variable.

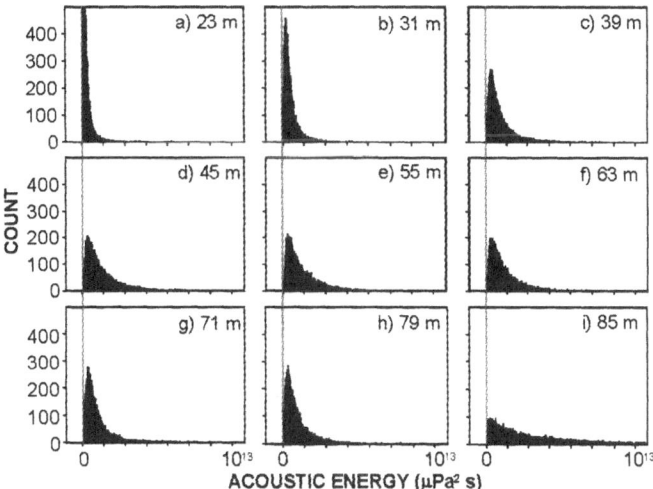

Figure 3. Histograms of the 224 Hz broadband energies measured over 63 hours (21,593 data points) at nine water depths. Histogram bin size is $10^{10}\ \mu Pa^2 s$.

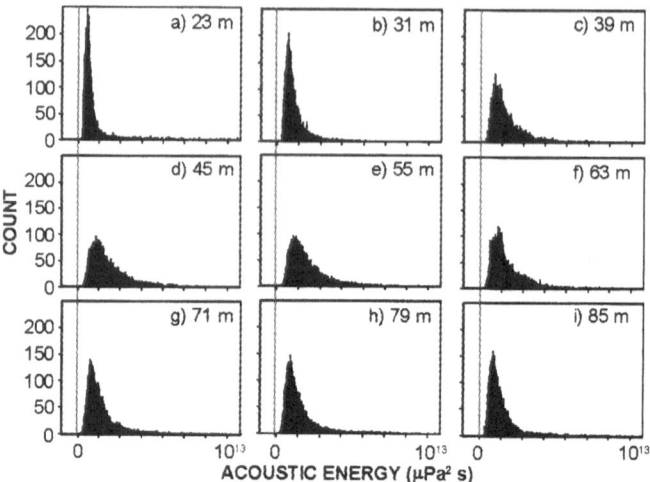

Figure 4. Histograms of the 400 Hz broadband energies measured over 63 hours (13,245 data points) at nine water depths. Histogram bin size is $10^{10}\ \mu Pa^2 s$.

3   Functional form of the histograms

The study of wave propagation in random media and the response of acoustic receiver arrays to signal variability requires quantification of the acoustic signal statistics in simple functional forms. In addition, it is necessary to characterize the stationarity of the signal statistics.

The scintillation index of narrowband signal intensity for a shallow water channel has recently been predicted to increase exponentially as a function of range, with the signal exhibiting a lognormal PDF for the saturated scattering case [3–6]. This situation contrasts with deep-water random media propagation, which is characterized as having three narrowband acoustic scattering régimes [7–9]. The first is the case where weak scattering or very short ranges are involved. In this case acoustic fluctuations are dominated by phase variability and described by the lognormal PDF. The scintillation index is much less than 1. The second régime is the partially saturated case where the scintillation index is greater than 1 due to focusing of the acoustic field and the signal amplitude varies considerably as phase variability creates strong interferences of the multipath components. Current theory does not provide a closed-form prediction for the intensity PDF in this régime. The third régime is the saturated scattering case where strong scattering or long ranges are involved and, over the statistical ensemble, the multipath components add completely incoherently. In this saturated regime the narrowband intensity statistics (with each sample having two degrees of freedom) are described by the exponential PDF, for which the scintillation index has value of 1.

Although we only present shallow water data for one range (and thus can not address the range dependence properties of the probability distribution functions) we have attempted to quantify the functional form of the measured probability distributions at the 42 km range using the Kolmogorov-Smirnov Test to compare predicted PDFs to measured PDFs and provide an estimate of the probability that the measured acoustic energy PDFs are either exponential or lognormal. The Kolmogorov-Smirnov Test has been applied to both the narrow and broadband 224 and 400 Hz data. It was calculated using acoustic data from receivers from 23 to 85 meters depth at 2 m intervals, over 63 contiguous 1-hour periods.

The narrowband Kolmogorov-Smirnov Probability (Fig. 6) for both the 224 and 400 Hz data sets indicates that the functional form of the probability distribution is depth dependent and variable in time. The test for the narrowband 400 Hz data shows frequent occurrences of the exponential distribution of energy, consistent with classical saturated scattering theory. The reason for the infrequency of a good fit of the 224 Hz narrowband data to the exponential distribution (in spite of its histogram's appearance) is likely due to the small number of multipath components at this frequency.

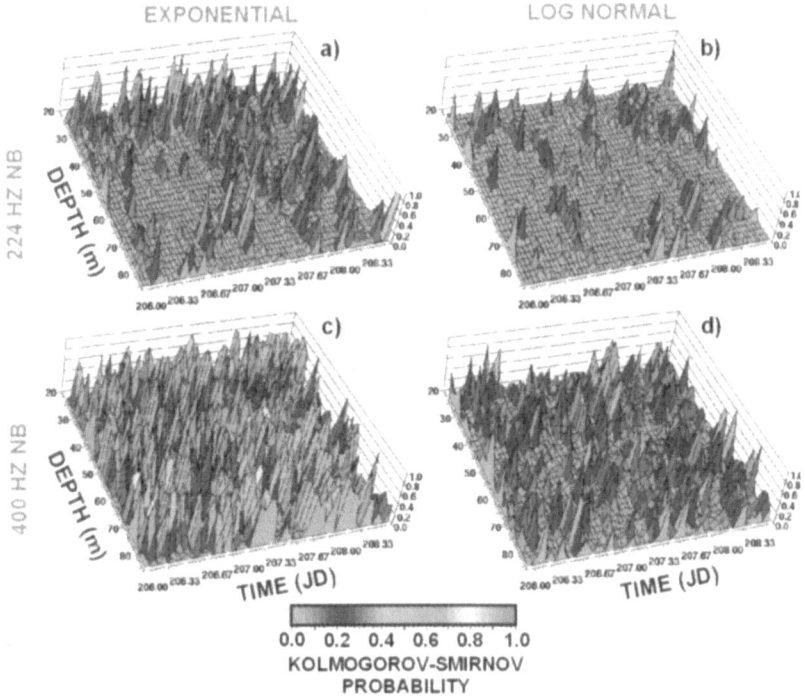

Figure 5. Probability as a function of time and depth that the 224 Hz and 400 Hz narrowband energies have an Exponential or Lognormal probability distribution function. Probabilities were calculated every 2 meters from 23 to 85 meters water depth, over 63 contiguous 1-hour periods.

The broadband Kolmogorov-Smirnov Probability (Fig. 7) is also both depth and time dependent and shows that the acoustic energy PDF is unlikely to be exponential in nature, a result that can be anticipated by noting that the broadband acoustic signals have much more than two degrees of freedom. There are several times and depths at which the number of degrees of freedom appears to be large enough that the PDF fits the lognormal distribution (particularly at 400 Hz), however there are also many times and depths where neither exponential nor lognormal distributions accurately describes the acoustic energy PDF.

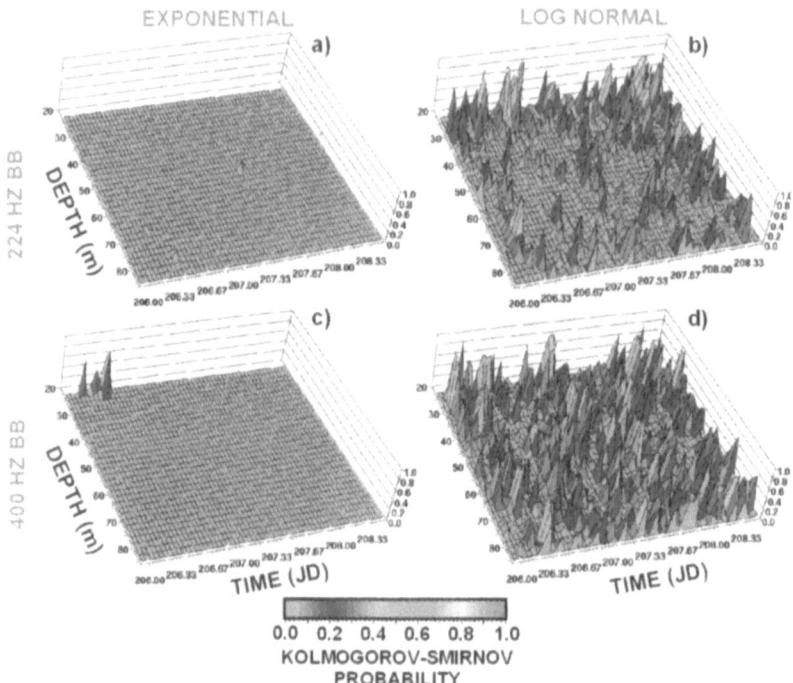

Figure 6. Probability as a function of time and depth that the 224 Hz and 400 Hz broadband energies have an exponential or lognormal probability distribution function. Probabilities were calculated every 2 meters from 23 to 85 meters water depth, over 63 contiguous 1-hour periods.

The Kolmogorov-Smirnov tests indicate that the probability distribution function for both the narrow and broadband 224 and 400 Hz acoustic intensity distributions are non-stationary and depth dependent. At this point in time we do not have a strong individual correlation between the probability distribution variability and the each of the different types of fluid processes that are randomizing the sound speed profile.

4   Scintillation index

The scintillation index (SI) is often used to characterize acoustic fluctuations. The narrowband scintillation indices measured over 1-hour periods (Fig. 8, c and d) show a strong time and depth dependence with values varying from 0.5 to 3.0. The larger values are apparently associated with signals whose fluctuations are controlled by phase modulations of partially coherent components as mentioned above. The narrowband 400 Hz data shows a slightly larger SI than the 224 Hz data.

For a Chi-squared energy distribution having N degrees of freedom, the scintillation index assumes a value of 2/N. The measured broadband scintillation indices (Fig. 8, a and b) have values varying from 0.2 to 1.0 with the 400 Hz SI slightly smaller than the 224 Hz SI, a result consistent with the observation of a richer resolved multipath structure in the 400 Hz matched filter results. Careful observation of the narrowband scintillation index variability shows an apparent 12 hr. cycle between small and large SI values, particularly on the 400 Hz deep receivers. Finette [10] has also seen a periodicity in SI in his numerical simulations of narrowband acoustic signal propagation through sound speed fields that are perturbed by linear and nonlinear internal waves.

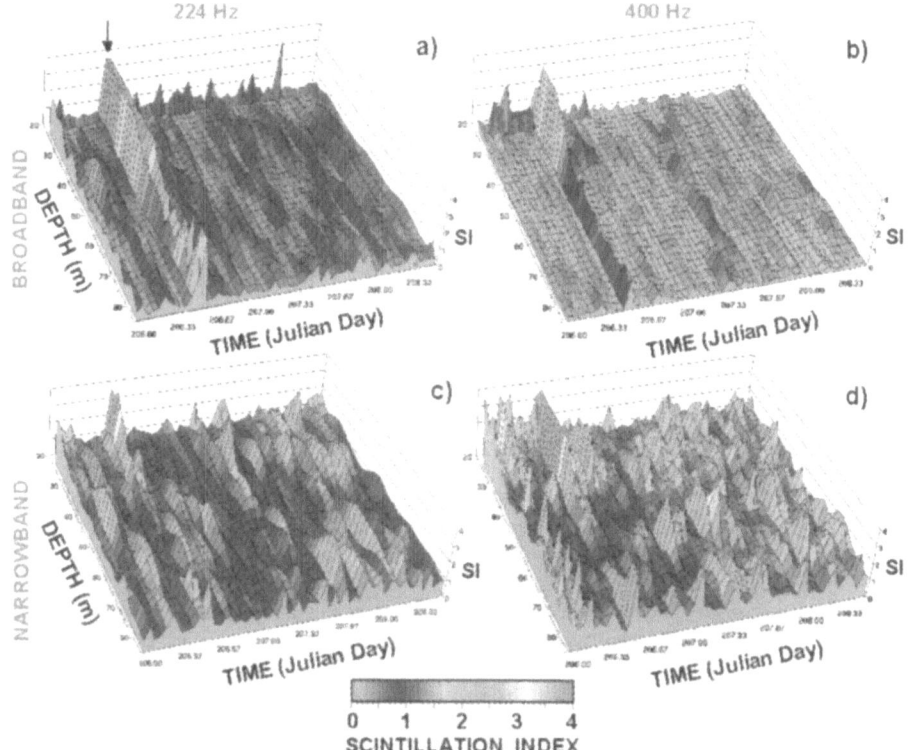

Figure 7. Scintillation Index as a function of time and water depth. Scintillation indices were calculated every 2 meters from 23 to 85 meters water depth, over 63 contiguous 1-hour periods. Large scintillation index value (arrow) is due to noise interference from a small boat and illustrates some of the dangers of measuring acoustic signal statistics in the real ocean.

5 Conclusion

At a source receiver range of 42 km, the SWARM 95 224 and 400 Hz narrowband and broadband acoustic signal intensity probability distribution functions are strongly non-stationary and depth dependent. Kolmogorov-Smirnov Tests show that the 1-hr averaged narrowband energy probability distributions can be better approximated by an exponential than a lognormal distribution, a sign that the range in the experiment was too short to observe the statistical signal properties predicted by Creamer [3]. This conclusion is reinforced by the direct evaluation of the narrowband scintillation index. Broadband signal statistics are not exponentially distributed and, in general, poorly described by the lognormal distribution. Scintillation indices measured for the broadband signals are indicative that the energy estimates have ~4 (at 224 Hz) to ~10 (at 400 Hz) statistical degrees of freedom. The larger number of degrees of freedom at the higher frequency is consistent with the observed multipath structure and the more frequent occurrences of the 400 Hz broadband energy approximating a lognormal distribution.

Acknowledgements

This work was supported by the Office of Naval Research. The SWARM 95 experiment was a major multi-disciplinary endeavor and would not have been successful without the contributions of many individuals. We thank the other members of the SWARM GROUP: J. Apel, M. Badiey, J. Berkson, K.P. Bongiovanni, J. Bouthillette, E. Carey, C. Chiu, T. Duda, C. Eck, S. Finette, R. Headrick, J. Irish, J. Kemp, A. Newhall, J. Presig, B. Racine, S. Rosenblad, S.A. Shaw, D. Taube, D. Tielbuerger, A. Turgut, K. von der Heydt and W. Witzell.

References

1. J. Apel, M. Badiey, J. Berkson, K. P. Bongiovanni, J. Bouthillette, E. Carey, C. Chiu, T. Duda, C. Eck, S. Finette, R. Headrick, J. Irish, J. Kemp, J. Lynch, A. Newhall, M. Orr, B. Pasewark, J. Presig, B. Racine, S. Rosenblad, S. A. Shaw, D. Taube, D. Tielbuerger, A. Turgut, K. von der Heydt, W. Witzell and S. Wolf, An overview of the 1995 SWARM shallow water internal wave acoustic scattering experiment, *IEEE J. Oceanic Eng.* **22**(3), 465–500 (July 1997).
2. B.F. Cron and W.R. Schumacher, Theoretical and experimental study of underwater sound reverberation, *J. Acoust. Soc. Am.* **33**(7), 881–888 (1961).
3. D.B. Creamer, Scintillating shallow water waveguides, *J. Acoust. Soc. Am.* **99**(5), 2825–2838 (1996).
4. D. Tielbuerger, S. Finette and S.N. Wolf, Acoustic propagation through an internal wave field bounded by a shallow water waveguide, *J. Acoust. Soc. Am.* **101**(2), 789–808 (1997).
5. X. Tang and F.D. Tappert, Effects of internal waves on sound pulse propagation in the Straits of Florida, *IEEE J. Oceanic Eng.* **22**(2), 245–255 (1997).
6. N.C. Makris, The effect of saturated transmission scintillation on ocean acoustic intensity measurements, *J. Acoust. Soc. Am.* **100**(2), 769–783 (1996).
7. S.M. Flatte, Wave propagation through random media: Contributions from ocean acoustics, *Proceedings of the IEEE* **1**(11), (November 1983).
8. B.J. Uscinski, *Elements of Wave Propagation in Random Media* (McGraw-Hill International Book Company, 1977).
9. A. Ishimaru, *Wave Propagation and Scattering in Random Media,* Vol. 2 (Academic Press, Inc., 1986), Chap. 20.
10. S. Finette, M.H. Orr, A. Turgut, J.R. Apel, M. Badiey, C.-S. Chui, R.H. Headrick, J.N. Kemp, J.F. Lynch, A.E. Newhall, K. von der Heydt, B.H Pasewark, S.N. Wolf and D. Tielbuerger, Acoustic field variability induced by time-evolving internal wave fields, *J. Acoust. Soc. Am.* **108**(3), 957–972 (2000).

COMBINATION OF ACOUSTICS
WITH HIGH RESOLUTION OCEANOGRAPHY

JÜRGEN SELLSCHOPP

Forschungsanstalt der Bundeswehr für Wasserschall und Geophysik
Klausdorfer Weg 2-24, 24148 Kiel, Germany
E-mail: jsellschopp@bwb.org

PETER NIELSEN

SACLANT Undersea Research Centre, Viale San Bartolomeo 400, 19138 La Spezia, Italy
E-mail: nielsen@saclantc.nato.int

MARTIN SIDERIUS

Science Applications International Corporation, La Jolla, CA 92037,USA
E-mail: sideriust@saic.com

The variability of underwater sound observed on a moving platform in littoral waters is a combination of the effects of temporal and spatial environmental variations, which are intermingled and obscured by the change of the platform position. Sound transmission experiments over a fixed range are an approach to separate different sources of environmental impact on acoustic propagation. Even in a fixed geometry time variability experiment, the most demanding observations are for the spatial distributions of controlling parameters. On a short time scale, the variation of spatial distributions is responsible for acoustic variability, while mean, range averaged conditions may stay unchanged. The instrumentation for monitoring the ocean environment during the ASCOT 01 acoustic trial included moored instruments for the measurement of temperature and current profiles, tidal and surface waves. By means of wave dispersion relations, moored records of ocean variability may be transformed into guessed spatial fine structure. But critical results such as correct slope statistics of sound channel boundaries and iso-velocity surfaces cannot be guaranteed. A 40-sensor CTD chain was continuously towed parallel to the acoustic range. The measured 2-dimensional sound velocity field has 2 m vertical and 4 m horizontal resolution. A primitive model, which propagates rays through measured high resolution sound velocity fields is able to explain observed multipath arrival structures on a vertical hydrophone array.

1 Introduction

High quality predictions of the underwater sound channel are desired, since they are the essential input to reliable modeling of the conditions for sonar detection and underwater communication. Environmental parameters of primary concern are the sound velocity profile, the depth and composition of the seabed and the sea state. By these parameters, the ocean volume and the boundaries of the littoral sound channel are covered. Realistically none of these parameters can be prophesied on scales of kilometer or hour orders. The best possible approach, to allow for a realistic environment in acoustic

N.G. Pace and F.B. Jensen (eds.), Impact of Littoral Environmental Variability on Acoustic Predictions and Sonar Performance, 19-26.

model calculations, is the separation between predictable average conditions and deterministically unpredictable environmental variability, which might however be well described by a statistical parameter such as the surface wave spectrum. Whether small scale environmental variability is treated adequately in acoustic modeling cannot be decided from the comparison of sound measurements with model results, if there are free adjustable parameters in the model, which can be tuned until the results match. If it is impossible to measure all parameters, which have an impact on acoustics, there is always a danger that a parameter with substantial influence is overlooked and a wrong parameter adjusted instead. This becomes especially obvious in reverse modeling, where the restriction on a limited set of tractable parameters forces all deviations of the environment from an assumed predefined state to act into the selected parameter subspace. Consequentially inversely modeled parameters may not be accurate. The range dependence of the results from inverse modeling of bottom parameters [1] is a hint that the treatment of environmental variability in the model is inappropriate. A description of the ambience as accurate as possible, deduced from extensive measurements, is required for a better connection of acoustic variability and uncertainty to ocean variability. Whether or not the variability of a certain environmental parameter can be finally neglected, can only be decided after sensitivity studies have been made with realistic data applied to appropriate models.

Small scale fluctuations have a much higher impact on the littoral sound channel than in the open ocean. In a shallow sound channel, multiple interactions with the surface and bottom boundaries influence the energy and direction of the propagated sound. In contrast to stationary bottom scatterers, the rough surface introduces time dependence into acoustic signals. But also in a stratified ocean under summer conditions, when the sound does not interact with the sea surface in motion, time variability of the range dependent sound velocity structure in the water column is able to significantly impact transmissions between fixed locations. Because of the difficulty of measuring the internal variability, its effect on acoustic signals is harder to access than that of surface waves.

In order to separate influences of different kinds, it is advantageous to perform experiments with a fixed source and receiver, which would at least keep the range constant and the bottom unchanged. Measurements of acoustic variability should be complemented by measurements of environmental variables, so that the assumptions introduced into acoustic models are minimized.

2   ASCOT 01 experimental setup

The acronym ASCOT is nicely translated into "Acoustic Scenario Connected with an Oceanography Trial", although it originally stands for "Assessment of Skill for Coastal Ocean Transients". The NATO Research Vessel *ALLIANCE* spent three weeks in June 2001 for a collaborative effort with Harvard University [2] in Massachusetts Bay and half of the Gulf of Maine. Data from an initial ocean survey were used to initiate the primitive equation forecast model of the Harvard Ocean Prediction System (HOPS). A limited amount of data was acquired for assimilation during the forecast period. The predictive skill experiment was concluded with a verification survey. After the first phase, there was an opportunity to embed ocean acoustics experiments of four days duration, interrupted by one day for logistics. The experimental setup was similar to the fixed range experiments on Adventure Bank (Strait of Sicily) two years before [1]. The compulsory second ship was hired from the University of New Hampshire. A site with suitable conditions, which are defined by proximity to the coast, no interference with protected ocean areas and an approximately constant water depth of about 100 m over 10 km range, was identified 30 nautical miles north of Cape Cod, 20 miles east of Cape Ann (Fig. 1).

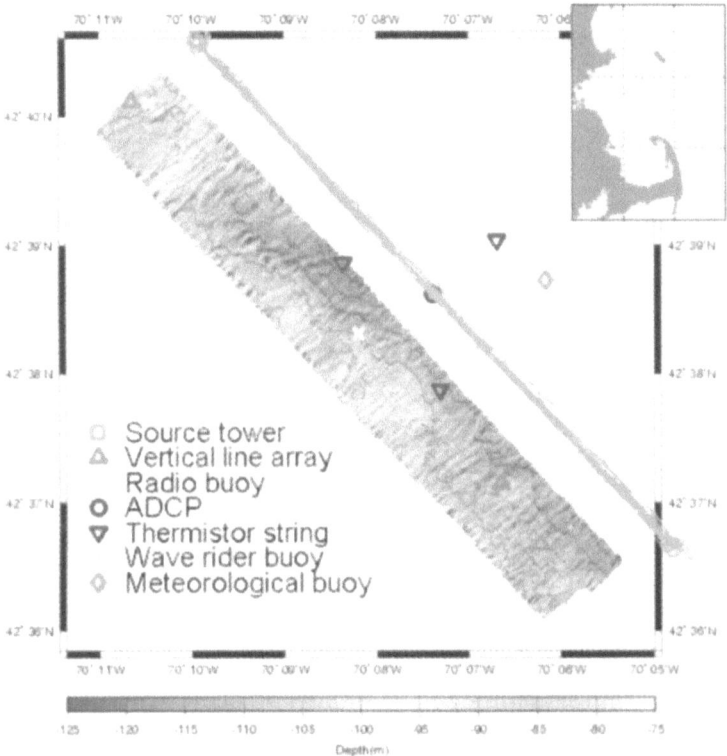

Figure 1. Bathymetry, mooring positions and tow track of the acoustic test site.

The bottom was hard and rougher than expected. Soundings were taken every 4 m on a straight line. The bottom contours of Fig. 1 were obtained by swath mapping with 20 m horizontal resolution. At the SE end of the site, a frame was lowered to the ground with a sound source mounted on top. A vertical line array with 64 hydrophones was deployed at distances 5, 2, 10 and 0.78 km from the source tower, each deployment lasting for one day. A suite of instruments was available for environmental monitoring. Three thermistor strings, each equipped with 11 thermistors with 5 m vertical separation, were deployed at the corners of an equilateral triangle with side length 2.35 km. While one standard thermistor string was sampled every 2 minutes, two strings had response times and sample rates of 10 seconds. An upward looking acoustic Doppler current profiler in the center of the triangle was used with 5 minutes averaging interval and sample rate. Sea surface roughness due to wind waves was continuously measured by a wave rider buoy. A meteorological buoy measured wind close to the surface and undistorted by the ship. A tide gauge, which was combined with the ADCP, was out of range. Qualitatively the measurements compare well with tides in Boston, which they precede by less than 30 minutes.

In addition to the moored instruments, a CTD chain [3] was used for the direct measurement of spatial structures. It was deployed from *ALLIANCE* and handed over to the *GULF CHALLENGER* for continuous coverage of a 10-km track parallel to the acoustic range. The CTD chain configuration involved 40 packages for the measurement of conductivity, temperature and depth (CTD), distributed over the 70-m aperture between surface float and depressor (Fig. 2). With standard scanning cycles of 2 seconds, the spatial sampling mesh width was 3.3 m horizontally and 1.8 m vertically.

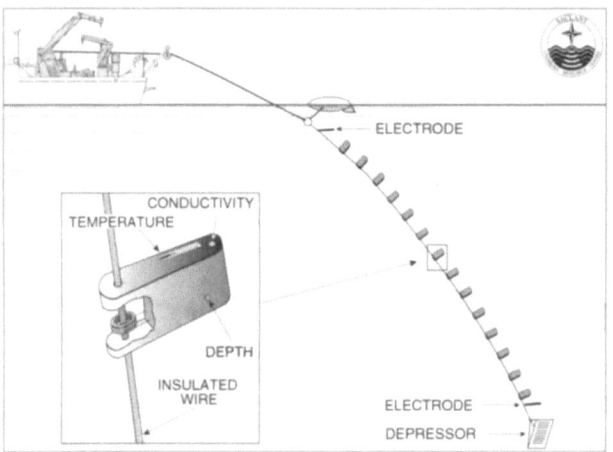

Figure 2. Sketch of the towed CTD chain. Sensors are inductively coupled to the towing cable.

3   Environmental variability and its impact on acoustics

In the course of a tidal cycle, the surface level changes by 2 m. The bottom moored thermistor strings feel the rise and fall of the thermocline with the tide. The temperature contours from low-pass filtered time series of the thermistor strings are highly correlated with, but not strictly parallel to surface elevations (Fig. 3). Acoustic energy propagating with low grazing angles would feel the thermocline position rather than the ocean surface. At 10 km range, the first four or five arrivals (see Fig. 4 of Nielsen *et al.* [4]) are from eigenrays, which do not touch the surface, but are influenced by sub-surface tidal signatures.

Tidal currents are approximately aligned with the acoustic track. They have maximum strength 25 cm/s and are nearly independent of depth. The long axis of the tidal ellipse is 2.5 km. Integrated over the 5 days of acoustic experiments, there was a dislocation of water masses to the SE by 10 km. Water mass properties were however sufficiently independent from horizontal coordinates, with the consequence, that flushing did not play a role.

Tidal waves alter the sound channel as a whole, but do not change its character in general. Higher components of the wave spectrum introduce range dependence to the sound propagation problem. Original (unfiltered) records from the moored thermistor strings contain signatures of internal waves with periods down to a few minutes, which would not be correctly sampled using standard instruments having response times of 2 minutes (Fig. 4). Records at different depths are well correlated. By inspection we find no indication of internal wave modes other than the lowest. From the density profile of the water column, the dispersion relation of internal waves was solved numerically [5]. A first mode oscillation of 1 (10) cycle(s) per hour is related to 0.6 (8.5) wave lengths per km. Phase velocities of these short internal waves are only slightly higher than the background current. The frequency spectrum of internal waves is therefore heavily affected by frequencies of encounter.

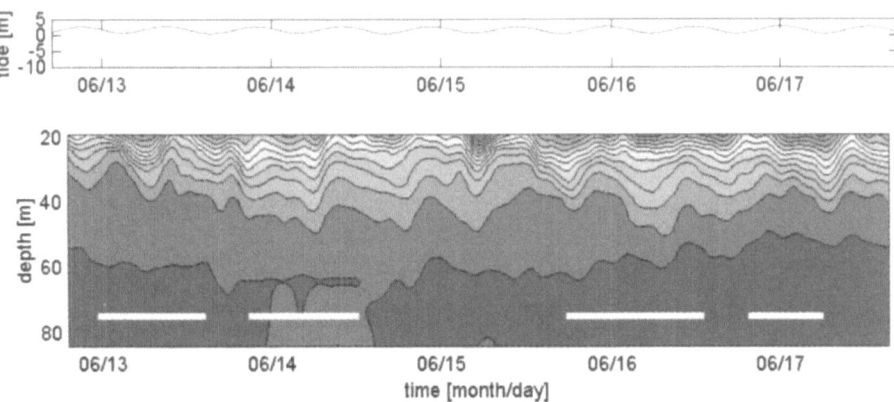

Figure 3. Tidal surface elevation (upper) and filtered temperature distribution (4°C – 10°C) from the eastern thermistor string. White bars indicate the duration of the acoustic experiments at 5, 2, 10 and 0.78 km range.

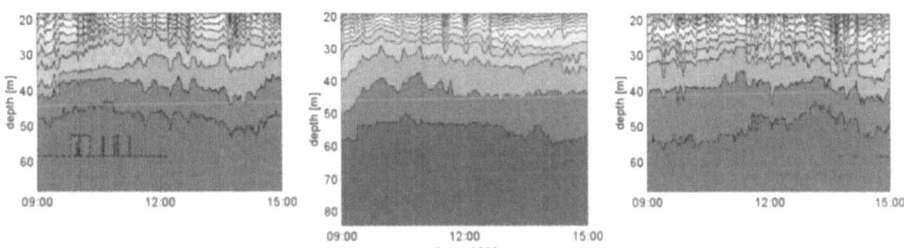

Figure 4. Temperature contours from original records of the western (left), southern (mid) and eastern (right) thermistor strings. While with 10 s sampling rate, two strings resolve high frequency internal waves, the third with 2 minutes under-samples.

For deterministic modeling of acoustic transmissions through a fluctuating ocean, the time and space dependent sound velocity field of the ensonified area must be known. Since it is impossible to obtain three or four dimensional measurements of ocean parameters with sufficient resolution, artificial representations of the sound velocity field are used instead with the correct fluctuation statistics in anticipation. Under favorable conditions and reasonable assumptions such as independence from the azimuth angle, a local frequency spectrum of internal waves may be converted into a horizontal wave number spectrum, which in turn is used for the realization of a spatially fluctuating sound velocity field. Because of the tidal currents, the moored measurements during ASCOT01 cannot be transformed into wave number spectra.

Spatial variability, directly measured by means of the towed CTD chain was analyzed for the wave number spectrum of contour displacements (Fig. 5). Low wave numbers, which would reflect repetitions of the 10 km track were excluded by high pass filtering. This also removes potential contamination by temporal effects such as tides.

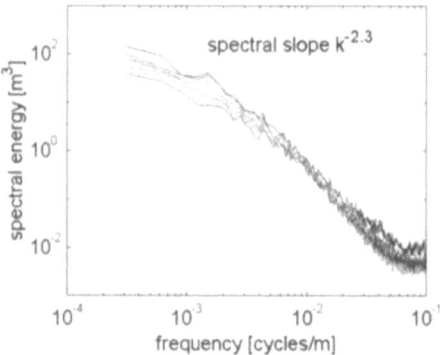

Figure 5. Wave number spectra of the vertical displacement of high pass filtered temperature contour lines from 6 to 12 °C. Note that the spectral energy increases with depth. The blue curve is for 6°C.

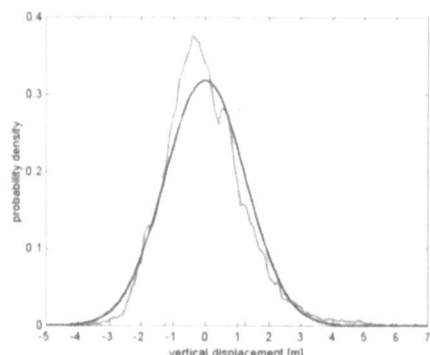

Figure 6. Distribution of vertical displacements of the high pass filtered 10°C contour. 6 to 12°C distributions look qualitatively the same. The standard deviation increases with depth (decreasing temperature). A Gaussian standard distribution is shown for comparison.

The spatial temperature records of the towed sections show pronounced downward excursions of the contour lines similar to those in Fig. 4. The asymmetry of internal fluctuations is reflected by the distributions of contour line depth. Throughout the thermocline, they deviate significantly from a Gaussian normal distribution (Fig. 6).

A primitive ray tracing routine was used for the simulation of acoustic pulses propagating through a real fluctuating sound velocity field. Measured sound velocity (by CTD chain temperature and salinity) was interpolated to a 1 x 1 m grid in the vertical plane spanning between source and receiver. The routine calculates sound velocity gradients, grazing angle, vertical position and time increments at each horizontal grid point. Two realizations are displayed in Fig. 7. Bottom topography was taken from echo sounder recordings and the sound velocity field from a parallel tow track. Only the rays of a narrow beam are displayed, which arrive at a receiver 40 m in depth.

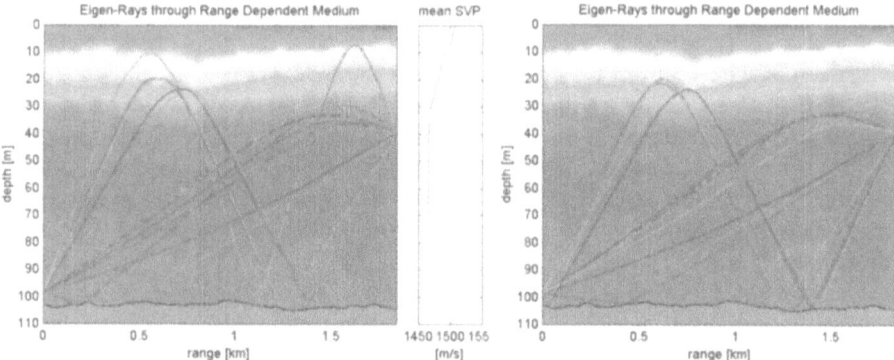

Figure 7. Multipath propagation through a realistic ocean. A hardly noticed horizontal shift of the sound velocity field causes a significant change of the pattern of eigenrays.

The simple relation between the initial ray elevation angle and the number of turning points, which exists in range independent environments with smooth boundaries, gets lost by range dependence. Different rays of the same class can coexist, appear and vanish by small changes in the sound velocity field. Small differences in travel time cause frequency dependent fading. Figure 8 shows pulse arrivals in the water column at the position of the receiving array for a range independent and for a sample range dependent situation.

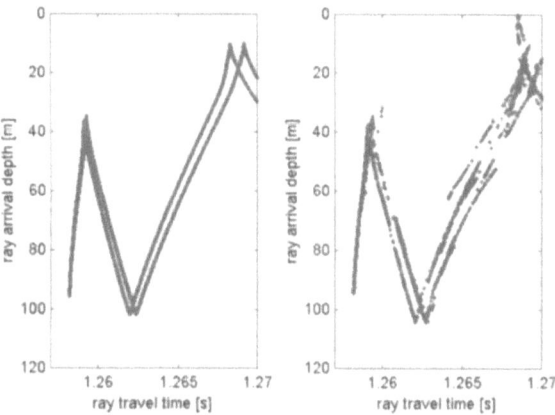

Figure 8. First few arrivals from multipath propagation over the 2 km test range. Left: Range independent. Right: Realistic range dependent example.

Bottom roughness amplifies the effect of fine structure ocean variability. Slightly modified refraction in the water column may shift the point of contact with the bottom to a facet with different slope, so that differential adjustment of the eigenray path will not bring the reflected ray back to the desired direction. With bottom roughness single eigenrays die and come to life much easier than with a smooth bottom (see Fig. 7). Because multipath arrival patterns become less stable, matched field processing is a challenging task in this environment.

4 Conclusion

Nielsen *et al.* [4] have shown that irrespective of a similar experimental setup the results of the ASCOT01 acoustic experiments differ largely from the ADVENT99 results. The matched field correlation of an early received signal with subsequent signals drops off very fast, even for low frequencies. While source localization by common matched field processing failed in the ASCOT01 environment, Siderius *et al.* [6] successfully used a more robust processor, which correlates signal envelopes separately for each hydrophone.

The unexpected resistance of the ASCOT01 environment against standard data processing, as far as can be judged at present, is due to high bottom reflectivity, bottom roughness and higher sound speed variability than during the ADVENT99 experiments [1]. The large amount and quality of environmental data, which were collected together with acoustics offers the opportunity to study the processes in detail, which are responsible for coherence loss and signal degradation in a relatively benign littoral ocean area.

Acknowledgements

The experiment was carried out as part of the SACLANTCEN Programme Of Work. We thank crews and cruise participants for their outstanding dedication to the task. FWG sponsored a considerable part of data processing including the development of processing tools.

References

1. M. Siderius, P.L. Nielsen, J. Sellschopp, M. Snellen and D. Simons, Experimental study of geo-acoustic inversion uncertainty due to ocean sound-speed fluctuations, *J. Acoust. Soc. Am.* **110**(2), 769–781 (2001).
2. A.R. Robinson *et al.*, http://www.deas.harvard.edu/~leslie/ASCOT01/ (2001).
3. J. Sellschopp, A towed CTD chain for two dimensional high resolution hydrography, *Deep Sea Research* **44**(1), 145–163 (1997).
4. P.L. Nielsen, M. Siderius and J. Sellschopp, Broadband acoustic signal variability in two "typical" shallow-water regions. In *Impact of Littoral Environmental Variability on Acoustic Predictions and Sonar Performance*, edited by N.G. Pace and F.B. Jensen (Kluwer, The Netherlands, 2002) pp. 237–244.
5. R. Evans, Program WAVE, http://oalib.saic.com/Other/wave/wave.zip (2001).
6. M. Siderius, P. Nielsen and J. Sellschopp, Source localization in a highly variable shallow water environment: Results from ASCOT-01. In *Impact of Littoral Environmental Variability on Acoustic Predictions and Sonar Performance*, edited by N.G. Pace and F.B. Jensen (Kluwer, The Netherlands, 2002) pp. 425–432.

EFFECT OF HURRICANE MICHAEL ON THE UNDERWATER ACOUSTIC ENVIRONMENT OF THE SCOTIAN SHELF

D. HUTT, J. OSLER AND D. ELLIS

DRDC Atlantic, Dartmouth, Nova Scotia, Canada B2Y 3Z7
E-mail: daniel.hutt@drdc-rddc.gc.ca

In October 2000 DRDC Atlantic carried out a detailed characterization of the shallow water environment in a 150 by 170 km area of the Scotian Shelf. The study area was centered at 44 deg. N, 61 deg. W and had an average water depth of 70 m. In addition to oceanographic moorings, two rapid environmental assessment surveys of water temperature profiles were made from Canadian maritime patrol aircraft which dropped 72 air-expendable baththermographs (AXBTs) in an 8 by 9 grid with 16 km nominal spacing. Between the AXBT surveys on Oct. 14 and 21, hurricane Michael passed over the study area. The AXBT surveys and satellite-derived sea surface temperature imagery show that passage of the hurricane cooled surface waters and changed the thickness of the mixed layer by up to 10 m. The effect of the environmental change on acoustical propagation in the 20 Hz to 10 kHz band was estimated by calculating broadband transmission loss with the PROLOS normal modes model using sound speed profiles measured before and after the hurricane and using climatological profiles.

1 Introduction

The complexity of continental shelf ocean processes and the spatial variability of seabed characteristics, results in an environment in which it is difficult to predict acoustic propagation. The increased military interest in littoral zones in recent years has led to a requirement to understand and model this environment. As a first step, we have begun development of a high-resolution oceanographic model of the Scotian Shelf. The model, based on the Princeton Ocean Model, will eventually assimilate remotely sensed sea surface temperature data [1]. Work is also underway to examine other types of remote sensing data such as ocean color and radar backscatter, for their potential to provide information about the underwater acoustical environment [2].

This paper describes ocean data obtained during cruise Q255 of DRDC's research vessel CFAV *Quest* on the Scotian Shelf in Oct. 2000. The goal of the cruise was to obtain *in situ* oceanographic data and remotely sensed sea surface data which would be used to test the ocean model. It happened that the most powerful hurricane of the Canadian 2000 season, Michael, passed over our study area providing a rare opportunity to study the impact of such a disturbance on shelf oceanography. Although underwater acoustical propagation measurements were not made during the trial, the measured water temperature profiles are interpreted here in terms of modelled acoustical propagation. It was found that the passage of hurricane Michael caused mixing of the water column,

N.G. Pace and F.B. Jensen (eds.), Impact of Littoral Environmental Variability on Acoustic Predictions and Sonar Performance, 27-34.

reducing the temperature of the mixed layer by 1°C and affected the depth of the thermocline by nearly 10 m.

2   Scotian Shelf environment

Figure 1 shows the continental shelf area southeast of Nova Scotia, known as the Scotian Shelf. Bounded to the northeast by the Laurentian Channel and in the southwest by the Gulf of Maine, the Scotian Shelf has a range of seabed types with clay in basins and sand or gravel on the shallow banks. An overview of the geoacoustic and oceanographic environment of the Scotian Shelf is given by Osler [3]. The average depth in the Q255 study area, shown in Fig. 1, is 70 m and much of the seabed in the southern half of the area is composed of Sable Island Sand.

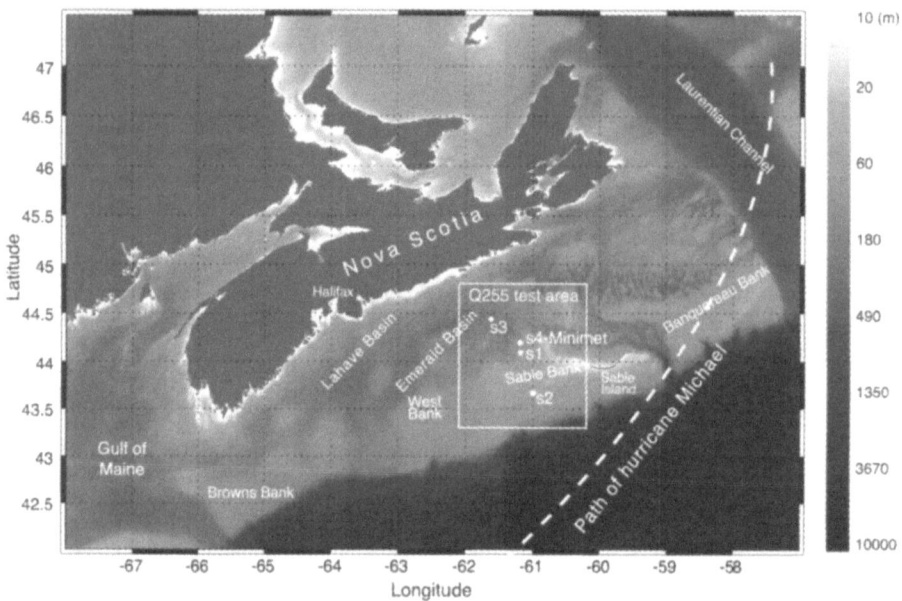

Figure 1. Scotian Shelf, southeast of Nova Scotia and the Q255 study area.

3   Environmental characterization

The locations of oceanographic moorings (sites s1 to s4) are shown in Fig. 1 with the instruments listed in Table 1. Self-locating Datum Marker Buoys (SLDMB) which drift with the surface current were used to measure currents during the course of the trial. Results from the SLDMB measurements can be found in Hutt *et al.* [4]. The Minimet buoy is a weather station equipped with standard meteorological sensors. For the purposes of this paper, discussion of ocean data is limited to the two rapid environmental assessment (REA) surveys that took place on Oct. 14 and 21, 2000. During each survey, maritime patrol aircraft (MPA) deployed 72 AXBTs (aircraft-deployed expendable bathy-thermographs) in an 8 by 9 grid with 16 km nominal spacing which covered the entire test area shown in Fig. 1.

Table 1.   Measurement sites for Q255, Oct. 2000.

Site	Latitude	Longitude	Depth (m)	Description
1	44°05.0 N	61°10.0 W	70	Geoacoustic experiments
2	43°40.0 N	61°00.0 W	60	S4 current meters SLDMBs (Oct. 14)
3	44°24.0 N	61°37.0 W	180	S4 current meters SLDMBs (Oct. 21)
4	44°10.0 N	61°10.0 W	65	ADCP S4 Current meter Minimet buoy Triaxys wave buoy SLDMBs (Oct. 14 and 21)

4   Hurricane Michael

4.1   Synoptic Overview of Hurricane Michael

Hurricanes which travel northward along the east coast of North America are a common occurrence in late summer and autumn. Their energy is sustained by circulation of the warm humid air over the Gulf Stream. Hurricanes typically lose energy quickly once they pass north of the Gulf Stream where they are affected by cooler water. However, in some cases hurricanes may reintensify due to acceleration by mid-latitude winds. Hurricane Michael, which passed over the Scotian Shelf on Oct. 19, 2000 was an example of such a "baroclinically enhanced" extra-tropical hurricane.

Formed in response to an upper-level low moving over a stationary surface front north of the Bahamas on Oct. 12, Michael attained hurricane status on Oct. 17. It began to move north on Oct. 18 and experienced significant intensification on Oct. 19 as it passed over the Scotian Shelf. It reached the south coast of Newfoundland late on Oct. 19 with its tropical core still intact. With maximum sustained winds of 140 km/h and barometric pressure of 966 mb at its center, Michael was the most intense hurricane to pass eastern Canada in 2000. The rapid passage of Michael across the Scotian Shelf is shown in the sequence of GOES East images in Fig. 2. Between 08:15 and 19:15 UTC on Oct. 19, 2000, Michael travelled about 1000 km with an average speed of 90 km/hr.

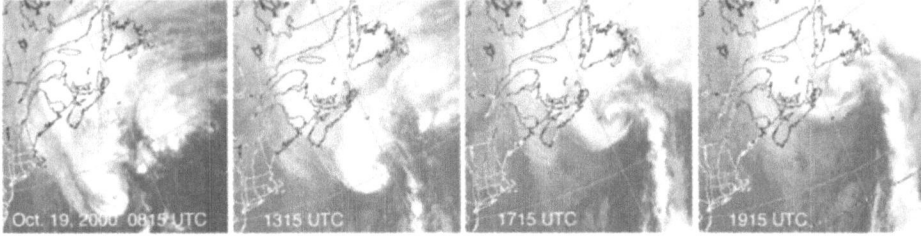

Figure 2. Progression of hurricane Michael across Scotian on Oct. 19, 2000.

4.2   Effect of Hurricane Michael on Test Area

The GOES imagery show that the closest the center of the hurricane was to the Minimet buoy, moored at site 4, was 75–100 km to the southeast at approximately 16:00 UTC on Oct. 19. Figure 3 shows times series of several of the meteorological parameters measured by Minimet sensors. The passage of the eye of the hurricane is easily seen in the figure as a deep trough of pressure late on Oct. 19. The characteristic lull in wind speed and reversal of wind direction as the center of the hurricane passes the buoy is also evident.

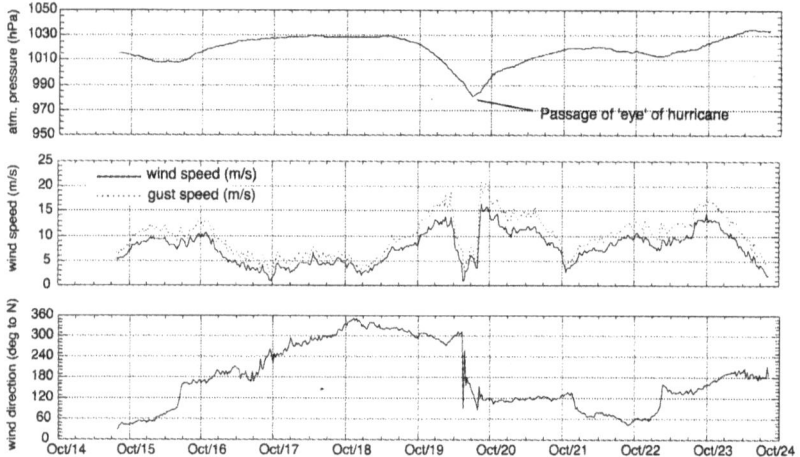

Figure 3. Sea surface meteorology during passage of hurricane Michael.

Although hurricane Michael passed over the Scotian Shelf in less than six hours, it was intense enough to effect changes in the shelf oceanography. Images of sea surface temperature (SST) from the NOAA-14 AVHRR sensor before and after the hurricane indicate that the surface of the water was cooled by an average of 1.2 °C (std. dev. 0.2 °C) in the study area. Figure 4 shows the SST images of Oct. 14 and 22, 2000 which were the last clear day before and the first clear day after passage of the hurricane.

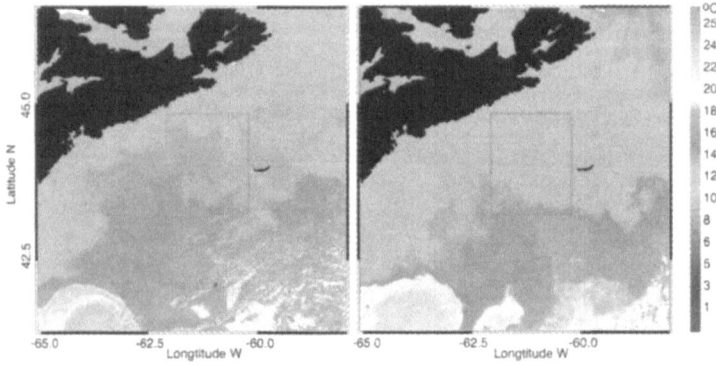

Figure 4. Sea surface temperature from NOAA-14 AVHRR imagery on Oct. 14, 2000, left (before hurricane) and on Oct. 22, 2000, right (after passage of hurricane).

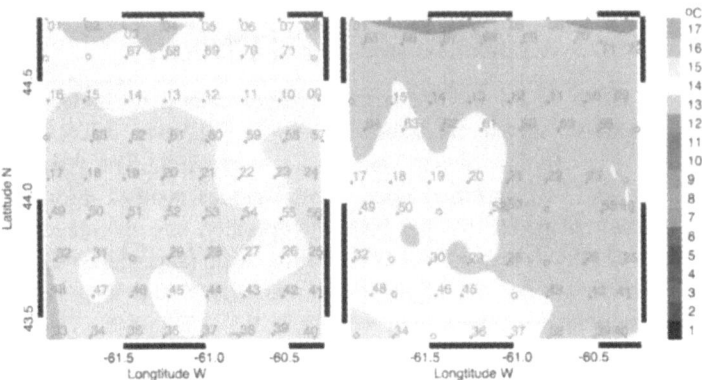

Figure 5. Sea surface temperature from REA surveys on Oct. 14, 2000 (left) and Oct. 21, 2000 (right), before and after passage of hurricane. AXBT drop sites shown as numbered red dots.

Data from the REA surveys on Oct. 14 and Oct. 21 show a similar 1°C drop in surface temperature caused by the hurricane. The uppermost temperature measurement of the AXBTs, made at a depth of 1.5 m, are shown as temperature contour plots in Fig. 5. The water column temperature structure before and after Michael is shown in transects from west to east through the middle of the two REA surveys in Fig. 6. The 1 °C drop in surface temperature between the surveys can be seen to extend throughout the mixed layer. In the deeper west side of the transect, there was a slight warming of water below the thermocline after the hurricane which is evidence that mixing occurred throughout the entire water column. Before Michael, the thermocline was at a constant depth of 32 m across the entire transect. After the hurricane, there is much more variability in the depth of the thermocline. Near AXBTs 20 and 21, the thermocline is 8 m lower after the hurricane and near AXBTs 22 and 23 a new thermocline has appeared at approximately 20m whereas before Michael the entire water column was isothermal in that area.

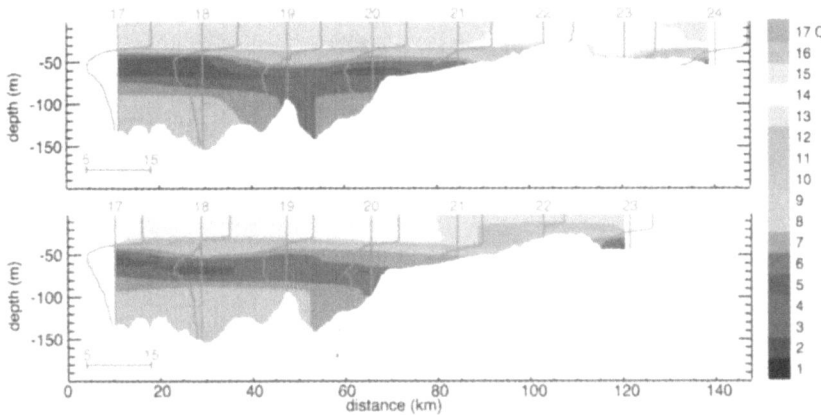

Figure 6. Water temperature profiles from REA surveys on Oct. 14 (top) and Oct. 21, 2000 (bottom). Transects are from west to east. Red vertical lines represent 10 °C.

5   Acoustical modelling

5.1   Broadband Transmission Loss Modelling with PROLOS

To examine the effect of the environmental changes caused by hurricane Michael on broadband propagation, incoherent transmission loss (TL) was calculated from 20 to 10000 Hz at discrete frequencies in one-octave increments. The propagation model employed, PROLOS [5,6], is based upon normal modes acoustic propagation theory. PROLOS is a research model that has been incorporated into the allied environmental support system (AESS). It can model propagation with range-dependent sound speed profiles, bathymetry, and sediment geoacoustic parameters and includes losses due to seabed and sea surface roughness. The results in this paper were calculated using a geoacoustic model for Sable Island Bank [3] with 50 m of sand, overlying 50 m of glacial till, and then a sedimentary rock half space. The roughness of the seabed and surface were specified as 0.01 and 0.3 m rms respectively.

5.2   Impact of Hurricane Michael on Acoustical Propagation

The incoherent transmission loss for range independent propagation was compared for *in situ* sound speed profiles before and after the passage of hurricane Michael and for a climatological profile for the month of October. The *in situ* sound speed profilesc were derived from the AXBT temperature profiles at drop site 21 shown in Figs. 5 and 6. The climatology profile, prepared as part of the ocean model project [1], is a simple depth average of a compilation of historical data. This approach yields a thermocline that is much less pronounced than the *in-situ* profiles but is typical of climatology-based sound speed profiles often used operationally for sonar performance prediction. The sound speed profiles used for the TL calculations are shown in Fig. 7.

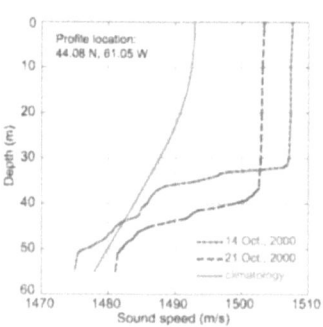

Figure 7. Sound speed profiles from AXBT site 21 before and after hurricane Michael, and sound speed profile from climatology.

Results of the PROLOS propagation modelling are shown in Figs. 8 and 9 as TL as a function of frequency and range for two different cases. In the first case (Fig. 8), the source and receiver are both located below the thermocline at a depth of 50 m. The result is that most of the energy is trapped between the thermocline and the seabed. There are higher losses above 1 kHz due to surface scatter and below 200 Hz due to penetration into the seabed. This gives rise to an optimal propagation frequency of approximately

500 Hz. The TL predicted using the climatological sound speed profile is in reasonable agreement with the TL calculated using both of the measured profiles. Thus, below the thermocline, hurricane Michael had negligible effect on the propagation.

For the second case (Fig. 9), the source is below the thermocline at a depth of 50 m but the receiver is above the thermocline at a depth of 30 m. Here, the effects of hurricane Michael and inadequacies in the climatology become evident. After the hurricane, losses are higher for the *in situ* profile, but they are restricted to frequencies above 500 Hz. The hurricane has not had a dramatic effect because the basic character of the profiles has been retained, that is a well mixed surface layer, a sharp thermocline, and a downward refracting sound speed profile between the thermocline and the seabed. There are however significant differences in TL for frequencies above 200 Hz when the predictions using the climatology are examined. The weaker thermocline in the climatology allows considerably more energy to reach the receiver whereas the TL for the *in situ* profiles is much higher.

The inadequacy of the climatology in this case underscores the requirement for a more suitable method of representing the structure of the sound speed profile in climatology and the requirement for nowcast and forecast capabilities of an ocean model on the Scotian Shelf.

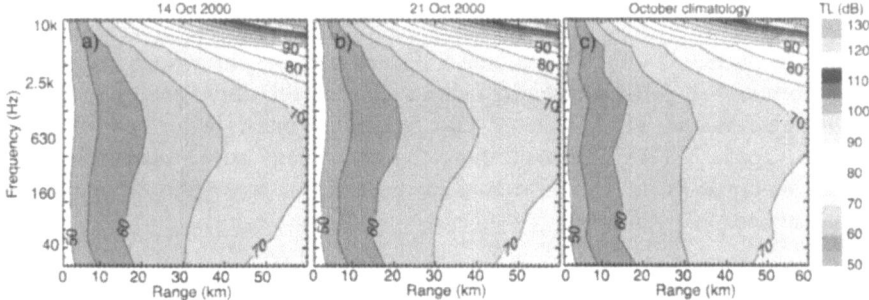

Figure 8. Broad-band transmission loss with source and receiver both at 50 m (below thermocline). a) before hurricane, b) after hurricane, c) climatology.

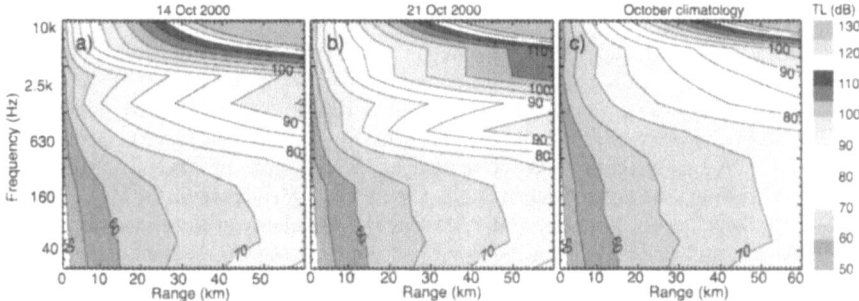

Figure 9. Broad-band transmission loss with source at 50 m (below thermocline) and receiver at 30 m (above thermocline), a) before hurricane, b) after hurricane, c) climatology.

6   Summary

We presented results of an environmental survey of part of the Scotian Shelf carried out in Oct. 2000. During the trial, hurricane Michael passed over the study area, affording a unique opportunity to observe the effect of an intense storm on shelf oceanography and on the acoustical propagation environment. The passage of the hurricane across the Scotian Shelf took only 6 h but it was intense enough to lower the temperature of the mixed layer by more than 1°C and to disturb the level of the thermocline by nearly 10m.

To estimate the effect the hurricane had on acoustical propagation, measured water temperature profiles from one location were use to calculate broadband TL in the 20 Hz to 10 kHz band with the PROLOS normal modes model. It was found that with the source and receiver both located below the thermocline, most of the energy remained trapped between the thermocline and the seabed with an optimal propagation frequency near 500 Hz. TL calculations for the same source-receiver geometry but using a sound speed profile based on climatology gave similar results. The conclusion was that the hurricane had little or no impact on the propagation environment below the thermocline and that use of a sound speed profile based on climatology was adequate for that case.

A second case was then examined with the source below the thermocline but with the receiver above. Here, it was found that although propagation losses were somewhat greater after the hurricane, the effect was not dramatic because the basic shape of the profiles was not altered by the hurricane. However, there were significant differences between TL calculated with the measured SSPs and with the climatological profile. The weaker thermocline in the climatology allows considerably more energy to reach the receiver whereas the TL for the *in-situ* profiles is much higher, particularly for frequencies above 200 Hz. This result shows that climatology can be a poor substitute for accurate, timely SSPs for some situations, particularly for propagation above or across the thermocline.

Acknowledgements

We thank Jim Abraham and Chris Fogarty of Meteorological Services Canada for synoptic analysis of hurricane Michael.

References

1. Bobanovic, J. and Thompson, K., Estimating Three Dimensional Properties of the Coastal Ocean from Remotely Sensed. DRDC contractor report (in press 2002).
2. Hutt, D., Acoustical oceanography and satellite remote sensing, *Backscatter Magazine* published by Alliance for Marine Remote Sensing, Autumn issue (2001), pp. 32–34.
3. Osler, J., A geo-acoustic and oceanographic description of several shallow water experimental sites on the Scotian Shelf. DREA Tech. Memorandum 94/216 (1994).
4. Hutt, D., Stockhausen, J., Osler, J. and Mosher, D., Capability of Radarsat-1 for estimation of ocean surface current on the Scotian Shelf, *Proc. IGARSS02*, Toronto, Canada (2002).
5. Ellis, D., A two-ended shooting technique for calculating normal modes in underwater acoustic propagation. DREA Report 85/105 (1985).
6. Deveau, T., Enhancement to the PROLOS normal-mode acoustic propagation model. DREA Contractor Report 89/442 (1989).

HIGH-FREQUENCY ACOUSTIC PROPAGATION IN THE PRESENCE OF OCEANOGRAPHIC VARIABILITY

M. BADIEY, K. WONG AND L. LENAIN

College of Marine Studies, University of Delaware, Newark DE 19716, USA
E-mail: Badiey@udel.edu

Broadband mid-to-high frequency (0.6–18 kHz) acoustic wave propagation in shallow coastal waters (< 20 m) is influenced by a variety of oceanographic conditions. Physical parameters such as temperature and salinity as well as hydrodynamic parameters such as surface waves, tide and current can influence amplitude and travel time of signal transmissions. In this paper a unique set of simultaneous ocean and acoustic observations that reveal interesting temporal behavior of the acoustic signal and its correlation with environmental variability are presented. The temporal variations in salinity, including those induced by the semi-diurnal tides and a northerly wind event, are accurately predicted by using the measured acoustic signals and temperature profile.

1   Introduction

Environmental variability, especially in shallow coastal regions, can cause amplitude and phase variations in acoustic signal propagation over different time and space scales. The cause-and-effect relationship between fluctuations in the ocean environment and those in the received acoustic signal needs to be better understood. Although studies over the past few years have made significant progress, more work needs to be done to fully capture the complexity and variability in shallow-water environment.

In 1997 a high-frequency broadband acoustic propagation experiment was conducted at a very shallow site in the Delaware Bay. The location of the experiment (shown in Fig. 1) was chosen in an area that observations could be conducted unobstructed for long periods. During the same period, a parallel research in characterizing the oceanography of the bay was conducted at the same location. The combined results from these two concurrent observations have provided an interesting opportunity in which oceanographic features such as salinity or temperature fronts could be observed directly by using the propagation of the acoustic signals.

In this paper we first present a description of the oceanography in this shallow-water acoustic waveguide. Then the acoustic wave propagation is qualitatively related to controlling factors of the environment, in particular to the frequently observed salinity features resulting from fresh water input in coastal regions. The frequency range of probe signals was 0.6–18.0 kHz which overlaps with that of the underwater acoustic communications signals.

N.G. Pace and F.B. Jensen (eds.), Impact of Littoral Environmental Variability on Acoustic Predictions and Sonar Performance, 35-42.

Figure 1. Map of the experiment location.

2   Shallow water oceanographic features observed in Delaware Bay

Delaware Bay is a major coastal plain estuary located on the east coast of the United States. Previous studies have shown that the bay is forced by a combination of tide, wind, and river induced motion [1]. The spatial and temporal variability associated with these mechanisms can have significant implications for the transmission of acoustic signals in the bay, thus providing an opportunity to field concurrent and calibrated oceanographic characterization and acoustic propagation tests. In this section, we provide a description of oceanographic features observed during a one-week period in September 1997.

Figure 2 shows the temporal variability in wind speed and direction as well as the vertical profiles of current, salinity, and temperature at the study site. For the sake of brevity, only the longitudinal component of the current will be considered here. The temporal variation in the observed current is dominated by the semi-diurnal lunar (M2) tide. The amplitude of the M2 current (about 45 cm/s) decreases slightly with depth, but the phase of the M2 current shows little variation with depth, indicating barotropic motion from surface to bottom. The diurnal tides are also present in the Delaware Bay, but the amplitude of the principal diurnal solar tide (K1) is about an order of magnitude weaker than that of M2.

In addition to the tidal variability, the current also exhibits non-tidal variability which operates over several-day time scales. The non-tidal current has a standard deviation of only about 1.2 cm/s, even though its magnitude may exceed +/-3 cm/s at times. The non-tidal current is largely forced by winds acting either over the surface of the bay or over the continental shelf adjacent to the bay. The record-mean distribution of the current profile shows a two-layer circulation, with outflow at the surface and inflow in the lower layer.

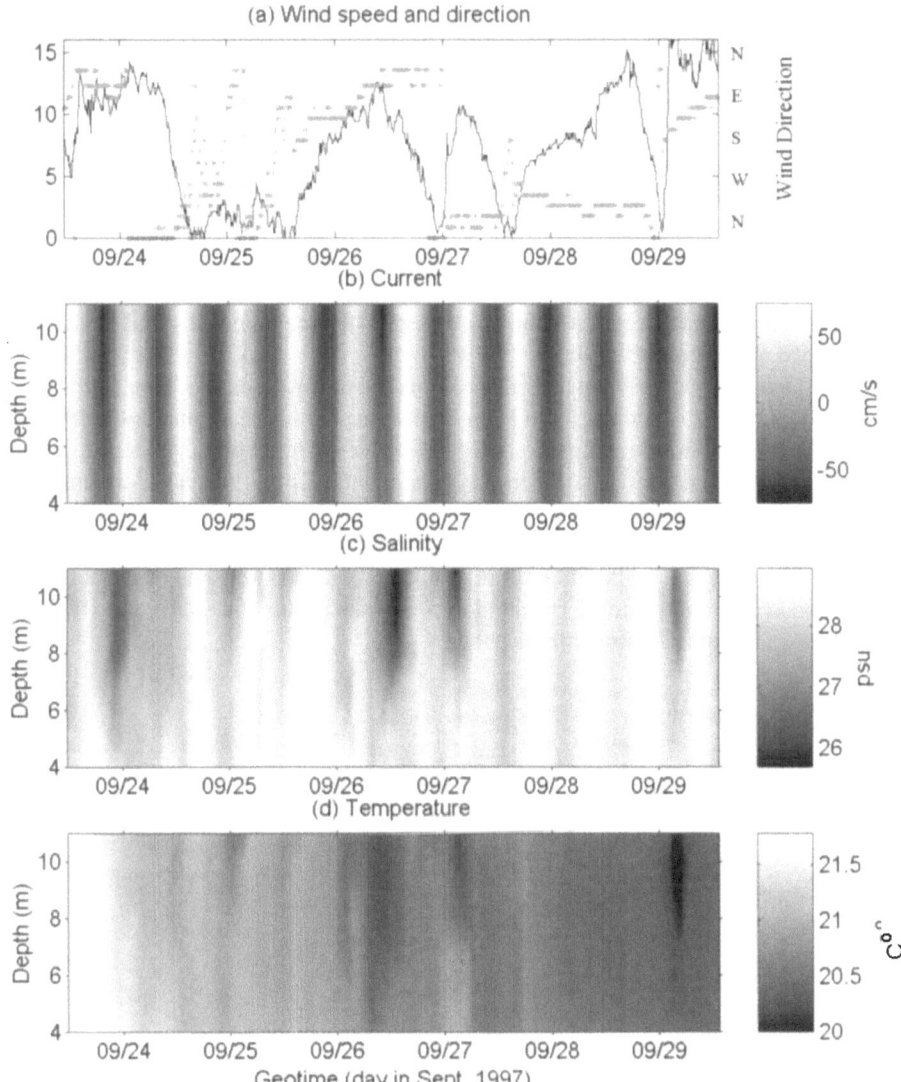

Figure 2. Oceanographic measurements in Delaware Bay during September 23–29, 1997; (a) Wind speed (solid line) and direction (dotted line) measured above the surface, (b) current profile, (c) salinity profile, (d) temperature profile.

Even though the magnitude of the mean flow and the non-tidal current variability is small, the low frequency process is important to the long-term transport and distribution of waterborne material in the bay.

The temporal variability in the current structure has a profound influence on the salinity distribution. The mean salinity distribution shows that the lower bay was weakly stratified during the study period, with a surface to bottom salinity difference of only 0.6 practical salinity units (psu). Similar to the current, the salinity also shows tidal and non-tidal variability on top of the mean distribution. Again the M2 tide dominated the salinity variability, and the semi-diurnal variation in salinity is 90 degrees out of phase with that in current, indicating that tidal advection plays a significant role in determining the salinity structure. The quadrature phase between current and salinity indicates that salinity is lowest at the time of slack water after ebb, and the highest salinity value is found at slack water after flood. The amplitude of the M2 salinity variation is largest near the surface (0.5 psu) and smallest near the bottom (0.12 psu). This indicates that the water column becomes most stratified at slack after ebb, and the water column is least stratified at slack after flood.

This ebb-flood asymmetry in the degree of stratification may be in part explained by the effect of tidal straining [2]. Given the larger tidal current amplitude near the surface than near the bottom, the upper part of the water column will experience a greater tidal excursion than the lower part of the water column over a tidal cycle. Assuming that the background longitudinal salinity gradient remains constant during the tidal cycle, the differential tidal excursion between the upper and lower parts of the water column will tend to intensify the surface to bottom salinity difference at the end of the ebb cycle. The reverse occurs at the end of the flood cycle, as the differential advection will reduce the surface to bottom salinity difference then.

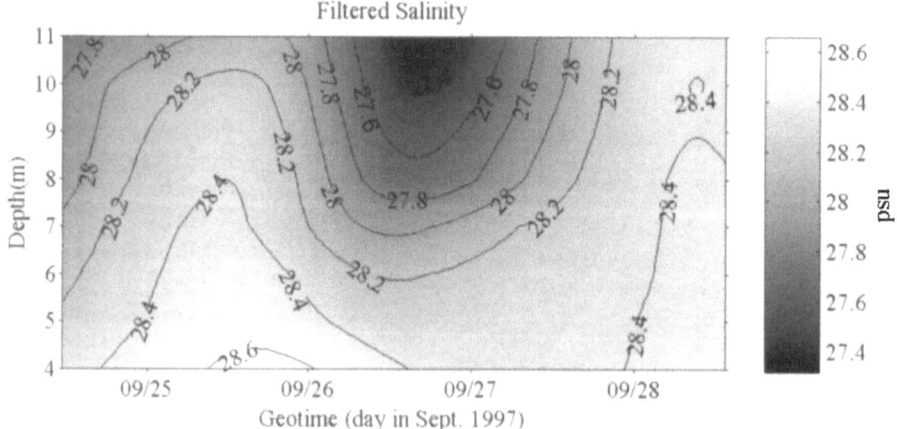

Figure 3. Non-tidal salinity variation for the week of experiment.

The data shows that salinity can undergo substantial non-tidal variation of up to 0.8 psu over 2 to 3-day time scales. This salinity variation is closely associated with wind events, such as the one on September 26 (see Figs. 2 and 3). On that day the wind was primarily from the north, and this wind caused a non-tidal advection of low salinity water from the upper part of the estuary into the lower bay. The temporal variations in salinity, both at the tidal and non-tidal time scales, can have a significant impact on the transmission of acoustic signals in the lower bay.

3   High frequency acoustic wave propagation experiment

A description of the oceanography of the experiment location was given in the previous section. To capture the physical parameter variability in the changing water column a one-week acoustic observation was conducted. Two fixed tripods, each having an acoustic source and three receiving hydrophones, were placed in 15 m of water separated by 387 m. The source was located 3.125 m above the sea floor and transmitted chirp signals over the frequency range of 0.6–18.0 kHz. The three receiving hydrophones were located at 0.33, 1.33 and 2.18 m above the sea floor, respectively.

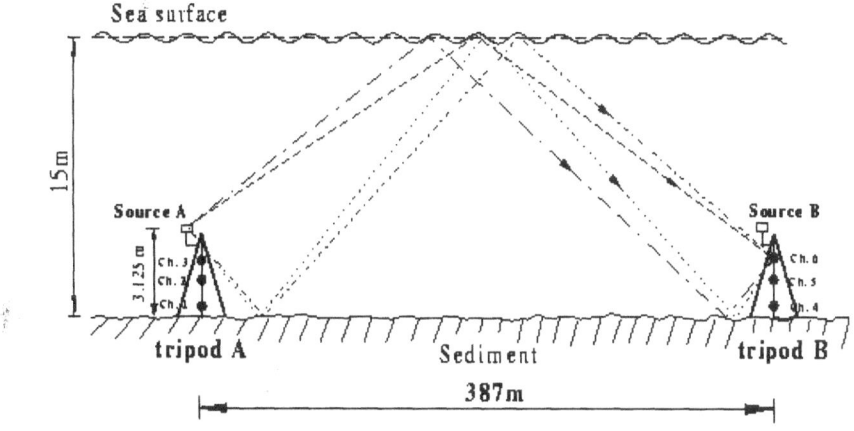

Figure 4.   Acoustic experiment set up.

To sample different scales of temporal variability in the experiment two different pulse transmission rates were used. In the first case a chirp signal was transmitted every 0.345 seconds for a five second period and repeated every ten minutes for the entire week. In the second case the same chirp signal was transmitted every 0.345 seconds for a 40-second interval and then repeated every hour for the entire duration of the experiment. In both cases the signal was received locally (by the three hydrophones attached to the same tripod as the acoustic source), as well as by the three remotely mounted hydrophone receivers (located 387 meters away). The 5-second sampling, repeated every ten minutes, was useful for studying the minutes-to-hour acoustic fluctuations due to sound speed changes in the water column, while the 40-second sampling, repeated hourly, was useful for analyzing the fast acoustic fluctuations due to surface waves. It is also noted that in both sampling cases each received signal has sufficient time to clear before the next signal arrives so that overlap between adjacent signals did not occur.

The acoustic signal following different ray paths arrives in groups of echos. The first group consisting of the energy arriving from direct and single bottom bounce paths (which arrive so close in time as to interfere and form a single peak); the second group

consisting of energy following four rays having one surface bounce. The third group
consisting of energy from four rays with two surface bounces; and so forth [4,5]. The
first group represents acoustic energy that has not interacted with the sea surface, while
all other groups consist of arrivals from rays that have one or multiple interactions with
the sea surface. Fluctuations of the first group can be correlated with variations in the
ocean current and the sound speed while fluctuations of later arriving groups also include
the effects of variations of the tide height and sea surface roughness. The detailed
reference for the group dynamics and the other analysis is found in [4,5].

For analysis here, we focus our attention on one of the ray paths that has only
interacted once with the sea surface. We are able to separate the path by beamforming at
the receiver array. Figure 5 shows for a specific geotime the beamformed signal for first
two groups of arrivals. In this figure, there is a clear and distinct arrival at approximately
$0°$ in arrival angle, corresponding to the first group having no sea-surface interaction.
After that, there are four dominant and distinct arrivals corresponding to the four ray
paths that comprise a group of arrivals, all having only a single sea-surface interaction.
We consider one of these ray paths that interacted with sea surface once and arrived
earlier than the rest (designated by dashed line in Fig. 5). Tracking this ray over the
period of observation, provides a depth averaged sound intensity that has traveled
through the upper water column.

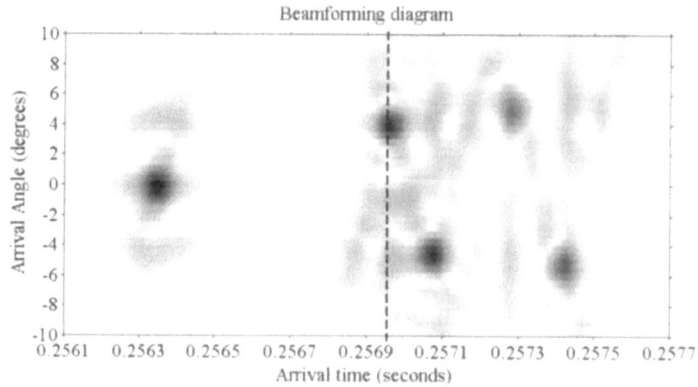

Figure 5. Acoustic pressure beamforming for a specific geotime.

4   Methodology

In this section a brief explanation of the methodology that was used to acoustically track
the aforementioned salinity feature is provided. Transmitted pulses are first beamformed
at the vertical hydrophone receiver array [5]. Then using the selected ray path and a
known source-receiver geometry, one can calculate the sound speed as $C_{calc} = R / t_a$
where R is the travel distance of the ray and t_a is the ray arrival time. The travel distance
of the ray is related to the source-receiver depth, distance, and the water depth that are all
known accurately in this experiment. C_{calc} is a depth-averaged sound speed across the

water column from the source to the sea-surface. Based on an empirical formula to calculate the sound speed from salinity and temperature profiles [3] the following equation can provide the salinity if sound speed and temperature are known

$$S(C_{calc},T)=35.0+\frac{C_{calc}-1449-4.6T+0.055T^2-0.00029T^3}{1.34-0.01T} \quad (1)$$

where the salinity S (psu), the sound speed C_{calc} (m/s), and the temperature, T(°Celsius) are all time dependent quantities. Using (1) the salinity is obtained for the entire week of the experiment from the acoustic propagation given the measured temperature profile. Figure 6 shows the comparison between the acoustically obtained values of the salinity and the direct measured values using CTD (see Fig. 2.c.) in geophysical time (geotime) for the entire week. A running averaged window can enhance the comparison results by eliminating the tidal effects (Fig. 6.b).

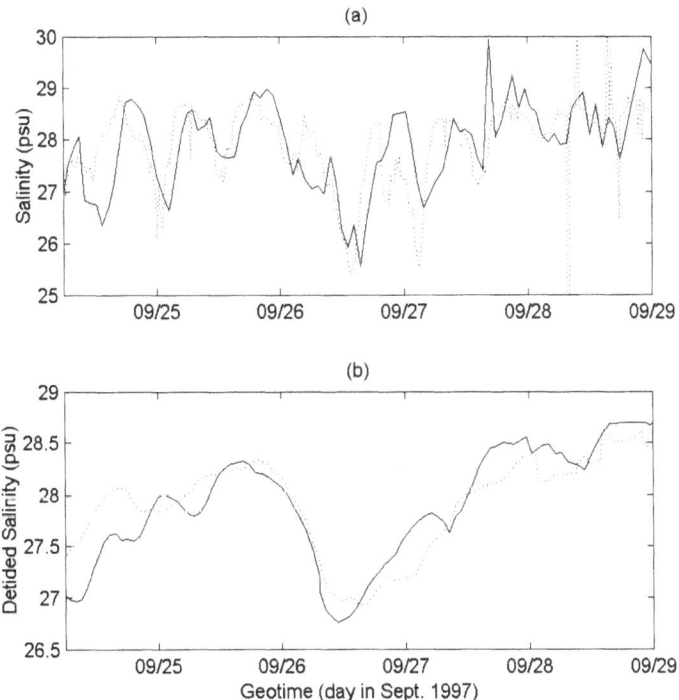

Figure 6. Comparison between predicted and measured salinity, (a) Raw data, (b) Filtered data to outline the non-tidal variations of the salinity. The solid line represents acoustically predicted value of salinity and the dotted line shows the measured value 2 m below the sea surface.

5   Summary

Concurrent oceanographic and acoustic observations were conducted in shallow water region of Delaware Bay. The purpose of these tests was to understand the correlation between the oceanographic features and the high frequency acoustic wave propagation.

Results show a direct (cause and effect) relationship between salinity and temperature changes with acoustic wave propagation in shallow waters. Separating a single ray path by beamforming technique the temporal variations of the salinity are detectable from the measured temperature and acoustic transmissions.

Acknowledgements

Authors wish to thank the crew of R/V Cape Henlopen and Captain Art Sundberg for his contribution in the experiment. Steve Forsythe helped with signal processing. This research was jointly sponsored by the Office of Naval Research and Sea Grant.

References

1. Wong, K.-C. and Garvine, R.W., Observations of wind-induced, subtidal variability in the Delaware estuary, *J. Geophys. Res.* **89**, 10589–10597 (1984).
2. Simpson, J.H., Brown, J., Matthews, J. and Allen, G., Tidal straining, density currents, and stirring in the control of estuarine stratification, *Estuaries* **13**, 125–132 (1990).
3. Mackenzie, K.V., Nine-term equation for sound speed in the oceans, *J. Acoust. Soc. Am.* **70**, 807–812 (1981).
4. Badiey, M., Simmen, J. and Forsythe, S., Frequency dependence of broadband acoustic propagation in coastal environment, *J. Acoust. Soc. Am.* **101**, No. 6 (1997).
5. Badiey, M., Mu, Y., Simmen, J. and Forsythe, S., Signal variability in shallow-water sound channels, *IEEE Journal* **25**, No. 4, 492–500 (2000).

INSTRUMENTED TOW CABLE MEASUREMENTS OF TEMPERATURE VARIABILITY OF THE WATER COLUMN

ANTHONY A. RUFFA AND MICHAEL T. SUNDVIK

Naval Undersea Warfare Center Division, 1176 Howell Street, Newport RI 02841, USA
E-mail: ruffaaa@npt.nuwc.navy.mil; sundvikmt@npt.nuwc.navy.mil

The Instrumented Tow Cable (ITC) measures the temperature variability of the water column with a spatial resolution of ½ meter along the cable and a temporal resolution on the order of 100 seconds. The ITC is a modification of a conventional steel-armored tow cable, involving the replacement of three outer steel armor wires with stainless steel tubes containing optical fibers. The overall cable diameter is unchanged, and there is otherwise no significant impact to the mechanical properties of the cable. The ITC survived all standard mechanical ruggedness tests, including long stroke cyclic bending over a 46-inch diameter sheave under tensions up to 22,500 lb. Temperature sensing is derived from Raman scattering effects in the optical fibers. In lake tests, the standard deviation against XBTs was found to be approximately 0.3 °C for repeated runs. Sea test data from the shelf-slope front south of New England mapped the location of the front within 150 m of the sea surface and showed indications of internal waves.

1   Introduction

The Office of Naval Research is funding the development of an environmentally adaptive upgrade to U.S. Navy active sonar systems under the EA89 program. This will lead to a capability to both measure and adapt to environmental variability. A key component of this effort is the Instrumented Tow Cable (ITC), an otherwise conventional steel-armored cable integrated with optical fibers, equipping it to measure the temperature of the water column with an order of magnitude improvement in both spatial and temporal resolution.

The ITC was originally implemented in the LBVDS (Lightweight Broadband Variable Depth Sonar) tow cable, a steel armored cable having a diameter of 1.6 inches. Three of the outer steel armor wires were replaced with stainless steel tubes, each containing an optical fiber. The fiber temperature is measured by processing scattered laser pulses utilizing Raman scattering effects. The optical fibers survived long stroke cyclic bending sheave mechanical tests up to 22,500 lb tension (over a 46" diameter sheave). The system supports temperature measurements every 100 seconds in time, and every ½ meter along the cable. In lake test measurements [1], the standard deviation against XBT data was 0.3 °C.

The purpose of this paper is to show the potential of the ITC for assessing environmental variability. This will primarily be done by showing evidence of internal wave activity taken from ITC sea test data.

N.G. Pace and F.B. Jensen (eds.), Impact of Littoral Environmental Variability on Acoustic Predictions and Sonar Performance, 43-48.

2   Description of sea tests

Although there are several versions of the ITC, the two sea tests discussed here were both conducted with the LBVDS tow cable, shown in Fig. 1. The first sea test took place on June 26–28, 2000 on the U.S. continental slope, South of New England (in the vicinity of 40 deg N, 71 deg W) in water depths of 200 m to 850 m. Site selection was based on previously well-sampled acoustic propagation and water column studies using instrumented tow body technologies and moorings as a part of the PRIMER experiments, conducted in 1996 by a team of researchers from WHOI, URI, and the Naval Postgraduate School [2].

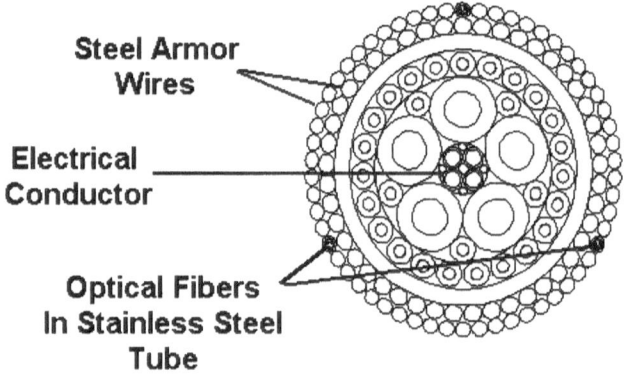

Figure 1. Cross-section of armored cable showing location of tubes containing optical temperature measurement fibers.

The cable was towed at a nominal speed of 5 knots for 24 hours, while the ITC system automatically collected temperature data every 2.5 minutes (with some gaps). During the same 24-hour period, a T-10 XBT measurement was taken approximately every fifteen minutes to provide a comparison.

The second sea test was conducted in April-May 2001 near the Hudson Canyon. Here the ITC was towed and was also deployed from a stationary ship. The data set consisted of four separate events, a shallow water tow at 3 knots along the 80-m bathymetric contour, two separate overnight deployments while the ship was at mooring, and a 7 knot tow further offshore in the vicinity of the shelf break front. Figure 2 shows the location of these measurements.

3   Data conditioning

For both sea tests, ITC and XBT data were compared as a check on the accuracy of the ITC data. ITC measurements are different from XBT measurements in two fundamental ways: (1) each ITC measurement is averaged over 100 seconds along an inclined tow cable while the ship is transiting (while XBT measurements are taken along a vertical line); and (2) the error from one ITC measurement cell to the next is independent, in contrast to XBT data [3]. The latter difference means that spatial averaging of ITC measurements will reduce the scatter or jitter. An 11-point smoothing in depth was performed, reducing the standard deviation of the error (Fig. 2). This is practical because the ½-meter spatial resolution is more than needed: at 5 knots, and with slightly over 300 meters tow cable deployed, the critical angle of the tow cable allows the water column to be sampled to a depth of 120 m. Thus, 600 measurements are made of the 120 m depth profile, or one measurement every 20 cm in depth.

Calibration of the ITC bias error was not complete at the time of the sea tests, requiring a 1.5 °C adjustment of the ITC temperature to remove the offset. This is also evident in Fig. 3

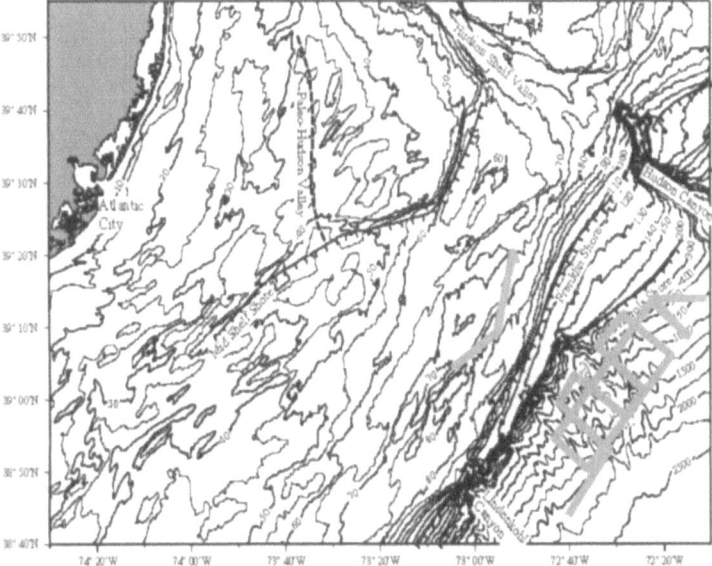

Figure 2. Location of measurements from second sea test conducted in April-May 2001. Orange lines indicate the location of the towing vessel's tracks. Moorings where measurements occurred overnight were at either end of the shallow track. The yellow box indicates the location of the New Jersey STRATAFORM geophysical survey region where very high-resolution bathymetry is available [5].

4   Results

Figures 4 and 5 are temperature plots from the first and second sea test, respectively. In both plots, the cable depth varied with tow speed, and was determined by matching appropriate ITC and XBT temperature profiles (there was no depth sensor).

Both plots appear to show strong evidence of internal waves. In Fig. 4, internal wave packets are seen in the data when the vessel was over the shelf waters, as variations in the depth of the isotherms with a period of about 10 minutes, where the dip in the isotherm is sharp. The time that the internal wave activity was most evident was also in agreement with one of the portions of the tidal cycles where solitons have previously been observed for that area [4].

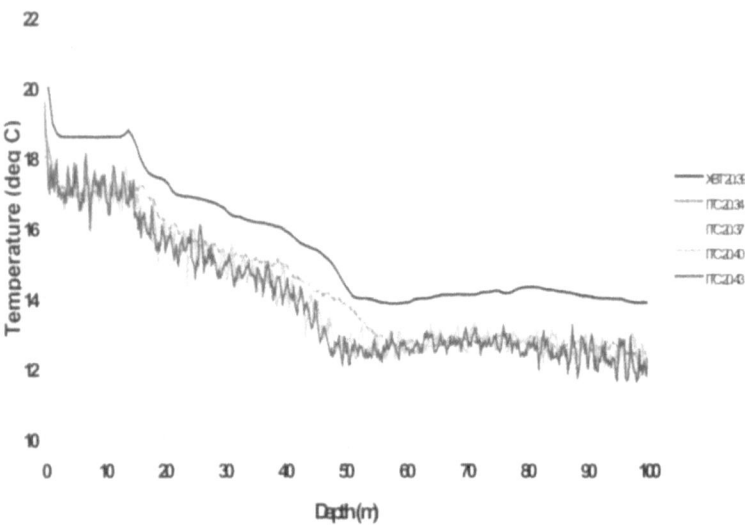

Figure 3. Comparison of XBT to ITC measurements collected nearly at the same time. Raw ITC measurements are at the bottom, the XBT at the top. As a result of these comparisons, an 11-point depth smoothing and an offset of 1.5 °C was added to the ITC measurements.

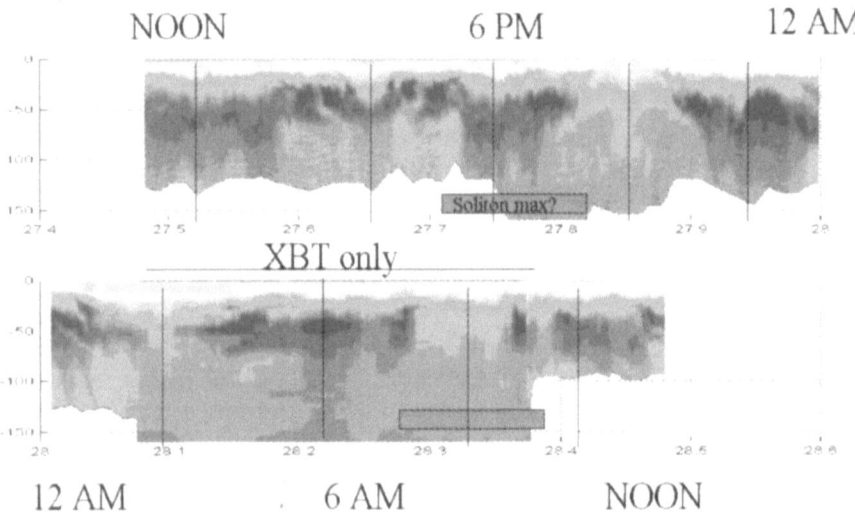

Figure 4. False color image plot of temperature from first sea test. Oscillations in the upper water column imply the presence of solitons at approximately 6 pm.

Figure 5 shows a false color plot from the second sea test. These temperature data comprise the whole of the shallow tow along the 80-meter bathymetric contour. The depth of the tow was inferred strictly from the speed of the tow, assuming a critical angle (cable was assumed to follow a straight line in the water). Therefore fluctuations in the depth of the bottom of the cable are due to variations in estimates of the ship's speed. Therefore, some error is induced by the uncertainties in ship speed as well as fluctuations in the water column itself. The average tow speed was 3 knots. Fluctuations in the depth of the thermocline are evident in the data, on the order of 1 to 5 meters in amplitude. These fluctuations are most likely temporal in nature, and are of much higher frequency than tidal fluctuations. They are most like produced by internal wave activity.

Figure 6 shows measurements taken while at anchor. These measurements have no uncertainty in the depth of the tow, as the cable was suspended vertically while on a mooring, and the currents and winds were small. (< 1 knot current, winds < 10 knots). They depict only temporal variability. Note the strong similarity in amplitude and frequency of these fluctuations to that of the slow tow. The measurements are a candidate for estimating the internal wave field at frequencies below the half the sampling frequency of the temperature measurement, or up to 1/120 Hz.

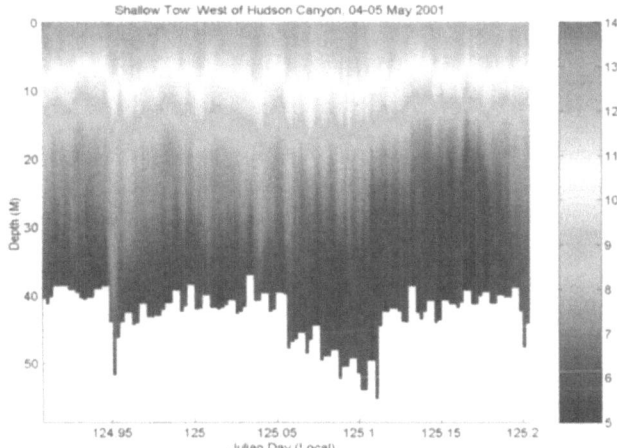

Figure 5. False color plot of temperature from second sea test. Oscillations implying presence of internal wave activity are evident.

5 Summary

The Instrumented Tow Cable has demonstrated strong evidence of measuring the temperature variability due to internal waves of the water column both in South of New England in the area of the shelf-slope front, and in the Hudson Canyon area. The ITC accomplished this as a straightforward upgrade to conventional cables commonly used by the Navy for towed systems, and all measurements can be made during ship transit. Thus, the ITC has the potential to provide a capability for directly measuring internal waves whenever the ship is in appropriate locations.

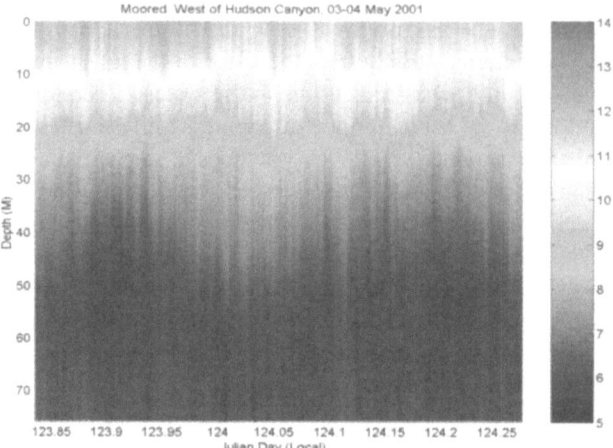

Figure 6. False color image of temperature data collected during R/V Endeavor Cruise EN-353a (second sea test) while at mooring. Colors representing degrees Celcius are shown at right side. Fluctuations are similar to those observed during the shallow tow of Fig. 5. These fluctuations are temporal indications of internal wave activity in the area.

Acknowledgement

This work is supported by the Office of Naval Research (ONR 321SS, Mr. Ken Dial) under the Environmentally Adaptive Sonar Technology and Lightweight Broadband Variable Depth Sonar Programs. The authors wish to acknowledge the helpful technical discussion with Dr. Norman Toplosky (NUWC Division Newport) in modeling the critical tow angle to obtain estimates of depth for the measurement cells. The assistance of Walter Paul (WHOI) and Jessica Mary Donnelly (MIT), and the seamanship of the crew of F/V Nobska, are also gratefully acknowledged. For the second sea test, the authors gratefully acknowledge the assistance of Darren Blier, Alyssa Cosmo, Jason Bard, the Officers and Crew of R/V Endeavor, and the data analysis and display work of Ronald Regnier.

References

1. Ruffa, A.A. and Bard, J.A.,   An instrumented tow cable for near real-time temperature measurement, *Sea Technology* **41**, 38–43 (November 2000).
2. Lynch, J., von der Heydt, K., Eck, C., Peters, D., Chiu, C.-S., Smith, K. and Miller, J., Acoustics portion of the New England shelfbreak front PRIMER experiment, http://acoustics.whoi.edu/AO/topics/Primer/Primer.html (1996).
3. Boyd, J.D. and Linzell, R.S. The temperature and depth accuracy of Sippican T-5 XBTs,   *J. Atmos. Oceanic Technol.* **4**, 128–136 (1993).
4. Brown, W.S. and Moody, J.A., Tides. In *Georges Bank* edited by Backus, R. H. and Bourne, D.W. (MIT Press, Cambridge, MA, 1987) pp. 100–107.
5. Goff, J.A., Swift, D.J.P., Duncan, C.S., Mayer, L.A. and Hughes-Clarke, J., High-resolution swath sonar investigation of sand ridge, dune and ribbon morphology in the offshore environment of the New Jersey margin, *Marine Geology* **161**, 307–337 (1999).

MESOSCALE – SMALL SCALE OCEANIC VARIABILITY EFFECTS ON UNDERWATER ACOUSTIC SIGNAL PROPAGATION

EMANUEL COELHO

SACLANT Undersea Research Centre, Viale S. Bartolomeo 400, 19038 La Spezia, Italy
E-mail: coelho@saclantc.nato.int

Naval Forces standard procedures use single sound speed profile measurements for Active Sonar Detection Range (ADR) or Active Sonar Counter Detection Range (CDR) estimates. These profile measurements are usually tasked to one ship or aircraft and support centre, which then provides regular reports, disseminated throughout the force. Rapid Environmental Assessment (REA) methodologies have been developed towards the optimization of the ADR and CDR estimation by providing oceanographic forecast data, consistent with the real conditions in the operation area. These methodologies include assimilation of the available oceanographic data into the numerical models that provide the snapshot sound speed cross-sections, which feed transmission loss models. Although the available ocean forecast schemes include a broad range of scales, they usually cannot account accurately for high frequency ocean phenomena (mesoscale to small scale). Furthermore, operationally available numerical tools cannot account accurately for 3D, non-hydrostatic phenomena, like short internal waves. To overcome this uncertainty at present, extensive oceanographic data collection is required, which is very expensive and likely it will not be feasible to obtain during a crisis scenario. The present work aims to contribute for the development of a system that can complement Naval Forces and REA procedures, by estimating ADR and CDR uncertainty due to local oceanic variability. An "Around the Ship Modeling System" is proposed to assess locally the initial phase uncertainty of the freely propagating modes and to include the effects of non-hydrostatic phenomena. This system may produce locally more accurate oceanographic field estimates and the evaluation of uncertainty error bounds on the sound speed profile estimates. These results can then be used on transmission loss Models for more robust ADR and CDR evaluation. An example is outlined based on a scale analysis on the non-linear internal waves regime of an area regularly used for Naval Exercises and a methodology is proposed for the generation of the ensemble of possible sound speed cross sections, which can be used for transmission loss variability assessment.

1 Introduction

In coastal regions the wind, tidal currents, river outflow and instabilities can force oceanographic phenomena affecting the propagation of acoustical signals at active sonar frequencies. Characteristic lengths range from meters (e.g. non-linear internal waves) to several kilometers (e.g. internal tides and mesoscale structures) with typical periods going from minutes to days. Furthermore, surface and internal mixed layers can be established, and significantly change, in the range of minutes [1], developing a homogeneous well-mixed water column and erasing all the other structures.

49

N.G. Pace and F.B. Jensen (eds.), Impact of Littoral Environmental Variability on Acoustic Predictions and Sonar Performance, 49-54.
© 2002 *Kluwer Academic Publishers.*

In deeper waters typical length and temporal scales can be expected to be larger. Though internal waves can exist, they can be expected to be less energetic and mainly forced by mixed layers depth changes and wind curl time variability. However, time dependence on the forcing mechanisms will promote an energy transfer directly towards inertial scales [2]. Therefore, inertial oscillations and quasi-inertial waves are likely to be present in any dynamical system and through non-linear interactions one can expect an energy transfer to higher frequencies and smaller scales.

In this paper, one example is introduced integrating the effects on tidally forced internal waves and non-linear high frequency internal waves in an area normally used for NATO Naval Exercises (e.g. Linked Seas 2000 and Strong Resolve 1998). Rapid Environmental Assessment (REA) strategies have been developed towards the optimization of the ADR and CDR estimation by providing forecast data, consistent with the real conditions in the operation area [3,4]. Although the used forecast schemes include all ranges of scales, they cannot account accurately for all high frequency (mesoscale to small-scale) oceanographic phenomena. For an area with high variability, like the one mentioned in the example below, this means that, though the models can provide fair energy budgets for the several scales of the oceanographic phenomena, they cannot produce accurate snapshots of the sound-speed profiles to be used into the transmission loss models.

To a certain extent this is due to an uncertainty on the initial phase of the forcing functions and to the freely propagating inertia-gravity waves resulting from the time variability of the forcing functions. Furthermore, the used numerical models are hydrostatic and therefore cannot account for steep topography effects and small-scale internal waves. To overcome this uncertainty at present, extensive oceanographic data collection is required and complex 3D models needed to be adapted. These actions will not be feasible to perform during a crisis scenario in operational and tactical timeframes. The main objective of this paper is therefore to introduce methodologies, concepts and hypothesis, which can guide future research on these topics. The proposed data analysis procedures are presented in Sect. 2. They follow the concept of "Around-The-Ship" ocean modeling, adapted and updated within tactical timeframes. Under this concept, available data from forecast models and actual observed local data is used to run simple process or feature models and fast relocatable high-resolution small domain ocean models. The results of these schemes are then used to extrapolate the local ship or other platform observations to a surrounding area within a range up to typically 10–20 km. As a process-modeling example, in Sect. 3 a test case for non-linear internal waves will be analyzed and a scale analysis performed using analytical solutions. In Sect. 4 the implementation example is discussed and concluding remarks presented.

2   Ocean variability models (NATO Tactical Ocean Modeling System)

The NATO Tactical Ocean Modeling System (NTOMS) flowchart is shown in Fig. 1. The main goal of the system is to estimate coherent oceanic fields directly from local (platform based) observations, taking into account historical data, regional models analysis and forecasts and remote sensing data. It acts as an interface between local observations and the regional ocean models, such that once new measurements are made they are directly used to update the estimates of surrounding coherent fields, that can be used for assimilation into the regional models instead of the point observations. The system intends to cover spatial scales from meters to tenths kilometers and temporal scales from minutes up to 24 hours. Depending on the scenarios, several modules and increased complexity can be set.

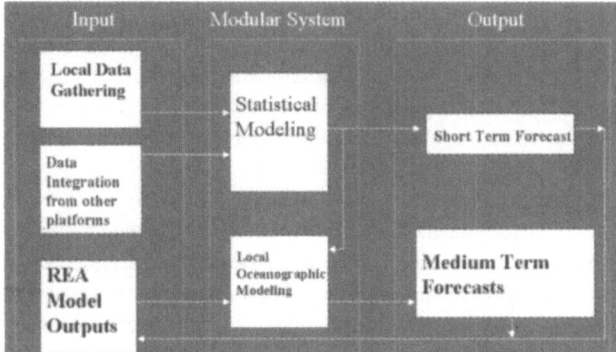

Figure 1 - The NATO Tactical Ocean Modeling flowchart.

One can expect that the coherent fields surrounding the platform will not be consistently observed. Therefore, the first step of the system consist on a statistical modeling approach which will allow the detection and classification of these coherent structures by adaptively implementing simple analytical solutions and evaluating its correlation with the along-track observed data.

Traditional oceanic forecast systems provide daily or 12 h forecasts for the next 2–4 days. As mentioned above, the model spatial and temporal resolution can be adapted accordingly to the high frequency phenomena, but free propagating waves and locally forced high frequency waves are not well predicted due to uncertainty in the initial and boundary conditions. Therefore, depending on the strategy, the model outputs should be interpreted as mean fields between the sequential forecasts or as representing one ensemble of possible realizations of the fields, within the considered period.

The second step of NTOMS aims to overcome this limitation, by producing local "Around-the-Ship", less accurate, shorter term predictions that are consistent with the along-track ship observations. These short-term forecasts are based on initial guesses feeding simple feature modeling or fast high performance relocatable models from which the predicted fields are estimated. The complexity and the feature models to be used, depend on each scenario and require an expert interpretation of the observed fields.

For the purpose of this paper, one case will be addressed. Non-linear internal wave and wind forced normal modes modules will be used for local ocean variability estimation based on observed fields and forcing conditions in an area frequently used for Naval Exercises.

3 Non-linear internal wave test case

The occurrence of non-linear internal waves is not easily predicted. Furthermore, the models that are now available for REA follow the hydrostatic approximation and therefore cannot accurately account for these features.

These waves can be responsible for significant variability at the thermocline levels, forcing isopycnic displacements of the order of 10s of meters, in the range of minutes. Surface signatures of these features can be seen through satellite SAR images [5,6] and significant differences in repeated profile measurements can suggest their occurrence [7].

Using appropriate scale analysis one can assess if these features are likely or not in a certain area and based on the background stratification and the measurement of one property like wavelength through SAR imagery, it can be possible to estimate their range of amplitudes and phase velocities. Furthermore, if they can be correlated with topographical features, one can estimate the variability using historical remote sensing and *in-situ* data, even if the local conditions do not allow for SAR imagery interpretation.

However, an accurate nowcast and forecast of these features usually requires dedicated assets and observations that are not easily performed from organic ships and available modeling is still not applicable operationally [8], though some preliminary implementations of operational schemes have been attempted in some areas.

In this work, a non-linear mode internal wave model is used, following the procedure described in Ostrovky *et al.* [9]. This model is set to run based on initial profiles estimated through XBT's or other similar instruments and then uses wavelength information interpreted either from actual or historical SAR imagery. As an alternative method, differences between consecutive or historical profiles at the same location can be used as estimates of the maximum amplitude of the non-linear internal wave oscillations. Also, it is assumed tidal currents are known.

Using this methodology, sound speed cross-sections are then estimated between two consecutive points crossing the internal wave field. These estimates can be produced for different initial phase and relative orientation and other relevant degrees of freedom determined by the geometry of the area and available data.

As an implementation example, data from the INTIFANTE'00 cruise is used, see Refs. [10,11]. During this campaign off the west coast of Portugal, simultaneous measurements of shipborn profiling systems and moorings were made superimposed to SAR imagery as described in [6]. In Fig. 2 one can see the NLIW signatures and the cross-section observed through a yo-yo CTD from the NRP *"D. Carlos"*.

 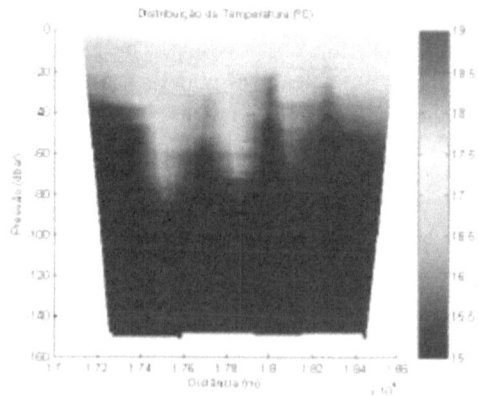

Figure 2.   Along track temperature profiles observed during the INTIFANTE'00 cruise. The profiles where observed using a yo-yo CTD from the NRP *"D. Carlos"*, while crossing the train of Non-Linear Internal Waves (NLIW), as shown in the image on the left side.

In [10] there was shown evidence of NLIW in the area slightly conditioned by the spring-neap tide cycle. Furthermore, the NLIW showed 0-70% likelihood at this site depending on the spring-neap cycle and overall is about 30% likely to occur.

Furthermore, observed temperature profiles showed vertical displacement of about 50 m to 35 m during the passage of non-linear internal waves in periods of about 3 h. The first empirical mode dominated the motion with the first eigenvalue ranging from 70% to 90%. Also, using the method developed by Inall, et al. [12], consistent transport estimates between theoretical (two layer theory) and observed (ADCP) showed maximum values ranging from 10 m²/s to 2.5 m²/s with the dispersion term not always being negligible, balancing non-linear terms, suggesting KDV theory is applicable

In Fig. 3 we have one realization using a 2-layer nonlinear internal wave model by fixing wavelength and number of waves in the train, as it can be determined by the SAR images described in [6]. This realization can be assumed to be a possible outcome of a random variable and used for transmission loss variability computations, as described above. In this example, the amplitudes of the components and the initial relative phase of each component determine the degrees of freedom.

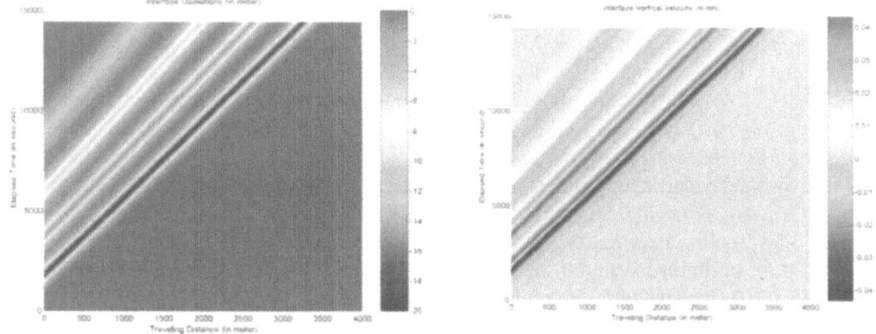

Figure 3. Realization from internal wave model.

4   Concluding remarks

At the present time the available ocean forecast tools do not allow for accurate small scale, short-term predictions. This paper proposes a methodology which, when implemented, might complement the available REA products for small areas and within short time periods. In order to develop this concept and demonstrate the improvements the NATO Tactical Modeling System (NTOMS) was presented. This system will integrate several modules starting from on-track data acquisition and processing capabilities, to small domain "Around-the-Ship", short-term, relocatable-domain numerical modeling. These tasks will be part of a SACLANTCEN project named NTOMS, starting 2003.

Acknowledgements

The author wishes to thank the Instituto Hidrografico – Portuguese Navy and the teams of the Universidade do Algarve, ENEIA and DERA, participating in the INTIFANTE00 and INTIFANTE01 for their cooperation.

References

1. Coelho, E.F. and Stanton, T.P., Tidal mixing over steep topography. In *Proc. 2nd International Conference in Air-Sea Interaction and on Meteorology and Oceanography of the Coastal Zone*, edited by A.M.S., Lisbon, Portugal, 1994.
2. Gill, A.E., *Atmosphere-Ocean Dynamics* (Academic Press, N.Y., 1982).
3. Sellschopp, J., Exercise Linked Seas 2000, SACLANTCEN CD-ROM, 2000.
4. Vitorino, J. and Monteiro, M.J., Swordfish 2001 - A study of the oceanographic conditions off Cape São Vicente (SW Portugal) using data assimilation models. 3ª assembleia Luso-Espanhola Geofisica, Valencia, Spain, 2002.
5. Jeans, D.R.G. and Sherwin, T.J., The evolution and energetics of large amplitude nonlinear internal waves on the Portuguese Shelf, *J. Marine Res.* **59**, 327–353 (2001).
6. Small, J. and Dovey, P., INTIFANTE99 oceanographic data report. Defense Evaluation and Research Agency (DERA) Report, 1999, 58 pp.
7. Rodrigues, O., Jesus, S., Stephan, Y., Coelho, E.F., Demoulin, X. and Porter, M.B., Nonlinear soliton interaction with acoustic signals: Focusing effects, *J. Comp. Acoust.* **8** (2), 347–363 (2000).
8. Lonzano, C.J., Robinson, A.R., Arango, H.G., Gandopadhyay, A., Sloan, Q., Haley, P., Anderson, L. and Leslie, W., An interdisciplinary ocean prediction system: assimilation strategies and structured data models. In *Modern Approaches to Data Assimilation in Ocean Modelling*, edited by P. Malanotte-Rizzoli (Elsevier Oceanography Series, Elsevier Science, The Netherlands, 1996) pp. 413–452.
9. Ostrovsky, L.A. and Stepanyants, Y.A., Do solitons exist in the ocean? *Rev. Geophys.* **27**, (Aug. 1989).
10. Coelho, E.F., Clemente, C., Quaresma, L., Beja, J., Caldas, J. and Marreiros, M., Submarine canyons near-inertial ocean dynamics – Recent work on the Nazare and Setubal systems. In *Canyons Workshop*, Sitges-Barcelona, April 2002.
11. Jesus, S.M, Coelho, E., Onofre, J., Picco, P., Soares, C. and Lopes, C., The INTIFANTE'00 sea trial: preliminary source localization and ocean tomography data analysis. In *Proc. MTS/IEEE Oceans 2001*, Honolulu, Hawaii, USA, 2001.
12. Inall, M.E., Shapiro, G.I. and Sherwin, T.J., Mass transport by non-linear internal waves on the Malin Shelf, *Cont. Shelf Res.* **21**, 1449–1472 (2001).

SPATIAL COHERENCE OF SIGNALS FORWARD SCATTERED FROM THE SEA SURFACE IN THE EAST CHINA SEA

PETER H. DAHL

Applied Physics Laboratory, University of Washington, Seattle, USA
Email: dahl@apl.washington.edu

Measurements of sea surface forward scattering, wind speed, and directional wave spectra made in 100 m of water in the East China Sea are discussed. The experiment was part of the Asian Seas International Acoustics Experiment (ASIAEX) conducted in May and June 2001. Signals were received at ranges near 500 m on two vertical line arrays that were co-located but separated in depth by 25 m. Estimates of the vertical spatial coherence along these arrays as a function of frequency, path geometry, and sea surface environmental conditions are compared with a model for spatial coherence. The model is based on identifying the probability density function that describes vertical angular spread at the receiver position, and requires computation of the sea surface bistatic cross section, done here with the small slope approximation. Model results for the equivalent horizontal spatial coherence are also presented. Forward scattering from the sea surface represents an important channel through which sound energy is transmitted, and spatial coherence determines in part the performance of imaging and communication systems that utilize the sea surface bounce path.

1 Introduction

Sound interaction with the sea surface is often a defining feature in shallow water, multipath propagation. With each interaction with the sea surface, the sound field may be further spread in time, frequency, and angle. Utilization of the sea surface multipath for detection, imaging, and communication purposes depends greatly on the nature of this spreading.

Here, we present new results from an experiment conducted in the East China Sea to measure the vertical spatial coherence of O(10)-kHz sound that has been forward scattered from the sea surface. Vertical spatial coherence relates, via Fourier transform, to the vertical angular spread imparted by sea surface forward bistatic scattering. To interpret the field measurements, we use a modeling approach that has been used previously to predict field measurements of horizontal spatial coherence, requiring computation of the sea surface bistatic cross section [1,2]; this approach is readily extended to modeling vertical coherence. Thus, here we shall compare measured and modeled estimates of vertical coherence and also present model curves of horizontal coherence for the same conditions and geometry.

N.G. Pace and F.B. Jensen (eds.), Impact of Littoral Environmental Variability on Acoustic Predictions and Sonar Performance, 55-62.
© 2002 *Kluwer Academic Publishers.*

2   Field experiment and measurements

The experiment was conducted from 29 May to 9 June 2001 in the East China Sea, off the Chinese continental margin 350 km east of Shanghai (29.65°N, 126.82°E) in waters nominally 100 m deep. The experiment was part of the Asian Seas International Acoustics Experiment (ASIAEX) field program for 2001, and as such consisted of a multi-tasked program of ocean acoustic and supporting environmental measurements conducted from the U.S. R/V *Melville* and the Chinese R/Vs *Shi Yan 2* and *Shi Yan 3* [3].

The vertical spatial coherence measurements were made from the R/V *Melville* using two, co-located vertical line arrays each of length 1 m, one at depth 26 m and the other at depth 52 m, referenced to the top element. The autonomous Moored Receiving Array (MORAY, Fig. 1) was deployed at a nominal range of 500 m from the acoustic source, with precise range varying somewhat because current conditions influenced the source position. Each array consisted of four elements (ITC 1042) with top-to-bottom element separation equal to 13 cm, 30 cm, and 60 cm. Data recorded on MORAY were sampled at 50 kHz and sent back to the *Melville* through an RF modem.

The acoustic source was deployed off the stern of the *Melville* at a depth of either 25 m or 50 m. Depending on frequency, one of three transducers (ITC 2010, ITC 1007, and ITC 2044) was engaged in transmission. Every 100 s, a sequence of 7 transmit pulses were sent, consisting of CW pulses of length 2 and 3 ms and center frequency between 2 and 20 kHz, plus various FM pulse forms. The sequence was repeated 20 times to obtain an ensemble of 20 pings of each pulse type for a given measurement set. Upon completion of a measurement set, the process was repeated. Two continuous measurement periods, each nominally 24 h, were carried out in order to capture environmental effects in the data. Here, we present results for three measurement sets representing differing conditions in terms of both sea state and geometry, and from these sets, the measurements made with CW pulses centered at 8 kHz and 20 kHz using the ITC 1007 source. For data interpretation both the source and receiving elements are assumed to be approximately omnidirectional.

The wind speed was measured continually using *Melville's* IMET station, and the local sea state was measured using a 0.9-m diameter TRIAXYS directional wave buoy. The buoy measured wave height variance spectra in 0.005-Hz bins from 0.3 Hz to 0.64 Hz, and in 3-degree directional bins, with spectra estimated every 0.5 h based on a 20-min averaging time. The buoy operated from 29 May to 8 June within 500 m of *Melville's* position, itself maintained by dynamic positioning, and data were sent back to the *Melville* via RF modem link. Figure 2 shows the surface waveheight spectra corresponding to the three measurement sets. The legend lists the nominal time of each measurement in UTC, rms wave height H, wind speed U, rms wave slope SL, and principal direction of the waves (direction from).

The sound speed profile was monitored with frequent CTD casts made from both the *Melville* and the nearby Chinese research vessel *Shi Yang 3*. Figure 3 (left side) shows an averaged sound speed profile representing the conditions in effect at the time of set 7, along with two individual profiles, taken before and after set 7. In this case the single profiles vary little from the mean profile. In general, however, the sound speed versus depth variation in the East China Sea is linked to the tides during late spring and summer months, and the sound speed for the upper mixed layer (nominal depth 30 m) often varied between 1525 and 1530 m/s.

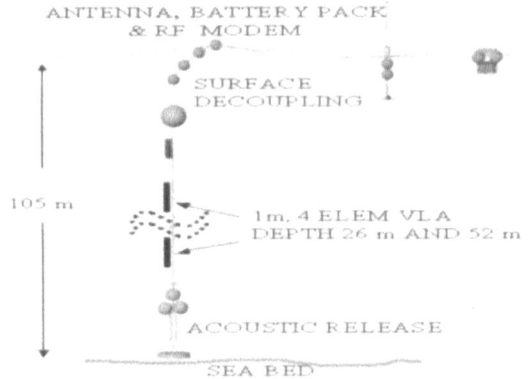

Figure 1. Diagram of MORAY vertical line array system.

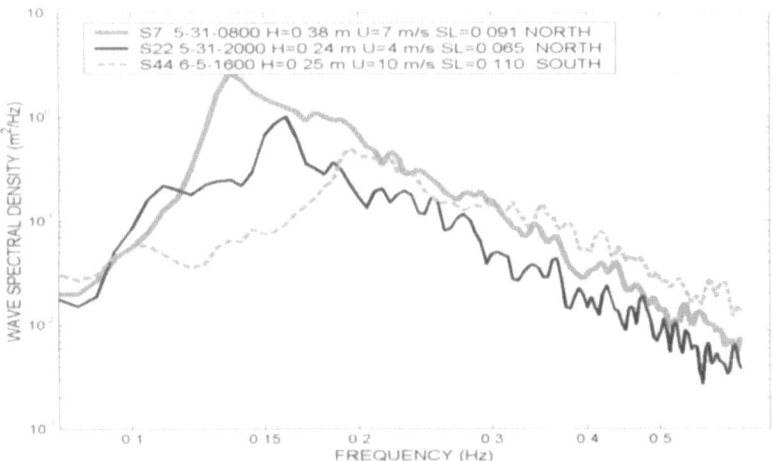

Figure 2. Surface waveheight spectra corresponding to the three measurement sets.

Propagation conditions for set 7 are shown by the ray diagram (Fig. 3, right) computed using the averaged sound speed profile (Fig. 3, left). The multipath structure of the 105-m deep channel is resolvable with the 3-ms CW pulse used in set 7, and four rays (direct, surface, bottom, surface-bottom) are shown in the ensemble average of 8-kHz data (Fig. 4). Surface-interacting paths typically displayed fluctuations consistent with a Rayleigh distribution in amplitude, e.g., the standard deviation of the log of intensity was close to 5.6 dB. In general, six ray arrivals were observed in the data for the set-7 geometry (i.e., were sufficiently energetic to be observed above the noise), and the gross channel impulse time was about 80 ms. Grazing angles for the bottom and surface-bottom bounce arrivals shown in Fig. 4 are 16.3° and 22.3°, respectively, which likely brackets the bottom critical angle. Of main in interest in this paper is the single surface bounce path, shown by the dashed line in the ray diagram of Fig. 3.

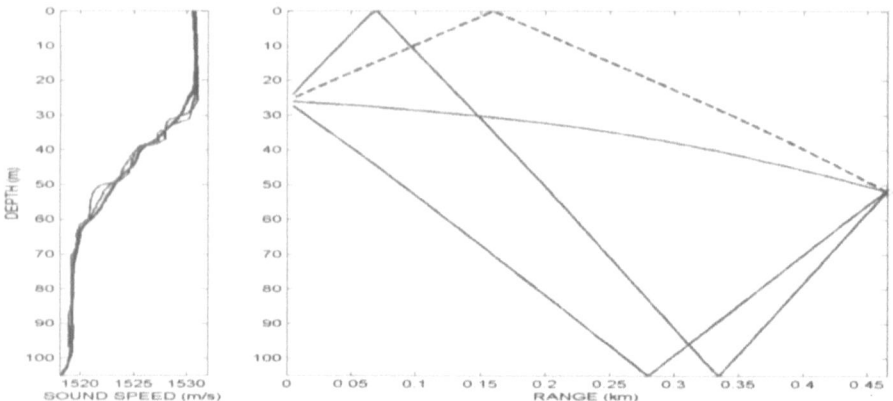

Figure 3. Left: average sound speed profile (thick line), and two sound speed profiles from single CTD casts (thin lines) taken before and after set 7. Right: ray diagram for set 7; source is on the left side at depth 26 m, and receiver on the right side at depth 52 m.

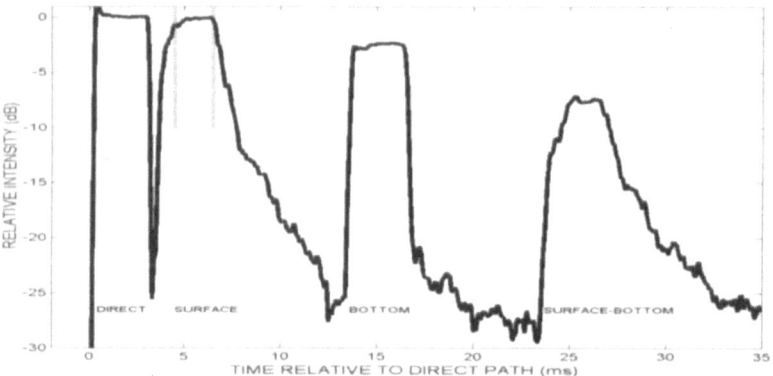

Figure 4. Ensemble averaged intensity (in dB arbitrary units) after leading edge alignment for the 8-kHz data from set 7. Vertical lines show time window within which coherence is estimated.

We estimate spatial coherence in the surface bounce path using

$$\hat{\Gamma}_{ij} = \frac{\left\langle e_i e_j^* \right\rangle}{\sqrt{\left\langle e_i e_i^* \right\rangle \left\langle e_j e_j^* \right\rangle}} \tag{1}$$

where e_i is the time-dependent complex signal (proportional to pressure) for the element i of the vertical line array. The brackets in Eq. (1) represent both a time average over the time window defined by the two vertical lines in Fig. 4 (i.e., a cross correlation of the two sample functions at zero time lag), and an ensemble average over the 20-ping ensemble. We assume that spatial coherence is stationary along the 1-m vertical array, and thus is a function only of element separation and not element position; for the four-element array, there are six non-zero spatial separations. For an approximation of the standard deviation of the coherence magnitude estimate $\left|\hat{\Gamma}_{ij}\right|$, we

use $(1-\left|\hat{\Gamma}_{ij}\right|^2)/\sqrt{n}$, with n being the number pings [4]. From numerical (bootstrap) simulation, we conclude that this same expression, multiplied by $\sqrt{2}$, can be used as an approximation for the standard deviation of the estimates for the *real* and *imaginary* parts of $\hat{\Gamma}_{ij}$.

3 Model for spatial coherence

Our model for vertical and horizontal spatial coherence is based on identifying the probability density functions (PDF) that describe the angular spread at the receiver position in vertical and horizontal arrival angle; these being $P_v(\theta_v)$ for vertical arrival angle, and $P_h(\theta_h)$ for horizontal arrival angle. The PDFs are constructed by summing the scattered intensities associated with discrete vertical and horizontal arrival angles, and normalizing the result. For a given patch of sea surface, the scattered intensity depends on the sea surface bistatic cross section, computed here with the small slope approximation [1,2].

Required to compute the bistatic cross section is an estimate of sea-surface spatial correlation function as derived from the sea-surface wavenumber spectrum. For the latter, we use data from the wave buoy for surface wavenumbers up to about 1.5 rad/m, and for the rms waveheight (Fig. 2). To fill in for higher wavenumbers not sensed by the buoy, we use a model [5] for the directional wave spectrum as a function of fetch and wind speed, which is run for fully-developed conditions in view of the O(100)-km fetch in the East China Sea. Note that we are presently evaluating new approaches to incorporate the raw directional spreading data from the wave buoy into a directional wavenumber spectrum, and ultimately into a two-dimensional sea-surface spatial correlation function. Thus, the results presented here will be based on a directionally-averaged wavenumber spectrum. This approximation has been demonstrated to be a reasonable one for frequencies of O(10) kHz [1].

The model for vertical and horizontal spatial coherence as function of wavenumber times receiver separation, or kd, is obtained upon taking the Fourier transform of the relevant PDF, equivalent to computing the characteristic function. For example, the model for vertical coherence is obtained by numerical evaluation of

$$\Gamma(kd) = \int_{-\infty}^{+\infty} P(\theta_v)\, e^{ikd\theta_v}\, d\theta_v \; . \tag{2}$$

4 Results and discussion

Figure 5 shows the theoretical PDFs for vertical arrival angle at 20 kHz corresponding to the three measurement sets. The mean value of each PDF is shown by the vertical line, and the standard deviation is given in the legend. Unlike the PDF for horizontal arrival angle, the one for vertical arrival angle is asymmetric with respect to the mean value with a positive coefficient of skewness, the value of which is between 2 and 3 for the PDFs in Fig. 5.

Figure 6 shows model curves for the absolute value (upper plots) and real part (lower plots) of vertical coherence plotted against element spacing normalized by wavelength, compared with measured values at 8 kHz and 20 kHz from the three measurement sets. (To reduce the complexity of Fig. 6, only the real part of coherence is displayed, as correspondence between model and data for the imaginary part was similar.) The plots are arranged from left to right with increasing wind speed (4 m/s, 7 m/s and 10 m/s). The effects of refraction are included in the model results; however, these tend to be small. The largest effect is in set 22 (upper left plot), where, for comparison, we have also plotted a 20-kHz model curve based on iso-velocity conditions. The downward refracting conditions tend to slightly compress the set of arrival angles at the receiver, thereby slightly increasing the vertical coherence.

In terms of the absolute value of vertical coherence, there is reasonable agreement between the model and data, although when estimates fall below about 0.5 they are encumbered with a high variance and in future work we will attempt to group data sets measured under similar conditions to reduce this variance. Estimates made at 8 kHz also stay slightly above those made at 20 kHz and become closer to those made at 20 kHz for increasing wind speed, two features which are in accord with the model.

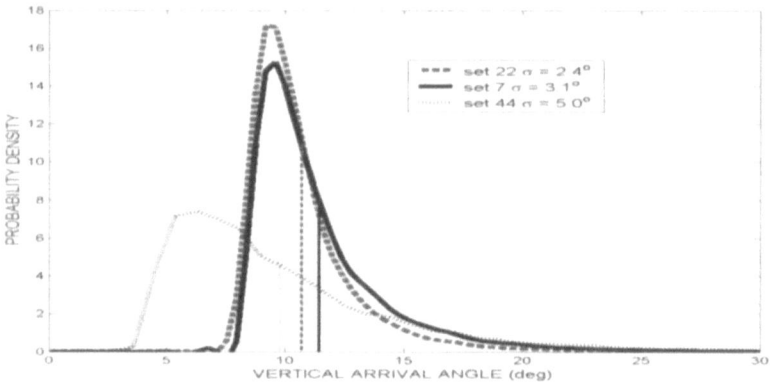

Figure 5. Model PDFs for the vertical arrival angle corresponding to the three measurement sets made at 20 kHz. Vertical lines mark location of the mean value, and the standard deviation of each PDF is given in legend. The PDFs are defined with arrival angle expressed in radians.

The standard deviations σ for the model PDFs relate to the absolute value of coherence as $\left|\Gamma(kd^*)\right| \approx 0.75$, where $kd^* = 1/\sigma$. Thus, we define the normalized receiver separation at which the absolute value of coherence reaches 0.75 as a *characteristic scale* for vertical coherence length, and its inverse gives the equivalent for vertical angular spread. The horizontal line at 0.75 across the upper plots (Fig. 6) indicates a reasonable consistency between the model and data for this key descriptor of angular spreading.

The upper plots (Fig. 6) also show horizontal coherence modeled for the same conditions and geometry (the two higher-coherence curves in each plot, without data points). In all cases, the characteristic scale for horizontal coherence length is about 5 times greater than that for vertical coherence. As discussed in [2], the characteristic scales for vertical and horizontal coherence have a geometric dependence in addition to their dependence on the sea surface environment and acoustic frequency. Specifically, vertical angular spread goes approximately as $cos(\theta_g) / (1+RD/SD)$ and horizontal angular spread goes approximately as $sin(\theta_g) /(1+RD/SD)$ where SD and RD are, respectively, source depth and receiver depth, and θ_g is a characteristic grazing angle. Thus, in terms of coherence length scales, the ratio of horizontal to vertical scale is approximately $cot(\theta_g)$, or about 5 when $\bar{\theta}$ from the PDFs is used as θ_g.

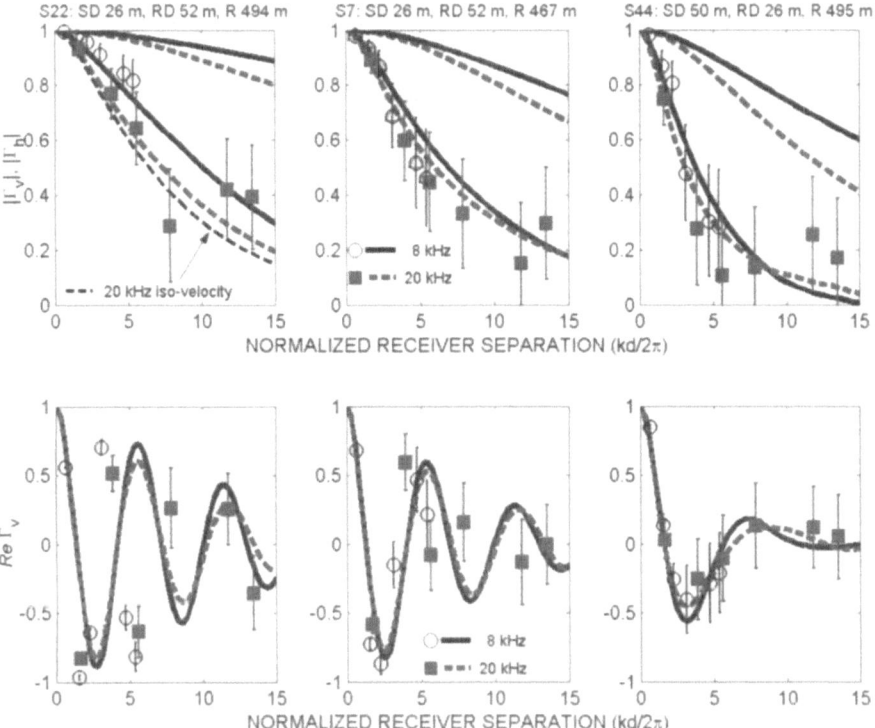

Figure 6. Comparison of measured values and model curves for the absolute value (upper plots) and real part (lower plots) of vertical spatial coherence versus normalized receiver separation at a frequency of 8 kHz (circle, solid line) and 20 kHz (square, dashed line). Results from the three measurement sets are displayed from left to right with increasing wind speed (4 m/s, 7 m/s, 10 m/s), and key geometric variables are listed in the upper plot corresponding to each set. The horizontal line across the upper plots intersects the point at which the absolute value of coherence reaches 0.75. The three upper plots also show model curves for the absolute value of horizontal coherence for the same conditions, geometry, and frequencies. The upper left plot shows an additional model result for 20 kHz based on iso-velocity conditions (thin, dashed line).

In terms of the phase of vertical coherence, the real and imaginary parts of vertical coherence are strongly influenced by the non-zero mean value of $P_v(\theta_v)$, or $\bar{\theta}_v$, an angle that is slightly greater than the nominal specular grazing angle. Were $P_v(\theta_v)$ to be symmetric about its mean value, then the real part of coherence would be $cos(kd\overline{\theta}_v)$ modulated by its absolute value. The PDF's skewness as seen Fig. 5 breaks this simple relation. However, as both the cosine and the PDF are smooth functions near $\overline{\theta}_v$, we can approximate the real part of the vertical coherence function up to and including the first zero-crossing by,

$$Re\,\Gamma \approx cos(kd\overline{\theta}_v)\left(1-(kd\sigma)^2/2\right) \qquad (3)$$

and the first zero-crossing for the real part is given by $kd\overline{\theta}_v = \pi/2$. In terms of this comparison, both model and data in the lower plots of Fig. 6 are in reasonable agreement. Model-data agreement for the entire real part of vertical coherence is very good for set 44 (right), is less satisfactory for set 7 (center), and is marginal for set 22 (left). The reason for the poor agreement with the real part of the coherence data for set 22 is unknown. Array tilt is a possible but unlikely cause. However, directivity in the sea surface waves may be influencing higher moments of the PDF for vertical arrival angle, and we shall be investigating this effect in future work.

The rms long-wave slope of the sea surface, or SL, is a key ocean environmental parameter governing the predictability of spatial coherence in the surface multipath. Estimates of SL from the wave buoy alone (Fig. 2) are necessarily low as they are based on limited sea surface wavenumber support. Still, the buoy estimates provide a useful correlate to the *effective SL* for surface forward scattering [2]. The increase in coherence between set 7 and set 22 was in large part due to the reduction in SL. However, the decrease in coherence between set 7 and set 44 was in part due to the change in grazing angle. The downward-refracting sound channel can increase vertical coherence by way of compressing the set of arrival angles, although this effect cannot be confirmed with the data analyzed thus far, owing to the high variance in the coherence estimates.

Acknowledgements

This study was funded by the Office of Naval Research Code 321 Ocean Acoustics Program via Contract No. N00039-91-C-0072. I wish to thank Chris Eggen of APL-UW for programming assistance, and Prof. Halvor Hobæk and the University of Bergen, Department of Physics, for their hospitality extended to me during my stay in Bergen during which much of this paper was written.

References

1. Dahl, P.H., On bistatic sea surface scattering: Field measurements and modeling, *J. Acoust. Soc. Am.* **105**, 2155–2169 (1999).
2. Dahl, P.H., High-frequency forward scattering from the sea surface: The characteristic scales of time and angle spreading, *IEEE J. Oceanic Eng.* **26**, 141–151 (2001).
3. Dahl, P.H., ASIAEX, East China Sea, cruise report of the activities of the R/V *Melville* 29 May to 9 June, 2001. APL-UW TM 7-01, July 2001.
4. Kendall, M.G. and Stuart, A., *The Advanced Theory of Statistics*, Vol. 1, 2nd ed. (Charles Griffen & Company, London, 1963) p. 236.
5. Plant, W.J., A stochastic, multiscale model of microwave backscatter from the ocean, *J. Geophys. Res.* (in press, 2002).

VARIABILITY IN HIGH FREQUENCY ACOUSTIC BACKSCATTERING IN THE WATER COLUMN

A.C. LAVERY, T.K. STANTON AND P.H. WIEBE

Woods Hole Oceanographic Institution, 98 Water Street, Woods Hole MA 02543, USA
E-mail: alavery@whoi.edu

High-frequency acoustic backscattering in the water column is highly variable in both space and time. We present selected results from a program designed to address the origin of this variability. There are many naturally occurring processes in the water column, of both physical and biological origin, that give rise to acoustic backscattering. The naturally occurring spatial and temporal variability of these physical and biological processes contribute significantly to variability in acoustic backscatter. In addition, there is uncertainty associated with identifying and obtaining high-resolution information of the physical and biological parameters that contribute to volume scattering. Uncertainty in predicting volume scattering also arises from possible inaccuracies of the scattering models, as well as variability due to speckle. Emphasis is given here to identifying the model parameters with the highest degree of uncertainty.

1 Introduction

High-frequency acoustic scattering instruments can be used to rapidly survey large regions of the ocean interior. The resultant data provide high-resolution synoptic information regarding the spatial and temporal distribution of the physical and biological processes that give rise to scattering (e.g., suspended sediments, bubbles, microstructure, zooplankton, and fish). It is generally observed that the scattering is highly variable in both space and time. One source of variability is the inherent speckle that arises from the intrinsic randomness caused by summing multiple echoes with random phases. Another significant source of variability arises from the spatial and temporal variability of the physical and biological properties of the water-column. This naturally occurring variability can lead to correspondingly large uncertainties in predicting volume scattering. Furthermore, there is also uncertainty associated with 1) identifying all the processes that give rise to volume scattering, 2) a general lack of accurate high-resolution information of the physical and biological parameters that contribute to volume scattering, and 3) the possible inaccuracy of the models (and associated input parameters) available for predicting volume scattering.

In this paper, we present a selection of data from a decade-long program that addresses many of the issues that lead to uncertainty in predicting acoustic volume backscattering (S_V). This program has involved significant model development, laboratory measurements, and field measurements [1]. A key component to this program is the towed sensor platform BIOMAPER-II (Bio-Optical Multi-frequency Acoustic Physical and Environmental Recorder) designed to acquire high-resolution multi-

N.G. Pace and F.B. Jensen (eds.), Impact of Littoral Environmental Variability on Acoustic Predictions and Sonar Performance, 63-70.

frequency (43, 120, 200, 420, and 1000 kHz) acoustic backscattering data together with biological and environmental information to assist in predicting volume scattering [2]. There are two identical sets of transducers mounted on BIOMAPER-II, one set facing upwards and the other set facing downwards, designed so that full coverage of the water column is possible even with the platform at depth. This system can be towed at constant depth or undulated up and down (tow-yo fashion) through the water column. A video plankton recorder (VPR) is mounted on BIOMAPER-II in order to obtain high-resolution video images of small biological organisms present in the water column. Valuable information regarding the orientation, size, and distribution of different organisms (relative to the acoustics) can be obtained from the VPR [3,4]. Depth, temperature, and conductivity, are also measured continuously (at 1/4 Hz). Together with MOCNESS [5] net tows to acquire detailed information on species composition and size, and CTD (conductivity, temperature, and depth) profiles, it possible to achieve a high level of ground truthing, particularly for the biological component of the water column. We present data and model predictions for a shallow water coastal region: the Gulf of Maine and the waters over Georges Bank (off Cape Cod, USA).

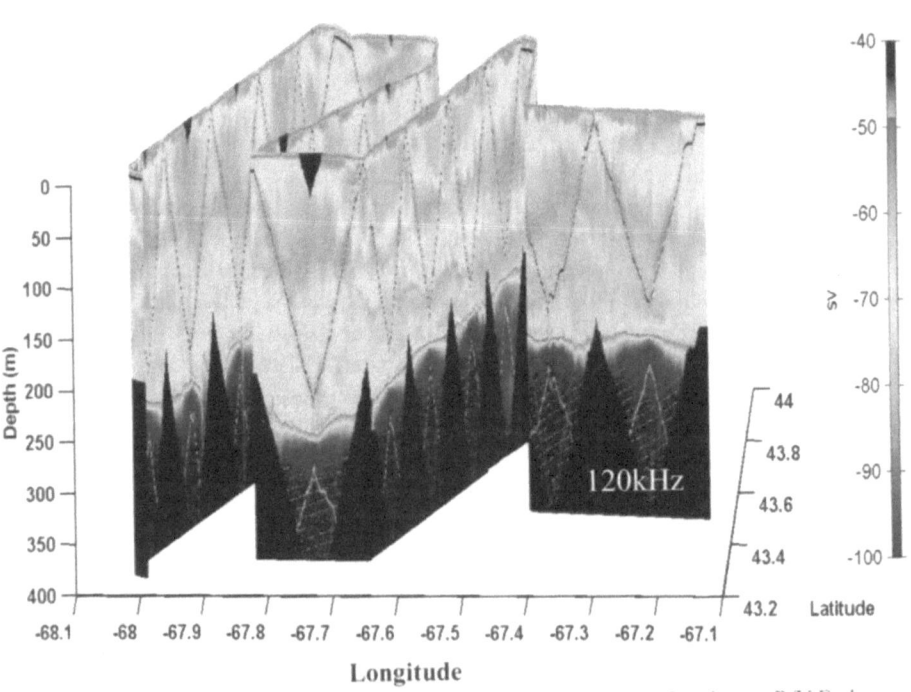

Figure 1. Acoustic scattering at 120 kHz in Jordan Basin in the Gulf of Maine on R/V Endeavor cruise 331 (December 1999). The BIOMAPER-II tow-yo track is apparent.

2   Spatial and temporal variability of the physical and biological processes in the water column

The naturally occurring spatial and temporal variability of the physical and biological processes in the water column can give rise to very significant levels of variability in volume backscattering data. For example, vertical variability at the spatial scale of a few hundred meters and on the temporal scale of a day occurs due to the vertical migration of zooplankton (Fig. 1). Superimposed on this is horizontal variability due to the size of zooplankton patches, which can extend up to many kilometers [6,7]. Variability close to the surface arises from scattering from bubbles due to breaking waves, with vertical scales of a few tens of meters. In addition, we have observed elevated scattering levels at depths that correspond to the location of the thermocline. Physical processes such as internal waves can also give rise to elevated scattering levels (Fig. 2), with vertical scales set by the amplitude of the wave and temporal scales set by the period of the internal wave. It remains uncertain if the elevated scattering levels observed in the vicinity of internal waves (and other physical processes) is a result of scattering from the physical process itself (due to changes in the acoustic impedance), or biological organisms acting as passive tracers. In fact, identifying all the possible processes that give rise to scattering is a challenging problem (Fig. 3). As a consequence of the complex nature and variable spatial and temporal scales of these processes, in order to interpret acoustics data collected with BIOMAPER-II we have gathered significant quantities of high-resolution ground truthing information with the VPR, MOCNESS, and CTD casts.

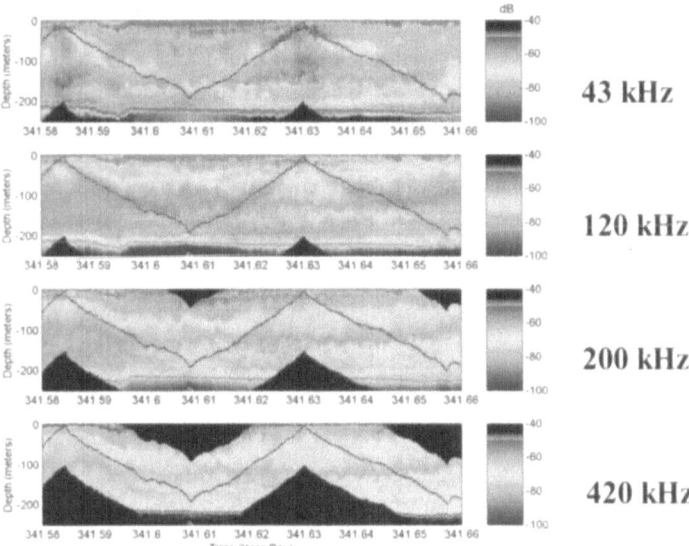

Figure 2. Acoustic scattering versus time (year-day) at 43, 120, 200, and 420 kHz, from a section in Jordan Basin. An internal wave can be seen at approximately 100 m, around the depth of the thermocline. There are two layers of elevated scattering associated with the internal wave.

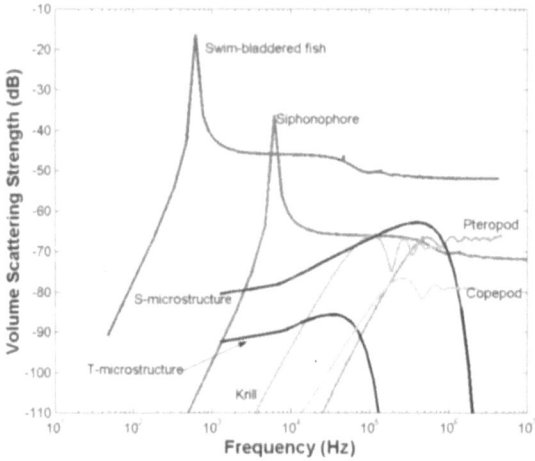

Figure 3. Comparison of model predictions for a number of physical and biological sources of scattering. The contribution to scattering from gas-bearing organisms (or bubbles) is expected to be very significant over a broad frequency range, but particularly at frequencies close to the resonance frequency.

3   Model accuracy and uncertainty in model parameters

Acoustic scattering models, in combination with appropriate ground-truthing information, are critical to the interpretation of scattering data. Uncertainty regarding the accuracy of the models can be assessed by comparison of model predictions to measurements performed in controlled laboratory experiments. The scattering models we have developed for zooplankton over the last decade have become increasingly more accurate, and have been rigorously tested in controlled laboratory experiments.

To understand the variability in acoustic scattering in the water column, it is also necessary to identify the model input parameters with the highest degree of uncertainty. Scattering from zooplankton is highly complex and depends on parameters such as the shape, size, orientation, material properties, and acoustic frequency. Since zooplankton communities are typically very diverse, in order to simplify model development they have been categorized into three groups according to their general scattering characteristics [8]: fluid-like (e.g. euphausiids and copepods), gas-bearing (e.g. siphonophores), and elastic-shelled (e.g. pteropods). Fueled by the naturally high abundances and general importance of certain species of fluid-like zooplankton, much of the modeling effort has been directed towards animals in this category [9]. Many of these models make simplifying assumptions regarding the body shape and size. To address this, we are currently in the process of developing scattering models for a number of animals typically found in the water column that make use of high-resolution computerized tomography (CT) to ascertain the shape and size of the body exterior (Fig. 4). We have compared the predictions of a scattering model, based on the distorted wave Born approximation (DWBA) with 3D CT measurements of animal shape as input, for decapod shrimp (which are weak-scatterers with fluid-like material properties) to measurements of live individual (and aggregations of) decapod shrimp, with reasonable success (Fig. 5). We have found that the target strength of an individual animal on a ping-by-ping basis depends very strongly on the angle of orientation. For decapod shrimp, our scattering model reproduces the data at broadside scattering better than at angles close to end-on incidence. Typically, volume scattering averages over many animals with many different orientations, reducing the effects due to the acute dependence on angle of orientation. However, to accurately model volume scattering it is still necessary to obtain information on the distribution of animal orientations in the water column, for example, through the use of the VPR. For fluid-like zooplankton, we have found that the scattering is also highly dependent on the material properties. Changes of only a few percent in the sound speed and density contrasts can lead to changes in volume scattering strength of up to 15 dB [10]. In addition, there is scant information available as to the 3D distribution of material properties within the body interior. Uncertainties in animal orientation and material properties are the leading source of uncertainties in predicting volume scattering for fluid-like zooplankton. For fish, uncertainties in the shape and orientation of the swim-bladder lead to significant uncertainty in predictions of volume scattering strengths.

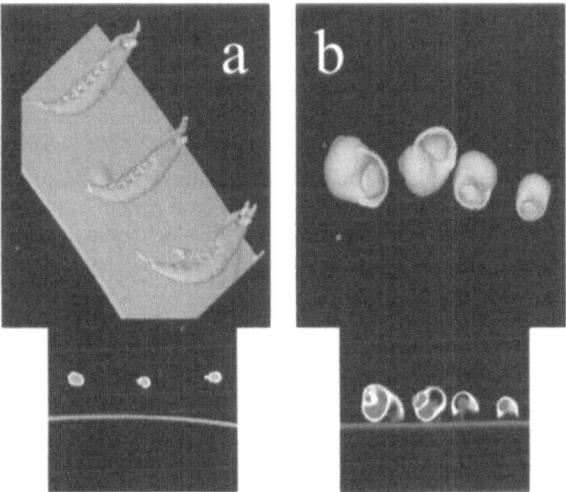

Figure 4. High-resolution measurements of animal shape obtained from CT scans: (a) fluid-like Antarctic krill and (b) elastic-shelled periwinkles. The top images in each panel show the 3D reconstruction of the outer boundary and the bottom images show representative cross-sectional slices of the animals.

We have also developed an acoustic scattering model, and performed laboratory experiments, for scattering from turbulent microstructure [11]. Our model includes contributions from fluctuations in the both the index of refraction and density. The input parameters for the model include: 1) the temperature spectrum, the salinity spectrum, and their co-spectrum, 2) the dissipation rates of turbulent kinetic temperature, ε, and temperature variance, χ, and 3) the acoustic frequency. We have found that our scattering model depends very sensitively on the values of ε and χ, which are difficult to measure without a microstructure profiler. We expect uncertainties in these parameters to significantly affect our predictions of volume scattering. It is also expected that other types of microstructure, such as salt fingers, may give rise to scattering.

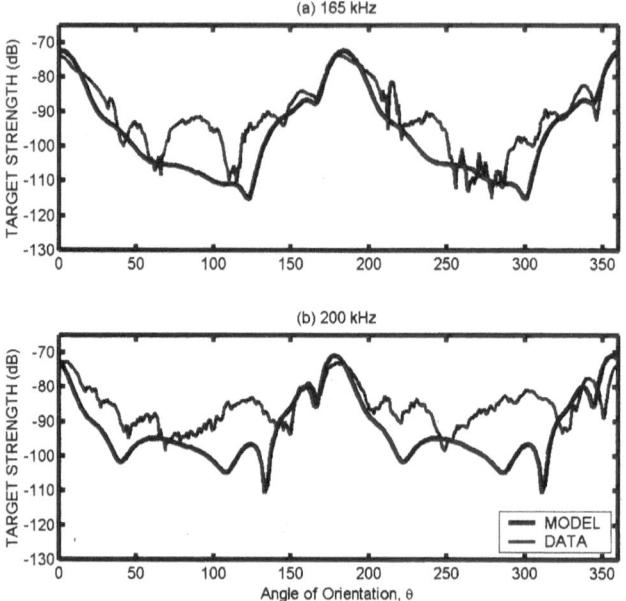

Figure 5. Comparison of model predictions and laboratory measurements, as a function of orientation, for backscattering from live individual decapod shrimp at (a) 165 kHz and (b) 200 kHz. The thin solid line corresponds to data, and the thick solid line to the DWBA-based scattering model that uses high-resolution 3D CT measurements of animal shape [12].

4 Comparison of model predictions and acoustic scattering data

We have used the data obtained from MOCNESS tows and CTD profiles to predict acoustic volume scattering. BIOMAPER-II is typically towed at the surface during MOCNESS or CTD casts, so it is possible to compare the acoustics data to model predictions with the model input parameters and acoustics data collected almost coincidentally in space and time. Scattering predictions using the models we have developed for microstructure and fluid-like and elastic-shelled zooplankton are compared to the acoustics data in Fig. 6 for a MOCNESS tow performed in Jordan Basin. The well-known fluid-sphere solution to the wave equation was used to predict scattering from gas-bearing zooplankton. The acoustics data at each depth were averaged over the time period it takes to perform the net tow. Detailed analysis of the net tow revealed that small copepods were numerically the most abundant. However, the contribution to scattering from siphonophore gas-inclusions, called pneumatophores, is predicted to dominate above microstructure and all other zooplankton contributions combined.

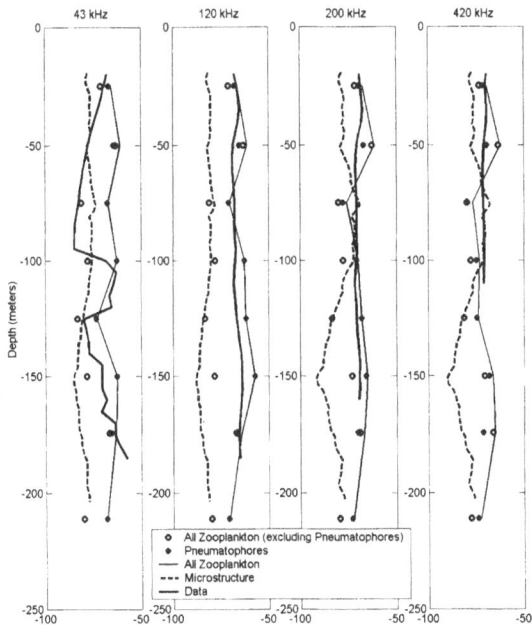

Figure 6: Comparison of model predictions and data at 43, 120, 200, and 420 kHz for a MOCNESS tow performed in Jordan Basin in the Gulf of Maine.

5 Synthesis

We have presented a selection of data that illustrate the high degree of variability in high-frequency acoustic backscattering in the water column. We have found that in order to understand the origin of the variability it is necessary to acquire significant quantities of high-resolution ground-truthing information about the physical and biological processes that occur in the water column, at relevant spatial and temporal scales. We are currently developing a new generation of scattering models for a selected number of important fluid-like and shelled zooplankton that make use of high-resolution measurements of animal shape and size obtained from CT scans. These models are used for the interpretation of the scattering data. We have found that for fluid-like zooplankton, uncertainties in the orientation and material properties give rise to the largest uncertainty in predicting volume scattering. We have also developed a scattering model for turbulent oceanic microstructure that includes fluctuations in both the density and index of refraction. For microstructure, lack of high-resolution information on dissipation rates of turbulent kinetic energy and temperature variance are the largest cause of uncertainty in predicting volume scattering. We have compared model predictions to acoustics data obtained at selected locations in the Gulf of Maine, and are currently working on mapping the basin-wide contribution to scattering from zooplankton versus microstructure.

Acknowledgements

We thank Mark Benfield, Chuck Greene, and Joe Warren for their invaluable assistance in collecting, analyzing, and interpreting, the acoustics and VPR data. We also thank Nancy Copley for analyzing the MOCNESS data. This research was supported by the United States Office of Naval Research (ONR), National Science Foundation (NSF), National Oceanic and Atmospheric Administration (NOAA), and Woods Hole Oceanographic Institution (WHOI). This is Woods Hole Oceanographic Institution Contribution Number 10703.

References

1. Stanton, T.K., From acoustic scattering models of zooplankton to acoustic surveys of large regions. In *Proc. IEEE Colloquium on Recent Advances in Sonar Applied to Biological Oceanography* (London, UK, 1998).
2. Wiebe, P.H., Stanton T.K., Greene, C.H., Benfield, M.C., Sosik, H.M., Austin, T., Warren, J.D. and Hammar, T., BIOMAPER-II: an integrated instrument platform for coupled biological and physical measurements in coastal and oceanic regimes, *IEEE J. Oceanic Eng.* (in press 2002).
3. Benfield, M.C., Davis, C.S. and Gallager, S.M., Estimating the *in situ* orientation of *Calanus finmarchicus* on Georges Bank using the Video Plankton Recorder, *Plankton Biol. Ecol.* **47**, 69–72 (2000).
4. Benfield, M.C., Lavery, A.C., Stanton, T.K., Wiebe, P.H. and Greene, C.H., Distributions of physonect Siphonulae in the Gulf of Maine and their potential as important sources of acoustic scattering, *J. Limnology and Oceanography* (submitted 2002).
5. Wiebe, P.H., Morton, A.W., Bradley, A.M., Backus, R.H., Craddock, J.E., Cowles, T.J., Barber, V.A. and Flierl, G.R., New developments in the MOCNESS, an apparatus for sampling zooplankton and micronekton, *Marine Biology* **87**, 313–323 (1985).
6. Wiebe, P.H. and Greene, C.H., The use of high frequency acoustics in the study of zooplankton spatial and temporal patterns. In *Proc. NIPR Symp. Polar Biol.* 7, 133–157 (1994).
7. Wiebe, P.H., Stanton, T.K., Benfield, M.C., Mountain, D.G. and Greene, C.H., High-frequency acoustic volume backscattering in the Georges Bank coastal region and its interpretation using scattering models, *IEEE J. Oceanic Eng.* **22**, 445–464 (1997).
8. Stanton, T.K., Wiebe, P.H., Chu, D., Benfield, M.C., Scanlon, L., Martin, L. and Eastwood, R.L., On acoustic estimates of zooplankton biomass, *ICES J. Mar. Sci.* **51**, 505–512 (1994).
9. Stanton, T.K. and Chu, D., Review and recommendations for the modeling of acoustic scattering by fluid-like elongated zooplankton: euphausiids and copepods, *ICES J. Mar. Sci.* **57**, 793–807 (2000).
10. Chu, D., Wiebe, P.H. and Copley, N., Inference of zooplankton material properties from acoustic and resistivity measurements, *ICES J. Mar. Sci.* **57**, 1128–1142 (2000).
11. Lavery, A.C., Schmitt, R.W. and Stanton, T.K., High-frequency acoustic scattering from turbulent oceanic microstructure: the importance of density fluctuations, (in preparation).
12. Lavery, A.C., Stanton, T.K., McGehee, D.E. and Chu D., Three-dimensional modeling of acoustic backscattering from fluid-like zooplankton, *J. Acoust. Soc. Am.* **111**, 1197–1210 (2002).

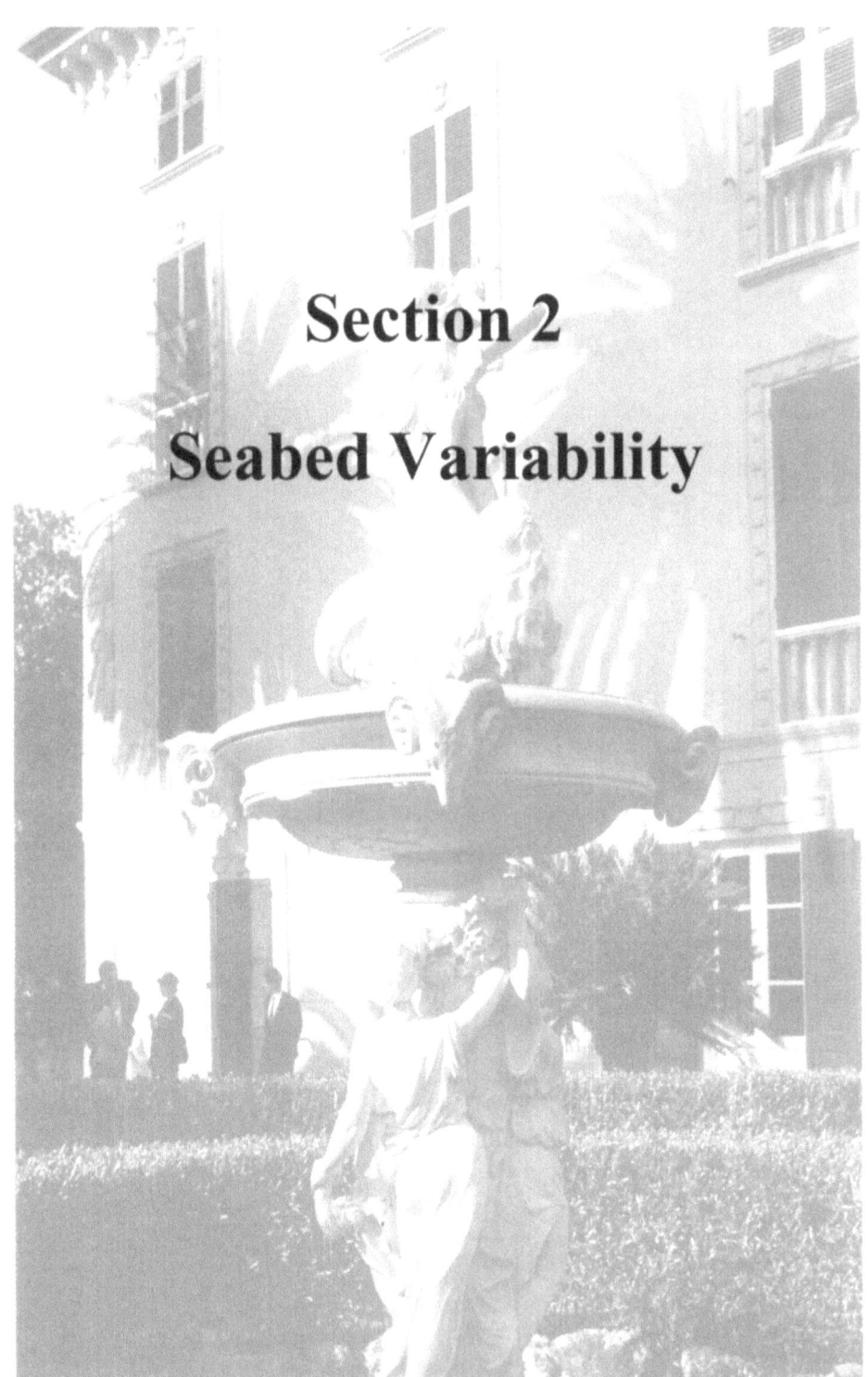

Section 2

Seabed Variability

INTRA- AND INTER-REGIONAL GEOACOUSTIC VARIABILITY IN THE LITTORAL

CHARLES W. HOLLAND

Pennsylvania State University, Applied Research Lab, PO Box 30, State College, PA 16804
E-mail: cwh10@psu.edu

Spatial variability of seabed geoacoustic properties generates uncertainty in the prediction of sonar system performance in littoral regions. In order to investigate geoacoustic variability within a region and variability between two regions, extensive geoacoustic and acoustic measurements were conducted in two areas in the Italian littoral. While it is not surprising that significant geoacoustic and acoustic variability is observed in each region, the surprising result is the marked similarity between the two regions. Geoacoustic properties were quite consistent when compared at commensurate water depths even with inter-region separation in excess of 800 km. Not only was the variability comparable between the two regions, but the geoacoustic regimes themselves were similar. The similarities have important implications for extrapolation of sparse geoacoustic data.

1   Introduction

Active and passive sonar systems that operate in littoral regions must contend with significant spatial variability of the ocean acoustic environment. An important and often dominating component of the variability is the seabed, or the geoacoustic properties. For geoacoustic point (e.g., cores) or line (e.g., propagation loss) measurements the concomitant question is: "how do the geoacoustic properties vary a few meters, kilometers, or tens of kilometers away from the measurement point?" In the extreme, the question becomes, "how can geoacoustic properties be estimated or extrapolated from very distant (hundreds of km) measurements, i.e., beyond the boundaries of the physiographic region?"

These two questions, the former termed intra-regional variability and the latter inter-regional variability, are addressed here from a measurements-based approach. Two study regions were selected: one on the Tuscany shelf and one in the Straits of Sicily (see Fig. 1). Each region has dimensions of order 50x50 km and the regions are separated by ~800 km. Limitations of a measurement-based approach are that state-of-the-art measurement techniques under-resolve some of the scales of geoacoustic variability and provide estimates of variability only for specific regions. However, the strength of a measurements-based approach is that very little is understood about inter-regional and intra-regional geoacoustic variability. Thus, measurements are badly needed to begin to bound the scales of variability and to identify potential techniques for describing and predicting the variability.

N.G. Pace and F.B. Jensen (eds.), Impact of Littoral Environmental Variability on Acoustic Predictions and Sonar Performance, 73-82.

The observation and description of seabed variability is cast with the underlying motive of understanding the impact of geoacoustic variability on acoustic variability. Thus, while it may be true that within 1 meter of a core, the sediment microscopic and macroscopic structure is variable, that variance may or may not have impact on sonar performance. For this study, the focus is on geoacoustic variability that has significant 0–5000 Hz range.

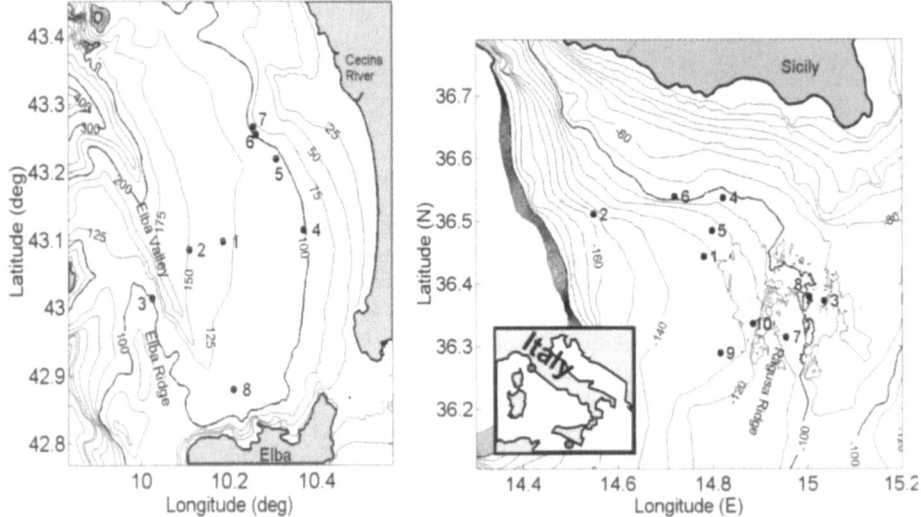

Figure 1. North Elba and Malta Plateau regions with site locations; depths are in meters. Seismic survey tracks are in gray.

2   Methodology

Extensive geologic/geophysical measurements (e.g., cores, grab samples, seismic reflection, swath bathymetry, and seafloor photography) as well as acoustic measurements (e.g., local seabed reflection/scattering and transmission loss/ reverberation) were conducted in both regions. Although each measurement type is useful for probing seafloor variability, the key measurements for this study are the local reflection [1] and scattering [2] measurements which provide much higher vertical and lateral resolution of the geoacoustic properties (of order $O(0.1)$ and $O(100)$ m respectively) than traditional propagation/reverberation measurements. Site locations were based on seismic data; at each site a suite of the above-listed measurements were conducted.

Generally, geoacoustic variability is expected to be greatest in a direction perpendicular to the isobaths. Thus, sites selected for this study lie roughly along a line perpendicular from the coast: North Elba (NE) sites are 5,1,2,3. Sites were chosen for the Malta Plateau (MP) at commensurate water depths, Sites 4,1,2,7, since geoacoustic properties may be correlated with water depth.

Geoacoustic variability is manifest over a continuum of scales. In order to describe variability it is useful to define distinct scales that capture the critical aspects of the environment; for purposes of this study they are: geoacoustic regimes, sedimentary classes, and seabed features. Within a region, geoacoustic regimes are defined as areas in which the physical mechanisms that govern reflection and scattering are similar. The angular and frequency dependence of the reflection/scattering may be spatially variable within a regime, but the variability is expected to be continuous rather than abrupt. Variability within a regime generally arises from geometric factors, e.g., variability in layer thicknesses. The boundaries between geoacoustic regimes may or may not be abrupt. Within a geoacoustic regime, sediment classes are defined as distinct sedimentary units. Seabed features are defined as 1–100 m scale discrete sedimentary objects that may produce sonar clutter.

3   North Elba geoacoustic variability

The Tuscan shelf was created (along with the Northern Apennines and the Adriatic Sea) from the collision of the Adria microplate and the Corsica-Sardinia Massif [3]. In the northern part of the shelf (north of Elba Island), the Elba valley running northwest from Elba Island separates two distinct regions, the basin to the east dominated by fine-grained sediments, and the Elba Ridge to the west, which exhibits coarser grained sand. At water depths between 115–130 m swath bathymetry reveals erosional features perpendicular to the isobaths (~0.5 m rms height, O(500) m spacing) apparently formed during the last sea level transgression. Near the southern edge of the region, are parallel ribbon fields (~0.5 m rms height) oriented east-northeast. Generally, seafloor and sub-bottom layer slopes are very small (less than 1°), with the exception being the Elba Ridge flank, which has slopes of ~15–20°.

The frequency and angular dependence of the reflection loss (defined as -20 log $|R|$, where R is the complex reflection coefficient) for various sites in NE is shown in Fig. 2 along with the associated water depth. The salient feature in the data is the critical angle, θ_c, defined as the "knee" where the reflection loss rapidly increases from near zero. At grazing angles below θ_c, the reflection coefficient is small and normal modes/rays propagate efficiently. The critical angle is a crucial indicator for long-range propagation; generally the higher θ_c, the better the propagation. For a homogenous half-space, the reflection coefficient is independent of frequency. However, θ_c can be frequency dependent due to several factors including: velocity dispersion (e.g., [4]), layering, and sound speed gradients. For velocity dispersion, theory predicts that θ_c should increase with increasing frequency. Sediment layering produces resonant effects in angle and frequency; a single layer produces nulls such as that observed at MP Site 1. Sediment sound speed gradients are generally positive and generate an apparent θ_c inversely proportional to frequency because the loss along refracted paths is proportional to frequency. An example of this is NE Site 3, where θ_c changes from ~30° at 500 Hz to ~18° at 5000 Hz.

The reflection data of Fig. 2 along with the associated raw time series were analyzed following the technique described in [1] to produce *in-situ* sediment geoacoustic properties, i.e., velocity, density, and attenuation. Of these, the controlling factor in shallow water propagation is generally the velocity, which is shown in Fig. 3 as an indicator of geoacoustic variability. The sound speed of the water near the seabed interface was nearly constant between sites and is shown to clarify the relative sound speed between the water and the seabed.

The predominant (i.e., covering the vast majority of the region) geoacoustic regime, Regime I, in NE is fine-scale layering (see Fig. 3 Sites 1 and 2 and Fig. 4b) consisting of a background mud matrix with interstitial high-velocity layers composed of sand with shell and coral fragments (see Fig. 4c,e). This layering structure was created by high frequency glacio-eustatic sea-level changes during the Pleistocene era [5]. Even though the interstitial layers are thin (of order tens of centimeters), the layers have a measurable

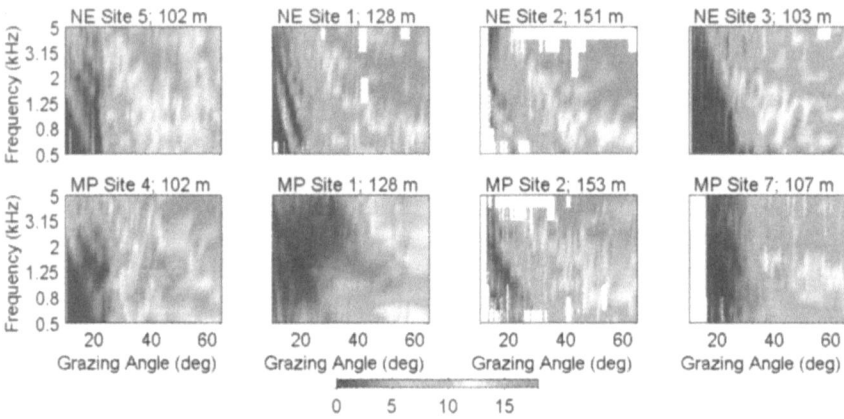

Figure 2. Reflection loss (dB) at various sites in North Elba (NE) and the Malta Plateau (MP).

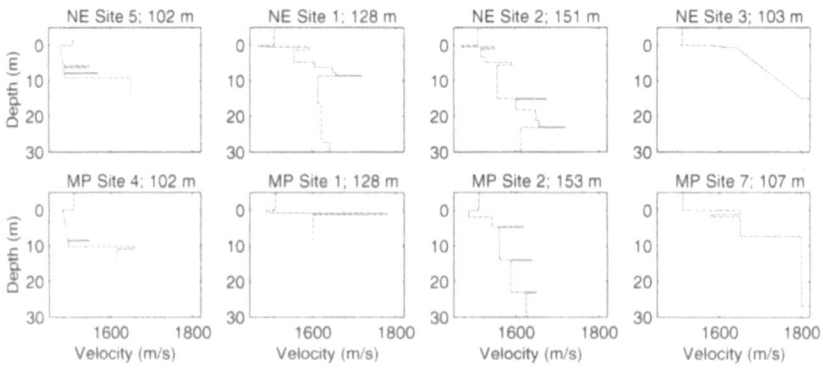

Figure 3. Sediment sound speed variability at the two study regions (see Fig. 1 for locations).

effect on seabed reflection above ~500 Hz [1]. At lower frequencies, reflection/propagation is dominated by paths that refract through the sediment. The intercalating layers are the dominant scattering mechanism across the entire frequency range of interest. At 1800 Hz and below, scattering is apparent from the intercalating layer at 25 m sub-bottom [2]. The variability of this geoacoustic regime is continuous in the sense that the dominant reflection/scattering mechanism (the intercalating shelly layers) have non-parallel dips (see Fig. 4b), however, the geoacoustic properties of each layer unit are constant with at least to lateral offsets of 5 km [6] and perhaps even to much longer scales. The geoacoustic variability within in this regime produces a minimal impact on the acoustic variability; i.e., there is relatively little variability in the reflection coefficient (i.e., compare NE Sites 1, 2 of Fig. 2; Site 8, not shown is also similar) and scattering strength [7]. The eastern boundary to this regime is abrupt and well-defined by the Elba Ridge; the boundary to the east is not abrupt, but continuous. Analysis of the reflection coefficient at Site 8 indicates that this regime persists to the southern part of the region to at least the 110 m contour.

Geoacoustic regime II is defined as the band between about 110–75 m in the northeastern part of the region where the uppermost silty clay layer rapidly increases to thicknesses of 10–15 m. In this zone, the dominant regime switches from fine-scale layering to a few thicker/higher impedance layers, which eventually wedge out to a silty-clay/basement contact at ~80 m depth contour. The basement is the flanks of the Secche di Vada Ridge which according to [3] is a Late Cretaceous flysch (largely sandstone). In this regime, the reflection/ scattering is dominated by the high impedance layer at the base of the thick silty-clay layer (see NE Site 5, Fig. 3) and at the shallowest depths, the basement. The sediment thins over the basement contact to a 1 meter or less in the northeast and southeast corners. The composition of the basement in the southeast is unknown. The relatively few sites in the regime makes it hard to estimate geoacoustic and acoustic variability in regime II, nevertheless the variability in regime II appears to be largely governed by the presence/absence of the basement contact and geometric effects (especially the thickness of the uppermost silty-clay layer). The reflection at Site 4 is quite similar to that at Site 5, with indication of refraction at the lowest frequencies.

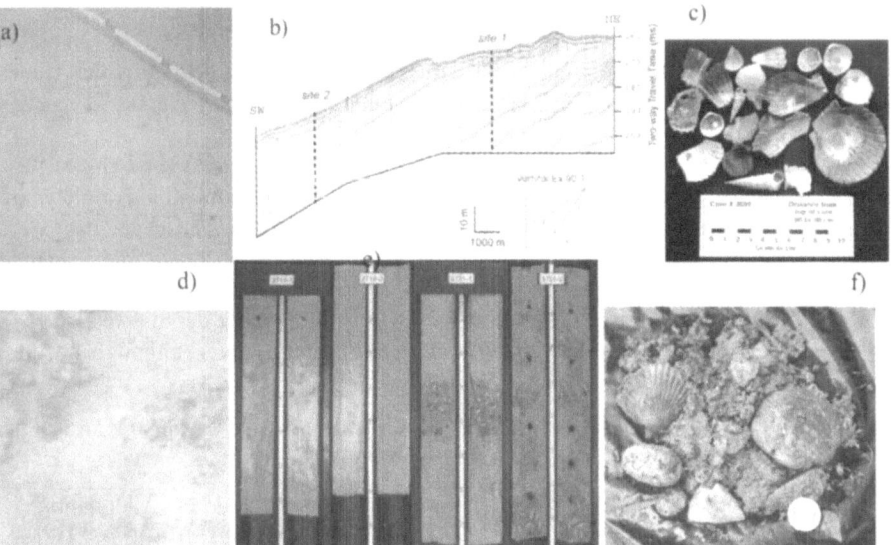

Figure 4. Images from top left to bottom right: a) seafloor at NE Site 5, black tape marks spaced 10 cm apart; b) NE Sites 1 and 2 seismic reflection data; c) NE Site 2 recovered shell material from a 10 cm section of core; d) seafloor on Ragusa Ridge showing rock outcrop in upper left, scale is about 30x40 cm; e) split cores south of MP Site 1 showing intercalating shelly layers, core lengths are about 80 cm; f) MP Site 1 shell and cobble in core nose, coin is 2.5 cm diameter.

Geoacoustic regime III is (an ostensibly homogeneous) sand, with a thin (0.5 m) sandy silt cover. Due to the paucity of data in this regime, very little is known about its lateral variability.

While there is considerable complexity in the sediment fabrics found in NE, there are distinct sediment classes that can be identified. These classes with their associated geoacoustic properties are given in Table 1. Silty-clay sediment completely blankets the seafloor east of the Elba Ridge. Acoustic measurements, core data, and seafloor photography indicate that the surficial properties of this layer are spatially uniform. Observations show a characteristic distribution of small (cm scale) holes from biologics (Fig. 4a). The sound speed gradients in this upper layer increase as a function of distance from the source (the Cecina River at ~43.3° N); e.g., at Site 5 the gradient is 1.5 s^{-1} at Site 4, 7 s^{-1} and at Site 2 at least 30 s^{-1}. The density gradient does not appear to vary with distance from the source and is ~0.1 $g/m^3/m$ over the first meter, with the upper 10 cm gradient being substantially higher. Sand covers the remaining ~10% of the region, surficial sand being found only on the Elba Ridge. Consolidated sediment (rock) within the upper ten meters of sediment is found in the northeast and southeast corner of the region. There are probably magmatic rock outcrops on the Elba Ridge but these have not been sampled.

Numerous shells, shell and coral fragments (Fig. 4c) were found in core and grab samples with sizes ranging up to the order of the core barrel (10 cm). Shell/coral fragments tend to exist in greatest number in layers that have a sand matrix, and are found much less frequently in the mud layers. The variability in the geoacoustic properties of these layers is quite high, depending on the volume fraction and size distribution of the shell and coral fragments. Between the shelly sand layers over much of the region there are mud layers. The sound speed gradient in the mud can be approximated by: $c(z) = c_o \left[(1+\beta)(1+2g_o z/(c_o(1+\beta)))^{1/2} - \beta \right]$ where c_o is the interface sound velocity, β the curvature and g_o the initial gradient [1]. Table 1 provides values for g_o and β, which differ from [1] because the mud and silty-clay classes have been separated in this study.

The primary features observed in the region are gas, and buried sand ridges. Gas charged sediments appear to be sprinkled throughout the region although they are generally deeper than 10m sub-bottom; pockmarks (presumably from escaped gas) were observed in the southeastern corner [5]. Lowstand coastal ridges observed in the seismic data are buried 3-5m below water-sediment interface, about 3–4 m high, 100–200 m in width, 0.5–2 km in length and ~3 km spacing in 100–130 m water depth (see also [5]). The composition of the ridges is probably coarse sand and gravel, although no core samples were obtained.

Table 1. Geoacoustic properties in North Elba region, see text for details. No entry means that the property could not be reliably estimated; cs is shear velocity.

	Velocity (m/s)	Vel grad (s^{-1})	Density (g/cm^3)	Den grad ($g/cm^3/m$)	Attenuation (dB/m/kHz)
Silty-clay	1475	1.5 +	1.3	0.1	0.01
Mud	1500	5, β=-0.9	1.5	via vel.[8]	0.015
Sand	1640	10	1.8	--	0.3
Shelly sand	1650±100	--	1.9±0.2	--	0.1
sandstone	cs: 1600	--	--	--	--

4   Malta Plateau intra- and inter-regional geoacoustic variability

The Malta Plateau occupies the northern edge of the North African passive continental margin and is a submerged section of the Hyblean Plateau of mainland Sicily [9]. The region (see Fig. 1) is divided by the Ragusa Ridge, roughly defined by depths shallower than 110 m, which forms a spine ~20 km wide between Sicily and Malta. The area west of the ridge is blanketed with a silty-clay sediment discharged from the Irminio River and other small streams from the flanks of Monte Lauro.

The seabed reflection coefficient on the Malta Plateau (Fig. 2) bears a striking resemblance with that at North Elba at corresponding water depths. Note the similarities between MP Site 4 and NE Site 5, also the strong similarity between MP Site 2 and NE Site 2, as well as the similarity between MP Site 7 and NE Site 3. Other than MP/NE Site 1, the comparison of sites within a region shows substantially more variability than from region to region at the same water depth.

Since the reflection coefficient (unlike propagation or reverberation) is independent of water depth and sound speed profile in the water column, similarity in reflection means similarity in sediment structure. This is borne out by examining the sediment sound speed of Fig. 3, which show numerous similarities. The sediment regimes of MP, in fact, closely parallel those in NE. Geoacoustic regime I occupies a large area in MP, dominated by fine-scale layering (see MP Site 2 Fig. 3 and Fig. 4e) of similar character to that in NE, i.e., shelly sand layers interspersed in mud layers.

Geoacoustic regime II is also analogous to NE. It exists from the 110 m contour shoreward composed of a silty-clay layer, thickening shoreward, over highly reflective layers that thin to a silty-clay/basement contact in several places. As an example of this regime, at MP Site 4, there is an 8-m layer of low velocity sediment (1480 m/s; 1.32 g/cm^3) that corresponds remarkably well to the 7.4 m thick layer at NE Site 5 (1477 m/s; 1.32 g/cm^3). The uncertainties in the velocity and density estimates are ±4 m/s and ±0.04 g/cm^3 respectively. Both sites have at least one intercalating high-speed layer, however, in both regions the strata at a depth of ~10m sub-bottom controls the reflection/scattering. Differences between the regions include the complexity of the basement topography and perhaps the basement geoacoustic properties (believed to be limestone, although geoacoustic properties have not been measured). Lineated basement outcrops occur both shallower than 110 m water depth and along the westward and eastward boundaries of the ridge. The outcrops are up to 2 km wide along the western edge of the ridge.

Geoacoustic regime III, on the Ragusa Ridge, is composed of ponded sand between small-scale rocky outcrops, 1–100 m in width and 1–8 m in height (see Fig. 4d). The thickness of the ponded sand thins towards the parallel escarpments which are ~12 km apart. At Site 7, the consolidated basement is not shallower than 44 m sub-bottom. The sand has a comparable interface velocity to that on the Elba Ridge, but MP Site 7 does not show the same strong frequency dependence of θ_c as NE Site 3 (see Fig. 3), indicating a smaller velocity gradient. In fact, the velocity gradient is small enough that it appears to play a negligible role and thus is not included in the geoacoustic model. A thin low velocity layer causes the high loss peak in the data from 40–65°.

The ostensible exception to inter-regional similarities is between MP and NE Site 1. At MP Site 1, a high velocity sub-bottom layer existed, composed of gravel and cobble sized stones and cemented sediments (Fig. 4f) that dominate the reflection/scattering processes [10]. This layer appears to be associated with a buried river channel. During the last glacial period, when the coastline was at the 80–90 m depth, the Irminio River was capable of transporting cobbles up to 25 cm in diameter [11]. The extent of the buried river channel network is believed to be relatively small and is probably better characterized as a feature rather than a geoacoustic regime.

The sediment classes closely parallel those found in NE. Sediment geoacoustic properties for each class are provided in Table 2.

Features observed in the Malta Plateau region include: gas plumes, small-scale rock outcrops (e.g., Fig. 4d), buried carbonate banks, and buried river channels. Two types of gas plumes were observed: large plumes, 100–300m in lateral extent rising within ~5 m of the seafloor, and small-scale plumes, 5–20 m in lateral extent and rising to seafloor interface. Both kinds of gas plumes tend to occur in clusters of multiple plumes. These plumes have been positively correlated with clutter events on low frequency active sonar in this area. Carbonate structures are found in the northwest part of the region [9]. They appear to be highly reflective and are found about 10 m below the seafloor, but may not be an important factor for producing clutter below 1 kHz. Buried river channels in the northern sector of the region have been associated with clutter events in 600 Hz sonar systems [12]. However, an alternative explanation in that area may be small-scale gas plumes, which because of their size may easily be missed in seismic surveys with coarse line spacing.

Table 2. Geoacoustic properties in the Malta Plateau region.

	Velocity (m/s)	Vel grad (s^{-1})	Density (g/cm^3)	Den grad ($g/cm^3/m$)	Attenuation (dB/m/kHz)
Silty-clay	1480	1.5 +	1.3	0.1	0.01
Mud	1500	15, -0.99	1.5	Via vel.[8]	0.015
Sand	1650	--	1.8	--	--
Shelly sand	1650±100	--	1.9±.2	--	0.1
Cobble	1780	--	1.85	--	0.2

5   Summary and Conclusions

Geoacoustic variability was examined in and between two regions in the Italian littoral. Within a region the variability was described by geoacoustic regimes, sediment classes, and sedimentary features. The predominant regime in both regions is characterized by a mud host material with a sound speed gradient larger than Hamilton [13] would predict and thin intercalating layers of sand mixed with shell and coral fragments. In this regime, the variability in layer geometry appears to have a minor effect on the acoustic response (i.e., reflection and scattering) of the seabed, simplifying the level of detail required for sonar performance prediction requirements.

In geoacoustic regime II, the sound speed gradients in the uppermost layer appear to be a function of distance from the sediment (riverine) source. This relationship may be general to littoral environments, and to the author's knowledge has not been heretofore published. Critical factors required to predict these gradients need to be identified and suitable models developed. There is tantalizing evidence that the sound speed gradients in the deeper layers may also be a function of identifiable and predictable factors. Note that at MP Site 2, the sound speed gradient is significantly higher than that at NE Site 2; it is conjectured that there is an underlying and perhaps predictable relationship between the gradients and the location on the shelf.

Between the two regions, remarkable similarities were observed. The surficial silty-clay sediment is uniform across large areas of each region and has almost identical properties between the two regions though the regions are separated by ~800 km. The similarities between the two regions go deeper. Each region has a similar layering structure (mud host with intercalating sandy-shelly layers) and concomitant reflection characteristics over a broad part of the region. Each region also has a broad area around the 100 m depth contour where the surficial silty-clay layer deepens to O(10) m in thickness. Significant differences in reflectivity were observed between the two regions at about the 130 m depth contour. Nearby core and seismic data suggest that the two regions are similar at these water depths, but that in the Malta Plateau, the reflection measurement sampled a feature (i.e., buried river channel) rather than the predominant regime.

The inter-regional similarities raise numerous questions: are these inter-regional similarities expected around the entire littoral Italian zone?, in other parts of the Mediterranean? Are these inter-regional similarities predictable and if so, at which level: the geoacoustic regimes? their boundaries? sediment classes? Is intra-regional variability predictable at some level? These questions are important because acoustic models will always be faced with insufficient geoacoustic data. The ability to extrapolate geoacoustic measurements from region-to-region could provide an important advance for sonar performance prediction. Geologic/geophysical models (e.g.[14]) may provide the framework for addressing many of these issues.

While the existing measurement techniques were capable of generally defining the geoacoustic regimes and general bounds, description of the variability within a regime is quite poor, i.e., only a few measurements (in some cases only 1) exist within a regime. By hosting the reflection and scattering measurement techniques on an AUV, rapid detailed surveys within a regime could be conducted, thus resolving much finer scales of variability that impact sonar system employment.

Acknowledgements

This research was sponsored by both the NATO SACLANT Undersea Research Centre and the Office of Naval Research. The author gratefully acknowledges the captain, officers, crew and the outstanding scientific staff aboard the R/V Alliance for their skill and dedication that contributed significantly to the success of the measurement program. Core photos were taken by Adolf Legner.

References

1. Holland, C.W. and Osler J., High-resolution geoacoustic inversion in shallow water: A joint time and frequency domain technique, *J. Acoust. Soc. Am.* **107**, 1263–1279 (2000).
2. Holland, C.W., Hollett, R. and Troiano L., A measurement technique for bottom scattering in shallow water, *J. Acoust. Soc. Am.* **108**, 997–1011 (2000).
3. Pascucci, S., Merlini, S. and Martini P., Seismic stratigraphy of the Miocene-Pleistocene sedimentary basins of the Northen Tyrrhenian Sea and western Tuscany, *Basin Res.* 11, 337–356 (1999).

4. Biot, M.A., Generalized theory of acoustic propagation in porous dissipative media, *J. Acoust. Soc. Am.* **34**, 1254–1264 (1962).

5. Brizzolari, E., Chiocci, F.L., Orlando, L. and Sacchi, L., Pleistocene evolution of the Tuscan shelf from Piombino headland to San Vincenzo (Tyrrhenian Sea, Italy), *Giornale di Geologia* **5312**, 11–30 (1991).

6. Holland, C.W., Regional extension of geoacoustic data. In *Proc. 5th European Conference on Underwater Acoustics*, edited by P. Chevret and M.E. Zakharia, Lyon, France (2000) pp. 793–798.

7. Holland, C.W., Spatial variability of shallow water bottom scattering: a case study. In *Proc. 5th European Conference on Underwater Acoustics*, edited by P. Chevret and M.E. Zakharia, Lyon, France (2000) pp. 1141–1146.

8. Bachman, R.T., Acoustic and physical property relationships in marine sediments, *J. Acoust. Soc. Am.* **78**, 616–621 (1985).

9. Osler, J. and Algan, O., A high resolution seismic sequence analysis of the Malta Plateau, SACLANTCEN SR-311, 1999.

10. Holland, C.W., Shallow water coupled scattering and reflection measurements, *IEEE J. of Oceanic Eng.* (in press 2002).

11. Amore, C., and Randazzo, G., First data on the coastal dynamics and the sedimentary characteristics of the area influenced by the River Irminio basin, *Catena* **30**, 357–368 (1997).

12. Max, M.D., Portunato, N., and Murdoch, G., Sub-seafloor buried reflectors imaged by low frequency active sonar, SACLANTCEN SM-306 (1996).

13. Hamilton, E.L., Geoacoustic modeling of the seafloor, *J. Acoust. Soc. Am.* **68**, 1313–1339 (1980).

14. Syvitski, J.P.M., Pratson, L. and O'Grady, D., Stratigraphic predictions of continental margins for the Navy. In: *Numerical Experiments in Stratigraphy: Recent Advances in Stratigraphic and Computer Simulations,* edited by J.W. Harbaugh, L.W. Whatney, E. Rankay, R. Slingerland, R. Goldstein and E. Franseen, SEPM special publication, No. 62 (1999) pp. 219–236.

ACOUSTIC AND *IN-SITU* TECHNIQUES FOR MEASURING THE SPATIAL VARIABILITY OF SEABED GEOACOUSTIC PARAMETERS IN LITTORAL ENVIRONMENTS

JOHN C. OSLER, PAUL C. HINES AND MARK V. TREVORROW

Defence R&D Canada – Atlantic, P.O. Box 1012, Dartmouth, Nova Scotia B2Y 3Z7, Canada
E-mail: john.osler@drdc-rddc.gc.ca

Geoacoustic properties of the seabed are required for accurate modeling of acoustic propagation, and hence sonar performance prediction. Characterizing acoustic interaction with the seabed is particularly important in shallower water environments as the propagation typically involves extensive interaction with the sea surface and seabed. DRDC Atlantic is developing acoustic and *in-situ* techniques for seabed classification and measuring geoacoustic parameters. The acoustic technique uses normal incidence acoustic returns, in the 1 to 10 kHz band, from the seabed and sub-bottom. The transition from interface to volume scattering depends upon frequency and sediment type and can be used to distinguish the composition of near surface marine sediments. An experimental methodology has been developed using a vertical line array of receivers and a downward-looking superdirective projector array. The technique is also being adapted for use with commercial normal incidence sub-bottom profilers. *In-situ* measurements are being made using the DRDC Atlantic free fall cone penetrometer probe (FFCPT). It has been developed to measure acceleration and dynamic sediment porewater pressure as a function of depth of penetration into the seafloor. It also records hydrostatic pressure and optical backscatter for detection of the mudline. This combination of sensors permits the direct application of geotechnical analysis methods and parametric-based correlations already long established in engineering practice. The FFCPT provides two independent means of calculating the undrained shear strength, as well as other engineering variables that are used to identify the sediment grain size characteristics. The probe has a modular design allowing additional sensor payloads to be integrated, the first of which uses resistivity as a means to determine sediment bulk density. Experimental results, using the acoustic and in-situ techniques, will be presented from two joint US-SACLANTCEN-CAN sea-trials in 2001 at the New Jersey Strataform area and the Scotian Shelf.

1 Introduction

DRDC Atlantic is developing acoustic and *in situ* techniques to determine geoacoustic properties of the seabed for accurate modeling of acoustic propagation. The emphasis of this research is on continental shelf water depths at frequencies that are relevant to tactical and low frequency active sonars. The surficial geology of the Eastern Canadian and American continental margins is tied to the geological processes associated with sea-level changes during periods of glaciation, and in the case of the Scotian Shelf, with the direct effects of glaciers themselves. This has led to high spatial variability in the

N.G. Pace and F.B. Jensen (eds.), Impact of Littoral Environmental Variability on Acoustic Predictions and Sonar Performance, 83-90.

composition and physical properties of the surficial sediments and is the motivation to develop the techniques described in this paper.

The acoustic approach extends the capabilities of an existing classification technique to lower frequencies, into the band of interest for ASW sonars. It is being adapted for use on vessels with commercial seismic profiling systems and for active sonobuoys. These could transit or drift through an area of operational interest as part of a rapid environmental assessment scenario. The *in-situ* approach uses a probe to measure geotechnical and geoacoustic properties of the seabed. These mechanical properties are useful for marine engineering, mine burial prediction, and permit a sediment classification based on grain size. One geoacoustic sensor module has been developed for the *in-situ* method and it measures resistvity as a means to determine porosity. It has been noted [1,2] that porosity is an effective parameter, for first order geoacoustic characterization of the seabed, to which the more traditional quantities of sound speed, density, and attenuation may be directly or empirically related.

2 Acoustic technique

Heald [3] has developed a surficial sediment classification technique using the first and second normal incidence acoustic seabed returns. It is typically used with mono-static echo sounder systems that operate at several 10s of kHz. At these frequencies, bottom roughness at the seabed interface is the dominant scattering mechanism. Following Pace *et al.* [4], the first seabed return is treated as a far-field, mono-static, arrival whose intensity is inversely proportional to the mean square slope—a roughness parameter. The second seabed return is treated as a near-field bi-static arrival, with a virtual source above the water surface, whose intensity is proportional to the Rayleigh reflection coefficient. The intensity of the first and second seabed returns are integrated over time to yield energy, E_1 and E_2 respectively, and the ratio E_1/E_2 is used to classify different sediment types [3].

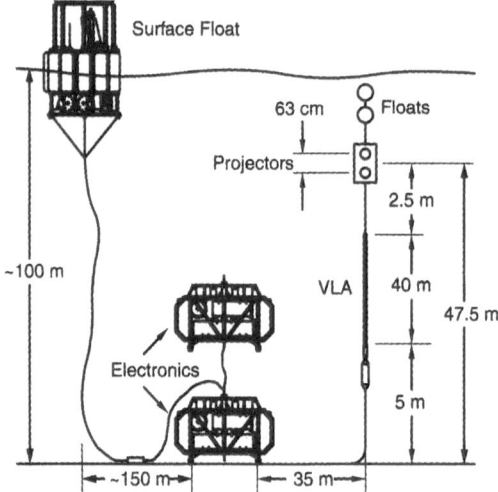

Figure 1. Schematic diagram of the DRDC Atlantic Underwater Acoustic Target deployment.

The technique in [3] is based on a Helmholtz-Kirchhoff model with gaussian roughness and ignores any coherent reflections and seabed penetration. Hines and Heald [5] have extended this theory to lower frequency (1 to 10 kHz range) by including volume scatter from coherent energy penetrating into the seabed. Scatter from the interface initially dominates the intensity of the seabed return, however because of its rapid decay, the return is subsequently dominated by volume scatter. The transition between surface and volume scatter, and their respective rates of decay, form the basis of the technique to classify near surface sediments. The intensity time series are a function of roughness parameters, frequency, and the composition of the surficial sediment layer. Wide bandwidth, or multi-frequency systems, may exploit the frequency dependence and provide further discrimination regarding seabed composition.

Experiments were conducted using the DRDC Atlantic Underwater Acoustic Target (UAT). The UAT is a ship-launched echo-repeater (Fig. 1) with a 15 hydrophone vertical line array (VLA). Immediately above the VLA, there is a pair of ITC 1007B 16 cm diameter spherical projectors whose acoustic centers are separated by 62.5 cm. The projectors are controlled individually but may be used in tandem with user specified time and phase delays. The power and electronics modules for the UAT are housed in two space frames that were displaced as far as possible horizontally from the VLA to minimize their influence on the scattering measurements.

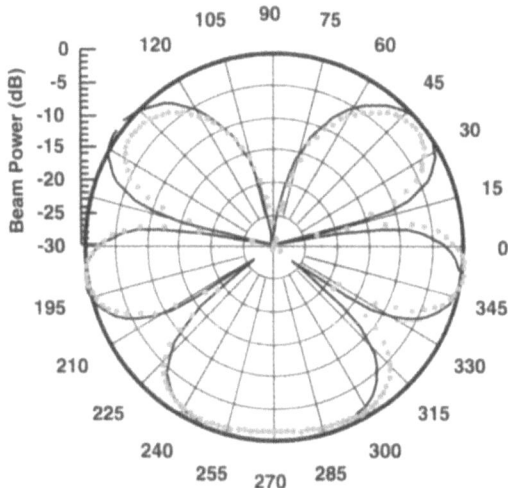

Figure 2. Theoretical (solid line) and measured (magenta dots) beam patterns for the steady state radiation from a pair of transducers separated by $5\lambda/4$ and driven in phase quadrature.

Acoustic energy that propagates from the source and is reflected by the sea surface gives rise to a series of arrivals that are unwanted and can mask the seabed reflected arrivals that are of interest. Their amplitude was reduced by creating a directional source, with the main lobe pointed at the seabed, by time delaying and phase inverting the output from the upper projector such that cancellation occurs above the transducers and reinforcement occurs below (Fig. 2). Frequencies were selected for transducer separations of $5\lambda/4$, $9\lambda/4$, and $13\lambda/4$ (approximately 3, 5.3, and 7.7 kHz) respectively. In open ocean experimentation, this two element projector array has reduced the amplitude of the surface reflection by approximately 10 dB. The reduction was less than that achieved during calibration tests, but further refinements should improve the performance.

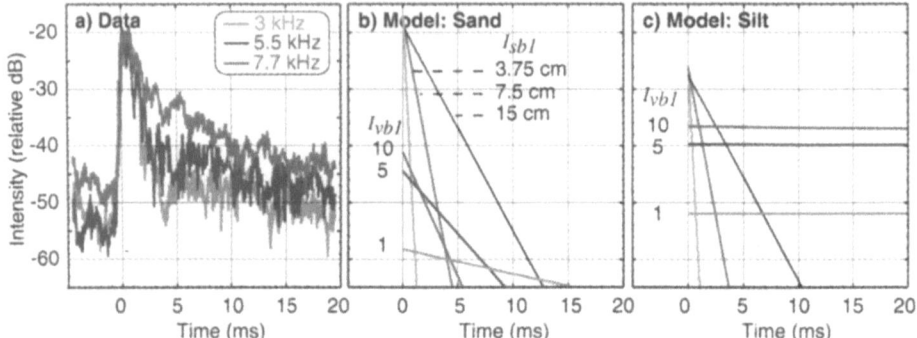

Figure 3. (a) Median intensity time series of the first seabed reflection of a 1 ms pulse arriving on the uppermost hydrophone; and model predictions for (b) sand and (c) silt seabeds of interface scatter, I_{sb1}, and of volume scatter, I_{vb1}, after [5].

Measured data and model predictions [5] for the intensity versus time decay of the first seabed return are presented in Fig. 3 for a pulse vertically incident on the seabed. The surface scatter intensity, I_{sb1}, is calculated for three values of surface roughness, 3.75, 7.5, and 15 cm, and a fixed correlation length of 1.5 m. The volume scatter intensity, I_{vb1}, is calculated at 1, 5, and 10 kHz using the roughness parameters specified in [5] and physical parameters in [6]. The model intensity curves begin (time zero) when the pulse that is incident on the seabed transitions to an annulus from a spot. This is also when the peak in I_{sb1} and I_{vb1} is expected to occur. The predicted intensity of the volume and surface scatter have been superimposed to illustrate that the initial level and its decay are dominated by surface scatter. At some later point in time, the intensity becomes dominated by the decay of the volume scatter. This transition between interface and volume scatter is a function of seabed type, seabed roughness, and frequency and may be exploited to classify near surface sediments. It should be possible to invert for the limited number of geoacoustic parameters in this simple yet effective seabed model and compare the results with normal incidence [7] and wide angle inversion techniques that yield high resolution geoacoustic models [8].

The median intensity of two hundred first seabed returns received on the uppermost hydrophone of the UAT at three discrete frequencies are displayed in Fig. 3a. Though the decay of the experimental data has a considerable amount of structure, perhaps due to some layering in the seabed, the overall behaviour compares favourably with the model predictions. That is, the initial decay is most rapid at the lowest frequency followed by a sharp transition to a slowly decaying volume scattering regime. At the two higher frequencies, the intensity levels at the onset of volume scattering are progressively higher and decay more rapidly than the lowest frequency. The measured intensity curves are consistent with theoretical predictions for a seabed composed of material that is intermediate between the sand and silt cases presented. (Note that the data have yet to be calibrated so its vertical axis is relative). These results were collected in the vicinity of the AMCOR 6010 site [9] on the New Jersey Strataform area. The geoacoustic seabed properties have been measured by independent means and are summarized in [9]. The seabed is described as having "near surface layering" with the first twenty five metres being a "sandy-silty-clay" layer. The average compressional sound speed in the first 5 m is 1560 m/s and reaches 1830 m/s at 30 m depth.

3 *In-situ* technique

The free fall cone penetrometer test (FFCPT) is a free fall probe that has been developed to measure mechanical (or geotechnical) and geoacoustic properties of the seabed. A diverse range of military and civilian applications for the probe are envisioned. Commercial applications may include pipeline survey work and support of dredging or marine contaminant disposal operations. The FFCPT requires less logistical support than the direct-pushed cone penetrometer test (CPT) and there is no hydraulic platform on the seabed to potentially disturb the seabed material being measured. It can conduct surveys quickly and may ultimately be integrated with a fully automated free fall winch that operates on a moving vessel. Military applications include support for: mine countermeasures by providing ground truth for seabed classification systems that are used to predict mine burial and the effectiveness of mine hunting sonars; and for shallow water active sonar by making measurements from which the geo-acoustic properties of the seabed, as required for propagation and reverberation modeling, may be derived.

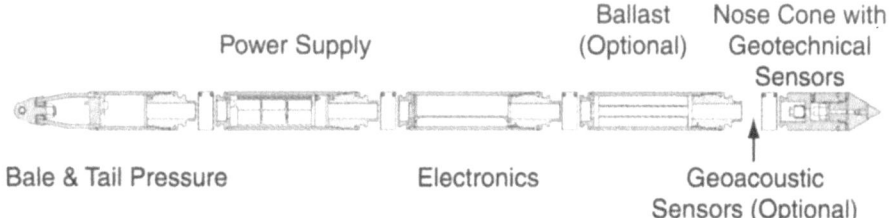

Figure 4. Schematic diagram of the DRDC Atlantic FFCPT. Optional sensor modules and ballast may be inserted between the nose cone and the electronics module.

The basic FFCPT consists of a nose cone instrumented with geotechnical sensors, power supply, electronics, and tail pressure sensor (Fig. 4). As the probe penetrates into the seafloor nose first, it measures acceleration and dynamic sediment porewater pressure as a function of depth. It also records hydrostatic pressure in the water, has an optical backscatter sensor to detect the mudline, and allows additional ballast and geoacoustic sensor modules to be integrated. This combination of sensors permit the direct application of geotechnical analysis methods and parametric-based correlations established in engineering practice [10]. The DRDC Atlantic FFCPT has been developed in collaboration with Brooke Ocean Technology (BOT) Ltd. and Christian Situ Geoscience (CSG) Inc. (both in Dartmouth, Nova Scotia). It incorporates the basic sensor suite from their earlier 11.43 cm (4.5 inch) O.D. prototype into a modular 8.89 cm (3.5 inch) O.D. design (Fig. 4). The first geoacoustic module being developed measures resistivity as a means to determine sediment bulk density. Future modules under consideration include: a linear actuator to measure shear rigidity (when the probe is at rest); and transducers for the direct measurement of compressional sound speed. The ability to combine geotechnical and geoacoustic sensors in a single probe, as demonstrated in [11], allows a more complete characterization of the seabed than that provided by acceleration-based penetrometers [12,13].

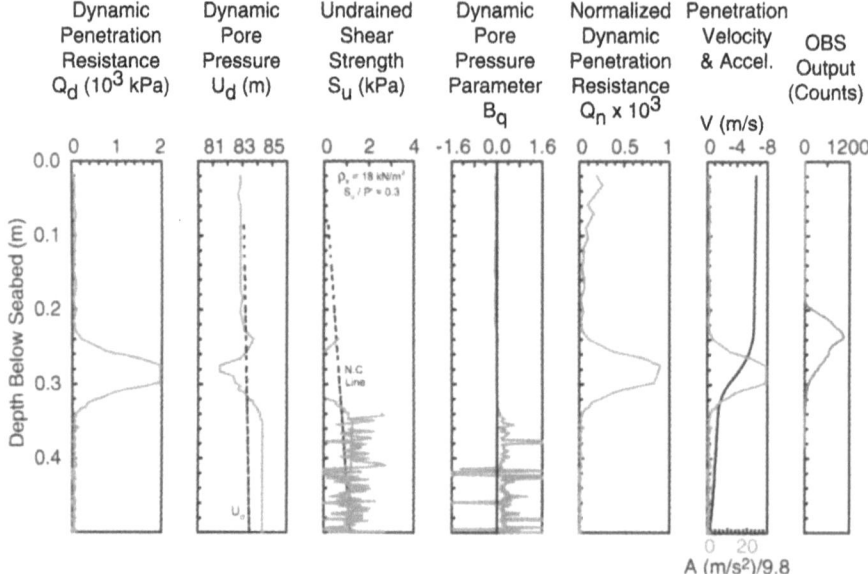

Figure 5. Material parameters versus depth for a deployment of the BOT FFCPT near NJ-1 on the ONR Strataform. Analysis provided by Christian Situ Geoscience Inc.

The resistivity module has been developed by BOT and ConeTec (Vancouver, British Columbia) using the design principles of a static resistivity CPT system [14]. The excitation signal is typically set to generate a current-switched AC sinusoidal wave, at a frequency of about 1 kHz. The primary controlling factor for resistivity is the conductivity of the pore water phase; the conductivity of the mineral grains is a secondary factor. Given a knowledge of the pore water chemistry, it is possible to make a second-order evaluation of bulk density or porosity. If the sediment profile composition is known in detail through an independent test (*e.g.* Fig. 6) then it is possible to account for vertical variability in the grain conductivity. Extending the capabilities of the resistivity module to permit measurements during penetration is under consideration [15]. The high rate of initial penetration poses several technical challenges, such as an excitation rate of several hundred kHz, that must be addressed.

The penetration displacement history is obtained from double integration of the recorded acceleration and uses the optical sensor to define the moment of impact (particularly helpful on softer bottoms). The FFCPT provides two independent means of calculating the undrained shear strength. One using the dynamic penetration resistance as measured using the accelerometer sensors and another using the dynamic porewater pressure response as measured by the sediment porewater pressure sensor (Fig. 5). The FFCPT utilizes a standard CPT-based sediment classification chart [10] as part of the interpretation algorithm. There are accepted empirical relationships between the dynamic pore pressure parameter, the normalized dynamic penetration resistance and sediment grain size characteristics (Fig. 6).

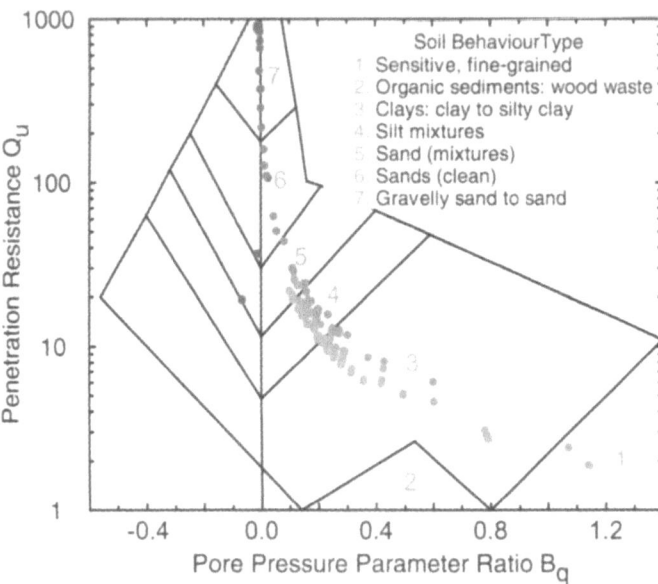

Figure 6. Sediment classification chart [10] using the penetration resistance and pore pressure parameter data in Fig. 5. Dots are color coded from violet at the surface to red at depth.

The 11.43 cm (4.5 inch) prototype was deployed at several locations near NJ-1 (39°14.949N, 72°51.798W) on the ONR Strataform area during the US-SACLANTCEN-CAN Boundary Characterization 2001 sea-trial. It has measured the geotechnical properties of a pervasive near surface sand layer that is approximately 10 cm thick and lies between clay and silt layers (Fig. 5). In several instances, the 11.43 cm O.D. prototype was not able to pierce through this sand layer. This limitation in its performance supported the decision to adopt a smaller diameter for the DRDC Atlantic FFCPT. An O.D. of 8.89 cm (3.5 inch) was selected as it represents the smallest diameter in which it is presently feasible to house the electronics and power supply. The 40% reduction in the cross-sectional area of the DRDC Atlantic FFPCT has led to higher terminal velocities for its descent through the water column, up to 10.5 m/s versus 6 m/s for the prototype, and deeper penetration into the seabed, in excess of 2 m, in soft sediments that include thin layers of shells or sand.

4   Conclusions

DRDC Atlantic is developing acoustic and *in-situ* techniques for seabed classification and to determine the geoacoustic parameters required by models that predict the performance of mine hunting and shallow water active sonars. Particular emphasis is being placed on techniques that can be employed by moving vessels and/or air deployed buoys to characterize the spatial variability of the seabed. An experiment to test the predictions of a seabed and sub-seabed sediment scattering model has been conducted. Initial results are consistent with theoretical predictions, demonstrating the expected frequency dependent transition between interface and volume scattering, and form the basis of a classification and inversion technique in the 1 to 10 kHz band. The *in-situ* technique employs a free fall probe that combines geotechnical and geoacoustic sensors.

Sediment grain size is determined using accepted empirical relationships between the dynamic pore pressure parameter and the normalized dynamic penetration resistance. Porosity is measured using electrical resistivity and can be related to the more traditional acoustic properties of sound speed, density, and attenuation.

Acknowledgements

The authors would like to thank the officers and crew of *CFAV Quest* for their assistance in the conduct of sea-trials, and Brooke Ocean Technology and Christian Situ Geosciences Inc for analysis of FFCPT data.

References

1. Richardson, M.D. and Briggs, K., On the use of acoustic impedance to determine sediment properties. In *Proc. Inst. Acoustics* **15**, 15–23 (1993).
2. Prior, M.K. and Marks, S.G., Deduction of seabed type using in-water acoustic measurements. In *Proc. Inst. Acoustics* **23**, 74–82 (2001).
3. Heald, G.J., High frequency seabed scattering and sediment discrimination. In *Proc. Inst. Acoustics* **23**, 258–267 (2001).
4. Pace, N.G., Al-Hamdani, Z. and Thorne, P.D., The range dependence of normal incidence acoustic backscatter from a rough surface, *J. Acoust. Soc. Amer.* **77**, 101–112 (1985).
5. Hines, P.C. and Heald, G.J., Seabed classification using normal incidence backscatter measurements in the 1–10 kHz frequency band. In *Proc. Inst. Acoust.* **23**, 42–50 (2001).
6. Ivakin, A.N., Models for seafloor roughness and volume scattering. In *Proc. Oceans '98*, 518–521 (1998).
7. Turgut, A., Determination of sub-bottom sediment properties and their spatial distributions from chirp sonar data, *J. Acoust. Soc. Am.* **99**, 2451 (1996).
8. Holland, C.W. and Osler, J., High resolution geoacoustic inversion in shallow water: a joint time and frequency domain technique, *J. Acoust. Soc. Am.* **107**, 1263–1279 (2000).
9. Carey, W.M., Doutt, J., Evans, R.B. and Dillman, L.M., Shallow-water sound transmission measurements on the New Jersey Continental Shelf, *IEEE J. of Ocean. Eng.* **20**, 321–336 (1995).
10. Robertson, P.K., Soil classification by the cone penetration test, *Can. Geotech. J.* **27**, 151–158 (1990).
11. Lambert, D.N., Submersible mounted *in situ* geotechnical instrumentation, *Geo-Marine Letters* **2**, 209–214, (1982).
12. Stoll, R.D. and Akal, T., XBP-A tool for rapid assessment of seabed properties, *Sea Technology* **40**, 47–51, (1999).
13. Poeckert, R.H., Preston, J.M., Miller, T., Relega, R. and Eastgaard, A., A seabed penetrometer. DREA Technical Memorandum 97/233 (1997) 24 pp.
14. Campanella, R.G. and Weemees, I., Development and use of an electrical resistivity cone for groundwater contamination studies, *Can Geotech. J.* **27**, 557–567 (1990).
15. Rosenberger, A., Weidelt, P., Spindeldreher, C., Heesemann, B. and Villinger, H., Design and application of a new free fall *in situ* resistivity probe for marine deep water sediments, *Marine Geology* **160**, 327–337 (1999).

MEASUREMENTS OF BOTTOM VARIABILITY DURING SWAT NEW JERSEY SHELF EXPERIMENTS

A. TURGUT

Naval Research Laboratory, Washington DC 20375, USA
E-mail: turgut@wave.nrl.navy.mil

D. LAVOIE

Office of Naval Research, VA 22217, USA

D. J. WALTER AND W.B. SAWYER

Naval Research Laboratory, Stennis Space Center, MS 39529, USA

Chirp sonar and vibracore measurements of bottom variability from scales of centimeters to tens of meters have been conducted during the recent Shallow Water Acoustic Technology (SWAT) experiments at the New Jersey Shelf. Chirp sonar (2–12 kHz) data were used to invert bottom geoacoustic properties with sub-meter resolution. Sediment properties such as density, porosity and sound-speed profiles are inverted from coherent acoustic returns at several sites with well-characterized subbottom reflectors. Also, centimeter-resolution geaoacoustic properties were measured from several vibracores collected along the survey tracks. Chirp sonar inversion results compare favorably with those of co-located sediment core measurements. Both deterministic and stochastic features are deduced from the sonar surveys and co-located sediment core data. Effects of measured bottom spatial variability on the broadband acoustic propagation are also studied by using a Parabolic Equation (PE) model. It was concluded that, in addition to the oceanographic variability, the seafloor and subbottom spatial variability might further spatially decorrelate the acoustic signals propagating in shallow waters.

1 Introduction

In shallow waters, the performance of both active and passive sonar systems is strongly influenced by interaction of acoustic energy with the seabed. Proper knowledge of certain bottom geoacoustic properties such as compressional wave speed, attenuation, and density structure is needed for the accurate performance prediction of most sonar systems. The structure of geoacoustic parameters in the sediment is rather complex so that a deterministic description of such a field is almost impossible for a given site considering the length scales of interest (a few centimeters to tens of meters). Stochastic description and parameterization of the field in terms of its statistical properties might be more useful for an acoustic propagation/scattering prediction model. The structure of a sediment geoacoustic parameter is described by partitioning the field into a deterministic part and a stochastic part. The deterministic part represents site-specific large-scale features of a given geological province. The stochastic part represents small-scale sound-

N.G. Pace and F.B. Jensen (eds.), Impact of Littoral Environmental Variability on Acoustic Predictions and Sonar Performance, 91-98.

speed structure that is modeled as a zero-mean quasi-stationary random process. Second-order statistics of the field are described by a 3-D power spectrum. Several forms of spectral representations of volume inhomogeneities were proposed by previous researchers with certain spectral parameters to be estimated from available sediment core samples (see [1] for references).

2   Deterministic and stochastic description of New Jersey Shelf

Core data and chirp sonar inversion results are used to obtain both deterministic and stochastic structure of the New Jersey shelf. Similar to previous studies (see e.g., [2,3]), several deterministic subbottom features such as the marine/non-marine erosional subsurface "R", a network of buried river channels as well as several seafloor features such as erosional channels, sand ridges, and iceberg scours are well characterized by the chirp sonar surveys.

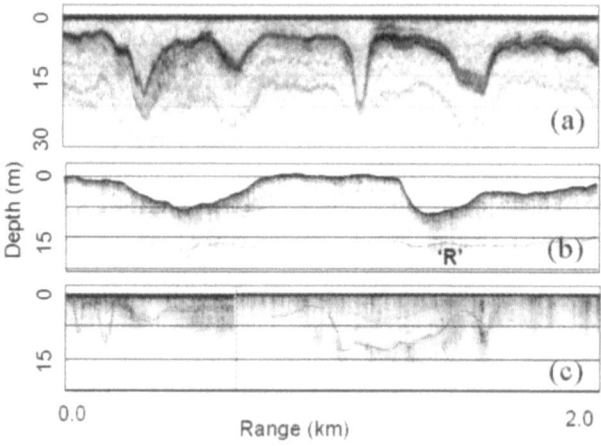

Figure 1. Spatial scales of a) internal solitary waves, b) erosional channels, and c) buried river channels observed at the New Jersey Shelf.

Figure 1 compares the typical spatial scales of buried river channels and erosional channels (geologic features) with an Internal Solitary Wave (ISW) packet (oceanographic feature) observed in the same area. Notice the similarities in the spatial scales between these geologic and oceanographic features that might influence acoustic wave propagation in the area. Chirp sonar inversions and vibracore measurements provided the high-resolution geoacoustic description of upper sediment layers as deep as 30 m. Figure 2 shows a comparison of sound-speed profiles obtained from vibracores and chirp sonar inversion at core sites H5 and H5A (Figs. 2a and 2b - see Fig. 2e for locations). The comparison was limited to the top few meters (maximum vibracore lengths), and the agreement is satisfactory. Notice also that both results show a sharp stiff-clay/sand layer interfaces verified by down-core photographic images (see Figs. 2c an 2d). A statistical modeling approach was used to characterize sediment volume inhomogeneities and bathymetry variation using an anisotropic von Kármán spectrum.

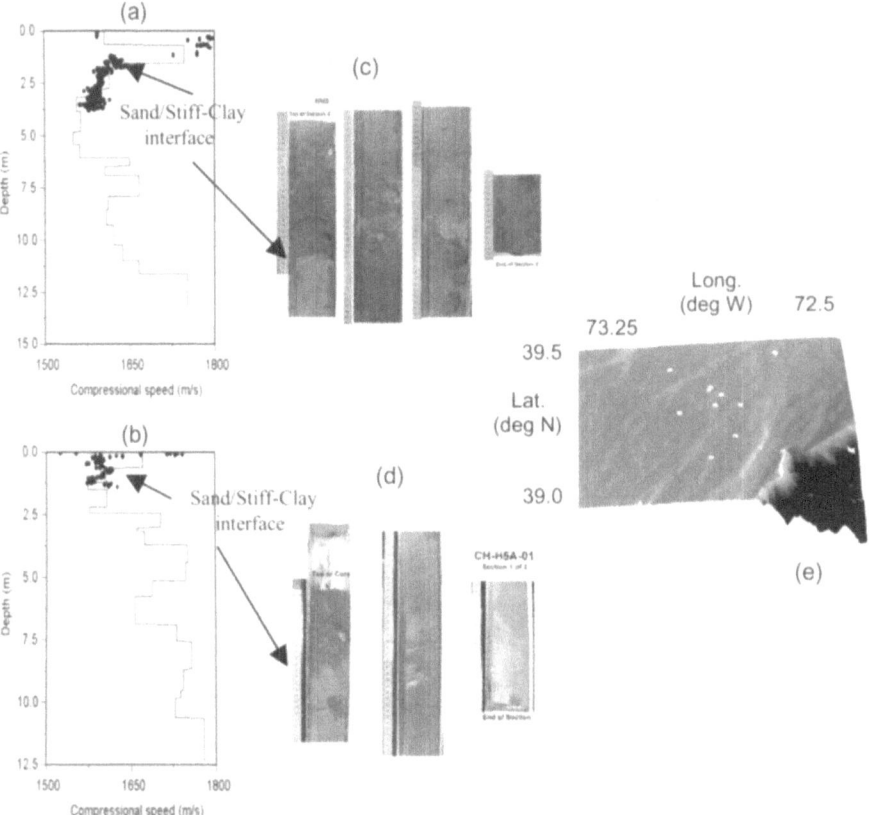

Figure 2. Comparison of sound-speed profiles obtained from vibracores and chirp sonar inversion at core sites a,c) H5, and b,d) H5A. Core locations are shown in (e).

The typical spectral parameters such as the spatial scale factors, variance, and the spectral exponent are obtained from the chirp sonar inversion results and vibracore measurements. The 3-D sediment sound-speed or density inhomogeneities can be described by an ellipsoidal von Kármán spectrum as

$$S(\xi) = \mu a_x a_y a_z \pi^{-3/2} \frac{\Gamma(m)}{\Gamma(m-3/2)} (1 + a_x^2 \xi_x^2 + a_y^2 \xi_y^2 + a_z^2 \xi_z^2)^{-m}, \qquad m > 3/2, \qquad (1)$$

where μ is the variance, a_x, a_y, and, a_z are spatial scale factors, Γ is the Gamma function, m is the spectral exponent, and ξ_x, ξ_y, ξ_z are the components of the wavenumber vector ξ. The 2-D and 1-D wavenumber spectra can be obtained by integrating the above spectrum over one and two components of wavenumber, respectively [1]. In Fig. 3a, 1-D form of the above spectrum was fitted to the spectra obtained from chirp sonar inversions and vibracore measurements of sound-speed profiles at core sites H5, H5A, and H10. Statistical parameters for the spectrum were estimated as $\mu = 0.002$, $a_z = 1.5$ m, and $m = 1.9$. Figure 3b shows the wavenumber spectrum of cross-shelf and along-shelf bathymetry calculated from the hull-mounted chirp sonar survey tracks. Statistical parameters obtained for the surface roughness were $\mu = 1.5$, $a_x = 5000$ m, $a_y = 1500$ m, and $m = 2.1$.

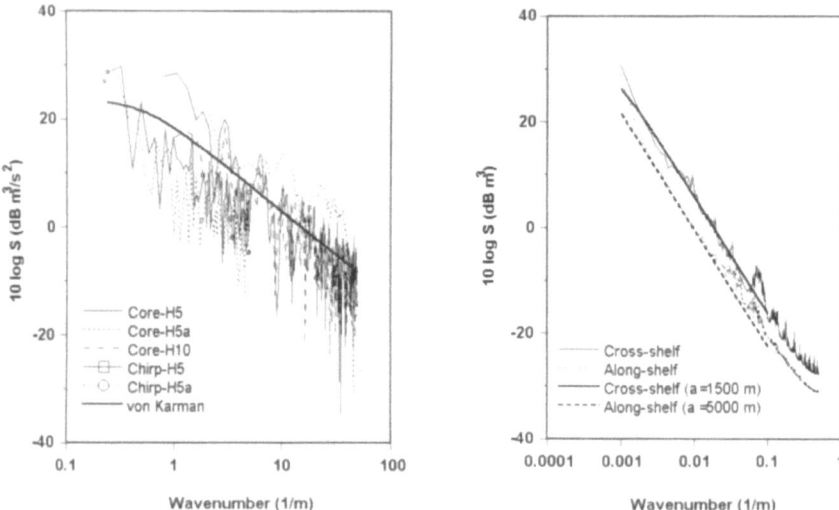

Figure 3. One-dimensional wavenumber spectra of a) sound-speed profiles obtained from chirp sonar and vibracore measurements, and b) seafloor roughness spectra obtained form chirp sonar surveys. 1-D von Kármán model predictions are also superimposed.

3   Numerical simulations for broadband acoustic propagation over deterministic and stochastic bottoms

In this section, both 2-D and 3-D numerical simulations are performed to study the coupling and refraction of the acoustic modes when they propagate through deterministic features such as erosion channels, buried river channels, and ISWs (see Fig. 4a). The vertical displacement of the deterministic features is described by $\eta(r) = A\text{sech}^2[(r-r_s)/L]$,, where A is the amplitude, r and r_s are the range variables, and L is the horizontal scale factor. Broadband acoustic propagation over stochastic bottoms is also studied using the statistical parameters of the seafloor roughness and sediment volume inhomogeneities presented in the previous section. As an example, a realization of both seafloor and sediment volume are depicted in Figs. 4b and 4c, respectively. First, we calculate the frequency and horizontal-scale dependency of mode coupling by propagating individual modes through a hyperbolic-secant range-dependency using a PE routine [4]. Figures 5a, 5b and 5c show the coupling of the first three modes as a function of the frequency and horizontal-scale factor, L, of the hyperbolic-secant function representing an ISW, an erosional channel, and a buried river channel, respectively. In general, coupling is stronger for the modes 2 and 3 in all three cases and slight differences can be detected in the frequency and horizontal-scale dependency. A relatively stronger coupling is observed in the ISW case. Also, mode coupling is stronger for the erosional channel case than that of buried river channel case.

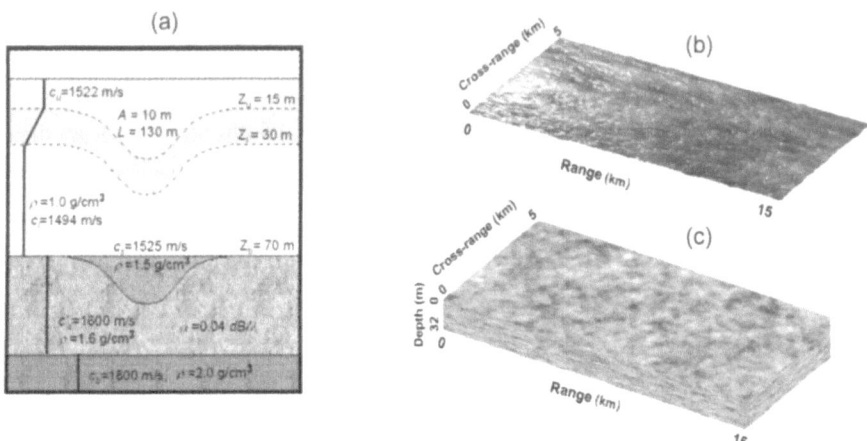

Figure 4. a) Deterministic description of an erosion channel, a buried river channel, and an ISW by a hyperbolic secant function. Stochastic description of b) seafloor morphology (rms height = 1.5 m), and c) sediment volume inhomogeneities by a von Kármán statistical model (gray scale covers from –6% to +6% variability).

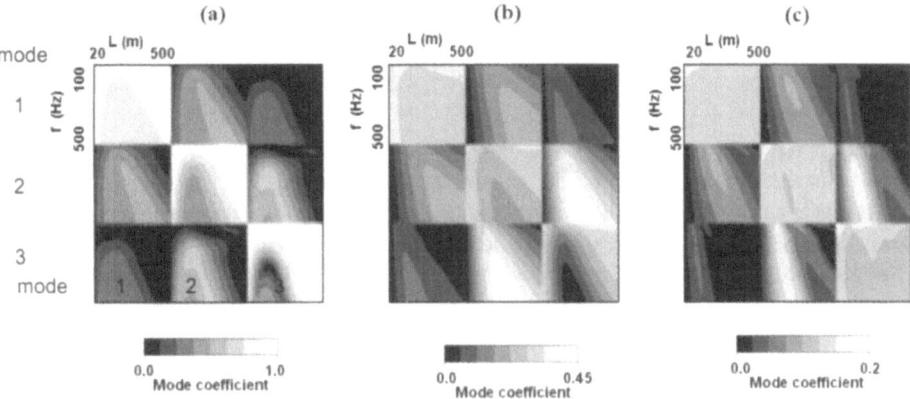

Figure 5. Frequency and horizontal-scale dependency of mode coupling for acoustic propagation, a) through an ISW, b) over an erosion channel, and c) over a buried river channel. Coupling of first three modes is displayed. The columns indicate starting-field mode number and the rows indicate the coupled-mode components. For display purposes, the panels on the main diagonals in (b) and (c) are subtracted by 0.65 and 0.85, respectively.

An adiabatic-mode PE modeling approach [5] was used to study the horizontal refraction of broadband acoustic signals propagating over deterministic and stochastic bottoms. The main difference between our modeling approach and that of Ref. [5] is that we used the PE formulation in a Cartesian coordinate system instead of a cylindrical or spherical coordinate system. Figures 6a, 6b, and 6c show the refraction of 70–190 Hz broadband signals obliquely propagating through a hyperbolic-secant range dependency of an ISW, an erosional channel, and a buried river channel, respectively. The azimuthal angle between acoustic propagation vector and the straight hyperbolic-secant disturbance is 8 deg., A = 10 m, and L = 130 m. The waterfall plots show the amplitude of pressure field received at 15 km range and 40 m water depth. The total cross-range extension is 5 km and the source is placed at 2.5 km cross-range distance. The 8-deg. angle corresponds to the cross-range distances of the maximum of the straight hyperbolic-secant disturbance being at rs = 3 km at the source range (0 km), and rs = 1.1 km at the receiver range (15 km). For the ISW case, the amplitude of the pressure field is larger at the leading edge of the disturbance, indicating a relatively stronger refraction. Also, similar to the mode coupling results, mode refraction is stronger for the erosional channel case than for the buried river channel case (note the smaller pressure amplitude in the disturbed field).

In Figs. 6d, 6e, and 6f, pressure amplitudes at 15 km range and 40 m water depth are compared for the environments with no range dependency, with bathymetric variability, and with sediment sound-speed and density variability, respectively. The pressure amplitudes in Figs. 6e and 6f are calculated for single realizations of bathymetry and sediment volume using the von Kármán spectra with the statistical parameters given in the previous section ($a_x = a_y = 1000$ m is assumed for the sediment variability). To quantify the effects of the environmental variability on pulse propagation, we define a spatial correlation coefficient for the transient pressure amplitude as [6]

$$\rho(\varsigma) = \frac{\left| \int p(t;0) p(t;\varsigma) dt \right|}{\left(\int \left| p(t;0) \right|^2 dt \right)^{\frac{1}{2}} \left(\int \left| p(t;\varsigma) \right|^2 dt \right)^{\frac{1}{2}}} \tag{2}$$

where t is the time and ς is the cross-range distance. The integrals in the above equation are evaluated over the width of the calculated pulses. In Fig. 6g, the spatial correlation functions are calculated for the deterministic cases along a 600 m aperture horizontal line array located at 15 km range, 40 m depth, and between 2.2 and 2.8 km cross-range distances (reference position is at 2.5 km).

For reference, the correlation function for the range independent case is also plotted since the source-receiver distance is different for the each element of the array. We note that the spatial decorrelation is stronger for the ISW case since the bottom interacting higher order modes are attenuated at the 15 km range. In Fig 6h, the spatial correlation functions for the stochastic bottoms are compared with range independent case. Note that bathymetric variability introduces stronger spatial decorrelation than does the sediment volume variability. In a recent numerical study, weak spatial decorrelation of CW signals are also observed in the case of acoustic propagation through diffuse background internal wave field [7].

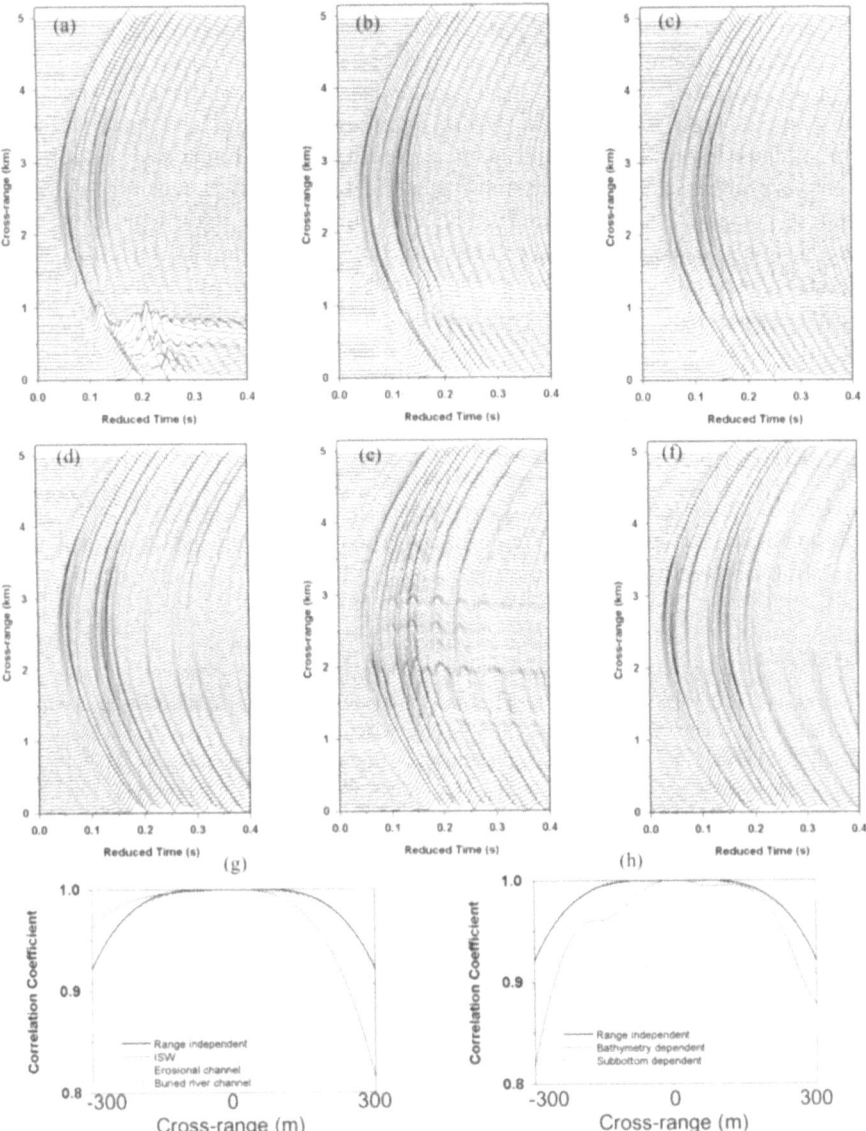

Figure 6. Transient pressure amplitudes calculated at 15km range and 40 m depth (source depth = 20 m) depicting horizontal refraction effects for deterministic and stochastic environments. A hyperbolic-secant disturbance is used to represent (a) ISW, b) erosional channel, and c) buried river channel. The transient pressure amplitudes for d) range independent, e) bathmetric variability, and f) sediment volume variability case. Corresponding spatial decorrelations reference to the mid-point (at 2.5 km cross-range) are shown in (g) and (h).

4   Discussion

In this study, we established a procedure for obtaining deterministic and stochastic bottom properties from vibracore and chirp sonar reflection measurements. Seafloor morphology and 1-D (vertical) sediment inhomogeneities are statistically characterized by using several data sets from the SWAT New Jersey Shelf experiments. Further analysis of chirp sonar inversions is underway to estimate horizontal scale factor of the sediment variability. Vibracore measurements and chirp sonar inversions of sound-speed profiles are compared at core sites H5 and H5A and a good agreement is observed. Numerical simulations of horizontal refraction of broadband signal indicated that several deterministic features, observed during the recent SWAT experiments, might introduce significant mode coupling and horizontal refraction. Also, in addition to oceanographic effects, measured levels of bottom variability might further decrease the spatial correlation of the low-frequency acoustic signals in shallow waters.

Acknowledgments

This work was supported by the Office of Naval Research. We thank Keith Ludwig of the USGS and Bruce Pasewark, Chad Vaughan, and Allen Reed of the NRL for participating in the SWAT vibracoring/chirp sonar experiment.

References

1. Turgut, A., Inversion of bottom/subbottom statistical parameters from acoustic backscatter data, *J. Acoust. Soc. Am.* **102**(2), 833–852 (1997).
2. Goff, J.A., Swift, D.J.P., Duncan, C.S., Mayer, A.M. and Hughes-Clarke, J., High resolution swath sonar investigation of sand ridge, dune, and ribbon morphology in the offshore environment of the New Jersey margin, *Marine Geology* **161**, 307–337 (1999).
3. Duncan, C.S., Goff, J.A., Austin, J.A., Jr., Fulthorpe, C.S., Tracking the sea-level cycle: seafloor morphology and shallow stratigraphy of the latest Quaternary New Jersey middle continental shelf, *Marine Geology* **170**, 395–421 (2000).
4. Duda, T.F. and Preisig, J.C., Coupled acoustic mode propagation through continental-shelf internal solitary waves, *IEEE J. Ocean. Eng.*, **24**, 256–269 (1997).
5. Collins, M.D., Adiabatic mode parabolic equation, *J. Acoust. Soc. Am.* **94**(4), 2269–2278 (1993).
6. Rouseff, D., Turgut, A., Wolf, S.N., Finette, S., Orr, M.H., Pasewark, B.H., Apel, J.R., Badiey, M., Chiu, C.-S., Headrick, R.H., Lynch, J.F., Kemp, J.N., Newhall, A.E., von der Heydt, K. and Tielbuerger, D., Coherence of acoustic modes propagating through shallow water internal waves, *J. Acoust. Soc. Am.* **111**(4), 1655–1666 (2002).
7. Oba, R. and Finette, S., Acoustic propagation through anisotropic internal fields: Transmission loss, cross-range coherence, and horizontal refraction, *J. Acoust. Soc. Am.* **111**(2), 769–784 (2002).

MAPPING SEABED VARIABILITY USING COMBINED ECHOSOUNDER AND XBPS FOR SONAR PERFORMANCE PREDICTION

K.M. KELLY AND G.J. HEALD

QinetiQ, Winfrith Technology Centre, Dorchester, Dorset, UK
E-mail: kmkelly@qinetiq.com, gcheald@qinetiq.com

The acoustic experiment, Cerberus, took place in the SouthWest Approaches to the English Channel in August 2001. During this experiment geoacoustic information was obtained using a combination of Expendable Bottom Penetrometers (XBPs) and the echosounder seabed classification systems, EchoPlus and Roxann. The results of combining these geophysical techniques mapped both variations in sediment type and the presence of large bathymetric features in the region. This improved understanding of the seabed type and variability will lead to improved acoustic prediction. Combining these techniques provides a simple and effective method for gathering seabed variability data for input to acoustic models. This paper investigates the practicability of using XBPs for ground truthing the echosounder systems and compares the results from the three systems. The effect of the seabed variability on sonar performance prediction is discussed.

1 Introduction

During August and September 2001 an Active Sonar experiment (Cerberus) took place in the Southwest Approaches. Environmental data was obtained during the experiment in support of the sonar operations. This included a geophysical dataset from the outer shelf of the Southwest Approaches continental shelf region.

The geophysical dataset obtained included data from two sediment discrimination systems, EchoPlus and Roxann, and from expendable Bottom Penetrometers (XBPs). The two discrimination systems use the first and second echoes from the echosounder which are related to roughness and hardness respectively. Background on these systems can be found in Chivers [1] and the theory of the first and second echo for sediment discrimination is given by Heald [2]. These systems give an indication of seabed spatial variability. The XBPs were cone penetrometers deployed using the ship's Expendable bathythermograph (XBT) launcher. They measured deceleration (g) and seabed penetration (cm). These parameters can be related to different sediment characteristices and hence to seabed type [3].

The seabed in this region is composed mainly of Quaternary sands and gravels which overlie the Pleistocene clayey sands of the Upper Little Sole Formation [4]. These Quaternary deposits are formed into a series of large tidal sand ridges, which are the main bathymetric feature of the outer shelf. They form ridges up to 60 m high, 200 m long and 15 km spacing and are relict features which formed during the Late Devensian

N.G. Pace and F.B. Jensen (eds.), Impact of Littoral Environmental Variability on Acoustic Predictions and Sonar Performance, 99-106.

lowstand of the ice age, when sea levels were lower and tidal forces over the shelf were stronger [5]. They are now in a state of decay.

After the Ice Age, as the sea levels rose, the ridges advanced landward until they reached the part of the shelf where sediment supply was insufficient to maintain their growth. Their landward extension stopped at about the 120 m isobath. Seabed reworking during this period formed a sandy gravelly lag pavement. The present seabed consists of partly mobile sediments which are swept by tides and wave induced currents into a series of beforms which range in size from minor sand ripples to tidal sand ridges. Figure 1 is the British Geological Survey (BGS) surface sediment map for the region.

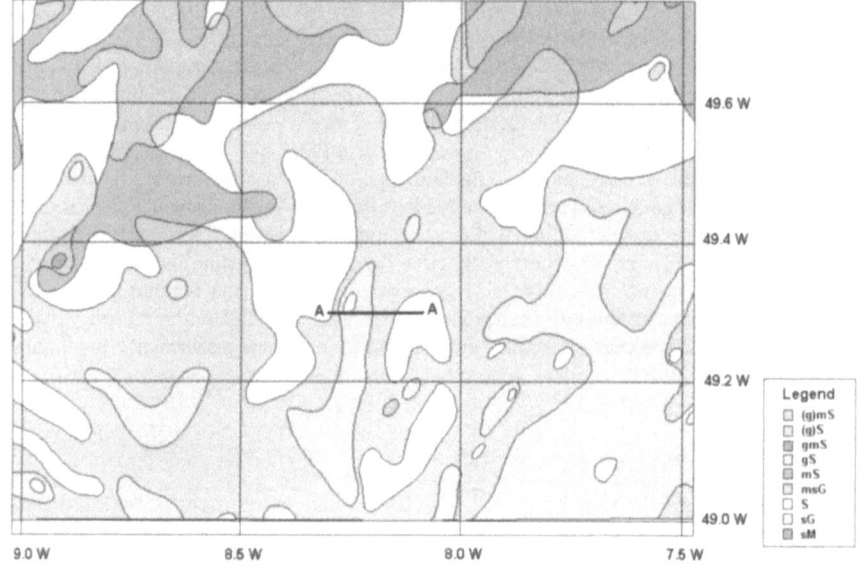

Figure 1. BGS surface sediment map for the survey area.

2 Geophysical datasets

2.1 Echoplus

EchoPlus is a seabed discrimination system that can be installed on any vessel with a single beam echosounder. It is hardwired into the ship's echosounder and uses the echosounder returns from the first and second echoes to determine the roughness and hardness of the seabed. Mapping the variations in the roughness and hardness outputs gives an indication of seabed variability and allows classification into discrete regions.

The first echo consists of a leading edge, which is caused by the initial reflection and scattering from the seabed, and a trailing edge, which is dominated by scattering from the seabed and possibly reflections from the sub-bottom. The analysis window integrates the trailing edge of this first echo to determine a measure of the seabed roughness. The rougher the seabed the more energy will be scattered back to the transducer and the more energy will appear in the integrated analysis window. The second echo includes contributions from seabed, sea surface and sub-bottom. This echo is used to determine the hardness of the seabed. The harder the seabed the stronger the nearfield scattering will be and the more energy appears in the analysis window [1].

EchoPlus was hardwired into the MV BREMEN echosounder which was an Elac LAZ 72 echosounder which was operating at 30 kHz. The system operated well throughout the exercise and a good quality dataset was obtained.

2.2   Roxann

Roxann works on the same principle as that already described for EchoPlus, using the tail of the first echo (E1) and the second echo (E2) to give an indication of roughness and hardness [4]. In this case Roxann used a Simrad echosounder operating at 200 kHz.

Normally the echosounder transducer is deployed over the side of the survey ship attached to a metal pole, but during Cerberus, due to the size of the ship, this could not be achieved. There was also no suitable moon pool available on the ship which could be used for tranducer deployment. Therefore a towed body arrangement was set up. This was by no means ideal since the length of cable required resulted in a reduced signal to noise ratio, meaning that the data obtained was very noisy. Also the system expects the transducer to be just below the surface, but with a towed body it was deployed somewhat deeper. This meant that the time delay until the second echo, calculated by the system using the water depth was slightly inaccurate resulting in a poor return from the second echo. This could be overcome by changing the timing based on the position of the transducer in the water column.

2.3   XBPs

Expendable Bottom Pentrometers (XBPs) are expendable seabed probes for measuring the in-situ physical properties of seabed sediments. They use the same technology as the widely used expendable bathythermograph (XBT), except that an accelerometer replaces the usual thermistor and special electronics and computer logic are employed to sense and record several different aspects of impact and penetration into the seabed.

When the probe impacts with the seabed the measured deceleration is proportional to the resistive force exerted by the sediment. This depends on the undrained shear strength of the sediment. After the probes downward motion stops there is a period of heavily damped oscillatory motion with no further net penetration, particularly in stiffer granular sediments. Whilst this motion is occurring the probe is behaving in a way analogous to a mass on an elastic foundation. This portion of the record can be analysed to obtain an estimate off the dynamic shear modulus which in turn can be used to calculate shear wave velocity [6].

There are much higher peak forces and rates of deceleration with very little penetration in areas where the surface sediment is sand. In softer fine grained sediments where porosites are typically greater than 50% the probe penetrates to a significant depth. The time required for full penetration is larger and the maximum deceleration is much smaller.

3   Comparison of geophysical datasets

Deploying the two echosounder systems simultaneously during Cerberus meant that a comparison could be made between the two systems. They use essentially the same processing techniques, the trailing edge of the first echo being used to give an indication of seabed roughness, and the second echo being used to give seabed hardness. The main differences between the two systems are their operating frequencies and the mode of deployment used.

The Roxann E1 signal indicated several regions of increased seabed roughness but the E2 signal remained consistently low, only indicating increased seabed hardness on two occasions. Both coincided with Roxann transects across a large patch of gravelly sand and sandy gravel.

Figure 2 shows both the Roxann and EchoPlus data for one of the transects of the gravel patch, section A-A Fig. 1. The Roxann E1 signal is giving a strong return associated with the gravelly sand patch. E2 shows a similar trend although the signal is weaker. The EchoPlus hardness signal is very similar to the Roxann E1 signal. There is no variation in roughness. This shows that both systems are detecting the same seabed features.

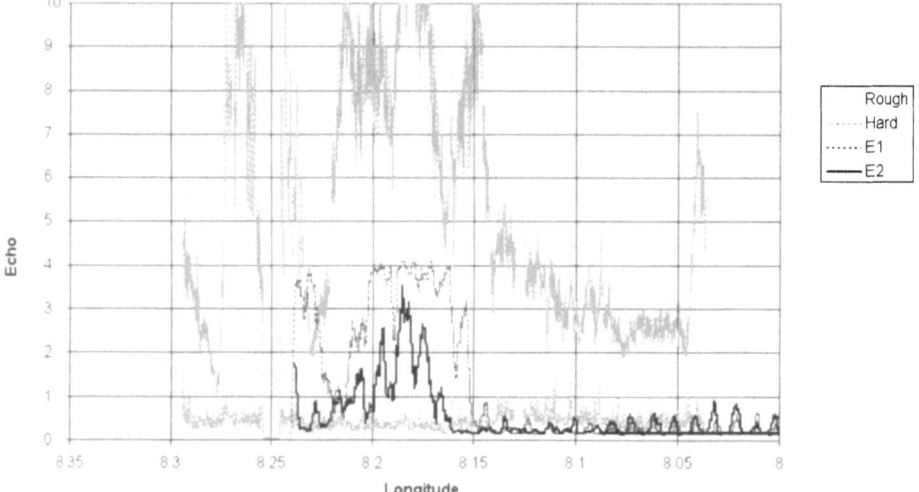

Figure 2. EchoPlus, roughness and hardness and Roxann E1 and E2 for section A-A (see Fig. 1).

Both the EchoPlus and Roxann systems provide information on seabed variability, but since the signal can be strongly affected by other environmental factors, such as the interaction of the second echo with the sea surface, repeated tracks under different environmental conditions can give different results. As a result the same roughness/hardness relationships will not always apply and some form of ground truth will be necessary in order to use these systems for actual seabed classification. They will however consistently show relative discrimination levels. Normally the ground truthing uses cores or grabs to identify the sediment present at the seabed. However this involves the ship stopping which is not always practical. The possibility of using XBPs for ground truth was therefore investigated.

Figure 3 is a plot for all the XBP data showing the deceleration curves. Although all the XBP deployments were identified by the XBP software as seabed type I, sand and gravel, there are three distinct curve types visible in the data. The first type shows a maximum deceleration of greater than 200 g. This probe was deployed on the top of

Haddock Bank and is likely to be representative of coarse sand or gravel. The second type corresponds to most of the remaining XBP deployments. Decelerations are between 100 and 200 *g*. These deceleration curves are representative of sands which makes up most of the seabed in this area. The final type shows decelerations lower than 100 *g* and taking place over a greater time spread. They also show a distinct double deceleration, a small initial deceleration being followed by the main deceleration. This preliminary deceleration is probably caused by a disturbance of the sediment by the downforce of water just ahead of the probe. These probes were deployed in a region to the west of the trials area where muddy sands are more prevalent on the seabed.

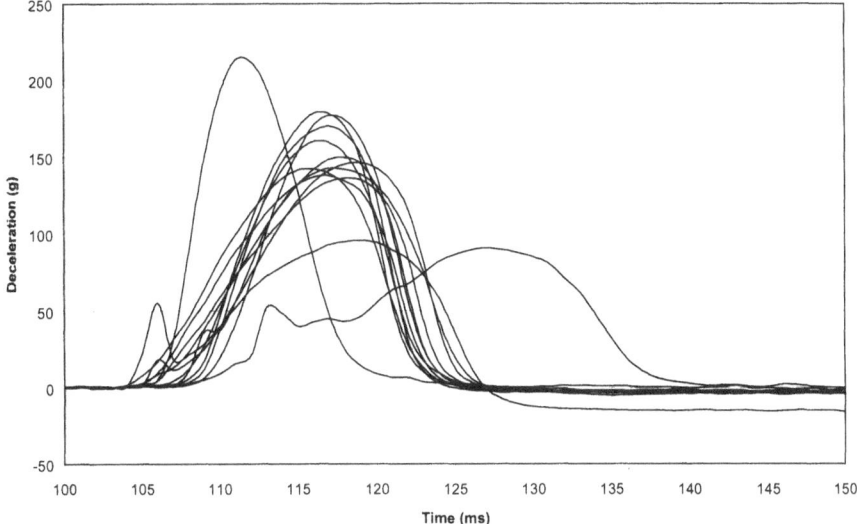

Figure 3. Deceleration curves for XBP dataset.

Figure 4 is a plot of penetration against deceleration for the XBP data. In general the two are related, higher decelerations being associated with less penetration, although there is some variability within the sands. The groupings of the deployments into the three seabed types identified are shown.

The definitions of gravel, sand and muddy sand determined from the XBP data were then used to identify the corresponding EchoPlus data distribution. Figure 5 shows roughness/hardness plots of EchoPlus data from a 1-min window surrounding 3 typical XBP deployment, one for each seabed type.

The data shows distinct variations in the roughness/hardness relationships for the different seabed types. Sands appear to give lower hardness values that both muddy sand and gravel. The main difference between the muddy sand and the gravel is that roughness is lower for the muddy sands and the data forms a distinct cluster. The roughness/hardness relationship for the gravels shows a greater spread of data values. As a result, when an attempt was made to define seabed type using this raw data, it was difficult to distinguish between the muddy sand and gravel as a result of the spread of values associated with the gravel. However the regions of sand could be clearly identified, see Fig. 6. This type of acoustic signal "overlay" is not uncommon in these types of dataset and emphasizes the need for good ground truth.

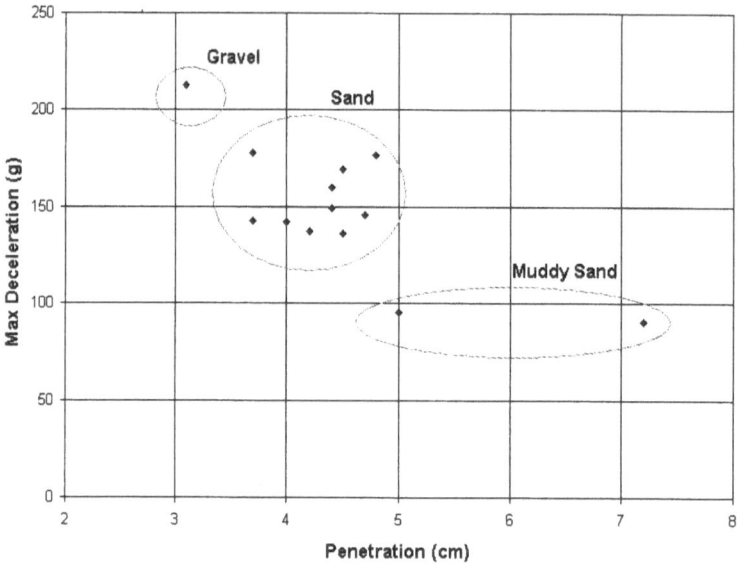

Figure 4. Plot of penetration against deceleration for the XBP dataset showing the distribution of the different seabed types.

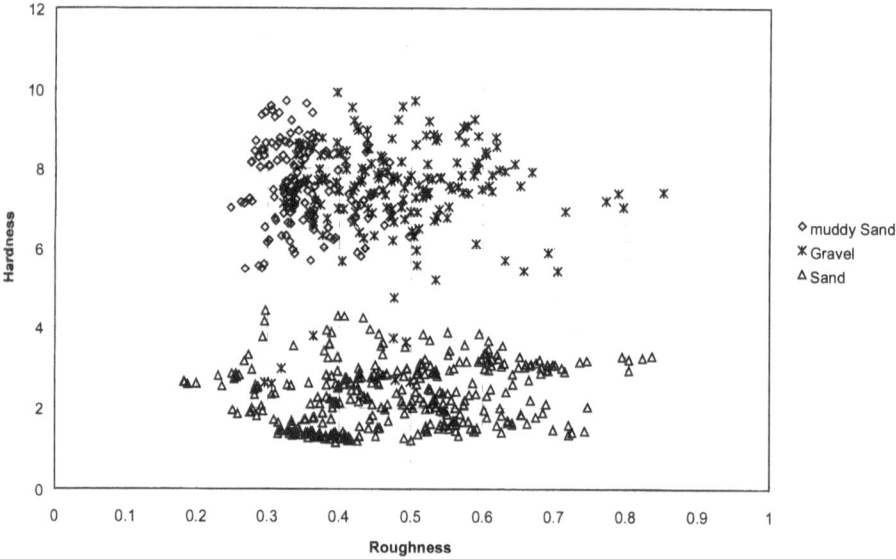

Figure 5. Relationship of roughness and hardness for the three different seabed types.

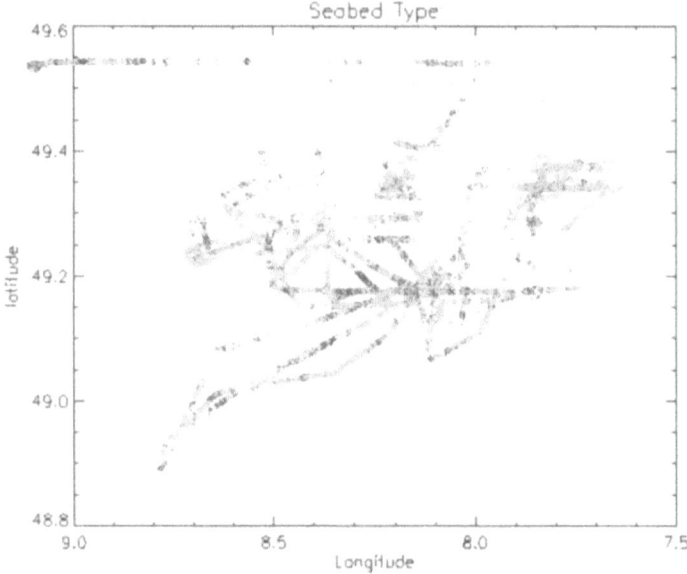

Figure 6. Seabed type map based on the seabed type distributions defined in Fig. 5.

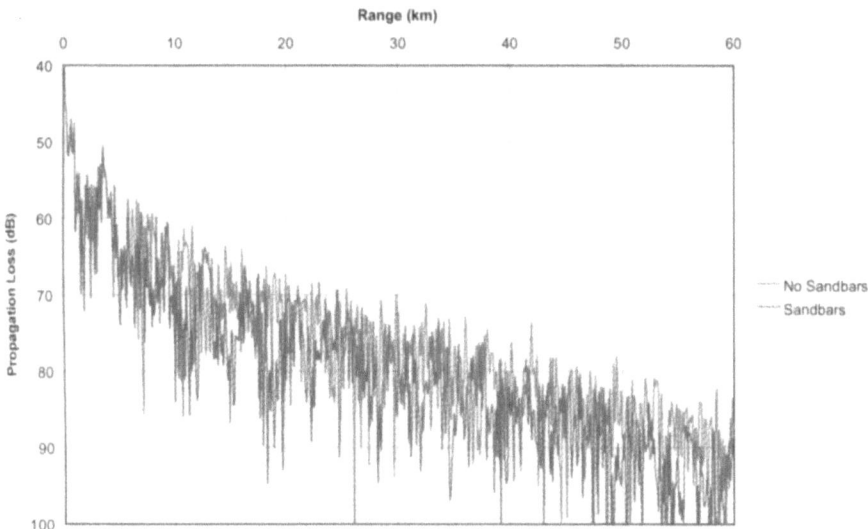

Figure 7. Propagation loss at 3.5 kHz for the Cerberus environment, with and without sandbars present.

4   Acoustic predictions

The geophysical data obtained during Cerberus indicated that the seabed in the survey area was relatively consistent in composition, ranging from muddy sands to gravels with most of the region consisting of sands and gravelly sands. Model predictions indicated that these variations did not significantly affect propagation loss. Neither did the inclusion of the sandbars, as is illustrated in Fig. 7. Propagation loss predictions for this environment, both with and without sandbars present only differs by a few dB.

5   Conclusions

The two echosounder based seabed discriminators, EchoPlus and Roxann, are useful survey tools for assessing seabed variability. In order to attempt to categorize the seabed types present some form of ground truth is required. This paper investigates the potential for using XBPs for such a purpose. The results would suggest that this is a potentially viable technique although there are some problems with acoustic overlay when relating the two types of data. The results indicate that the effects of seabed variability in the Cerberus area are not significant for sonar performance prediction.

References

1. Chivers, R.C., Emerson, N.C. and Burns, D., New acoustic processing for underway surveying, *The Hydrographic Journal* **56**, 9–17 (1990).
2. Heald, G.J. and Pace, N.G., Implications of a bi-static treatment for the second echo from a normal incidence sonar. In *Proc. 3rd European Conference on Underwater Acoustics*, edited by J.S. Papadakis, Crete, Greece (1996) pp. 649–654.
3. Hamilton, L.J., Mulhearn, P.J. and Poeckert, R., Comparison of RoxAnn and QTC-View acoustic bottom classification system performance for the Cairns area, Great Barrier Reef, Australia, *Cont. Shelf Res.* (1999).
4. Evans, C.D.R., United Kingdom offshore regional report: The geology of the western English Channel and its western approaches (London: HMSO for the British Geological Survey, 1990).
5. Belderson, R.H., Pingree, R.D. and Griffiths, D.K., Low sea-level tidal origin of Celtic Sea sandbanks – evidence from numerical modelling of M2 tidal streams, *Marine Geology* **73**, 99–108 (1986).
6. Stoll, R.D. and Akal, T., XBP-tool for rapid assessment of seabed sediment properties, *Sea Techology*, 47–51 (Feb. 1999).

VARIABILITY OF SHEAR WAVE SPEED AND ATTENUATION IN SURFICIAL MARINE SEDIMENTS

MICHAEL D. RICHARDSON

Marine Geosciences Division, Naval Research Laboratory, SSC, MS 39529-5004, USA
E-mail: mike.richardson@nrlssc.navy.mil

Values of shear wave speed and attenuation in surficial marine sediments (upper 30 cm) are summarized and then correlated with easily measured sediment physical properties (porosity, bulk density, and mean grain size). Shear wave speed ranges from a low of 5 m/s in soft silty-clays to a high of 150 m/s in hard pack fine sands. Shear wave speed increases with decreasing porosity, increasing bulk density, and increasing mean grain size. Strong gradients in shear wave speed in the upper meter of sediment, related to increased effective stress (overburden pressure) and vertical gradients in sediment properties complicate theses predictive relationships. Shear wave attenuation follows the opposite trends as shear wave speed, increasing with increasing porosity, decreasing mean grain size and decreasing bulk density.

1 Introduction

Knowledge of sediment geoacoustic properties is of fundamental importance to marine environmental, military, and engineering applications. For instance, geoacoustic properties are used to predict the stability of marine slopes, sediment consolidation behavior, strength of marine foundations, liquefaction potential, mine burial, and scattering of acoustic energy into and from the seafloor. *In situ* measurement of compressional and shear wave speed and attenuation in marine sediments is a well-developed technology [1,2]. Values of shear wave speed and attenuation measured in a variety of silicilastic and carbonate sediments over the past 14 years are compiled and new regressions developed between sediment physical and geoacoustic properties are presented.

2 Methods

Shear wave speed and attenuation were measured using several different versions of the *In situ* Sediment geoAcoustic Measurement System (ISSAMS) [1,2]. The compiled data include previously published studies near La Speizia, Italy during 1988 [1,3,13], in the Adriatic Sea during 1989 [1,8], in Eckernförde Bay, Baltic Sea during 1993, 1995, and 1997 [7,12], off Boca Raton, Florida in 1994 [8], in the West Sound off Orcacs Island in Puget Sound, Washington during 1995 [10], offshore the Eel River, along the Northern California Coast in 1996 [4,5], in the northeastern Gulf of Mexico, near Panama City during 1989, 1993, 1998 [7,8] and near Ft Walton Beach during 1998 [9,11], and in the lower Florida Keys during 1994–1997 [6]. The data base also includes unpublished data from the North Sea (1997) and unpublished data collected

N.G. Pace and F.B. Jensen (eds.), Impact of Littoral Environmental Variability on Acoustic Predictions and Sonar Performance, 107-114.

along the lower Florida Peninsula during 1998. Sediments range from very soft, low-porosity mud in Eckernförde Bay to fine-to-medium well-sorted sands in the northeastern Gulf of Mexico. A few sites near La Spezia and in the northeastern Gulf of Mexico were poorly sorted coarse sands. Carbonate sediments in the Florida Keys and along the lower Florida Peninsula, were generally poorly sorted with a variety of coarser reef debris and shell fragments mixed with finer grained calcareous algae. Calcareous sediments have been shown to slightly different values of sediment geoacoustic properties than silicilastic sediment of the same physical properties [6].

2.1   Geoacoustic Measurements

Values of shear and compressional wave speed and attenuation were measured using several different versions of the In Situ sediment geoAcoustic Measurement System (ISSAMS) [1,2,8,13]. Pulse techniques that utilize time-of-flight and amplitude measurements between pairs of compressional and shear wave probes are used to measure speed and attenuation. The geoacoustic probes are driven into the sediment using a variety of platforms including diver-deployed, mechanically-operated, and hydraulically-operated systems [8]. Although electro-mechanical methodologies of data acquisition and probe deployment have evolved over the years, actual geoacoustic probe design and signal processing has changed little. Compressional wave transmit and receive probes are identical radial-poled ceramic cylinders mounted on a hollow stainless steel tubes. The ceramic cylinders are potted in polyurethane resin to electrically isolate the probes from seawater and for protection during insertion. Shear wave transmit and receive probes are bimorph ceramic benders, potted in a stainless steel ring with silicone rubber to allow unrestricted bender movement. A thin coating of much harder polyurethane resin holds the ceramics in place and provides a tough coating to protect the ceramics during insertion.

Compressional and shear wave speed and attenuation are typically measured over pathlengths ranging from of 30 to 100 cm at depths of 5–30 cm below the sediment-water interface. For compressional wave measurements, transmit pulses are driven utilizing 38 to 58 kHz pulsed sine waves and time delays and voltages are used to determine values of speed and attenuation. Actual values of compressional speed and attenuation are calculated by comparison of received signals transmitted through the sediment with those transmitted through seawater overlying the sediments. Shear speed is measured as time of flight between probes driven at 100 to 2000 Hz. Recently; techniques based on transposition have been developed to measure compressional and shear attenuation without standards [5]. Even more recent improvements to compressional wave measurement techniques, based on comparison of signals of paired receivers [8], have not yet been applied to these data sets but would probably change values of shear wave speed and attenuation very little.

2.2   Sediments

Sediment cores were carefully collected by either by divers in shallow water or from box core samples in deeper water. After cores are acoustically logged, sediments were sectioned at 2-cm intervals and water content was determined by water loss of samples dried in an oven at 105°C for 24 h. Porosity and bulk density were calculated from values of water content and values of grain density measured with a Quantachrome Ultrapycnometer. The size distribution of gravel- and sand-sized particles was determined by dry-sieving and a Micromeritics sedigraph or pipette analysis was used to determine the silt- and clay-sized fractions. Grain size distributions are described by the graphical methods of Folk [7]. Great care was exercised to minimize disturbance on sediment cores and physical properties are measured soon as physically possible. The original data for most of the data sets analyzed here can be found in the references [1–13].

3   Data compilation

Shear wave speeds together with sediment physical properties (porosity, bulk density and mean grain size) were measured at approximately 100 sites during that last 14 years. In most cases the original data has been published as part of regional and local geological and geophysical studies, high-frequency acoustic experiments, site surveys, or papers on development and/or improvements to ISSAMS or data analysis techniques. Several papers have provided summaries of past in situ measurements [6,8]. This paper is confined to comparisons of values of shear wave speed and attenuation to sediment physical properties.

In most cases shear wave speed was measured on multiple deployments of ISSAMS within 25-m of the same location. Shear wave speed and attenuation were compiled from multiple depths (10–30 cm) below the sediment water interface and between multiple pairs of transducers. A typical deployment would yield 12 independent values shear wave speed, and 3 independent values of shear wave attenuation. The measurement frequency (70–2000 Hz) is dependent of transducer loading with lower frequencies used in softer muddy sediments and higher frequencies used in sandy sediments. Typically, 1000 Hz is used for most fine-to-medium sands and 100–300 Hz is used in soft muddy sediments. Shear wave speed was measured at 100 sites but shear wave attenuation has been only been compiled for 18 silicilastic sites. Work proceeds slowly on the reanalysis waveforms to determine shear wave attenuation. Usually multiple cores were collected are each site and sediment physical properties measured at 2-cm interval down cores.

4   Regressions

Near surface gradients of shear wave speed are common, especially in sandy sediment (Fig.1), These gradients are primarily in response to increases in mean effective stress, rather than changes in sediment physical properties [3]. The empirical relationships developed in this paper do not account for the effects of these gradients and can best be considered averages for the upper 30 cm of sediments. Corrections can be easily applied to the relationship between shear wave speed and physical properties which will account for these gradients. Scatter diagrams with empirical fits for shear wave speed and attenuation compared to sediment physical properties are presented in Figs. 2 and 3.

As expected from past compilations of shear wave speed [6,8,13] shear wave speed is highest in sandy sediments with high density and low porosity. The regressions for shear wave speeds with porosity ($R^2 = 0.73$) and density ($R^2 = 0.85$) were improved when restricted to silicilastic sediments ($R^2 = 0.897$ and 0.92 respectively). Regressions

of shear wave speed with mean grain size were not improved with the deletion of
carbonate sediments.   Shear wave speeds for the poorly-sorted carbonate sediment
found in the lower Florida Keys were higher than shear wave speeds in silicilastic
sediments, given the same porosity and density, and accounted for most of the
differences between sediment types. Very little evidence of cementation was evident in
TEM micrographs of these sediments suggesting other causes for the higher rigidity or
shear wave speed [6]. It was suggested that high intraparticle porosity (10–15%) in
lower Florida Keys sediments is sufficient to account for the difference higher shear
wave speeds given porosity or bulk density and yet preserve the shear wave mean grain
size relationships. For now it is suggested that the two different predictive regressions
be used for silicilastic and carbonate sediments or grain size be used to predict shear
wave speed. The first extensive compilation of shear wave attenuation and sediment
physical properties appears to yield some useful relationships. Shear wave attenuation
follows the opposite trends as shear wave speed, increasing with increasing porosity,
decreasing mean grain size and decreasing bulk density. Although the percentage of the
data accounted for by these linear regressions in not high (Fig. 3) the trends appear
convincing.

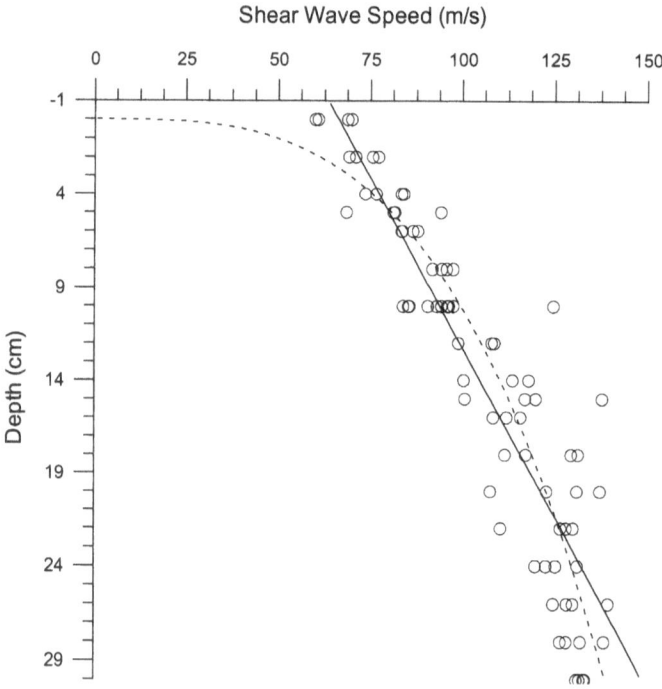

Figure 1. Gradients in shear wave speed at a well-sorted fine sand site (C1) in the North Sea.
Depth regressions include a linear fit (Vs = 72.9 = 2.3D) and the traditional power law fit
(Vs = 56.8D$^{0.25}$).

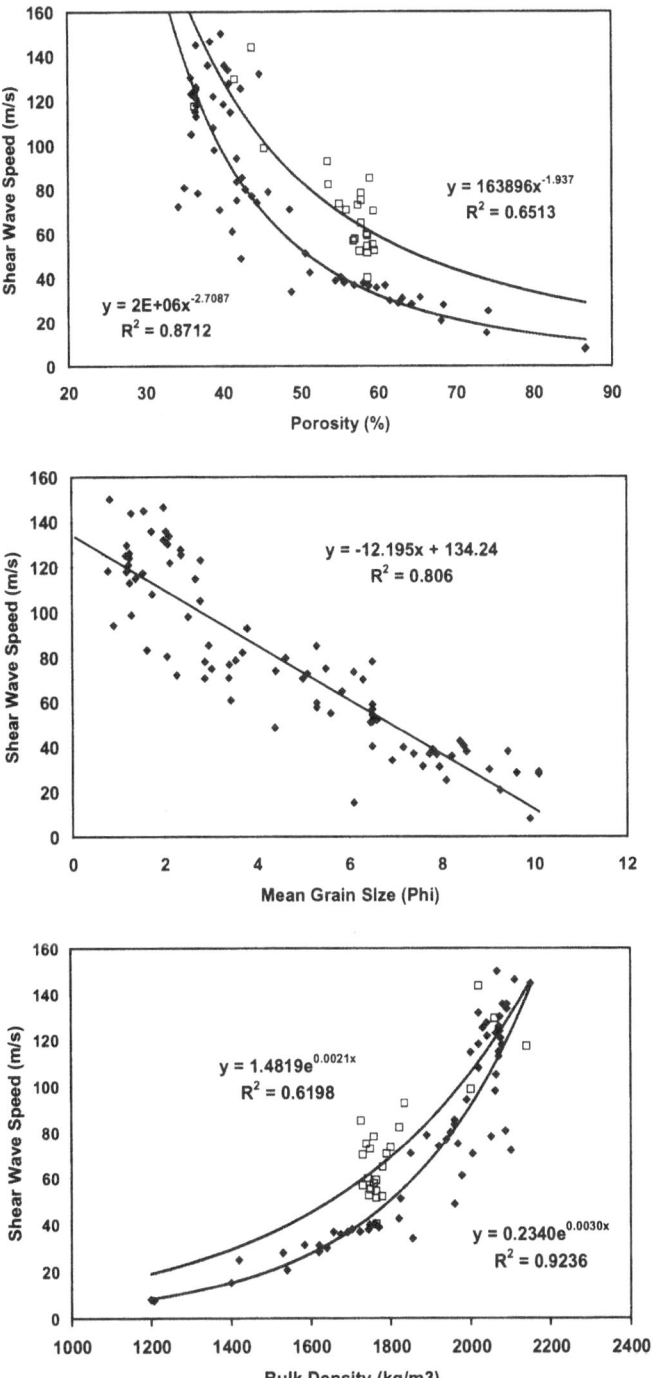

Figure 2. Empirical relationships between shear wave speed (m s^{-1}) and sediment physical properties (porosity, mean grain size and bulk density).

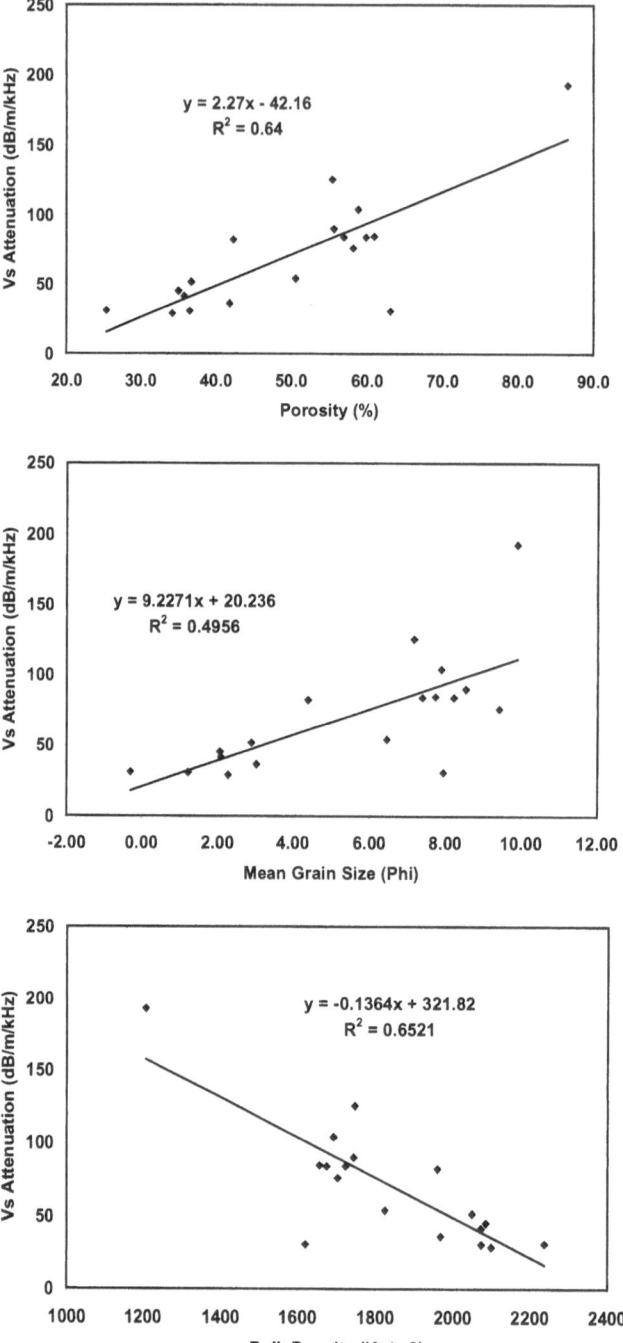

Figure 3. Empirical relationships between shear wave attenuation (dB m^{-1} kHz^{-1}) and sediment physical properties (porosity, mean grain size and bulk density).

Acknowledgements

ISSAMS benefited from the electro-mechanical expertise of Enrico Muzi and Bruno Miaschi (SACLANTCEN) during early development and from engineering expertise of Sean Griffin and Frances Grosz (Omni Technologies) for its current remotely-operated, hydraulic configuration. Kevin Briggs (NRL), Briano Tonarelli (SACLANTCEN) and Fedra Turgutcan (SACLANTCEN) provided most of the sediment characterization. This work was supported by NRL Program Element 601153N, Herb Eppert, Program Manager and is NRL contribution NRL/PP/7430-02-0003.

References

1. A. Barbagelata, M.D. Richardson, B. Miaschi, E. Muzi, P. Guerrini, L. Troiano and T. Akal, ISSAMS: An *in situ* sediment geoacoustic measurement system. In *Shear Waves in Marine Sediments,* edited by J.M. Hovem, M.D. Richardson and R.D. Stoll (Kluwer Academic Publishers, Dordrecht, The Netherlands, 1991) pp. 305–312.

2. S.F. Griffin, F.B. Grosz and M.D. Richardson, ISSAMS: A remote *in situ* sediment acoustic measurement system, *Sea Technology* **37**, 19–22 (1996).

3. M.D. Richardson, E. Muzi, B. Miaschi and F. Turgutcan, Shear wave gradients in near-surface marine sediment. In *Shear Waves in Marine Sediments,* edited by J.M. Hovem, M.D. Richardson and R.D. Stoll (Kluwer Academic Publishers, Dordrecht, The Netherlands, 1991) pp. 295–304.

4. M.D. Richardson, K.B. Briggs, S.J. Bentley, D.J. Walter, T.H. Orsi, Biological and hydrodynamic effects on physical and acoustic properties of surficial sediments off the Eel River, Northern California, *Marine Geology* **182** (Dec. 2001).

5. M.D. Richardson, Attenuation of shear waves in near-surface sediments. In *High-Frequency Acoustics in Shallow Water,* edited by N.G. Pace, E. Pouliquen, O. Bergem and A.P. Lyons. SACLANTCEN Conference Proceedings CP-45, La Spezia, Italy (1997) pp. 451–457.

6. M.D. Richardson, D.M. Lavoie and K.B. Briggs, Geoacoustic and physical properties of carbonate sediments of the Lower Florida Keys, *Geo-Marine Letters* **17**, 316–324 (1997).

7. M.D. Richardson and K.B. Briggs, In-situ and laboratory geoacoustic measurements in soft mud and hard-packed sand sediments: Implications for high-frequency acoustic propagation and scattering, *Geo-Marine Letters* **16**, 196–203 (1996).

8. M.D. Richardson, In-situ, shallow-water sediments geoacoustic properties. In *Shallow-Water Acoustics,* edited by R. Zang and J. Zhou (China Ocean Press, Beijing, 1997) pp. 163–170.

9. M.J. Buckingham and M.D. Richardson, On tone-burst measurements of sound speed and attenuation in sandy marine sediments, *IEEE J. Oceanic Eng.* (in press 2002).

10. R.F.L. Self, P. A'Hearn, P.A. Jumars, D.R. Jackson, M.D. Richardson and K.B. Briggs, Effects of macrofauna on acoustic backscatter from the seabed: Field manipulations in West Sound, Orcas Island, WA, USA, *J. Marine Res.* **59**, 991–1020 (2001).

11. M.D. Richardson, K.B. Briggs, D.L. Bibee, P.A. Jumars, W.B. Sawyer, D.B. Albert, R.H. Bennett, T.K. Berger, M.J. Buckingham, N.P. Chotiros, P.H. Dahl, N.T. Dewitt, P. Fleischer, R. Flood, C.F. Greenlaw, D.V. Holliday, M.H. Hulbert, M.P. Hutnak, P.D. Jackson, J.S. Jaffe, H.P. Johnson, D.L. Lavoie, A.P. Lyons, C.S. Martens, D.E. McGehee, K.D. Moore, T.H. Orsi, J.N. Piper, R.I. Ray, A.H. Reed, R.F.L. Self, J.L Schmidt, S.G. Schock, F. Simonet, R.D. Stoll, D.J. Tang, D.E. Thistle, E.I. Thorsos, D.J. Walter and R.A. Wheatcroft, An overview of SAX99: Environmental considerations. *IEEE J. Oceanic Eng.* **26**, 26–53 (2001).

12. R.H. Wilkens and M.D. Richardson, The influence of gas bubbles on sediment acoustic properties: In situ, laboratory and theoretical results from Eckernförde Bay, Baltic Sea, Germany, *Cont. Shelf Res.* **18**, 1859–1892 (1998).

13. M.D. Richardson, E. Muzi, L. Troiano and B. Miaschi, Sediment shear waves: A comparison of in situ and laboratory measurements. In *Microstructure of Fine Grained Sediments,* edited by R.H. Bennett, W.R. Bryant and M.H. Hurlbert (Springer-Verlag, New York, 1990) Chap. 44, pp. 403–415.

IN-SITU DETERMINATION OF THE VARIABILITY OF SEAFLOOR ACOUSTIC PROPERTIES: AN EXAMPLE FROM THE ONR GEOCLUTTER AREA

LARRY A. MAYER AND BARBARA J. KRAFT

Center for Coastal and Ocean Mapping, University of New Hampshire, Durham N.H. 03824, USA
E-mail: lmayer@unh.edu

PETER SIMPKIN

IKB Technologies Ltd., 1220 Hammonds Plains Rd., Bedford N.S. B3B 1B4 Canada
E-mail: ikb@seistek.com

PAUL LAVOIE, ERIC JABS AND ERIC LYNSKEY

Center for Coastal and Ocean Mapping, University of New Hampshire, Durham N.H. 03824, USA
E-mail: lmayer@unh.edu

In support of the US ONR-sponsored Geoclutter program, we have developed, built, and deployed a relatively inexpensive, robust, small-ship-deployable device (**ISSAP** – In situ **S**ound **S**peed and **A**ttenuation **P**robe) for rapidly measuring sound speed and attenuation in near-surface sediments. We have demonstrated its ability to make reliable and precise measurements (+/– 1–2 m/s for sound speed, < +/– 1 dB/m for attenuation). We have found that in the Geoclutter area the sound speed varies on the order of 200–300 m/s over spatial scales of 10's of kms and the attenuation (at 65 kHz) varies on the order of 60 dB/m. On scales of less than one kilometer, the sound speed can vary by more than 100 m/s and attenuation by approximately 25 dB/m. On the sub-meter scale, much of the seafloor is relatively homogeneous but some areas show sound speed variation of approximately 50 m/s and attenuation variation on the order of 25 dB/m. These variations are probably related to the presence of large clasts or shells in the measured path.

1 Introduction

With growing pressure to operate in shallow waters, navies around the world are being faced with the challenges of understanding the complex acoustic environment of near-coastal regions. With this in mind, the U.S. Office of Naval Research has undertaken a series of research programs aimed at gaining a better knowledge of both the ocean and seafloor environments in shallow water settings. Amongst these is the Geoclutter program, whose long-term goal is to understand the causes and implications of geologic clutter (reverberation) in a geologically well-characterized shallow-water environment. The field area selected for the Geoclutter program is the mid-outer continental shelf off New Jersey, USA (Fig. 1). The New Jersey margin was chosen for the Geoclutter study because the bathymetry and portions of the shallow subsurface of this area had already been mapped in detail as part of an earlier ONR program aimed at understanding the

N.G. Pace and F.B. Jensen (eds.), Impact of Littoral Environmental Variability on Acoustic Predictions and Sonar Performance, 115-122.

origin of subsurface stratigraphy on continental margins (STRATAFORM; [1,2]).   In
addition to multibeam bathymetry, 'calibrated' backscatter data (at 95 kHz from the
multibeam sonar) was also collected as part of the STRATAFORM program.

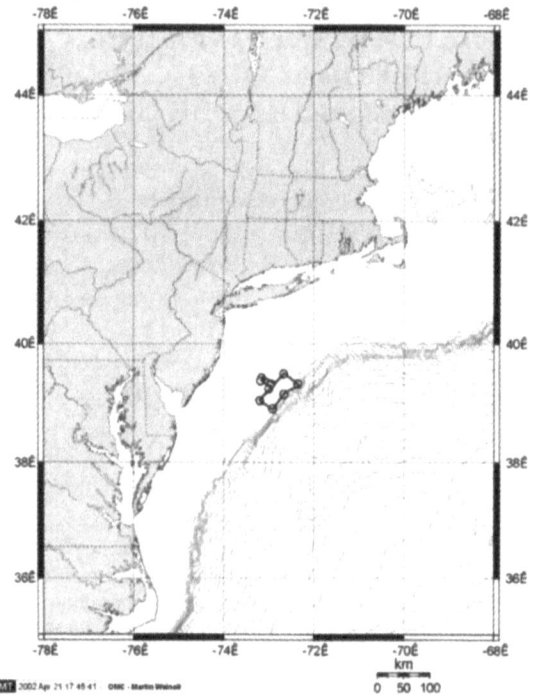

Figure 1. Location map for Geoclutter field area on the continental shelf off New Jersey, USA.
Survey area extends from approximately 50 m to about 150 m water depth and covers an area of
approximately 1300 sq. km.

The overall scientific objectives of the Geoclutter program are: 1) to understand,
characterize, and predict lateral and vertical, naturally-occurring heterogeneities that
may produce discrete acoustic returns at low grazing angles (i.e., "geologic clutter") and
then; 2) to conduct precise acoustic reverberation experiments at this site to understand,
characterize, and potentially mitigate the geologic clutter.   In order to meet these
objectives and to properly implement acoustic models for the Geoclutter area, we need
to know, or predict, the key acoustic and physical properties throughout the volume of
interest (i.e., grain size, density, sound speed, attenuation). The properties of the near-
surface seafloor sediments are particularly important.   A possible approach to this
problem is to use the 95-kHz multibeam backscatter data collected in the region, which
may provide information on seafloor sediment properties. The relationship between
backscatter and sediment properties remains ambiguous however, and cannot yet be
used as a direct and quantitative predictor of seafloor properties.   Understanding the
relationship between the multibeam backscatter and the properties of the seafloor is a
sub-theme of our Geoclutter research program.

We thus fall back on more traditional means of sampling and laboratory
measurements to obtain the needed seafloor property data.   Given the coarse-grained,
sandy nature of the sediment in the region we were concerned that laboratory

measurements of certain properties (in particular sound speed and attenuation) on core samples may not reflect *in situ* values as sandy sediments tend to de-water very quickly. We thus chose to develop a simple and relatively inexpensive device designed to measure, *in situ*, the spatial variability of sound speed and attenuation in near-surface sediments at the Geoclutter site. The *in situ* measurements would then be combined with the data collected from cores (by investigators from the Universities of Texas and Delaware) to better understand the variability of *in situ* sediment physical and acoustic properties in the Geoclutter area.

2 Approach

The *in situ* measurement of seafloor acoustic properties is not a new concept. Hamilton pioneered the collection of *in situ* acoustic property data in 1956 using diver-deployed probes in shallow water [3]. This was followed by deep-water measurements from the bathyscaph *Trieste* in 1962 [4] and the *Deepstar* 4000 submersible in 1966–70 [5]. Other investigators used divers or small-frames in very shallow water or remotely determined *in situ* properties using reflection techniques (see summary in [5]). More recently sophisticated platforms have been built to insert a range of sensors into the seafloor (e.g. [6,7]). Our objective was to design and deploy a relatively inexpensive, robust, small-ship-deployable device that would specifically address the question of the spatial variability of seafloor sediment properties by rapidly making multiple measurements of sound speed and attenuation in near-surface sediments. Our concept was to design an instrument that was like a box-corer and that could rapidly make multiple measurements of *in situ* properties by simply "pogo-ing" on the bottom and thus cover a relatively large area of the seafloor in a short period of time.

The system we designed uses four 2.54 cm (diameter) by 30 cm long probes that are inserted 15 cm into the seafloor by 250 kg of reaction weight attached by armored coaxial cable to a free-swinging inner frame within a protective outer tripod. This design, in combination with articulated tripod feet, allows the probes to be inserted vertically on slopes up to 20 degrees (Fig. 2). The transducer probes are mounted on an inner frame assembly through precision-machined Delrin^{TM} collars designed to decouple the acoustic signals from the frame. The probes operate at frequencies of either 65 or 100 kHz (all measurements reported here are at 65 kHz). Probe separation can be adjusted in 10 cm increments from 10 to 60 cm. For this study the probes were arranged in a square pattern with nominal path separations of 20 and 30 cm (Fig. 2). An onboard computer and topside electronics control the paths selected and the number of measurements per path. A typical deployment involves measurements across five paths including both long (30 cm) and short (20 cm) paths. In addition to the acoustic probes, the ISSAP also has a color video camera that provides imagery of the seafloor and the probes as they penetrate, a 65 kHz altimeter to independently monitor height off the bottom, and temperature, pressure, pitch, roll, and heading sensors to monitor the stability and orientation of the platform. Finally, a bottom sense switch provides yet another indication of the platform's height above the bottom (Figs. 2 and 3).

The system is lowered to the bottom on a coaxial cable until the altimeter, bottom sense switch, and camera indicate proximity to the bottom. When the bottom is in sight, a bottom-water measurement cycle is initiated with a short (40 microsecond) pulse transmitted from one of the probes and received by another. Ten measurements are for a total of 150 measurements in a measurement cycle.

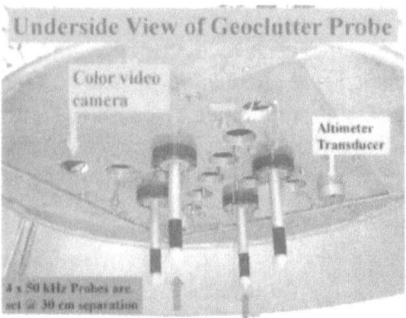

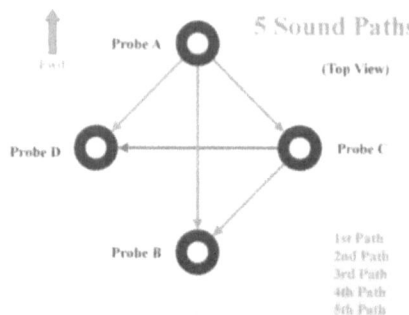

Figure 2. Underside view of ISSAP showing orientation of probes (left top). Also shown are color video camera and light, as well as the 65 kHz altimeter. The altimeter is used to determine height of instrument off the bottom as well as to provide a vertically incident return from the seafloor. On top right is diagram showing 5 paths used for sound speed and attenuation measurement. To left is photo of tripod and probe assembly. Tripod is designed to allow a vertical insertion of the probes on slopes up to 20 degrees.

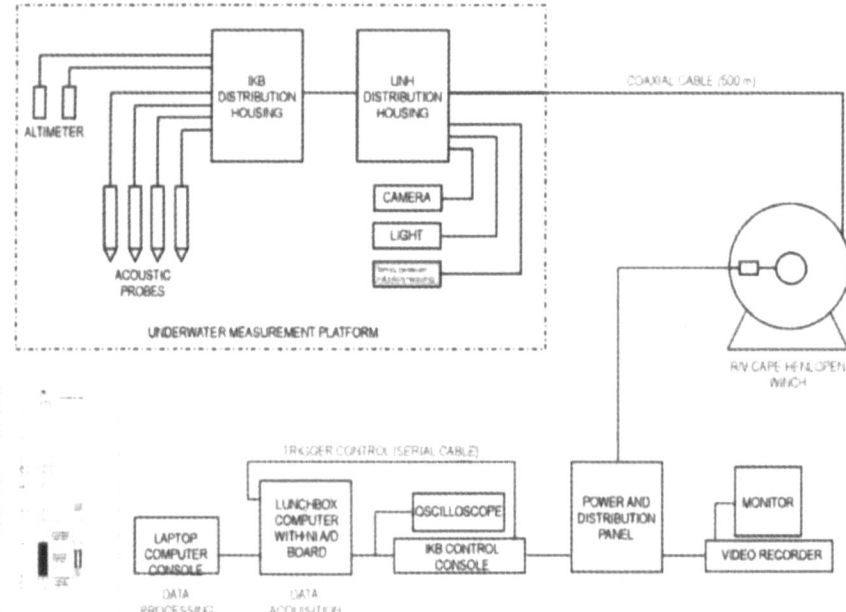

Figure 3. Block diagram of ISSAP system as installed on the R/V Cape Henlopen. Detail of probe is presented in insert on bottom left.

Upon completion of the bottom-water measurement cycle the system is lowered into the seafloor where two measurement cycles of 150 measurements each over the 5 paths are made in the sediment. When both sediment measurement cycles are complete, the system is pulled out of the seafloor and another bottom-water measurement cycle is completed. A sampling station thus typically consists of two bottom-water cycles and two sediment cycles with a total of 600 independent measurements of acoustic travel time over 5 independent paths with different separations. Each measurement cycle takes less than one minute; sampling an entire station thus takes on the order of five minutes to complete.

The transmit and receive pulse for each measurement is sent up the coax and digitized at 2 MHz on the topside acquisition computer and sent to a processing computer (Fig. 3). An entire measurement cycle (150 measurements) results in approximately 75 Mbytes of data; a typical station (2 bottom-water and two sediment cycle) produces about 300 Mbytes of data. The fundamental measurement is that of the travel time (time-of-flight) between the transmit and received pulse. Travel times are determined by several methods and converted to sound-speed through a calibration process. The details of the analytical procedures are presented in Kraft *et al.* [8]. There are two levels of calibration available. The most precise involves collecting data in distilled water at a known temperature and using the well-established variation in sound speed with temperature to precisely determine the separation of each pair of transducers. This is done at the beginning of the cruise, at the end of the cruise, and several times during the cruise. We also carry out an ongoing calibration by measuring the speed of sound in seawater (at known temperature) before and after each penetration into the seafloor. Bottom-water calibrations also allow us to determine if the insertion of the probes into the bottom resulted in a change in their relative path length.

Along with measurement of time-of-flight (and thus sound speed) we can also compare the digitized sediment and water path pulses in order to measure sediment attenuation. Several approaches have been used to measure attenuation. The relative amplitude of the received waveforms over the different path-lengths is one indication of attenuation as is the spectral ratio (or difference) between the seawater received waveform and the sediment received waveform. We also use the filter correlation technique of Courtney and Mayer [9] that was developed especially for short time series of the type we are measuring. The details of attenuation processing are presented in Kraft *et al.* [8].

3   Results and discussion

The ISSAP was deployed in the Geoclutter area off New Jersey on the *R/V Cape Henlopen* between 30 July and 5 August, 2001. The system performed flawlessly recovering water column and sediment data at 99 stations selected to represent a range of seafloor backscatter types over an area of approximately 1300 sq. km. (Fig. 4). More than 40 gigabytes of digital data were collected (representing 58,200 individual measurements) as well as more than 20 hours of video.

With rare exception, the waveforms recovered from the ISSAP were remarkably clean allowing not only an unambiguous measurement of time-of-flight but also for the for calculation of attenuation (see Kraft *et al.* [8]). Most importantly, the tremendous redundancy of our measurements at each station (typically 300 measurements in the sediment and 300 measurements in the water column) allow us to put well-grounded

confidence limits on our measurements and thus understand the true local variability of
sound speed and attenuation in the Geoclutter area.

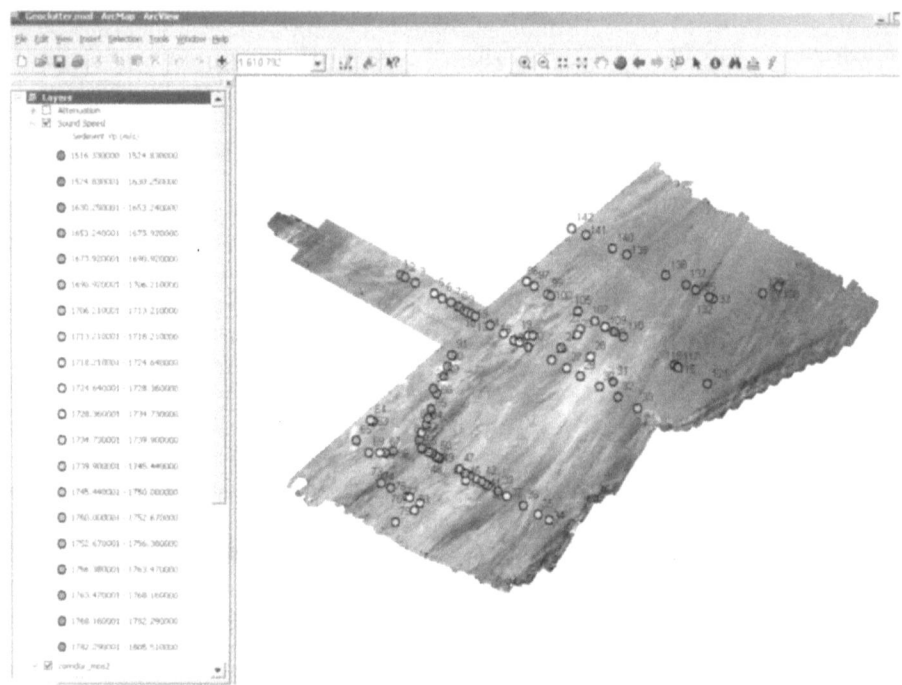

Figure 4. Location of 99 ISSAP stations superimposed on gray-scale display of average
backscatter at 45 degrees collected with the 95 kHz multibeam sonar. High backscatter is
represented by light color, low backscatter by dark color. Backscatter range is less than 20 dB.
Values of sound speed are color coded as in table on left. Range is from 1524 to 1801 m/s. Water
depths range from 50 m on the eastern limit to 150 m on the western limit of the area.

Distilled water calibration runs resulted in a standard deviation of .354 m/s
indicating high precision. At each station, the 300 measurements of seawater sound
speed showed a standard deviation of less than 1 m/s, again indicating high precision.
Across the entire area the mean speed of sound in seawater was 1500.8 m/s with a range
of less than 10 m/s, within the expected change due to variations in bottom-water
properties. These values indicate that the system geometry remained constant and the
timing precise throughout our operations. In contrast to the consistency of the water
column, real and substantial variations in seafloor sound speed and attenuation were
measured (Table 1). The system was deployed in sediments ranging from muddy, silty
sands, to gravels and shell hash deposits with a video record of each deployment
providing an indication of the degree of penetration of the probes as well as the nature
of the surface sediment. In addition, a grab sample was collected at each station; grain
size analyses have been made by scientists from the University of Texas and are
introduced in Kraft et al. [8].

Table 1. Range of values measured at all 99 stations @ 65 kHz.

Property	Min	Max	Mean	Mean SD[1]	Min SD[2]	Max SD[3]
Spd (m/s)	1524.4	1801.4	1726.6	12.0	1.3	29.0
Att(dB/m)	10.0	71.3	34.2	6.3	0.8	19.7

1 Mean SD = the average of the 99 standard deviations measured at each station.
2 Min SD = the minimum of all of the 99 station standard deviations.
3 Max SD = the maximum of all the 99 station standard deviations.

As can be seen from Table 1, within the approximately 1300 sq. km of the Geoclutter area, there is a substantial range of both sound speed and attenuation (277 m/s for sound speed and 61.3 dB/m for attenuation). This is not surprising for a dynamic area of this size. What is more surprising and important to those charged with modeling acoustic propagation in this regime is the range of variability on a much smaller spatial scale. If we look at areas where stations are close to each other, we see that the range of values for stations less than 1 km apart can be as much as half the overall range in the Geoclutter region. For example the difference in sound speed between station 44 and 45 (less than 1 km apart) is 118 m/s while the difference in attenuation is 24.2 dB/m (Fig. 6). While there is a clear difference in backscatter between these two sites (Fig. 6), and some relationship between backscatter, sound speed and attenuation, the relationship is complex and difficult to generalize (see Kraft et al. [8]).

More intriguing is the scale of variability seen at an even smaller spatial scale. If we look at the variability at each individual station we see that for any given path length (20 or 30 cm) the standard deviation for the 60 measurements made across any given path rarely exceeds 1 m/s. Thus, once again we are assured of the precision of our measurements. If we look at the variation amongst paths at a given station, we see that the majority of stations show relatively small variation along the different paths. Typically the path-to-path standard deviation for sound speed at these stations is 4–5 m/s (attenuation ~2–3 dB/m) and thus the seafloor is quite homogenous over these scales. However, for a subset of stations the standard deviation of the sound speeds are on the order of 20–25 m/s with a range of ~50m/s (attenuation ~8–10 dB/m with a range of ~25dB/m) indicating a significant amount of variability even over a spatial scale of 20–30 cm. In these cases, it is typically one of the paths that has consistently higher values of sound speed and attenuation and probably represents the presence of a large clast (gravel or shell) in the path.

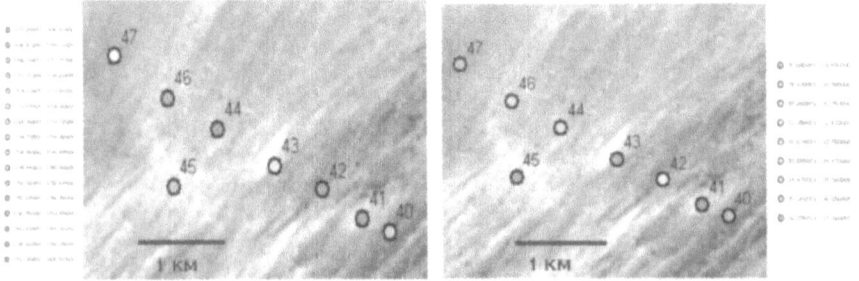

Figure 6. Close-up of sound speed (left) and attenuation (right) at a series of stations superimposed on multibeam sonar backscatter data.

4   Conclusions

We have designed a simple and relatively inexpensive device to rapidly measure, *in situ,* the spatial variability of sound speed and attenuation in near-surface sediments and have demonstrated its ability to make reliable and precise measurements (+/– 1–2 m/s for sound speed, <+/– 1 dB/m for attenuation).   We have deployed this system in the Geoclutter field area on the New Jersey margin and found that the sound speed varies on the order of 200–300 m/s over spatial scales of 10's of kms and the attenuation (at 65 kHz) varies on the order of 60 dB/m.   On spatial scales of less than one kilometer, the sound speed can vary by more than 100 m/s and attenuation by approximately 25 dB/m. On the sub-meter scale, much of the seafloor is relatively homogeneous but some areas show sound speed variations of approximately 50 m/s and attenuation variations on the order of 25 dB/m.   These variations are probably related to the presence of large clasts or shells in the measured path.   The relationship of changes in the acoustic properties to other physical properties, and particularly the relationship of the measured sound speed and attenuation to remotely measured backscatter, are areas of ongoing research.

Acknowledgements

We gratefully acknowledge the support of the Office of Naval Research Grant Number N00014-00-1-0821 under the direction of Roy Wilkens and Dawn Lavoie.   We also thank the captain and crew of the *R/V Cape Henlopen*, who quickly and skillfully adjusted to the deployment of a strange instrument.

References

1. Mayer, L.A., Hughes-Clarke, J.E., Goff, J.A., Schuur, C.L. and Swift, D.J.P., Multibeam sonar bathymetry and imagery from the New Jersey continental margin: Preliminary results, *EOS, Trans. AGU* **77**, F329 (1996).
2. Goff, J.A., Swift, D.J.P., Duncan, C.S., Mayer, L.A. and Hughes-Clarke, J.E., High-resolution swath sonar investigation of sand ridge, dune and ribbon morphology in the offshore environment of the New Jersey margin, *Marine Geology* **161**, 307–337 (1999).
3. Hamilton, E.L., Low-sound velocities in high porosity sediments, *J. Acoust. Soc. Am.* **28**, 16–19 (1956)
4. Hamilton, E.L., Sediment sound velocity measurements made in-situ from the bathyscaph TRIESTE, *J. Geophys. Res.* **68**, 5991–5998 (1963).
5. Hamilton, E.L., Compressional wave attenuation in marine sediments, *Geophysics* **37**, 620–646 (1972).
6. Griffin, S.R., Grosz, F.B. and Richardson, M.R., In-situ sediment/geoacoustic measurement system, *Sea Technology*, 19–22 (April 1996).
7. Best, A.I., Roberts, J.A. and Somers, M.L., A new instrument for making in-situ acoustic and geotechnical measurements in seafloor sediments, *Underwater Technology* **23**, 123–132 (1998).
8. Kraft, B.J, Mayer, L.A., Simpkin, P., Lavoie, P., Jabs, E., Lynskey, E. and Goff, J., Calculation of in-situ acoustic wave properties. In *Impact of Littoral Environmental Variability on Acoustic Predictions and Sonar Performance*, edited by N.G. Pace and F.B. Jensen (Kluwer, The Netherlands, 2002) pp. 123–130.
9. Courtney, R.C., and Mayer, L.A., Calculation of acoustic parameters by a filter-correlation method, *J. Acoust. Soc. Am.* **93**, 1145–1154 (1993).

CALCULATION OF *IN SITU* ACOUSTIC WAVE PROPERTIES IN MARINE SEDIMENTS

B.J. KRAFT, L.A. MAYER, P. SIMPKIN , P. LAVOIE, E. JABS AND E. LYNSKEY
Center for Coastal and Ocean Mapping, University of New Hampshire, Durham NH 03824, USA
E-mail: bjkraft@cisunix.unh.edu

J.A. GOFF
University of Texas Institute for Geophysics, Austin TX 78759, USA

The importance of estimating compressional wave properties in saturated marine sediments is well known in geophysics and underwater acoustics. As part of the ONR sponsored Geoclutter program, *in situ* acoustic measurements were obtained using ISSAP (In situ Sound Speed and Attenuation Probe), a device developed and built by the Center for Coastal and Ocean Mapping (CCOM). The location of the Geoclutter field area is the mid-outer continental shelf off New Jersey. Over 30 gigabytes of seawater and surficial sediment data was collected at 99 station locations selected to represent a range of seafloor backscatter types. At each station, the ISSAP device recorded waveform data across five acoustic paths with nominal probe spacing of 20 or 30 cm. The transmit/receive probes were arranged in a square pattern and operated at a nominal frequency of 65 kHz. The recorded waveforms were processed for sound speed using two methods, cross-correlation and envelope detection, and compared. The waveforms were also processed for sediment attenuation using the filter-correlation method. Results show considerable variability in the acoustic properties at the same and nearby seafloor locations.

1 Introduction

The earliest published work involving *in situ* measurement of acoustic properties was performed using diver-deployed probes in shallow water [1], a difficult and time-intensive task. During the late 1960's, a deep-diving submersible [2] was used to measure the attenuation of compressional acoustic waves in deep-water, but again divers were left to deploy probes in shallow waters and only a small number of stations were measured. A complete summary of the early research is contained in [3]. Recently, *in situ* measurements have been obtained with sophisticated platforms capable of obtaining multiple, rapid measurements of near surface values of sediment geoacoustic and geotechnical properties [4, 5]. Current propagation models predicting the interaction of acoustic waves with the seafloor are limited by a lack of data correctly depicting the spatial variability of the seafloor. To expand present understanding of acoustic wave propagation mechanisms in marine sediments, it is imperative to obtain abundant and high-resolution measurements in their natural setting.

N.G. Pace and F.B. Jensen (eds.), Impact of Littoral Environmental Variability on Acoustic Predictions and Sonar Performance, 123-130.

2   Experiment description

As part of the ONR Geoclutter program the Center for Coastal and Ocean Mapping designed, built, and deployed ISSAP, a geoacoustic measurement system capable of measuring surficial sediment compressional wave velocity and attenuation. ISSAP was constructed of aluminum and stainless steel, weighed approximately 275 kg, had a height of 1.5 m, and a 9.4 m square footprint. It had two principal parts; an outer frame that acted as a guide for an inner frame assembly. The outer frame consisted of a protective tripod reinforced with a tapered skirt. Articulated tripod feet allowed for vertical probe insertion on slopes up to 20 degrees. Included in the inner frame assembly were a load bearing box beam structure, a 0.36 m square aluminum platform, and a guard ring slightly larger than 1 m in diameter. Mounted on the inner frame platform were two pressure housings for electronics, a color video camera and light, and a Jasco Research UWINSTRU, which measured platform heading, pitch, and roll, depth, and bottom water temperature. The transducer probes were mounted to the underside of the platform with Delrin™ precision machined collars (Fig. 1) designed to minimize travel of the acoustic signal through the ISSAP frame and displacement of the probes during insertion. Multiple locations were available for probe placement. Acoustic path lengths were adjustable in 10 cm increments from 10 to 60 cm.

The ISSAP instrument used four transducer probes arranged in a square pattern

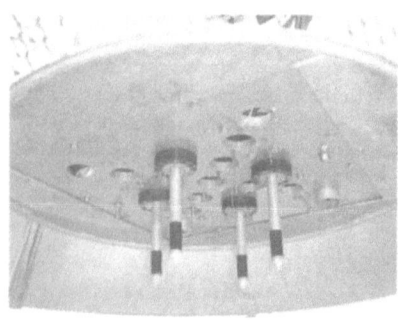

Figure 1. Arrangement of transducer probes on ISSAP and method of mounting.

giving approximate acoustic path lengths of 30 cm and 20 cm. The active elements were piezoelectric ceramic cylinders with diameter and length of 2.54 cm. Overall probe length was about 30 cm which allowed for up to 20 cm insertion into the sediment. The active zone of the transducer was located at a maximum insertion depth of 15 cm. The transducers were used to transmit and receive, and operated at a frequency of 65 kHz. Sensitivity and response between transducer pairs at different angles was approximately equal.

Five acoustic paths were used to measure compressional speed of sound and attenuation; two long paths (30 cm) and three short paths (20 cm). A 40 µs pulse was generated at a repetition rate of 30 Hz. The acoustic signal detected by the receive transducer was amplified and combined with the transmitter gate pulse to generate a composite signal (see Fig. 2). The gain mode was set to LOW (0 dB) for seawater measurements and HIGH (12 dB) for most sediment measurements. The composite signal was sampled at a frequency of 2 MHz with a National Instruments PCI-6110E A/D data acquisition board. The composite sampled waveform contained all information necessary to calculate the time-of-flight of the acoustic pulse. A distilled water calibration procedure was performed to compensate for fixed system delays. For a complete description of the ISSAP instrument see Mayer *et al.* [6].

Acoustic measurements with ISSAP, and sediment samples from the seafloor (grab and a few slow-core), were collected at 99 station locations over an area approximately 1300 square km. At each location, the ISSAP instrument was lowered to a height ~10 m

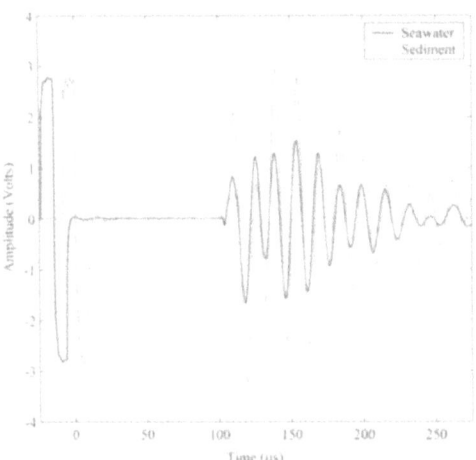

Figure 2. Composite waveforms included the trigger and received waveform. A typical amplified sediment waveform (12 dB) and a seawater waveform are shown following the cross correlation step.

above the seafloor. A measurement cycle (150 measurements) was obtained in seawater for calibration purposes and for use in the attenuation calculation. Using the real-time video as a guide, ISSAP was lowered and the transducers inserted in the sediment. Two sediment measurement cycles were obtained for a total of 300 acoustic measurements. ISSAP was removed from the sediment, raised to a height of ~10 m above the seafloor and a second seawater measurement cycle obtained. Due to failure of the UWINSTRU A/D board, the bottom water temperature and depth were not measured for most stations. Data obtained at 2 stations could not be processed.

3   Data processing

3.1   Sound Speed

3.1.1   Cross-Correlation

The relative time delay between two signals may be estimated from the peak of the cross-correlation of the two signals [7]. The cross-correlation function is estimated as a function of correlation lag and has its maximum at a lag equal to the time delay. First, the time-of-flight in seawater was calculated as the elapsed time between two zero-crossings of the seawater sampled waveform (see Fig. 2); the zero-crossing with negative slope from the trigger and the first zero-crossing with positive slope on the received waveform. Least-squares regressions were performed to resolve the zero crossing between samples. After flight times in seawater were calculated, cross-correlation of the sediment waveform with the seawater waveform was performed to estimate the relative time delay. The cross correlation was limited to the first half cycle of each waveform to prevent a tendency (or shift) in correlation lag towards secondary multipath arrivals. The time-of-flight of the acoustic pulse in the sediment was determined as the difference of the seawater time-of-flight and the time delay estimate from the cross-correlation. Shown in Fig. 2 are typical results obtained after performing the cross-correlation.

A disadvantage of the cross-correlation technique was that determination of the peak correlation (lag resolution) was limited by the sampling frequency of the received waveforms. Estimation of the peak correlation was improved by performing a cubic least squares regression. In some sediment types, loading on the transducers produced waveforms with slow rise times. In these instances, the cross-correlation technique may underestimate the sediment sound speed. To address these problems the sound speeds were also calculated using envelope detection. This method additionally considers phase differences between the seawater and sediment waveforms.

3.1.2   Envelope Detection

This method required a filtering step to remove an artifact on the leading edge of the seawater waveforms possibly due to a high-frequency mode of the piezoelectric crystal. The seawater waveforms were low-pass filtered using a 5th-order Butterworth digital filter with a 3 dB dropoff at the cutoff frequency of 90 kHz. To compensate for the effect of the filter delay on the seawater waveforms, the sediment waveforms were also filtered using the same filter. The waveform envelopes were determined from the magnitude of the discrete analytic signal. By definition, an analytic signal is a complex signal that has the original signal in the real part and the Hilbert transform of the original signal, a 90° phase shifted version of the original signal, in the imaginary part.

The time-of-flight (in seawater and sediment) was calculated as the elapsed time between the trigger zero crossing with negative slope and the zero-crossing of (extracted from) the leading edge of the envelope. Waveform envelopes were normalized by the first detected peak and the zero crossing determined by performing a linear least square regression to all samples between 0.4 and 0.8 of the normalized amplitude. A second calibration step was needed to determine the effective acoustic path lengths using this zero crossing.

3.2   *Attenuation*

Attenuation in marine sediments may be estimated using either time or spectral domain methods. Spectral domain methods are difficult to use when secondary reflections of the transmitted pulse are located within the sampling window of the first arrival pulse. In this instance, short window lengths in the time domain are required to separate the secondary arrival from the first arrival, which greatly reduces the spectral resolution. To overcome the limitations of the spectral methods, a filter-correlation method was proposed by Courtney and Mayer [8]. This method showed accurate estimates of attenuation parameters could be obtained even when window lengths in the time domain were reduced to minimize the effects of secondary reflections. Additionally, the filter correlation method may be used to estimate attenuation as a function of frequency by filtering a broadband signal with bandpass filters over a range of passbands, and cross-correlating the filtered attenuated signal with a filtered reference signal.

Although a narrow band signal was used in this experiment, bandpass filtering the received waveforms aided the identification of secondary reflections. No attempt was made to estimate attenuation as a function of frequency. All waveforms were filtered using an 8th-order digital Butterworth filter with passband from 52 to 82 kHz and a 3 dB dropoff at the passband edges. The envelopes of the bandpass filtered waveforms were determined from the magnitude of the analytic waveform as described above. To account for phase differences between the seawater and sediment arrivals, it was important to perform the cross-correlation using the envelopes of the waveforms, derived from the Hilbert transform, and not the waveforms themselves. After the

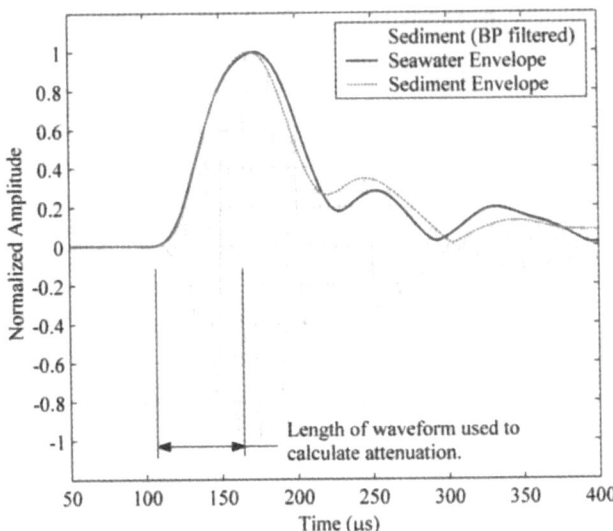

Figure 3. Received waveform in sediment after filtering with Butterworth bandpass filter. Also shown are the envelopes of the sediment and seawater filtered waveforms following the normalization and cross-correlation. Evident is a secondary multipath arrival at approximately 200 μs.

waveforms were correctly aligned the sediment attenuation (in dB/m) relative to the bottom water was calculated using

$$\alpha = -20\log_{10}\left(\frac{A_s}{A_w}\right) \cdot \frac{1}{\Delta L} \qquad (1)$$

where A_s represents the *rms* energy in the sediment waveform, A_w represents the *rms* energy in the seawater waveform, and ΔL is the measured physical path length between the transmit and receive transducer probes. The number of samples used to calculate the *rms* energy was determined by the length of the sediment waveform and included all samples from the leading edge of the envelope to the zero-crossing prior to the first peak of the sediment envelope (see Fig. 3).

An algorithm based on the deviation between the sediment and seawater envelopes was used to help identify waveforms corrupted by early secondary reflections. The sediment envelopes were cross-correlated with the seawater envelopes and normalized by the first (arrival) peak of the envelope. The seawater envelope was shifted in time by the correlation lag and the deviation between the seawater and sediment envelopes calculated. Mean deviations were determined for all samples along the leading edge of the sediment envelope. For example, the mean deviation for the nth sample was determined as the average of the n-1, n, and n+1 sample deviations. The mean deviations were summed over the same range of samples used in the attenuation calculation. In some instances, the sum of the mean deviations was a poor predictor of secondary reflections and visual inspection of the waveforms was required.

Attenuation results of paths associated with possibly corrupted waveforms were flagged and not included in the average attenuation for that station. Also, waveforms that were clipped from insufficient dynamic range on the A/D data acquisition board were excluded.

Table 1. Summary of sound speed results calculated with the cross-correlation and envelope detection methods and attenuation results (at f = 65 kHz) determined using the filter-correlation method. The average of the standard deviations measured at each station is represented by σ (MEAN).

Cross-Correlation	MIN	MEAN	MAX
Seawater, V_p (m/s)	1493.9	1500.8	1508.7
σ (m/s)	0.6	1.1	1.5
Sediment, V_p (m/s)	1516.3	1718.1	1810.9
σ (m/s)	0.7	11.5	29.0
Velocity Ratio	1.009	1.145	1.209
Envelope Detection	MIN	MEAN	MAX
Seawater, V_p (m/s)	1493.4	1500.8	1508.6
σ (m/s)	0.4	0.8	1.3
Sediment, V_p (m/s)	1524.4	1726.6	1801.4
σ (m/s)	1.3	12.0	29.0
Velocity Ratio	1.014	1.150	1.203
Filter Correlation	MIN	MEAN	MAX
Attenuation, α (dB/m)	9.9	34.2	71.3
σ (dB/m)	0.8	6.3	19.7

4   Discussion of results

A summary of the sound speed results using both methods and the attenuation results using the filter-correlation method is shown in Table 1. The two methods used to calculate sound speed produced consistent results in seawater. The envelope detection method produced an increase in sediment sound speed at 84 stations, with the average increase equal to 10.8 m/s (maximum increase of 25.8 m/s and minimum increase of 0.1 m/s). A total of 13 station locations experienced a decrease in sediment sound speed, with the average decrease equal to 6.1 m/s (maximum decrease of 14.9 m/s and minimum decrease of 0.4 m/s). This method resulted in an overall average increase in sediment sound speed of 8.5 m/s for all stations. A discussion of sound speed and attenuation results and their relationship to the spatial variability of the seafloor is included in Mayer et al. [6].

Most often, the geotechnical parameter expected to exhibit some correlation with the attenuation coefficient, $k = \alpha / f$ (in units of dB/m·kHz) is the mean grain diameter. Sediment core and grab samples were collected by investigators at the Universities of

Texas and Delaware and a preliminary comparison of attenuation and grain size distribution performed (see [9] for a description of a similar grain size analysis). Comparison with published results was somewhat difficult in that published results primarily extend over a broad range of mean grain sizes [3, 10].

Station data, including grain size distributions, were sorted into descending order based on the average attenuation coefficient and divided into groups representing 0.1 dB/m·kHz decreases in attenuation. An average grain size distribution (in % fraction of sample based on weight) representing each group was determined using the grain size distributions of each station in the group. The results of this averaging process are shown in Fig. 4, where each series plotted represents the averaged grain size distribution for a group of stations with similar attenuation coefficient.

Although preliminary, the averaged grain size distributions shown do present interesting results. The group of stations with the highest attenuation coefficients (k = 0.8–0.9 dB/m·kHz), had the largest weight percent of fine sand (.175 to .25 mm) as well as a higher percentage of coarse grains with diameters greater than 4 mm. The group of

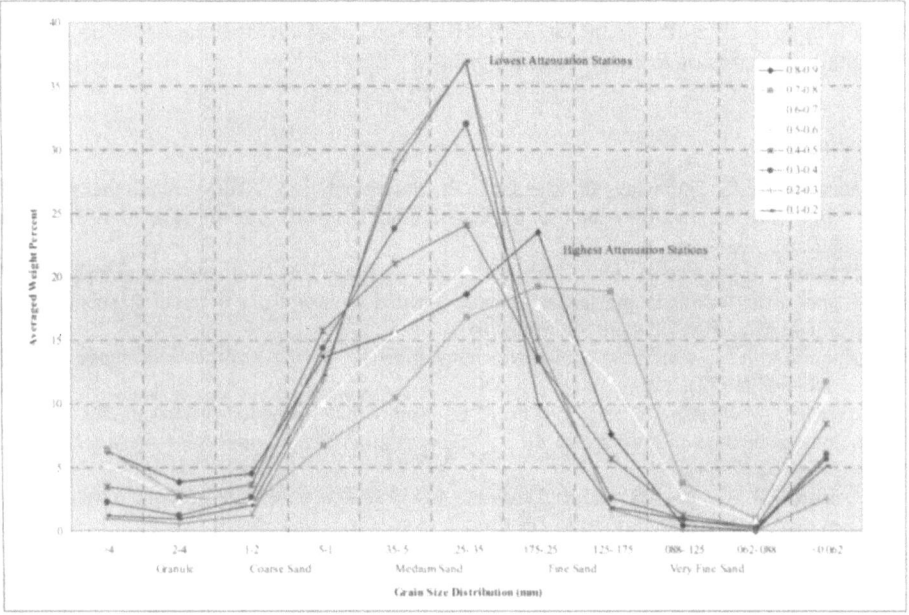

Figure 4. Averaged grain size distributions based on weight percent for each group of stations corresponding to the averaged attenuation coefficient for each station.

stations with slightly lower attenuation coefficients (k = 0.7–0.8 dB/m·kHz), had the second largest weight percent of fine sand, the highest weight percent of grains with diameters < 4 mm, and the highest weight percent of fine grained sediments, those with diameter < 0.062 mm. In contrast, the stations with the lowest attention coefficients had the highest weight percent of medium sand (.25 to .35 mm). These stations had grain size distributions indicating relatively, well-sorted sediments (mostly homogeneous, medium grained sands), while the high-attenuation stations contained a mixture of course and fine grained sediments.

5   Conclusions

ISSAP, a geoacoustic *in-situ* measurement system was used to rapidly and accurately measure compressional wave speed and attenuation in surficial marine sediments. An interesting comparison of attenuation with averaged grain size distributions was introduced. This work is in process and a more complete analysis will be presented at a later time. Continuing research will also explore the relationship between measured acoustic properties and remotely measured backscatter.

Acknowledgements

This research was supported by the Office of Naval Research Grant Number N00014-00-1-0821 under the direction of Roy Wilkens and Dawn Lavoie. The first author is supported by Tyco International Ltd. through the Tyco Postdoctoral Fellowship in Ocean Mapping. We also thank Dr. Lloyd Huff for many useful suggestions as well as the captain and crew of *R/V Cape Henlopen*.

References

1. Hamilton, E.L., Shumway, G., Menard H.W. and Shipek, C.J., Acoustic and other physical properties of shallow water sediments off San Diego, *J. Acoust. Soc. Am.* **28**, 1–15 (1956).
2. Hamilton, E.L., Bucker, H.P., Keir, D.L. and Whitney, J.A., Velocities of compressional and shear waves in marine sediments detected *in situ* from a research submersible, *J. Geophys. Res.* **75**, 4039–4049 (1970).
3. Hamilton, E.L., Compressional-wave attenuation in marine sediments, *Geophysics* **37**, 620–646 (1972).
4. Best, A.I., Roberts, J.A. and Somers, M.L., A new instrument for making in situ acoustic and geotechnical measurements in seafloor sediments, *Underwater Technology* **23**, 123–132 (1998).
5. Griffin, S.R., Grosz, F.B. and Richardson, M.D., In situ sediment geoacoustic measurement system, *Sea Technology*, April (1996).
6. Mayer, L.A., Kraft, B.J., Simpkin, P., Lavoie, P., Jabs, E. and Lynskey, E., *In situ* determination of the variability of seafloor acoustic properties: an example from the ONR GEOCLUTTER area. In *Impact of Littoral Environmental Variability on Acoustic Predictions and Sonar Performance*, edited by N.G. Pace and F.B. Jensen (Kluwer Academic, The Netherlands, 2002) pp. 115–122.
7. Vaseghi, S.V., *Advanced Digital Signal Processing and Noise Reduction,* 2nd ed. (John Wiley & Sons, Ltd., 2000) pp. 62–64.
8. Courtney, R.C. and Mayer, L.A., Calculation of acoustic parameters by a filter-correlation method, *J. Acoust. Soc. Am.* **93**, 1145–1154 (1993).
9. Goff, J.A., Olson, H.C. and Duncan, C.S., Correlation of side-scan backscatter intensity with grain-size distribution of shelf sediments, New Jersey margin, *Geo-Marine Letters*, **20**, 43–49 (2000).
10. Hamilton, E.L. and Bachman, R.T., Sound velocity and related properties of marine sediments, *J. Acoust. Soc. Am.* **72**, 1891–1904 (1982).

SUB-BOTTOM VARIABILITY CHARACTERIZATION
USING SURFACE ACOUSTIC WAVES

MANELL E. ZAKHARIA

French Naval Academy
IRENAV BP 600 F-29240 BREST NAVAL, France
E-mail: zakharia@ecole-navale.fr

The study of both acoustic propagation and reverberation require a "realistic" information on bottom and sub-bottom properties, especially in shallow water environment. Several methods have been proposed to study the sub-bottom structure that are often faced to a "blurring" effect due to the strong echo corresponding to the water-sediment interface. We propose a new approach that, unlike conventional ones, relies on this interface as an "information carrier" and make use of the Stoneley-Scholte surface acoustic waves, SSW. After describing briefly the properties of SSW, we will use their velocity dispersion to recover the main properties of the sediment: density profile, profile of shear and compression waves velocity. The investigation of the reflection and refraction properties of SSW show that they can be treated as "conventional" ones. A new concept of "surface wave sonar" for detecting embedded objects as well as sediment inhomogeneities is thus possible. Several illustrations will be given displaying results issued from tank experiment on scaled mock-ups (scale of about a hundred).

1   Introduction

Prediction models and accurate investigation of acoustic fields require detailed information on the bottom and sub-bottom variability mainly in the low frequency or /and shallow water cases. Several approaches have been used to estimate the bottom topography and the properties of the top few meters of sediment. They commonly use conventional sonar (or seismic) systems towed in the water column and exploiting information issued from the reflection and refraction of an incident P wave (emitted in the water column). Most of these systems are faced with a "blurring" effect due to the strong echo from the water-sediment interface. Several methods have been developed to overcome this problem by using low-frequency high-resolution arrays (synthetic aperture or/and parametric arrays for detecting embedded objects, for instance). The approach we use in this paper is a complementary one based on the use of the seabed interface as a "carrier of information" or a "waveguide". The corresponding system is a sonar system towed on the bottom and using the surface acoustic waves for sediment characterization and object detection (Fig. 1). Although such an approach can be found in some marine seismic devices, conventional sonar processing is seldom used [1].

N.G. Pace and F.B. Jensen (eds.), Impact of Littoral Environmental Variability on Acoustic Predictions and Sonar Performance, 131-138.
© 2002 *Kluwer Academic Publishers.*

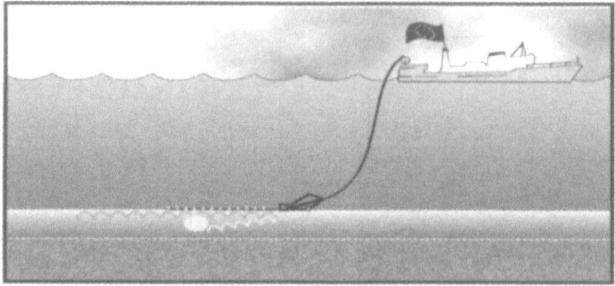

Figure 1. Surface wave sonar concept.

2   Velocity dispersion of SSW

The Stoneley-Scholte Waves, SSW, are heterogeneous dispersive waves guided along the interface and possessing a vertical exponential decrease in both the water column and the sediment [2]. They are relevant waves for sediment characterization for several reasons:

1. Their group velocity is highly dependant on the shear velocity in the sediment.
2. Their penetration into the sediment is about a wavelength (a few meters).
3. They are not affected by geometrical spreading loss (guided waves); nevertheless they are evanescent ones (exponential vertical decrease of amplitude) and have to be generated at the interface and detected very close to it.

Several modes could be generated [3]. In the work described, only the first-order mode has been used (for transmitter simplicity considerations). Its energy distribution in the water and the sediment depends on the impedance contrast. Two major cases can be distinguished. For both cases, the major properties are summarized in Table 1:

1. Soft bottom: c_{ss} (z) < c_w, the shear velocity in the sediment c_{ss} (z) is lower than the sound velocity in the water c_w. The energy is mostly concentrated in the sediment but enough energy is present in the water for characterizing bottom properties. This is the most common case for "conventional" sediments.
2. Hard bottom: c_{ss} (z) > c_w. SSW energy is mostly concentrated in the water. Although the generation and detection is easier, the bottom variability is of lower interest (rocky bottom).

Table 1.   Main properties of SSW [4].

	Soft bottom	Hard bottom
Energy concentration	in the sediment	in the water
Penetration depth	one λ in the sediment one λ in the water	one λ in the sediment several λ in the water
Velocity	dispersion equation approximated by 0.8 c_{ss}	dispersion equation no simple approximation

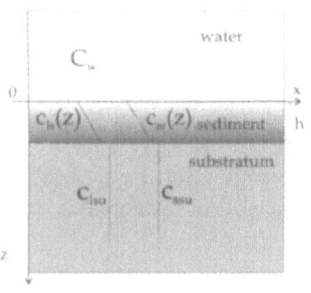
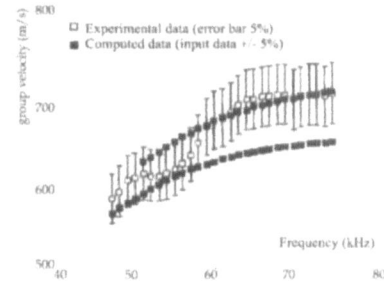

a- experiment geometry   b- dispersion law (experimental results)

Figure 2. Velocity dispersion of SSW.

As the penetration depth of SSW depends on the frequency, each frequency bin carries some information on a corresponding depth layer; when using wideband signals, several propagation depth are investigated and the group velocity dispersion of the SSW depends on the velocity profile in the sediment. The group velocity can be easily measured on a time-frequency representation of wideband transmitted signals [5].

Figure 2b shows a typical example of velocity dispersion of SSW. Full squares represent theoretical data and hollow ones experimental results. The sediment is modeled by 1-cm PVC layer with constant density and linear profiles of velocity (Fig. 2a): c_l (z)= 2090 + 101z, cs (z) = 778 + 20 z; (c in m/s and z in mm).

The velocity profiles have been measured (with an accuracy of 5%) and the experimental profiles have been introduced in a numerical model for predicting velocity dispersion. Bias of ± 5% have been added to the data in order to take into account the input error. The two model outputs (with ± 5% error) have been plotted as full squares. An experiment was conducted using the PVC layer over a marble substrate [6]. A shear wedge generator and a point receiver were used. For two positions of the receiver, the wideband signals were analyzed using the Wigner-Ville time-frequency distribution [5] from which, the group delay was extracted and the velocity dispersion computed and plotted (hollow squares, with an error bar of 5%). Figure 2b shows that the match between the predicted and the computed data is better than a few percents.

3   Inversion

The inverse problem consists in determining the sediment properties from SSW properties. We proposed a new method based on the use of an artificial neural network ANN [5,7]. For computation time reasons, we show results only for a simple benchmark including a single uniform layer. In this case, the number of parameters to recover is reduced to a few: thickness, shear velocity, compression velocity and density (for N layers, it would be 4N parameters). The method can be summarized as follows:

1. Several simulations of a benchmark case are realized for various input data values; a few frequency bins of the dispersion curve are used (typically 7).

2. A random data selection (frequency bins and geoacoustical properties) is used for training an artificial neural network (ANN).

3. The ANN input is then reduced to the frequency bins (no more geoacoustical data) and the network is asked to classify the input vector. While doing the classification, an error minimization is achieved and the values corresponding to the minimal error are extracted and compared to the actual ones. The operation is repeated several times to reduce the influence of training set.

In order to solve the inverse problem, very loose *a priori* information needs to be used (as opposed to methods such as conjugate gradient). An example is presented in Table 2.

Table 2. A priori information for solving the inverse problem

Material	Thickness	Shear velocity	Compression velocity	density
	m	m/s	km/s	
Sediment	$3 \leq h \leq 19$	$140 \leq c_{ss} \leq 460$	$1.7 \leq c_{ls} \leq 2.9$	$1.4 \leq \rho \leq 2.2$
Substratum	semi ∞	3850	6.3	2.7

Only five values were used for each parameter leading to 625 realizations of the direct problem. Half of the data (randomly selected) were used for training the ANN and the other half for inversion. The results of data inversion (simulation) are given in Table 3.

Table 3. Error on the inversion of sediment properties.

parameter	Thickness	Shear velocity	Compression velocity	Density
error	$\varepsilon_h < 5\%$	$\varepsilon_{cs} < 2\%$	$\varepsilon_{cl} < 18\%$	$\varepsilon_d < 17\%$

For tank experiments, a similar approach was applied: several simulated realizations of the direct problem were used for ANN training and the ANN tried to classify data issued from the experiment (estimated group velocity). Similar accuracy was obtained: shear velocity: $\varepsilon < 0.5\%$, thickness: $\varepsilon < 11\%$, other parameters: $\varepsilon < 20\%$.

These results show the relevance of SSW for describing the sediment properties and detecting their small variations. In order to investigate the performance of a SSW based sonar system, we have studied the reflection and refraction properties of these waves.

4   Reflection and refraction of SSW

In the case of solid-solid interface (such as the presence of a rock in a sediment), the reflection and transmission coefficients of SSW were investigated. As SSW are evanescent waves (two components), two hypotheses have been found in the literature (theoretical work) concerning the continuity conditions at the interface: continuity of each component of SSW or continuity of their resultant. Several experiments were run using various materials possessing impedance contrast close to the one encountered in the seabed. Figure 3 shows the reflection of an SSW in the case of a PEHD/Polyester interface (two bonded solids) with the following properties:

- PEHD: $\rho = 920 \pm 50$ kg/m^3, $c_l = 2000 \pm 30$ m/s, $c_t = 1000 \pm 30$ m/s, $c_s = 760 \pm 30$ m/s
- Polyester: $\rho = 1120 \pm 50$ kg/m^3, $c_l = 2480 \pm 30$ m/s, $c_t = 1230 \pm 30$ m/s, $c_s = 950 \pm 30$ m/s

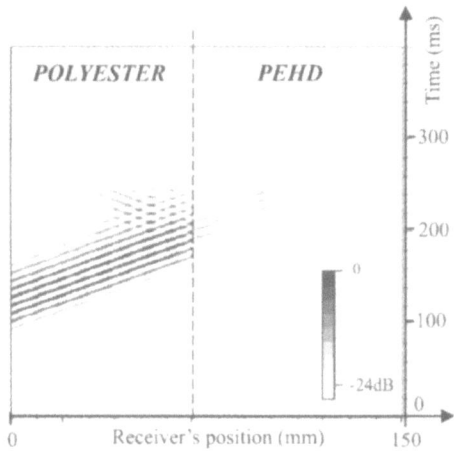

Figure 3. Reflection and refraction of SSW at solid-solid interface. Vertical: time axis, horizontal: receiver position (for a normal incidence transmitter).

From experiments, we have found that the second hypothesis (continuity of the resultant) matches better the results and that, at oblique incidence, SSW follow laws similar to the Snell-Descartes ones [8]:

1. SSW is reflected as a SSW (with same velocity).
2. SSW is transmitted as another SSW (with a SSW velocity corresponding to the second medium).
3. A critical angle was observed (similarity with the case of compression waves).
4. No other waves (mode conversion) or components could be observed.

Thanks to these properties, the propagation of the SSW can be interpreted in a simple and conventional manner. Even for 3D geometry, one can easily talk about reflection, diffraction, directivity pattern, ray description, SSW beam forming, SSW tomography, etc.

5   SAW sonar

The first sonar application we have investigated is the detection of embedded objects using a surface acoustic wave sonar approach (Fig. 1). Several tank experiments have been carried out on buried targets in a homogeneous resin [9]. The impedance contrast has been chosen to be as close as possible to real conditions as shown in Table 4.

Table 4.   Impedance contrast at sea and in tank.

SEA	TANK
Sediment: $Z = 2.5$ Mrayls	Polyester resin: $Z = 3$ Mrayls
Rocks: $Z = 14$ Mrayls	Metal: $Z = 46$ Mrayls
Impedance contrast: 5.6	Impedance contrast: 8

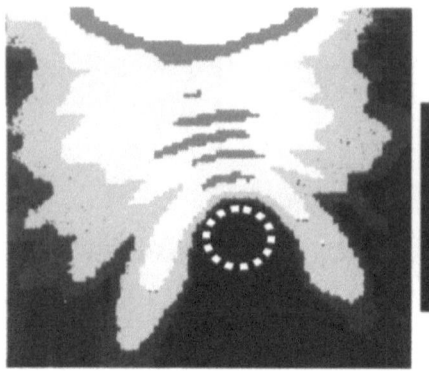 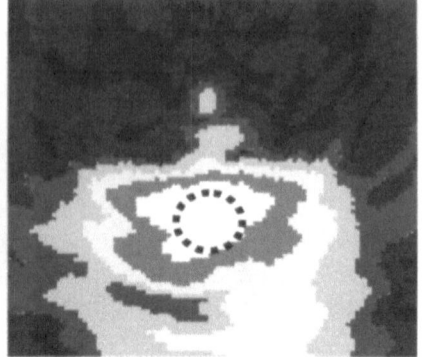

Figure 4. Energy scattering by a buried   Figure 5.   SSW sector scanning sonar.
sphere. Shadow effect similar to sidescan   Imaging of a buried solid sphere.
sonar case. Sphere (16 mm); frequency 100   Sphere (16 mm); frequency 100 kHz,
kHz, λ_s: 10 mm. Scale: 10 x 10 cm. 2 λ_s: 10 mm. Scale: 10 x 10 cm. 3 dB/color
dB/color

SSW transmission was achieved using a periodic excitation of the water-resin interface
(comb transducer centered around 100 kHz, scale factor: about 100). SSWs were clearly
identified by their velocity and by their vertical exponential decrease. Figures 4 and 5
show experimental results for a sphere of 16 mm (about 1.5 λ_s). In Fig. 4, for a given
transmitter position, the surface was finely scanned and the SSW energy on the interface
was computed. In this figure, one can see from top to bottom:

1. The incoming wave loosing some energy while propagating; the source position
    is 10 cm above the top of the figure.

2. An interference zone (between the incident signal and the sphere-reflected echo).

3. A shadow zone after the sphere, similar to the one encountered in sidescan sonar.

Similar results were obtained with several other embedded objects [9]. In all cases, the
presence of an object was clearly defined by SSW energy scattering even for a small
sphere (diameter = 0.4 mm ≈ 0.4 λ) where shadow contrast was better than 10 dB.

Figure 5 shows an example of experimental results for a monostatic sonar
configuration. A sector scanning approach was used:

1. A transmitter with low directivity index (aperture ≈ 60°, range to target: 10 cm).

2. A fixed receiving array of hydrophones (broadside configuration at 2 cm from
    the target, array length: 8 cm ≈ 8 λ).

3. A digital beamforming for sector scanning (Hamming array shading).

Figure 5 clearly shows the possibility of using SSW in sector scanning surface wave
sonar. The 3 dB resolution (white color) is comparable to the target size (extended on
each side, due to the limited array resolution). Comparable results were obtained for
various targets in other geometrical configurations for both broadside and endfire arrays
(endfire arrays are easier to tow and to handle at sea).

6   Tomographic reconstruction of SAW velocity

Several applications such as the survey of hazardous areas in offshore (continental slope stability), the survey of harbor sedimentology, the prediction of acoustical propagation, etc. require a permanent survey tool. The tool could be sitting on the bottom and monitoring the changes in some sub-bottom properties [8,10]. For such a purpose, we have investigated the ability of SSW to provide insight description of the seabed from remote distance. A transmission tomography approach was developed and applied to the characterization of changes of sediment properties. A tomographic experiment has been conducted in a tank using a cylindrical inclusion in a homogeneous resin. It simulates the effect of gas migration through the sediment (gas hydrates in hazardous areas). The cylindrical inclusion of polyester was inserted in a homogeneous resin (PEHD). The properties of the SSW waves in both cases are the following:

- Polyester: $c_s = 960 \pm 30$ m/s, $Z_s = 1.07 \pm 0.07$ Mrayls
- PEHD: $c_s = 760 \pm 30$ m/s, $Z_s = 0.7 \pm 0.07$ Mrayls

Both transmitter and receiver were moved along a circle in angular steps of 10° for the transmitter and 5° for the receiver. Conventional transmission tomography algorithms were developed based on the measurement of time of flight. Figure 6 shows an example of reconstructing velocity values in the mock-up using the back-propagation method. The error on the substrate velocity is very low (2%). Although the error on the inclusion is higher (18%), the shape of the cylindrical insert is quite well reconstructed. Other methods (such as SIRT) provided comparable results.

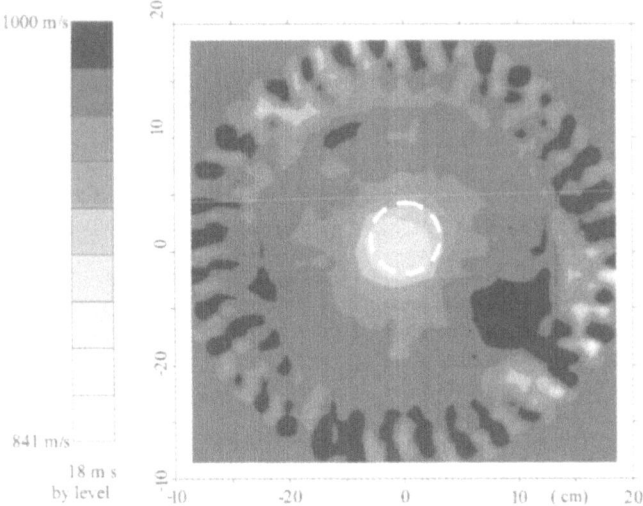

Figure 6. Tomographic reconstruction of a cylindrical inclusion in a homogeneous substrate.

7 Conclusions

This paper shows that, like conventional p or s waves, SSW can be used for detailed description of the seabed: velocity profile estimation, sector scanning sonar, detection of embedded objects, tomographic reconstruction of sediment inhomogeneities, etc. The concepts presented have been illustrated by several tank experiments showing the feasibility of surface acoustic wave sonar. Effort is presently put on the development of an efficient single mode generator at sea, using preferably contact-less devices.

Acknowledgements

This work was supported by the European Commission and the French Ministry of transport. The author thanks Jérôme Guilbot (Total-Fina-Elf), Edouard Mouton (Sage-Geodia) and Emmanuelle Chauvet for their contribution.

References

1. Stoll, R.D. and Batiste, E., New tools for studying seafloor geotechnical and geoacoustic properties, *J. Acoust. Soc. Am.* **96** (5), (1994).
2. Ewing, W.M., Jardetzky, W.S. and Press F., *Elastic Waves in Layered Media* (McGraw-Hill, New York, 1957).
3. Rauch, D., On the role of bottom interface waves in ocean seismo-acoustics: a review. In *Ocean Seismo-Acoustics,* edited by T. Akal and J.M. Berkson (Plenum, New York, 1986) pp. 623–641.
4. Guilbot, J., Caractérisation acoustique des fonds sédimentaires marins par étude de la dispersion de célérité des ondes d'interface de type Stoneley-Scholte, Ph.D. Thesis, Sep. 1994, Lyon, 256 pp.
5. Zakharia, M.E. and Chevret, P., Neural network approach for inverting velocity dispersion; application to sediment and to sonar target characterization, *Inverse Problems* **16**, 1963–1708 (2000).
6. Guilbot, J. and Zakharia, M.E., Tank experiments on a sediment small-scale model. Shear wave velocity profile inversion via Stoneley-Scholte waves. In *Proc. 2nd European Conference on Underwater Acoustics,* edited by L. Bjørnø (European Commission, Brussels, 1994) pp. 979–984.
7. Guilbot, J. and Magand, M., Determination of the geoacoustical parameters of a sedimentary layer from surface acoustic waves: a neural network approach. In *Full Field Inversion Methods in Ocean and Seismic Acoustics*, edited by O. Diachok, A. Caiti, P. Gerstoft and H. Schmidt (Kluwer Academic Publishers, 1995) pp. 171–176.
8. Mouton, E., Détection des instabilité des sédiments marins par des précurseurs acoustiques, Ph.D. Thesis, Oct. 2000, Lyon, 180 pp.
9. Zakharia, M.E. and Châtillon, J., Interaction of interface waves with a buried object. In *Proc. 3rd European Conference on Underwater Acoustics,* edited by J.S. Papadakis (European Commission, Brussels, 1996) pp. 39–44.
10. Mouton, E. and Zakharia, M.E., Reconstruction of sediment inhomogeneities using surface wave tomography. In *Proc. 5th European Conference on Underwater Acoustics*, edited by M.E. Zakharia, P. Chevret and P. Dubail (European Commission, Brussels, 2000) pp. 245–250.

THE INFLUENCE OF NOISE AND COHERENCE FLUCTUATIONS ON A NEW GEO-ACOUSTIC INVERSION TECHNIQUE

C.H. HARRISON

SACLANT Undersea Research Centre, Viale S. Bartolomeo 400, 19138 La Spezia, Italy.
E-mail: harrison@saclantc.nato.int

Distributed noise sources at the sea surface (wind, rain, waves) and distant shipping provide an ideal plane wave spectrum for investigating geo-acoustic properties. Using the fact that the vertical noise directionality is closely related to the bottom properties, a technique has recently been established theoretically and experimentally [C.H. Harrison, D.G. Simons, "Geoacoustic Inversion of Ambient Noise: A Simple Method." Submitted to JASA] for deducing geo-acoustic properties from the noise directionality. Theory suggests that the simple ratio of up-to-down beam-steered power is, in fact, the power reflection coefficient of the seabed – potentially a function of angle and frequency. Experiments at six sites in the Mediterranean and on the New Jersey Shelf with a 64-element vertical array have shown that stable solutions can be obtained in a few minutes. In preparation to extend the technique to the use of a vertical synthetic aperture the sptial and temporal variability of the coherence has been investigated by selecting pairs of hydrophones still from the VLA. Surprisingly the coherence for a pair is quite unstable on a period of 10 seconds while the reflection loss derived from the entire array remains reasonably stable. The paper discusses reasons for the fluctuations and implications for the inversion technique when synthesising an aperture.

1 Introduction

A new method of deriving bottom reflection loss from ambient noise [1–3] has been tried at five Mediterranean sites (during ADVENT99 and MAPEX2000bis) and most recently at one site on the New Jersey Shelf (during Boundary2001). It can be shown by ray or flux arguments [1,2,4] that if the wind provides a sheet source (with or without distant shipping) the ratio of down-going to up-going noise at a given angle to the horizontal is simply the bottom power reflection coefficient. This results from the peculiarity of sheet sources that the propagation does not undergo the usual geometric spreading with range. Instead there is a geometric series of noise contributions resulting from the surface interaction once per ray cycle. The downward beam has exactly the same geometric series as the upward beam but with one extra bottom reflection near the receiver. Therefore a VLA can measure reflection loss almost directly as a function of angle and frequency. From this measurement one can then deduce the geoacoustic parameters (if necessary!) by employing a model of plane wave reflection loss, such as OASR [5] or more simply, the formulae in [6] for multiple fluid layers with an underlying solid bottom. The only significant effect of refraction is to modify angles according to Snell's law and the sound speeds at the bottom and receiver. A numerical demonstration of the fidelity of the theory was given in [1,7], and it was shown in [7,8] that the measured properties are within about a water depth of the receiver.

In this paper we take the experimental results from two Mediterranean sites and the New Jersey Shelf site to demonstrate geographic variability. Then at the first site we look at the influence of spatial and temporal variability on the deduced reflection loss. One aim is to investigate the feasibility of using a synthetic aperture rather than a full VLA, and we will find that variability has a dramatically different impact in the two cases.

2   Geographic variation of reflection loss

If the method really works we naturally expect to see changes from site to site since bottom properties are known to vary. Here we demonstrate geographic changes with examples from three out of the six sites visited to date. The full array method is deliberately insensitive to non-uniformity of noise source distribution.

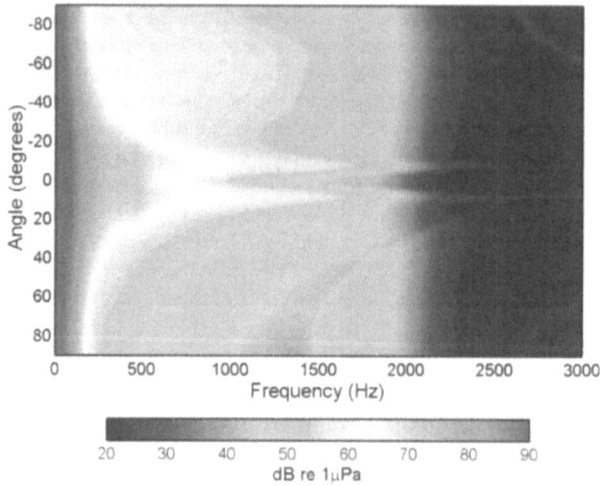

Figure 1. Sicily sand VLA beam response.

One of the sites visited in November 2000 during MAPEX2000bis was south of Sicily at the northern end of the Ragusa Ridge, where the bottom was thought to be sandy. The vertical beam response of the central 32 elements (0.5m spacing) of the 62m VLA is shown in Fig. 1. The "noise-notch" is evidence of the strong downward refraction. The "up-to-down" ratio is shown in Fig. 2. Interpreting this as reflection loss we immediately see the classical interference fringes [6] (indeed these can be seen in the lower part of Fig. 1 when we know what to look for).

By inspection we can deduce several things, the most obvious of which is the critical angle (about 30°). In this example the fringes are regular in frequency (notwithstanding the experimental artefacts caused by grating lobes at high frequencies and broad beams at low frequencies [1]) which means there must be only two dominant reflectors or boundaries. Also the layer separation is directly related to the fringe separation (see later). The strength of the reflection and the depth of modulation of the fringes provide information about the combined speed and density in the two layers. Volume absorption in the sediment will certainly affect the low loss region to the left of the critical angle, but experimentally this region is rather unreliable since the up-to-down ratio is so close to unity. Inversions and goodness of fit are discussed in [1,7].

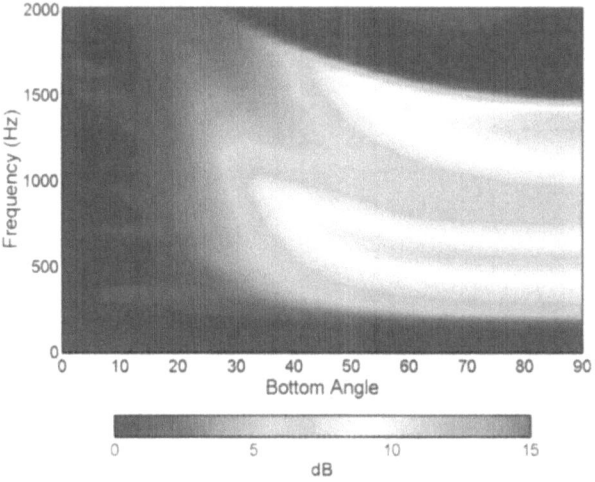

Figure 2. Sicily sand: Experimentally deduced reflection loss.

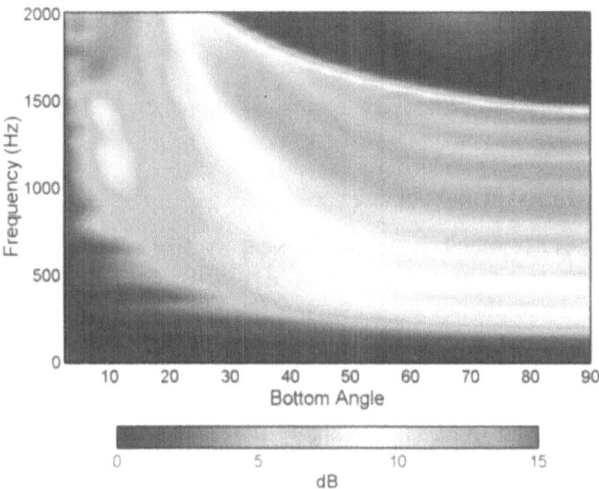

Figure 3. Elba mud: Experimentally deduced reflection loss.

A contrasting site in the same experiment was 'lossy mud' east of Elba. This resulted in the reflection loss plot of Fig. 3. Here we see evidence of higher losses at low angles, i.e. an absence of a clear critical angle. Also we see two superimposed fringe patterns, one fine – with seven or eight loss maxima visible at, say 90°, and one coarse – with only one maximum and one minimum visible. It is immediately obvious that one cannot use a simple three layer (i.e. two boundary) model to explain this data. Indeed the pattern suggests three boundaries, two of which are wide apart and two of which are close together. In fact this can be matched very well with a three boundary calculation [1,7]; furthermore the deduced layer spacing agrees with independent experiments in the vicinity.

At the New Jersey Shelf site (39°19'N 72°33'W), where we expect a 'stiff clay', we see in Fig. 4 a clear critical angle again (about 30°) but subjectively either many complicated fringes or none at all. This suggests many equally weakly reflecting layers, effectively a half-space.

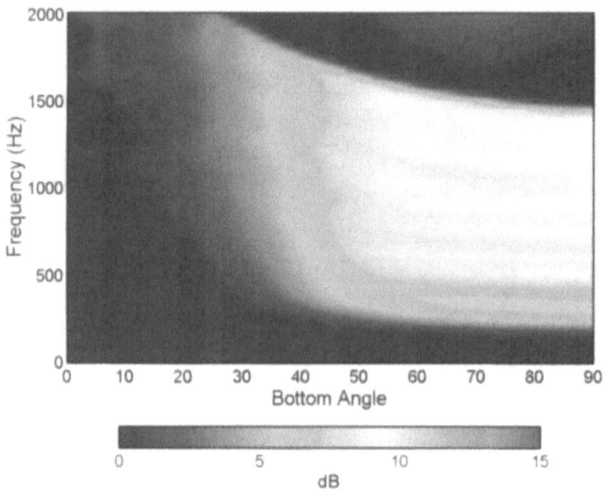

Figure 4. New Jersey Shelf clay: Experimentally deduced reflection loss.

3   Spatial variability of coherence

Spatial variability is a crucial issue for vertical synthetic aperture processing. By definition we build a cross-spectral density (CSD) matrix for the synthetic aperture by taking real pairs of hydrophones at real depths. For the sake of economy we keep one hydrophone fixed, so we rely on the coherence estimates being only dependent on separation and independent of depth (i.e. the CSD matrix must be Toeplitz). However normal mode theory tells us to expect non-Toeplitz behaviour especially within a few wavelengths of boundaries. In the current applications where water depths are about 100 m and frequencies are between 100 Hz and 2000 Hz we expect only residual effects. There are various experimental ways of demonstrating this. One is to plot the CSD matrix as a colour contour surface; we expect to see bands parallel to the diagonals and no variation along the diagonals. Another way is to use each row of the CSD matrix to plot a graph of coherence *vs* separation; a spread indicates deviation from Toeplitz. A more enlightening way, in the current context, is to compare full array reflection loss with a forced Toeplitz array reflection loss. "Forced Toeplitz" means that we use instead a Toeplitz array whose diagonals are the average values along the diagonals of the true

CSD matrix. In practice it is difficult to see any difference at all in these reflection loss plots, which means the Toeplitz assumption is good enough for this application.

4   Temporal variability of coherence, beam response, and deduced reflection loss

Using the full array this geoacoustic inversion technique is remarkably insensitive to variations in noise directionality. However, temporal variability becomes important when we build the CSD matrix from the coherence between hydrophone pairs taken at different times. In this section we try out several processing options solely to investigate behaviour. It is hoped that the requirement for averaging can be moderated by future processing schemes.

4.1   Full Array Processing (Unmodified CSD Matrix)

Figure 2 shows the average of 20 batches of 10-second files (3 minutes' worth of data). The noise is sampled at 6kHz, enabling several hundred 128-point FFTs to be averaged during each 10 seconds. In fact the solution is fairly well converged after *only one* 10-second file, as Fig. 5 shows.

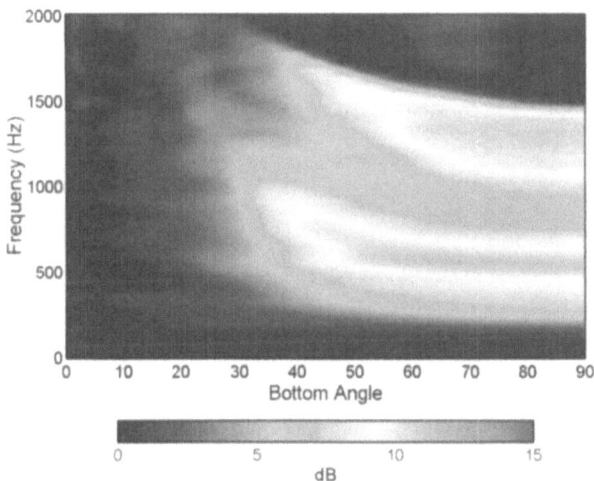

Figure 5. Experimental reflection loss deduced from a single 10 second file, *cf* the 20 file average shown in Fig. 2.

4.2   "Forced Toeplitz" Processing (Toeplitz Array built from Complete Array)

Our first option is to build a Toeplitz array from the full array for one time. However, as noted in Section 3 there is hardly any difference between this and the original. The next possibility is to take each diagonal (whilst retaining Hermitian symmetry) from a different 10-second file. Naturally we need 32 files to construct each new array. Figure 6 shows that even allowing for this, after 7×32 files (exhausting the file supply!), we still have relatively poor convergence compared with Figs. 2 and 5. So even with the spatial averaging (along the diagonals) combining coherence measurements from different times introduces instability.

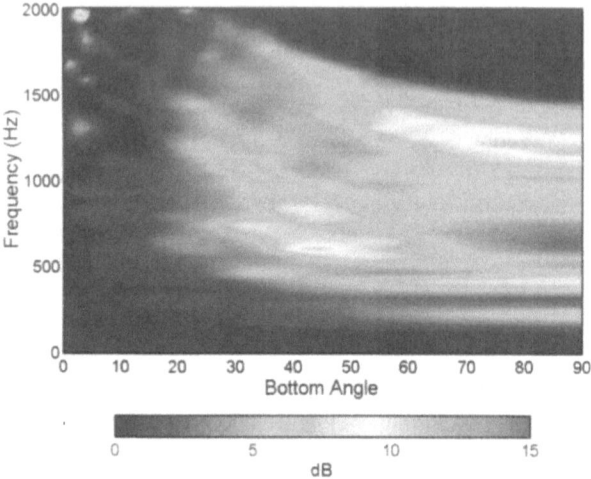

Figure 6. Reflection loss deduced via the average beam response (over 224 (=7×32) 10-second files) from a Toeplitz CSD matrix constructed from averages along diagonals taken from separate 10-second files.

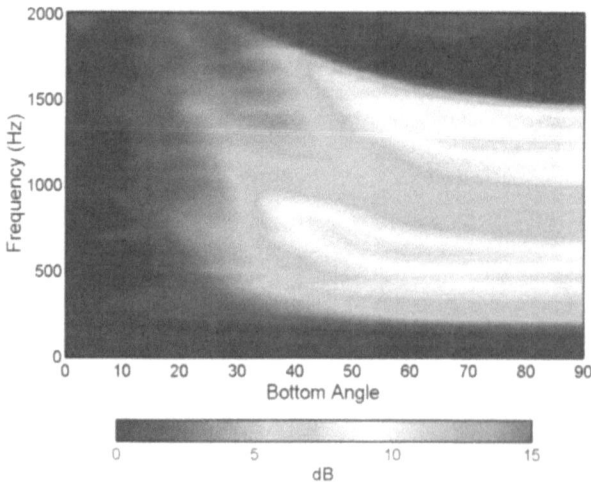

Figure 7. Reflection loss deduced via the average beam response (over 30 10-second files) from a Toeplitz CSD matrix constructed from simultaneous hydrophone pairs 32 and n.

4.3   Hydrophone Pair Processing (Toeplitz Array built from Hydrophones 1 and n)

If we now select hydrophone pairs (1 and n, up to 32) from the full array for one time we find that the response is poor to start with but converges reasonably well after 30 or so 10-second files (Fig. 7). Thus taking hydrophones pairs is worse than taking the full array, but remembering that here we have only N independent coherences whereas the full array has $(N^2-N)/2$, it is not surprising that we need a further N averages (~ 30) to achieve similar stability.

Finally if we take each hydrophone pair from a different 10-second file we find very slow convergence. The supply of files is exhausted long before we achieve convergence (Fig. 8).

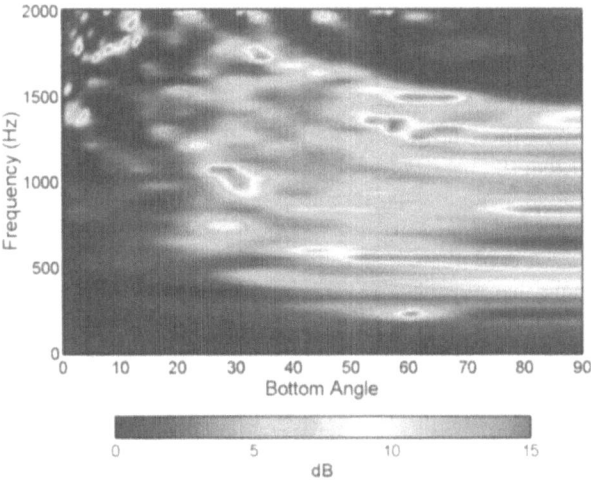

Figure 8. Reflection loss deduced via the average beam response (over 192 (=6×32) 10-second files) from a Toeplitz CSD matrix constructed with hydrophone pairs from separate 10-second files.

4.4   Explanation of these Anomalies

So why is there such a big discrepancy between the required averaging time for the full array and for the various synthetic apertures? Obviously to build a n-element array takes at least n times longer, but the above results demonstrate an extra factor. Imagine a small array near the seabed with "point splashes" at the surface. Each splash (with beam-forming and up-down comparison) gives straight away an estimate of R at that ray angle and the frequencies of the broad-band splash. No averaging is required at all, we just need to fill in each angle and frequency. Therefore we have to wait for sounds to have arrived from all the required directions, but we don't need to wait for any evening out or convergence process. It's rather like scribbling on a piece of paper pressed on a coin; the image of the head appears wherever you scribble – scribbling harder doesn't help.

In contrast when we take hydrophone pairs we cannot do any steering at all until we have a complete set of C_{nm}. Since each C_{nm} originates at different times they each have been generated by a different splash (potentially a different angle and frequency). The resulting correlation matrix consists of a set of numbers that are all incorrect in as far as they are not the average for a sheet. To a certain extent there will be some averaging when we perform the $\Sigma\Sigma C_{nm}$ process to get the beam response, however it is not at all obvious that the average of 'wrong' numbers converges to the right answer! What is true, though, is that we certainly have to wait for the 'filling-in' time in order to more or less cover all angles. In addition now, however, we need to ensure that we have the same angle distribution (i.e. source spatial distribution) for each n,m and this requires some time for the average to settle down. In short, the reason for the discrepancy is that the original full array method is self-compensating and builds its reflection loss picture from estimates of reflection loss (that need no averaging) rather than from coherence (which certainly does need averaging).

5   Conclusions

A technique has been established for determining bottom reflection loss and hence geoacoustic parameters from noise directionality. It has already been shown that the theory is sound and that the measurement is local to the receiver array. In order to test the feasibility of using a synthetic aperture array instead of a full vertical array we have investigated the dependence on variability. Geographic, spatial (water column), and temporal variability were addressed. It is found that the averaging time for synthetic aperture is much more than n times the time for a n-element full array. This surprising finding is explained by the self-compensating nature (and therefore rapid convergence) of the full array technique rather than any anomalies with synthetic aperture.

Acknowledgements

The authors would like to thank the Captain and crew of *RV Alliance* and the chief scientists on the three cruises Jürgen Sellschopp (ADVENT99), Martin Siderius (MAPEX2000bis), Charles Holland (BOUNDARY2000) for making time available for these experiments.

References

1. Harrison, C.H. and Simons, D.G., Geoacoustic inversion of ambient noise: a simple method, *J. Acoust. Soc. Am.* (submitted Nov. 2001).
2. Harrison, C.H. and Simons, D.G., Geoacoustic inversion of ambient noise: a simple method. In *Proc. Inst. of Acoust. Conf. on Acoustical Oceanography*, Southampton, UK (April 2000).
3. Aredov, A.A. and Furduev, A.V., Angular and frequency dependencies of the bottom reflection coefficient from the anisotropic characteristics of a noise field, *Acoustical Physics* **40**, 176–180 (1994).
4. Harrison, C.H., Noise directionality for surface sources in range-dependent environments, *J. Acoust. Soc. Am.* **102**, 2655–2662 (1997).
5. Schmidt, H., OASES user's guide and reference manual. Dept of Ocean Engineering, MIT (1999).
6. Jensen, F.B., Kuperman, W.A., Porter, M.B. and Schmidt, H., *Computational Ocean Acoustics* (AIP Press, New York, 1994) pp. 46, 50, 54.
7. Harrison, C.H. and Baldacci, A., Bottom reflection properties by inversion of ambient noise. In *Proc. 6th European Conf. on Underwater Acoustics*, Gdansk, Poland (2002).
8. Harrison, C.H. and Baldacci, A., Simulated geoacoustic inversion of ambient noise, *J. Acoust. Soc. Am.* (in preparation).

ESTIMATING SHALLOW WATER BOTTOM GEO-ACOUSTIC PARAMETERS USING AMBIENT NOISE

DAJUN TANG

Applied Physics Laboratory, University of Washington, 1013 NE 40th St., Seattle WA 98105, USA
E-mail: djtang@apl.washington.edu

Knowing bottom geo-acoustic parameters is of great importance for using sonar systems effectively in shallow waters. In this paper, ambient noise data recorded on a vertical hydrophone array taken in the frequency range of 1000 to 3000 Hz were used. Forward modeling and model/data comparison show that the energy ratio of down-looking and up-looking beams, after proper average over time and frequency, is the energy reflection coefficient of the bottom. From the reflection coefficient, critical parameters of the sediments, the sound speed, density and attenuation coefficient, are obtained. Core data taken at the experimental site support the inversion results.

1 Introduction

This paper is motivated by the desire of devising a practical and reliable way to estimate sediment geo-acoustic parameters in shallow water. In shallow water environments, sound propagation is dominated by modes corresponding to small grazing angles, as such, the sound field is greatly influenced by the presence of sediments, especially the surficial layer of sediments. Therefore, knowing the geo-acoustic parameters of the surficial sediments is of crucial importance for improving sonar performances in shallow water regions. Direct measurements of reflection loss is difficult and impractical, since at least one pair of well separated source and receiver are needed, and such scheme only provides reflection coefficient at one particular grazing angle. In addition, at small grazing angles this approach is prohibitively challenging because the presence of shallow water boundaries.

In this paper, we present a method of estimating key sediment parameters using ambient noise recorded on a vertical line array. The parameters obtained this way are the compressional sound speed, the density, and the attenuation coefficient. The approach has the following advantages:

1. Needs only a single measuring station with a vertical array.

2. Is passive.

3. Provides data over wide frequency band.

4. With a moving vertical array, provides potential for large area survey.

5. Needs no knowledge of the noise sources.

6. Has potential to be applied to range-dependent environments since the array is sensitive only to local modes.

The idea of taking advantage of the presence of ambient noise to measure bottom properties is not new. For examples, Deane and Buckingham [1], Buckingham and

147

N.G. Pace and F.B. Jensen (eds.), Impact of Littoral Environmental Variability on Acoustic Predictions and Sonar Performance, 147-154.
© 2002 *Kluwer Academic Publishers.*

Carbone [2], and Harrison [3] presented models and model/data comparisons of vertical coherence of noise and provided a basis for using coherence to invert for bottom properties. Extensive numerical evaluations of such models are also available [5]. Carbone *et al.* [4] used that approach to estimate both the compressional and shear wave speeds. While the vertical coherence is simple to measure, for this method to be applicable, certain conditions on the noise source have to be met, which in many environments are not the case.

A team of Russian scientists led by Furduev has reported their work on using ambient noise to estimate bottom relection coefficient in deep water [6–8]. Their approach is based on ray theory and uses a vertical line array to measure up- and down-looking beams, and from which estimates the bottom energy reflection coefficient as a function of angle and frequency.

Our approach in the present paper is very similar to that used by the Russian scientists. However, since we work in a shallow water area where modal interference is strong, no simple analytical results can be obtained. For our experiment scenario, numerical analysis shows that the energy ratio of the down- and up-looking beams of noise is indeed the energy reflection coefficient of the bottom, provided that averaging over frequency as well as time is performed. At small grazing angles, the noise comes from distant sources; near normal incidence (large grazing angle), the noise in our case comes from the ship from which the vertical array is deployed; there is no appreciable noise in the mid-grazing angles. The energy reflection coefficient at small grazing angles provides information on sediment sound speed, whereas that at large grazing angles gives information on sediment density. Information concerning bottom attenuation coefficient are found in both grazing angle regions.

We will in the following two sections present experimental data and modeling results, respectively, and conclude by discussions.

2 Experiment

During May and June, 2001 in the East China Sea, as part of the Office of Naval Research sponsored ASIAEX experiment, noise over the band of 500 to 5000 Hz was recorded on a 31 element vertical array, which was deployed from the research ship the Melville. The experiment was conducted in an area where many small fishing boats are within visible range, and shipping and wind noise are also present. In addition, the Melville was performing dynamic positioning, therefore engine noise from the Melville is a major source to be considered. The water depth in the experiment site is 105 m. The sound speed profile in the water column corresponding to the time of the noise measurements was measured from CTD casts and is given in Fig. 1. It is a typical summer time profile with a thermocline extending to a depth of 30 m. The sound speed below the thermocline (deeper than 70 m) is essentially a constant, where the vertical line array is deployed. The element spacing is 21.43 cm, the sampling rate is 12,000 Hz, and the noise band recorded is 500 to 5000 Hz. Segments of noise data, each 0.5 s long, are taken every 5 s on each of the 31 elements of the array. The data segments are bandpass filtered and Fourier transformed. At each frequency bin from 1000 to 3000 Hz, beams are formed using all elements and a Hanning window. The beam angle ranges from $-90°$ (up-looking) to $+90°$ (down-looking) with one degree increments. The square of the absolute

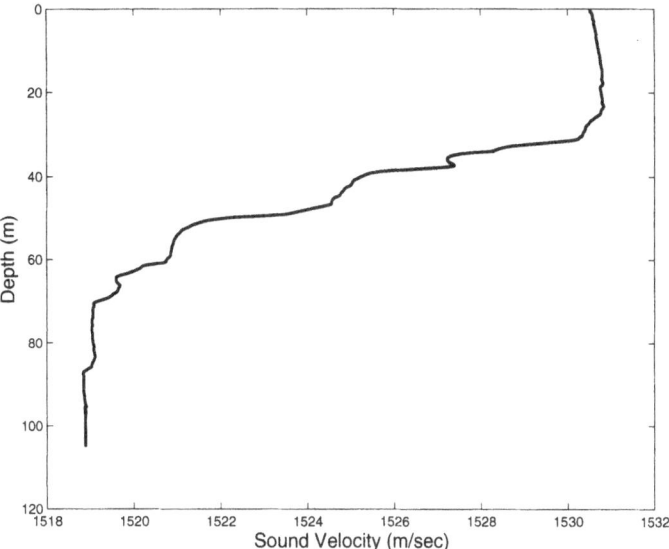

Figure 1. Sound speed profile obtained from CTD measurements.

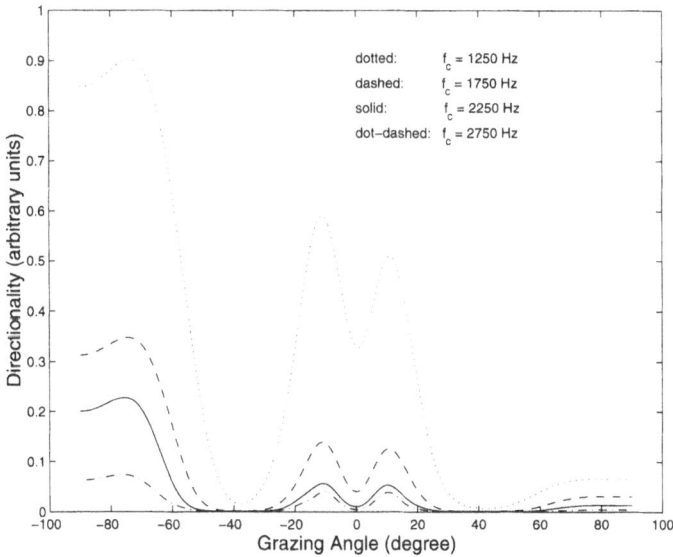

Figure 2. Directionality from data.

value of the beams are termed Directionality. Figure 2 gives the directionality that was first averaged over 50 data segments in the same frequency bins, and was then averaged over neighboring frequencies of 500 Hz. In the figure, directionality is given for four center frequencies. The values of the directionality from different frequencies show the relative strength of the noise field over the frequency band, with decreasing strength versus increasing frequency. The directionality of all four frequencies show similar features: (1) A minimum at zero grazing angle, caused by the fact that the noise sources are all near

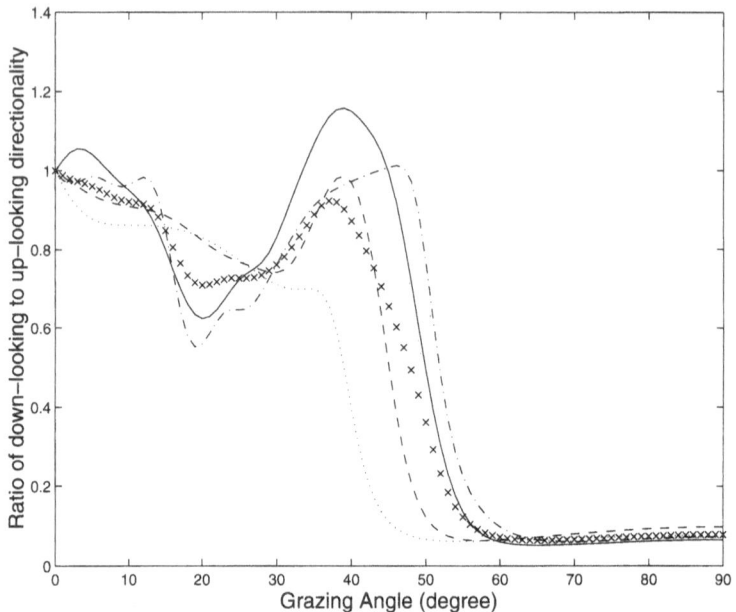

Figure 3. Ratio of down-looking directionality to the up-looking directionality from data. The 4 curves correspond to frequencies as given in Fig. 2. The crosses are the average of result from all four frequency bins.

the sea surface and the thermocline minimizes the excitation of modes with very small grazing angles. This will be further explained in the next section. (2) Within 20° there is a peak on either side of the minimum, with the one corresponding to up-looking beams larger than the one corresponding to the down-looking beams. These beams are associated with noise trapped in the waveguide and propagated from long distance to the array. (3) The large peak on the left of the figure is due to the self noise from the Melville, and the small peak on the right is the reflected beam by the bottom of the self noise. The ratio of the down-looking directionality to that of the corresponding up-looking one is given in Fig. 3. We will show in the next section that this ratio corresponds to the energy reflection coefficient of the bottom.

3   Modeling

We assume that the noise filed with frequency f on an array element at depth z can be modeled as the summation of a large number of un-correlated points sources of the following modal form:

$$p_j(z) = a_j \sum_n \phi_n(z)\phi_n(z_j)e^{ik_n r_j}/\sqrt{r_j}$$
$$= a_j \sum_n \phi_n(z_j)e^{ik_n r_j}/\sqrt{r_j}(e^{-ik_{zn}z} + V_n e^{i\psi_n}e^{ik_{zn}z}), \qquad (1)$$

where a_j is the source strength, z_j and r_j are the depth and range of the source, k_n is the complex eigenvalue of mode n of the waveguide, and k_{zn} is the vertical wavenumber of

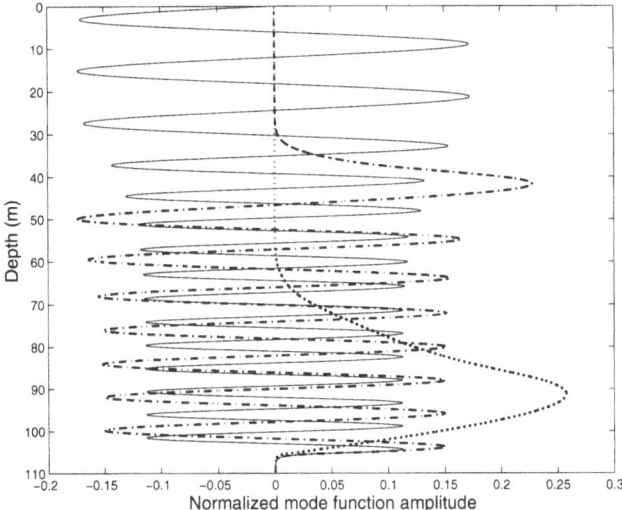

Figure 4. Modes number 1 (dotted), 15 (dot-dashed) and 30 (solid) at 2000 Hz. Notice that modes 1 and 15 have low amplitudes near the surface

mode n. The quantity V_n is the bottom reflection coefficient at the particular frequency associated with an angle related to the eigenvalue, and ψ_n is simply a phase factor accumulated through one cycle in the water column for mode n. This expression is true only when the receiver array is in a region where sound speed profile is a constant. In the expression ϕ_n is the depth-dependent mode function. Figure 4 shows three mode functions versus depth and demonstrates that lower order modes have little excitation if the source is near the surface.

To form a beam at direction θ, we have

$$B_j(\theta) = a_j \sum_n \phi_n(z_j)e^{ik_n r_j}[b(k\sin\theta - k_{zn}) + V_n e^{i\psi_n}b(k\sin\theta + k_{zn})], \qquad (2)$$

where k is the wavenumber at depth z_j and b is the array response of a beam pointed at angle θ. Sum over all sources, and average over realizations, we obtain the following expression for the directionality:

$$< |B(\theta)|^2 > \approx \sum_j |a_j|^2 \sum_n \sum_{n'} \phi_n(z_j)\phi_{n'}^*(z_j)e^{i(k_n - k_{n'})r_j}[b(k\sin\theta - k_{zn}) +$$

$$V_n\, e^{i\psi_n}b(k\sin\theta + k_{zn})][b(k\sin\theta - k_{zn'}) + V_{n'}e^{i\psi_n}b(k\sin\theta + k_{zn'})]^*(3)$$

where the star represents complex conjugate and $< ... >$ represents averaging over realizations. If the array has an ideal response: $b(\theta) = \delta(\theta)$, then

$$< |B(\theta_n)|^2 > = \sum_j |a_j|^2 \sum_n |\phi_n(z_j)|^2,$$

$$< |B(-\theta_n)|^2 > = \sum_j |a_j|^2 \sum_n |\phi_n(z_j)|^2|V_n|^2, \qquad (4)$$

and the energy reflection coefficient $|V_n|^2$ can be obtained by the ratio of the two beams. However, for a finite-length array such as the one used, we do not have a close form result

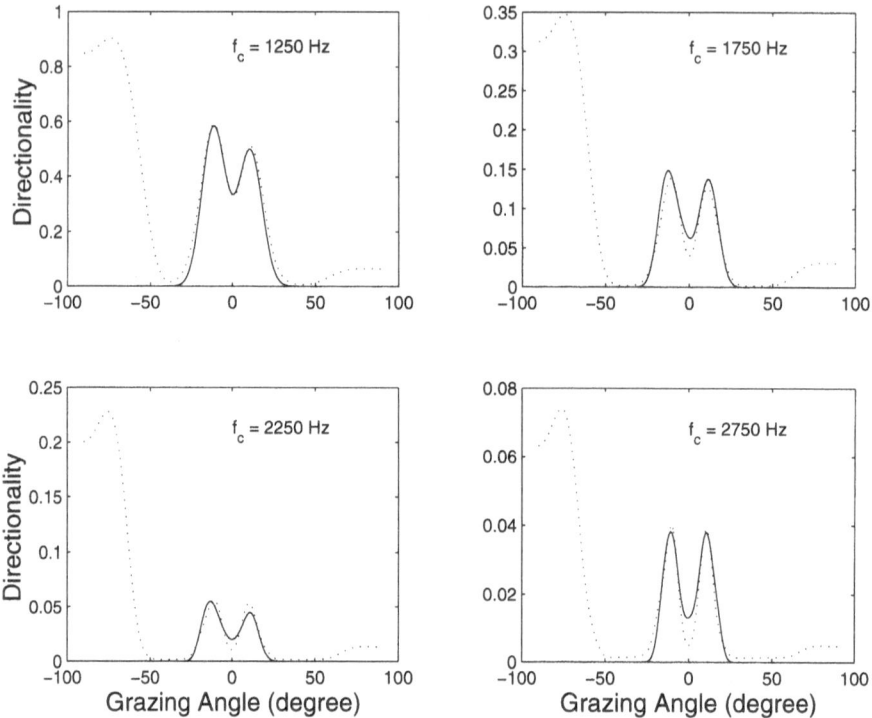

Figure 5. Model/data comparison of beam directionality. The dotted curves are data, solid curves
are model results.

showing that the ratio of the two terms in Eq. (4) is the energy reflection coefficient. Here
we used KrakenC [9], a normal mode code, to simulate the array response to the noise
field. For a given frequency, we assume that there are 200 independent point sources
randomly distributed near the sea surface. The ranges of the sources are 100 to 5000 m.
The depth of the sources is also random and ranges from 0.1 to 5.0 m. The contribution
of the 200 sources are summed on each element on the array and beamforming was
performed at all angles. This constitutes one realization. Repeating this for 100 time,
and an average is obtained. The same procedure was performed for many frequencies,
with 50 Hz increments. Further, the beam output from neighboring frequency bins (500
Hz total) was also averaged. By comparing the simulation results to those from data, and
changing bottom parameters in the normal mode code, we arrive at a set of "optimal"
results (eyeball fit). The results are given in Fig. 5. Note that the model results are
scaled to fit the experiment result. The set of parameters used in the Kraken code for
this optimal case are: sound speed is 1600 m/s, density is 1.78 g/cm^3, and the attenuation
coefficient is 0.11 dB/m kHz. The energy reflection coefficient based on these parameters
is given in Fig. 6, along with simulation results and results from data.

Cores were taken [10] in the experiment site. Analysis of the core show that the
surficial sediment has a sound speed of 1600±10 m/s from 11 out 14 cores taken. This is
consistent with our result from noise analysis. So far density and attenuation coefficient
analysis from cores are not available.

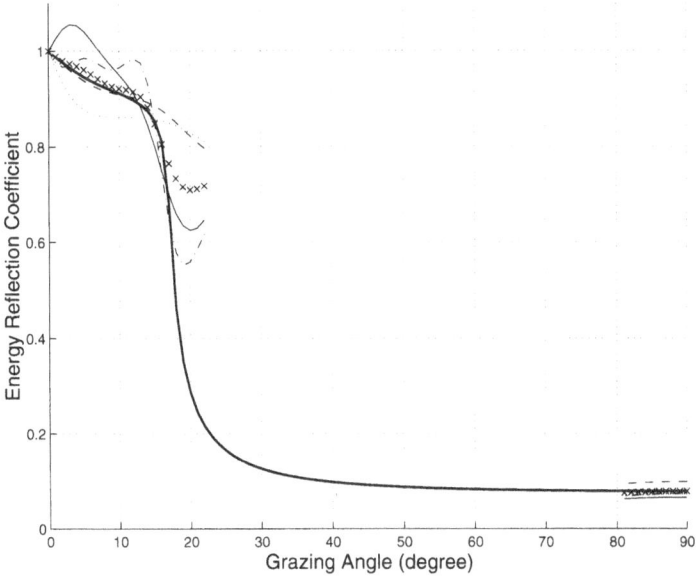

Figure 6. Energy reflection coefficient from data and model. The solid curve is the theoretical energy reflection coefficient. The circles are results from simulations averaged over realizations and all frequency bins. Others are the same as those in Fig. 3.

4   Discussion

Summarizing the results from data analysis and simulations, we found that by averaging over time segments (for data) and realizations (for simulations), we obtained approximate energy reflection coefficient with considerable fluctuations. By further averaging over a wide band of frequencies, the ratio of the down- and up-looking directionality converges to the true energy reflection coefficient.

What needs to be done to further validate this approach is to conduct a detailed statistical analysis through simulations. A set of criteria should be established to guide field data processing.

Since the vertical array response is sensitive only to local geo-acoustic conditions, this approach can potentially be used to conduct surveys in shallow water with changing sediment composition.

Another intriguing possibility is to use the data in mid-angles, where there is no real noise source, to estimate bi-static bottom and surface scattering coefficients.

Acknowledgements

This work was supported by the U.S. Office of Naval Research, Code 321OA.

References

1. Deane, G.B., Buckingham, M.J. and Tindle, C.T., Vertical coherence of ambient noise in shallow water overlying a fluid seabed, *J. Acoust. Soc. Am.* **102**, 3413–3424 (1997).

2. Buckingham, M.J. and Carbone, N.M., Source depth and the spatial coherence of ambient noise in the ocean, *J. Acoust. Soc. Am.* **102**, 2637–2644 (1997).

3. Harrison, C.H., Formulas for ambient noise level and coherence, *J. Acoust. Soc. Am.* **99**, 2055–2066 (1996).

4. Carbone, N.M., Deane, G.B. and Buckingham, M.J., Estimating the compressional and shear wave speeds of a shallow water seabed from the vertical coherence of ambient noise in the water column, *J. Acoust. Soc. Am.* **103**, 801–813 (1998).

5. Harrison, C.H., Brind, R. and Cowley, A., Computation of noise directionality, coherence and array response in range dependent media with CANARY, *J. Comp. Acoust.* **103**, 801–813 (1998).

6. Aredov, A.A., Okhimenko, N.N. and Ferduev, A.V., Anisotropy of the noise field in the ocean (experiment and calculations), *Sov. Phys. Acoust.* **9**, 327–345 (2001).

7. Aredov, A.A. and Ferduev, A.V., Relation of underwater noise level to wind speed and the dimensions of the volume of water determining the noise field, *Izvestiya, Atmospheric and Oceanic Physics* **15**(1), 58–62 (1979).

8. Aredov, A.A. and Ferduev, A.V., Angular and frequency dependencies of the bottom reflection coefficient from the anisotropic characteristics of a noise field, *Acoustical Physics* **40**(2), 176–180 (1994).

9. Porter, M.B. and Reiss, E.L., A numerical method for bottom interacting ocean acoustic normal modes, *J. Acoust. Soc. Am.* **77**, 1760–1767 (1985).

10. Miller, J.H., Data to be published by Miller of the University of Rhode Island, (Private communication).

EFFECT OF ENVIRONMENTAL VARIABILITY ON MODEL-BASED SIGNAL PROCESSING: REVIEW OF EXPERIMENTAL RESULTS IN THE MEDITERRANEAN

JEAN-PIERRE HERMAND

Dept of Optics and Acoustics, Université Libre de Bruxelles,
ave F.-D. Roosevelt 50, CP194/5, B-1050 Brussels, Belgium
E-mail: jhermand@ulb.ac.be

Over the last decade experiments were conducted in the Mediterranean for localising controlled sound sources and for deducing bottom geo-acoustic properties from the measurement of the acoustic-channel impulse response over a broad frequency range. Inversion of the large time-bandwidth-product signals transmitted to probe the medium was performed using a coherent receiver, the model-based matched filter. The reference channels incorporate Green's function models for partially known or hypothesised environmental conditions and source parameters (range, depth and Doppler). The search terminated when most of the time-spread energy on single or multiple elements of the receive array was recombined coherently into a single peak. Although performance limitations were imposed by environmental and modelling uncertainties the model-based processor was always seen as intrinsically robust to the acoustic-signal variability encountered in our experiments. Substantial processing gains were obtained in most situations depending on the time dispersion, spatial diversity, predictability and coherence of the specific acoustic channel. Most importantly, correct and stable inversion results were possible even from a sparse but representative set of hydrologic data.

1 Introduction

Ocean-duct and shallow-water transmission channels are characterised by time dispersion and frequency-selective fading of the acoustic signal. The performance of a conventional matched filter (MF) receiver can be drastically improved if the reference channel compensates for the amplitude and phase distortion occurred in the time dispersive medium [1]. Source-location and environmental parameters were determined by varying the reference channel of the model-based matched filter (MBMF) receiver [2, 3]. It was shown that the use of large time-bandwidth(TW)-product signals for probing the medium was important not only to improve resolution performance but also to reduce propagation fading.

This paper focuses on the effects of environmental variability and other modelling uncertainties upon MBMF inversion results and the role of the transmit frequency bandwidth and spatial receive aperture in limiting these effects.

In Sect. 2 a theory is proposed to explain the robustness of a multichannel MBMF receiver to environmental and geometric variability, observed in various experiments. Section 3 reviews source localisation results achieved with a towed source and horizontal receive array (HRA). Section 4 reviews geo-acoustic inversion results obtained with a fixed source and single and multiple elements of a vertical receive array (VRA). Section 5 concludes the paper.

155

N.G. Pace and F.B. Jensen (eds.), Impact of Littoral Environmental Variability on Acoustic Predictions and Sonar Performance, 155-162.

2   Uncertainty in model-based matched filter processing

Let us define frequency integrals of the Green's function $G(\omega, r)$ of a stationary medium,

$$\alpha(r) \triangleq \left| \frac{1}{\Delta\omega} \int_\Pi G(\omega, r) d\omega \right|^2 \quad \text{and} \quad \beta(r) \triangleq \frac{1}{\Delta\omega} \int_\Pi |G(\omega, r)|^2 d\omega \tag{1}$$

where r is the radius vector between the source and a receiver and Π is the frequency range of radiation. For a signal with centre frequency ω_0 and large time-bandwidth product $\Delta t \Delta\omega$ such that $\Delta t (\Delta\omega)^2 / \omega_0 \gg 2\pi$, the coefficients $\alpha(r)$ and $\beta(r)$ determine to the output peak signal-to-noise ratio (SNR) $\rho(r)$ of the MF and optimum receivers,

$$\rho_{mf}(r) = 2\mathcal{E}/N_0 \, \alpha(r) \quad < \quad \rho_{opt}(r) = 2\mathcal{E}/N_0 \, \beta(r) \tag{2}$$

where $\mathcal{E}$ is the source signal energy and N_0 is the white noise power spectral density at the receiver input. The upper bound on $\alpha(r)$ is equal to $\beta(r)$ and their ratio

$$\gamma(r) = \alpha(r)/\beta(r) \quad \leq \quad 1 \tag{3}$$

is a measure of the degree of *time dispersion* in the real propagation channel.

The output peak SNR of a MBMF processor which partially compensates for the dispersion distortion of the received signal is given by [1]

$$\rho_{mb}(r) = 2\mathcal{E}/N_0 \, \beta(r) \, \max_\xi |\mathcal{R}_{GG_M}(\xi, r)|^2 \quad < \quad \rho_{opt}(r) \tag{4}$$

where $\mathcal{R}_{GG_M}$ is a measure of the degree of *similarity* between the real sound field and the model field $G_M(\omega, r)$,

$$\mathcal{R}_{GG_M}(\xi, r) \triangleq \frac{\int_\Pi G(\omega, r) G_M^\dagger(\omega, r) \exp(j\omega\xi) d\omega}{\left[\int_\Pi |G(\omega, r)|^2 d\omega \int_\Pi |G_M(\omega, r)|^2 d\omega \right]^{1/2}} \quad < \quad 1 \tag{5}$$

and ξ is a phase variable which includes the mean travel time.

Since transmission over most ocean channels involves propagation through a random time-varying medium, the integrands in Eq. (1) must be averaged over an ensemble of realisations of the channel transfer function,

$$\overline{\alpha}(r) = \frac{1}{(\Delta\omega)^2} \iint_\Pi \text{Re}\left[\mathcal{R}_G(\omega_1, \omega_2; r) \right] d\omega_1 d\omega_2 \quad \text{and} \quad \overline{\beta}(r) = \frac{1}{\Delta\omega} \int_\Pi \mathcal{R}_G(\omega, r) d\omega \tag{6}$$

where $\mathcal{R}_G \triangleq \langle G(\omega_1, r) G^\dagger(\omega_2, r) \rangle$ is the complex-valued two-frequency correlation function [4] and $\langle \cdot \rangle$ denotes ensemble averaging. $\mathcal{R}_G$ is a measure of the correlation between the *fading* at different frequencies.

To quantify the overall effect of environmental and modelling uncertainties on MBMF performance, Eq. (5) is replaced by the combined correlation coefficient [5]

$$\mathcal{K}_{GG_M}(\Omega, R, \tau) \triangleq \frac{\int_Q \langle \int_\Pi GG_M^\dagger \exp(j\omega\xi) d\omega \rangle dQ}{\left[\int_Q \langle \int_\Pi |G|^2 d\omega \rangle dQ \int_Q \langle \int_\Pi |G_M|^2 d\omega \rangle dQ \right]^{1/2}} \quad \leq \quad 1 \tag{7}$$

where $G \triangleq G(\omega, r, t)$, $G_M \triangleq G_M(\omega + \Omega, r + R, t + \tau)$ and $dQ = drdt$. The function combines averaging of the channel transfer functions over a certain space-time domain Q and ensemble averaging. The intervals along the coordinate axes Ω_{det}, ρ_{det} and τ_{det} that characterise a high degree of *determinacy* (predictability) of a specific acoustic channel are determined from the condition that the above quantity decays to $1/2$.

An important property of Eq. (7) is that the intervals of deterministic behaviour can be greater than the corresponding correlation radii of the acoustic field itself [6, 7]. The frequency band Ω_{det} can be much larger than the coherence band Ω_{coh} of the field. Also the time τ_{det} can be much larger than the characteristic fluctuation time of the environmental parameters that affect the acoustic transmission. Consequently, the MBMF processing technique can be less sensitive to uncertainty in knowledge about the environmental conditions, and about the source and receiver configuration, than expected when the above intervals are considered separately. Therefore, as will be demonstrated by the experimental results in the next sections, compensation of distortion in the received signal can remain effective even in a randomly inhomogeneous dispersive channel for which $\Omega_{det} \supset \Pi \supset \Omega_{coh}$. Moreover, a MBMF receiver implemented as a bank of filters matched to a number of anticipated channels is expected to be robust, to some extent, to the uncertainty upon the environment. From Eqs. (3), (4) and (7), the MBMF processing gain relative to the MF equals the degree of time dispersion in the real channel $[1/\gamma(r)]$ weighted by the degree of determinacy between the real and model fields,

$$\Gamma(r) \triangleq \max_{\xi, Q} \mathcal{K}_{GG_M}(\Omega, R, \tau)/\gamma(r). \qquad (8)$$

3 Source localisation in ocean ducts

The WEST SARDINIA 89-90 sea trials took place under winter conditions at a 2.8-km deep water site west of Sardinia. Doppler-resolvent, large TW-product, linearly frequency modulated (LFM) signals were transmitted from a controlled source to a 64-element HRA, towed at constant depths. The signal bandwidth was varied over one decade 50–500 Hz around a centre frequency 590 Hz. The medium Green's function was modelled using a range-independent (RI), generalised ray theory model. The environmental input was a single sound speed profile (SSP) calculated from a measured near-surface temperature profile (XBT), a deeper depth archival temperature data and salinity winter mean data.

In the fixed range experiment the source and receiver were towed near the axis of a refractive duct [1]. Amplitude and phase scintillations of the transmitted signals caused a structural change in the duct multipath-arrival structure from one signal to the next with deep fading occurring occasionally for the smaller signal bandwidths. This was due to small variations in range and tow depths and to slow spatial and temporal fluctuations in duct sound speed. Figure 1(a) compares the MBMF gain [Γ, Eq. (8)] to the estimated energy spreading loss (ESL) [γ, Eq. (3)] for each of the received transmission. Most importantly, for every transmission, the MBMF consistently gathered the time-spread signal energy back into a single output peak. The gain, varying from 0.5 dB to 3 dB, was proportional to the time- and bandwidth-dependent ESL measured at the MF output. These calculations included only the time-dispersed, ducted R-path arrivals through eigenray filtering. Moreover, it was also possible to isolate in depth and range the source that emitted the acoustic signal and the depth of the receive array [Fig. 1(b)]. The overall

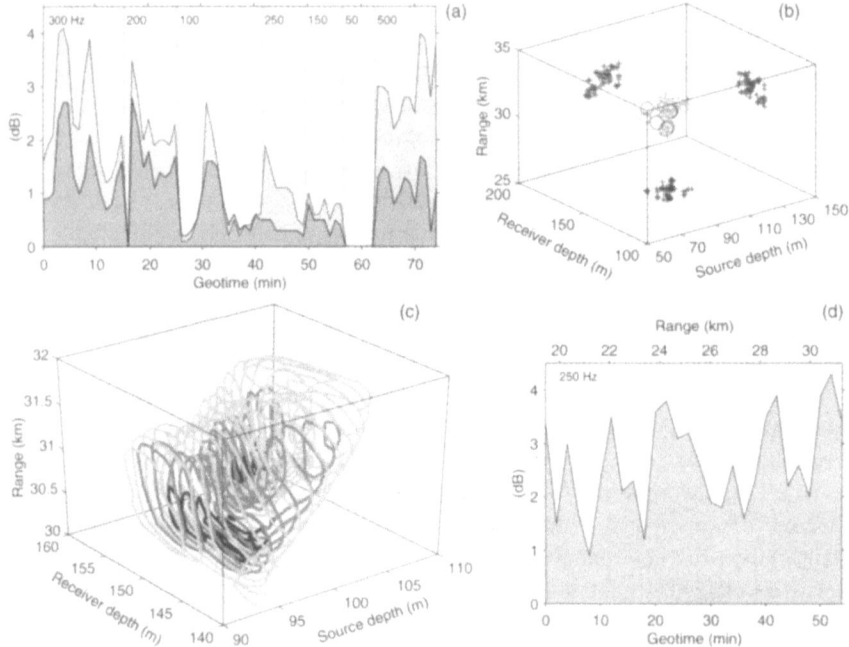

Figure 1. Effect of environmental variability upon MBMF source localisation. (a) Time history of MBMF processing gain relative to the MF (dark grey) and ESL (light grey) over 1.2-h period. Signal bandwidths and corresponding time intervals of transmission are indicated. January 24, 1989. (b) Scatter of the source-receiver depths and range estimates for all transmissions. The axes limits correspond to the search intervals. (c) Contour slices of MBMF gain. light grey: 1 dB; grey: 2 dB and black: 3 dB. (d) Same as (a) for opening range over 1-h period. February 6, 1990. West Sardinia.

effect of environmental, modelling and geometrical uncertainty is represented by the ambiguity volume in Fig. 1(c). The centroid was very close to the nominal range and depths of the source and receiver: 31.3 km, 100 m and 150 m, respectively.

In the opening range experiment the source and receiver were towed near the surface. The transmitted signals were strongly time-dispersed by propagation, through a near-surface duct and were Doppler-shifted (time-expanded). Here, surface interaction and near-surface variability played an important role. Most of the energy was conveyed by low-grazing angle paths with narrow dispersion in elevation angle. Hence there was no significant down Doppler shift and spread. Because of the Lloyd-mirror effect, the multipath interference structure was highly sensitive to small variation in relative depth of source and receiver. MBMF processing improved the peak SNR by 1–4 dB in spite of the moderate sea state (3–4) and the sensitivity to geometry [Fig. 1(d)]. The gain resulted from phased recombination of a continuum of R-path and RSR-path arrivals which were not resolved by the MF. Range and relative velocity were correctly determined by a multichannel MBMF [2]. Comparison with MF cross-ambiguity functions showed that large spurious peaks built up away from the true time delay and Doppler shift [8] were suppressed by the MBMF processor. These experimental results demonstrate the relevance of the combined correlation coefficient Eq. (7) and its properties discussed in Sect. 2.

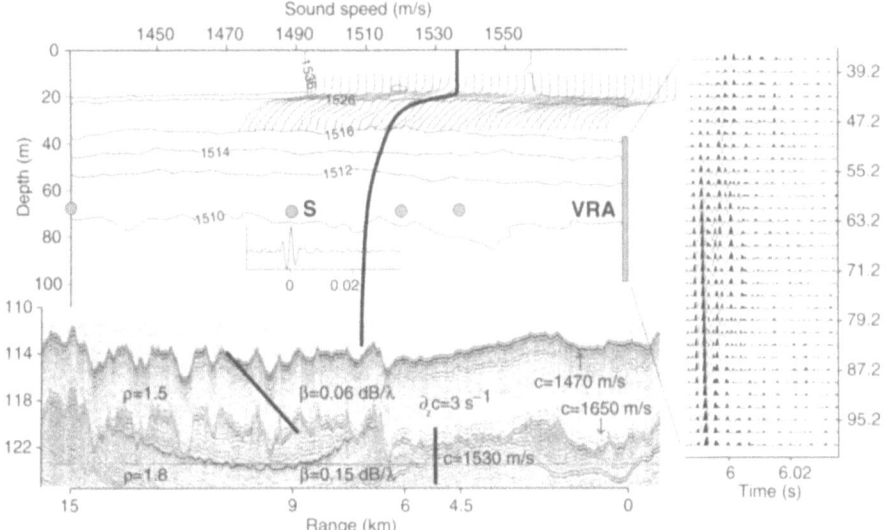

Figure 2. Geometry and RD environmental model of the YELLOW SHARK 94 experiment. *Top panel:* Hydrologic variability. Source (S,○) and VRA positions. *Bottom:* Geo-acoustic inversion results [3]. *Right:* Snapshot of channel impulse responses measured on the 32-element VRA for the source at 9-km range. *Middle:* Model-based time reversal mirror. Temporal focus. September 10-11, 1994. Northwest of Formiche di Grosseto islands, off the west coast of Italy.

4   Geo-acoustic inversion in shallow waters

The YELLOW SHARK 94 sea trials were carried out in a shallow water, muddy bottom site, off the west coast of Italy (Fig. 2). A multitone signal and a LFM signal with the same frequency band 200–800 Hz were transmitted from a mid-depth acoustic projector bottom-moored at four ranges to a 32-element VRA [9, 3].

Because of the perfectly static configuration of the acoustic moorings the observed acoustic variability was due entirely to the ocean volume: essentially, the time-varying thickness of the mixed layer and fluctuating sound-speed structure in the thermocline. All frequencies were affected with larger and shorter-term variations at the higher frequencies.

In the experimental setup where both source and VRA were positioned below a well-developed thermocline, the depth of the resulting propagation channel and associated sound-speed gradient were important. For each range, the ocean SSP input to the forward model was the ensemble average of all profiles measured along the transect during the acoustic transmissions. The time variation of the effective channel depth were accounted for by varying both the range-averaged mixed-layer and water depths. The top depth of the thermocline was controlled by translating the whole SSP in the vertical but maintaining the bottom-water sound speed constant.

4.1   Multitone Matched-Field Results

The effect of water-column variability upon acoustic parameter estimation and the importance of signal bandwidth were investigated by inverting separately each of 38 observations of the multitone pressure field of 12-s duration spaced by 2 min. For each of these,

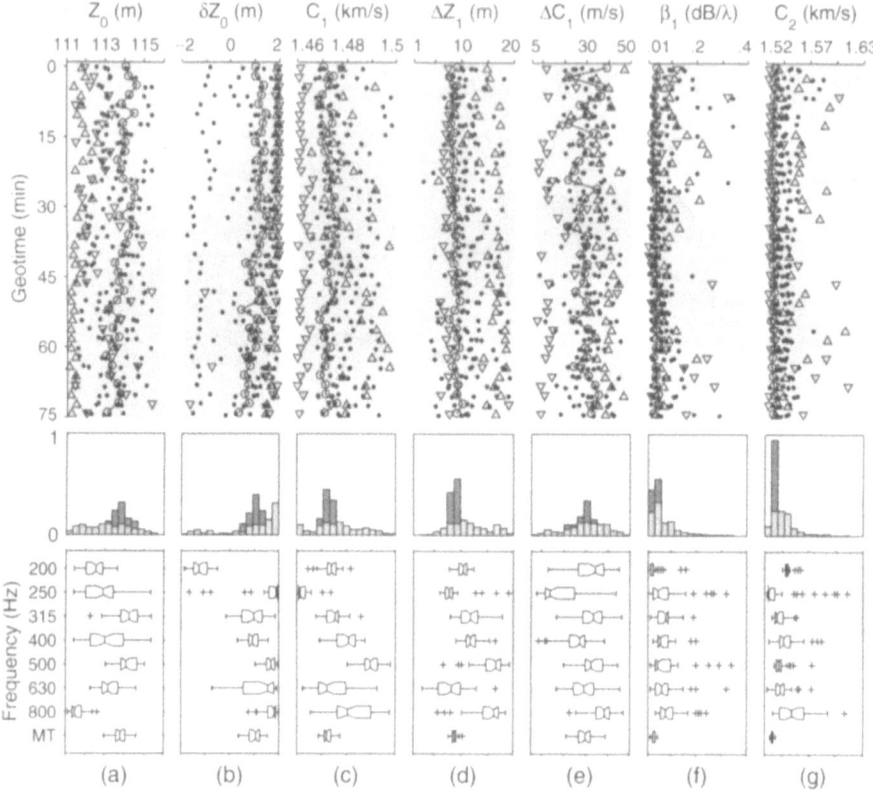

Figure 3. Effect of water-column variability upon matched-field geo-acoustic inversion. *Top panels:* Time history of (a) water depth, (b) thermocline depth, (c) sediment p-wave speed, (d) thickness, (e) speed gradient, (f) attenuation and (g) basement speed estimates. The curve and symbols are for inversions based on: 200–800-Hz seven tones (solid line), 250-Hz tone (∇), 800-Hz tone ($\triangle$) and other individual tones (•). The horizontal scales correspond to the search intervals. *Middle:* Normalised histograms of multitone (black) and concatenated single-tone (grey) inversion results. *Bottom:* Statistics for the single-tone and multitone (MT) inversions. vertical lines of the box: lower quartile, median, and upper quartile values. horizontal lines: extreme data extent within $1.5\times$ interquartile range. +: outliers.

genetic-algorithm optimisation was performed for each tone individually and for all tones jointly.

Figure 3 shows the time histories and statistics of two water-column and five bottom parameters for the 9-km range. The single-tone bottom estimates were widely scattered over the search intervals with all parameters being highly sensitive to water-column variability. The effect on each parameter was strongly frequency-dependent. In comparison, the broad-band, multitone (MT) processing provided remarkable stability and correct bottom characterisation. For all parameters, the bias and variance of the MT estimates were much smaller than the ones of the individual tones. Concatenating single-tone results did not improve the estimation. Related experimental results are discussed in [9–11].

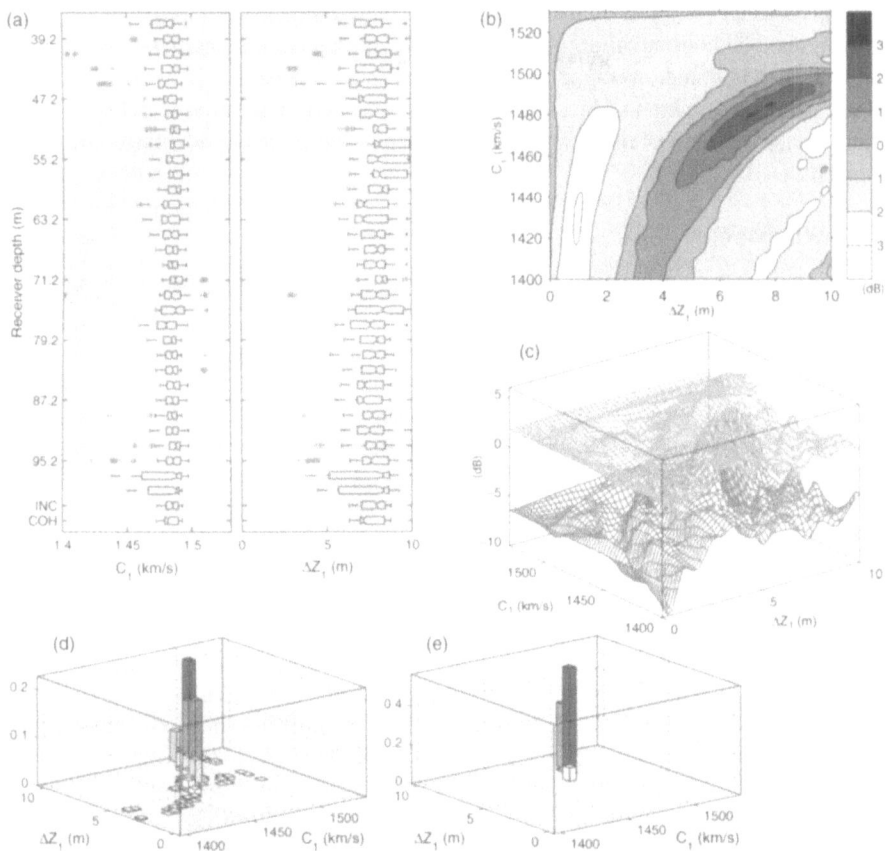

Figure 4. Effect of water-column and bottom variability upon **MBMF** geo-acoustic inversion. (a) Statistics of sediment-speed and layer-thickness results for each receiver depth, all ranges included (see Fig. 3 caption). **MBMF** speed-thickness ambiguity function for the 9-km range: (b) mean and (c) extreme values. Normalised histograms of (d) concatenated single-depth results and (e) space-coherent results, all ranges included.

4.2 Model-Based Matched Filter Results

The effects of both water-column and bottom variability upon **MBMF** processing were quantified for single and multiple array-signals. All **LFM** transmissions received at the 4 ranges and 32 depths were included in the analysis (5472 signals). A simplified **RI** model of the environment was used where only the two most range-dependent (**RD**) bottom parameters were varied: the depth-average sound speed and thickness of the upper clay layer. The other parameters were fixed to the baseline values obtained with a six-parameter model [3].

The temporal variability of bottom parameters inverted from single **VRA**-elements was remarkably small because of the frequency diversity useage as for the above MT matched-field results [Fig. 4(a,d)]. Greater variability was observed near the waveguide boundaries. Space incoherent (**INC**) and coherent (**COH**) processing of all **VRA** signals further reduced the effect of hydrologic variability by exploiting spatial diversity (a,e).

COH processing over a few elements, optimally distributed in depth, provided similar stability. Actual RD bottom variability is revealed by prominent histogram peaks (d,e) which correspond to validated, average values over the different ranges. The two-parameter ambiguity features (b) were stable in spite of the processing gain variations due to acoustic signal variability (c). Detailed analysis of the depth dependency and sparse array results are given in [3].

5 Conclusions

The experimental results reviewed in this paper show that model-based matched filter and matched-field processing over a broad signal-frequency range are effective in reducing the detrimental effect of environmental and modelling uncertainties. For the encountered conditions of oceanographic variability, the peak signal-to-noise ratio for the model-based processor was always larger than the conventional. Correct and stable results of source localisation and seabed geo-acoustic characterisation were possible even from a sparse set of hydrologic data.

Acknowledgements

The preparation of this article was supported by the Royal Netherlands Navy. The data were collected by the SACLANT Undersea Research Centre.

References

1. Hermand, J.-P. and Roderick, W.I., Acoustic model-based matched filter processing for fading time-dispersive ocean channels: Theory and experiment, *IEEE J. Oceanic Eng.* **18**, 447–465 (1993).
2. Hermand, J.-P., Model-based matched filter processing for delay-Doppler measurement in a multipath dispersive ocean channel. In *Proc. IEEE/OES Oceans'93*, 306–311 (1993).
3. Hermand, J.-P., Broad-band geoacoustic inversion in shallow water from waveguide impulse response measurements on a single hydrophone: Theory and experimental results, *IEEE J. Oceanic Eng.* **24**, 41–66 (1999).
4. Kennedy, R., *Fading Dispersive Channels* (Wiley, New York, 1969).
5. Kravtsov, Yu.A. and Petnikov, V.G., Partial determinacy of wavefields, *Soviet Physics Doklady* **30**, 1039–1040 (1986).
6. Bogusch, R.L. *et al.*, Frequency-selective scintillation effects and decision feedback equalization in high data-rate satellite links, *Proc. IEEE* **71**, 754–767 (1983).
7. Gindler, I.V. and Kravtsov, Yu.A., Degree of determinacy of a wave field and coherent processing of wideband signals, *Soviet Physics Acoustics* **34**, 142–144 (1988).
8. Hermand, J.-P. and Roderick, W.I., Delay-Doppler resolution performance of large time-bandwidth product linear FM signals in a multipath ocean environment, *J. Acoust. Soc. Am.* **84**, 1709–1727 (1988).
9. Hermand, J.-P. and Gerstoft, P., Inversion of broadband multitone acoustic from the Yellow Shark summer experiments, *IEEE J. Oceanic Eng.* **21**, 324–346 (1996).
10. Siderius, M. and Hermand, J.-P., Yellow Shark Spring 1995: Inversion results from sparse broadband acoustic measurements over a highly range dependent soft clay layer, *J. Acoust. Soc. Am.* **106**, 637–651 (1999).
11. Dosso, S.E. and Nielsen, P.L., Quantifying uncertainty in geoacoustic inversion. II. Application to broadband, shallow-water data, *J. Acoust. Soc. Am.* **111**, 754–767 (2002).

RAPID GEOACOUSTIC CHARACTERIZATION FOR LIMITING ENVIRONMENTAL UNCERTAINTY FOR SONAR SYSTEM PERFORMANCE PREDICTION

KEVIN D. HEANEY AND HENRY COX

ORINCON Defense, 4350 N. Fairfax Dr. Suite 400, Arlington VA 22203, USA
E-mail: kheaney@east.orincon.com

Ocean acoustic propagation and sonar performance in shallow water environments are dominated by interactions with the seafloor and is therefore sensitive to the geo-acoustic properties of the sediment. The goal of this research is improve sonar performance prediction by estimating the environment and determining the sensitivity to uncertainty. An approach is presented that links the observed acoustic signals of a sonar system to the environmental characterization and then, via simulation, to the environmental sensitivity. Relevant observables are extracted from data taken from a tactical sonar system (passive towed array, bi-static active, etc). These observables are taken from the striation patterns (time spread, slope) and the received level vs. range curves for surface ships of opportunity. The geo-acoustic parameters are estimated using the Rapid Geo-acoustic Characterization (RGC) algorithm that matches the observables to a parametric model of the sediment based upon Hamilton's equations. Once a baseline geo-acoustic model is determined, simulation studies are used to examine the sensitivity of the acoustic observables to variations in the environment. The resulting performance prediction curves can then computed with relevant confidence intervals. The approach reduces the mismatch by estimating the geo-acoustic environment and captures and communicates the uncertainty in performance prediction to the end user.

1 Introduction

The difficulty in predicting sonar performance is that acoustic propagation is sensitive to environmental variables and these variables generally are not well sampled in standard navy databases. To restore confidence in, and the utility of, sonar performance algorithms, a method for determining the environmental parameters *in situ* must be developed. Experience in the ocean acoustics research community has shown that accurate geo-acoustic inversions are possible in controlled experiments. The technical challenge addressed in this paper is to determine whether a surface ship of opportunity could be used as a source for covert environmental acoustic calibration. The robust Rapid Geo-acoustic Characterization (RGC) algorithm was developed and applied to a data set from a surface ship in the Gulf of Mexico. The algorithm successfully determines a geo-acoustic profile that permits the reproduction of the relevant acoustic propagation. We present here the signal processing, parameter extraction, and inversion algorithm used to extract a geo-acoustic profile that reproduces the key features of the

N.G. Pace and F.B. Jensen (eds.), Impact of Littoral Environmental Variability on Acoustic Predictions and Sonar Performance, 163-170.

acoustic propagation. *We stress that the RGC algorithm is seeking to rapidly match the acoustic propagation and is not primarily concerned with an accurate representation of the geo-acoustic profile.* Accurate geo-acoustic inversions can subsequently be performed using Full Search algorithms based upon simulated annealing [1].

Much work has been done in the area of environmental sensitivity and uncertainty. In general, acoustic modelers look at how the acoustic field varies with the changing of database values, and tactical navy personnel look at the variability in sonar performance as a function of position and time. We seek a systematic approach to the environmental uncertainty problem of sonar performance prediction that incorporates acoustic modeling and system performance on real data.

In Section 2 we outline the geo-acoustic inversion approach to define a baseline environment. In this section the data analysis procedure is presented for a bottom mounted horizontal line array as well as the parameter extraction and inversion technique for rapidly estimating a geo-acoustic profile. Section 3 concludes with how this information could be used in-situ to provide accurate representations of the effect of environmental uncertainty on the sonar performance.

2   Rapid geoacoustic characterization algorithm

2.1   Experimental Data

In 1999, the Applied Research Laboratory – University of Texas (ARL-UT) conducted an ocean acoustics experiment off the south coast of the Florida panhandle. A small portion of this experiment was run with the intention of performing geo-acoustic inversions from a surface ship of opportunity. This dataset provides an excellent opportunity for the development and demonstration of inversion techniques based upon surface ships of opportunity. A 534-m 52-element array was deployed on the bottom.

The data analysis procedure is as follows:

- Beamform to improve SNR and estimate number of interferers.
- Compute track-beam spectra for various sub-apertures of the array.
- Use the ship GPS and convert spectra to range/frequency data for geo-acoustic inversion.

After beamforming with an array length (17 phones) that maximizes SNR while limiting signal rejection due to small beamwidths, we fuse the acoustic (scissorgram) and the position (GPS) data into a single data file that will permit geo-acoustic characterization. The range-frequency data set is shown in Fig. 1. The striations resulting in coherent multipaths propagation are clearly visible. These patterns are ubiquitous to surface ship passes in shallow water environments.

Range/Frequency Tracked Spectrogram

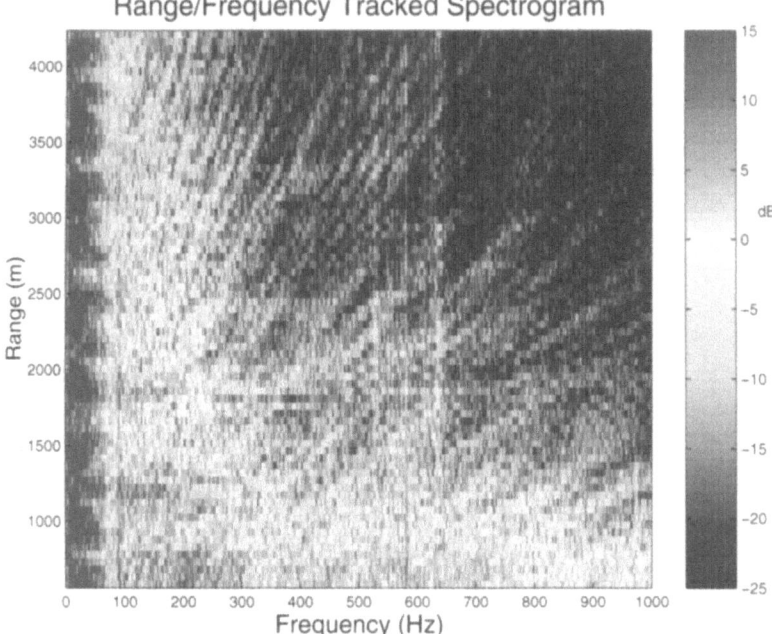

Figure 1. Range/frequency curve from beamformed acoustic data and position data.

2.2   Robust Observables for Inversion

Three observables are chosen to perform the environmental characterization.   These features are chosen because they are robust, easily measured and sensitive to the bottom parameters.   The *time-spread (Δt)* between the fastest and slowest propagating rays is the reciprocal of the spacing in frequency of the striations (Δf).   It is related to the critical angle of the bottom and depends on the sound speed in the sediment.   The second observable can be either the striation spacing in range or the slope of the striation patterns.   The slope of the striations is related to the waveguide-invariant β. It is a measure of how important refraction in the bottom is.   The final observable, α, is the slope of the TL vs. range curve (after correcting for cylindrical spreading). α is a measure of the overall attenuation.   For very hard bottoms, α is nearly zero.

2.2.1   Time Spread/Critical Angle

The striations visible in Fig. 1 are the result of constructive and destructive interference of the acoustic multipaths.   Looking at the interference pattern in the frequency domain leads to determination of the time spread of an acoustic pulse and therefore the spread in group velocities of the propagating energy.   At a range of 3 km and a frequency of 300 Hz, the frequency spacing is computed. A frequency spacing of 31 Hz is associated with a time spread in the acoustics of 32 m/s.   This is quite a small time spread and can easily be associated with a soft-bottom.

2.2.2   Striation Slope (β)

The slope of the striations is easily obtained from the display in Fig. 2. The slope of the striations is an indication of the angle spread between the lowest and highest angle acoustic energy and is therefore another robust feature that depends upon the geo-acoustic parameters as well as the water column sound speed. Along a ridge of constant intensity, the relative phases or phase differences between components are nearly constant. Consider two interfering components with travel times T_1 and T_2 and phases ϕ_1 and ϕ_2, respectively. Let ϕ_{12} be the phase difference, $\phi_2 - \phi_1$. Then

$$\phi_{12} = 2\pi f (T_2 - T_1) + \varepsilon_2 - \varepsilon_1 \tag{1}$$

where ε_1 and ε_2 are the phase changes associated with boundary interactions and are constant (unless on a caustic)

$$d\phi_{12} = 2\pi f \left[\frac{1}{v_2} - \frac{1}{v_1} \right] dr + 2\pi (T_2 - T_1) df \tag{2}$$

Applying the stationary phase condition $d\phi_{12} = 0$ yields the following relationship that is satisfied along a ridge on constant intensity

$$\frac{df}{f} = - \frac{\left[\frac{1}{v_2} - \frac{1}{v_1} \right]}{T_2 - T_1} dr = - \frac{\left[\frac{1}{v_2} - \frac{1}{v_1} \right]}{\left[\frac{T_2 - T_1}{r} \right]} \frac{dr}{r} \tag{3}$$

or

$$\frac{df}{f} = \beta_{12} \frac{dr}{r} \tag{4}$$

where β_{12} is the non-dimensional parameter:

$$\beta_{12} = - \frac{\left[\frac{1}{v_2} - \frac{1}{v_1} \right]}{\frac{T_2 - T_1}{r}} \tag{5}$$

Equation (5) is very general. No assumptions about range independence of the environment, or details of the sound speed profile, have been made. β_{12} depends on both local properties: the difference in phase slowness of the two components at the field point (r, z) and the travel time difference that involves the accumulated effects of propagation from the source to the field point. The quantity $\overline{\mu}_i = r/T_i$ is the average horizontal propagation speed or average group velocity of the ith component. Thus, Eq. (5) can be written as

$$\beta_{12} = -\frac{\left[\dfrac{1}{v_2} - \dfrac{1}{v_1}\right]}{\left[\dfrac{1}{\overline{u}_2} - \dfrac{1}{\overline{u}_1}\right]} \qquad (6)$$

A number of important simplifications occur. In range-independent environments, the group velocities can replace the range-averaged group velocities. In general β_{ij} depends on which components are interacting. There are two situations in which β_{ij} is nearly invariant with regard to the components that are involved. These are the iso-speed sound profile that is often a useful approximation in shallow water and steep angles, and the duct with a linear sound speed profile.

2.2.3 Transmission Loss Attenuation (α)

We are interested in matching the general characteristics of propagation and are not concerned (at this point) about matching the locations of the peaks and valleys. With received level measurements made over various ranges, the band-averaged transmission loss is computed. This reduces the high spatial frequency variability. Using a bandwidth of 60 Hz, the received levels for 200 and 500 Hz are shown in Fig. 3. It should be noted at this point that the hydrophones have not been calibrated. The more rapid attenuation with range of the higher frequency is consistent with a soft bottom.

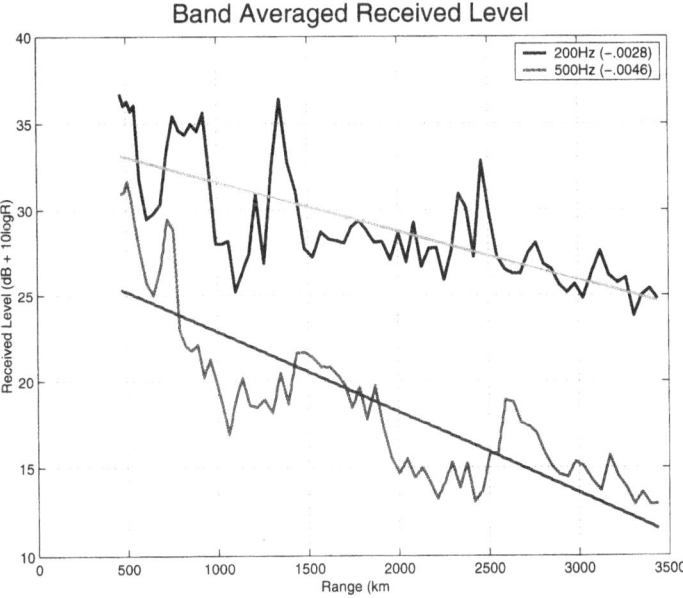

Figure 2. Band-averaged received level (RL) with cylindrical spreading taken out. Received levels matched with linear fits with *alpha* coefficients (200 Hz = -2.8 dB/km, 500Hz = -4.6 dB/km).

2.3   Forward Computation for RGC

To perform the Rapid Geo-acoustic Characterization, a simple geo-acoustic model that can represent a large variety of sediments and is consistent with our knowledge of geo-acoustics is needed. To this end, a simple two-layer sediment model is used. The sediment layer is considered to be of thickness H, overlaying a basement (or acoustic half-space). The sediment layer is inferred as having a unimodal particle size distribution with mean grain diameter ϕ, with associated geo-acoustic parameters ascribed by regression formulas presented in Hamilton and Bachman [2,3].

- Density:$\rho = (22.85 - \phi)/10.275$
- Velocity ratio:$(Cw/Cp) = 1.180 - 0.034 \phi + 0.0013 \phi^2$
- Sediment Phase Speed:$Cp(z)$:
- Sand:$\phi < 3.25$: $Cp(z) = Cp(0) * (20z)^{0.015}$
- Silt:$\phi > 5.75$: $Cp(z) = Cp(0) + 0.712 z$
- Mixtures:$3.25 < \phi < 5.75$: use silt factor: $\alpha = (\phi - 3.25)/(5.75 - 3.25)$
- $Cp(z) = \alpha\, Cp(silt) + (1 - \alpha)\, Cp(sand)$

where z is the depth of the sediment in meters, measured from the water sediment interface. The attenuation was chosen as a linear fit from 1.0 dB/Wavelength for sand to .05 for fine silt. It has a linear dependence on frequency, which may be suspect. The parameters for the basement are $Cp = 2000$ m/s, $\rho = 2.0$, $\alpha = 0.1$. For sediment thickness greater than 5 m, there is little dependence of the acoustics on these parameters. It must be stated here that we are not after the exact parameters of the bottom.

   For a given environmental estimate, a CW normal mode solution is generated and the striation spacing (slope and time-spread) are computed as well as the incoherent Transmission Loss (equivalent to the band-averaged). We now have a way of rapidly mapping parameters of a simple model to the observables. An exhaustive search of the predicted observables is performed for the two parameters that define the geo-acoustic model (ϕ, H). This is done for each range and frequency. We can add to the list of observables that we search over; however, this leads to the problem of finding a best fit in a large multidimensional search space.

2.4   Inversion Results

To determine the goodness of fit of a particular sediment estimate, we take the frequency mean of the square of the difference between the predicted and data observations. Each observable is normalized by it's mean and the and then the global cost function is determined by summing across the normalized observables. This yields a single cost function value for each sediment estimate. The results for each observable and for the final cost function are shown in Fig. 3. The total cost function is plotted in the lower right, revealing a global minimum at $H = 14$ m and $Cp = 1535$ m/s ($\phi = 2.6$ corresponding to a thin soft sand sediment).

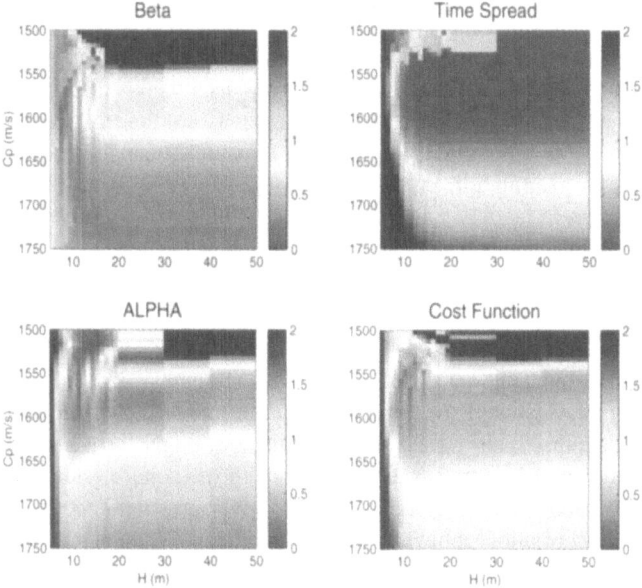

Figure 3. Cost functions for three observables and sum.

Once the RGC geo-acoustic model is chosen, a synthetic broadband TL curve is computed and used to generate a striation pattern that subsequently is compared to the data. Before comparing with the data however, the received level (RL) must be converted to TL, requiring an estimate of both the source level (SL) and the source spectrum. From historical work on ambient surface noise, a simple $(1/f^2)$ fall-off for the spectrum of the surface ship is used. The results are shown in Fig. 4.

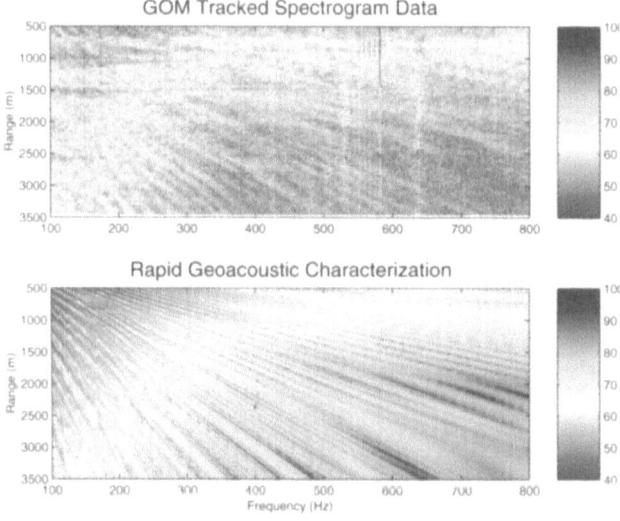

Figure 4. GOM beamformed data and RGC solution TL striations.

This result shows that the gross features of the data (slope, frequency dependence, spacing, TL) have been well matched. This is particularly encouraging because only data at 1 range (3 km) and 4 frequencies (200–500 Hz) were used and there is good agreement at higher frequencies and other ranges. There are places where the simulated field has much higher spatial frequencies than the data. This may be due to poor range resolution in the data (owing to the time it takes to do an FFT) or to a mismatch in the environment. This is conceded as a limitation of the RGC solution, but it is not a requirement for accurate use of TDAs in tactical sonar situations. A full global optimization will lead to higher precision results.

3   System approach

We have shown that by using data from a passing surface ship on a horizontal line array (similar to many navy systems) we can generate an estimate of the acoustic propagation for a particular region. This data has been taken through the sensor and therefore contains much information about what is important to the sonar system. To understand the system performance variability as a function of environmental uncertainty we take this estimate (and the information about sensitivity) as a starting point. Acoustic modeling with various perturbations to the environment can now be done to examine sensitivity and determine the robustness of the acoustic performance prediction. Communicating this uncertainty to the end user is a primary goal of this research.

References

1. Collins, M.D. and Kuperman, W.A., Simulated annealing applied to the geo-acoustic inversion problem, *J. Acoust. Soc. Am.* **98** (2002).
2. Hamilton, E.L., Geoacoustic modeling of the seafloor, *J. Acoust. Soc. Am.* **68** (1980).
3. Bachman, R.T., Parameterization of geoacoustic properties, *J. Acoust. Soc. Am.* **85** (1989).

ENVIRONMENTAL UNCERTAINTY IN ACOUSTIC INVERSION

STAN E. DOSSO AND MICHAEL J. WILMUT

School of Earth and Ocean Sciences, University of Victoria, Victoria B.C. Canada
E-mail: sdosso@uvic.ca

Acoustic processing for source localization is often limited by uncertainties in the physical properties of the ocean environment, such as the seabed geoacoustic parameters and water-column sound-speed structure. Uncertainties in environmental parameters can be the result of measurement uncertainties and of spatial and temporal variability. Quantifying environmental uncertainties and examining how they are transfered into uncertainties for source localization represents an important and challenging problem. In this paper, Bayesian inference theory is applied to estimate the uncertainties of geoacoustic inversion in the form of marginal probability distributions for the seabed parameters. In addition, a Bayesian form of focalization is developed which incorporates uncertain environmental parameters into the source localization, and quantifies the resulting localization uncertainty in terms of the 2-D marginal probability distribution for source range and depth (i.e., a probability ambiguity surface). Localization uncertainties are examined as a function of the uncertainties in geoacoustic parameters and ocean sound-speed profile.

1 Introduction

Acoustic fields propagating in a shallow-water ocean environment are affected by the location of the acoustic source and by physical properties of the environment (water column and seabed). Environmental properties are often not well known, due to a lack of data, measurement uncertainties, and spatial and temporal variability. Considerable effort has been applied in recent years to invert ocean acoustic fields for source location and for environmental properties, particularly geoacoustic parameters. Although these two inverse problems are often treated independently, in practical applications uncertainty always exists in environmental parameters and degrades the ability to perform localization. The problems are further linked as geoacoustic inversion is often applied to provide improved environmental information for subsequent source localizations. The goals of this paper are to quantify the uncertainties of geoacoustic inversion and source localization, and to examine localization uncertainties as a function of the uncertainties in geoacoustic parameters and ocean sound-speed profile (SSP). As acoustic inverse problems are strongly nonlinear, the analysis is carried out using Bayesian inference theory.

In Bayesian inversion [1], the solution is characterized by its posterior probability density (PPD). The PPD combines prior information for the unknown model parameters with the information from an observed data set expressed in terms of a likelihood function. To interpret the multi-dimensional PPD, the state of information about model parameters is typically quantified in terms of moments of the PPD, such as marginal probability distributions. Estimating these moments involves computing multi-dimensional integrals

171

N.G. Pace and F.B. Jensen (eds.), Impact of Littoral Environmental Variability on Acoustic Predictions and Sonar Performance, 171-178.

of the PPD, which is usually carried out using sampling methods. Obtaining efficient, unbiased sampling and verifying that the integral estimates have converged are important issues in Bayesian inversion. For instance, Monte Carlo integration is based on sampling at random from a uniform distribution over the parameter space. However, if the integrand is concentrated in localized regions of the space, many of these models will not contribute significantly to the integral. The method of importance sampling draws samples from regions that contribute most to the integral, providing more efficient sampling. In particular, if the sample of models can be drawn from the PPD itself, the integral evaluation is straightforward and efficient. This can be accomplished using Gibbs sampling (GS) methods, such as Metropolis and heat-bath sampling, which also form the basis for simulated annealing inversion.

A Bayesian formalism was applied to geoacoustic inversion by Gerstoft and Mecklenbräuker [2], who used a genetic-algorithms based approach to sample the PPD. Dosso [3] developed a fast Gibbs sampler (FGS) approach to Bayesian inversion, which provides an efficient, unbiased sampling of the PPD with a rigorous convergence criterion. The FGS algorithm was compared to standard GS and Monte Carlo integration for benchmark test cases and found to produce identical results in orders of magnitude less computation time. FGS analysis has been applied to a broadband, shallow-water geoacoustic survey [4]. The marginal distributions obtained for measured data agreed well with those of synthetic test cases, illustrating that simulations can provide a meaningful evaluation of practical cases. The marginals were found to have simple, smooth forms that facilitate straightforward comparisons of the information content for different cases.

To address environmental uncertainty in source localization, Collins and Kuperman [5] included environmental parameters as additional unknowns in a simulated-annealing inversion for the optimal source location, an approach known as focalization. Shorey et al. [6] applied a Bayesian formulation to this problem, referred to as the optimal uncertain field processor (OUFP), which estimates the source location by integrating over uncertain environmental parameters. As their goal was to estimate the maximum-probability source location rather than characterize the localization uncertainty, intensive sampling was not required, and Monte Carlo integration for a fixed number of samples was applied.

2 Theory

This section briefly summarizes the GS approach to Bayesian acoustic inversion; a more complete treatment of Bayesian theory is given in [1]. Let $\mathbf{m}$ and $\mathbf{d}$ represent model and data vectors, respectively, with elements m_i and d_i considered to be random variables. Bayes' rule for conditional probabilities leads to

$$P(\mathbf{m}|\mathbf{d}) \propto L(\mathbf{d}|\mathbf{m})\, P(\mathbf{m}), \tag{1}$$

where $P(\mathbf{m}|\mathbf{d})$ represents the PPD, $L(\mathbf{d}|\mathbf{m})$ is the likelihood function, and $P(\mathbf{m})$ is the prior distribution. In this paper, the prior is taken to be a uniform distribution over pre-defined lower and upper bounds for each parameter. The likelihood function is determined by the form of the data and errors: for Gaussian errors $L(\mathbf{d}|\mathbf{m}) \propto \exp\left[-E(\mathbf{m})\right]$, representing a Gibbs distribution, where $E(\mathbf{m})$ is the appropriate error function. For multi-frequency acoustic data $\mathbf{d} = \{\mathbf{d}_f, f = 1, F\}$ due to a source with unknown spectrum and random errors assumed to be uncorrelated over frequency and space with standard deviation σ_f at the fth frequency, the error function can be written [3]

$$E(\mathbf{m}) = \sum_{f=1}^{F} (1 - B_f(\mathbf{m})) \, |\mathbf{d}_f|^2/\sigma_f^2. \tag{2}$$

In (2), $B_f(\mathbf{m})$ represents the (normalized) Bartlett processor defined

$$B_f(\mathbf{m}) = \frac{|\mathbf{d}_f^\dagger \, \mathbf{d}_f(\mathbf{m})|^2}{|\mathbf{d}_f|^2 \, |\mathbf{d}_f(\mathbf{m})|^2}, \tag{3}$$

where $\mathbf{d}_f(\mathbf{m})$ is the replica acoustic field computed for model $\mathbf{m}$. The normalized PPD can thus be written

$$P(\mathbf{m}|\mathbf{d}) = \frac{\exp\left[-E(\mathbf{m})\right] P(\mathbf{m})}{\int_{\mathbf{m}'} \exp\left[-E(\mathbf{m}')\right] P(\mathbf{m}') \, d\mathbf{m}'}, \tag{4}$$

where the integration spans the model space.

To interpret the multi-dimensional PPD requires computation of its integral properties. Here, parameter uncertainties are considered in terms of marginal probability distributions. The 1-D marginal distribution for parameter m_i is defined

$$P(m_i|\mathbf{d}) = \int_{\mathbf{m}'} \delta(m_i' - m_i) \, P(\mathbf{m}'|\mathbf{d}) \, d\mathbf{m}', \tag{5}$$

where δ is the Dirac delta function; higher-dimensional marginal distributions are defined in a similar manner. Multi-dimensional PPD integrals such as (5) are typically evaluated using sampling methods. This paper applies the GS methods of Metropolis and heat-bath sampling which asymptotically sample from the PPD itself, providing a straightforward and efficient evaluation of integrals such as (5).

In Metropolis sampling, each parameter of the model is perturbed in turn, with perturbations accepted if a uniform random number ξ drawn from the interval $[0, 1]$ satisfies

$$\xi \leq \exp\left[-\Delta E/T\right], \tag{6}$$

where T represents a control parameter referred to as temperature. In heat-bath sampling, a discretized 1-D distribution is formed for each parameter m_i in turn by sampling $\exp\left[-E(\mathbf{m})/T\right]$ across the parameter bounds while keeping all other parameters fixed. A new value for m_i is then drawn at random from this distribution. Markov-chain analysis [1] indicates that, in the limit of a large number of perturbations, both the Metropolis and heat-bath methods provide an unbiased sampling of the Gibbs distribution

$$P_G(\mathbf{m}) = \frac{\exp\left[-E(\mathbf{m})/T\right]}{\sum_{\mathbf{m}'} \exp\left[-E(\mathbf{m}')/T\right]}, \tag{7}$$

which is proportional to the PPD (4) for uniform priors. In simulated annealing inversion, T is reduced during sampling to ultimately obtain the model that minimizes E and hence maximizes the PPD. In Bayesian analysis, intensive sampling is carried out at the fixed temperature $T = 1$ to sample from the PPD (4) and evaluate integrals such as (5). However, cooling from a high temperature to $T = 1$ prior to accumulating the sample is recommended to initiate the sampling in a high-probability region of the space.

A fast Gibbs sampling (FGS) algorithm for geoacoustic inversion based on Metropolis sampling has recently been developed [3]. In FGS, the efficiency of standard Metropolis

GS is improved by sampling in a transformed parameter space rotated to minimize inter-parameter correlations, and by adaptively determining appropriate perturbation sizes for sampling individual parameters. This approach is applied here for geoacoustic inversion.

While Metropolis sampling has proved effective for geoacoustic inversion, for acoustic localization with uncertain environmental parameters (focalization) the heat bath method can provide important advantages. The major challenge in localization involves the potentially large number of isolated, locally-optimal solutions dispersed over source range and depth (r and z), referred to as side-lobes on an ambiguity surface. As a result, independent 1-D sampling in r and in z can be very inefficient since the probability of jumping between locally-optimal solutions not aligned with the parameter axes is small. However, 2-D sampling in the r-z plane precludes this difficulty. To accomplish this within the heat bath method, a range-depth probability surface is generated and taken to be equivalent to the results of intensive 2-D sampling of these parameters. Hence, 2-D marginal probability distributions can be formed by summing the r-z probability distributions for a large number of realizations of the environmental parameters, generated using GS. Note that this is equivalent to averaging r-z probability surfaces over environmental parameters sampled from the PPD, while the OUFP method [6] averages over parameters sampled from the prior distribution. For range-independent or adiabatically range-dependent problems, r-z probability surfaces can be computed efficiently since the modal properties need be evaluated only once.

The final component required in GS is a rigorous convergence criterion. Convergence is judged here in terms of the difference between two two independent samples of models collected in parallel with the GS algorithm. For geoacoustic inversion, the criterion adopted requires the maximum difference between the 1-D cumulative marginal distributions for all parameters be suitably small (less than 0.1). For source localization in an uncertain environment, an additional requirement is included that the difference between the two 2-D cumulative marginal distribution in r and z be less than 0.02. In each case, the final sample for integration is taken to be the union of the two independent samples.

3   Inversion examples

This section illustrates the Bayesian analysis of geoacoustic inversion and source localization in an uncertain environment with realistic synthetic examples. The range-independent environment considered in both cases is given in Fig. 1, which shows the environmental parameters consisting of the ocean SSP c_1 (at surface), c_2 (at 10-m depth) and c_3 (at seabed, 100-m depth), sediment and basement sound speeds, c_s and c_b, sediment thickness, h, and seabed density and attenuation, ρ and α (constant over sediment and basement). Acoustic fields due to a source at range r and depth z are measured at a 19-sensor vertical line array (VLA) which extends from 0–90 m depth.

The first example consists of inversion for the seabed geoacoustic parameters with the SSP known exactly and source range and depth known to within small corrections (typical assumptions for geoacoustic inversion). The data consist of acoustic-field measurements at 50, 100, and 200 Hz, with complex Gaussian errors added to the data to achieve a signal-to-noise ratio of SNR = 10 dB at each frequency. The marginal probability distributions for all parameters were computed using the FGS algorithm and are shown in Fig. 2(a) (top set of distributions). This figure also indicates the true values and search

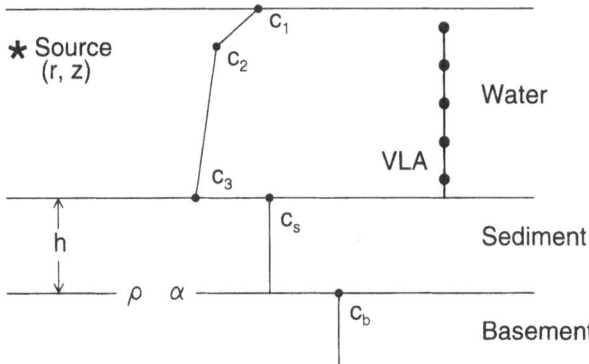

Figure 1. Schematic diagram of ocean environment for geoacoustic inversion and source localization examples. Water depth is 100 m. Parameters are described in text.

bounds for each parameter. Wide bounds are applied for the geoacoustic parameters so that the data (not the prior information) constrain the solution. The marginal probability distributions are unimodal and approximately symmetric, with smooth, simple forms. Figure 2(a) shows that the sediment properties, h and c_s, are well determined, while c_b, ρ, and α are less well determined but still constrained by the data. The source range r is poorly determined within the small correction bounds, while the source depth z is reasonably well determined. Figure 2(b) shows the marginal distributions determined when the SSP parameters are considered unknown within a 10 m/s interval. The SSP parameters themselves are almost completely undetermined via inversion; however, the effect on the other parameters is minimal (distributions for c_s and z are slightly wider). This indicates that uncertainty in the SSP, possibly due to spatial or temporal variability, does not preclude good results for geoacoustic inversion, provided this uncertainty is

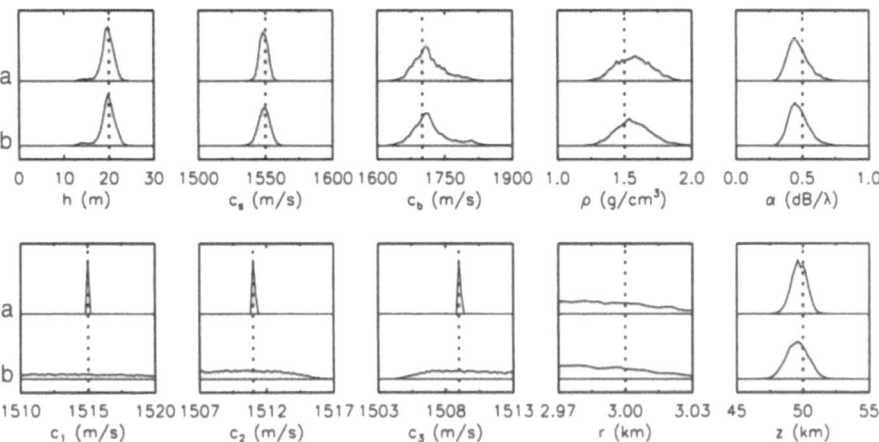

Figure 2. Marginal probability distributions from geoacoustic inversion. Upper set of distributions (a) shows results with known SSP; lower set (b) shows results with unknown SSP. Dotted lines indicate true parameter values, and the range of abscissa values indicates the parameter search bounds.

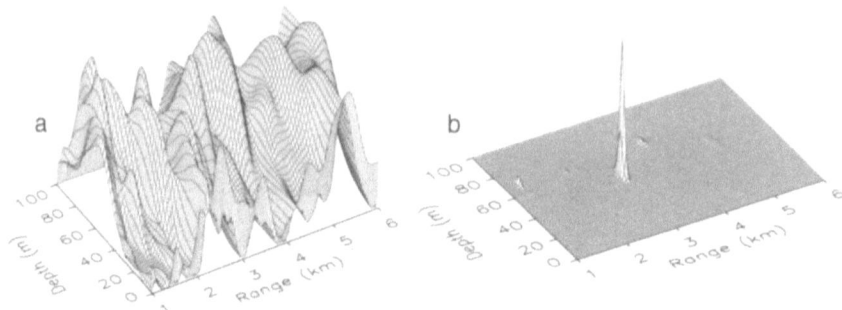

Figure 3. (a) Ambiguity surface for Bartlett match. (b) Probability ambiguity surface. Environmental parameters are set to their true values and SNR = 5 dB. Distributions are normalized to a common maximum value.

accounted for in the inversion procedure. It is important to note that this result does not imply that a good knowledge of the SSP is irrelevant in geoacoustic inversion. Holding the water-column parameters fixed at incorrect values within the prior bounds can lead to poor inversion results. This point is examined later in the paper.

The next example involves source localization in an uncertain environment. The environmental parameters are as given in Fig. 1, and the data consist of acoustic fields due to a source at $(r, z) = (3 \text{ km}, 50 \text{ m})$ at a frequency of 100 Hz with SNR = 5 dB. The inversion results are considered in terms of the 2-D marginal probability distribution over r and z, referred to as a probability ambiguity surface (PAS). The PAS is compared to a standard ambiguity surface (AS) for the case of exact environmental parameters (i.e., no integration required) in Fig. 3. The AS consists of a plot of the Bartlett processor $B(\mathbf{m})$, defined by (3), at all points of a grid from 1–6 km range and 0–100 m depth. The PAS in this case consists of a plot of $P(\mathbf{m}|\mathbf{d}) \propto \exp[B(\mathbf{m}) |\mathbf{d}|^2/\sigma^2]$ over the same grid. The AS and PAS have maxima at the same point; however, the exponentiation of the PAS stretches the surface by an amount determined by the ratio of the squared data magnitude to its variance. The AS indicates the fit to the data vs. r and z; the PAS indicates the probability that particular (r, z) points correspond to the source position. The probability that the source is within a particular r-z region can be determined by integrating (summing) P over the region.

Figure 4 shows PAS, computed using heat-bath sampling, for a variety of states of enviromental information including wide and narrow geoacoustic bounds and known and unknown SSP (10 m/s uncertainty), as described in the figure caption. The wide geoacoustic bounds correspond to the parameter search bounds indicated in Fig. 2, i.e., the available information prior to geoacoustic inversion. The narrow geoacoustic bounds correspond to 95% maximum-probability intervals for the parameters determined from the geoacoustic inversion results (i.e., the marginal distributions in Fig. 2). For comparison, Fig. 4(a) shows the PAS for the exact environment. The integrated probability for an acceptable region defined to be within 250-m range and 5-m depth of the true source location is $\bar{P} = 0.88$. Figure 4(b) shows that poor geoacoustic/SSP information significantly degrades the PAS, with a wider main peak and substantial sidelobes. The highest peak is at the correct source location, but the integrated probability for the acceptable region is just $\bar{P} = 0.23$. Applying narrow geoacoustic bounds in Fig. 4(c) improves the

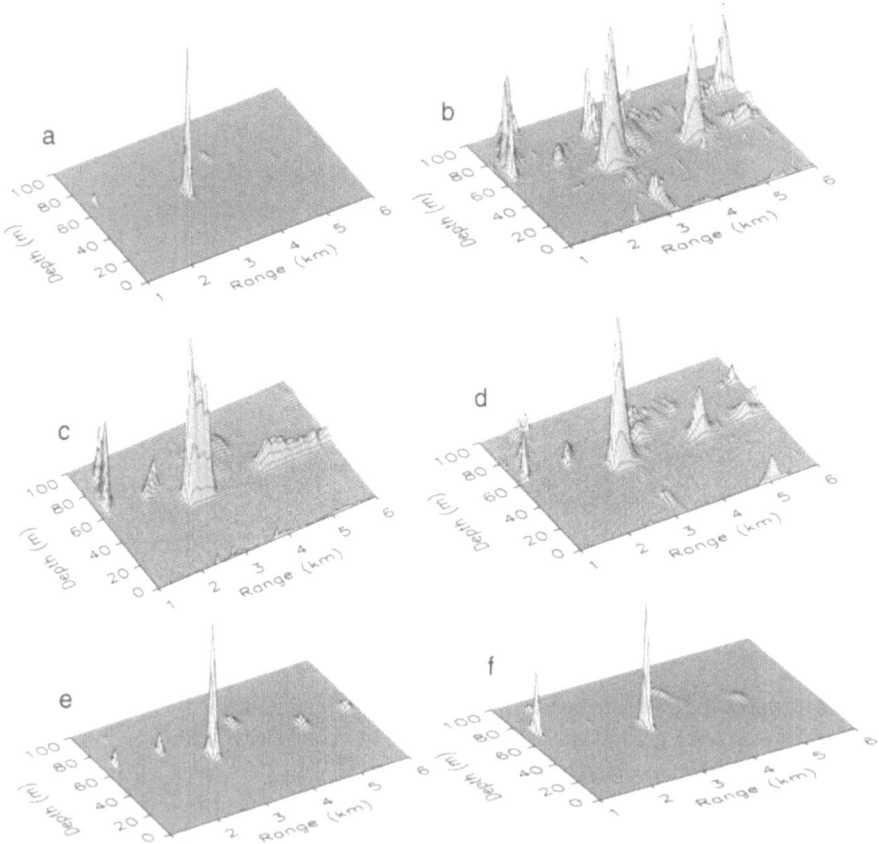

Figure 4. PAS for source localization example. Panel (a) shows PAS for exact enviromental parameters. The remaining panels show PAS for: (b) wide geoacoustic bounds and unknown SSP; (c) narrow geoacoustic bounds and unknown SSP; (d) wide geoacoustic bounds and known SSP; (e) narrow geoacoustic bounds and known SSP; and (f) exact geacoustic parameters and unknown SSP.

PAS substantially, with $\bar{P} = 0.47$. Figure 4(d) shows similar resuts for wide geoacoustic bounds and known SSP, with $\bar{P} = 0.45$. Figure 4(e) and (f) show good results for narrow geoacoustic bounds and known SSP and for exact geoacoustics and unknown SSP, with $\bar{P} = 0.70$ and 0.65, respectively.

The PAS shown in Fig. 4 quantify the state of available information for source localization under differing states of environmental information. The PAS indicate the probability that the source is located in a particular r-z interval, providing meaningful uncertainties for source localization. The reasonably good localization results obtained for most cases in Fig. 4 does not imply that environmental uncertainties are unimportant or can be be ignored. Rather, the results indicate how Bayesian analysis takes these uncertainties into account. To illustrate this, Fig. 5 shows PAS computed for geoacoustic parameters drawn at random from the wide prior bounds (SSP is exactly known). Very poor localization results in these cases. Similar results were obtained for the other cases in Fig. 4(b)–(d).

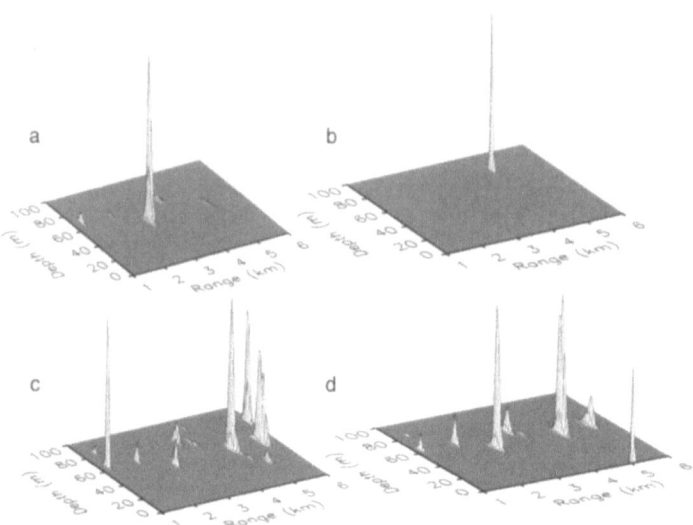

Figure 5. Panel (a) shows the PAS for exact environment; panels (b)–(d) show PAS for geoacoustic parameters drawn at random from wide bounds (exact SSP).

4   Summary

This paper applied Bayesian analysis to examine the available information content in ocean acoustic inverse problems. The dependence of geoacoustic inversion on knowledge of the ocean SSP and of source localization on SSP and geoacoustic parameters was considered. The computation of 2-D marginal probability distributions in range and depth (PAS) quantifies uncertainty in localization, and effectively addresses the effects of environmental uncertainty. This analysis was carried out efficiently using a variant of heat-bath GS.

Acknowledgements

We thank Neil Frazer for suggesting the heat bath method for this problem.

References

1. Sen, M.K. and Stoffa, P.L., *Global Optimization Methods in Geophysical Inversion* (Elsevier, Amsterdam, 1995).
2. Gerstoft, P. and Mecklenbräuker, C.F., Ocean acoustic inversion with estimation of *a posteriori* probability distributions, *J. Acoust. Soc. Am.* **104**, 808–819 (1998).
3. Dosso, S.E., Quantifying uncertainties in geoacoustic inversion I: A fast Gibbs sampler approach, *J. Acoust. Soc. Am.* **111**, 129–142 (2001).
4. Dosso, S.E. and Nielsen, P.L., Quantifying uncertainties in geoacoustic inversion II: Application to a broadband shallow-water experiment, *J. Acoust. Soc. Am.* **111**, 143–159 (2002).
5. Collins, M.D. and Kuperman, W.A., Focalization: environmental focusing and source localization, *J. Acoust. Soc. Am.* **90**, 1410–1422 (1991).
6. Shorey, J.A., Nolte, L.W., and Krolik, J.L., Computationally efficient Monte Carlo estimation algorithms for matched field processing in uncertain ocean environments, *J. Comp. Acoust.* **2**, 285–314 (1994).

MEASURING THE AZIMUTHAL VARIABILITY OF ACOUSTIC BACKSCATTER FROM LITTORAL SEABEDS

PAUL C. HINES, JOHN C. OSLER AND DARCY J. MACDOUGALD

Defence Research Establishment Atlantic, P.O. Box 1012, Dartmouth, NS, Canada, B2Y 3Z7

E-mail: paul.hines@drdc-rddc.gc.ca

The acoustic backscattering strength of the seabed has been demonstrated to be one of the key inputs required in sonar performance prediction (SPP) models. Extending range independent SPP models to full 3-dimensional – or even the simpler Nx2-dimensional models – requires measurements of the azimuthal variability of the backscattering strength. DREA's Wide Band Sonar (WBS) system, which consists of a parametric transmitter and a superdirective receiver, is ideally suited to make these measurements. In May 2001 the system was used as part of a joint US-SACLANTCEN-CA experiment referred to as Boundary Characterization 2001. The system collected acoustic backscatter data as a function of azimuth at two shallow water sites off the east coast of North America. In addition, backscattering strength measurements were made at several sub-critical grazing angles. In this paper the attributes of the WBS are outlined and some of the results from the experiment are presented.

1   Introduction

Sonar performance prediction (SPP) models require several environmental inputs in order to estimate detection ranges. For sonars operating in littoral waters in the low kHz band, key inputs include bottom loss, sea surface loss, backscattering strength (BSS) of the seabed and the sea surface, ambient noise, and the sound-speed profile for the water column. Reasonable estimates of sea surface scatter, forward loss, and ambient noise can be obtained using the wind speed [1–3], and bathythermographs are routinely used to obtain the sound-speed profile. In fact, advances in satellite sensing show potential for *remotely* estimating both the wind speed and sound-speed profile [2]. By contrast, estimating those sonar parameters related to the seabed poses a greater problem. For example, there is no easily measured correlate from which to infer the bottom backscattering strength in the way one uses wind speed to estimate surface scatter. In fact, the relative importance of seabed roughness, sediment type, and grain size to bottom scattering strength is still widely debated. As a result, many sonar performance prediction models rely on data bases to estimate seabed scatter. Often, these data bases are sparsely sampled spatially or use broadly defined regional descriptors such as "silt" or "rock" to estimate the scattering strength. Measurements of seabed scatter are critical to provide ground truth validation for these data bases. Measurements should also unravel the underlying physics of scattering which will improve ones ability to predict scatter given geo-technical information about the seabed.

N.G. Pace and F.B. Jensen (eds.), Impact of Littoral Environmental Variability on Acoustic Predictions and Sonar Performance, 179-186.

Scattering strength estimates for littoral seabeds have been extracted from reverberation experiments [e.g. 3]; however, this technique requires careful interpretation because scatter from several grazing angles having undergone different numbers of boundary interactions arrive simultaneously. We require direct measurements of seabed scattering – preferably unhindered by vertically directed sidelobes which generate unwelcome fathometer returns. Furthermore, if one hopes to extend range independent SPP models to full 3-dimensional (or even the simpler Nx2-dimensional models) measurements of the azimuthal variability of the backscattering strength are required. In shallow water, conventional sonars are poorly suited for this task at low kiloHertz frequencies. This is because the sonar's size is roughly proportional to its acoustic wavelength and to project a narrow acoustic beam, the sonar must be many wavelengths long. This in turn places the seabed in the near-field of the array which adds an extra layer of complexity to data interpretation. Additionally, strong returns on array sidelobes can mask weak on-axis returns resulting in erroneous estimates of the directional dependence of the scatter.

To address these issues the Defence Research and Development Canada (DRDC) Atlantic (formerly DREA) has developed a wide-band active sonar (WBS) with which to interrogate the seabed and quantify its geo-acoustic properties. The WBS employs a parametric transmitter to make these measurements. The parametric transmitter offers several advantages for the current experimental requirement. First and foremost, due to the nature of signal generation in the parametric array, no sidelobes are formed. This feature avoids extraneous boundary interactions when making measurements in shallow water. Second, the beamwidth of the parametric array is extremely narrow relative to the transmitter aperture. In the present case a square transducer measuring 41 cm on a side yields horizontal and vertical beam widths of approximately 3° across the difference frequency band. This coupled with the absence of sidelobes enables accurate measurements of the azimuthal variability of the backscattering strength. Third, one can obtain a wider bandwidth than that obtained using a conventional source. In the present case a bandwidth of 1 kHz to 10 kHz is realized from a single transducer. The wide bandwidth means that a single transducer can be used in place of a suite of transducers, each of which may require separate power and tuning circuitry.

In May 2001 the system was used as part of a joint US-SACLANTCEN-CA experiment referred to as Boundary Characterization 2001. The system collected acoustic backscatter data as a function of azimuth and grazing angle at two shallow water sites off the east coast of North America – one near New Jersey and one near Nova Scotia. Measurements were made at grazing angles ranging from 30° down to 2.5°, at frequencies of 4 and 8 kHz. Following the introduction, the attributes of the WBS are outlined and a sample of the measurements of backscatter variability are presented.

2   The Wide Band Sonar (WBS)

Defence Research Establishment Atlantic has developed a wide-band sonar (WBS) for collecting environmental acoustic data in the open ocean [4]. A schematic of the system is shown in Fig. 1. The primary acoustic sensor suite consists of a parametric array transmitter (PATS) and a superdirective endfire line array receiver (SIREM).

Mechanical steering of these arrays is remotely controlled from the research ship via an RF radio link fixed to the system's surface float. This minimizes the risk of acoustic interference from the ship and prevents ship motion from compromising its stability. In addition to acoustic sensors, the system is instrumented with a range of non-acoustic sensors to assist in the evaluation of the data. The non-acoustic sensors include depth, tilt, roll, and heading sensors to monitor the position and direction of the parametric array, as well as accelerometers to monitor platform vibration. The sonar head can be panned through 360° azimuth. The system can be configured to be either bottom-tethered (Fig. 1a) or bottom-mounted (Fig. 1b). In bottom-tethered mode the sonar support arm can be rotated 180° vertically so that the sonar can be positioned above or below the space frame. This enables measurements through 4π steradians. In this configuration, platform stability is achieved by de-coupling the space frame from the surface float through a weighted cable and streaming the space frame into the prevailing shear current. In bottom-mounted mode a remote command is sent from the ship to the surface float to flood the sub-surface floats on the space frame following deployment. This causes the system to descend slowly to the seabed. Once the floats are fully flooded the system weight in water is approximately 4500 N. This offers an extremely stable platform that permits coherent averaging of multiple pings. This is used in low SNR conditions such as measurements of backscattering strength at very shallow grazing angles. Prior to recovery, compressed air is forced into the sub-surface floats to evacuate the water. When the system is bottom mounted, physical constraints limit the range of vertical angles to -30° from the horizontal up to +90°; however, the head is still able to rotate 360° in azimuth. The experiments discussed in this document were all performed with the system in bottom-mounted mode.

The parametric transmitter – the advantages of which have been highlighted already – consists of nine square ceramic piston transducers arranged in a 3-by-3 grid. At full power, the primary source level is 242 dB re 1μPa@1m at 100 kHz. Pulse duration can range from 50 μs to 250 ms. The main disadvantage in employing a parametric array is that the conversion of energy from the primary frequencies to the difference frequency is a second order process, and therefore inefficient. Table 1 lists the difference frequency source level (SL_d) referred to 1 m range measured at a selection of difference frequencies.

Table 1. Parametric array source levels measured at several difference frequencies.

f_d (kHz)	SL_d (dB//1μPa^2)
1.0	166
2.0	172
4.0	182
8.0	188

To complement the compactness of the parametric transmitter, a six-channel superdirective hydrophone line array (SIREM) was developed as the principal receiver. SIREM is mounted on the parametric array head and turns with the array. (Recall Fig. 1). A superdirective array relies on computing pressure gradients and as such requires inter-element spacings that are a small fraction of an acoustic wavelength. Thus, by its very nature, a superdirective array is much more compact than a conventional array.   The principal disadvantage with a superdirective array is its susceptibility to incoherent noise such as sensor self-noise and inter-channel phase errors. That is to say, the very process of taking the difference between the acoustic signals at two sensors means that the signal to noise ratio (SNR) must degrade. In practice the array yields gains of up to 15 dB across the sonar's frequency band from an array aperture only 0.8 m long.

The system also has a tri-axial intensity array known as SIRA (Sound Intensity Receiver Array) fixed to the spaceframe. This secondary receiver is used to measure ambient noise directionality. The superdirective array weights can then be optimized for the specific ambient noise field. (SIRA can also be used to localize the platform during bistatic experiments. It's bearing accuracy is better than ±0.5°.) Note that all data reported in this paper were collected using a single hydrophone from SIREM.

In the following section, an experiment to measure the scattering strength of the seabed as a function of azimuth and grazing angle is described. Particular emphasis will be placed on the variability of the scattered energy as a function of azimuth.

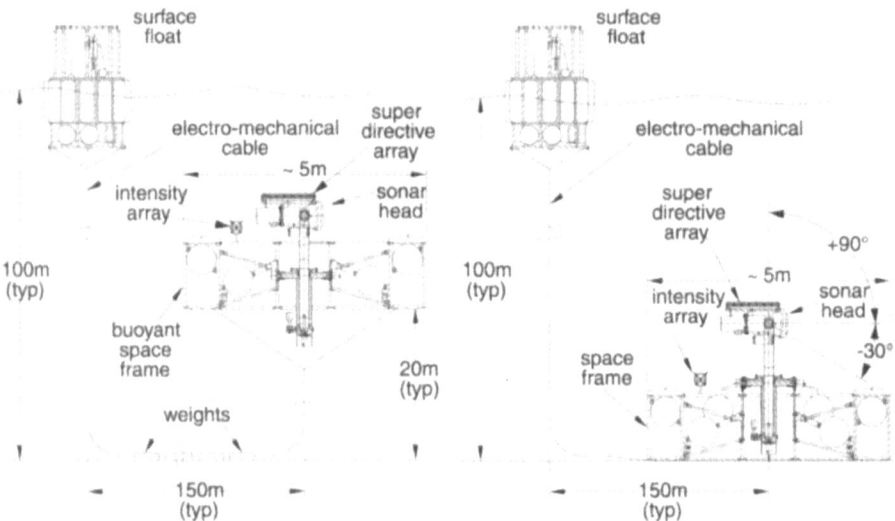

Figure 1. Schematic of Wide Band Sonar bottom-tethered (left), bottom-mounted (right). Note that the pivot point for the sonar support arm is different for the two configurations.

3   The scattering experiments

A series of experiments to measure the dependence of seabed backscatter on azimuth and grazing angle was performed at two shallow water sites on North America's Eastern Seaboard – one site known as Strataform is located off the New Jersey coast and the second site is about 100 nmi. southeast of Halifax, Nova Scotia on Canada's Scotian

Shelf. The sites were chosen primarily because of the availability of supporting geo-technical measurements to assist in data interpretation and modeling [5, 6].

Figure 2 contains a sketch of the geometry for the measurements of azimuthal scattering. The array was pointed at a grazing angle in the range of 5° to 15°. At each azimuthal angle, a series of 50 pulses was transmitted by the WBS at 4 and 8 kHz and

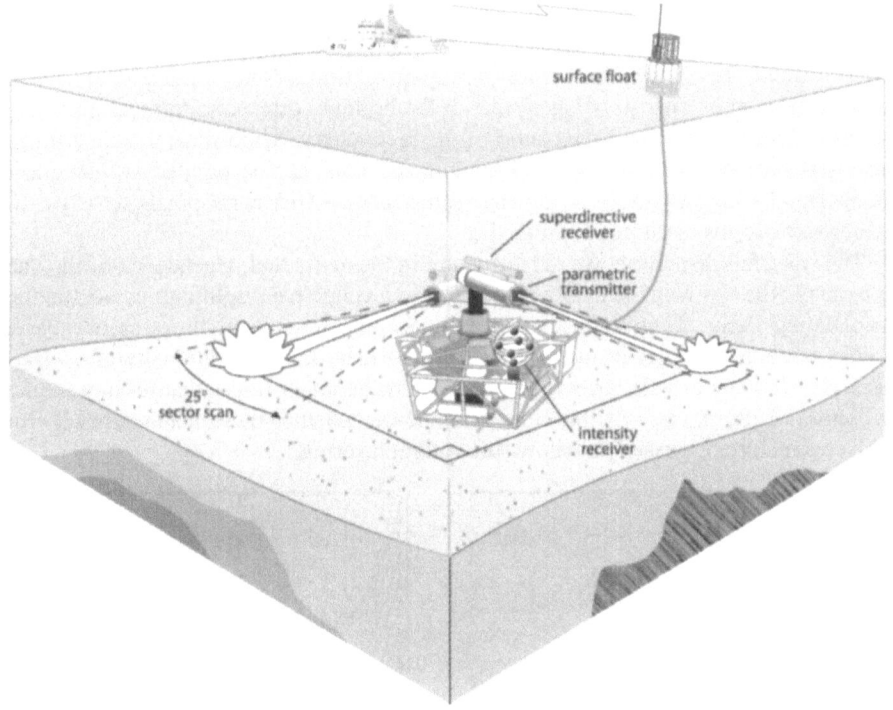

Figure 2. Geometry for measurements of the azimuthal variability of acoustic scattering.

acoustic backscatter from the seabed was recorded on SIREM. A pulse duration of 2 ms was used with a pulse repetition frequency (PRF) of 4 per second. The parametric array transmitter was rotated approximately 1.5° in azimuth and the sequence was repeated. This sector scan covered approximately $\Delta\theta = 25°$. Time constraints did not permit a 360° sector scan; as a compromise, the parametric array was rotated 90° in azimuth relative to the center of the sector scan and the experiment was repeated for 2 azimuthal angles separated by 3°. This procedure was adopted to allow an examination of the small scale transition of backscatter across azimuth, while at the same time providing an opportunity to observe any drastic variations that might only show up with a substantial change in azimuth. Backscattering strength was measured at grazing angles ranging from 20° down to 2.5° at the midpoint and the extrema of the sector scan. In the following section, samples of the azimuthal variability and the grazing dependence of the backscatter are presented.

4 Results and discussion

Figure 3 shows waterfall displays of the azimuthal dependence of the backscattered energy vs. time as measured on a single hydrophone in SIREM. The data are from the Strataform site at 4 kHz (left) and 8 kHz (right). The top (dashed) trace is the average of the 17 angles used in the sector scan. The 17 solid lines below it correspond to the data at each azimuthal angle. Each trace in the waterfall represents the coherent average of 50 pulses at a single azimuth. These data were taken with the parametric array pointed at 10° grazing. This corresponds to a two-way travel time of approximately 20 ms at the center of the beam. The two dashed lines at the bottom correspond to measurements at azimuths of 90°and 93° from the center of the sector scan. Unfortunately, experimental constraints required that for these measurements the sonar be pointed at 13° grazing (rather than 10° as was used during the sector scan). This corresponds to a two-way travel time of 15 ms to the center of the beam.

The waterfall display shows rich structure in the azimuthal dependence of the data. For example the arrow in the 8 kHz data points to a single peak splitting in two and then consolidating back into a single peak. This occurs a couple of times as one passes through azimuth. Similar features occur in the 4 kHz data. Additionally at 4 kHz, a sharp ridge (denoted by the arrow) can be seen in the upper traces that disappears near the bottom of the figure. At this time it is unclear whether these features result from interface structure or possibly shallow sub-bottom layering.

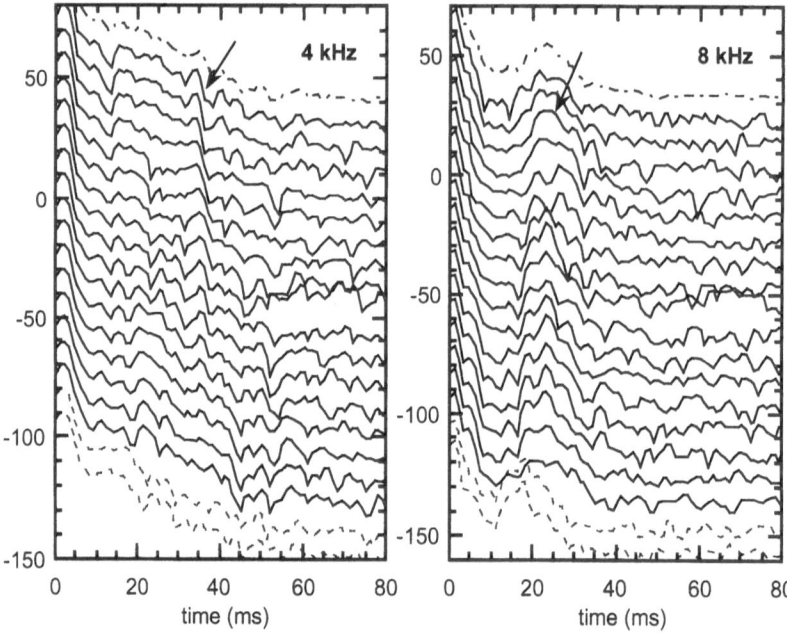

Figure 3. Waterfall display of azimuthal dependence of backscattered energy vs. time for the Strataform site taken at 4 kHz (left) and 8 kHz (right). The topmost (dashed) trace is the average for the entire sector scan. The two dashed lines at the bottom are measurements taken at azimuths 90° and 93° relative to the center of the sector scan (see text).

The backscattered energy was measured with the acoustic axis centered at slant angles of 5°, 7.5°, 10°, 15°, and 20°. To prevent discrete features from biasing the results, measurements were made at several azimuthal angles and the rms average was computed. The difference frequency source level at the seabed was estimated using a combination of calibration data and model evaluation [7]. Correcting the source level for the parametric array beam pattern and then converting time of flight to grazing angle allowed a range of grazing angles to be spanned for each slant angle. This procedure allowed for some overlap in the curves thereby increasing confidence in the results.

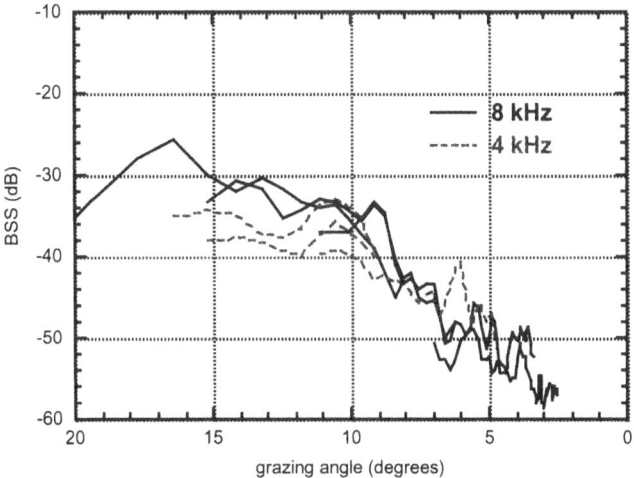

Figure 4. Backscattering strength vs. grazing angle at 4 and 8 kHz for the Strataform site.

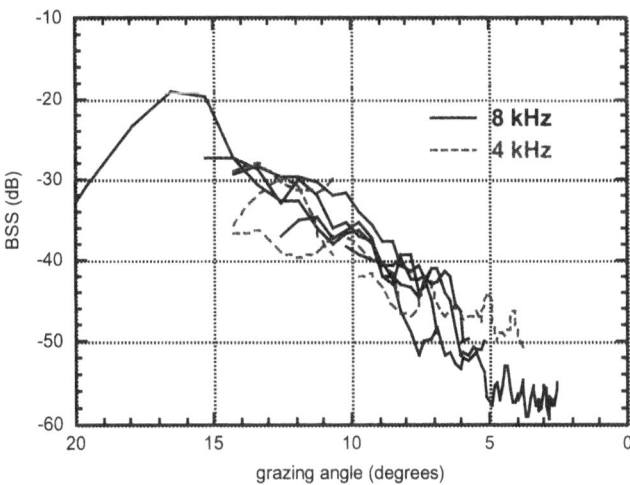

Figure 5. Backscattering strength vs. grazing angle at 4 and 8 kHz for the Scotian Shelf site.

Figure 4 shows the BSS at 4 and 8 kHz for the Strataform site plotted as a function of grazing angle. There appears to be little, if any, frequency dependence in the scattering strength. The slope of the 8 kHz data appears to be slightly steeper than for the 4 kHz data; however, this may result from interference from the noise floor at 4 kHz at the shallow angles. Figure 5 shows the BSS at 4 and 8 kHz for the Scotian Shelf site plotted as a function of grazing angle. As for Strataform, the BSS appears to be independent of frequency within the statistical accuracy of the measurements.

5 Concluding remarks

The Wide Band Sonar is an effective tool for examining seabed acoustics. It permits a wide range of experimental geometries, minimizes the risk of acoustic interference from the ship, and prevents ship motion from compromising array stability. It's narrow beam width makes the system particularly well suited to measure the variability of acoustic scattering from the seabed in addition to making direct (rather than inferred) measurements of the backscattering strength of the seabed.

In this paper, the azimuthal variability of the scattered intensity and the grazing angle dependence of the seabed backscattering strength are reported for frequencies of 4 and 8 kHz. The measurements were made at two shallow water sites referred to as Strataform and Scotian Shelf. Initial results show substantial azimuthal variability of seabed backscatter. At both sites, the backscattering strength measurements appear to be independent of frequency within the statistical accuracy of the data.

References

1. M. Nicholas, P.M. Ogdan and F.T. Erskine, Improved empirical descriptions for acoustic surface backscatter in the ocean, *J. Oceanic. Eng.* **23**, 81–95 (1998).
2. D. Hutt, Acoustical oceanography and satellite remote sensing, *Backscatter Magazine* published by Alliance for Marine Remote Sensing, Autumn issue, pp. 32–34 (2001).
3. D.D. Ellis, J.R. Preston, R. Hollett and J. Sellschopp, Analysis of towed array reverberation data from 160 to 4000 Hz during Rapid Response 97. SACLANTCEN SR-280 (2000).
4. P.C. Hines, W.C. Risley and M.P. O'Connor, A wide-band sonar for underwater acoustics measurements in shallow water. In *Proc. Oceans '98*, Nice, France (1998).
5. J.A. Goff, D.J.P. Swift, C.S. Duncan, L.A. Mayer and J. Hughes-Clarke, High-resolution swath sonar investigation of sand ridge, dune and ribbon morphology in the offshore environment of the New Jersey Margin, *Marine Geology* **161**, 307–337 (1999).
6. J.C. Osler, Geoacoustic characterization of Boundary 2001 experimental locations. In *Proc. Acoustic Interaction with the Seabed*, Conference held at DREA, Canada, 2–3 Oct. 2001, DREA TM-2001-185.
7. R.H. Mellen and M.B. Moffett, A numerical method for calculating the nearfield of a parametric acoustic source, *J. Acoust. Soc. Am.* **63**, 1622–1624 (1978).

BACKSCATTER FROM ELASTIC OCEAN BOTTOMS: USING THE SMALL SLOPE MODEL TO ASSESS ACOUSTICAL VARIABILITY AND UNCERTAINTY

ROBERT F. GRAGG, RAYMOND J. SOUKUP AND ROGER C. GAUSS

Naval Research Laboratory, Washington DC 20375-5350, USA
E-mail: robert.gragg@nrl.navy.mil

The scattering strength of the ocean bottom as a function of angle and frequency is a fundamental input for predicting the performance of active sonar systems, particularly in littoral waters. The small-slope formulation for scattering from the rough water/bottom interface is by now well established both as a physical theory and as a numerical algorithm [Gragg *et al.*, JASA **110** (2001)]. In this work, a data-model comparison is used to address the following questions. How well do the predictions of this elastic theory agree with data measured at sea? How sensitively does the theoretical prediction depend on the set of input parameters that characterize (a) the geoacoustics of the bottom material and (b) the spectrum of the surface roughness? For a given littoral area, how much needs to be known about these parameters for ASW/MCM purposes? How much of that could be estimated by remote (e.g., acoustic inversion) methods?

1 Introduction

We first extract the bottom scattering strength (in the 2.0–3.5 kHz band) from LWAD [1] data sets taken at a pair of nearby littoral sites that have a bare limestone bottom. The processing techniques complement those of Holland *et al.* [2], with special emphasis on squeezing the widest possible range of grazing angles out of a relatively simple system (omni source and VLA receiver). We then use small-slope theory to model the scattering, given a minimal set of input parameters that specify the bottom's roughness and geoacoustic properties (including shear). Finally, we adjust these inputs, under the control of a simplex/annealing search algorithm, to maximize the data-model agreement across frequency and grazing angle (f, θ). These geoacoustic inversions yield information on the importance of each of the parameters in scattering and on their site-to-site variability.

The water/sediment interface is the dominant scattering mechanism often enough that a solid understanding of it is essential—especially for sand or rock because these have significant ranges of sub-critical θ that are important in sonar operation. CST [3] measurements in the 100–1000 Hz band have supported a variety of f and θ dependences, illustrating the need for an improved physical model. The small-slope theory of scattering from rough interfaces has recently been used to develop such a physics-based formulation for elastic bottoms [4]. This model typically predicts θ dependences that are considerably more complex than the familiar $\sin^n \theta$ empirical descriptions (e.g. in Ref. [5]).

Our procedure relies on data with multiple frequencies and a wide range of sub-critical angles to get enough information to invert for parameters that are rarely measured in-situ;

N.G. Pace and F.B. Jensen (eds.), Impact of Littoral Environmental Variability on Acoustic Predictions and Sonar Performance, 187-194.
© 2002 *Kluwer Academic Publishers.*

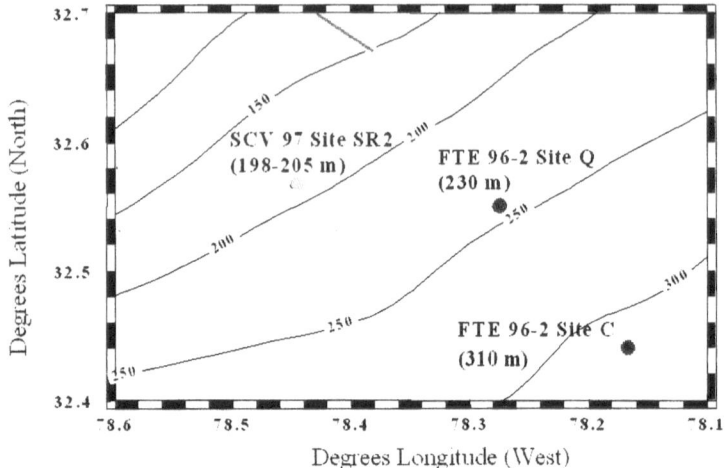

Figure 1. Test sites described in the text.

e.g. the bottom roughness spectrum. We use LWAD data from the FTE 96-2 experiment [1, 6, 7]—specifically from Sites Q and C, Fig. 1. MKS units are tacitly used throughout this article; e.g., "$c = 1500$" with the "m/s" that specifies the units left implicit.

2   Acquiring and processing the data

Both experiments were essentially monostatic. (The source and receiver were mounted only 3 m apart on a common cable.) At each site, deployments were made in *shallow*, *middle* and *deep* configurations (source depths 35 m, 60 or 65 m, and 105 m respectively). The receiver (a 9-phone VLA cut for 3750 Hz) provided ten beams, five of which are relevant here. These were steered at 90°, 51°, 34°, 19° and 6° below horizontal, and are designated 0 through 4, respectively. At each frequency, sets of 12–15 gated 10-ms CW waveforms (separated by 3 s) were transmitted. Analysis of longer (50-ms) pulses from these experiments has already been reported [1, 8]. Our use of these shorter signals is supported by the weak frequency dependence seen in the scattering data (~ 1 dB/kHz across the 240 Hz analysis band).

After beamforming, power spectra were obtained using Fourier transforms of length equal to the ping duration, the successive transforms proceeding to the end of the time series with 90% overlap. A frequency band representing the total energy about the zero-Doppler peak was selected and a time series was formed for each ping using only the energy in that band. The pings' direct arrivals were then temporally aligned and then averaged (before conversion to dB), producing a single reverberation curve for each beam and frequency bin. Integration over the zero-Doppler spectral peak yielded the total returned power over time and beam. Transmission loss terms to and from the scattering patch were obtained from the geometric spreading loss along each ray path. The computed beam pattern and ray trace were used to calculate the scattering patch area. With these inputs, bottom scattering strength was calculated from the sonar equation as a function of beam, f and θ.

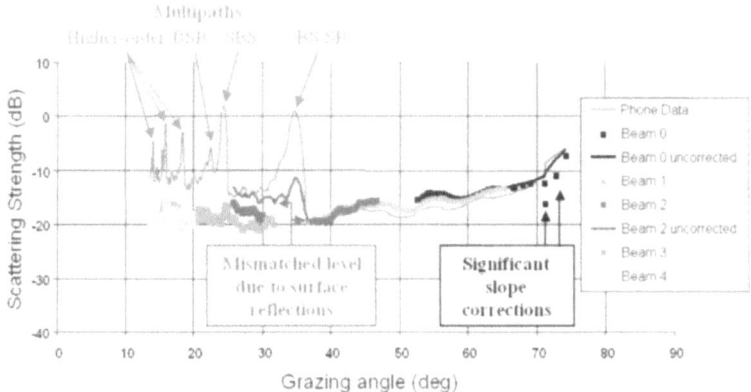

Figure 2. Scattering strength vs grazing angle at 3 kHz for a deep deployment at Site Q. The two types of corrections mentioned in the text are illustrated. Multipath designations B, SB, BSB etc. correspond to Fig. 3.

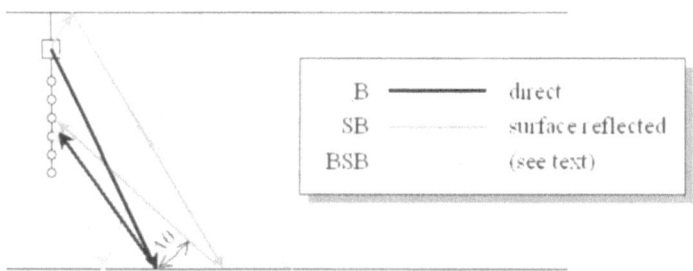

Figure 3. Sketch of multipaths that produce ambiguous returns. (See Fig. 2.)

As a final step, the data were processed to correct for (a) multipaths (as described in detail in Ref. [9]) and (b) reverberation decay over the pulse length. Figure 2 illustrates the situation, with multipaths labeled according to their boundary interactions (bottom and surface are abbreviated B and S). These are most prominent in the "Phone Data" curve, which comes from a single hydrophone. Two examples of uncorrected data are included for illustration. There is a sizable disparity between the uncorrected levels for beam 2 on either side of the BS/SB arrival. However, after correction for multipath effects, these levels agree and are consistent with beam 3. For beam 0, reverberation slope corrections become significant at high θ.

3 Geoacoustic inversion

The geoacoustic parameters of the problem are the densities and sound speeds of the two media. For the water, these are ρ_w and c_w (both real); for the bottom, they are the density ρ_b and the complex-valued compressional and shear speeds, c_p and c_s. We assume the water/bottom interface to have an isotropic, power-law, roughness spectrum of the form $S(k) = w_2/(h_0 k)^{\gamma_2}$ with $2 < \gamma_2 < 4$ (which corresponds to a fractal dimension between 1 and 2). We take the essentially arbitrary [4] reference length h_0 to be 1. This leaves ρ_w, c_w, w_2, γ_2, ρ_b, c_p, and c_s as model inputs. Before inverting for their values, we impose two conditions. We fix the parameters whose values are in no real doubt, $\rho_w = 1000$ (nominal sea water) and $c_w = 1487$ (from in-situ measurement), and vary the others within appropriate bounds. We also examine the data to identify a grazing angle $\hat{\theta}$ that appears to mark the onset of sub-bottom scattering, and then limit our data/model comparisons to $\theta < \hat{\theta}$. Inversion is then a matter of quantifying the data-model deviation across all the experimental frequencies and allowed grazing angles by devising a cost function Φ, and then searching the appropriate region of the parameter space for the minimum of $\Phi(w_2, \gamma_2, \rho_b, c_p, c_s)$. All that remains is to specify the cost function, the search algorithm, and the search region. As with most inversion techniques, some uniqueness problems are to be expected. One should anticipate fairly precise evaluations for the parameters that strongly affect scattering, but only rough estimates for the rest.

We examined several cost functions based on the calculated theory-minus-data deviations over the experimental θ and f ranges. The simplest choice, an RMS average deviation (in decibels), performed relatively poorly. We concluded that this was due to (i) the high concentration of data points at low angles (the median data angle is only about $20°$), and (ii) an unexplained ripple that persists in some of the data at higher angles. To counter these effects, we inserted a bin-averaging step in which the data points are assigned to bins of width $\Delta\theta \approx 4°$, bin values are computed by averaging (of the signed deviations, not their absolute values), and finally the bin values are RMS averaged to form the cost. This proved an effective remedy, and was adopted as our final working definition for Φ.

Since the parameter space is seven-dimensional, some algorithm more efficient than an exhaustive search was called for. We chose the "amebsa" algorithm—essentially a fast downhill simplex method that efficiently negotiates narrow valleys in parameter space and is augmented with simulated annealing to prevent trapping by local minima [10, 11]. Given a suitable empirically chosen cooling rate, the technique usually "freezes" into a final state of near-minimum Φ well within 50 temperature steps.

We chose the parameter search region based, as far as possible, on archival records for the geoacoustics of limestone and on roughness data from rocky sea floors. For convenience, we first changed from using the shear speed $Re(c_s)$ itself as a search parameter, and used the compressional/shear speed ratio $\xi = Re(c_p)/\,Re(c_s)$ instead. This does not materially affect the operation of the search algorithm and is more convenient in two respects. Hamilton has concluded from his analysis of a large collection of data sets dealing with saturated marine limestone that, although $Re(c_p)$ and $Re(c_s)$ vary considerably, their ratio is to be found in the interval $|\xi - 1.90| < 0.06$, "within 95% confidence limits" [12]. One can relax this empirical statistical rule somewhat, allowing a larger variance about the mean value ($\langle\xi\rangle = 1.90$) by taking $|\xi - 1.90| < n \times 0.06$ with $n > 1$. We use $n = 5$, and are thus dealing with the range $1.6 < \xi < 2.2$ (Table 1). (Since $n < 12$, this also automatically respects the physical requirement [13] $\xi > 2/\sqrt{3}$.) The limits on $Re(c_p)$ come from the observation that the compressional critical angle in our data appears to lie within the range $70° < \theta_p < 73°$. The low and high values for ρ_b embody the range reported in Hamilton's references [14, 15]. The bounds on γ_2 and w_2 reflect our experience with seafloor spectra. Reference [16] reports $\gamma_2 \approx 2.64$ for a scarp of the Mid-Atlantic Ridge (MAR). The range in Table 1 includes this value. The range of w_2

Table 1. Inversion parameters and their high and low values (MKS units).

	w_2	γ_2	ξ	$Re(c_p)$	ρ_b	$Im(c_p)$	$Im(c_s)$
low	0.0003	2.4	1.6	4348	2400	-300	-900
high	0.0009	3.0	2.2	5086	2800	-5	-5

is chosen to undershoot the MAR value [16], $w_2 \approx 0.0021$, because the present bottom is expected to have lower relief. The high values of $Im(c_p)$ and $Im(c_s)$ correspond to attenuations reported for pure, homogeneous, water-saturated limestone samples (Table I of Ref. [14]). The low values (corresponding to compressional and shear attenuations of approximately 0.5 and 7 dB/m/kHz, respectively) are essentially guesswork inspired by the saprolitic (weathered-in-place) nature of the bottom [17].

4 Results

Figure 4 plots the data at all four experimental frequencies for site C along with a simulation curve produced using the optimal environmental parameters from the inversion. The curve is black where the data-model fit is calculated ($\min(\theta) < \theta < \hat{\theta} = 70°$) and gray elsewhere. The critical angles θ_p, θ_s are calculated from the optimal values of $Re(c_p), Re(c_s)$. As the frequency increases, the scattering strength, σ, in the figure rises by $\Delta\sigma \approx 3$dB at small θ, consistent with the dependence $\sigma \propto f^{4-\gamma_2}$ to be expected from small-slope computations [4]. The fit is quite good: $\Phi \approx 0.5$ dB and the maximum absolute data/model deviation is only $\Delta \approx 4$ dB. There are large slope corrections near θ_p because the first derivative of the reverberation time series changes abruptly there. (In measuring the slope at such a discontinuity, we obtain not a consistent value among different frequencies and sites but rather an artifact that indicates that a change in a scattering mechanism has affected the derivative of the time series.) The marked data-model divergence as θ exceeds θ_p suggests the onset of significant sub-bottom volume scattering at that point. The optimal values of the parameters $w_2, \gamma_2, \rho_b, Re(c_p), Im(c_p), Re(c_s), Im(c_s)$ are listed in the figure along with the associated values of Φ and Δ. These numbers are simply the raw output of the algorithm presented without any regard for significant figures. "$\gamma_2 = 2.8258$," for example, does *not* mean that we have determined the spectral exponent to four decimal places. We have obtained very similar theoretical curves even using subsets of the allowed angles, e.g. with $\hat{\theta} = 50°$. However, attempts to invert using still smaller subsets, e.g. with $\hat{\theta} < 45°$ or using only angles in the range $45° < \theta < 70°$, have had little success. Evidently, inversion in this environment requires data over a fairly broad range of angles and at least a modest bandwidth. Otherwise, the system point seems too easily lured into physically dubious regions of the parameter space.

As noted above, inversion results are rarely, if ever, 100% repeatable. To investigate this tendency, we first ran a series of fifty independent inversions for Site C. The outcome, in Fig. 5, provides a look at how successfully our inversion process determines that site's parameters. The results are displayed as • symbols in separate vertical panels (one for each of the seven parameters), with the fifty repetitions arranged along the horizontal. In each panel, the solid horizontal line marks the parameter's mean value, and the dashed lines are one standard deviation away. (The third panel also has dotted lines to mark Hamilton's nominal interval for limestone.) It is clear from this that the spectral parameters w_2 and

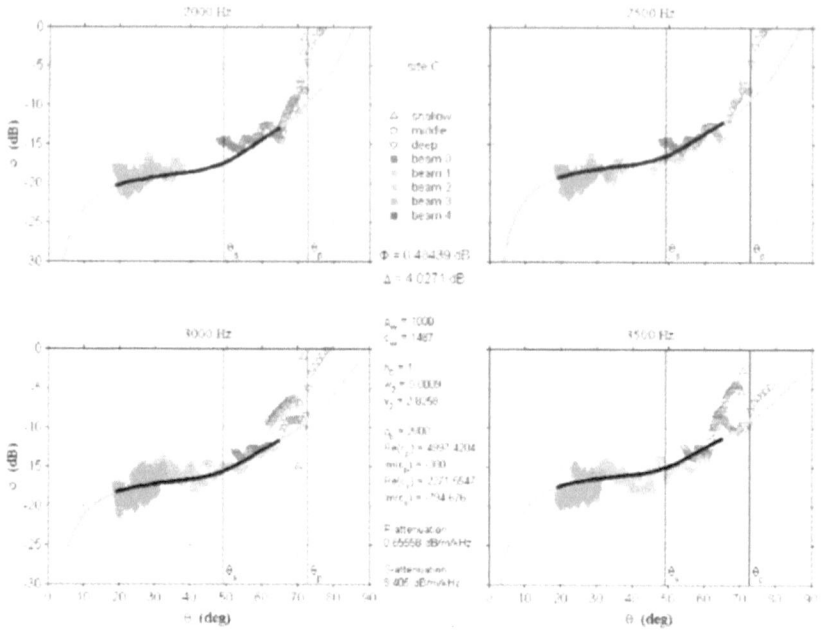

Figure 4. Site C data (symbols) and associated inversion result (curves).

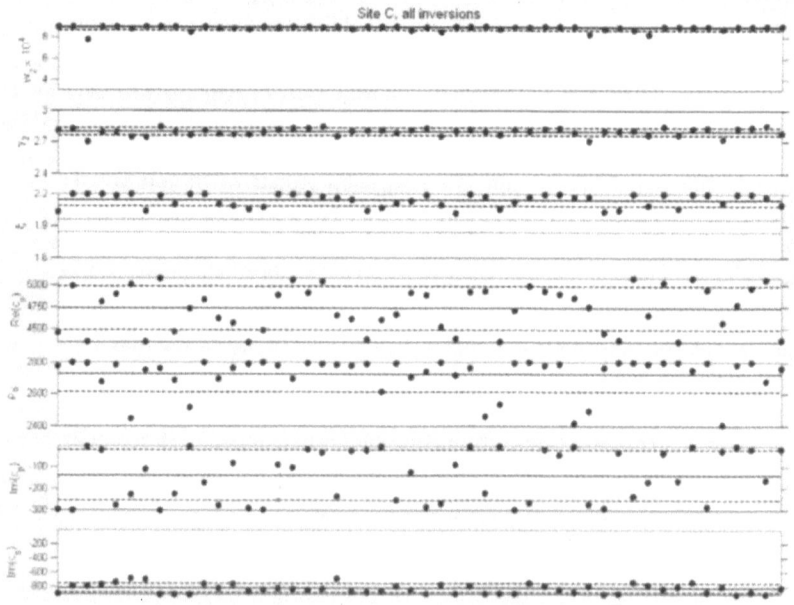

Figure 5. Results of a series of 50 independent inversions for Site C.

γ_2 are both well determined (though in retrospect we might have extended the search range for w_2 a bit higher). So is $Im(c_s)$, which means that the shear attenuation is definitely substantial (presumably due to the saprolitic nature of the rock). The ξ ratio is also well determined, though definitely somewhat above Hamilton's range. The remaining three—ρ_b and the real and imaginary parts of c_p—are not well determined. They are not among the principal parameters that determine the scattering strength of this bottom (for 2.0 kHz$\leq f \leq$3.5 kHz and $\theta < \hat{\theta}$, at least).

Similar computations were also done for Site Q, with a similar outcome: w_2, γ_2, ξ, and $Im(c_s)$ were well determined, but ρ_b, $Re(c_p)$, and $Im(c_p)$ were not. However, the results (summarized in Table 2) do exhibit some site-to-site variability. In particular, the spectral strength w_2 more than doubles over the 10 miles between the two sites.

Table 2. Summary of inversion results for the two sites.

		$w_2 \times 10^4$	γ_2	ξ	$Re(c_p)$	ρ_b	$Im(c_p)$	$Im(c_s)$
Site Q	mean	4.11	2.59	2.04	4691	2630	-172	-721
	std. dev	0.34	0.05	0.11	271	139	92	90
Site C	mean	8.86	2.80	2.14	4733	2723	-141	-825
	std. dev.	0.24	0.04	0.06	252	113	116	63

5 Discussion

We have been able to invert acoustic backscatter measurements for the essential parameters in interface scattering from sea floors. One key to the success of this effort has been the use of data covering a wide range of grazing angles and a moderate frequency band. Of equal importance has been the availability of a physical theory of scattering (namely, small-slope) that allows both large roughness on the interface and elasticity in the bottom. The physical basis of such scattering models supports extrapolation in angle and frequency and also provides a foundation for relating geophysical variability to acoustic variability.

Our results suggest that the geoacoustic parameters that are important for scattering in typical ASW scenarios could well be evaluated in-situ by inverse methods. Site-to-site variability is an important consideration here; however, preliminary indications from our geographically very sparse data sets indicate that the variability problem may not be too severe. (The spectral strength w_2 changes by only about 3 dB over the 10 miles between Sites Q and C, for example.) Extending the technique to the MCM realm, where bottom penetration is more important, would probably require the inclusion of a volume scattering module in the scattering algorithm (a step which we have under active consideration).

Acknowledgements

This work was supported by the Office of Naval Research.

References

1. Soukup, R.J. and Ogden, P.M., Bottom backscattering measured off the South Carolina coast during Littoral Warfare Advanced Development Focused Technology Experiment 96-2. NRL Memorandum Report 7140-97-7905, Washington, D.C., 1997.

2. Holland, C.W., Hollett, R. and Troiano, L., Measurement technique for bottom scattering in shallow water, *J. Acoust. Soc. Am.* **108**, 997–1011 (2000).

3. Holland, C.W., Ogden, P.M., Sundvik, M.T. and Dicus R., Critical Sea Test bottom interaction overview. In *Critical Sea Test White Paper SPAWAR CST/LLFA-WP-EVA-46* (Space and Naval Warfare Systems Command, Arlington VA, September 1996).

4. Gragg, R.F., Wurmser, D. and Gauss, R.C., Small-slope scattering from rough elastic ocean floors: General theory and computational algorithm, *J. Acoust. Soc Am.* **110**, 2878–2901 (2001).

5. McCammon, D.F., Low grazing angle bottom scattering strength: Survey of unclassified measurements and models and recommendations for LFA use, *JUA(USN)* **43**, 33–47 (1993).

6. Kerr, G.A., Bucca, P.J., Fulford, J.K. and Snyder, S.W., Environmental variability during the Littoral Warfare Advanced Development Sponsored Focused Technology Experiment (FTE 96-2). NRL Memorandum Report 7182-97-8056, Stennis Space Center, MS, 1997.

7. Thompson, C.H., Nero, R.W. and Love, R.H., Volume scattering on Littoral Warfare Advanced Development Focused Technology Experiment 96-2, NRL Memorandum Report 7174-97-8061, Stennis Space Center, MS, 1997.

8. Soukup, R.J., Bottom backscattering measured off the Carolina coast during the Littoral Warfare Advanced Development System Concept Validation Experiment 97 (LWAD SCV 97). NRL Report 7140-98-9885, Washington, D.C., 1998.

9. Soukup, R.J. and Gragg, R.F., Backscatter from a limestone seafloor at 2–3.5 kHz: Measurements and model validation, *J. Acoust. Soc. Am.* (submitted 2002).

10. Press, W.H., Flannery, B.P., Teukolsky, S.A. and Vetterling, W.T., *Numerical Recipes*, 2nd ed. (Cambridge University Press, Cambridge, 1986).

11. Press, W.H. and Teukolsky, S.A., Simulated annealing operations over continuous space, *Computers in Physics*, Jul/Aug, 426–429 (1991).

12. Hamilton, E.L., V_p/V_s and Poisson's ratios in marine sediments and rocks, *J. Acoust Soc. Am.* **66**, 272–280 (1979), Table III.

13. Landau, L.D. and Lifschitz, E.M., *Theory of Elasticity*, 3rd ed. (Pergamon, New York, 1986) p. 10.

14. Peselnick, L. and Zietez, I., Internal friction of fine-grained limestones at ultrasonic frequencies, *Geophysics* **24**, 285–296 (1959).

15. Peselnick, L., Elastic coefficients of Solenhofen limestone and their dependence upon density and saturation, *Geophysical Res.* **67**, 4441–4448 (1962), Table I.

16. Gragg, R.F. and Wurmser, D., Scattering from rough elastic ocean floors: Small-slope theory and experimental data. In *Proceedings of the Environmentally Adaptive Sonar Technologies (EAST) Peer Review*, 8–11 February 2000, Austin, TX (Office of Naval Research, Code 321, Arlington, VA, 2000).

17. Fulford, J.K., Naval Research Laboratory, Stennis Space Center, MS (personal communication).

SPATIAL AND TEMPORAL VARIABILITY IN BOTTOM ROUGHNESS: IMPLICATIONS TO HIGH FREQUENCY SUBCRITICAL PENETRATION AND BACKSCATTER

KEVIN L. WILLIAMS, DARRELL R. JACKSON, ERIC I. THORSOS
AND DAJUN TANG

Applied Physics Laboratory, College of Ocean and Fishery Sciences,
University of Washington, 1013 N.E. 40th St., Seattle, WA 98105, USA
E-mail: williams@apl.washington.edu

KEVIN B. BRIGGS

Marine Geosciences Division, Naval Research Laboratory,
Stennis Space Center, MS 39529, USA
E-mail: kevin.briggs@nrlssc.navy.mil

Quantitative prediction of high frequency, low grazing angle penetration into, and scattering from, sand sediments requires knowledge of the roughness of the water/sand interface. Since the sediment roughness evolves due to hydrodynamic and biological processes, concurrent, co-located measurement of roughness and acoustic penetration/backscattering is essential for testing acoustic models or using such models to determine the likelihood of buried target detection. Here, we examine both roughness and acoustic measurements carried out during a six week sediment acoustics experiment in 1999 (SAX99). A ripple field was present throughout the experimental period but changed wavelength and orientation as a result of a storm event (i.e., the ripple field was temporally non-stationary). The predicted impact of this change in the ripple field on acoustic penetration at shallow grazing angles is presented. The small-scale roughness important for backscattering was measured at several locations near to, but not co-located with, acoustic backscattering measurements. These measurements indicate roughness changes with location. The effect of this spatial non-stationarity on tests of alternative backscattering models is discussed. Finally, simple sonar equation predictions of high frequency, low grazing angle, buried mine detection are made using various combinations of interface roughness conditions.

1 Introduction

A sediment acoustics experiment (SAX99) was carried out to examine scattering from, penetration into, and propagation within a sand sediment at high frequencies [1,2]. One of the goals was to characterize the sediment throughout the course of the experiment [2]. It could be anticipated that sediment interface roughness would be an important factor for understanding at least some aspects of sediment acoustics. Therefore several means were used to characterize this roughness [2–5]. As shown in Sect. 2, roughness measurements indicate that the sediment interface changed both temporally and spatially.

N.G. Pace and F.B. Jensen (eds.), Impact of Littoral Environmental Variability on Acoustic Predictions and Sonar Performance, 195-202.
© 2002 *Kluwer Academic Publishers.*

Two groups carried out acoustic penetration experiments during SAX99 [6,7]. Experimental data and associated modeling [6] showed that the sediment ripple structure present at the site was the primary sediment feature leading to penetration at grazing angles less than the critical angle (approximately 30°), so called "subcritical" penetration. Other SAX99 experiments quantified backscattering to an in-water source/receiver. Backscatter models [8] were able to predict the scattering levels to reasonable accuracy using the measured sediment roughness. However, the spatial changes of measured roughness makes discrimination between models difficult since higher precision is required. The effects of temporal and spatial changes in interface roughness on penetration and scattering are presented in Sects. 3 and 4.

Detection of a buried target at shallow grazing angles (long ranges) is examined in Sect. 5 for different sediment ripple wavelengths and rms heights. The signal-to-noise ratio (SNR) of a buried target is shown to be very sensitive to these two parameters. Also, the spatial variability of the ripple and small scale roughness directly affects the uncertainty in the predicted SNR.

2   Sediment interface roughness

The sand sediment interface roughness at SAX99 was a superposition of a ripple field that can be modeled with a narrow Gaussian spectrum centered on the mean ripple wave vector and smaller scale roughness with a power law spectrum. These two components are discussed separately.

2.1   Sand Ripples

During SAX99 an autonomous 300 kHz sonar system was deployed that created backscattering images of the sediment interface scattering (see discussion of XBAMS in [1] for details) every 90 minutes from Oct. 6th to Oct. 30th. A storm occurred on Oct 8th that caused a significant change in the ripple field (water depth was about 19 m). Figure 1 shows images of the backscattering (in terms of the Lambert parameter [1]) from the sediment interface generated using scans 36 h apart. The ripple field is clearly seen in both images. The storm changed both the wavelength and direction of the ripples. Spectral analysis of the images in Fig. 1 indicates that, before the storm, the ripple wavelengths averaged 75 cm, but after the storm they averaged 50 cm.

For the period after the storm (when the majority of SAX99 experiments were carried out) a bottom mounted system, IMP for In situ Measurement of Porosity [2,3], allowed measurement of the rms ripple height as well as the power law portion of the roughness spectrum. Figure 2 shows an IMP measurement of a 1 m x 0.15 m section of the sediment interface taken on Oct 18th as well the one-dimensional roughness spectrum derived from that measurement. The rms ripple height was approximately 1 cm. Similar results were found using a bottom mounted optical system [5]. IMP was not deployed before the storm, but a few stereo photographs were taken that indicate that before the storm the rms roughness was slightly higher (an rms height of at least 1.3 cm).

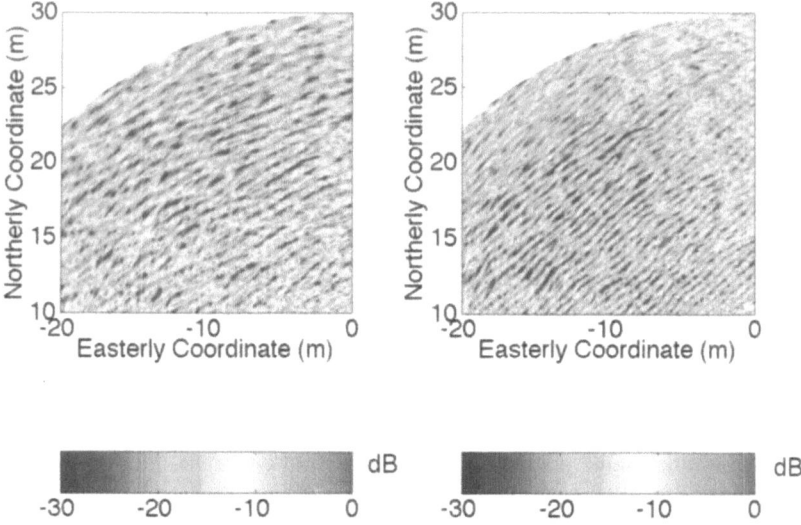

Figure 1. Images of sediment scattering before (left) and after (right) a storm event that caused a change in ripple wavelength and orientation.

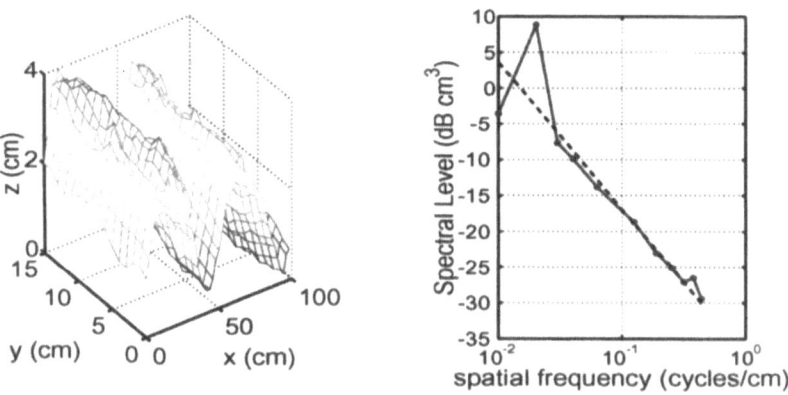

Figure 2. Measurement of a 1 m x 0.15 m section of the ripple field present during SAX99 (left), and one-dimensional spectrum derived from the measurement (right). The enhanced spectral level at the ripple wavelength is evident.

2.2   Power Law Interface Roughness

Measurements of the interface roughness with stereo photography were used to examine spatial variability [3,4]. A two-power law best fit [3] to measured 1-D bottom roughness spectra averaged from four different locations within the SAX99 area is shown in Fig. 3 (heavy dashed line).

In [3] the roughness profiles from which the best fit line was derived were grouped into four sets of 3. Each set of 3 profiles was generated from a single pair of stereo photographs. The 95% confidence limits associated with each set are presented in [3]. The confidence limits from two of the four sets are plotted in Fig. 3. The confidence limits do not overlap over most of the spatial frequency range implying that the roughness realizations at the two locations are not from the same population, i.e., the roughness is spatially nonstationary. This nonstationarity has important ramifications on backscatter model/data comparisons. In particular, estimates of the uncertainty in the roughness spectral level at a site of backscattering measurements must account for this nonstationarity, and these uncertainties need to be used in determining the bounds on model predictions.

For the model/data comparisons discussed in Sect. 4 the model uncertainty bounds were estimated by examining the 95% confidence limits in Fig. 3 relative to the best fit line. When plotting model predictions for comparison with data, the model uncertainty was determined by calculating the dB difference between the best fit line and the most extreme upper and lower 95% confidence curve at the spatial frequency appropriate for the acoustic frequency being examined. (Note that Fig. 3 includes vertical dashed lines indicating the spatial frequencies associated with $10°$ grazing angle backscattering at 20 and 40 kHz.) The uncertainties for scattering at 20 kHz are +2.4 and -2.7 dB and at 40 kHz are +0.7 and -2 dB.

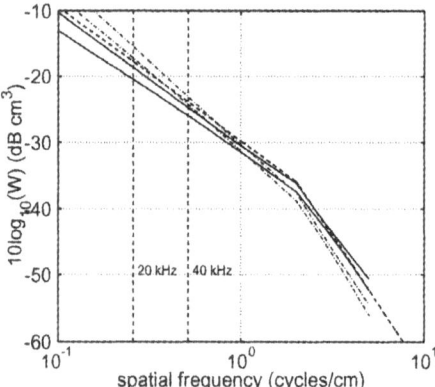

Figure 3. The (slanted) heavy dashed line is the best fit to measured bottom roughness spectra averaged from four different locations within the SAX99 experimental area. The pairs of solid and dash-dot lines indicate the 95% confidence limits of spectra measured at two locations.

3   Signals received via sub-critical penetration at high frequencies

In [6] a simple diffraction theory model using first order perturbation theory and a sinusoidal ripple was developed as a starting point to examine subcritical penetration. It was shown that this model captures much of the behavior seen in penetration experiments. Equation (16) of [6] allows calculation of the transmission across the water/sand interface and propagation to a buried target. Reciprocity allows the same expression to handle transmission of the buried target signal back across the interface to an in-water receiver (to obtain the reciprocal transmission one must multiply the result from Eq. (16) of [6] by the ratio of the water density to the sediment density).

Figure 4 shows the result of a sonar equation calculation for the signal received from a focusing sphere (diameter equal to 25.4 cm) buried such that the top of the sphere is 6.4 cm below the mean water/sand interface. The sand acoustic parameters are given in [6]. The sphere is assumed to have a target strength of -10 dB. A plane wave with a 0 dB re 1 μPa pressure level at the interface is incident at 10° grazing angle with the propagation direction normal to the ripple strike. The receiver is at a horizontal range of 25 meters and is 4.4 meters above the water/sediment interface (resulting in a 10° grazing angle as measured directly above the sphere at the sand/water interface).

The curves in Fig. 4 correspond to the following ripple parameters: (1) ripple wavelength equal to 75 cm, ripple rms height equal to 1 cm, (2) ripple wavelength equal to 50 cm, ripple rms height equal to 1 cm, (3) ripple wavelength equal to 75 cm, ripple rms height equal to 2 cm, (4) ripple wavelength equal to 50 cm, ripple rms height equal to 2 cm.

From Fig. 4 it is easy to see that if the rms ripple height doubles the received signal for a particular ripple wavelength increases by 12 dB. This is because the transmitted intensity across the water/sand interface (the square of Eq. (16) in [6]) is proportional to the second power of the rms height and the signal must traverse the interface twice to get to the receiver. Also, if one decreases the wavelength from 75 to 50 cm (as occurred because of the storm during SAX99) the peak in the received level for a particular rms ripple height increases in magnitude and moves to a higher frequency. An explanation for the frequency behavior can be found from a perturbation theory result used in the simple diffraction theory of [6]. First order perturbation theory predicts [6] that the angle of transmission into the sediment is given by

$$\cos\theta_2 = \frac{c_2}{c_1}(\cos\theta_1 - \frac{\lambda_1}{\lambda_r}) \qquad (1)$$

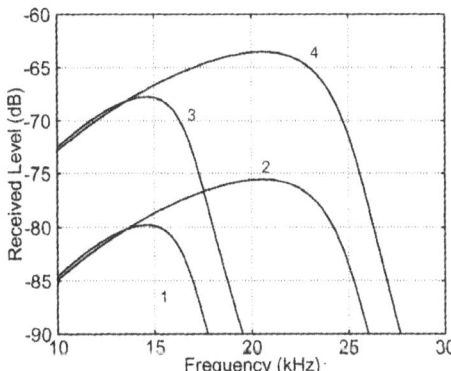

Figure 4. Received pressure levels from a buried sphere for a plane wave incident at a 10° grazing angle onto rippled sand/water interfaces (see text for details).

where c_2 is the sediment sound speed, c_1 is the water sound speed, θ_2 is the transmission grazing angle, θ_1 is the incident grazing angle, λ_1 is the acoustic wavelength in the water, λ_r is the wavelength of the ripple, and attenuation in the sediment has been ignored. Equation (1) yields a real transmission angle only if the right hand side is less than or equal to 1. This imposes a high frequency cutoff above which the acoustic energy is evanescent (θ_2 becomes imaginary) and the received level drops rapidly. A smaller λ_r leads to a higher cutoff frequency. This cutoff phenomenon was seen in SAX99 experimental data [6]. The salient point here is that changes in ripple characteristics (wavelength or height) can lead to vastly different buried target receive levels.

4   Testing backscattering models

Measured interface roughness was used in [8] to test fluid sediment and Biot sediment backscattering models. Results at 20 and 40 kHz are shown in Fig. 5. The dashed lines in each panel are predictions for the upper and lower bounds for the fluid sediment model and the solid lines are the upper and lower bounds for the Biot model. These bounds attempt to account for the spatial variability of roughness as discussed in Sect. 2.2. The asterisks and circles represent data from two different acoustic measurement systems [8]. The uncertainty due to the spatial changes in roughness is large enough that it makes discrimination between models difficult. However, data on sound speed dispersion [9] imply that the Biot model is more appropriate for the SAX99 sand. It has been shown [10] that a fluid model of the sediment with an effective density determined from the Biot model gives backscattering results equivalent to the Biot model. In the signal-to-noise calculations of the next section we have used this effective density model.

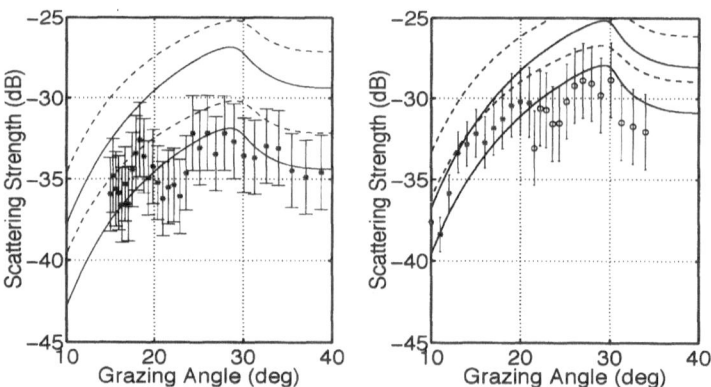

Figure 5. Backscattering from a sand sediment: left, 20 kHz; right, 40 kHz.

5   High frequency buried target detection at sub-critical angles

The variability of ripple characteristics and small scale roughness have a direct effect on the signal-to-noise ratios and their uncertainties for buried target detection using a high resolution sonar (e.g., a Synthetic Aperture Sonar (SAS)). The panels in Fig. 6 were calculated using target signal levels discussed in Sec. 3 and an ensonified area for backscattering of 0.84 m x 0.3 m. This ensonified area is based on more sophisticated simulations [11] carried out to determine the region of the sand/water interface that contributes significantly to sub-critical penetration for the geometries of interest; we believe this method leads to a more realistic estimate for SNR than if the sonar resolution cell is used directly to determine ensonified area [11].

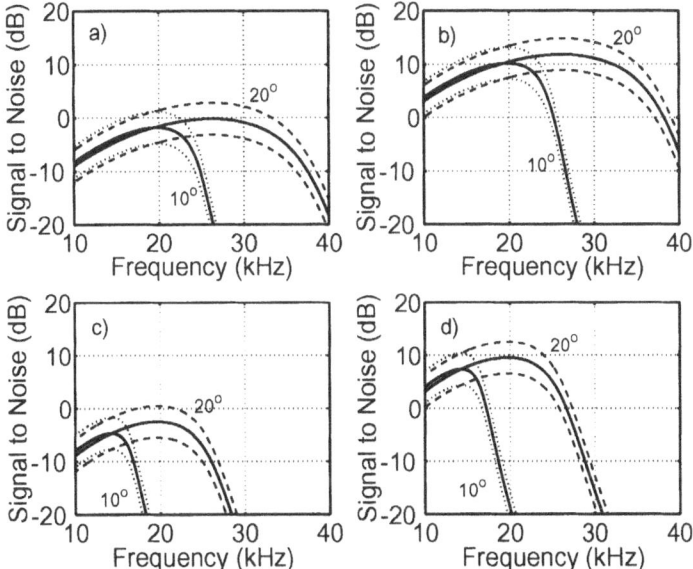

Figure 6. Signal-to-noise ratios for a buried target under differing interface conditions and for two different grazing angles. Ripple conditions: a) ripple wavelength: 50 cm, rms ripple height: 1 cm, b) ripple wavelength: 50 cm, rms ripple height: 2 cm, c) ripple wavelength: 75 cm, rms ripple height: 1 cm, d) ripple wavelength: 75 cm, rms ripple height: 2 cm. The heavy lines in each panel use the mean backscattering level predicted by the Biot model and the lighter lines represent uncertainty bounds based on spatial variability of roughness at scales important for backscattering at 20 kHz.

Comparing panel (a) with (b) or (c) with (d) demonstrates the sensitivity to ripple height. Likewise, comparing (a) with (c) or (b) with (d) indicates the importance of ripple wavelength. As an example, for a high resolution sonar operating at 20 kHz and using a $10°$ incident angle, the SNR could vary from +10 dB (panel b) to less than -20 dB (panel c) over the conditions used in Fig. 6. Viewed another way, panel (b) implies that detection of the buried sphere is possible down to $10°$ when the ripple wavelength is 50 cm but panel (d) implies detections only down to $20°$ when the ripple wavelength is 75 cm. Finally, the spatial variability of the power law roughness implies these predicted SNRs have an uncertainty of about +/- 3 dB (spatial variability in ripple height would increase this uncertainty further).

From an operational standpoint the sensitivity of detection to roughness conditions suggests the need to determine these parameters during mine hunting operations. Obtaining the ripple wavelength is relatively easy; indeed, SAS images collected during

SAX99 allowed ripple wavelength to be determined. Also, backscattering from the interface is directly obtainable from sonar images if the systems are absolutely calibrated. Obtaining the ripple rms height and the spatial variation of ripple height are more of a challenge. Current work involves use of high resolution backscattering images (e.g., Fig. 1) combined with backscatter modeling to determine the ripple slopes from which the ripple height could then be found.

Finally, though the first order perturbation theory used here suggests a quick reduction in signal above the cutoff frequency, both experimental evidence and theoretical modeling indicate that higher order perturbation effects can contribute to the transmitted signal and allow for continued penetration above the cutoff frequency. Further work is being carried out to investigate these effects.

Acknowledgements

This work was supported by the U.S. Office of Naval Research, Code 321OA.

References

1. Thorsos, E. I., *et al.,* An overview of SAX99: Acoustic measurements, *IEEE J. Oceanic Eng.* **26**, 4–25 (2001).
2. Richardson, M.D., *et al.,* An overview of SAX99: Environmental considerations, *IEEE J. Oceanic Eng.* **26**, 26–54 (2001).
3. Briggs, K.B., Tang, D. and Williams, K.L., Characterization of interface roughness of rippled sand off Fort Walton Beach, Florida, *IEEE J. Oceanic Eng.* (to be published 2002).
4. Lyons, A.P, Fox, W.L.J., Hasiotis, T. and Pouliquen, E., Characterization of the two-dimensional roughness of wave-rippled sea floors using digital photography, *IEEE J. Oceanic Eng.* (to be published 2002).
5. Moore, K.D. and Jules, J.S., Time-evolution of high-resolution topographic measurements of the seafloor using a 3D laser line scan mapping system, *IEEE J. Oceanic Eng.* (to be published 2002).
6. Jackson, D.R., Williams, K.L., Thorsos, E.I. and Kargl, S.G., High-frequency subcritical acoustic penetration into a sandy sediment, *IEEE J. Oceanic Eng.* (to be published 2002).
7. Chotiros, N.P., Smith, D.E and Piper, J.N., Refraction and scattering into a sandy ocean sediment in the 30 to 40 kHz band, *IEEE J. Oceanic Eng.* (to be published 2002).
8. Williams, K.L., Jackson, D.R., Thorsos, E.I., Tang, D. and Briggs, K.B., Acoustic backscattering experiments in a well characterized sand sediment: Data/model comparisons using sediment fluid and Biot models, *IEEE J. Oceanic Eng.* (to be published 2002).
9. Williams, K.L., Jackson, D.R., Thorsos, E.I. and Tang, D., Comparison of sound speed and attenuation measured in a sandy sediment to predictions based on the Biot theory of porous media, *IEEE J. Oceanic Eng.* (to be published 2002).
10. Williams, K.L., An effective density fluid model for acoustic propagation in sediments derived from Biot theory, *J. Acoust. Soc. Am.* **110**, 2276–2281 (2001).
11. Piper, J.E., Commander, K.W., Thorsos, E.I. and Williams, K.L., Detection of buried targets using a synthetic aperture sonar, *IEEE J. Oceanic Eng.* (to be published 2002).

VARIABILITY OF BOTTOM BACKSCATTERING STRENGTH IN THE 10–500 KHZ BAND AT SHALLOW GRAZING ANGLES

NICHOLAS P. CHOTIROS

Applied Research Laboratories, The University of Texas at Austin,
P.O. Box 8029, Austin TX 78713-8029, USA.
E-mail: chotiros@arlut.utexas.edu

Bottom backscatter is often the dominant component of reverberation for sonars operating in the band from 10 to 500 kHz in littoral waters. A grazing angle of 10° is at the intersection of the range of angles encountered by operational minehunting sonars and the angles reported in most experimental studies. Analysis of the measured values provides an indication of the processes involved. The effectiveness of models may be gauged by comparison with measured values.

1 Introduction

The variations in the first order statistic, that is the backscattering strength, are fundamental. The variability in question is the difference in backscattering strength between different areas of the seafloor that seem to be similar. The similarity may be quantified in several ways. The simplest and most popular metric is mean grain size. The expectation has been that areas with the same mean grain size should have the same backscattering strength at the same acoustic frequency, but the reality is more complicated. In this study, measured backscattering strength will be examined as a function of frequency and grain size, and from which simple deductions will be made. The data are compared with models to gauge their fidelity.

2 Bottom backscattering strength

Backscattering strength is defined as the mean backscattered intensity referenced to a unit distance, produced by a unit area of the bottom, in any direction, in response to an incident wave of unit intensity. It is often plotted as a function of grazing angle.

Most sonar applications, particularly in minehunting, involve grazing angles 10° and below, because of the need to maximize detection range within a limited water depth. Most measurements of backscattering strength reported in the literature are made at grazing angles 10° and above because of the difficulty in obtaining reliable measurements at smaller angles. Therefore, this angle lies at the intersection of the needs of the applications and the availability of measurements. The reported values [1–22] are shown in Fig. 1(a) and the key to the data source is shown in Fig 1(b).

The data points cover published backscattering strength values up to 1997. Although the collection is not completely up to date, the histogram adequately shows a mean and standard deviation. The global mean of –34 dB is significant, because it appears to be applicable across the band.

N.G. Pace and F.B. Jensen (eds.), Impact of Littoral Environmental Variability on Acoustic Predictions and
Sonar Performance, 203-210.

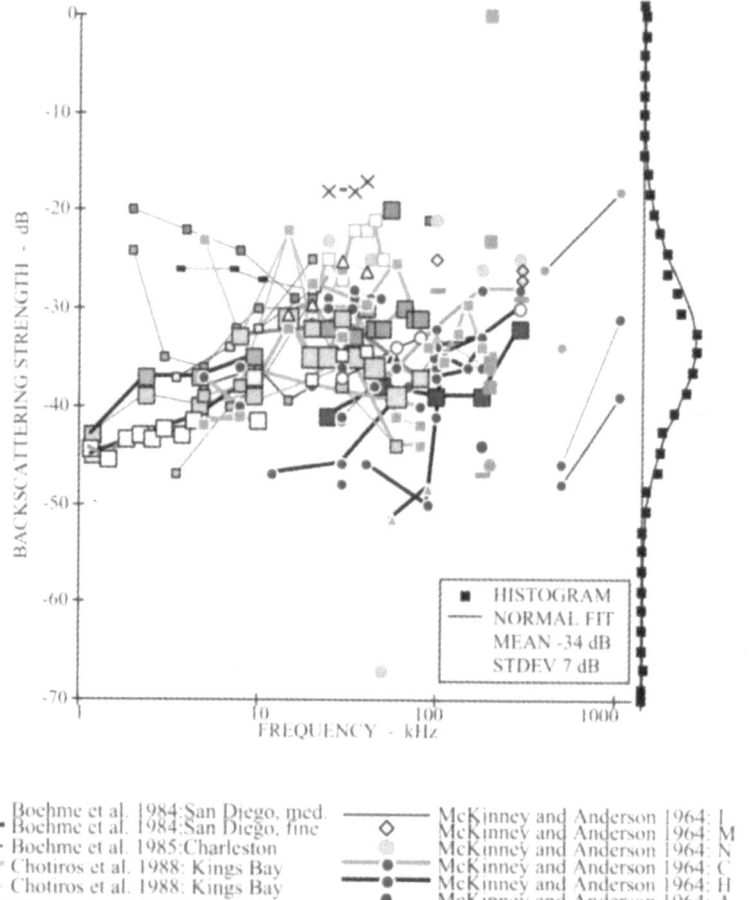

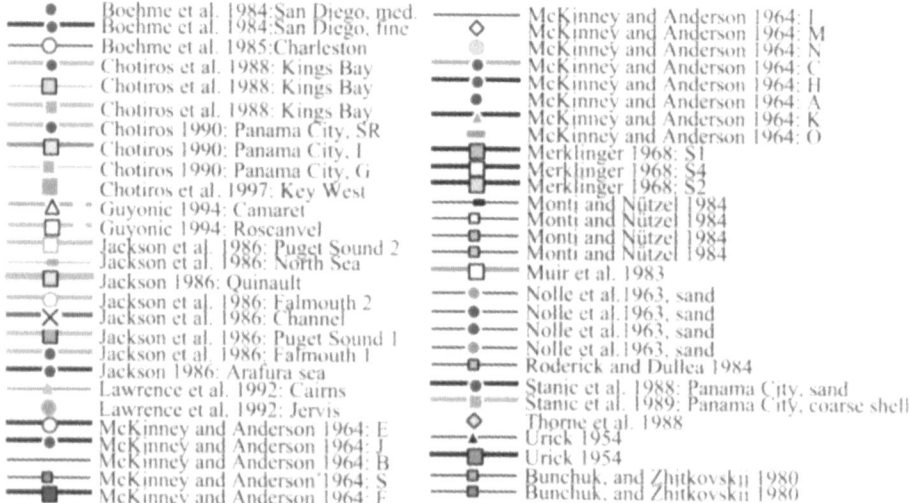

Figure 1. (a) Measured bottom backscattering strength at 10° grazing angle as a function of frequency, histogram and the best-fit normal distribution curve. (b) Key to data sources.

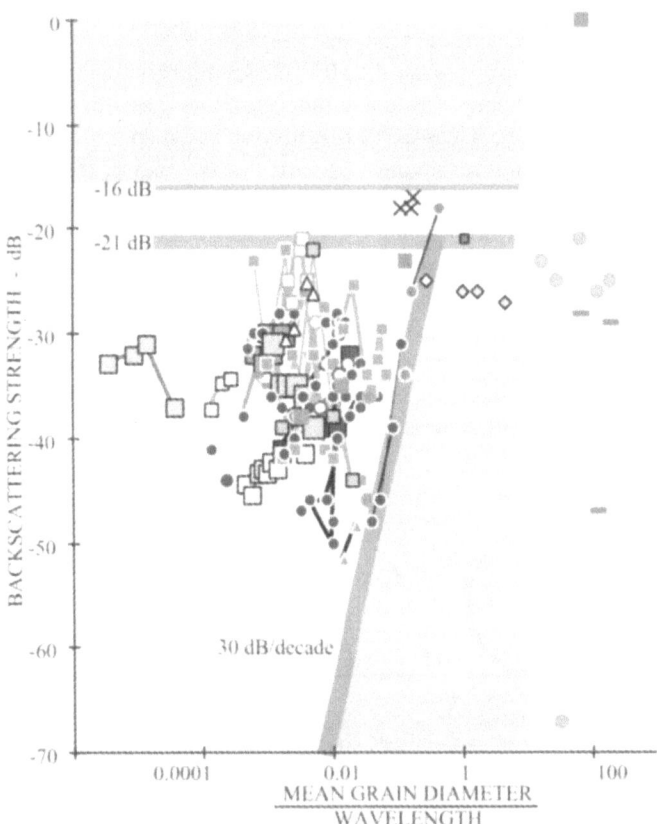

Figure 2. Values of bottom backscattering strength at 10° grazing angle as a function of normalized mean grain size.

There are measurement sites and frequency ranges in which backscattering strength increases with frequency, but they are counter-balanced by others that have the opposite trend. To show the trends, data points from the same site are connected. With very few exceptions, the values lie between –20 and –50 dB. A normal distribution curve with a standard deviation of 7 dB appears to fit the histogram.

To examine the connection with sediment properties, the backscattering strength was plotted against the mean grain size of the sediment, normalized by the acoustic wavelength in water, as shown in Fig. 2. Laboratory measurements [16], made with graded sands of various mean grain diameters and a flat interface, follow a power law with a slope of 30 dB per decade of normalized grain size. It appears to be a lower bound for all data points with normalized grain sizes less than 1, and it represents the intrinsic backscattering strength of the granular structure. Since all the other measurements, most of which were taken in situ, give values that are significantly higher, it must be concluded that other factors dominate the backscattering process in ocean sediments. These factors include perturbations caused by hydrodynamic forces and biological activity, which result in roughness of the interface and volume inhomogeneities. In many cases, biological activity causes sediment grains to be cemented together into larger pellets, giving the sediment a larger effective acoustic grain size [23].

At normalized grain sizes greater than 0.1, the scattering strength trend levels off at a saturation value. It is possible to estimate the saturation value for certain simple cases. With reference to Fig. 3(a), if Lambert's rule may be assumed, i.e. the average scattered signal intensity is isotropic in azimuth and varies as the sine of the elevation angle θ_2, then conservation of energy requires that the saturation value should not exceed −21 dB. Most of the data points at normalized grain sizes less than 10 appear to fall below this upper bound, with a few exceptions. Of those that exceeded this upper bound, one was a gravel sediment in the English Channel measured at frequencies between 20 and 40 kHz [8] and the other a sand sample in the laboratory at 1 MHz [16]. With reference to Fig. 3(b), if the average scattered intensity were isotropic in elevation and azimuth, then conservation of energy would limit the saturation value to −16 dB. This appears to cover the values of all data sets at normalized grain sizes up to 10.

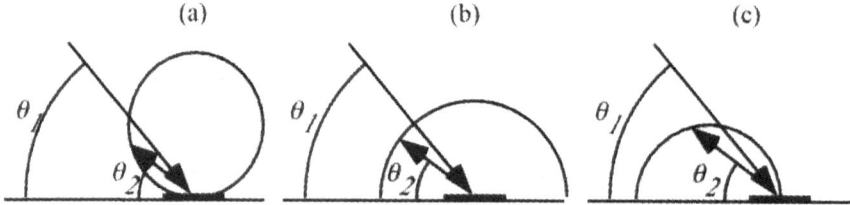

Figure 3. Illustration of (a) Lambert's rule, (b) isotropic and (c) directional scattering.

At normalized grain sizes greater than 10, the situation is complicated by outliers in both directions. A directional scattering surface may account for such a wide range of variations. With reference to Fig. 3(c), if the backscattered intensity followed a directional beam pattern that was directed back toward the sonar, then it is possible to obtain values in the region of 0 dB, which is consistent with the highest outlier. The process is one of specular reflection from a large perpendicular facet rather than scattering from a randomly rough surface. Conversely, if the sound beam made oblique angles to all of the facets, then the backscattered intensity would be extremely small and consistent with the lowest outliers. Both the high and low outliers were from areas containing solid rock [12] and coral reefs [4].

The data are separable into three regimes. In the first regime, at normalized grain sizes less than 0.1, the values lie between the isotropic scattering upper bound, and the 30 dB per decade lower bound. In the second regime, at normalized grain sizes between 0.1 and 10, the values lie between the isotropic scattering upper bound and an empirical lower bound of −30 dB. In the third regime, at normalized grain sizes above 10, the scatterers may be either large grains or facets. In the former, the upper and lower bounds are similar to that of the second regime. In the latter, the range of possible values will be very large depending on the alignment of the facets.

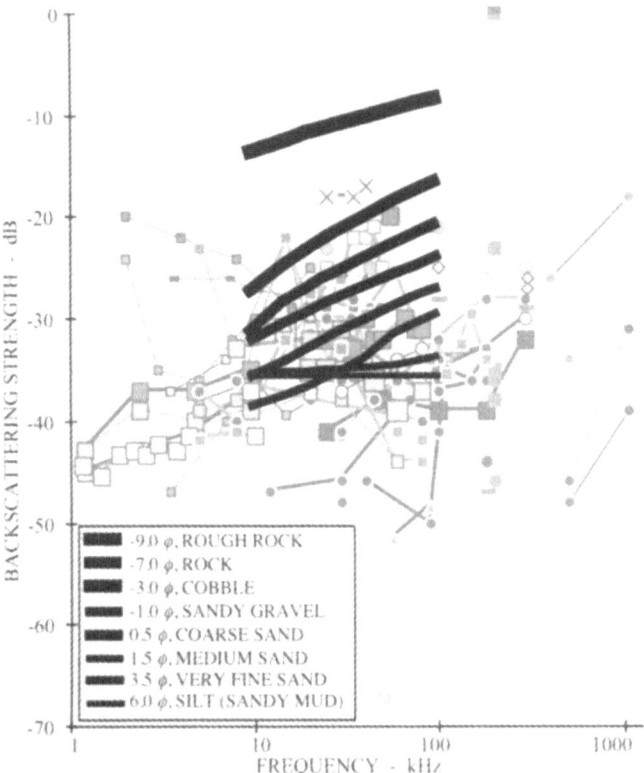

Figure 4. Comparison of generic bottom backscattering strength curves from APL-UW9407 with measured values as a function of frequency.

3   Comparison with APL/UW 9407

Since the collection of models in APL-UW9407 [24] has been accepted into the Oceanographic and Atmospheric Master Library (OAML), it is worth examining its bottom backscattering strength model predictions in the light of the data available. The model addresses the 10 to 100 kHz band and grazing angles greater than 8°. It is completely defined by six input parameters. In practice, one is often unable to obtain all six parameter values. For this eventuality, the authors have provided a set of 23 generic bottom types, and may be invoked by name (e.g. rock, gravel, or sand) or by a numerical mean grain size (ϕ). Each one contains a preprogrammed set of parameter values that were judged to be typical. For comparison, the backscattering strength curves from a broad sampling of generic bottom types are superimposed on the measured values as a function of frequency in Fig. 4. The generic curves fall on top of a large proportion of measured values, but they do not cover the lower range of measured values. The generic curves show backscattering strengths that are constant or increasing with frequency, but never decreasing with frequency.

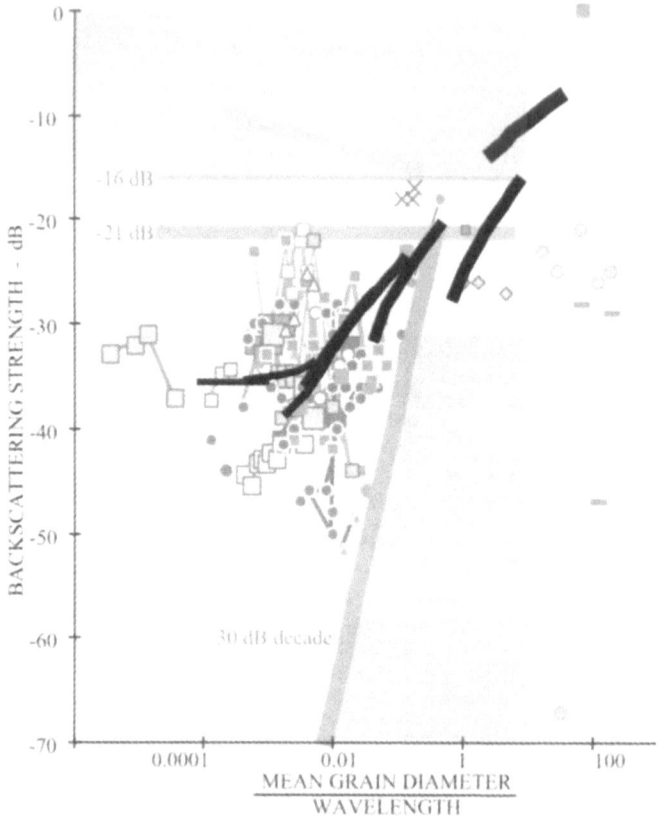

Figure 5. Comparison of generic bottom backscattering strength curves from APL-UW9407 with measured values as a function of normalized grain size.

In Fig. 5, the generic curves are superimposed on the measured values as a function of normalized grain size. For normalized grain sizes less than 0.1, the generic curves are clustered in a narrow region in the middle of the range of measured values. In this sense, the generic curves may be considered typical. However, they are clustered in a very narrow region, representing only a small subset of the trends manifested in the measured data. For example, none of the generic curves can represent the instances where the backscattering strength drops below –40 dB, rises above –25 dB, or where the backscattering strength decreases with normalized grain size, of which there are quite a few. It is evident that the generic curves represent only a small subset of the backscattering strength trends that are found in the database. For normalized grain sizes between 0.1 and 10, the generic curves lie between –30 and –16 dB in agreement with the data. For normalized grain sizes greater than 10, the generic curves continue their upward trend. As deduced earlier, the backscattering in this regime may be due to large grains or facets. The generic curves overestimate the range of values of the former, and are unable to track the wide variations of the latter.

4   Conclusions

The database of bottom backscattering strength values, at a grazing angle of 10°, was examined. The global average is –34 dB with a standard deviation of 7 dB. The values appeared to follow a normal distribution. Overall, no significant frequency dependent trends were discernible, but different frequency dependent trends exist at individual sites.

When plotted against normalized grain size, the data are separable into three distinct regimes. (1) At normalized grain sizes below 0.1, scattering is dominated by extrinsic features, such as interface roughness and inclusions. The values lie between the isotropic scattering upper bound, and the intrinsic scattering lower bound. (2) Between 0.1 and 10, scattering is dominated by the intrinsic scattering strength, and the values lie between the isotropic scattering upper bound, and an empirical lower bound. (3) Beyond 10, scattering may be due to large grains or facets. In the former, the values fall within the same bounds as (2), and in the latter the values have a very wide range of variation.

The generic bottom backscattering strength curves provided by APL-UW9407 were compared with the measured data. The generic curves are most successful in regime (2). In (3), the generic curves are unable to track the facet scattering process, and overestimate the backscattering strength of the large grain scattering process. In (1), the generic curves occupy a small region in the middle of the range of measured values. In this sense, the generic curves are typical, but they represent only a small subset of the measured data. For this reason, they are not suitable for inversion applications, in which a best-fit generic bottom type is inverted from measured reverberation data. For inversion purposes, the six input parameter values should be independently adjusted.

Acknowledgements

This work is sponsored by the Office of Naval Research (ONR), Code 321 OA, under the management of J. Simmen.

References

1. Boehme, H., Chotiros, N.P. and Churay, D.J., High frequency environmental acoustics: bottom backscattering results. Technical Report No. (ARL-TR-85-30), Applied Research Laboratories, The University of Texas at Austin (1985).
2. Boehme, H., Chotiros, N.P., Rolleigh, L.D., Pitt, S.P., Garcia, A.L., Goldsberry, T.G. and Lamb, R.A., Acoustic backscattering at low grazing angles from the ocean bottom, Part I. Bottom backscattering strength, *J. Acoust. Soc. Am.* **77**(3), 962–974 (1985).
3. Bunchuk, A.V. and Zhitkovskii, Yu., Sound scattering by the ocean bottom in shallow-water regions (review), *Sov. Phys. Acoust.* **26**(5), (1980).
4. Chotiros, N.P., Altenburg, R.A. and Piper, J.N., Analysis of acoustic backscatter in the vicinity of the Dry Tortugas, *Geo-Marine Letters* **17**, 325–334 (1997).
5. Chotiros, N.P. and Boehme, H., Analysis of bottom backscatter data from the Kings Bay Experiment. Technical Report ARL-TR-88-6, Applied Research Laboratories, The University of Texas at Austin (1988).
6. Chotiros, N.P., High frequency bottom backscattering: Panama City Experiment. Technical Report ARL-TR-90-22, Applied Research Laboratories, The University of Texas at Austin (1990).
7. Guyonic, S., SACLANT Undersea Research Centre MCM Workshop, 20–22 Sep. 1994.

8. Jackson, D.R., Baird, A.M., Crisp, J.J. and Thompson, P.A.G., High-frequency bottom backscatter measurements in shallow water, *J. Acoust. Soc. Am.* **80**(4), 1188–1199 (1986).

9. Jackson, D.R., High frequency bottom backscattering strength at the Quinault range. APL-UW-8-86, Applied Physics Laboratory, University of Washington (1986).

10. Jackson, D.R., High frequency bottom scattering in the Arafura sea. APL-UW-5-86, Applied Physics Laboratory, University of Washington (1986).

11. Lawrence, T.N., *et al.*, Acoustic bottom scattering measurements at high frequencies, presented at the Shallow Water Undersea Warfare Symposium, Naval Research Laboratory/Stennis Space Center, Feb. 1992.

12. McKinney, C.M. and Anderson, C.D., Measurement of backscattering of sound from the ocean bottom, *J. Acoust. Soc. Am.* **36**(1), 158–163 (1964).

13. Merklinger, H.M., Bottom reverberation measured with explosive charges fired deep in the ocean, *J. Acoust. Soc. Am.* **44**(2), 508–511 (1968).

14. Monti, J.M. and Nutzel, B., Acoustic scattering from the sea floor. Tech Doc. 7293, Naval Underwater Systems Center (1984).

15. Muir, T.G., Thompson, L.A., Shooter, J.A. and DeMary, T.E., Frequency response measurements on backscattering from a shallow sea floor using a parametric source. In *Ocean Reverberation*, edited by D.D. Ellis, J.R. Preston and H.G. Urban (Kluwer Academic Press, The Netherlands, 1993) pp. 91–96.

16. Nolle, A.W., Hoyer, W.A., Mifsud, J.F., Runyan, W.R. and Ward, M.B., Acoustic properties of water filled sands, *J. Acoust. Soc. Am.* **35**(9), 1394–1408 (1963).

17. Roderick, W.I. and Dullea, R.K., High resolution bottom backscatter measurements. Technical Document 7181, Naval Underwater Systems Center (1984).

18. Smailes, I.C., Bottom reverberation measurements at shallow grazing angles in the NE Atlantic and Mediterranean sea, *J. Acoust. Soc. Am.* **64**(5), 1482–1486 (1978).

19. Stanic, S., Briggs, K.B., Fleischer, P., Sawyer, W.B. and Ray, R.I., High-frequency acoustic backscattering from a coarse shell bottom, *J. Acoust. Soc. Am.* **85**(1), 125–136 (1989).

20. Stanic, S., Briggs, K., Fleisher, P., Ray, R. and Sawyer, W.B., Shallow water high frequency bottom scattering off Panama City, Florida, *J. Acoust. Soc. Am.* **83**, 2134–2144 (1988).

21. Thorne, P.D., Pace, N.G. and Al-Hamdani, Z.K.S., Laboratory measurements of backscattering from marine sediments, *J. Acoust. Soc. Am.* **84**(1), 303–309 (1988).

22. Urick, R.J., The backscattering strength of sound from a harbor bottom, *J. Acoust. Soc. Am.* **26**(2), 231–235 (1954).

23. D'Andrea, A.F., Lopez, G.R., Nitrouer, C.A. and Wright, L.D., Fecal pellets of Abra alba as traces of sediment movement in Eckernförde Bay. In *Proc. Workshop on Modeling Methane-Rich Sediments of Eckernförder Bay*, Eckernförder, 26–30 June 1995.

24. APL-UW high frequency ocean environmental acoustic models handbook. Technical Report APL-UW TR 9407, Applied Physics Laboratory, University of Washington (1994).

PREDICTING SCATTERED ENVELOPE STATISTICS OF PATCHY SEAFLOORS

ANTHONY P. LYONS AND DOUGLAS A. ABRAHAM

The Pennsylvania State Univ., Applied Research Lab., P.O. Box 30, State College, PA 16804
E-mail: apl2@psu.edu

ERIC POULIQUEN

SACLANT Undersea Research Centre, Viale San Bartolomeo 400, 19138 La Spezia, Italy
E-mail: pouliq@saclantc.nato.int

Local hydrodynamic or biological influences often produce seafloors in shallow water that consist of differing types of material. The scattering properties from the components of these kinds of seafloors may have a complicated relationship in terms of their frequency dependence and/or angular response. Consequently, this relationship directly influences the angular and frequency response of the scattered envelope distributions. The probability distribution function (PDF) for a scattering scenario such as this is not easy to obtain analytically. However, a recently developed model for a patchy seafloor with a single dominating component [1] allows for numerical analysis of the envelope PDF for more complicated seafloors through the use of Hankel transforms of the joint characteristic function (JCF) of the complex envelope. The JCF is straightforward to construct for complicated patchy seafloors. In this study, a direct link between environmental parameters and the envelope distributions of backscatter is developed. The influence of the relative scattering properties of the seafloor patches on the scattered envelope statistics will be examined in detail.

1 Introduction

The envelope distribution resulting from roughness scattering from a seafloor with uniform properties is expected to bc Raylcigh as the central limit theorem holds resulting in Gaussian reverberation. Local hydrodynamic or biological influences, however, often produce seafloors in shallow water that consist of several different types of material. Seagrass and shellfish are examples of scatterers that often do not exist uniformly on the seabed but are distributed in patches of varying density. Examples of heterogeneous seafloors are shown in the photographs displayed in Fig. 1. Patchiness in the scattering properties commonly found in shallow water suggests that the Rayleigh distribution model might not always be appropriate, especially when the area ensonified by the transmit and receive beams is not large enough to encompass enough of the patches of differing scattering strength. The acoustic expression of non-uniform seafloors manifests itself in the form of non-Rayleigh or heavy tailed envelope distribution functions and is termed clutter. Clutter affects the detection of targets in acoustic imagery by increasing the probability of false alarm. In this respect clutter is at least as important, if not more important, a problem to object detection than mean scattered levels in shallow water. To ameliorate the effect of clutter through adaptive systems or signal processing algorithms

N.G. Pace and F.B. Jensen (eds.), Impact of Littoral Environmental Variability on Acoustic Predictions and Sonar Performance, 211-218.

it will first be necessary to understand the properties and causes of non-Rayleigh scattered envelope statistics at high frequency.

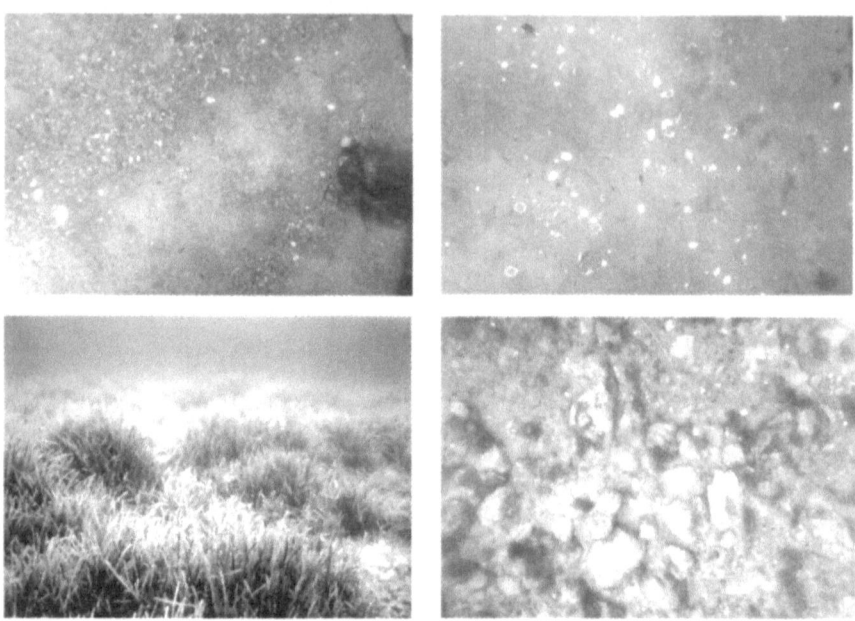

Figure 1. Examples of heterogeneous seafloors: clockwise from top left gravel, shell, rocks, *Posidonia oceanica* seagrass.

The statistical character of reverberation is a function of the sonar system parameters of beamwidth, bandwidth and frequency and environmental properties such as the number, size distribution, and scattering strength of patches. The relationship between the scattering strength of the patches and that of the surrounding seafloor can be a function of frequency and grazing angle and so directly influences the angular and frequency response of the scattered envelope distributions. In order to improve predictive capabilities for high-frequency acoustic systems operating in shallow water areas that have spatially heterogeneous seafloors it is imperative to link the scattered envelope distributions to seafloor scattering models. Unfortunately, there are very few physics-based models that link a description of the environment and sonar system to the probability density function of the clutter induced matched filter envelope output. Additionally, these models, which assume an infinite number of scatterers [2], do not retain the pertinent description of the environment that would allow prediction of how the envelope PDF changes as a function of sonar system parameters or is affected by varying environmental scattering mechanisms. A recently developed model for a seafloor consisting of a finite number of scattering patches [1] allows for numerical analysis of the envelope PDF for more complicated seafloors through the use of Hankel transforms of the joint characteristic function (JCF) of the complex envelope. The JCF is straightforward to construct for complicated patchy seafloors allowing a direct link between environmental parameters and the envelope distributions of backscatter to be developed. Using these concepts, the influence of the relative scattering properties of the seafloor patches on the angular and frequency response of the scattered envelope statistics will be examined in detail for several example scenarios in this paper.

2 Statistical model for scattering patches on a Gaussian background

Abraham and Lyons have recently developed a theoretical model for scattered envelope distributions resulting from interface scattering from a seafloor comprised of patches [1]. Diffuse scattering from the interface produces a Gaussian distributed return from each patch by virtue of the central limit theorem, with power proportional to the patch area and backscattering coefficient. Now assume that there are n such patches within a resolution cell of the sonar system. The clutter component of the complex envelope of the received signal in this resolution cell (i.e., after beamforming and matched filtering) may then be represented as

$$R = \sum_{i=1}^{n} \sqrt{A_i} Z_i \qquad (1)$$

where A_i is the area of the ith patch and Z_i is a zero-mean, complex, Gaussian random variable with variance σ^2. In the model of Abraham and Lyons it was assumed that the patches dominated the background scattering and were characterized as having an exponentially distributed area

$$A_i \sim f(a) = \frac{1}{\mu} e^{-\frac{a}{\mu}} \qquad (2)$$

where μ is the average patch area. Assuming that A_i and Z_i are independent of each other and the responses of the other patches, the JCF of R is shown to be [1]

$$\Phi_K(\omega, \gamma) = \frac{1}{\left[1 + \frac{1}{4} \mu \sigma^2 (\omega^2 + \gamma^2) \right]^n} \qquad (3)$$

This turns out to be of the same form as that for the JCF of the K-distribution [1] with shape parameter $\alpha = n$, the number of patches in a resolution cell, and scale parameter $\lambda = \mu \sigma^2$, the average area of each patch times the backscattered power per unit area of each patch. The resultant sum is non-Gaussian owing to the random power of each component even though each patch produces a Gaussian response.

Using the equivalence of the matched filter envelope for scattering from patches (with no contribution from background scattering) to a K-distribution, the complex envelope in the case including background scattering can be described in a manner analogous to adding a Gaussian target to K-distributed reverberation [3]. The complex envelope of the reverberation is now written as the sum of the contribution of the scattering patches and of the Gaussian background

$$R = Z_0 + \sum_{i=1}^{n} \sqrt{A_i} Z_i \tag{4}$$

The background seafloor scattering (Z_0) can be thought of being composed of a multitude of scatterers and is therefore complex Gaussian distributed with zero mean owing to the central limit theorem. The CDF for the combined patch and background-scattered envelope can be found by using the JCF of the complex envelope [3]. The advantage of using JCFs is that the JCF of the combined patch plus background scattering is the product of the individual JCFs. The JCF for the patch scattering is given by Eq. (3) and the JCF for the Gaussian background with power λ_0 would be

$$\Phi_B(a) = e^{-\frac{\lambda_0}{4}a} \tag{5}$$

and the JCF, Φ_A, for the combined background reverberation plus patch scattering, would be the product of Φ_K given by Eq. (3) and Φ_B given by Eq. (5). The cumulative distribution function (CDF) can be obtained from the JCF via a Hankel transform of order one,

$$F_Y(y) = \sqrt{y} H_1 \left\{ \sqrt{y}, \frac{1}{x} \Phi_A(x^2) \right\} \tag{6}$$

A useful indicator of the non-Rayleigh behavior of scattering is the scintillation index given by

$$\sigma_I^2 = \frac{\mu_2 - \mu_1^2}{\mu_1^2} \tag{7}$$

where μ_1 and μ_2 are the first and second moments of $Y = |R|^2$. Assuming that the form of the complex matched filter is described by Eq. (4), the first and second moments can easily be derived (b.d.) as $\mu_1 = \lambda_0 + \alpha\lambda$ and $\mu_2 = 2\mu_1^2 + 2\alpha\lambda^2$ yielding the scintillation index

$$\sigma_I^2 = 1 + \frac{2\alpha\lambda^2}{(\lambda_0 + \alpha\lambda)^2} \tag{8}$$

Experimentally determined values of high frequency reverberation scintillation index from Lyons and Abraham [4] will be compared later to predicted values.

3   Seafloor patch scattering models

The backscattered signal from a sonar resolution cell may consist of contributions from a number of surface features or patches. These include surface scattering from a variety of bare sediment types, such as sands or gravels, and volume scattering from various types of vegetation or shell distributions of varying density. In this section, backscattering coefficient models will be described for the individual components of scattering: rough sand or gravel with homogeneous roughness and impedance properties and dense, homogeneous shell distributions or *Posidonia oceanica* seagrass patches. We are interested in effects of the relative scattering from the different patches over the whole angular range for backscatter for frequencies from 10–100 kHz.

The model used in this study for scattering from a rough sand interface is a combination of a perturbation approximation surface scattering model for low grazing angle regimes and a Kirchhoff approximation for steep angles (i.e., near $\theta = 90^{\circ}$). As the models and the assumptions used in their derivation are discussed in Lyons, et al. [5], we will give only a brief description of the required model inputs here. The surface scattering model requires the statistical properties of the seafloor roughness, which is assumed to exhibit power-law spectra with isotropic statistics:

$$W(\mathbf{K}) = \Gamma K^{-\xi} \tag{9}$$

where $\mathbf{K}$ is a 2-D wavenumber vector with magnitude equal to the wavenumber K, Γ is the spectral strength, and ξ is the spectral exponent. The acoustic properties of density ratio, compressional velocity ration are also required for the interface scattering model as is the compressional loss parameter, which is related to the attenuation coefficient. This model will also be used for scattering from a gravel interface.

The model used to estimate scattering from a shell-covered seafloor is essentially that described in Stanton [6]. The scattering coefficient as a function of grazing angle for this model is given by

$$s_A(\theta) = F\left(\frac{\mathfrak{R}_{12}^2}{4\pi}\right) \sin^2\theta \tag{10}$$

where $\mathfrak{R}_{12}^2$ is the reflection coefficient and $\mathfrak{R}_{12}^2 = (\rho c - 1)/\rho c + 1)$, where ρ and c are the mass density and sound speed of the shell material, respectively, normalized by the corresponding quantities for the surrounding water. F is the packing factor and is equal to the fraction of the seafloor covered by shells. The angular dependence of seafloor scattering for shell-covered areas is included via Lambert's law in the model. High-frequency scattering data from a shell-covered seafloor presented in Lyons and Abraham [4] support the typical $\sin^2\theta$ dependence predicted by the model. The scattering coefficient for the shell-covered seafloor scattering model is frequency independent as long as the packing factor remains constant.

Due to a lack of understanding of the exact mechanisms causing scatter, backscatter from *Posidonia oceanica* seagrass will be estimated using a simple empirical model instead of a physical model, which is given by the expression:

$$s_B(\theta) = \psi \sin\theta \tag{11}$$

where ψ is the strength of the return at normal incidence. Implicit in this model is the assumption of uniform scatter in all directions. While this model is unrealistic for surfaces, it is often used for volume scattering from fish or bubbles (a probable mechanism for scattering from seagrasses). We also assume that the scattered level is independent of frequency. Evidence for the frequency independence of backscattering from *Posidonia oceanica* over the range of 20–100 kHz as well as the $\sin\theta$ dependence can be found in Lyons and Pouliquen [7].

4   Results and discussion

Scattering strength calculations based on the individual scattering models presented above are presented in Fig. 2. Parameters from Stanic *et al.* [8] are used as inputs to model sand interface scattering and from Stanic, et al. [9] for gravel interface scattering with the values of spectral strength and spectral exponent changed to 0.015 and 3.0 respectively (a statistically rougher seafloor than in reference [9]). The parameter ψ was set to 0.01 in the seagrass scattering model to match data presented in Lyons and Pouliquen [7] while the parameter combination $F\Re_{12}^{2}/4\pi$ was set to 0.03 in the shell-covered seafloor model to match data presented in Lyons and Abraham [4].

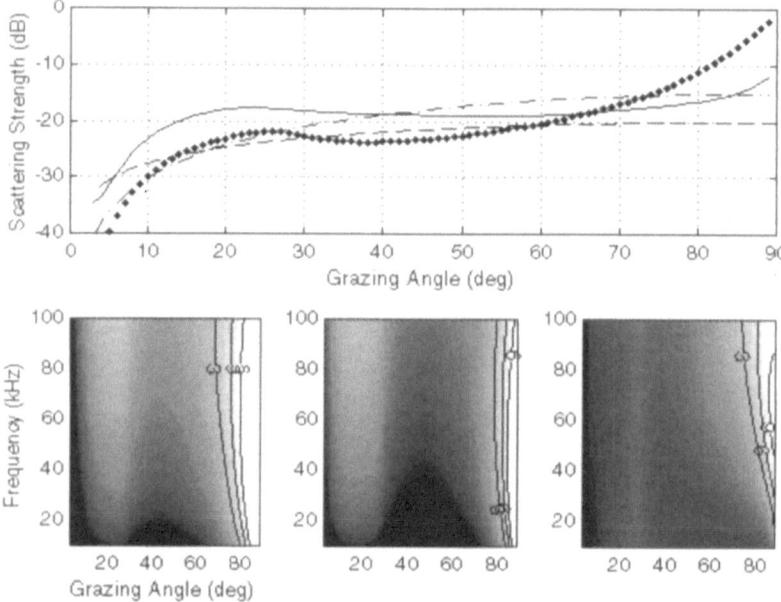

Figure 2. Top: Predictions of 80 kHz scattering strength for sand (diamonds), shell-covered (dash-dot line), gravel (solid line), and seagrass-covered (dashed line) seafloors. Bottom: Differences in scattering strength versus grazing angle and frequency between sand and: seagrass-covered (left), shell-covered (middle), and gravel (right) seafloors. Larger values are lighter on these plots and contour lines of 3, 6 and 9 dB are also shown.

For the examples used in the present study, the resolution cell consists of a combination of only two of the four possible surface types. In the following examples a sand bottom was chosen to be the scattering patch type. The bottom panels of Fig. 2 show the difference between predictions for sand scattering and each of the three other seafloor types. It is apparent that the seafloor appears acoustically patchy, i.e., has scattering from the patches that is significantly above the background, near normal incidence. Experimentally determined scintillation index values from 80 kHz scattering data for several non-homogeneous seafloors were presented in Lyons and Abraham [4] as a function of area. Figure 3 shows these values replotted as a function of grazing angle. The data show non-Rayleigh behavior near normal incidence (values greater than one are indicative of a non-Rayleigh envelope distribution). Scintillation index estimated using the predicted scattering strengths, the number of patches (a function of the ensonified area) at normal incidence equal to 2, and a patch to background area ratio of 1:4 are shown in Fig. 4. The angular dependence of scintillation index predicted using realistic parameter values is seen to agree well with experimental values.

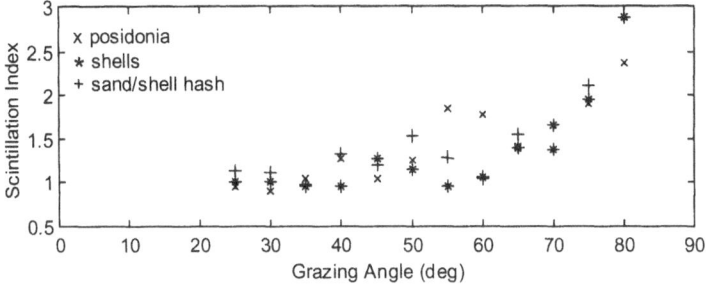

Figure 3. Experimental values of scintillation index versus grazing angle from [4].

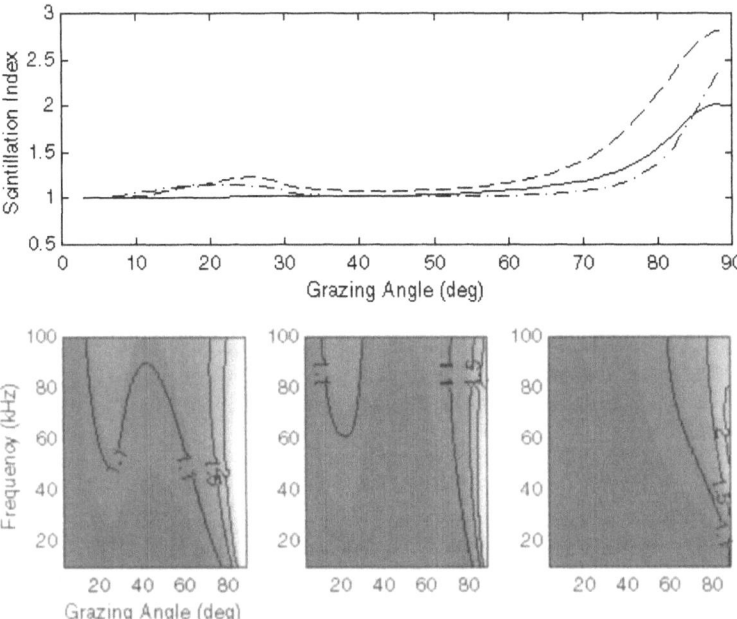

Figure 4. Top: Predictions of 80 kHz scintillation index versus grazing angle for sand patches on: shell-covered (dash-dot line), gravel (solid line), and seagrass-covered (dashed line) seafloors. Bottom: Scintillation index versus grazing angle and frequency for sand patches on: seagrass-covered (left), shell-covered (middle), and gravel (right) seafloors. Larger values are lighter on these plots and contour lines of 1.1, 1.5 and 2 are also shown.

5   Conclusions

In this paper a predictive model for the statistical distribution of clutter resulting from scattering from two different contributing seafloor types within the same resolution cell has been developed. The seafloors were modeled as being comprised of a finite number of homogeneous exponentially sized scattering patches (in contrast to the more traditional asymptotic derivation of the K-distribution) on a background (the area around the scattering patches) that was assumed to produce a Gaussian scattered return. Comparisons of scintillation index predictions made using the developed model and realistic input parameters for several example seafloor descriptions were compared with scintillation index estimated from high-frequency acoustic scattering data. The predicted scintillation index compared quite favorably with the level and angular dependence of experimental data. The strong peak seen in both the experimental and predicted scintillation index near normal incidence resulted solely from the angular dependence of the relative scattering strengths of the two contributing seafloor types within a resolution cell. The importance of the present work lies in the ability to link the clutter envelope distribution to measurable geo-acoustic properties in conjunction with sonar system parameters, providing the foundation necessary for solving several important problems related to the detection of targets in non-Rayleigh clutter. The direct link between system and environmental parameters and the statistical distribution of reverberation will allow: performance prediction for different systems based on seafloor properties, extrapolation of performance to other system/bandwidths, and optimization of system parameters such as bandwidth to local environment.

Acknowledgements

The Office of Naval Research sponsored this work (grant numbers N00014-01-1-0352 and N00014-02-1-0115).

References

1. Abraham, D.A. and Lyons, A.P., Novel physical interpretations of K-distributed reverberation, *IEEE J. Oceanic Eng.* (submitted 2001).
2. Jakeman, E. and Pusey, P.N., A model for non-Rayleigh sea echo, *IEEE Trans. Ant. and Prop.* **24**, 806–814 (1976).
3. Abraham, D.A., Signal Excess in K-distributed reverberation, *IEEE J. Oceanic Eng.* (submitted 2001).
4. Lyons, A.P. and Abraham, D.A., Statistical characterization of high-frequency shallow-water seafloor backscatter, *J. Acoust. Soc. Am.* **106**, 1307–1315 (1999).
5. Lyons, A.P., Anderson, A.L. and Dwan, F.S., Acoustic scattering from the seafloor: Modeling and data comparison, *J. Acoust. Soc. Am.* **95**, 2441–2451 (1994).
6. Stanton, T.K., On acoustic scattering by a shell-covered seafloor, *J. Acoust. Soc. Am.* **108**, 551–555 (2000).
7. Lyons, A.P. and Pouliquen E., Measurements of high-frequency acoustic scattering from seabed vegetation. In *Proc. 16th Int. Congress of Acoustics and the 135th Meeting of the Acoustical Society of America* (AIP, Woodbury, NY, 1998) pp. 1627–1628.
8. Stanic, S., Briggs, K.B., Fleischer, P., Ray, R.I. and Sawyer, W.B., Shallow-water high-frequency bottom scattering off Panama City, Florida, *J. Acoust. Soc. Am.* **83**, 2134–2144 (1988).
9. Stanic, S., Briggs, K.B., Fleischer, P., Sawyer, W.B. and Ray, R.I., High-frequency acoustic backscattering from a coarse shell ocean bottom, *J. Acoust. Soc. Am.* **85**, 125–136 (1988).

THE EFFECT OF SEABED BACKSCATTERING VARIABILITY ON THE PROBABILITY OF DETECTION AND ON THE PERFORMANCE OF SEABED CLASSIFICATION ALGORITHMS

E. POULIQUEN AND L. PAUTET

SACLANT Undersea Research Centre, Viale San Bartolomeo 400, 19138 La Spezia, Italy
E-mail: pouliq@saclantc.nato.int

A.P. LYONS

The Pennsylvania State University, Applied Research Lab., P.O. Box 30, State College, PA 16804
E-mail: apl2@psu.edu

As the scattering properties of the seafloor affect not only the mean angular dependence of backscatter but also cause a wide spread of scattered amplitudes, higher moment statistics are essential for applications such as target detection and seabed classification and characterization. The angular and frequency responses of the scattered amplitude distribution are caused primarily by water-sediment interface roughness, upper sediment heterogeneity, patchiness and discrete scatterers. To quantify the impact of seabed properties on higher moment statistics, scattered amplitudes were acquired at various sites displaying a large spectrum of seabed properties. Concurrent measurements of the seabed properties were conducted using two-dimensional stereo-photogrammetry, core and grab samples, videos and seabed penetrometers. To complement this experimental work, a temporal snapshot model based on the fourth order small slope approximation for interface scattering and on the small perturbation theory for the volume has been developed. Valid at high frequency and at all grazing angles, it allows the multiplication of scattering scenarios and will help to understand and quantify the impact of complicated seabeds on detection performance and on seabed classification algorithms. To illustrate high frequency acoustic variability, this paper presents selected experimental and simulated results.

1 Introduction

At high frequency, seabed backscattering is a major component of sonar reverberation. It is usually quantified in terms of averaged backscattering strength (BS) which relates directly to an averaged intensity scattered from a unit surface at a unit distance. For very large acoustic footprints compared to the wavelength, the BS is a meaningful and sufficient indicator for many detection and classification purposes. At high frequency where the acoustic footprint is often reduced and approaches the dimension of the acoustic wavelength, the problem becomes more complicated as scattering fluctuates greatly from place to place even in apparently large-scale stationary environments. Higher moment statistics and probability distribution functions (PDFs) provide enhanced information for detection and classification applications. The dominant mechanisms affecting the PDFs

N.G. Pace and F.B. Jensen (eds.), Impact of Littoral Environmental Variability on Acoustic Predictions and Sonar Performance, 219-226.

are of three kinds: 1) scattering caused by the intrinsic nature of the seabed (i.e., interface roughness, volume heterogenities, patchiness, discrete scatterers), 2) the sensing geometry (i.e., the beam pattern, the size of the scattering cell, the angle of incidence) and 3) the pulse shape, spectrum and duration. In some cases, other phenomena such as sea surface multipath and water column fluctuation may also significantly affect the PDF. To understand and quantify the impact of each of these phenomena, *in situ* measurements are necessary. Modelling of higher order statistics of high frequency backscatter is also required to allow a multiplication of scattering scenarios. Recent modeling work by Abraham and Lyons [1, 2] and Pouliquen *et al.* [3–5] offers two complementary approaches to predict higher moment statistics, treating patchiness with Joint Characteristic Functions (JCF) and using snapshot realizations, respectively. Processed signals acquired at sea and snapshot model outputs will be presented in this paper. The interpretation of the signals acquired is facilitated by *in situ* ground truth in terms of roughness (from stereo-photographs) and histograms of grain size (from gravity cores and grab samples), mainly. This paper discusses the effect of seabed scattering variability on the performance of classification algorithms based on normal incidence monobeam echosounding. Examples of measured and simulated ampitude PDFs and Probability of False Alarm (PFA) at off-normal incidence are also presented. It illustrates the highly variable nature of seabed acoustic backscatter at high frequency and outlines the dominant role imposed by seabed backscatter in target detection algorithm performance.

2   At sea measurement of seabed backscattering statistics

A prototype, hereafter refered to as ESP (Environmental Sensor Package) was designed and assembled by SACLANTCEN to acquire temporal backscattered signals from the seabed at various incident angles and frequencies. It was used at a number of sites near

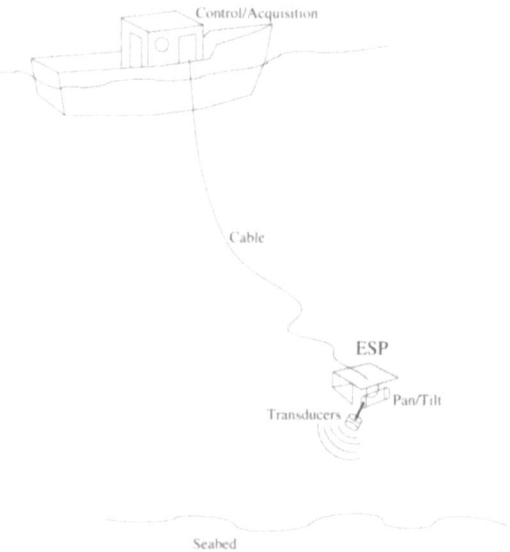

Figure 1. Experimental configuration of signal acquisition using ESP in a driting mode. An additional pinger provides the vertical distance between the source and the seabed.

Halifax, Canada, during the MAPLE'2001 experiment [6]. ESP is a high frequency/multi-incidence/multi-look system. Power, control commands and data collection are transmitted through a 200 m cable (Fig. 1). The acquired signals allow computation of higher moment statistics seafloor backscatter (i.e., mean levels of reverberation, variance, probability distribution, probability of false alarm) and also provide seabed type information when sounding at normal incidence. The system was designed to be operated in a drifting mode (geographicaly referenced), in a stationary position, or mounted on a tower. The latter allows the study of the effect of water column fluctuation on seabed backscattering variability (assuming ping-to-ping stationary of the seabed response) [7]. The drifting allows the study of the evolution of seabed response statistics as the environment changes. The stationary mode allows the quantification of acoustic variability for given seabeds assuming local stationarity and ergodicity. ESP was used at different depths (as low as 2 m above the seabed) and operated 3 different acoustic sources resonating at three different frequencies (50, 100 and 140 kHz). Thirteen positions were established on a large variety of seabeds with various levels of clutter often changing within less than a few metres. Several acquisitions were also made as ESP was drifting. High frequency multibeam systems, sidescan sonars, monobeam echosounders and advanced ground truth systems were also deployed at the sites. Dense gridding of sediment sampling (cores and grabs), expendable bottom penetrometers, videos and stereo-photos were used. Grabs allowed quantification of the spatial variability of the sediment grain size distribution, which provides information on acoustic impedance and heterogeneity affecting volume scattering. Gravity cores provided vertical information on density and compressional velocity with centimetric resolution at some locations. The core analysis (sound speed, density profile) was done immediately using the multi-sensor core logger mounted vertically in order to preserve the integrity of the upper structure of the core. Stereo-photographs [8] provided two-dimensional digital information on the water-sediment interface roughness which has an impact on interface scattering. One hundred pairs of stereo-photos (*e.g.*, Fig. 2), 40 grabs, and 7 cores were taken during the sea trial.

3 Examples of seabed acoustic backscattering variability

3.1 Variability at Normal Incidence

Signals from echosounders provide point-to-point bathymetry and are also used to remotely classify the seabed. These algorithms base their analysis on the shape and energy of narrow-band returned signals to provide segmentation and classification information of surveyed areas. The shape and energy of the returned echoes strongly depends on the interference structures produced by a heterogenous seabed volume bounded by a rough interface. The sounding geometry particularly around the specular direction may change from one ping to another. The source coordinates (x, y, z), pitch θ and roll ϕ work towards producing a snapshot of the temporal response from the seabed for a high frequency transmit pulse of duration of the order of milliseconds. This ping-to-ping variability affects the performance of classification algorithms and may provide unreliable/ambiguous classification. To quantify this variability, two typical classifying parameters were computed from signals acquired at-sea and simulated signals. As an example, 20000 time series with various acquisition geometries were recorded on a rippled sandy seabed shown in Fig. 2. The acoustic source movements were recorded. For normal incidence sound-

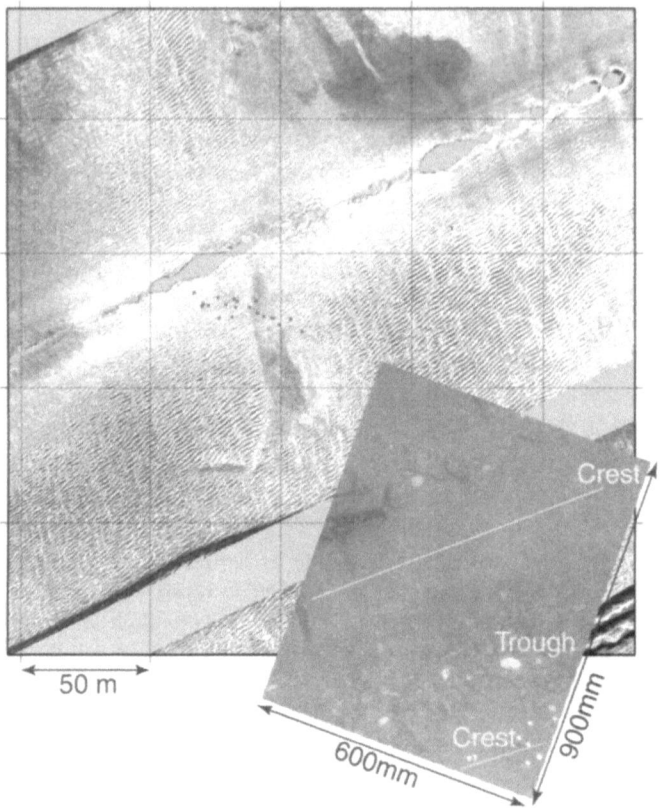

Figure 2. Example of a sidescan sonar image featuring a strong ripple field. The red dots correspond to the ESP acoustic acquisitions at a station. The yellow dots mark the locations where stereo-photographs were taken. One stereo-photograph shows the presence of a ripple field with coarser grains in the troughs and finer grains on the crests.

ings, the vertical distance between the seabed and the source showed a RMS variation of 20 cm around a mean value of 10.71 m. As the ship was anchored, the maximum horizontal translation during each acquisition series was less than 5 m (series of 1200 pings in average were recorded at a ping rate of 4.1 pings/s). Pitch and roll RMS variations were 0.26° and 1.50° around mean values of 0.69° and 0.19°, respectively. From each received amplitude envelope, two typical classifying parameters a_i and b_i are computed as follows:

$$a_i = C * \int_{t(\theta_0=0°)}^{t(\theta_1=40°)} s_i(t)dt \qquad b_i = \frac{1}{a_i} \int_{t(\theta_2=18°)}^{t(\theta_3=25°)} s_i(t)dt, \qquad (1)$$

with C being a normalizing factor depending on the acoustic source and i being the ping number. The amplitude envelope $s_i(t)$ is power corrected (*i.e.*, transmission loss and footprint size are removed) and time normalized so that the echo streching is independent of the source-seabed distance. The a_i parameter is mostly sensitive to the water/sediment impedance contrast (*i.e.*, "hardness") whereas the b_i parameter depends on both volume heterogeneity and interface roughness. Given the variability of the signals, the pings are often aligned in time and averaged to provide a more stable result:

$$A_i = \frac{1}{N} \sum_{j=i-N/2+1}^{i+N/2} a_j \qquad B_i = \frac{1}{N} \sum_{j=i-N/2+1}^{i+N/2} b_j, \qquad (2)$$

with N being the number of pings being averaged. The value of N is usually between 5 and 10. Figure 3 shows a high spread of a, b, A and B computed from the measured signals. Given the consistent texture of the sidescan sonar image (Fig. 2), stationarity of the seabed properties for each acquisition series is a reasonable assumption. Nevertheless, the non-averaged parameters a and b vary significantly from one ping to another. Even the averaged parameters A and B (with $N = 10$) display a large spread wich corresponds for this particular sediment type to uncertainty in roughness of about ±1.5 cm RMS and to relative acoustic impedance variation of about ±15%. Similar

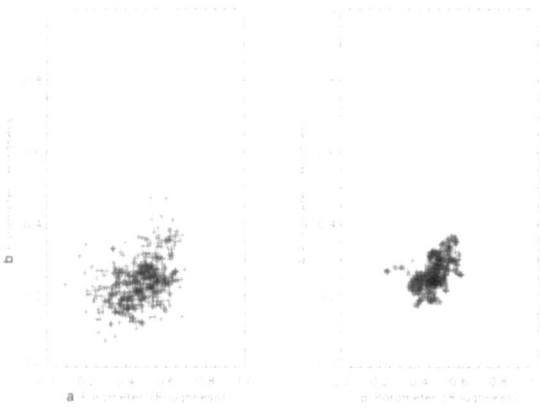

Figure 3. Example of classification parameter variability computed from signals acquired by ESP at Rose Bay (Fig. 2). The carrier frequency is 50 kHz, the beam pattern is 26°, the pulse length is 0.5 ms.

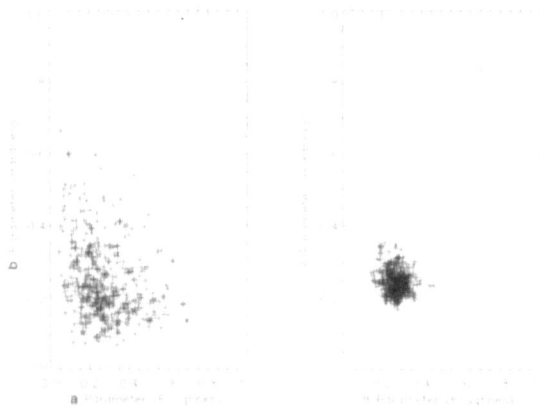

Figure 4. Example of classification parameter variability using BORIS-SSA. The acquisition geometry (source/sediment distance, roll and pitch) and the transmit pulse are identical to the ones of Fig. 3 but the geo-acoustic properties are different.

observations can be made (Fig. 4) from simulated data using the snapshot model BORIS-SSA [3–5]. The same geometry and source motion was used in the simulation. Volume scattering was not included because the heterogeneity structure of the sediment was not measured at this site. Horizontal isotropy of the water/sediment interface was assumed and the RMS roughness was estimated from non-processed stereo-photos ($\sigma \simeq 3$ cm). The spread of the non-averaged a and b parameters is greater than the measured ones. The averaging process to obtain parameters A and B is more effective as the vertical distance source-sediment is known exactly but also displays a high inherent variability. Variability observed on measured and simulated echoes illustrates the limits of classification algorithms based on monofrequency echosounder signals. Recent experiments have shown that multi-frequency sensing of the seabed is a way to improve the classification algorithm performance. As volume and interface backscatter is frequency dependent a multi-frequency approach provides orthogonal information that can significantly reduce ambiguity in the classification process [6].

3.2 Variability at Oblique Incidence

High frequency and low grazing angle backscattering from cluttered seabeds is an important factor limiting target detection. Compared to near normal incidence, the often abrupt changes of the reflection and transmission coefficients around critical angle and shadowing (*e.g.*, in Fig. 2) are additional physical effects producing complicated backscattered signals. With the intent of understanding the formation of probability density functions, off-normal incidence signals were also acquired by ESP at several sites displaying a large spectrum of seabed types with various levels of complexity. Acquisitions for different geometries and transmitted signals were made. Figure 5 displays an example of a sequence of acoustic envelopes backscattered from the same seabed (a mixture of silt and gravel). Despite apparent seabed stationarity, the signals are highly variable. In Fig. 6, their PDF and related PFA show a large spread of backscattering strength. A heavy tail of the PDF or the slow decay of the related probability of false alarm (PFA) *versus* detection threshold reveal a particular difficulty in detecting targets. The presence "patchiness" of shells and gravel may be the cause. Similarly, Fig. 7 shows a PDF and a PFA computed from synthetic signals obtained using BORIS-SSA [5]. The simulations were made on a on a large number of snapshots on the same interface having well defined statistical properties. The physical interface and volume properties, geometry of acquisition and the source characteristics can be chosen and allow a multiplication of scenarios. Measurements and simulations illustrate the high variability of the seabed response and the critical need to consider not only the mean backscatter as an indication of detection probabilities but also higher moments as well as the probability density function. The comparison between acquired and simulated statistics is promising and will be the object of further studies.

4 Summary

This paper presented a selection of measured and simulated signals with their related higher statistical characteristics represented in terms of PDF and PFA. At normal incidence, acoustic signal variability causes a large spread of classification parameters which reduces the classification algorithm performance. At oblique incidence, signal variability is such that higher moment statistics are essential for target detection.

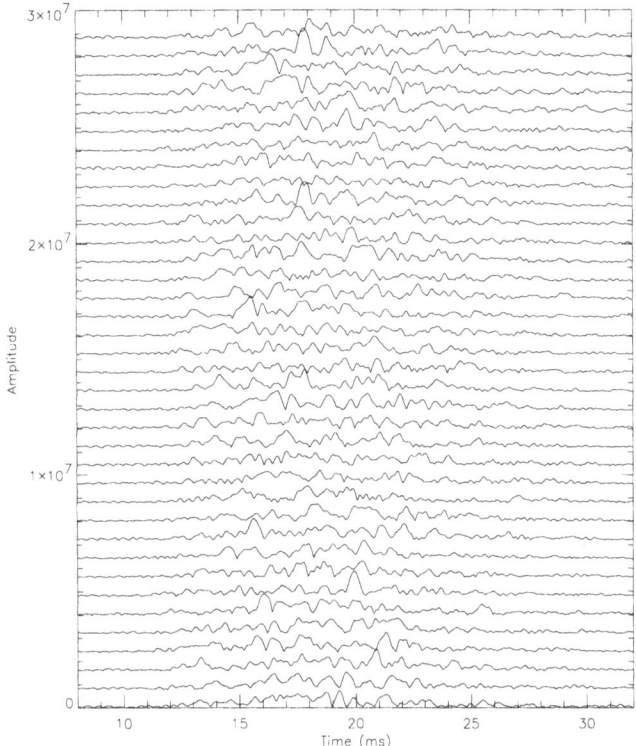

Figure 5. 36 successive amplitude envelopes acquired by ESP at 100 kHz over a silt+gravel area. Incident angle: $\theta = 60°$. Beam aperture is $16°$. Ping rate is 4.1 pings/s.

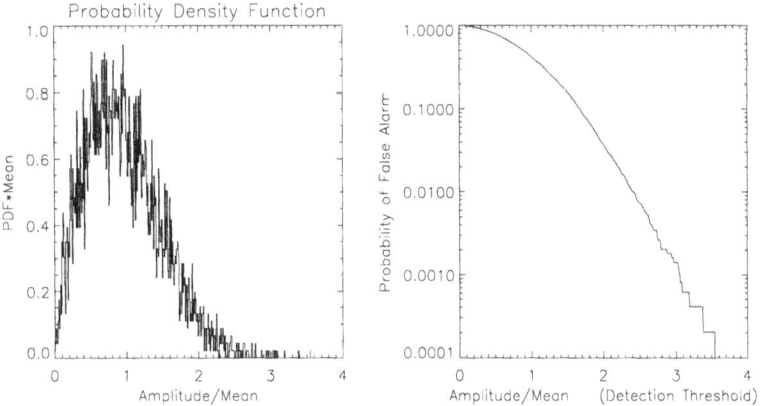

Figure 6. PDF and PFA obtained from a series of 1200 pings acquired by ESP at 100 kHz over a silt+gravel area. Incident angle: $\theta = 70°$. Beam aperture is $16°$. Ping rate is 4.1 pings/s.

Acknowledgements

We would like to thank everyone who took part to the MAPLE'2001 sea trial. Prototypes were designed and assembled by the SACLANTCEN Engineering & Technology

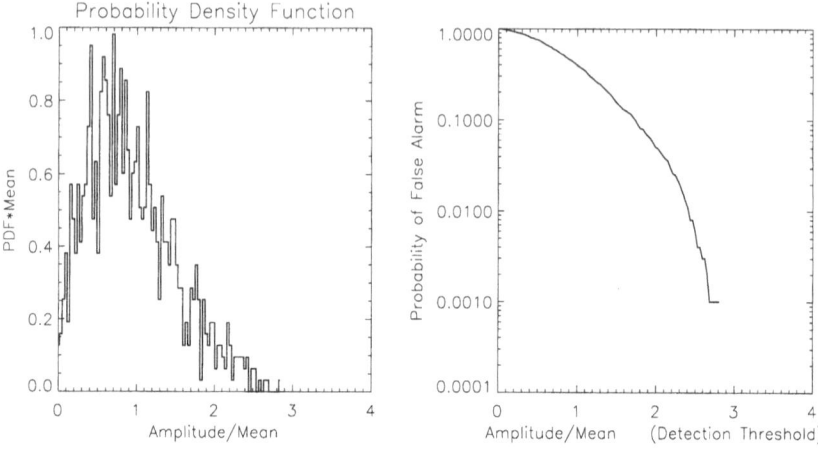

Figure 7. PDF and PFA obtained from time series simulated by BORIS-SSA. Only seabed interface scattering is considered. $f = 100$ kHz. The seabed is an isotropic filtered "power-law" power spectral density with an 3.2 exponent. Incident angle: $\theta = 50°$. Beam aperture is 16°.

Department. Special thanks to F. Cernich, A. Figoli, P. Franchi, P. Guerrini, R. Lombardi and P.A. Sletner. The experiment was conducted from the R/V *Alliance* and the CFAV *Quest*.

References

1. Abraham, D.A., Lyons, A.P., Novel physical interpretation of K-distributed reverberation, *IEEE J. Oceanic Eng.* (submitted 2001).

2. Lyons, A.P., Abraham, D.A. and Pouliquen, E., Prediction of scattered amplitude statistics of patchy seafloors. In *Impact of Littoral Environmental Variability on Acoustic Predictions and Sonar Performance,* edited by N.G. Pace and F.B. Jensen (Kluwer, The Netherlands, 2002) pp. 211–218.

3. Pouliquen, E., Bergem, O. and Pace, N.G., Time-evolution modeling of seafloor scatter. I. Concept, *J. Acoust. Soc. Am.* **105**(6), 3136–3141 (1999).

4. Bergem, O., Pouliquen, E., Canepa, G. and Pace N.G., Time-evolution modeling of seafloor scatter. II. Numerical and experimental evaluation, *J. Acoust. Soc. Am.* **105**(6), 3142–3150 (1999.)

5. Pautet, L., Pouliquen, E., Canepa, G., A study on ping-to-ping coherence of the seabed response, In *Impact of Littoral Environmental Variability on Acoustic Predictions and Sonar Performance,* edited by N.G. Pace and F.B. Jensen (Kluwer, The Netherlands, 2002) pp. 489–496.

6. Pouliquen, E., Trevorrow, M., Blondel, Ph., Canepa, G., Cernich, F. and Hollett, R., Multi-sensor analysis of the seabed in shallow water areas: overview of the Maple'2001 experiment. In *Proc. 6th European Conf. on Underwater Acoustics*, Gdansk, Poland, 2002.

7. Pautet, L. and Pouliquen, E., Experimental study of fluctuation in coherent backscattering. In *Proc. 6th European Conf. on Underwater Acoustics*, Gdansk, Poland, 2002.

8. Lyons, A.P., Fox, W.L.J., Hasiotis, T. and Pouliquen, E., Characterization of the two-dimensional roughness of shallow-water sandy seafloors, *IEEE J. Oceanic Eng.* (July 2002).

9. Orlowski, A., Application of multiple echoe measurements for evaluation of sea bottom type, *Oceanologia* **19**, 61–78 (1984).

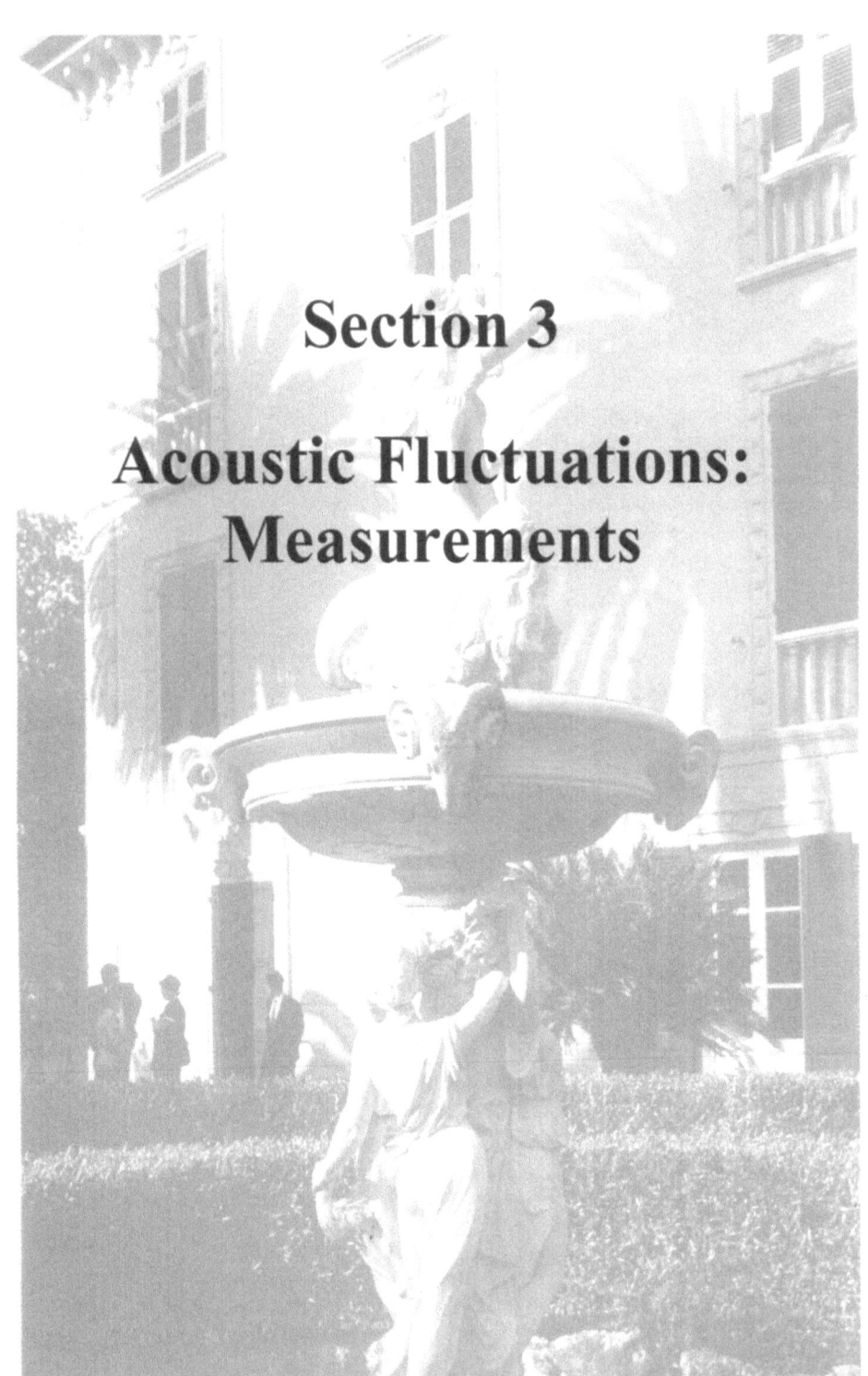

Section 3

Acoustic Fluctuations:
Measurements

EFFECTS OF ENVIRONMENTAL VARIABILITY ON ACOUSTIC PROPAGATION LOSS IN SHALLOW WATER

TUNCAY AKAL

TUBITAK Marmara Research Centre, Earth and Marine Science Research Institute
P.K. 21 Gebze, Kocaeli 41470, TURKEY
E-mail: tuncay.akal@posta.mam.gov.tr

Transmission loss as a sonar parameter describes the magnitude of acoustic energy loss of sound propagation in the ocean. Broadband (10 Hz – 300 kHz) propagation loss measurements conducted over the years in varying areas show the effect of the spatial and temporal variability on acoustic propagation for bottom-limited conditions. An example from the existing acoustic data obtained over the last three decades has been utilized to demonstrate the effects of environmental variability on propagation loss, and our ability to simulate the propagation conditions with existing propagation models are presented.

1   Introduction

Acoustic propagation in the ocean is influenced by many factors: the physical and chemical properties of the water column cause attenuation and refraction, while variations in seafloor properties and boundary roughness complicate reflections. Thus, any attempt to understand propagation of acoustic energy in shallow/coastal water requires an accurate knowledge of the propagation medium, especially sound speed structure of the water column, seafloor properties and sea surface roughness. The shallow/coastal water environment is often characterized by wind driven flow, and/or current flow interacting with bottom topography and increased variability in water properties as a result of enhanced mixing. The ocean sound speed structure varies both in time and space. There are various oceanographic processes like currents, internal waves, fronts, eddies and thermal changes that control the sound speed structure and thus the acoustic characteristics of this environment. In order to understand the range and time dependence of acoustic propagation over this broad frequency range in shallow/coastal waters, simultaneous environmental measurements are required.

As part of SACLANTCEN's research program over the last three decades acoustic data covering a broad frequency range (10 Hz – 100 kHz) were collected in order to adequately understand acoustic propagation, including penetration into the seabed, along with water column environmental characteristics. Transmission loss (TL) as a sonar parameter describes the magnitude of acoustic energy loss of sound propagation in the ocean. Data from a polar region are utilized to demonstrate the effects of a polar front on acoustic propagation, and our ability to simulate the propagation conditions in such a complex environment with existing propagation models are presented.

N.G. Pace and F.B. Jensen (eds.), Impact of Littoral Environmental Variability on Acoustic Predictions and Sonar Performance, 229-236.
© 2002 Kluwer Academic Publishers.

2   Experimental technique

The sound speed structure of the water column is known to be the controlling factor in acoustic propagation. Acoustic energy propagating through a shallow water channel interacts with the sea floor causing partitioning of waterborne energy into different types of seismic and acoustic waves. The propagation and attenuation of these waves observed in such an environment are strongly dependent on the physical characteristics of the sea bottom. Transmission loss representing the amount of energy lost along an acoustic propagation path (in range and time) carries the information relative to the environment through which the wave is propagating.

In all the TL data reported, explosive charges and transducers were used as sound sources and a vertical hydrophone string containing omnidirectional hydrophones was used to receive the transmitted acoustic signals. The experimental procedure used for the runs was the following: the receiving ship would launch a vertical hydrophone string suspended from a spar buoy with a digital radio link; this would then be drifted away from the ship in order to minimize the effect of ship-radiated noise and surface waves. The number of hydrophones used varied depending on the water depth, spaced in such a way as to cover most of the water column as shown in Fig. 1.

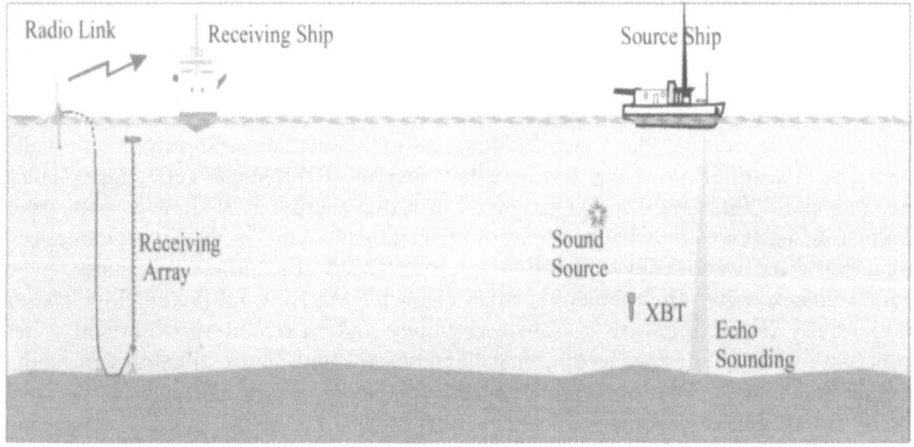

Figure 1. Experimental set-up for transmission loss measurements.

The source ship would then move away or towards the receiving ship at a fixed, predetermined course launching sound sources or towing a CW/FM source transmitting at regular intervals and set for different depths. The source ship would also launch XBTs and obtain echo soundings along the track for the calculation of the sound speed and bathymetric profiles.

The transmission losses reported are energy losses in dB with reference to a source level at 1 m. For those signals in which the noise could be considered stationary, the data have been corrected for noise by subtracting the measured noise energy of the same time duration observed prior to the signal arrival. The data for which this was not valid, or for which the signal-to-noise ratio was found to be less than one, have been omitted.

3   Effects of an oceanographic front in space and time

Areas where oceanographic fronts are present are characterized by a strong temporal and spatial variability due to the mixing process between two different water masses. Acoustic and environmental data collected in an Arctic front region are used to demonstrate effects of an oceanographic front on TL both in time and space.

The acoustic experiment was carried out in a very complicated environmental due to the presence of the polar oceanic front where the cold Arctic water meets the warmer Atlantic water. This boundary between two different water masses meanders with tidal currents and creates a very complex medium that varies in both space and time.

3.1   Spatial Variability of Transmission Loss across a Front

Figure 2 shows a cross section profile of the sound speed variation perpendicular to the front calculated from XBT recordings obtained by the source ship during TL measurements. The front can be identified approximately at a range of 12 to 14 km from the receiver position. As also shown in Fig. 2 is the bathymetry of the seafloor, which is characterized by a shallow shelf with water depths varying from 100 to 130 m, extending to 35 km distance from the receiver position, then the shelf drops from 130 to 340 m depth within 10 km distance, and a trough where the water depth remains 340 m. At depths less than 150 m the sea floor is covered by large-grained material (i.e. sand, gravel and boulders) with large patches of shells. As the water depth increases the bottom composition changes from sand to silty sand, then to sand-silt-clay in the deepest part.

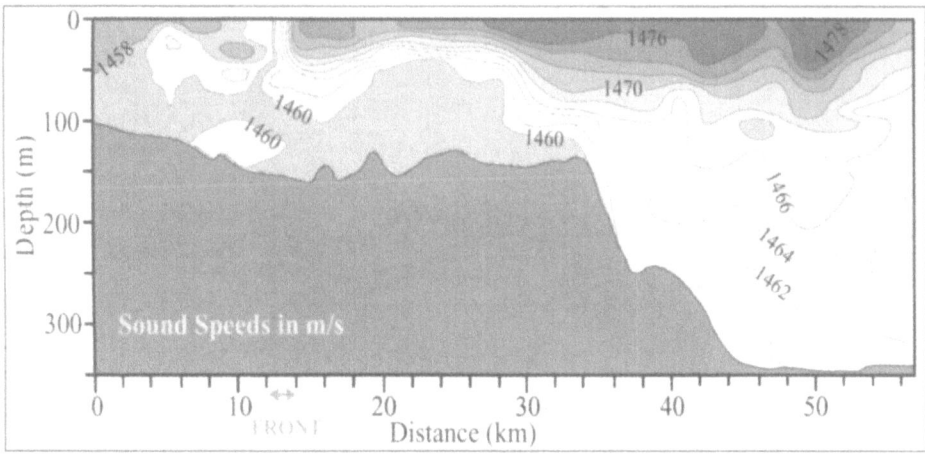

Figure 2. Sound speed cross section observed during acoustic measurements.

The range-dependent propagation path was perpendicular to the front with the receiving ship situated at the Arctic water side of the front. Figure 3 show the measured TL as a function of frequency and distance for a source at 25 m and a receiver at 50 m depth. Within the frequency band of the measurements, an optimum propagation frequency around 200 to 800 Hz is observed. This phenomenon is characteristic of shallow water propagation and it is caused by a low-frequency attenuation due to bottom interaction and attenuation of higherfrequencies due to scattering and volume absorption [1, 2]. This optimum frequency corresponds to propagation through the water column and the optimum frequency increases with bottom hardness and the associated presence of high velocity shear waves and decreases with increasing water depth due to diminishing bottom interaction.

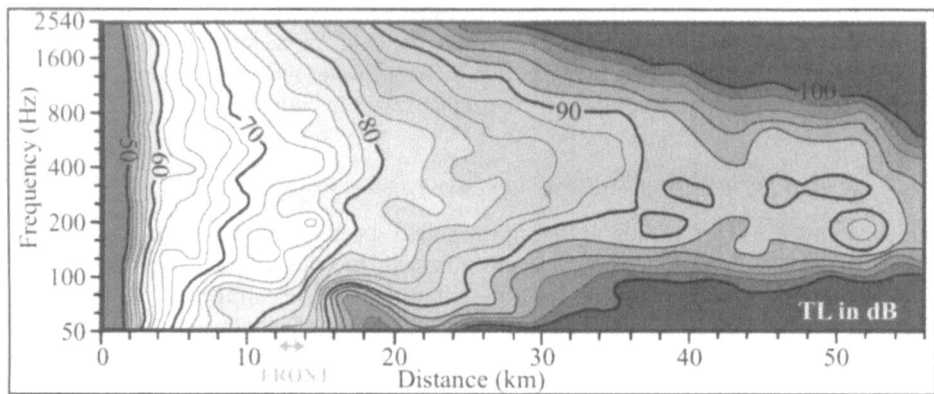

Figure 3. Measured TL as a function of frequency and distance for a source at 25 m and a receiver at 50 m depth.

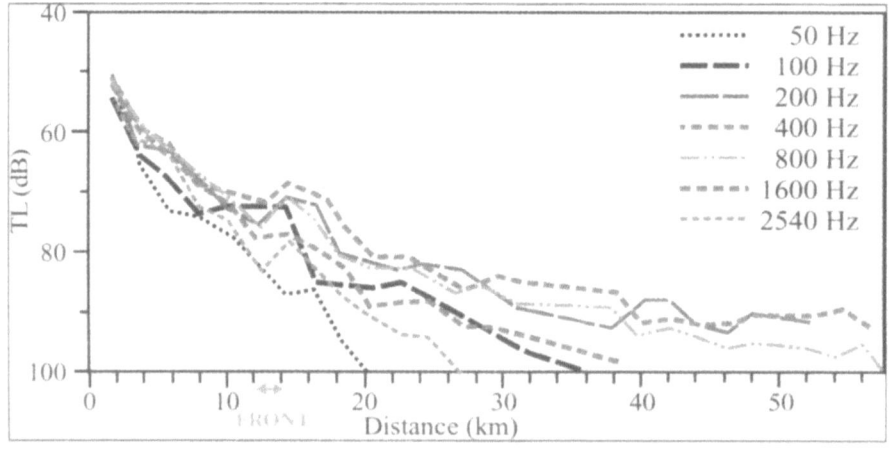

Figure 4. TL as a function of distance for selected frequencies.

Figure 4 shows the TL curves for selected frequencies. In this figure the effect of the front is clearly seen as a 5 to 6 dB decrease within the 100 to 800 Hz intermediate frequency band. This is due to a change in the water column sound-speed structure within the frontal area where the distribution of acoustic energy changes, causing less interaction with the seafloor.

Using the environmental data collected during the acoustic measurements, several different simulations models (RAM, PAREQ and C-SNAP) [3–5] have been used to model the TL (Fig. 5). Figure 6 is an example of the acoustic field simulated for this particular environmental data set using the RAM model for 400 and 1600 Hz. After the best agreement was found between simulation results and the measured data (Fig. 5), we simulated a second case by removing the front and making sound speed profiles at the receiver position range independent, but keeping the same bathymetric profile and bottom properties as the range dependent case, where the effect of the front can clearly be seen as being up to 15 to 18 dB on TL. The difference is shown in Fig. 7 where the frontal effects are absent. Both low and high-frequency acoustic energy distributions show completely different characteristics.

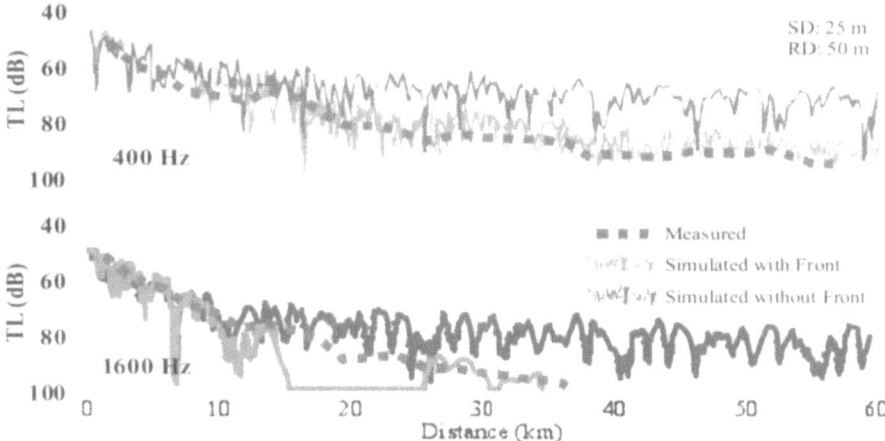

Figure 5. Measured data and comparison of simulations with and without the polar front.

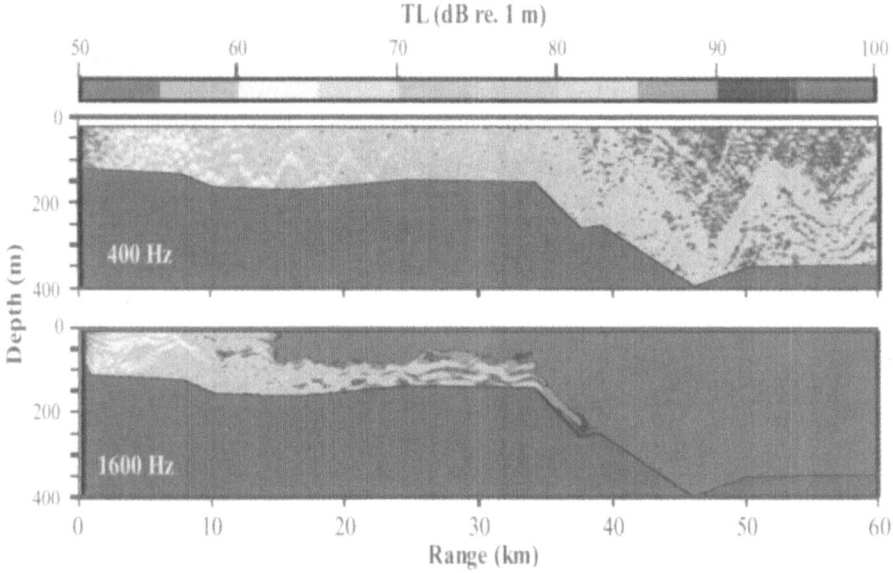

Figure 6. An example for the acoustic field simulated for 400 Hz and 1600 Hz through the front.

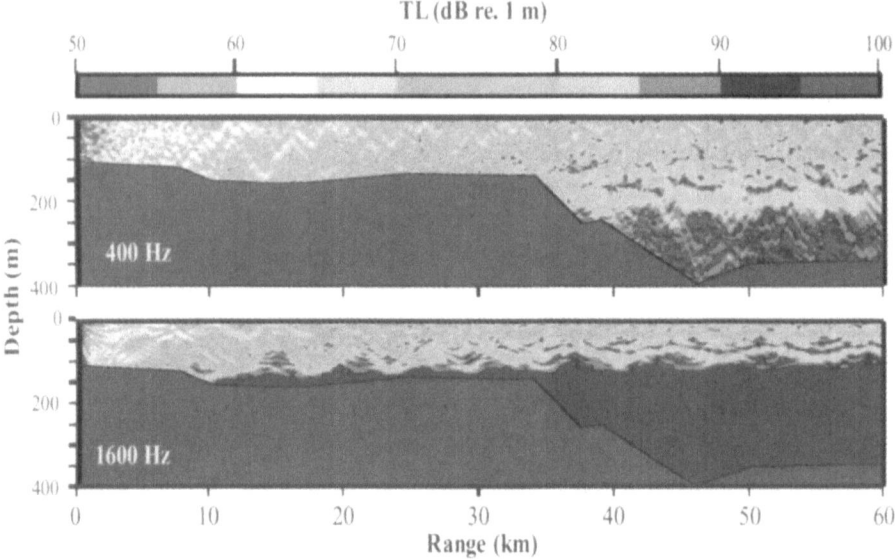

Figure 7. An example for the acoustic field simulated for 400 Hz and 1600 Hz without the front.

3.2   Temporal Variability of Transmission Loss across a Front

This experiment was conducted to study the temporal variability of the TL across the front between to positions over a 48 hour period.   As can be seen in Fig. 8 the environmental conditions did not remain constant during this period as shown by the temperature cross section at the receiver position.   On several occasions the front passed through the receiver position and clearly affected the propagation conditions.

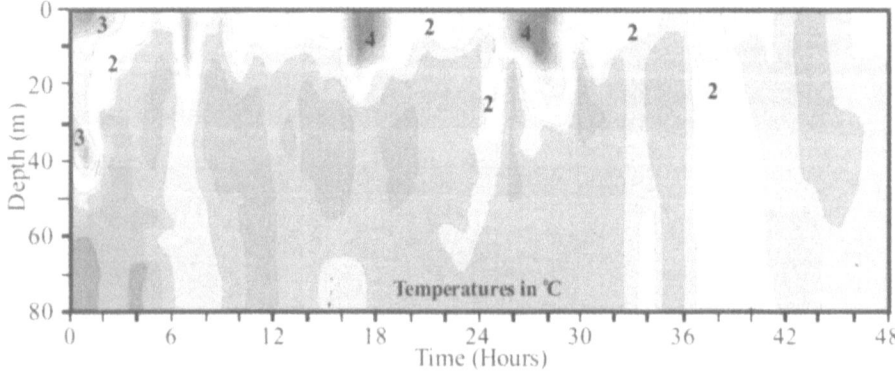

Figure 8. Temperature structure as a function of time at the receiver position.

As shown by the TL contours in Fig. 9 the maximum TL exhibits approximately 12 h oscillations. All the losses, especially for the intermediate frequencies, show marked oscillation with time, with variations of up to 8 to 12 dB. When the predicted tidal curves (Fig. 9) are superimposed on the loss contours, there is very good correlation between high tide and maximum TL. The temperature cross-section (Fig. 8) and the current measurements made at the receiver position, supported by XBTs taken by the source ship, strongly indicate that the best transmission was observed when the front was furthest to the south, coinciding with the current running south and low

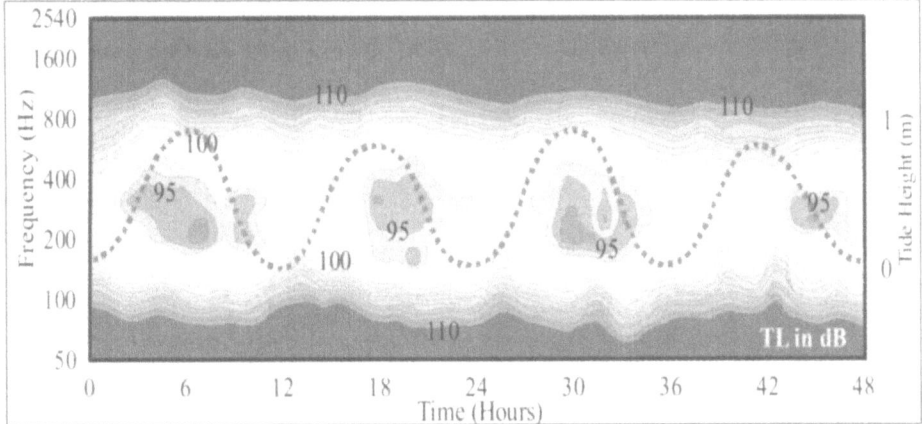

Figure 9. Predicted tidal curves and TL as a function of time.

tide, while the poorest propagation was observed with the front at its northern-most position. We are therefore dealing with two different situations as shown on Fig. 10. When the receiving ship was in isothermal water north of the front (polar water), an important part of the propagation in the shallow water will be under upward refracting conditions until the position of the front, but in the deeper water part, it will be under downward refracting conditions causing relatively small total losses. When the receiving ship remained in the Atlantic water south of the front, the propagation was under very strong downward refracting condition over the shallow water area causing higher losses due to increased bottom interaction.

4   Conclusions

Acoustic and environmental data collected in an Arctic polar front region are used to demonstrate effects of an oceanographic front on TL both in space and time. These areas are characterized by a strong temporal and spatial variability due to the mixing process between two different water masses. The results indicate that the polar front can have significant effect on acoustic propagation both in space and time. Range dependent data may have up to 15 to 18 dB different TL. The TL in the intermediate frequencies can oscillated up to 12 dB within a 48-h period. These oscillations are well correlated with the tide, which also control the position of the front and thereby changes the acoustic propagation characteristics.

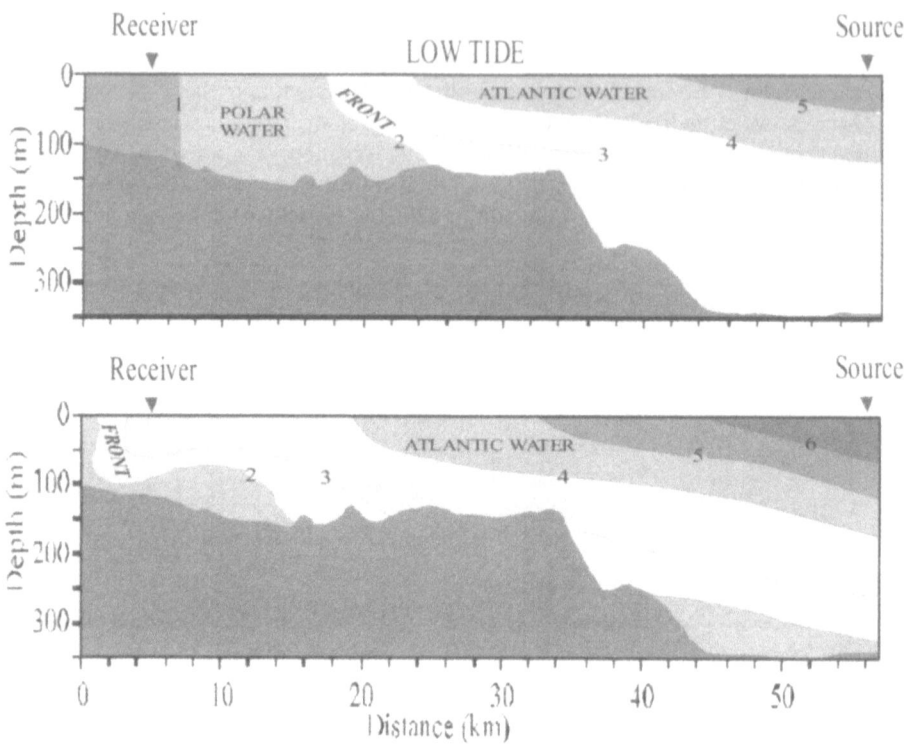

Figure 10. Simplified temperature structure during low and high tide periods.

Acknowledgements

I would like to thank G.F. Edelmann of MPL-Scripps/UCSD, M.G. Martinelli and C.M. Ferla of SACLANTCEN for their help in running the different propagation models.

References

1. Akal, T., Sea-floor effects on shallow-water acoustic propagation. In: *Bottom Interacting Ocean Acoustics*, edited by W.A. Kuperman and F.B. Jensen (Plenum Press, 1980) pp. 557–575.
2. Jensen, F.B. and Kuperman, W.A., Optimum frequency of propagation in shallow water environments, *J. Acoust. Soc. Am.* **73**, 813–819 (1983).
3. Collins, M.D., User's Guide for RAM. Naval Research Laboratory, Washington, DC.
4. Jensen, F.B. and Martinelli, M.G., The SACLANTCEN parabolic equation model (PAREQ). SACLANTCEN Undersea Research Centre, La Spezia, Italy (1985).
5. Ferla, C.M., Porter, M.B. and Jensen, F.B., C-SNAP: Coupled SACLANTCEN normal mode propagation loss model. Rep. SM-274, SACLANT Undersea Research Centre (1993).

BROADBAND ACOUSTIC SIGNAL VARIABILITY IN TWO "TYPICAL" SHALLOW-WATER REGIONS

PETER L. NIELSEN

SACLANT Undersea Research Centre, Viale San Bartolomeo 400, 19138 La Spezia, Italy
E-mail: nielsen@saclantc.nato.int

MARTIN SIDERIUS

Science Applications International Corporation, La Jolla, CA 92037, USA
E-mail: thomas.martin.siderius@saic.com

JÜRGEN SELLSCHOPP

FWG, Klausdorfer Weg 2–24, 24148 Kiel, Germany
E-mail: JuergenSellschopp@bwb.org

Successful sonar performance predictions in shallow-water regions are strongly dependent on accurate environmental information used as input to numerical acoustic prediction tools. The sea-surface and water-column properties vary with time and this
time variability of the ocean introduces fluctuations in received acoustic signals. The
lack of knowledge of the environmental changes results in uncertainty in predictions
of the acoustic propagation. SACLANTCEN has recently conducted two experiments
to quantify the impact of the time-varying ocean on broadband acoustic propagation in
"typical" shallow-water regions. Extensive oceanographic data were collected during
the acoustic transmissions. Broadband acoustic signals were transmitted every minute
over a fixed propagation path up to 18 h. The signals were received on a vertical array at
fixed ranges of 1 to 10 km from a moored source. The variability of the oceanographic
and acoustic data is presented for the two experimental areas. Numerical modelling of
the sound propagation using the measured environmental data is shown and compared
to the acoustic data. The possibility of predicting the received signals with an extensive
knowledge of the underwater environment is discussed.

1 Introduction

Sound propagation in the ocean depends strongly on the actual location. In shallow-
water regions the seabed properties are known as the key parameters that affect the
sound propagation. However, experimental data from repeated acoustic transmissions
over fixed propagation path in particular shallow-water regions shows significant impact
from the time-varying ocean on the sound propagation as variability in transmission
loss (TL) and signal arrival time. Prediction of sound propagation in shallow water is
generally performed by assuming a time-invariant ocean. This assumption is sufficient
for certain shallow-water regions as numerical modelling of the sound propagation has
been performed successfully by using "frozen" environmental inputs [1–3]. Experiments
conducted in particular shallow-water regions show significant variability in acoustic data

N.G. Pace and F.B. Jensen (eds.), Impact of Littoral Environmental Variability on Acoustic Predictions and
Sonar Performance, 237-244.

caused by changes in the oceanographic conditions with time [4–6]. Sudden increase in TL at particular frequencies, amplitude fading and arrival-time variability of received time series were detected during these experiments, and this variability in the acoustic data is most likely caused by the presence of internal waves. Successful prediction of sound propagation in these time-varying environments cannot be achieved without uncertainty unless detailed spatial and temporal information about the environment is available.

In May 1997 SACLANTCEN conducted the PROSIM'97 experiment south of the Elba Island, Mediterranean, in April/May 1999 the ADVENT'99 on Adventure Bank, Mediterranean, and the ASCOT'01 experiment in June 2001 off the coast of Massachusetts Bay, USA. Only data from ADVENT'99 and ASCOT'01 are presented in this paper. The ADVENT'99 experiment was conducted in very *benign* conditions on the Adventure Bank, Mediterranean with very weak tidal effects. Cores, seismic surveys and model-based geoacoustic inversion results [7] indicate a sandy-like sediment layer overlaying a harder sub-bottom. Acoustic Linear-Frequency-modulated (LFM) signals were transmitted every minute from a bottom-moored sound source for up to 18 h. The acoustic signals covered a frequency band from 200–3800 Hz, and the signals were received on a 64-element vertical array at 2, 5 and 10 km range (recover/deploy for each range). Extensive oceanographic data were collected during the acoustic transmissions to correlate changes in the environment with changes in the received acoustic data. In particular, a 49-element Conductivity-Temperature-Depth (CTD) chain was towed by ITNS Ciclope continuously along the 10-km track acquiring range- and depth varying sound-speed structures. Each of the CTD structures are separated by 1 h. The weather conditions were favorable during the acoustic transmissions with maximum significant sea surface wave height of 1.5 m [7].

The configuration of the ASCOT'01 experiment was similar to ADVENT'99. However, the location of the experiment was known *a priori* to have a more variable environment than for the ADVENT'99. Strong tidal effects are present and the moorings of the sound source and vertical array were close to the continental shelf. These conditions can create strong internal waves affecting the acoustic signals over time. The source, receiving array and signals were the same as in ADVENT'99 with source-receiver separations of 1, 2, 5 and 10 km. The bathymetry is more range dependent than for ADVENT'99 with changes up to 12 m within 2 km. The acoustic signals were transmitted every 30 s for up to 12 h. There was no seismic survey or corering performed along the propagation tracks limiting the knowledge of seabed properties and layering structure of the bottom. However, U.S. Geological Survey [8] has performed analysis on acoustic backscatter intensity measured in the ASCOT'01 area, and the result from this analysis shows rapidly changing sediment properties corresponding to a mixture of sand and gravel.

2   Oceanographic data

There are 3 environmental factors that are considered as main contributors affecting the fixed-path acoustic propagation over time: (1) tidal effects, (2) water-column sound-speed fluctuations, and (3) scattering from the bathymetry and seabed. The tidal effects during ADVENT'99 are considered negligible, as the tide in the Mediterranean is less than 0.5 m. Direct measurement of the tide was not performed during the ASCOT'01 experiment but the tidal stations Boston Harbor and Boston Light are located relatively close to the experimental area. The tide amplitude and phase are almost the same for Boston Harbor

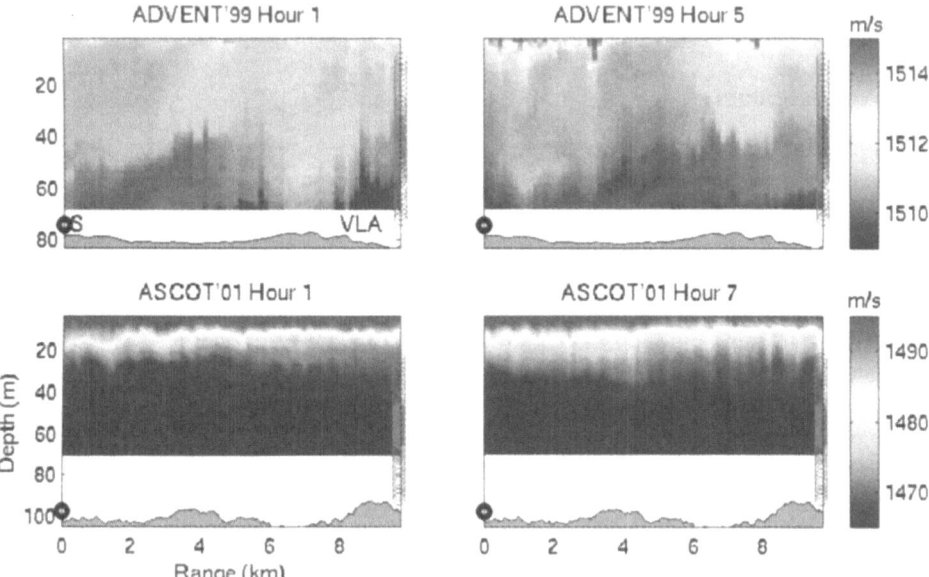

Figure 1. Sound-speed structures acquired during the ADVENT'99 (upper panels) and ASCOT'01 (lower panels) along the 10-km acoustic propagation track.

and Boston Light. Although the 2 stations are separated by only 20 km it is assumed that the measured tide is representing the tide at the ASCOT'01 site about 80 km from the Boston tidal stations. The water depth varies around ±1.2 m with a period of 12 h.

The time-, range- and depth dependent sound-speed structures along the propagation tracks were measured by a towed CTD-chain. A total of 18 sound-speed sections were acquired during the 18-h acoustic transmission along the 10-km for ADVENT'99. Only 7 sound-speed sections were acquired along the 10-km track during ASCOT'01 for a 10-h transmission period. Sound-speed profiles measured at different times along the 10-km acoustic track is shown in Fig. 1.

There is a clear difference in the sound-speed structures from ADVENT'99 (upper panel) and ASCOT'01 (lower panel). The water column of ADVENT'99 is almost iso-velocity with only a few m/s change in sound speed over depth. The sound-speed is also very weakly range-dependent with a tendency to divide the track into a low and high sound-speed region. The sound-speed structures from ASCOT'01 show typical downward refracting profiles along the track. The water column in this area is clearly more range dependent than ADVENT'99 with indications of soliton-like features in the upper part of the water column. The scattering of the acoustic field from the seabed may change in time as the sound-speed profiles change. This change of sound-speed alters the insonification of the seabed and may cause additional fluctuations in the received acoustic signals. The scattering characteristics of the seabed is considered range dependent while the tidal effect and water-column sound-speed are both time and range dependent. The bathymetry along the 10-km measured by a single-beam echo-sounder is shown together with the sound-speed structures in Fig. 1. The water depth changes by a few metres for ADVENT'99 along the 10-km track, while it changes up to 12 m within a couple of km in range and significant roughness is observed for ASCOT'01. The impact of the bathymetry changes on the acoustic propagation is stronger for ASCOT'01 than ADVENT'99.

3   Acoustic data

The acoustic data received on the moored VLA have been processed for transmission loss (TL) (low level) and for establishing ping-to-ping correlation (high level) used as measures to assess the acoustic fluctuation with transmission time. The arrival structure of the matched-filtered (MF) time-series across the VLA at 10 km and at 3 different transmission times is shown in Fig. 3 for ADVENT'99 (upper panels) and ASCOT'01 (lower panels). The first 2 figures in the upper and lower panel of Fig. 3 show signals separated by 1 min, and the last figure in the upper and lower panel is signals received 1 h later. The received signals are stable at 1-min separation but after 1 h clear changes in the arrival structure can be observed. Especially for the ASCOT'01 data changes in individual multi-path arrivals appear as a focusing-defocusing effect. The time-varying ocean causes these fluctuations of the acoustic signals. Note that the time dispersion of the ASCOT'01 data is significantly longer than for ADVENT'99 indicating a higher sound speed in the bottom at the ASCOT'01 area. In this case steeper and later arriving multi-paths are trapped in the water column caused by the higher critical angle of the bottom. The effect of the tide is only observable for the ASCOT'01 data as the tide in the Mediterranean is negligible. The tide alters the absolute arrival time of the multi-paths as the water depth changes. This change in arrival time is larger for the steep and late arrivals as these paths travel longer (or shorter) distances than the shallow paths before arriving at the VLA.

   TL is considered as a robust but low level processing of the acoustic field. The TL has been calculated for all received signals in both experiments as calibrated source signatures were available. The TL is averaged in a 10 Hz frequency band around the centre frequencies 250, 550 and 750 Hz. In addition, the signals are averaged over transmission time resulting in mean TL and standard deviation over depth for a 12-h transmission period (Fig. 2).

   In general, the TL is higher for the ADVENT'99 area (upper panel) for all frequencies than for ASCOT'01 (lower panel). However, the standard deviation of the TL is almost the same for both experiments and for all frequencies regardless the difference in the environmental conditions. The standard deviation of the TL also increases with increasing frequency as expected and the deviation reaches ±5 dB at 750 Hz. Correlation of time series to assess variability is a much stricter measure than TL. Matched-Field Correlation (MFC) is applied to illustrate signal similarity and how fast the signals degrade over transmission time. The MFC is using a standard Bartlett processor as the correlator between two signals [7]. One of the signals received early during the transmissions is denoted as a reference signal. This reference signal is correlated with all the subsequently received signals and normalized with the total energy in the two signals. The correlator has a value of 1 for two similar signals and 0 for totally un-correlated signals. The MFC is calculated for all transmitted signals in the frequency band from 200–800 Hz and for each discrete frequency obtained through the Fourier transform of the time series. The correlation is shown in Fig. 3 for 2, 5 and 10-km propagation range obtained during the ADVENT'99 (upper panels) and ASCOT'01 (lower panels).

   The ADVENT'99 data show a high correlation between the received signals for all

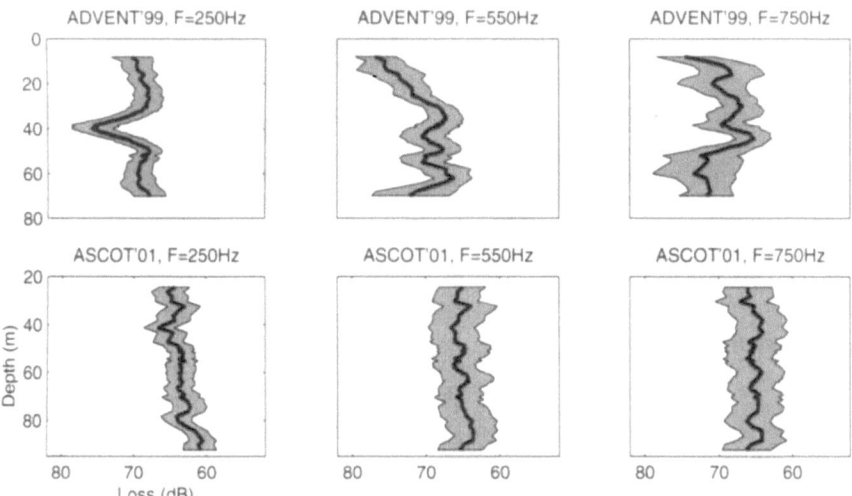

Figure 2. Time and frequency averaged TL from ADVENT'99 (upper panels) and ASCOT'01 (lower panels) received at the 10-km propagation range. The TL is averaged over a 10-Hz band around 3 centre frequencies of 250, 550 and 750 Hz. The solid line and black-shaded areas are the mean and standard deviation respectively over a 12-h transmission period.

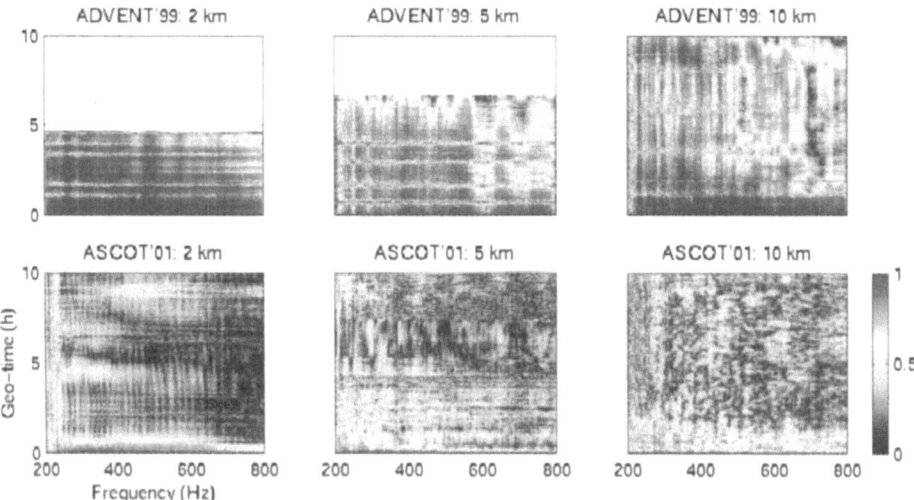

Figure 3. Matched-Field Correlation of an early received signal with the subsequently received signals from ADVENT'99 (upper panels) and ASCOT'01 (lower panels) for 10-h transmission time. The correlation is shown for 2, 5 and 10-km propagation range in the frequency band from 200 to 800 Hz.

ranges and for frequencies below 650 Hz during the entire transmission period. At higher frequencies and longer ranges the signals start to de-correlate within 1 h of transmission. This de-correlation is caused by changes in the environment, which have more impact at higher frequencies and longer ranges as the signals propagate through a larger amount of water mass. The de-correlation time for the received signals during the ASCOT'01

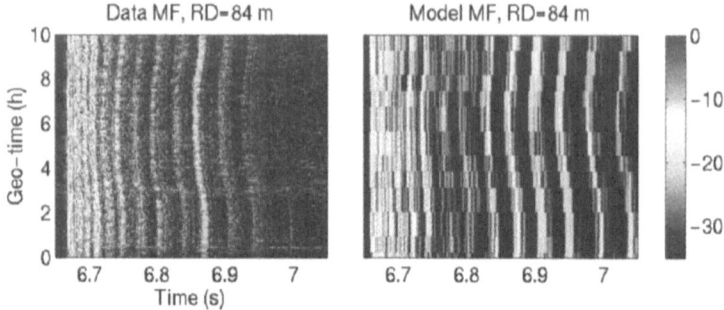

Figure 4. Data (left) and model (right) of the envelope of MF at 84-m depth received over a 10-h period.

experiments is much lower than for the ADVENT'99 data. The ASCOT'01 data de-correlate in less than 2 h regardless of propagation range but the de-correlation time is slightly longer at lower than at higher frequencies. At later transmission time, high correlation can be observed within a short period and for particular frequencies. The frequencies where high signal correlation appears depend on the time of transmission (5 and 10-km track in Fig. 3 lower panels). High correlation of the signals along the 2-km track is observed for a large frequency band after 5 and 9 h of transmission. This indicates significant changes in the environment for a period during the transmissions, and then the environment returns close to the initial condition. The MFC clearly shows that the ASCOT'01 environment is significantly more time and range varying than the ADVENT'99 environment, which has severe impact on broadband sound propagation in this region. Prediction of sound propagation in the ASCOT'01 environment is extremely difficult without a detailed spatial- and temporal description of the environment.

4   Model-data comparison

Fully range-dependent acoustic propagation modelling has been performed for both the ADVENT'99 and ASCOT'01 scenarios to assess the feasibility of predicting the acoustic signals by including the detailed measurements of the underwater environment in the modelling. The measured range-dependent bathymetry, tidal effects, sound-speed profiles varying in time and range, and geoacoustic properties from inversion of the acoustic data are used as input to the propagation model. High-fidelity geoacoustic inversion was achieved for the ADVENT'99 data, and an excellent agreement between acoustic modelling results and data was obtained in frequency, range and time [7]. The bottom properties for ASCOT'01 are known with less confidence than for ADVENT'99. Acceptable geoacoustic inversion results have only been achieved along the 2-km track. These bottom properties are assumed range-independent out to 10 km for the modelling purposes. The propagation model used is the coupled normal-mode model C-SNAP [9]. The model-data comparison of the tidal effect is shown in Fig. 4 as the MF signals received at 84-m depth over transmission time. There is good agreement between data and model with the correct model prediction of the tidal effect of the late multi-path arrivals. Note the slightly higher amplitude of the late arrivals in the modelling results compared to the data. This discrepancy in amplitude due to insufficient knowledge of the bottom properties.

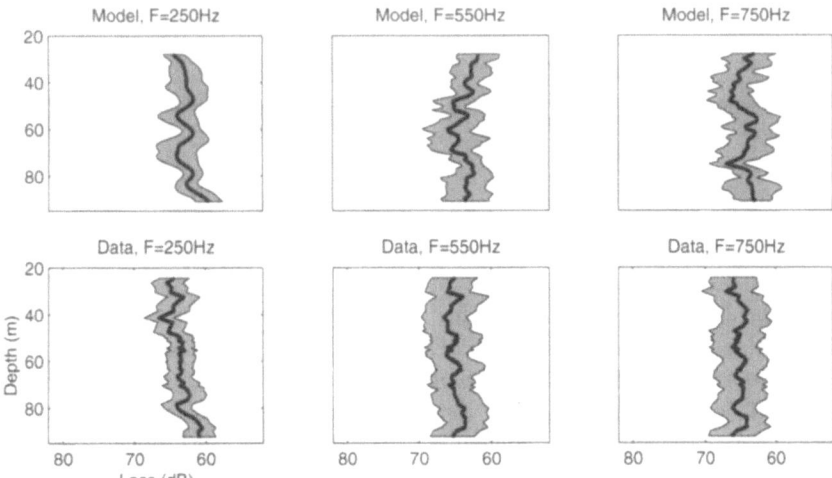

Figure 5. Model (upper panels) and data (lower panels) of frequency- and time averaged TL for centre frequencies of 250, 550 and 750 Hz received at a range of 10 km.

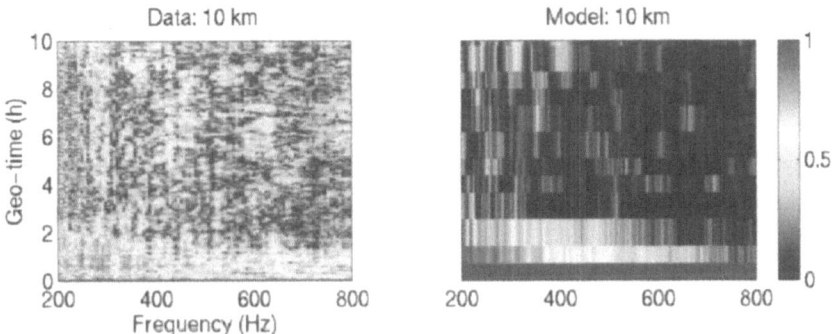

Figure 6. Data (left) and model (right) of MFC at a range of 10 km. The MFC is shown for a 10-h period in the frequency band from 200 to 800 Hz.

The modelled TL for centre frequencies of 250, 550 and 750 Hz is shown in Fig. 5 together with the experimental data at a range of 10 km. The same averaging in frequency and time is applied to the modelling results as for the data. There is a fairly good agreement between model and data of the mean TL levels but the modelling results are not completely matching the interference structure in depth. The standard deviation of the TL data is also captured by the propagation model by including the measured time- and range varying environmental properties.

The modelling of signal correlation over transmission time follows the same tendency as observed in the data (Fig. 6). The correlation time is 2 h for frequencies around 300 Hz, but the correlation time decreases as the frequency increases. These features in the modelling results can only be achieved if a good representation of the environment is available.

In general, the time- and range varying properties of the measured acoustic field for the ADVENT'99 and ASCOT'01 experimental sites are predictable if detailed information about the environment is available. The acoustic propagation model includes the main time- and range dependent features observed in the acoustic data. The uncertainty in predicting sound propagation in shallow water is mainly a question of predicting the state and changes in the environment accurately rather than the reliability of acoustic propagation models.

5 Conclusions

Oceanographic and acoustic data have been presented from the 2 shallow-water fixed propagation path experiments ADVENT'99 and ASCOT'01. The ADVENT'99 was conducted under *benign* conditions while the ASCOT'01 environment was known *a priori* to be hazardous for sound propagation. The ASCOT'01 environment is more time- and range dependent than ADVENT'99 which is reflected in the variability of the received acoustic signals. These results demonstrate the diversity of broadband sound propagation in two "typical" shallow-water regions. Successful prediction of sound propagation in the ADVENT'99 and ASCOT'01 region is achievable provided a detailed spatial and temporal environmental description is available. The uncertainty of predicting the sound propagation during ADVENT'99 and ASCOT'01 is not introduced by the acoustic propagation models but rather the prediction of the underwater environment.

Acknowledgments

The authors gratefully acknowledge the participants of ADVENT'99 and ASCOT'01 experiments. Special thanks to the Engineering Department at SACLANTCEN and crew on R/V *Alliance* for their efforts during the experiments.

References

1. Jensen, F.B., Comparison of transmission loss data for different shallow-water areas with theoretical results provided by a three-fluid normal-mode propagation model. Rep. CP-14, SACLANT Undersea Research Centre, La Spezia, Italy (1974) pp. 79–92.
2. Hermand, J.-P. and Gerstoft, P., Inversion of broadband multitone acoustic data from the Yellow Shark summer experiment, *IEEE J. Oceanic Eng.* **23**, 324–346 (1996).
3. Knobles, D.P., Westwood, E.K. and LeMond, J.E., Modal time-series structure in a shallow-water environment, *IEEE J. Oceanic Eng.* **23**, 188–202 (1998).
4. Zhou, J., Zhang, X. and Rogers, P.H., Resonant interaction of sound wave with internal solitons in the coastal zone, *J. Acoust. Soc. Am.* **90**, 2042–2054 (1991).
5. Lynch, J.F. *et al.*, Acoustic travel-time perturbations due to shallow-water internal waves and internal tides in the Barents Sea Polar Front: Theory and experiment, *J. Acoust. Soc. Am.* **99**, 803–821 (1996).
6. Apel, J.R. *et al.*, An overview of the 1995 SWARM shallow-water internal wave acoustic scattering experiment, *IEEE J. Oceanic Eng.* **22**, 465–500 (1997).
7. Siderius, M., Nielsen, P.L., Sellschopp, J., Snellen, M. and Simons, D., Experimental study of geo-acoustic inversion uncertainty due to ocean sound speed fluctuations, *J. Acoust. Soc. Am.* **110**, 769–781 (2001).
8. http:/pubs.usgs.gov/factsheet/fs78-98.
9. Ferla, M.C., Porter, M.B. and Jensen, F.B., C-SNAP: The Coupled SACLANTCEN normal-mode propagation loss model. Rep. SM-274, SACLANT Undersea Research Centre, La Spezia, Italy (1993).

VARIABILITY, COHERENCE AND PREDICTABILITY OF SHALLOW WATER ACOUSTIC PROPAGATION IN THE STRAITS OF FLORIDA

H.A. DEFERRARI, N.J. WILLIAMS AND H.B. NGUYEN

RSMAS – University of Miami, 4600 Rickenbacker Cswy, Miami FL 33149, USA
E-mail: hdeferrari@rsmas.miami.edu

Results of two shallow water propagation experiments are analyzed and compared with model predictions using observed environmental parameters as model inputs. The site of the experiments is off the coast of South Florida near Ft. Lauderdale nearby the future location of the planned Acoustic Observatory. Unique to the Florida Straits Propagation Experiments (FSPE) is an autonomous source that transmits broad band pulse-like signals at each of six center frequencies from 100 to 3200 Hz in octave steps. The transmissions last for 28 days and are received with a 32 element vertical array that is connected to shore by fiber-optic cable. Pulse arrivals along water born paths are identified by comparison with PE and normal mode model predictions. Three mode/ray groups of arrivals are identified: 1) RBR arrivals, which refract in the water column and interact with the bottom below the critical angle. These modes have low loss and nearly identical group velocities so that they coalesce to form a very intense focused arrival, 2) SRBR arrivals, that are spread in time and have increasing bottom angle with mode number and 3) numerous and mysterious late arrivals that couple with deep layers of the bottom and rapidly attenuate with higher frequency. The ocean environment, near the edge of the Florida Current, is highly variable with a saturated GM internal wave field and relatively large sub-inertial fluctuations from eddies and stream meanders. Sound speed fluctuations are generally 1 to 2 orders of magnitude larger than observed in the deep ocean. The bottom is composed of unconsolidated carbonate granules that have the density of sand and attenuation of fine sediment. Fluctuation statistics and coherence are computed and modeled in a parameter space of range, depth and frequency. The acoustic propagation, like the environment, is highly variable and complicated and many new interesting dependencies are revealed.

1 Introduction

The acoustic measurements reported here are from a 10-km propagation experiment conducted in Dec/Jan of year 1999–2000. The system installation and general features of the range site are described in a previous paper. The source and receiver arrays were situated along a nearly constant depth contour of 145 m as shown in Fig. 1. Two thermistor arrays were located symmetrically at ranges of 2.5 and 7.5 km from the source. The source is a multi-frequency and autonomous broadband transmitter that is moored and transmits for a period of one month under battery power. Several sets of transducers transmit m-sequence coded pulse trains at each of 6 carrier frequencies, $f_c = 100, 200,$

N.G. Pace and F.B. Jensen (eds.), Impact of Littoral Environmental Variability on Acoustic Predictions and Sonar Performance, 245-254.

400, 800, 1600 and 3200 Hz with a .25x f_c bandwidth. Each frequency was transmitted continuously for 1 h and then cycling to repeat sequence every six hours.

Data were processed following the SHARP methods of Birdsall and Metzger. The result is one pulse response per minute. The duration of sequence period is very nearly 2.55 s for each frequency and sample resolution is $1/f_c$ seconds. Once the data are averaged and pulse compressed they are stored on a server that can be access in either MATLAB of FORTRAN. The hourly pulse responses are readily access by specifying the transmission number (time), the frequency of the transmission and the hydrophone number (depth). For some of the time history of pulse response plots that follow the hourly samples are run together ignoring the five-hour gaps between samples.

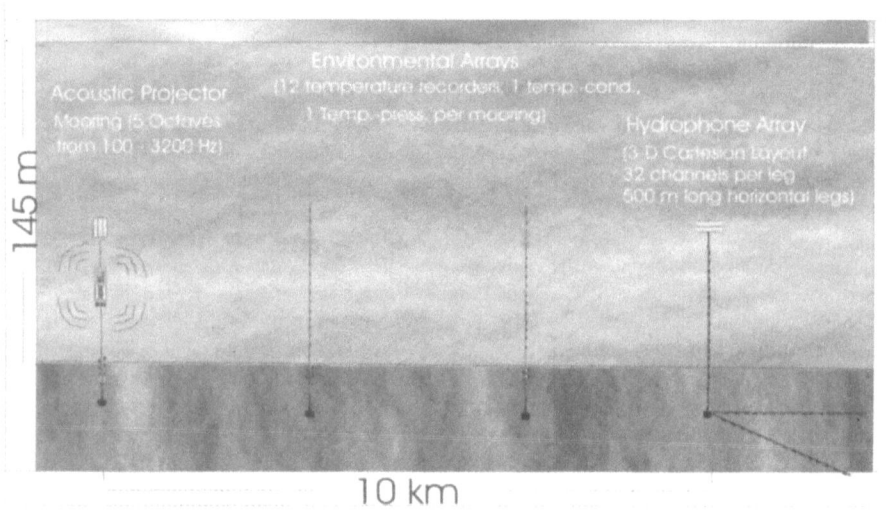

Figure 1. Experimental geometry of the 10-km Florida Straits propagation experiment.

Very energetic oceanography fluctuations and a highly variable sound speed field characterize the acoustic environment along the coast of south Florida. The mean sound speed profile is strongly downward refracting. The profile is typical of shelf areas shoreward of western boundary currents and comes about from the quasi-geostrophic balance of the current field. Sub-inertial fluctuations, with periods longer than the local inertial period of 25.6 h, result from meanders and eddies of the edges of the Florida current as well as coastal up- and down-welling produce very large variations in sound speed at inshore locations. Likewise internal waves and tides are energetic so that the overall sound speed variations are typically an order of magnitude greater than observed in the deep ocean.

The temperature data, Fig. 2, were collected during the acoustic experiment conducted Dec/1999 – Jan/2000 and are referenced to "experimental time". The computed sound speed profile is below. The 28-day long time series exhibits large slow fluctuations with roughly a fortnightly period. The temperature profile varies from an exceptionally strong thermocline to a nearly isothermal profile over the period. The cause is dynamical effects from the edges of the Florida Current (i.e. meanders and eddies). Perturbations to the temperature profile are evident over the internal wave band of frequencies although solitons are rare.

The total variability $< \Delta c / c >$ including the sub-inertial fluctuations, is a factor of 10 greater the typical internal wave fluctuations of Flatté for the deep ocean (Fig. 3). Comparing variations over the same IW band and including internal tidal contributions the Florida Straits site exhibits about twice the magnitude of the deep ocean.

The internal wave variability is hardly stationary. A burst of coherent wave trains, possibly a soliton, occurs around hour 200 and persists for two days with several 5–6 h. cycles. Likewise, the internal wave energy is greater during the time periods before and after the large changes in the mean profile. This observation is confirmed by the calculation of η^2 (Fig. 4) which approximates the internal wave potential energy

$$\eta' = T' / dT / dz, \tag{1}$$

where T' is the temperature perturbation over the internal waveband and dT/dz is the vertical temperature gradient. η is related to the potential energy of the internal waves by,

$$PE = (\rho / 2) N^2 \eta^2, \tag{2}$$

where ρ is the density and N is the Vaisala frequency.

Geo-acoustic properties of the bottom at the site of the experiments are not completely understood. The bottom sediment is thought to consist of unconsolidated carbonate covered with a veneer of finer sediment. The density of the carbonate is about the same as sand and the attenuation nearly that of fine sediment. Sound speed in the bottom is upward refracting but little is known about the sub-strata and sub-bottom at this site of these experiments. Further to the north, cores have been analyzed and bottom properties described by Monjo siting four studies as follows; "The sediment is 25 to 100 m thick composed of partially lithified sand or silty sand, made up of approximately 85% carbonate material."

The geo-acoustic model constructed by Monjo, when used with PE and normal mode propagation models predicted channel pulse responses in good agreement with measurements for previous experiments. A similar bottom and sediment model is used for the predictions that follow (Table 1).

Table 1. Geoacoustic model.

Velocity (m/s)	Gradient (1/s)	Density	Loss (dB/km/Hz)	Shear (m/s)	Shear Loss (dB/km/Hz)
1550	1.4	1.85	0.30	300	3.30

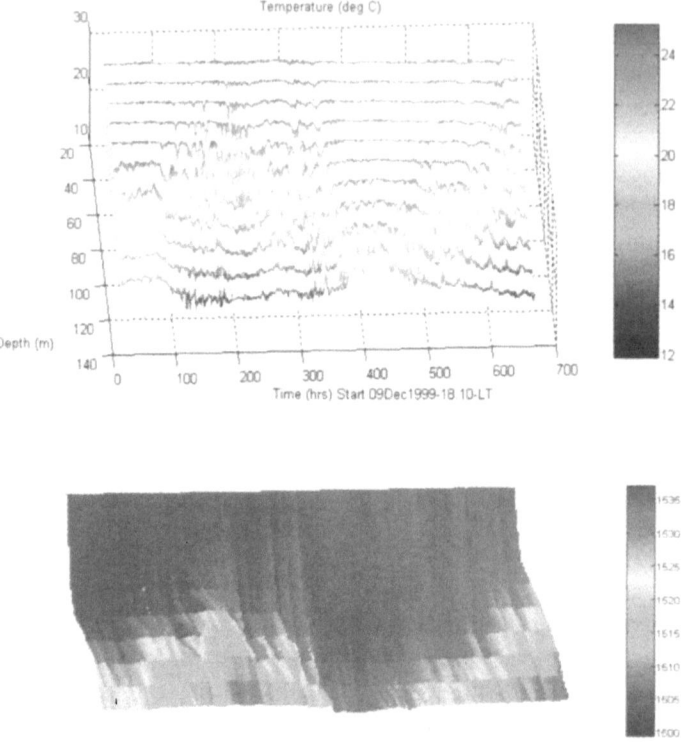

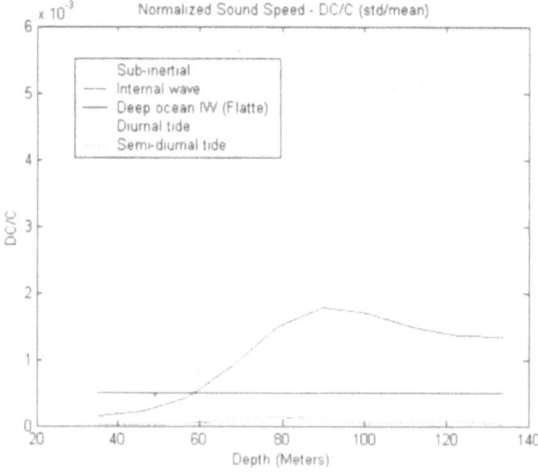

Figure 2. Temperature observations and the computed sound speed profiles.

Figure 3. Averaged normalized sound speed variations over various frequency bands.

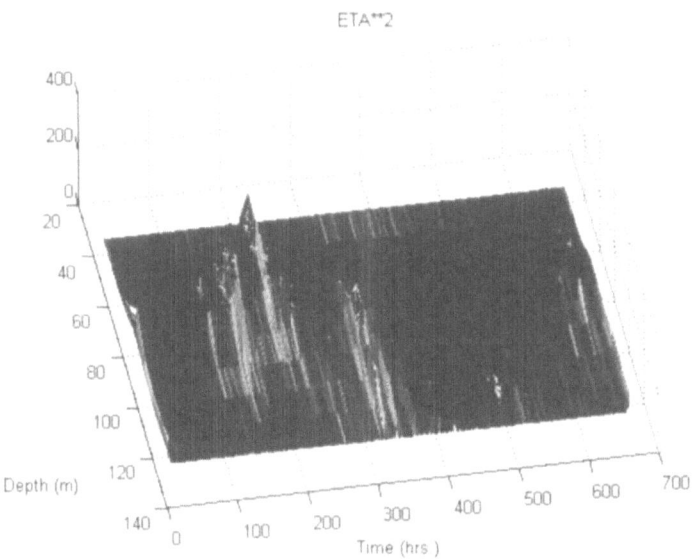

Figure 4. η^2 vs depth and time.

2 Analysis

The objective of the experiments is to study fluctuations, coherence and predictability in a parameter space of frequency, range of transmission and receiver depth. The computed measures are straightforward.

The coherency is computed as an averaged lagged product in space or time. The temporal coherence has three time variables: 1) T - the arrival time along the pulse, 2) t - the arrival time of the pulse in 1min intervals and 3) τ - the pulse to pulse lag time also in minutes.

$$COH(T,t,\tau) = < p(T,t)^* p(T,t+\tau) >^2 / < |p(T,t)| >^2 < |p(T,t+\tau)| >^2 . \qquad (3)$$

In this way we compute a coherency for every cycle of the received pulse response. For fluctuations we time gate arrivals, and compute intensity distributions and scintillation index SI is given by

$$SI = < I^2 > / < I >^2 -1 . \qquad (4)$$

The scintillation index can be computed for a particular arrival at several depths.

Predictability is studied by comparing measured pulse responses to the predicted pulse responses using propagation models with measured sound speed profiles as inputs. The models used are:

1. **PROSIM** Broadband Normal Mode Model , F. Bini-Verona, P.L. Nielsen and F.B. Jensen, SACLANTCEN SM-358 (based on ORCA model by E. Westwood *et al*).

2. **SNAP**: Saclantcen Normal mode Acoustic Propagation model, F.B. Jensen and M.C. Ferla (with SUPERSNAP solution engine by M.B. Porter and E.L. Reiss).
3. **MMPE**: Monterey-Miami Parabolic Equation model, K.B. Smith and F.D. Tappert.

3   Identifying arrival groups

We report on data collected during the first 14 days of the 10 km experiment. The sound speed profile varies from strongly downward refracting during the first few days to nearly isothermal profile during the two day. (Hrs 104 through 140, Fig. 2.) Figure 5 displays 6 1-hour long samples of the 800 Hz pulse reception. The blue gap between hourly records is 5 h in duration. Persistent arrivals result from two types of water borne paths, surface reflected – bottom reflected (SRBR) modes, comprising the early arrivals, and refracted – bottom reflected BRB modes that focus in time to form the single intense late arrival. This is consistent feature of the 200, 400 and 800 Hz pulse responses.

Both PE and normal mode models predict similar pulse responses at all frequencies. A PE prediction of the pulse response vs. range (Fig. 6) shows the SRBR arrivals fanning out in time with increasing range while the BRB group remains focused. Normal mode calculations of the group velocity (Fig. 7) show that the group velocity of the first 10 RBR modes is very nearly constant thus explaining the focusing. Figure 8 shows the 200 Hz pulse response for 14 days. The focused arrival persists until the sound speed gradient weakens during day 12.

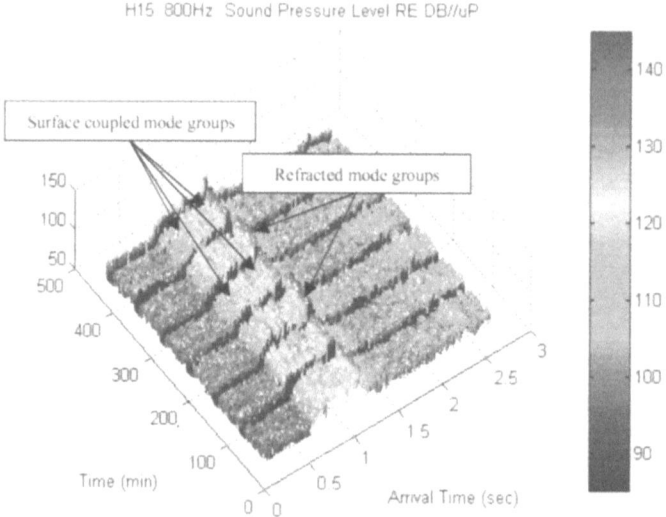

Figure 5. Pulse responses for the 800 Hz transmission showing SRBR and BRB mode groupings.

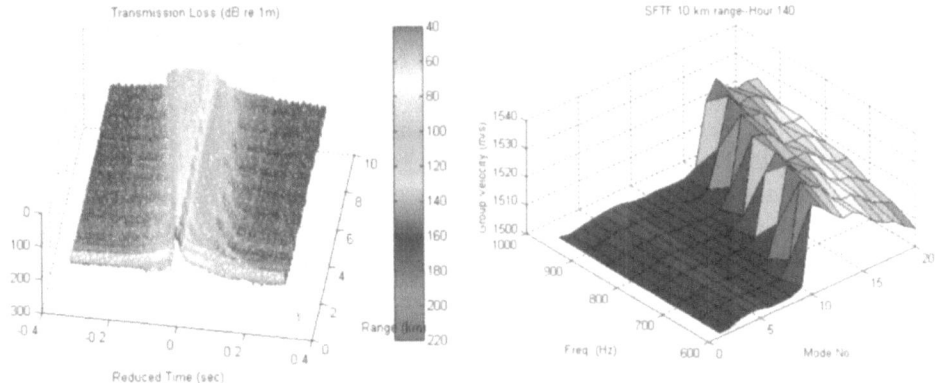

Figure 6. PE prediction of 800 Hz pulse vs. range.   Figure 7. Group velocity for 800 Hz modes.

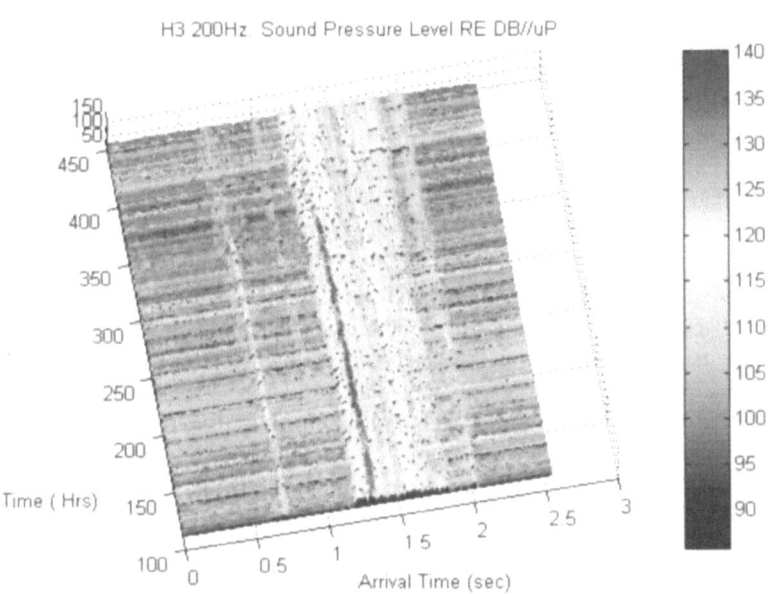

Figure 8. Pulse response for the 200 Hz. data fore 14 days. The BRB mode group focused arrival persists until the gradient weakens during hour 375.

4   Frequency dependence of intensity and coherence

Pulse response arrival patterns have been analyzed for an 11-day period during which time the sound speed profile remained strongly downward refracting. Beyond 11 days the profile became nearly iso-velocity and it was no longer possible to identify the same sets of arrivals. The discussion and conclusions that follow are based on some 44 plots like the example shown in Fig. 9. For five consecutive hours, pulse responses measured every minute and coherency as computed by Eq. (3) are shown for center frequencies of 200, 400, 800, 1600 and 3200 Hz. Some of the conclusions that follow are not clearly exemplified with this figure and the example is presented as a guide for discussion and not as proof. It is not practical to present all the data.

For all frequencies the waterborne paths (discussed above) arrive during the time interval of 1–1.2 s. The later arrivals that are more intense at lower frequencies are not yet identified. We suspect they come about from modes that travel deep into the bottom perhaps as much as 100 to 200 m. In fact, we can simulate the observed arrivals with normal mode models by admitting some layers deep in the sub-bottom. More information about the geo-acoustic properties is needed to resolve the issue.

The later arrivals are remarkably coherent and stable in time as one might expect if a significant portion of the propagation path is through the bottom. In the example, the arrival at time 1.45 s, for the 200 Hz pulse, is less intense but more coherent than any of the water born paths.

The late modes strip away with increasing frequency and are barely detectable for the 800 Hz signal. At first look, the mode stripping continues through the waterborne paths with the late arrivals attenuating more rapidly with increasing frequency. Mode stripping usually results from the higher order mode incurring more bottom loss because of steeper bottom angles. But here, the late arriving refracted modes have lower bottom angles than the earlier surface reflected modes.

Another loss mechanism may be at play. At 200 Hz the reception is dominated by the focused refracted arrival which is on average 15 dB higher than the individual SRBR arrivals. Likewise at 400 Hz. But the focused late arrival at 1600 Hz is about the same level as the SRBR's. Further, the RBR mode arrivals are always much less coherent in time generally decorrelating in less that half the time of SRBR mode arrivals that have nearly the same arrival time and bottom angle. The essential difference in the two mode types in that the BRB modes interact with stratified density fluctuations with near zero grazing. We suspect that there is a substantial loss associated with volume scattering near turning points. The lower the frequency the more lossless is the refraction. Model that use smooth sound speed profiles miss this effect and over predict the intensity of the focused BRB group.

Statistics of intensity fluctuations were computed for several receiver depth below and above the source depth. The RBR mode arrivals were time gated for these calculations. Results are summarized in Fig. 10. As the receiver depth approaches the source depth the intensity distribution looks more nearly Rayleigh; further away either above or below the distributions become more log–normal with increasing SI.

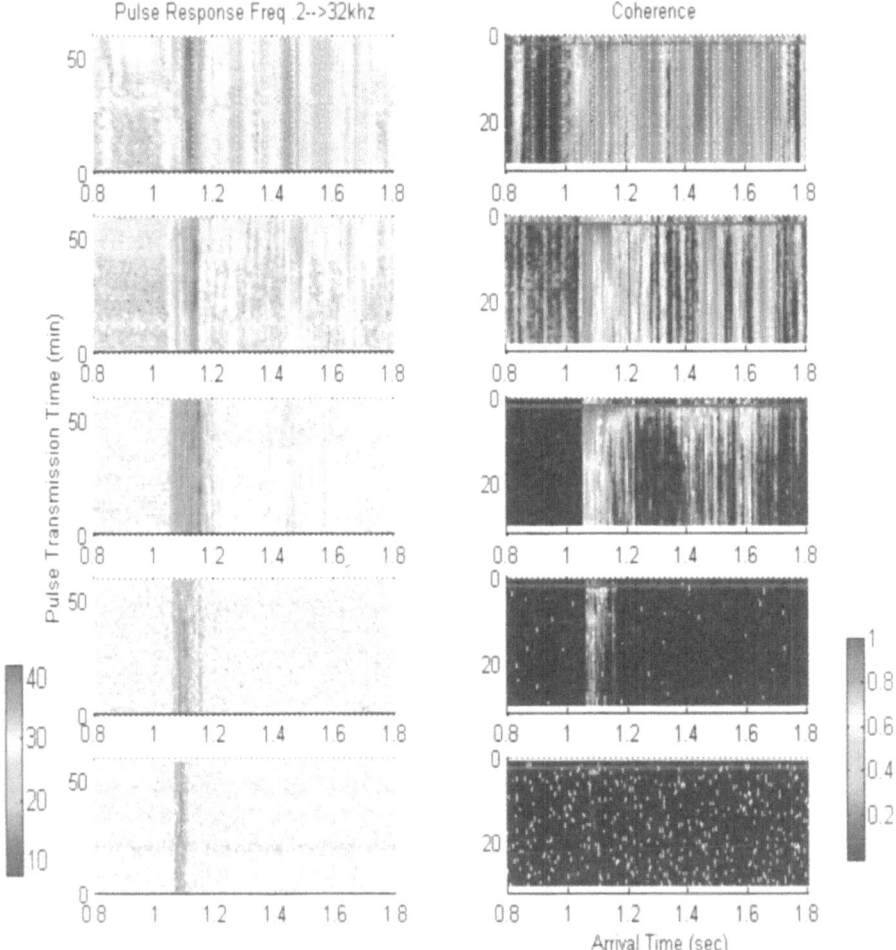

Figure 9. Left column: Pulse intensity (dB//up) vs arrival time for 200, 400, 800, 1600 and 3200 Hz transmission – 58 1min samples. Right column: Coherency for the corresponding pulse history.

Near the source depth all of the BRB modes have about equal amplitude and multipath interference dominates and hence saturated Rayleigh statistics. Further above or below one or two modes dominate and the distributions are more log–normal in appearance. We suspect, but have not yet been able to establish, that the distributions of fluctuation of the intensity of single modes will be log–normal.

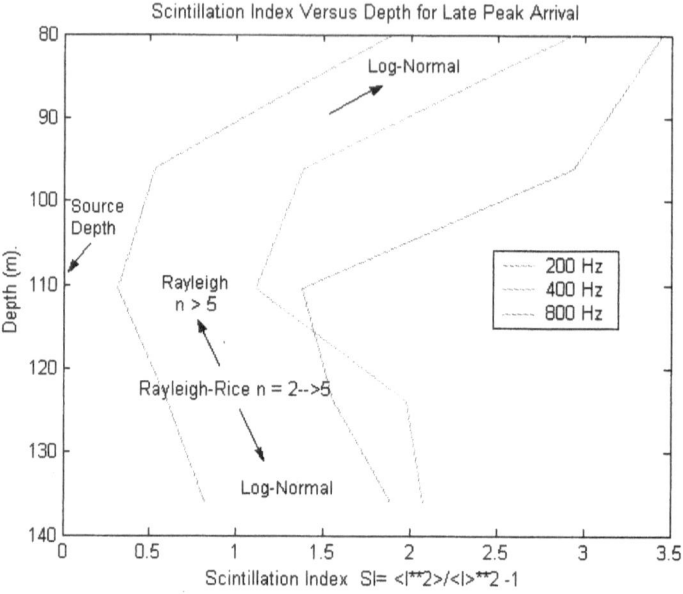

Figure 10. SI versus depth for BRB mode group arrival.

5   Summary

Low frequency pulse propagation is dominated by a single arrival consisting of several unresolved RBR modes – about 3 modes for the 200 Hz signals, 6 for the 400 and 12 for the 800. As much energy is carried in an SRBR group, they fan out in time and are readily resolved for the higher frequency measurements. In fact, arrivals associated individual modes are resolvable, stable and persistent for many hours even for the 3200 Hz signals. SRBR modes are generally more coherent and stable for all frequencies. There is a large observed transmission loss (10dB+) for RBR modes at higher frequencies that is not consistent with bottom loss. Models don't predict the loss and SBRB modes are immune. We hypothesize the loss results from volume scattering at low grazing near turning depths.

AMBIENT NOISE AND SIGNAL UNCERTAINTIES DURING THE SUMMER SHELFBREAK PRIMER EXERCISE

PHILIP ABBOT, CHARLES GEDNEY AND IRA DYER[1]
Ocean Acoustical Services and Instrumentation Systems, Inc.,
5 Militia Drive, Lexington, MA 02421, USA
E-mail: abbot@oasislex.com, gedney@oasislex.com, idyer@aol.com

CHING-SANG CHIU
Naval Postgraduate School, Monterey CA, 93943, USA
E-mail: chiu@nps.navy.mil

Uncertainties in noise level, and in signal level after long-range (42 km) acoustic shallow water transmissions, from a pulsed source, are determined from the summer shelfbreak PRIMER experiment. Fluctuations over the 10-day period are not stationary, but are rendered so by tracking the wandering of their means. Then narrow-sense stationary probability density functions are obtained of ambient noise and signal peak transmissions from a match-filter output with time-bandwidth product = 1. The data are centered at 400 Hz, in a 100 Hz bandwidth, and analyzed from three individual hydrophones of a vertical line array. The ambient noise fluctuations closely follow the phase-random Log-Rayleigh density, with standard deviation σ = 5.6 dB. Signal peak statistics are determined from demeaned 50 s samples. The signal statistics over an 8-h period are approximately similar, but not identical, to those over the entire 10-day period. The signal has narrower histograms ($\sigma \approx 0.8$ dB) than the noise. Log Chi-Square densities, fit to the signal histograms, suggest that about 30 equal intensity components contribute to the fluctuations, many more than can be attributed to the idealized modal structure in the shelfbreak duct. This suggests that either strong scattering affects the signal transmission, or the signal process is not fully phase-random.

1 Introduction

Realistic sonar performance predictions are served by understanding the causes of temporal and spatial fluctuations in the environment [1–3]. The 1996 summer shelfbreak PRIMER exercise [4–8] provided a high resolution environmental acoustics data set that we use to study the temporal variability of noise and acoustic transmissions. The intent is to evaluate and characterize fluctuations of the ambient noise and signals transmitted over a long range, in a shallow water, downward-refracting environment.

2 Background

The summer Shelfbreak PRIMER experiment was conducted from July 24 to August 2, 1996. The experimental site covered a 60 km square at the shelfbreak of the Middle

---

[1] Also, MIT Department of Ocean Engineering, Cambridge, MA 02139.

N.G. Pace and F.B. Jensen (eds.), Impact of Littoral Environmental Variability on Acoustic Predictions and Sonar Performance, 255-262.

Atlantic Bight, south of New England, as shown in Fig. 1. The site was selected because of the presence of complex oceanographic phenomena affecting sound propagation across the shelfbreak. These phenomena include the meandering shelfbreak front, and the nonlinear, large-amplitude internal tides and internal solitary waves.

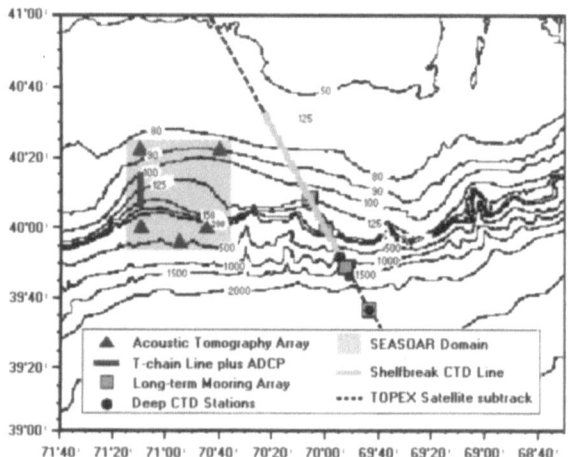

Figure 1. Shelfbreak PRIMER exercise area (triangles show source and receiver locations).

Figure 1 shows the locations of three acoustic sources and two receiver arrays. We consider the propagation along the western edge of the site (from the 400 Hz source at the southwest corner to the array at the northwest corner). The propagation range is 42 km, with water depths of 299 m and 85 m at the source and receiver, respectively. The source was at a depth of 294 m and the receiver array was deployed from 30.5 m to 83 m below the surface (16 hydrophones, 3.5 m spacing). In this paper, we consider signals and noise measured by hydrophones located at depths of 30.5, 55 and 83 m.

The source radiated a 5.11 s pseudo-random binary sequence which provided a pulse compression gain of 27 dB at the output of the matched filter. Pulses were repeated every 5.11 s, over a period of 5 min (resulting in 45–50 pulses per transmission cycle). After each 5 min transmission cycle, the source was shut off for 10 min (while other sources operated). This sequence was repeated for all 10 days of the exercise. The center frequency of the pulse was $f = 400$ Hz, with bandwidth $B = 100$ Hz, at a source level of about 180 dB re 1µPa at 1m. The received signals were processed using a matched filter and the output of this filter is used in the analysis presented here. The τB product of the processed pulses and noise samples is ≈ 1, that is, the temporal resolution $\tau \approx 10$ ms.

3   Analysis and discussion

The matched filter output received at the 30.5 m hydrophone for a typical transmitted pulse is shown in Fig. 2. The peak signal level is about 85 dB re 1µPa and arrives at about 28 s (source to receiver time delay). The signal then decays into the noise, which has peaks of about 70 to 75 dB re 1µPa. For the present analyses, each pulse is divided into two regions: an ambient noise region, with time delays greater than 29 s; and a signal plus noise region, with time delays less than 29 s.

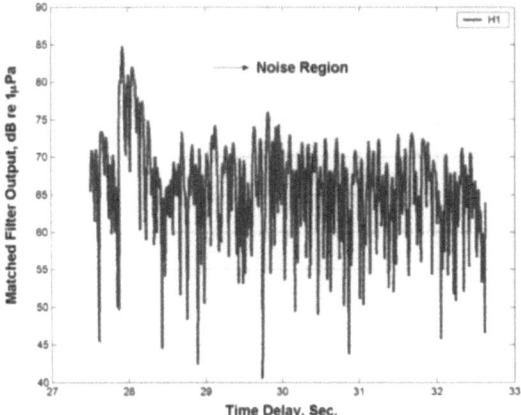

Figure 2. Typical matched filter output time series for a single 5.11-s transmission pulse at the 30.5-m hydrophone. Recorded on 8/1/96.

The mean (μ) and standard deviation (σ) of the matched filter output (in dB) for the ambient noise region is measured for each pulse within a transmission cycle. Figure 3 shows the variations in μ and σ computed in the noise region at the three hydrophones during a typical 5-min transmission cycle. This figure shows that the noise is spatially and temporally wide-sense non-stationary, with varying μ, but σ is relatively constant at about 5.5 dB. This behavior is also observed in the noise data measured at other hydrophones and at other times throughout the exercise. To render the noise narrow-sense stationary, each noise sample was corrected for μ at each hydrophone to give a zero-mean sample set. The resulting histogram, demeaned and aggregated for all 16 hydrophones, over the 5-min transmission cycle (48 pulses) is shown in Fig. 4 (series of dots, normalized to match the PDF). From Dyer [2] we may suppose that, for $\tau B \approx 1$, acoustic ambient noise can be represented as a phase-random process, comprised of one arrival from one distant ship, or another ambient source, resulting in the Log-Rayleigh probability density function (PDF), with standard deviation $\sigma = 5.6$ dB. Figure 4 compares the measured histogram with the Log-Rayleigh PDF, and supports the foregoing supposition (with $\sigma = 5.5$ dB). Similar results were obtained at all other times ($\sigma = 5.5$ to 5.6 dB), and the means wandered. The figure confirms that the phase-random single-component PDF is an excellent representation of the measured noise fluctuations.

The peak output of the matched filter (referred to as the "signal peak") as determined from the signal plus noise region was also recorded for each pulse within a given transmission cycle. Figure 5 shows the signal peak for the shallow hydrophone as measured over an 8-h period on July 24 1996 (0545 to 1345). The noise means are also shown in this figure for comparison. The figure shows the signal and noise at the matched filter output and the corresponding signal fluctuations that are of interest in the present study. In particular, the signal level varies by about 16 dB (from about 80 to 96 dB re 1 μPa) over the 8-h period. It is also interesting to note that the noise mean spreads by about 16 dB as well.

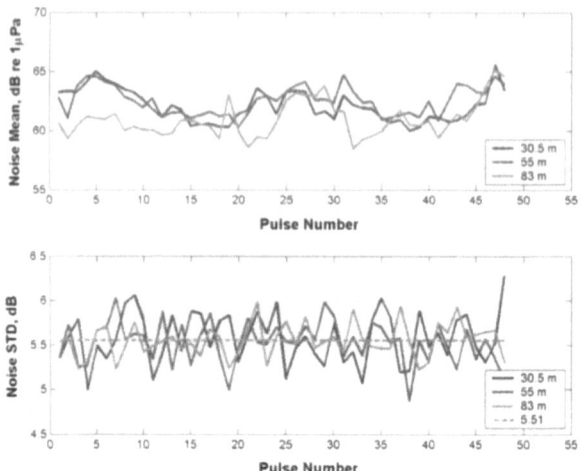

Figure 3. Variation of ambient noise μ and σ for hydrophones located at 30.5, 55 and 83 m depths during a typical 5-min transmission cycle. The time between pulses is 5.11 s. Recorded 8/1/96.

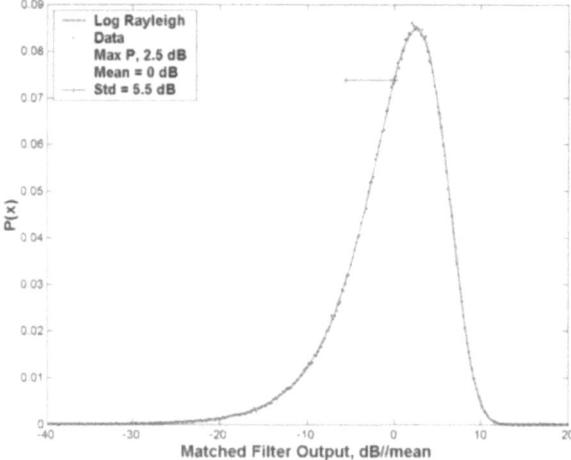

Figure 4. Comparison of the zero-mean noise PDF aggregated in depth over a 5-min transmission series with the single-component phase-random (Log-Rayleigh) PDF. The data points are the normalized height of the histogram level in 0.2 dB bins. Recorded on 8/1/96.

Since there is only one signal peak for each pulse, it is inappropriate to process the signal peaks the same way as for the noise. Rather, we demean the signal peaks every 10 pulses (50 s). We choose $N = 10$ because it is large enough to show fluctuations in the sample set (and it provides 4 separate groups of 10 samples within a given transmission cycle). Thus for each sample set of 10 pulses (50 s), the μ and σ are determined, the samples demeaned and aggregated, then the process is repeated for the next 10 pulses and so on, for a fixed observation interval (initially set to 8 h). The resulting 8-h histograms over the period as in Fig. 5, and for the three hydrophones, are given in Fig. 6(a)–(c), with the corresponding σ ≈ 0.87, 0.76, and 0.73 dB, respectively starting at the 30.5-m hydrophone. These are much smaller than those for the noise, and decrease slightly with increasing depth. Also shown in the figures are Log Chi-Square fits to the histograms, based on Dyer [2], with the supposition that the signal also is a phase-random process, but with more than one component. For simplicity, these components can be taken as equally intense, and then the fluctuation process would correspond to a number of independent phase-random transmission components, of order 25 to 36 (as shown in the figure). This greatly exceeds that which can be attributed to the idealized modal structure at the receivers in the shelfbreak duct, which suggests that either strong scattering affects the signal transmission, or the process is not fully phase-random. We need to explore these alternatives.

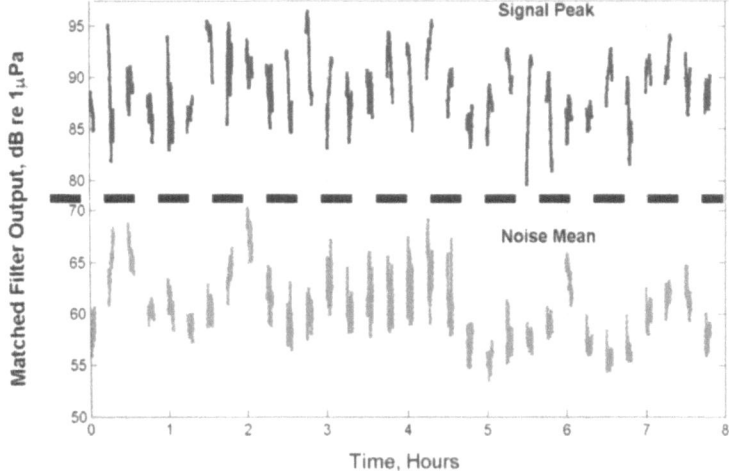

Figure 5. Signal peak output and mean noise level at the 30.5 m hydrophone, versus time for 8-h period, 7/24/96 (0545 to 1345). The blank periods correspond to the 10-min of each 15-min cycle when the source was off (other experiments had different sources transmitting during these times).

A histogram formed over the entire 10-day test (50 s sample size) is shown in   Fig. 6d) for the 83 m hydrophone. When compared to the one from the 8-h period on 7/24/96 (Fig. 6c), this histogram is similar in shape, with nearly identical σ. The K-S (Kolmogorov-Smirnov) 2-sample test indicates that the 8-h sample from 7/24/96 has the same PDF (with a 96 % probability) as the 10–day sample. Thus it appears that the fluctuations about the wandering means for an 8-h period adequately represents the fluctuations over the entire 10-day test period, while the means of course vary (see Fig. 8).

As an independent check of this observation, we observed the fluctuations from two other 8-h periods, one during 7/31/96, the other during 7/26/96. On 7/31/96, the signal peak histogram was similar to 7/24/96 (σ = 0.74 dB) with the K-S test indicating a 50% probability that it has the same distribution. Interestingly, the histogram from 7/26/96 was different, with σ = 0.86 dB, and the K-S test indicating a different PDF. Thus, the fluctuations on 7/26/96 were different relative to the 10-day sample set.

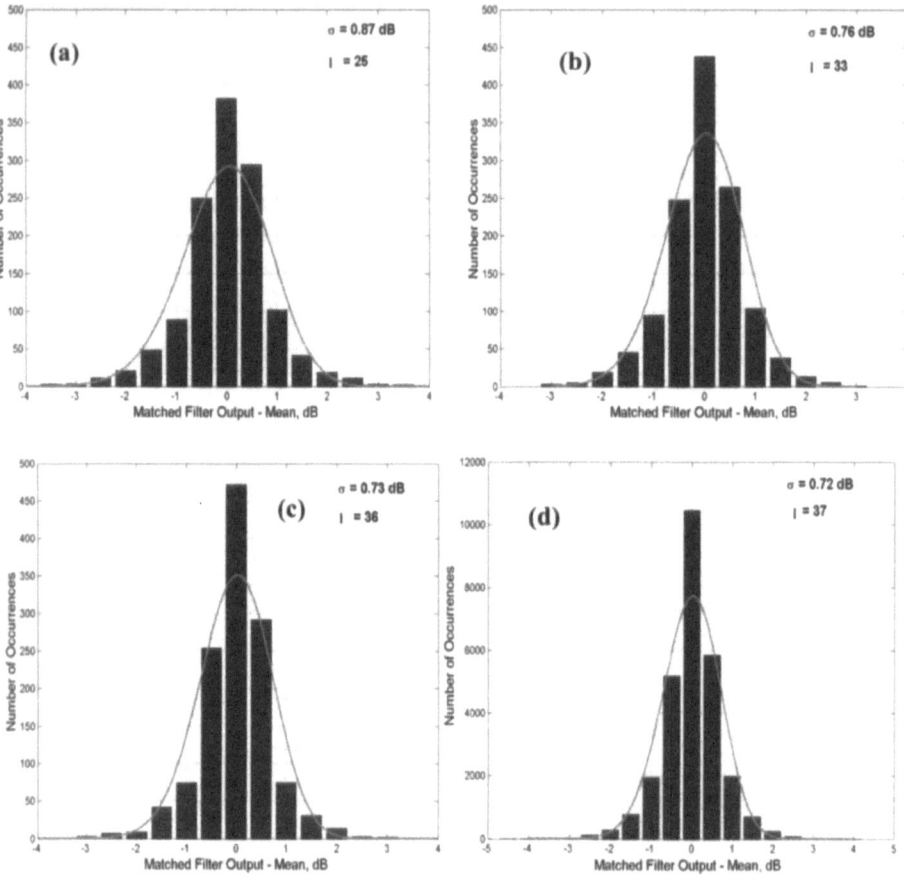

Figure 6. Histograms of demeaned signal peak (50 s sample size) and Log Chi-Square fit (using the l components as shown), during the 8-h period on 7/24/96 for hydrophones located at: a) 30.5 m, b) 55 m, c) 83 m, d) 10-day at 83 m hydrophone .

Figure 7 shows the demeaned σ plotted versus observation interval at the three hydrophones, for intervals of 1, 2, and 8 h, and 1, 2, 5 and 10 days. The figure shows that the σ tends toward a constant level of about 0.8 dB for intervals of 8 h and larger. Below 8 h, the σ are different and dependent on the observation interval, thus suggesting that the fluctuation mechanisms are different for the intervals below and above about 8 h.

In Fig. 8a, the histograms for the slowly wandering mean μ of the noise and signal at the 83 m hydrophone are shown for the 8-h period on 7/24/96. The noise μ varies considerably (with mean M = 59.5 dB re 1 μPa, and standard deviation Σ = 3.0 dB), very much like the noise data in Fig. 3 for a shorter observation time. The signal μ varies as well (M = 91.1 dB re 1 μPa and Σ = 3.1 dB). The wandering means from the 10-day period at the 83 m hydrophone are given in Fig. 8b. Over the 10-day period for the signal, M = 90.0 dB re 1 μPa and Σ = 4.0 dB. These figures show that the signal and noise wandering means have similar characteristics and that the environment is likely affecting both similarly.

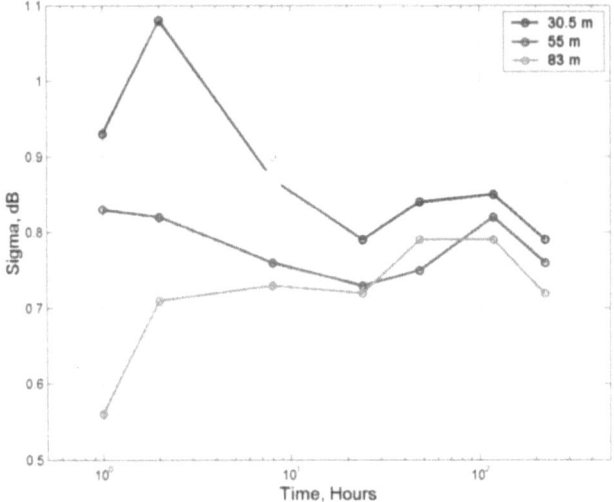

Figure 7. Measured σ versus observation interval (50 s pulse sample size) for hydrophones located at a) 30.5 m, b) 55 m and c) 83 m.

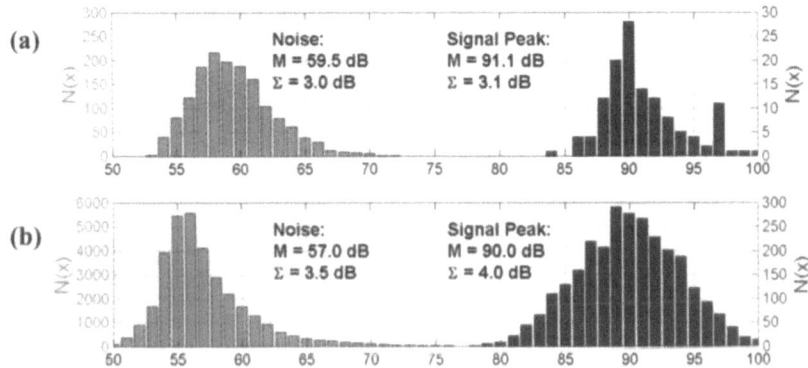

Figure 8. Histograms of noise and signal wandering means at the 83 m hydrophone, a) 8-h, 7/24/96, b) 10-day period.

4   Summary

By demeaning in small time windows, the summer PRIMER ambient noise and signal peak levels were rendered narrow-sense stationary, and the noise and signal histograms about their slowly wandering means were found. The noise fluctuations (about their wandering means) over the 10-day interval are closely phase-random, for they follow a Log-Rayleigh PDF, with $\sigma = 5.6$ dB ($\tau B \approx 1$). The signal peak fluctuations (also about their wandering means) are narrower than the noise with $\sigma \approx 0.8$ dB ($\tau B \approx 1$), for observation intervals of 8 h or larger. Fluctuations over an 8-h period are similar to those over the entire 10-day period, with one exception, suggesting a possible division at 8 h or less in the physical processes underlying the observed fluctuations.

We attempted unsuccessfully to fit signal histograms with an n-component phase-random process model [3] using estimates for the modal amplitudes. The data suggest that there is a significantly larger number of path arrivals contributing to the signals than predicted by the modal analysis. The possibility of micro-pathing in this environment, along with other possibilities, such as incomplete phase-randomness is left to future studies. Also, planned are fluctuation studies for observation intervals smaller than 8 h.

Acknowledgements

This work was sponsored by the Office of Naval Research. We also thank C. Miller (NPS) and C. Emerson (OASIS) for their assistance in processing the PRIMER data set.

References

1. Abbot, P. and Dyer, I., Sonar performance predictions based on environmental variability. In *Impact of Littoral Environmental Variability on Acoustic Predictions and Sonar Performance*, edited by N.G. Pace and F.B. Jensen (Kluwer Academic, The Netherlands, 2002) pp. 611–618.
2. Dyer, I., Statistics of sound propagation in the ocean, *J. Acoust. Soc. Amer.* **48**, 337–345 (1970).
3. Dyer, I., Statistics of distant shipping noise, *J. Acoust. Soc. Amer.* **53**, 564–570 (1973).
4. Lynch, J., Gawarkiewicz, G., Chiu, C., Pickart, R., Miller, J., Smith, K., Robinson, A., Brink, K., Beardsley, R., Sperry, B. and Potty, G., Shelfbreak PRIMER - An integrated acoustic and oceanographic field study in the mid-Atlantic Bight. In *Shallow-Water Acoustics*, edited by R. Zhang and J. Zhou (China Ocean Press, 1997) pp. 205–212.
5. Miller, C., Estimating the acoustic modal arrivals using signals transmitted from two sound sources to a vertical line hydrophone array in the 1996 Shelfbreak PRIMER Experiment. Naval Postgraduate School M.Sc. Thesis, 1998.
6. Colosi, J., Lynch, J., Beardsley, R., Gawarkiewicz, G, Chiu, C. and Scotti, A., Observations of nonlinear internal waves on the New England continental shelf during summer shelfbreak PRIMER, *J. Geophys. Res.* (in press 2002).
7. Chiu, C., Realistic simulation studies of acoustic signal coherence in the presence of an internal soliton wavepacket. In *Proc. IOS/WHOI/ONR Internal Solitary Wave Workshop*, Victoria, B.C., Canada, October 27–29, 1998.
8. Chiu, C., Lynch, J., Gawarkiewicz, G., Miller, C., Sperry, B. and Newhall, A., Measurement and analysis of the propagation of sound from the continental slope to the continental shelf, (in preparation).

VARIABILITY EFFECTS DUE TO SHALLOW SEDIMENT GAS IN ACOUSTIC PROPAGATION: A CASE STUDY FROM THE MALTA PLATEAU

K.M. KELLY

QinetiQ, Winfrith Technology Centre, Dorchester, Dorset, UK
E-mail: kmkelly@qinetiq.com

A comparison was made between measured and modelled propagation loss data from the Malta Plateau. The measured data was obtained during November 1999 at 3.5 kHz, with both source and receiver beneath the thermocline. This maximised the interaction of sound with the seabed. Propagation loss was modelled using the Synthetic Pulse Reception Model (SPUR). Different sources of geoacoustic data for input to SPUR were compared and gave very different predictions for propagation loss. The variability observed in the data was attributed to the presence of shallow gas in the seabed sediments. Once this had been taken into account, a good match between the measured and modelled propagation loss was obtained. These results illustrate the importance of good environmental characterisation for sonar performance prediction and highlight the significant effect that shallow gas can have on the geoacoustic properties of sediments and the resulting acoustic propagation.

1   Introduction

In order to predict propagation loss for sonar performance assessment, a number of models are available. One of the most recent to be developed is the Synthetic Pulse Reception Model (SPUR). This model allows complex range dependent environments to be modelled.

To test the accuracy of the model predictions, and the effect of improved geoacoustic input to these models, a comparison was made with measured sonar data. The Low Frequency Active Sonar trial, Mercury'99, which took place on the Malta Plateau in November 1999, provided an ideal propagation loss dataset for this study. Good supporting environmental data was obtained during the trial and the trials area has been geologically and geoacoustically well characterised by both seismic surveys and core measurements.

GEOSEIS is a geoacoustic database which provides an alternative generic approach for obtaining geoacoustic information from areas where other sources of data may be absent. This study compared existing geophysical data available for the area, with GEOSEIS predictions, to provide geoacoustic parameters for input to SPUR. The results show a good match between predicted propagation loss using the best GEOSEIS geoacoustic prediction available for the region and the measured propagation loss data.

The most significant factor affecting the geoacoustic properties of the Malta Plateau sediment was the presence of shallow gas. Once this had been taken into account that the best match between measured and modelled propagation loss was obtained.

N.G. Pace and F.B. Jensen (eds.), Impact of Littoral Environmental Variability on Acoustic Predictions and Sonar Performance, 263-270.

2   Geoacoustics of the Malta Plateau

The Malta Plateau is the shallow water region between Sicily and Malta. The central portion of the plateau has an almost constant depth of about 140 m. To the east the 200 m contour marks the top of the Malta Escarpment which plunges steeply into the Ionian basin. To the west the plateau deepens gently to a slope break at about 160 m.

The acoustic basement over the plateau is associated with the Messinian sea level lowstand, when the Malta Plateau underwent sub-aerial erosion. This basement consists of Miocene limestones and dolomites overlying an upper Cretaceous volcanic horizon at the top of a Cretaceous dolomite succession [1].

The central Malta Plateau is an area of significant Plio-Quaternary sedimentation. An 8 to 12 m thick layer of very soft fine grained sediments covers a horizontally bedded succession of unconsolidated sediments. There are six seismo-stratigraphic units, each representing a phase of sedimentation followed by a period of erosion, probably associated with a fall in sea level. These units are thickest in the central plateau, thin towards the basement highs and pinch out against the topographic high to the east, where there is a rough region of exposed rockhead. The seabed is almost flat with occasional pockmarks. These combined with the acoustic signature of the top unit in seismic sections, indicate the presence of shallow gas in the uppermost Plio-Quaternary sediments [1].

To the west of the plateau the base of the slope consists of slumped deposits derived from the plateau region. This slumping would appear to be recent and the margin may still be active [1].

The geoacoustic parameters for the Malta Plateau sediments, for propagation loss modelling, were obtained from GEOSEIS [2]. This is a geoacoustic database containing data on a wide variety of sediment and rock types. It can be used to provide generic geoacoustic parameters for any given sediment. The more detailed sedimentological information that can be provided the more representative the geoacoustic parameters are likely to be. GEOSEIS has the potential to provide realistic geoacoustic parameters for areas where there is little or no geoacoustic data available.

Sedimentological data was available for five cores from the Malta Plateau [3], listed in Table 1. Of the five cores 254 and 255 were sedimentologically very similar, composed mainly of silt with small fractions of clay and sand and 30–45% $CaCO_3$. 256 was composed of fine silty sand and 258 dominated by silt and clay, both with a slightly lower $CaCO_3$ content. Core 257 has been excluded since it was from a region containing debris deposits adjacent to one of the rocky outcrops and was therefore not representative of the plateau region [3].

Table 1. Sedimentological analysis of core data [1].

Core	% gravel	% sand	% silt	% clay	% $CaCO_3$
254	3.1	28	58.7	10.2	34.9
255	1.4	30.5	48.1	18.3	38.6
256	0.6	68.7	21.3	9.9	30.1
258	1.2	9.6	44.9	44.3	29.6
Average	1.575	34.2	43.25	20.675	33.3

The sedimentological analysis from the cores was used to derive geoacoustic parameters using the GEOSEIS algorithms. Two sets of parameters were obtained GEOSEIS 1, which used the average sedimentological composition for the cores, and GEOSEIS 2, which also took porosity into account. This was slightly lower at 50–55% for the cores, than the average porosity for similar sediments in GEOSEIS (63%). The parameters obtained from GEOSEIS are average values for sediments of similar composition within the database and have been corrected for frequency dependence using the Kramers-Kronig relationship derived from Kolsky [4];

$$Vpf = Vpm\left[1 + \frac{1}{\pi Qp}\ln\left(\frac{f(kHz)}{m(kHz)}\right)\right] (ms^{-1}) \qquad (1)$$

where

f	is the required frequency,
m	is the measurement frequency,
Qp	is the quality factor,
Vp	is the p-wave velocity.

The final factor, which was taken into account, was the presence of gas in the Plio-Quaternary sediments of the Malta Plateau, as indicated by the pockmarks and seismic signature [1]. The presence of even only a small amount of gas in sediments can have a very significant effect on sediment acoustic properties, with velocities being reduced by as much as 15–50% [5]. Domenico [6] showed that a velocity decrease of 36% occurred between 6% and 13% gas content, indicating that a small but critical amount of gas can significantly alter the acoustic properties of the sediment.

Density and attenuation are also affected by the addition of gas. Density is reduced as gas replaces the pore water so decreasing the density of the pore fluid. Since the average porosity of the sediment is known it was possible to calculate a reduced density for a 10% gas content. The equation used to calculate the density was:

$$\rho_{sediment} = \rho_{grain}(1-\varphi) + \rho_{water}\varphi \qquad (2)$$

where

$\rho_{sediment}$	is the density of the fluid saturated sediment,
ρ_{grain}	is the density of the sediment grains,
ρ_{water}	is the density of the pore water,
φ	is the fractional porosity.

GEOSEIS contains data on gassy sediments and this allowed predictions to be made for the sedimentological analysis (GEOSEIS 1) data with gas contents of 2% and 5%. Domenico [6] showed that after the initial velocity decrease as gas content increased to 13%, velocity then remained nearly constant regardless of how much more gas was added. Additional predictions were therefore made for a 10% content in both the sedimentological analysis prediction (GEOSEIS 1) and the prediction taking porosity into account (GEOSEIS 2) by reducing velocity by 30%.

There is little information on the effects of gas on attenuation, however it does appear that attenuation is significantly increased. Doubling the attenuation for 10% gas content gave a reasonable approximation, which was consistent with GEOSEIS

For the purpose of this study the Malta Plateau was modelled as a range independent environment. However, it should be noted that there is variability across the region in both sediment composition, as indicated by the differences in the sediment compositions of the cores, Table 1, and in the sediment gas content [1].

3   Acoustics Results

Propagation loss runs were carried out using the Parabolic Equation model, SPUR. This is an improved version of the Range-dependent Acoustic Model (RAM). RAM is based on the split-step Padé solution [7], which allows large range steps and is the most efficient PE algorithm developed to date.

The Malta Plateau provided a range independent acoustic scenario. The seabed was relatively flat and featureless with little geoacoustic variability. Expendable Bathythermographs (XBTs) deployed during the acoustic experiment showed only slight oceanographic variability with time. Wind speed and sea state remained constant and low, true wind speed rarely exceeding 3 m/s. Figure 1 shows the sound speed profile used in the modelling. This was an average for the XBT profiles obtained during the experiment. For a source depth of 70 m this profile provides a strongly seabed interactive environment.

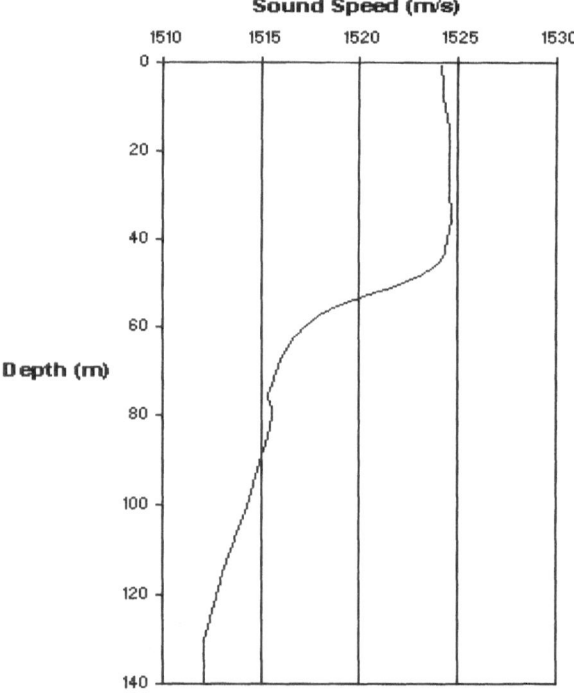

Figure 1. Averaged XBT profile used in the propagation loss modelling.

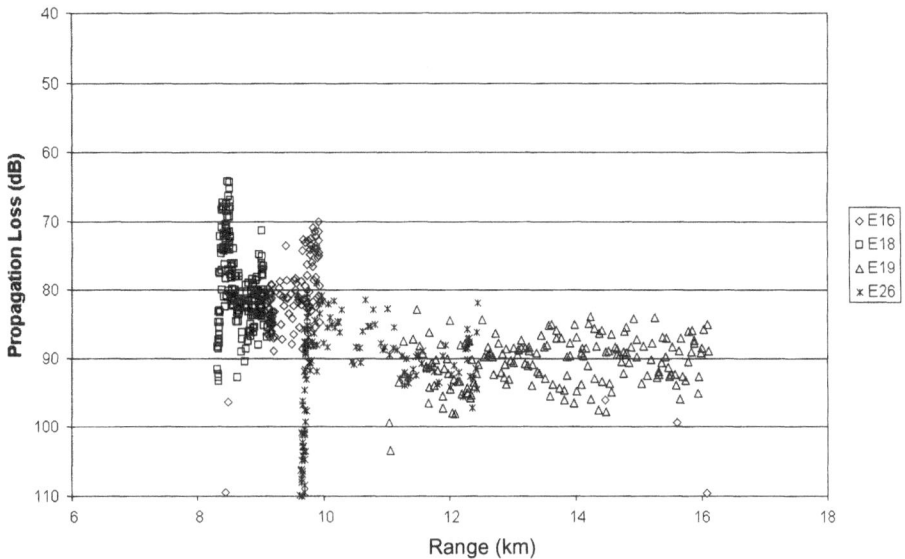

Figure 2. Measured propagation loss from four runs during Mercury'99.

Measured energy level data were obtained from four propagation loss runs carried out at 3.5 kHz during the Mercury'99 experiment. Propagation loss was calculated using a source level of 184 dB re 1μPa@1m. Combining these four runs provided a set of propagation loss data covering ranges of 7–16 km from the source, Fig. 2. This could be done because the environmental conditions did not vary significantly either within or between the different runs. The data could then be compared with the propagation loss predictions made using SPUR.

SPUR runs were carried out for the different seabed types listed in Table 2. The SPUR results and measured data were overlaid to allow a comparison to be made between the model predictions and real propagation loss for the range independent scenario on the Malta Plateau, see Figs. 3 and 4. Changing the geoacoustic properties of the seabed made a significant difference to the propagation loss predictions obtained.

Table 2. Summary of geoacoustic predictions for the Malta Plateau.

Sediment	Vp	Density g/cm^3	Attenuation dB/λ
GEOSEIS 1 (detailed sedimentology)	1433	1.626	1.01
2% Gas	1397	1.522	1.26
5% Gas	1408	1.610	1.40
10% Gas	1003	1.178	2.00
GEOSEIS 2 (including porosity)	1511	1.855	0.35
10% Gas	1118	1.548	0.70

The presence of gas in these sediments is a major factor affecting their geoacoustic properties. Figures 3 and 4 show the effects of varying the amount of sediment gas on the SPUR predictions for the GEOSEIS 1 and GEOSEIS 2 geoacoustics. For the GEOSEIS 1 prediction, Fig. 3, the measured propagation loss data can be seen to fall between the 5% and 10% sediment gas content predictions, with a difference of 20 dB between the no-gas and 10% gas scenarios at 15 km.

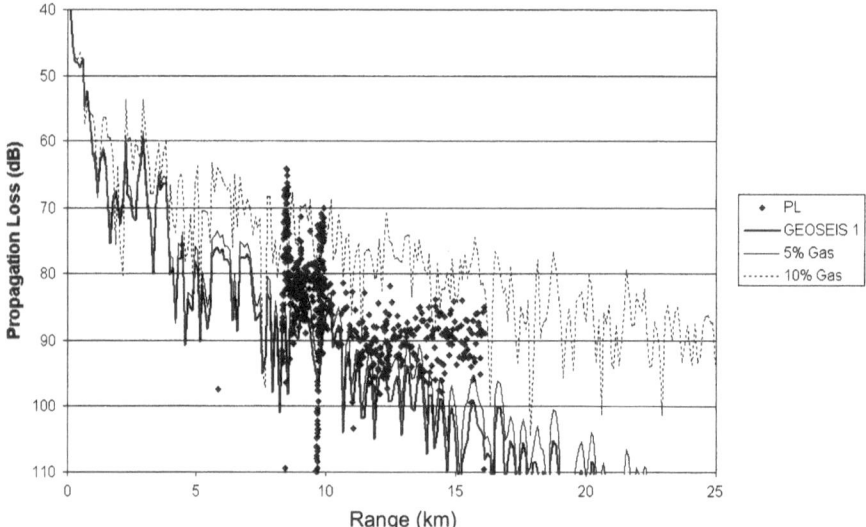

Figure 3. Effect of increasing gas content on the GEOSEIS 1 prediction, based on the sedimentological analysis of core data.

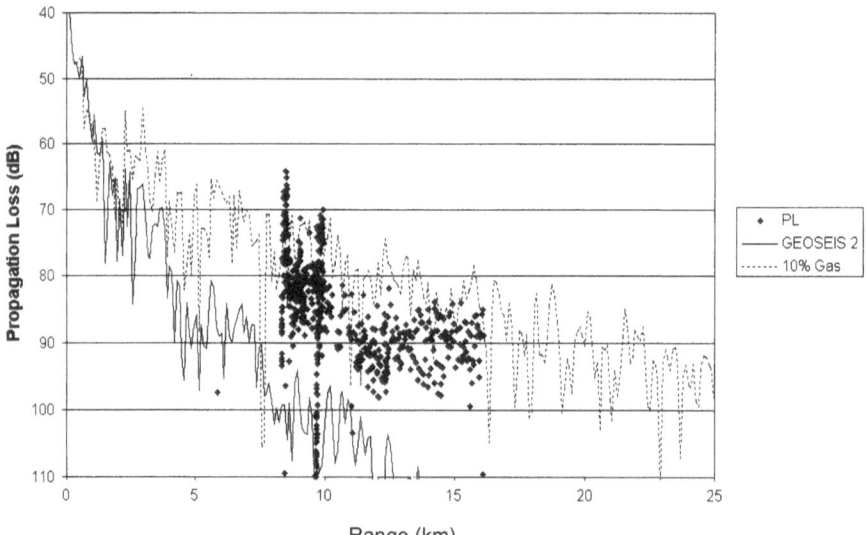

Figure 4. Effect of increasing the gas content on the GEOSEIS 2 prediction which takes porosity into account.

For the GEOSEIS 2 geoacoustics, which takes sediment porosity into account the measured data gives propagation loss values slightly higher than the SPUR predictions for a 10% sediment gas content with a difference of about 30 dB between no gas and 10% gas, see Fig. 4. GEOSEIS 2 with 10% gas is the best GEOSEIS geoacoustic prediction that could be made for the Malta Plateau, based on all the available sedimentological and geological information. The SPUR prediction using these parameters gives the closest match with the measured data.

Although it is often possible to recognise the presence of gas in seabed sediments, as a results of pockmarks or seismic signature, it is not possible to give an accurate estimate of how much gas is present. Since small quantities of gas can have a very marked acoustic effect this presents a serious shortcoming. Even in an area as well characterised as the Malta Plateau a quantitative estimate of gas content is purely speculative. However, the evidence would suggest that sediment gas content is relatively high, and the predictions for a 10% gas content, for the best GEOSEIS prediction, GEOSEIS 2, gives the closest match to the measured data. This result is within 2–3 dB of the measured data.

These results illustrate that for a well characterised, range independent environment, such as the Malta Plateau, a good match between measured and modelled propagation loss can be obtained. This is however dependent on good quality seabed information being available. In this case the issue of gas content is particularly important. Since there are no means of quantifying the sediment gas content the effects this has on the geoacoustic properties of the seabed sediments can only be estimates.

4 Conclusions

These results demonstrate the potential usefulness of GEOSEIS as a generic tool for providing geoacoustic data for input to acoustic models. Provided that a sedimentological breakdown can be provided for a particular seabed sediment, GEOSEIS should be able to provide sufficiently accurate geoacoustic parameters for use in propagation loss modelling.

The propagation loss predictions provided by SPUR will only be as good as the environmental data available for modelling. In shallow water regions, such as the Malta Plateau, a good understanding of the seabed is essential for sonar performance prediction. These results therefore highlight the importance of good geoacoustic information for areas of operational importance.

One particular geoacoustic problem highlighted by this study is the problem of modelling sediments that contain shallow gas. Shallow gas is a common phenomenon, present in many seabed sediments throughout the world. Even a small quantity of gas can have a dramatic effect on the geoacoustic properties of the seabed.

References

1. Max, M.D., Kristensen, A. and Michelozzi, E., Small-scale Plio-Quaternary sequence stratigraphy and shallow geology of the west-central Malta Plateau. In *Geological Development of the Sicilian-Tunisian Platform*, edited by M.D. Max and P. Colantoni, *UNESCO reports in Marine Science* **58**, 117–122 (1992).

2. McCann, C., McDermott, I., Grimbleby, L., Marks, S.G., McCann, D.M. and Hughes, B.C., The GEOSEIS database: a study of the acoustic properties of sediments and sedimentary rocks. In Proc. Oceanology International 94, Vol. 4 (1994).

3. Tonarelli, B., Turgutcan, F., Max, M.D. and Akal, T., Shallow sediment composition at four localities on the Sicilian Tunisian Platform. In *Geological Development of the Sicilian-Tunisian Platform*, edited by M.D. Max and P. Colantoni, *UNESCO reports in Marine Science* **58**, 123–128 (1992).

4. Kolsky, H., The propagation of stress pulses in viscoelastic solids, *Phys. Mag.* **1**, 693–710 (1956).

5. Edrington, T.S. and Calloway, T.M., Sound speed and attenuation measurements in gassy sediments in the Gulf of Mexico, *Geophysics* **49**, 297–299 (1984).

6. Domenico, S.N., Effect of brine-gas mixture on velocity in an unconsolidated sand reservoir, *Geophysics* **41**, 887–894 (1976).

7. Collins, M.D., A split step Padé solution for parabolic equation method, *J. Acoust. Soc. Am.* **93**, 1736–1742 (1993).

ACOUSTIC FLUCTUATIONS AND THEIR HARMONIC STRUCTURE

R. FIELD, J. NEWCOMB, J. SHOWALTER, J. GEORGE AND Z. HALLOCK

Naval Research Laboratory, Stennis Space Center, MS 39529, USA
E-mail: bob.field@nrlssc.navy.mil

Acoustic fluctuations are experimentally and theoretically shown to be harmonically related to the sound speed. The acoustic data analyzed here is shown to fluctuate predominately at the first and second harmonics of the M_2 tide. Theoretical results based on the Pekeris waveguide show that a large-scale, slowly fluctuating sound speed like the M_2 can cause more rapid fluctuations in the acoustic field in the form of acoustic harmonics. The theoretical results compare favorably with the acoustic data. The theoretical results also show the impact of the water column sound speed on the horizontal wave numbers and possibly on acoustic inversion methods that utilize them.

1 Introduction

From September 26 through November 2, 2000 the U. S. and Japan conducted a joint experiment in the New Jersey Bight off the U.S. East Coast. The experiment was conducted in three legs, the first of which investigated low frequency acoustic fluctuations in the presence of a shelf/slope front and internal waves. Legs II and III investigated acoustic scattering at 5.5 kHz and methods of acoustic inversion for sea floor properties, respectively. This paper focuses on a subset of the experimental results of Leg I.

Leg I was conducted from September 26 through October 5. During this leg, while the oceanographic measurements were underway, an autonomous, bottom-moored acoustic source continually transmitted signals in the 800 to 1250 Hz frequency range. The acoustic data was received on two acoustic sensors for a period of 24 hours. Toward the end of Leg I, a lower frequency acoustic source was suspended and transmitted 50, 125, 275 and 475 Hz signals simultaneously and continuously for twelve hours. Depending on frequency, the data shows 15–18 dB fluctuations over periods of 7–10 min. Of the four source/receiver geometries instrumented, only one is discussed here.

Section 2 lays out the experimental geometry for the acoustic track that was parallel to the shelf break and reviews the experimental acoustic and sound speed data. It is shown that the acoustic fluctuations are harmonically related to the M_2 tide. Section 3 lists a new set of fluctuation equations for the Pekeris waveguide. These equations, along with those of the rigid waveguide, were derived in a paper currently under review [1]. The equations derived for these simple waveguides predict the harmonic nature observed in the acoustic fluctuation measurements. Section 4 compares a "back-of-the-envelope", Pekeris waveguide simulation with the measured fluctuations. Since acoustic bottom inversion may utilize the horizontal wavenumber spectrum, the effects of sound speed

271

N.G. Pace and F.B. Jensen (eds.), Impact of Littoral Environmental Variability on Acoustic Predictions and Sonar Performance, 271-278.

dynamics in the water column on the horizontal wavenumber spectrum is shown in Sect. 5. The effects can be accounted for from the fluctuation equations listed in Sect. 3. Section 6 summarizes and concludes.

2   Experimental geometry and data

2.1   Experimental Geometry

Figure 1 shows the general location of the SWAT experiments and the specific location of one of the four source/receiver geometries used during the acoustic propagation experiment.

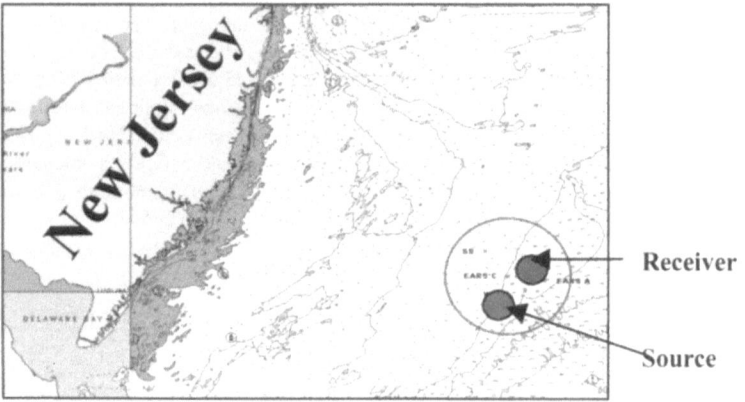

Figure 1. The general location of the SWAT experiments is shown in the open circle. One of the four source-receiver geometries is shown for the acoustic propagation part of SWAT.

2.2   Acoustic Data

The acoustic signals were transmitted from an autonomous, bottom-moored source. The water depth at the source position was 132 meters. The source was 9 meters off the sea floor making the source depth 123 meters. The ocean depth over most of the acoustic path and at the receiver was 139 meters. The receiver depth was 50 meters and its range from the source was 9715 meters. The transmitted signals were linear sweeps from 800 Hz to 1250 Hz. The duration of each sweep was 10 s for a sweep rate of 45 Hz/s. The sweeps were repeated every 20 s resulting in 10 s of ambient noise between each sweep. Because of limitations in the acoustic recording system, only sweep frequencies up to 1107 Hz were processed.

The data were spectrally processed using a contiguous series of non-overlapped 256 point Hann weighted FFTs resulting in a frequency bin width of 8.7 Hz and an effective noise bin width of 13 Hz. Thus, each FFT spanned a time period of only 0.115 s that corresponded to about 5 Hz of the received sweep. These processing parameters were chosen to ensure that each FFT (i.e., time period) would have the minimum possible number of bins containing a portion of the received sweep.

A time series for each frequency bin was created for frequencies from 800 Hz to 1107 Hz. Since the signal arrived at the recording system every 20 s, a time series of the received signal was created with an effective sampling rate of 4320 samples/day. Figure 2a displays the calibrated received level with the average removed at a frequency of 854.4922 Hz. The time series is 24-hours long and has been low pass filtered to 360 cycles per day (cpd) to correspond to the two-minute time sampling of the temperature and salinity data upon which the sound speed is constructed. Figure 2b displays its amplitude spectrum in cpd. From the spectrum, it can be seen that most of the high amplitude fluctuation energy is below 50 cpd.

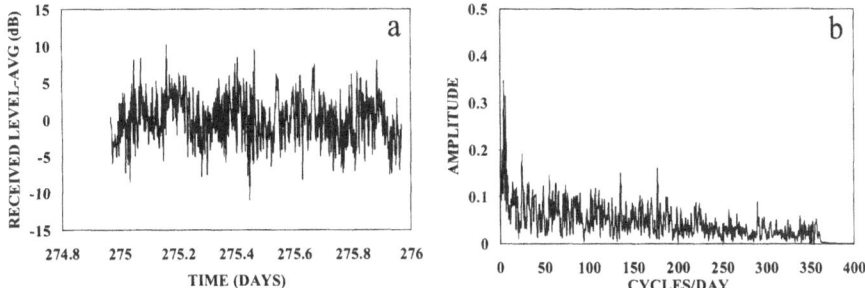

Figure 2. a) Acoustic received level at 854.4922 Hz with average removed over a 24-hour period. b) The amplitude spectrum of the acoustic received level time series.

Figure 3a shows the acoustic time series of Fig. 2a low pass filtered to 20 cpd and Fig. 3b displays the amplitude spectrum of Fig. 2b out to only 50 cpd. Figure 3b shows what appears to be fluctuations close to the M_2 tide, F_0, and its first and second harmonics, F_1 and F_2, respectively. To show that this is in fact the case, the sound speed data will be analyzed.

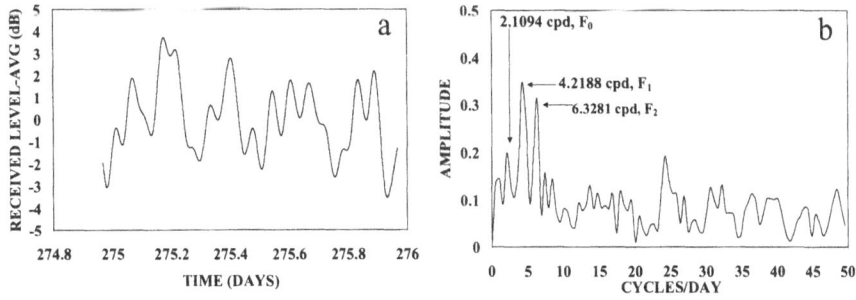

Figure 3. a) Acoustic time series low pass filtered to 20 cpd. b) The amplitude spectrum of the acoustic time series out to 50 cpd showing the M_2 tidal fluctuation and its first and second harmonics.

2.3   *Sound Speed Data*

Figure 4a shows a time series of sound velocity recorded just above the acoustic receiver at a depth of 45 meters. Figure 4b displays its amplitude spectrum. The two largest peaks in Fig. 4b are the Inertial Period and the semi-diurnal tide, M_2. The nominal value of the M_2 is 1.9322 cpd. For this relatively short time series the measured value is 1.9336 cpd.

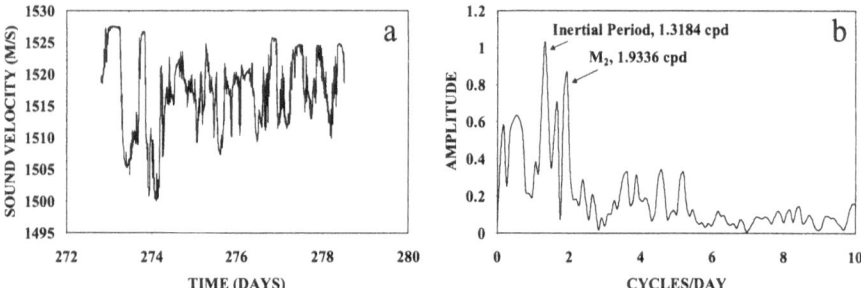

Figure 4.   a) Sound speed time series at 45 meters just above the acoustic receiver.   b) The amplitude spectrum of the sound speed time series showing the Inertial Period and the M_2 tide.

Figure 5a shows the sound speed time series of Fig. 4a band pass filtered to pass only the M_2 component of the spectrum. The vertical lines in the figure denote the time period during which the acoustic transmissions shown in Fig. 2a took place. During this time period the M_2 fluctuates about an average of approximately 1517.5 m/s with an amplitude of ± 1.625 m/s. Figure 5b displays the amplitude spectrum of the M_2 during the acoustic transmission time. Because of the short time window, the M_2 takes on its maximum value at 2.1094 cpd. The acoustic transmissions "see" the M_2 fluctuating at this rate, which is the value of F_0 shown in Fig. 3b. The highest amplitude acoustic fluctuations are at the first and second harmonics, F_1 and F_2, respectively, of the M_2.

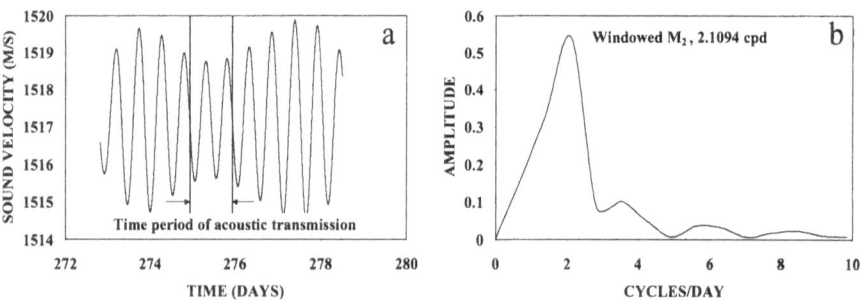

Figure 5.   a) Sound speed time series band-pass filtered to pass only the M_2.   b) The amplitude spectrum of the M_2 sound speed time series during the time of acoustic transmission.

3   Fluctuation equations for the Pekeris waveguide

In order to show that the measured acoustic fluctuations shown in Figs. 3a and 3b are theoretically expected, results from a paper currently under review will be given [1]. From Frisk [2] the expression for the complex acoustic pressure in a Pekeris waveguide is given by

$$p^P(r,z) = \frac{\sqrt{2\pi}e^{i\pi/4}}{\rho} \sum_{n=1}^{N_{max}} \Gamma_n^2 \sin(k_{zn}^P z_0)\sin(k_{zn}^P z)\frac{e^{ik_n^P r}}{\sqrt{k_n^P r}} \tag{1}$$

where Γ_n is the amplitude, k_{zn} and k_n are the vertical and horizontal wavenumbers, respectively, ρ is the density and r is the range. If the time dependent sound speed is written as

$$C(t) = C_0 + Ae(t)\cos(\Omega_0 t) \tag{2}$$

where C_0 is the average sound speed, A, is the amplitude of the sound speed fluctuation, $e(t)$ is a slow-varying envelope, $\Omega_0 = 2\pi F_0$ is the radial "frequency" and F_0 is the fundamental frequency of the sound speed variation, it can be shown that   Eq. (1) can be written in the time dependent form

$$p^P(r,z,t) = \frac{\sqrt{2\pi}e^{i\pi/4}}{\rho} \left\{ \begin{bmatrix} \sum_{n=1}^{N_{max}} \Gamma_{n0}^2 \frac{\sin(k_{zn0}^P z_0)\sin(k_{zn0}^P z)}{\sqrt{k_{n0}^P r}} e^{ik_{n0}^{P1} r} J_0(I_n^P) \end{bmatrix} + \\ \begin{bmatrix} 2\sum_{l=1}^{\infty}(-1)^l \cos(2l\Omega_0 t) \sum_{n=1}^{N_{max}} \Gamma_{n0}^2 \frac{\sin(k_{zn0}^P z_0)\sin(k_{zn0}^P z)}{\sqrt{k_{n0}^P r}} e^{ik_{n0}^{P1} r} J_{2l}(I_n^P) \end{bmatrix} - \\ \begin{bmatrix} i2\sum_{l=0}^{\infty}(-1)^l \cos[(2l+1)\Omega_0 t] \sum_{n=1}^{N_{max}} \Gamma_{n0}^2 \frac{\sin(k_{zn0}^P z_0)\sin(k_{zn0}^P z)}{\sqrt{k_{n0}^P r}} e^{ik_{n0}^{P1} r} J_{2l+1}(I_n^P) \end{bmatrix} \right\} \tag{3}$$

with

$$k_n^P r \equiv k_{n0}^{P1} r - I_n^P \cos\Omega_0 t \tag{4}$$

$$k_{n0}^{P1} \equiv k_{n0}^R + \left(k_{zn}^R - \frac{\varphi_{n0}}{4h} \right)\frac{\varphi_{n0}}{2hk_{n0}^R} \tag{5}$$

$$I_n^P \equiv \left[-\left(k_{zn}^R - \frac{\varphi_{n0}}{2h} \right)\frac{\varphi_{n0}'}{2h} + \frac{\omega^2}{C_0^3} \right]\frac{rAe(t)}{k_{n0}^R} \tag{6}$$

In Eq. (3), the J_n terms are Bessel functions of the first kind and Eq. (4) defines the time dependent horizontal wavenumbers. In Eq. (5), φ_{n0} is the bottom phase evaluated at the average sound speed, h is the water depth and k_{n0}^R is the horizontal wavenumber of the rigid waveguide evaluated at the average sound speed.   In Eq. (6), φ_{n0} is the first derivative of the bottom phase and ω is the acoustic frequency. From Eq. (3) it can be seen that the only allowed fluctuations in the acoustic field are those at the sound speed fundamental and its harmonics. Equation (2) can be generalized to a real sound speed field by expressing it as a series of sound speed components with different amplitudes and envelopes. For the real sound speed case, the acoustic field will fluctuate at the fundamental and the harmonics of each sound speed component.

4 "Back-of-the-envelope" calculations vs. data

This section compares time dependent Pekeris waveguide calculations with the data shown in Figs. 3a and 3b. It is meant to be only a "back-of-the-envelope" calculation that accounts for the low frequency part of the acoustic fluctuation spectrum. Accurate modeling would, of course, require a range and depth dependent model to account for higher frequency fluctuations found in the acoustic spectrum shown in Fig. 1b. For the Pekeris simulation the following parameters are chosen: water depth = 132 meters, bottom sound speed = 1800 m/s, bottom density = 2.0 g/cm^3 , range = 9715 meters, source depth = 123 meters, receiver depth = 50 meters and acoustic frequency = 854.4922 Hz. The water depth at the source was chosen. The bottom sound speed and density were chosen based on average coring values within the first 100 cm of the bottom sediment. The range, source depth, receiver depth and acoustic frequency are measured values. The Pekeris waveguide time series was computed with a "frozen ocean" approach. Equation (2) was used to generate a synthetic M_2 fluctuation in sound speed. From Figs. 5a and 5b, the following sound speed parameters for Eq. (2) were chosen: A = 1.625 m/s, C_0 = 1517.5 m/s and F_0 = 2.1094 cpd. The envelope function, $e(t)$, was set equal to one over the 24-hour period. The sampling used for the simulation was the same as the data, 4320 cpd (every 20 s) over a period of 24 h. The results are shown in Figs. 6a and 6b.

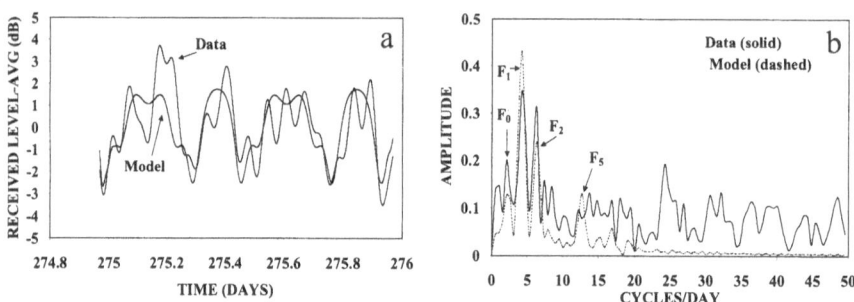

Figure 6. a) Acoustic time series over 24 hours. Solid line is data low pass filtered to 20 cpd. The dashed line is the Pekeris M_2 simulation at 2.1094 cpd. b) The amplitude spectrum of each time series. The M_2 fundamental and its harmonics that are labeled refer to the Pekeris simulation.

Figures 6a and 6b show that the Pekeris simulation gives a reasonable answer for the fluctuations caused by the M_2 tide. The simulation shows that the two largest acoustic fluctuation components are the first and second harmonics of the M_2 tide. In Fig. 6b the harmonics that are labeled refer to the model simulation.

5   Impact of water column dynamics on the horizontal wavenumber spectrum

Since acoustic bottom inversion is a major component of the SWAT collaboration, the Pekeris waveguide will be used to generate complex pressures as a function of range and time in order to determine how changes in the water column sound speed affect the horizontal wavenumber spectrum. The same Pekeris parameters will be used as in the previous simulation except the frequency is changed to 50 Hz and the source and receiver depths are 61 m (half the water depth). The procedure used to compute the horizontal wavenumber spectrum is to calculate complex pressures as a function of range for a constant sound velocity, then do a Hankel transform over the range aperture to get the horizontal wavenumber spectrum. This procedure is repeated over a 24 h time period where each Hankel transform is computed with a new but constant sound velocity over the range interval. The range aperture is calculated in 1-meter intervals from 5000 meters to 9046 meters.

Figure 7a shows the result of the calculation for a sound velocity that remains constant over the 24 h period. Four modes are found and the spectrum remains constant. Mode 1 has the largest amplitude. Figure 7b shows the same calculation using the sound velocity time series measured at a depth of 45 meters. This is the same sound velocity shown in Fig. 4a from day 274.8 to 275.8. The sound velocity time series is plotted to the left of Fig. 7b. The maximum sound velocity, labeled 1, is 1524.5 m/s and the minimum, labeled 2, is 1507.8 m/s. From Eqs. (4) and (6), it can be seen that the time dependent part of the wavenumber adds energy to the wavenumber spectrum and makes one wavenumber "bleed" into adjacent bins. Therefore, the water column will impose limits to the extent one can unambiguously interpret changes in the wavenumber spectrum in terms of changing geoacoustic parameters.

6   Summary and conclusions

It has been experimentally and theoretically shown that acoustic fluctuations are harmonically related to the sound speed. For a "monochromatic ocean signal", the only fluctuations allowed in the acoustic field are the sound speed fundamental and its harmonics. In general, the harmonic content is a function of the amplitude of the sound speed fluctuation, the acoustic frequency, the number of modes, the bottom phase and range. Because the time dependent horizontal wavenumber includes an additional term on the order of $\omega^2 A / C_0^3 k_{n0}^R$, each bin of the horizontal wavenumber spectrum can accumulate energy over time causing the amplitudes of the wavenumbers to "bleed" from one bin to another. This may pose significant problems for bottom inversion in cases where changes in the horizontal wavenumber spectrum are interpreted as changing geoacoustic conditions.

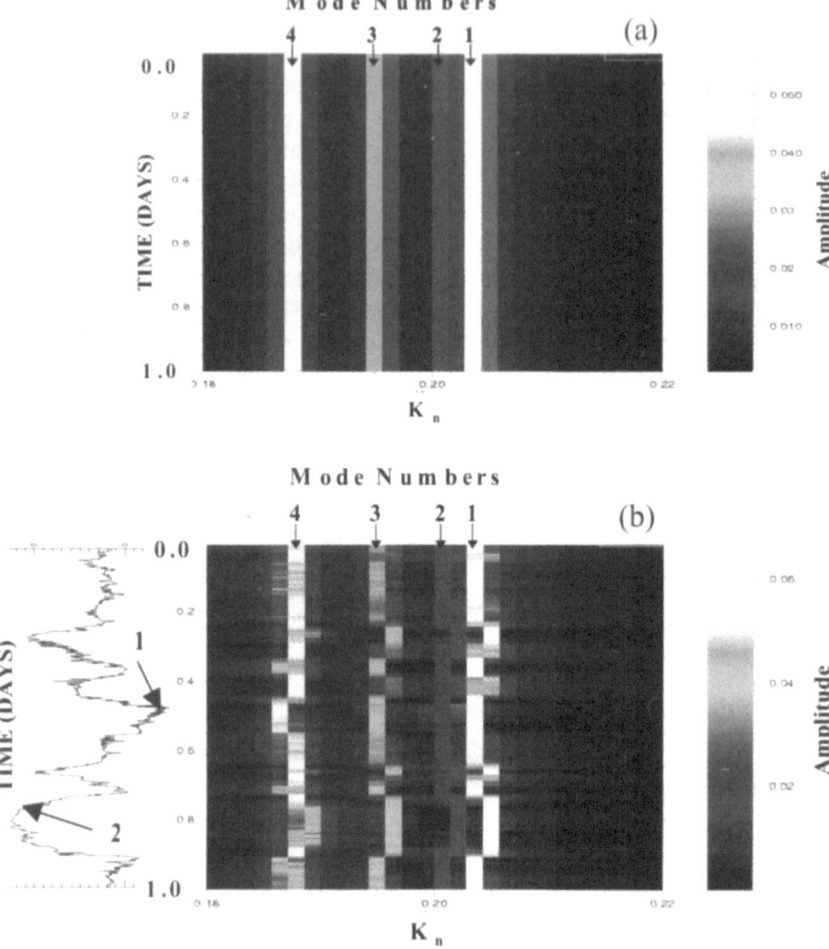

Figure 7. a) The horizontal wavenumber spectrum computed with a constant sound speed. The first four modes are shown. b) The horizontal wavenumber spectrum computed with the measured sound speed time series shown on the left.

Acknowledgements

This work is supported by the Office of Naval Research, Program Element PE 62435N.

References

1. Field, R.L. and George J., Acoustic fluctuations in simple shallow water waveguides, *J. Acoust. Soc. Am.* (submitted Jan. 2002).
2. Frisk, G.V., *Ocean and Seabed Acoustics: A Theory of Wave Propagation* (Prentice-Hall, Inc., New Jersey, 1994) Chap. 5.

GROUP AND PHASE SPEED ANALYSIS FOR PREDICTING AND MITIGATING THE EFFECTS OF FLUCTUATIONS

W.A. KUPERMAN, S. KIM, G.F. EDELMANN, W.S. HODGKISS AND H.C. SONG

Scripps Institution of Oceanography, University of California,
San Diego, La Jolla CA 92093-0701, USA
E-mail: [wak,seongil,geoff,wsh,hcsong]@mpl.ucsd.edu

T. AKAL*

SACLANT Undersea Research Centre, 19138 La Spezia, Italy
E-mail: tuncay.akal@posta.mam.gov.tr

The relationship between group and phase speed in an ocean waveguide is a robust descriptor of signal structure. Since different propagation paths or mode groups have characteristic group-phase speed relations (described by waveguide invariants), it turns out that this group-phase speed (g-p) structure is also useful for identifying the dominant fluctuating modal regions in, for example, shallow water internal wave fields. Since an acoustic time reversal mirror (TRM) works on a retransmission mechanism based on group speed structure ("last in, first out"), we will use TRM experimental results to illustrate the relation between acoustic fluctuations and g-p structure and to provide guidance for developing robust acoustic processing methods in fluctuating environments.

1 Introduction

It is well known that quantitative and qualitative information about sound propagation in a waveguide can be obtained by studying the modal (or ray) group speed (or cycle distance) and phase speed (or launch angle) relation which we term g-p structure. Much of this information is embedded in the waveguide invariant [1–4]. In this paper we relate the g-p structure to acoustic fluctuations, providing insight that can lead to signal processing methods that are robust in fluctuating environments. We will summarize the background theory and use results from recent time-reversal experiments (or phase conjugation in the frequency domain) [5–9] to demonstrate the existence of simple robust diagnostics that ultimately guide us to stabilizing processing methods. The signal processing we suggest is then demonstrated by simulation.

2 Group and phase speed structure in shallow water

For both reflection and refraction dominated propagation paths, phase speed increases with increasing launch angle. However, for reflection dominated paths, group speed decreases with increasing phase speeds whereas for refracting paths, group speed increases with

---

*Present address: TUBITAK-MAN, Marmara Research Center, Earth and Marine Sciences Research Institute, P.K.21 Gebze, Kocaeli 41470, Turkey.

N.G. Pace and F.B. Jensen (eds.), Impact of Littoral Environmental Variability on Acoustic Predictions and Sonar Performance, 279-286.

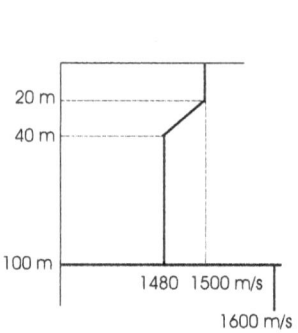

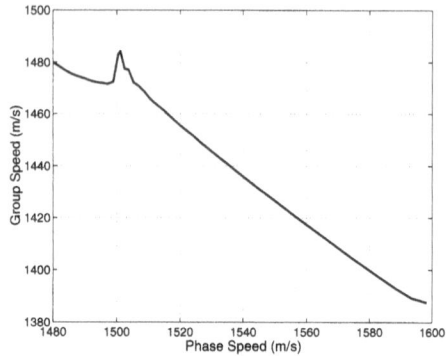

Figure 1. Group speed vs phase speed for an idealized summer profile (1000 Hz).

increasing phase speed. These relations are summarized by the waveguide invariant for mode groups defined as [1–4]

$$\overline{\beta}_{nm} = -\frac{\Delta \overline{s}_{g,mn}}{\Delta s_{p,mn}}. \tag{1}$$

Here $\Delta \overline{s}_{g,mn} = \overline{s}_{g,m} - \overline{s}_{g,n}$ and $\Delta s_{p,mn} = s_{p,m} - s_{p,n}$. The $\overline{s}_{g,m}$ and $s_{p,m}$ are the m-th mode group and phase slowness, respectively. The upper bar means the range-averaged value in a range-dependent ocean environment. By mode groups we mean to include cases in a waveguide where both refracting and reflecting modes are present (e.g., summer profile where low modes are refracted below thermocline and higher modes are reflected at both boundaries); the demarcation between the mode groups varies with frequency.

We use a range-independent waveguide for our baseline environment. The invariant β describes the shift in the waveguide interference structure with respect to frequency through the relation (ignore the second term on the r.h.s of the equation for the time being)

$$\frac{\delta r}{r} = \frac{1}{\beta}\frac{\delta \omega}{\omega} - \frac{\gamma}{\beta}\frac{\delta h}{h}. \tag{2}$$

For example the peaks in the interference pattern at angular frequency ω as a function of range r are shifted by δr when the frequency under consideration is shifted by $\delta \omega$. Clearly the sign of β is important; β is typically positive for reflecting environments and negative for refracting environments [1, 2, 4, 10]. One generalization to the waveguide invariant can be written down from Ref. [3] which is given by all the terms in Eq. (2), where h is a duct thickness confining a particular mode group. For an ideal waveguide, $\beta = 1$ and $\gamma = -2$. Thus, increasing the channel depth shifts the interference pattern out to increasing range, a phenomenon measured during tidal variations and theoretically predicted from a simple perturbation analysis [11]. Though the result is simple, if one has a fixed receiver, this ebb and flow of the single frequency interference pattern appears as an almost random fluctuation though a spectrogram (intensity vs ω and time) would show a distinct pattern. These type of fluctuations have also been addressed in terms of waveguide invariants both theoretically and experimentally, the latter using the frequency-time diversity of FM sweeps [12–15].

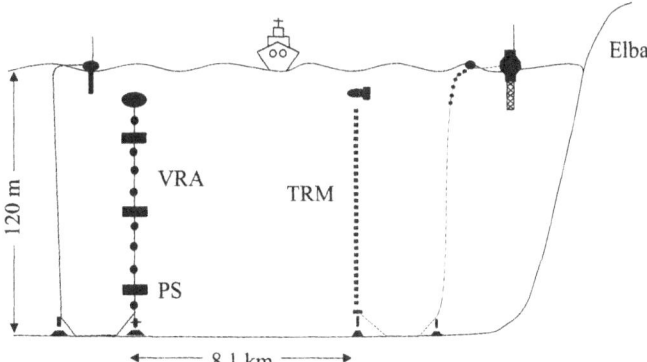

Figure 2. Schematic of TRM experiment. The prope source (PS) is attached to a vertical receive array (VRA) which sample the backpropagated pulse at the focal distance.

With simple group and phase speed arguments, we can identify two types of fluctuations:

1. The first can be thought of as the result of an ebb and flow of modes. If the modes stay within groups with the same invariant quantity, the flow is similar to the tidal variation example discussed above.

2. The other type is from a modal transition region [16] where the nature of a mode changes, for example, a case when a refracted mode changes to a reflected mode. This can happen for a summer profile where the next mode after the highest mode refracted below the thermocline is now surface reflected. Such a transition, would also occur for a given mode in the transition region by going from a higher to lower frequency (cutoff from refracting to reflecting ducts). By virtue of the generalized waveguide invariant above, this process would occur at a given frequency by a change in the refracting duct size as could be caused by internal waves.

Examine Fig. 1 which is a g-p phase speed plot of an idealized summer profile environment. It is where there is an abrupt change in the sound speed profile where internal waves have the maximum amplitude. For a ray turning around (mode is evanescent) in the phase speed region around 1500 m/s, we see a maximum variation in group speed; in this region there will be fluctuations between different modes (phase speeds) with the same group velocity. The fluctuation is between refracting and reflecting paths and therefore is in the region where β is changing sign (slope reversals in g-p curves). A time reversal experiment would emphasize these two phenomena since the first-in last-out processing would fluctuate between the spatial structure of the incoming field and one in which the modes are scrambed in the transition region (i.e., same modes transitioning to other group speeds and modes at the same group speed being redistributed about the local maximum).

3 Performance of a Time Reversal Mirror in fluctuating environments

We have already reported on a series of Time Reversal Mirror (TRM) experiments [5–9]. Figure 2 shows the basic geometry of these experiments. The probe source (PS) transmits a signal to the TRM where it is time reversed and retransmitted; a focus is produced at

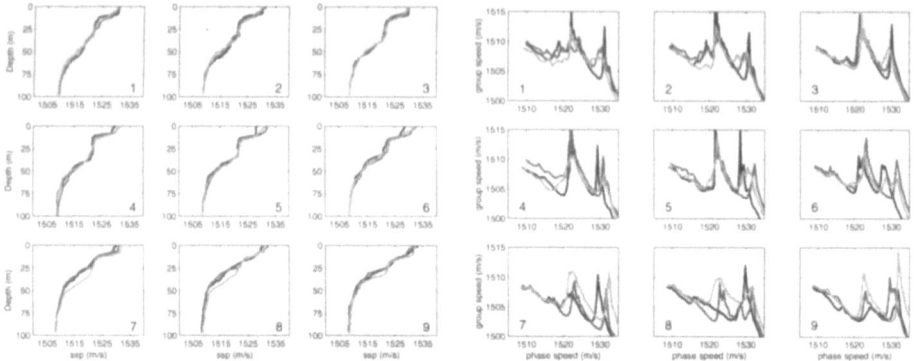

Figure 3. Sound speed profiles on three consecutive days at nine different positions north of Elba and corresponding g-p curves.

the PS if the forward and backpropagating environments are the same. Figure 3 shows the environments of the last experiment and the computed g-p curves. By the arguments above, one should be able to determine which environments would support time reversal (minimal fluctuation between forward and backpropagating time intervals). For example, station 1 shows a very unstable profile with many slope reversals whereas the station 9 is stable up to the phase speed of 1520 m/s corresponding to the maximum propagating mode. Indeed, the acoustic data verified this relation between stability and g-p curves to the point that by just measuring the sound speed profile and making the g-p computation, we could predict when time reversal would be robust or problematic.

Now we go to a case where we examine the fluctuating focus as a function of time. Figure 4 is an example of a single TRM result and Fig. 5 shows a sequence of TRM foci for which the same probe source signal at the TRM was retransmitted every minute for 50 min. In this particular one of many examples, the focus degrades considerably after 10 minutes but reappears after 40 min. What we would like is a signal processing method which stabilizes the focus.

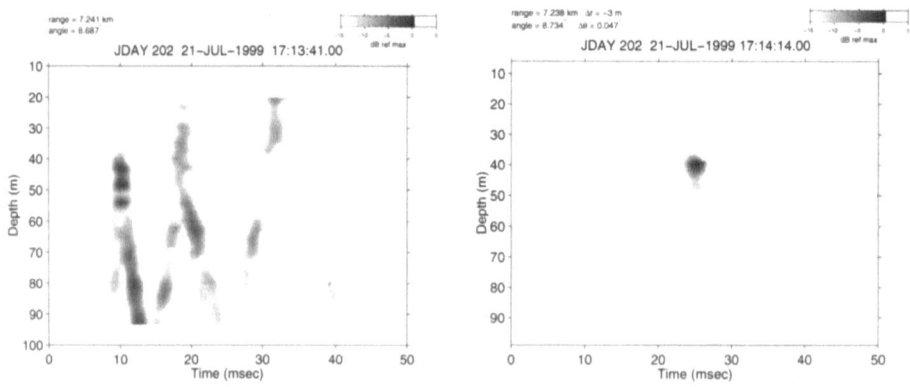

Figure 4. Multipath signal received by the TRM (left) and refocused signal at the original PS location (right).

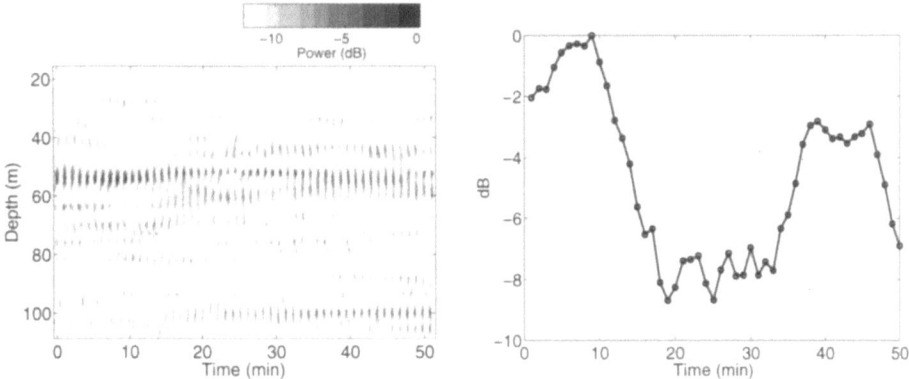

Figure 5. Measured time reversal stability with 3.5 kHz transmissions: sequence of TRM foci due to the same probe signal transmitted every minute for 50 min (left) and variation of the intensity level at the PS positions (right).

4 Mitigating fluctuation effects by adaptive processing

Phase conjugation and matched field processing (MFP) are closely related [17], the latter using a computer simulation (replicas) for backpropagation. For MFP, an assortmnt of robust algorithms have been proposed [18]. The sound speed perturbation method (MSC) [19] based on multi-constraint array processing methods (MLC) [20] appears promising and more or less uses replica vectors obtained from a singular value decomposition (SVD) of a large matrix representing an ensemble of environments. From Eq. (2), we can think of the frequency bins of a pulse as containing acoustic information representing different ranges and/or environmental realizations. The frequency-range shift has alread been verified by TRM experiments [7]. Hence, based on Ref. [19], we propose the following multi-frequency constraint (MFC) procedure [9]:

The first step is constructing a probe signal matrix by from probe signal vectors as

$$\mathbf{P} = [\mathbf{p}_1, \mathbf{p}_2, \cdots, \mathbf{p}_N], \tag{3}$$

where $\mathbf{P}$ is a $J \times N$ probe signal matrix at frequency ω and each element $\mathbf{p}_n$ is a $J \times 1$ signal vector. J is the number of transducers in the TRM and N is the number of probe signal vectors received by the TRM, each corresponding to a frequency component extracted from a broadband pulse

$$\mathbf{p}_n = G(\mathbf{r}_j; \mathbf{r}_s, t_1, \omega_n), \tag{4}$$

where ω_n is a frequency bin around ω. The performance of the adaptive method depends on how well signal matrix constructed by the multiple frequency bin vectors spans the ensemble of backpropagating environments as per the physics of Eq. (2).

Next, we adopt the arguments and procedure described in Ref. [19] to the frequency vectors hypothesized to span the environmental ensemble as estimated from the generalized waveguide invariant: If we take many frequency bins, the number of vectors spanning all possible wave-front perturbations can be large but the modal phase perturbations can be highly correlated among the modes as well as signal pings. The dimension actually is determined by the number of effective internal wave modes interacting with acoustic modes. Normally, internal waves can be represented by a few modes in shallow water [21]. The design of an efficient constraint space for the signal vector consists of selecting the minimum number of vectors that can best approximate the phase perturbation space. This order reduction can be achieved using the singular value decomposition of the signal matrix $\mathbf{P}$ with a rank K approximation,

$$\mathbf{P}(\omega) \simeq \mathbf{U}\Sigma\mathbf{V}^{\dagger}, \tag{5}$$

where $\dagger$ is the Hermition transpose, $\mathbf{U}$ is a $J \times K$ matrix whose columns are the left singular vectors, Σ is a $K \times K$ matrix whose diagonal elements are the singular values of $\mathbf{P}$, and $\mathbf{V}$ is a $N \times K$ matrix whose columns are the right singular vectors. Now the field vector $\mathbf{H}$ for backpropagation is obtained by the linear combination of left singular vectors,

$$\mathbf{H}(\omega) = \mathbf{U}(\omega)\mathbf{q}, \tag{6}$$

where $\mathbf{q}$ is a $K \times 1$ vector representing the coefficient used for the linear combination of the singular vectors. We expect the singular values to decrease rapidly with increasing number so that in many cases the first singular vector corresponding to the largest singular value is sufficient as a field vector for stable focusing. In this case $\mathbf{q} = [1, 0, \ldots, 0]^{T}$.

The final step is replacing $G(\mathbf{r}_j; \mathbf{r}_s, t_1, \omega)$ for the data on the TRM so that the adaptive time-reversed pressure field becomes

$$p_{tr}(\mathbf{r}, t_2, \omega) = \sum_j S^*(\omega)H^*(\mathbf{r}_j, \omega)G(\mathbf{r}; \mathbf{r}_j, t_2, \omega). \tag{7}$$

Here a new source spectrum $S(\omega)$ is inserted in the backpropagation which can have a different shape from the original source spectrum used for the probe source signal. This process should exhibit a stable focal structure since the field vector $H(\mathbf{r}_j, \omega)$ was designed to maintain high correlation with $G(\mathbf{r}; \mathbf{r}_j, t_2, \omega)$ for all possible wave-front perturbations. This leads to increased focal sizes and a stable focus since in most cases, higher order modes, and in particular, those representing rays that turn in the internal wave regions are de-emphasized.

This algorithm was motivated by the data in our TRM experiments but developed after the last experiment. Therefore we must use simulation to demonstrate its effectiveness. Figure 6 shows a simulation of the data from a TRM experiment exhibiting much of the same behavior of the data shown in Fig. 5. Internal wave simulations were based on Ref. [21, 22] and the broadband normal propagation utilized Ref. [23]. Figure 6 also shows an application of the MFC algorithm together with the two other algorithms, MSC and MLC of Refs. [19, 20] where the latter two methods require an additional (and impractical) large amount of data – multiple pings or multiple probe source locations [9]. The stabilization of the TRM focus is quite evident.

5 Conclusion

In this paper we reviewed how group and phase speed (g-p) structure as summarized by the generalized waveguide invariant relates environment to the space-time-frequency

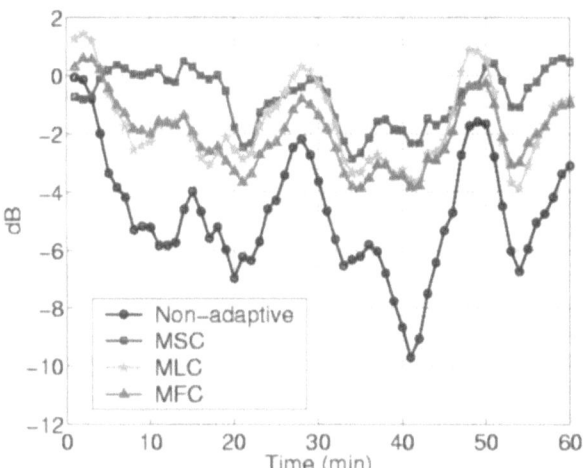

Figure 6. Simulated time-reversal focal strength using various adaptive methods: multiple sound speed constraints (MSC), multiple location constraints (MLC), and multiple frequency constraints (MFC).

structure of a propagating signal. We then showed that our stability results from TRM experiments were consistent with our interpretation of the g-p structure. Basically, in this picture, environmental fluctuations have acoustic trajectories in the multidimensional space-time-frequency space and we normally sense these fluctuations in one of the basis planes (e.g., space-time) – possibly and unintentionally maximizing the appearance and effect of fluctuations. Our understanding of frequency-range shift caused by the environment through the generalized waveguide invariant concept then provides guidance to construct an adaptive procedure to stabilize the focal structure of a TRM in a fluctating environment. The proposed processing was successfully demonstrated with simulation and the method, by virtue of its derivation, should also be applicaple to MFP.

Acknowledgements

This research was supported by the U. S. Office of Naval Research.

References

1. S.D. Chuprov, Interference structure of a sound field in a layered ocean. In *Acoustics of the Ocean*, edited by L.M. Brekhovskikh and I.B. Andreevoi (Nauka, Moscow, 1982) pp. 71–91.
2. L.M. Brekhovskikh and Y.P. Lysanov, *Fundamentals of Ocean Acoustics* (Springer-Verlag, 1991).
3. G.A. Grachev, Theory of acoustic field invariants in layered waveguide, *Acoust. Phys.* **39**(1), 33–35 (1993).
4. G.L. D'Spain and W.A. Kuperman, Application of waveguide invariants to analysis of spectrograms from shallow water environments that vary in range and azimuth, *J. Acoust. Soc. Am.* **106**, 2454–2468 (1999).

5. W.A. Kuperman, W.S. Hodgkiss, H.C. Song, T. Akal, C. Ferla and D. Jackson, Phase conjugation in the ocean: Experimental demonstration of an acoustic time-reversal mirror, *J. Acoust. Soc. Am.* **102**, 25–40 (1998).

6. W.S. Hodgkiss, H.C. Song, W.A. Kuperman, T. Akal, C. Ferla and D.R. Jackson, A long range and variable focus phase conjugation experiment in shallow water, *J. Acoust. Soc. Am.* **105**, 1597–1604 (1999).

7. H.C. Song, W.A. Kuperman and W.S. Hodgkiss, A time-reversal mirror with variable range focusing, *J. Acoust. Soc. Am.* **103**, 3234–3240 (1998).

8. S. Kim, G.F. Edelmann, W.A. Kuperman, W.S. Hodgkiss, H.C. Song and T. Akal, Spatial resolution of time-reversal arrays in shallow water, *J. Acoust. Soc. Am.* **110**, 820–829 (2001).

9. S. Kim, W.A. Kuperman, W.S. Hodgkiss, H.C. Song, G.F. Edelmann and T. Akal, Adaptive time reversal for robust focusing in the ocean, *J. Acoust. Soc. Am.* (submitted 2002).

10. G.L. D'Spain, J.J. Murray, W.S. Hodgkiss, N.O. Booth and P.W. Schey, Mirages in shallow water matched field processing, *J. Acoust.Soc. Am.* **105**, 3245–3265 (1999).

11. D.E. Weston and K.J. Stevens, Interference of wide-band sound in shallow water, *J. Sound and Vibration* **21**, 57–64 (1972).

12. V.M. Kuz'kin, The effect of variability of ocean stratification on a sound field interference structure, *Acoust. Phys.* **41**, 300–301 (1995).

13. V.M. Kuz'kin, A.V. Ogurtsov and V.G. Petnikov, The effect of hydrodynamic variability on frequency shifts of the interference pattern of a sound field in a shallow water sea, *Acoust. Phys.* **44**, 77–82 (1998).

14. V.M. Kuz'kin, Frequency shifts on the sound field interference pattern in a shallow water, *Acoust. Phys.* **45**, 224–229 (1998).

15. V.G. Petnikov and V.M. Kuz'kin, Shallow water variability and its manifestation in interference pattern of sound field. In *Ocean Acoustic Interference Phenomena and Signal Processing*, edited by W.A. Kuperman and G.L. D'Spain (AIP Press, New York, 2002).

16. W.A. Kuperman, G.L. D'Spain and K.D. Heaney, Long range source localization from single hydrophone spectrograms, *J. Acoust. Soc. Am.* **109**, 1934–1943 (2001).

17. W.A. Kuperman and D.R. Jackson, Ocean acoustics, matched-field processing and phase conjugation. In *Imaging of Complex Media with Acoustic and Seismic Waves,* in *Topics Appl. Phys.*, edited by M. Fink *et al.* (Springer-Verlag, Berlin, 2002) pp. 43–97.

18. A.B. Baggeroer, W.A. Kuperman and P.N. Mikhalevsky, An overview of matched field methods in ocean acoustics, *IEEE J. Oceanic Eng.* **18**, 401–424 (1993).

19. J.L. Krolik, Matched-field minimum variance beamforming in a random ocean channel, *J. Acoust. Soc. Am.* **92**, 1408–1419 (1992).

20. H. Schmidt, A.B. Baggeroer, W.A. Kuperman and E.K. Scheer, Environmentally tolerant beamforming for high-resolution matched field processing: Deterministic mismatch, *J. Acoust. Soc. Am.* **88**, 1851–1862 (1990).

21. T.C. Yang and K. Yoo, Internal wave spectrum in shallow water: Measurement and comparison with the Garrett-Munk model, *IEEE J. Ocean. Eng.* **24**, 333–345 (1999).

22. F.S. Henyey, D. Rouseff, J.M. Grochocinski, S.A. Reynolds, K.L. Williams and T.E. Ewart, Effects of internal waves and turbulence on a horizontal aperture sonar, *IEEE J. Oceanic Eng.* **22**, 270–280 (1997).

23. E.K. Westwood, C.T. Tindle and N.R. Chapman, A normal mode model for acousto-elastic environments, *J. Acoust. Soc. Am.* **100**, 3631–3645 (1996).

HIGH-FREQUENCY PROPAGATION FOR ACOUSTIC COMMUNICATIONS

MICHAEL B. PORTER, PAUL HURSKY AND MARTIN SIDERIUS

Science Applications International Corporation
1299 Prospect St, Suite 303, La Jolla, CA 92037, USA
E-mail: michael.b.porter@saic.com

VINCENT K. MCDONALD AND PAUL BAXLEY

SPAWARSSC
San Diego, CA 92152-6145, USA

In recent years there has been great progress in developing undersea wireless networks. The physical layer of these networks is generally an acoustic link operating in the 10–50 kHz band. Interestingly, our understanding of the acoustic propagation has not kept up with the elegant signaling schemes used to transmit the information. For instance, one common system uses adaptive equalizers to not only recombine the ocean multipath but to track the ocean dynamics. The tracking occurs on a millisecond time scale. Predicting system performance then requires an understanding of the transmission loss, the noise background, and the dynamics of the multipath. Here we use the term 'transmission loss' loosely, glossing over subtleties about the coherent and the incoherent field, which also play an important role in the system performance. This talk will summarize the issues and our knowledge about them, drawing upon results from an extensive set of sea tests conducted under the SignalEx program.

1 Introduction

It is not hard for today's cell phone user to appreciate the benefits of wireless connections. On land, 802.11a,b, and g are rapidly becoming established for wireless local area networks and Bluetooth™ for more specialized connections. In the ocean, this sort of technology is just emerging; however, the pay-off is arguably greater, given the costs of laying cable at sea.

One proposal under development would deploy a wide-scale network to monitor sewage outfall, plankton blooms, and nutrients, to observe the ocean ecosystem. The flexibility in terms of adapting the network configuration (including autonomous vehicles) to evolving features is a major benefit of the wireless structure. In the ocean, of course, wireless normally implies an acoustic rather than electromagnetic link.

On the human level this application presents another communication problem, namely that between the communications engineer and the underwater acoustician. The first group is generally only casually interested in the channel characteristics; the second group is only casually interested in the modulation schemes. Actually, the techniques from communication theory are very clever in finessing the effects of a varying

N.G. Pace and F.B. Jensen (eds.), Impact of Littoral Environmental Variability on Acoustic Predictions and Sonar Performance, 287-294.

multipath environment. However, we believe there is much to be gained for underwater acoustic communications in bridging these fields. That is the goal of this short paper.

2   Communications schemes

Before we can discuss the role of the acoustic channel, we should understand something about the existing communication schemes [1,2]. There are many variations on these themes but a high-level taxonomy breaks these into *non-coherent* and *coherent* schemes. We will discuss some of the more widely used approaches. In general, increased receiver complexity provides increased bandwidth efficiency, i.e. we can get more bits through per second if we work harder.

It is useful to remember that with sufficient SNR, there is no limit on the achievable bit rate (we assume a discrete-time, memory-less, additive Gaussian noise channel). To understand this, consider a pulse-amplitude transmission scheme where the amplitude of a pulse is given by a 32-bit integer. (We assume the SNR is such that the 32-bits are resolvable.) If we have a 1-kHz bandwidth, we can send a new pulse every millisecond, providing 32 kbits/s or 32-bits/Hz. Since the data rate is proportional to the number of bits we use to encode the pulse amplitude we can transmit as many bits as we want in the 1-kHz band.

Noise of course makes our ability to distinguish levels less reliable and as the SNR drops the bit-rate drops. Thus, the channel capacity, C, in bits/s is a function of bandwidth, W, and SNR:

$$C = W \log_2 (1 + SNR).$$

Higher data rates can therefore be achieved by increasing one or the other. Practically speaking, bandwidth efficiencies in a wide variety of systems tend to be O(1) bits/Hz.

2.1   MFSK (Multiple Frequency-Shift Keying)

Imagine playing a piano underwater. Obviously, the pattern of notes can be used to encode an arbitrary data stream. Of course, the multitude of echoes could make it difficult to hear clearly. Simply playing one chord and waiting for all the reverberation to die down before playing the next can solve this multipath problem. This is the basis for *Multiple Frequency-Shift Keying* and is the technique of the widely used Benthos modems. It is also the basis of the so-called Interoperability Standard proposed as a sort of Esperanto for acoustic modems (which in turn may speak their own language for more floral expression).

The details of one particular scheme may be of interest to the reader. We synthesize a piano with 128 keys spaced 40 Hz apart from 8 kHz up to 13.2 kHz. The upper and lower 4 tones are reserved for pilot tones that can be used to compensate for Doppler. Since 40-Hz tones are readily resolved in an FFT over a 25-ms time frame, we strike the notes for 25 ms. We then allow 0–200 ms of clearing time before striking the next chord. In MATLAB the demodulator is little more than a one-line call to generate the spectrogram of the received waveform as seen in Fig. 1. (These data are taken from a SignalEx [3–5] test conducted near Ship Island, Mississippi.) One small trick here is that '1 of 4' coding is used. This means that every 2 bits of the data stream are used to select within blocks of 4 tones. The decoder then decides which of the 4 tones is loudest rather than trying to address the SNR-dependent question of whether one specific tone has been struck.

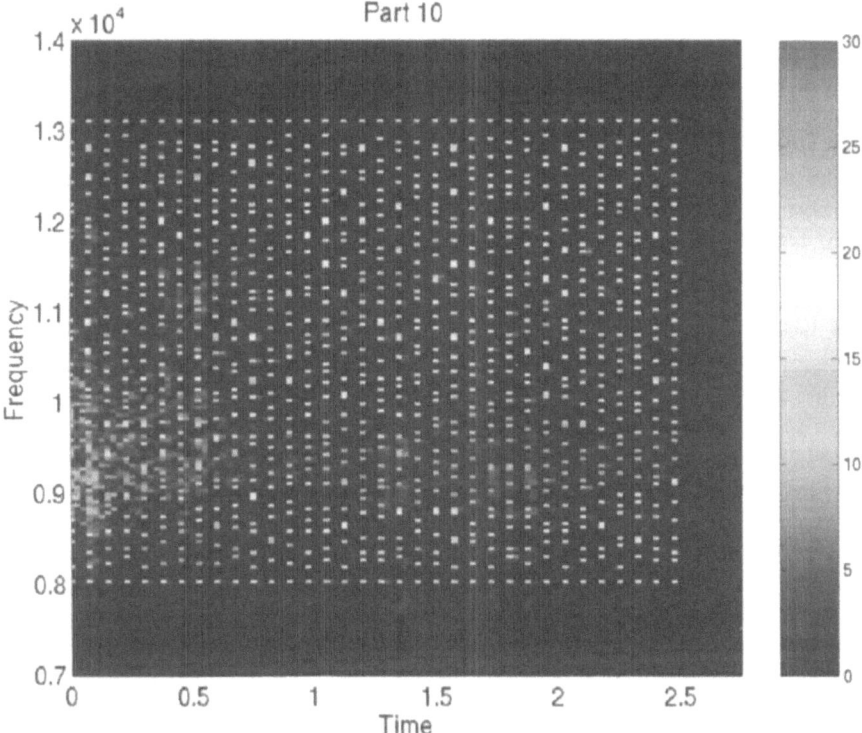

Figure 1. Spectrogram of an MFSK reception at a range of 5 km (SignalEx/Ship Island).

To get a feel for the data rates, consider that we have 120 tones available every 200 ms. Since we are using 1-of-4 coding, 30 tones are played in every chord and each tone encodes 2-bits. That implies 60 bits every 200 ms or 300 bps. It turns out that that amount of channel clearing time is seldom needed in shallow water and in Ship Island no clearing time at all was needed so a data rate of 2400 bps was reliably achieved. (You can see in Fig. 1 that there is little evidence of echoes.)

2.2   Direct-Sequence Spread Spectrum (CDMA)

The technology of Direct-Sequence Spread Spectrum (DSSS) or Code Division Multiple Access (CDMA) is embedded in the Telecommunications Industry Associate standard IS-95 and is widely used in cell phones. The technique begins with a repeating stream (a so-called *spreading code*) that looks like a random pattern of +/–1's. In the standard terminology these are called *chips*, a term that distinguishes them from the bits of the data stream. The chips flip (or do not flip) the phase of a sine wave or carrier and this is referred to as *Binary Phase-Shift Keying*.

For our testing we typically use a 12-kHz carrier with 4000 chips/s. This implies we flip (or more precisely, control) the sign 4000 times per second, and thereby generate a transmitted wave with energy predominantly in the 8–16 kHz band. (The bandwidth spreading is proportional to the chip rate.)

The process described so far is not allowing us to transmit data. To get our data stream in we simply multiply the bit stream of our data by the chip stream. (We are considering our bit stream as +/–1 so that multiplication makes sense.) Typically our bit stream will run at 200 bps, which of course is much slower than the 4000 chips/s. To summarize, the bits flip the sign of the chip stream 200 times/s; the chip stream flips the sign of the carrier 4000 times/s.

To recover the data on the receiver end, we correlate, i.e. matched-filter, the received time series with a BPSK signal encoded using the spreading code directly. The result is the carrier wave with phase determined by the bit stream. Note that where MFSK beats the multipath by waiting for the channel to clear, CDMA does so by the intrinsic orthogonality of the spreading code. There is an initial acquisition process to synch the receiver matched-filter to the frames of the data. However, once that is done a later (or earlier) arriving multipath has a piece of the spreading code that is de-correlated from the dominant path used in the acquisition. With that understanding we can see that as the multipath spread increases we will generally need to reduce the bit rate so that the time-window for each bit never sees an echo of a previous bit.

In a short space, it is difficult to do justice to the CDMA approach. It should be noted that the standard procedure also includes a RAKE receiver that decodes the received signal on a tapped-delay line and recombines the taps. This process provides a further benefit in a fading multipath environment. Multiple-access is handled by simply assigning a different spreading sequence for each user. For moving sources, a Doppler tracker is also typically required to account for drifts in the frame positions. Finally, the typical implementation is DPSK (Differential Phase Shift Keying) rather than BPSK. Benthos Corp. and Nautronix Corp. are currently testing DSSS/CDMA in their commercial modems.

2.3   Coherent Schemes with Adaptive Equalizers

Getting the highest data rates through a channel typically requires careful attention to channel equalization. On a superficial level one might imagine that for a channel that propagates lower frequencies better than high frequencies that you simply boost the high end at the receiver to recover the natural sound. The problem is actually far more difficult. The channel transfer function in the ocean is just the Fourier transform of the impulse response. Consequently it has a variation in the frequency domain that occurs on a scale proportional to the reciprocal of the multipath spread.

To take some round numbers, in the 8–16 kHz band we may see up to 200-ms spread in a shallow water channel. That implies a 5-Hz scale to the variation of the transfer function. Over an 8–16 kHz band one may then have a very large number of deep fades. The equalizer must compensate for these variations across frequency. It must also track the time variations of the transfer function. This is done through an adaptive equalizer, which is actually implemented in the time-domain as a tapped-delay line. The output of the adaptive equalizer is roughly speaking a version of the received waveform with the multipath effects eliminated.

3   Implications for high-frequency modeling

Numerous sea tests in the lower frequency band have established that acoustic propagation can usually be understood in terms of a sequence of distinct echoes. These echoes are typically fairly undistorted copies of the transmitted waveform. There are far fewer published results in the 8–16 kHz band and one may well-wonder if an echo off a rough bottom or clouds of bubbles near the surface leaves much of a signal. A representative sample shown in Fig. 2 illustrates that even out to ranges of 6 km, one may see a clear multipath structure. (The colorbar is intensity in dB with an arbitrary reference.) These data from the SignalEx/New England Shelf test were taken during a fairly rough sea state with wave heights of approximately 3 m.

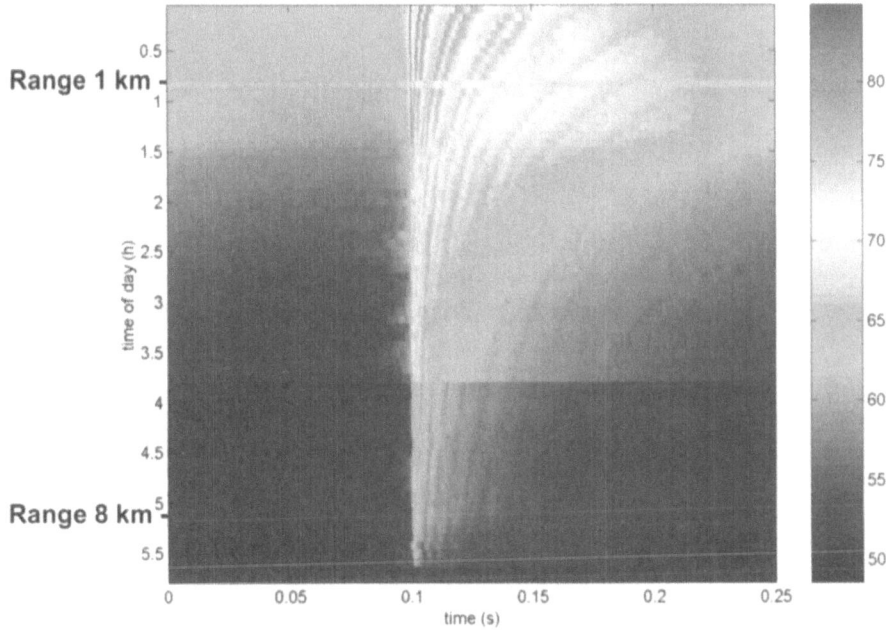

Figure 2. Variation of impulse response in a shallow-water channel (depth=45m) from SignalEx/New England Shelf. The response was measured as the source drifted out in range.

The first-order predictor of communications performance is simply SNR. In upward-refracting conditions the sole arrival is sometimes the surface reflection. A rough sea forms a murky acoustic mirror and simultaneously generates a lot of ambient noise. Network outages have been clearly correlated with wind speed.

The second key consideration is multipath spread. All of the above described modulation schemes are sensitive to this in some way. The MFSK approaches requires channel clearing time based on the spread; the DPSK scheme suffers from intersymbol interference; and the PSK approach with adaptive equalizers requires a tap-delay line accomodating the spread (and even so, will degrade in trying to equalize a more complicated channel). Of course, multipath spread can vary significantly from one environment to another. For instance, Fig. 3 shows a result taken during the

SignalEx/Ship Island test in very shallow water (< 5 m) and shows a multipath spread of about 2 ms. This type of channel is an easy one for most modulation schemes.

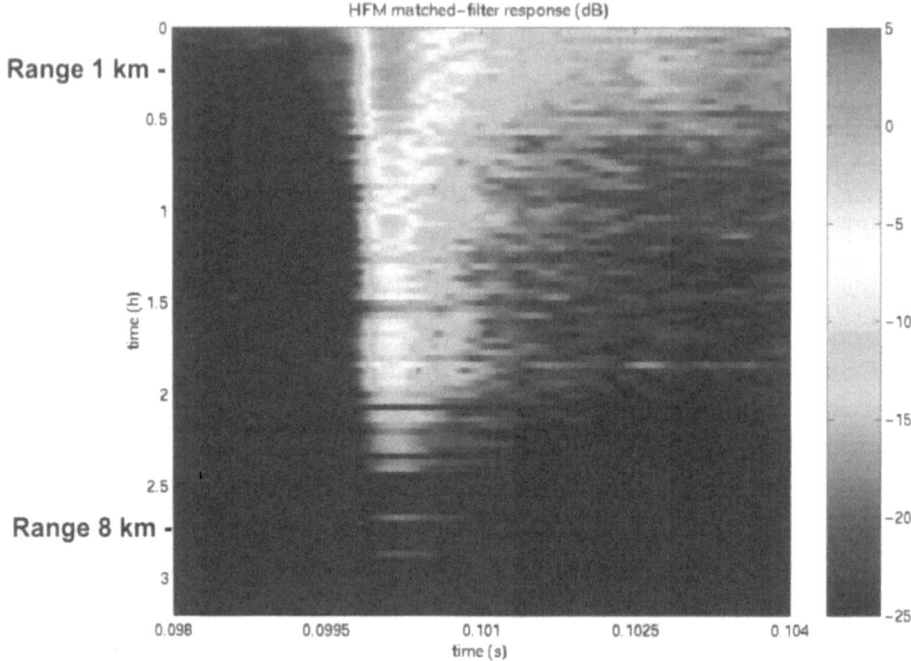

Figure 3. Variation of channel impulse response in a very shallow water (<5 m) site   (from SignalEx/Ship Island, Mississippi).

The final consideration is the variability in the channel impulse response. This is sometimes characterized in terms of Doppler spread and can be measured by transmitting a tone and then forming its spectrogram. An example is shown in Fig. 4 from ModemEx conducted off the coast of San Diego. Note that the 10-kHz tone has been both shifted and spread. The former is principally due to the fact that the source is moving; the latter probably represents fluctuations due primarily to swell. It is typically easy to compensate for Doppler shift.

Variability is obviously a factor for MFSK schemes in that it limits the narrowness of the individual tones. The effects on   other schemes are much more subtle. For instance, the adaptive equalizers are designed to track the dynamics of the multipath structure. To understand how they lose track, we must understand how the equalizer reacts to very short scale variations in the multipath structure. With a moving source, those variations can be significant on a 100 ms time scale.

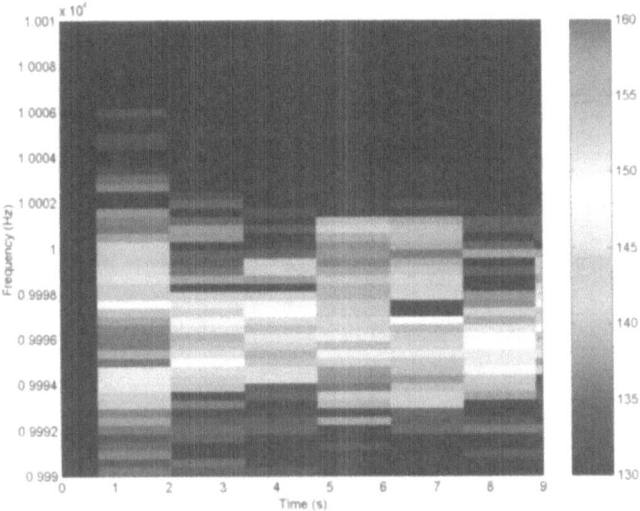

Figure 4. Doppler shift and spread of a 10-kHz tone (from SignalEx/Pilot Test San Diego).

Direct measurements of the time-dependent impulse response are also useful in characterizing the variability. The left panel in Fig. 5 shows such a result from the SignalEx/Point Loma test with chirps transmitted every 250 ms. We see that even at the 12-kHz carrier frequency the impulse response is very stable from ping to ping. Repeating the process 2 minutes later we again get a stable impulse response. However, comparing left and right panels we can clearly see the evolution in time. A fixed source/receiver geometry like this is clearly a much easier situation than one where there is relative source/receiver motion, e.g. in communicating to an AUV. Typically we find that with drifts of just 0.5 m/s the impulse response varies significantly between pings separated by 250 ms.

4   Summary

Communications engineers have developed clever schemes to mitigate the effects of a time-varying channel. The schemes have been optimized to yield the best performance subject to constraints on complexity. However, it is important to realize that understanding the channel characteristics is still a very important problem if we wish to predict the actual achievable performance. For instance, in deploying a network we have to select a spacing between nodes. That in turn will depend on our best guess of achievable data rates relative to bit-error rates.

The state of the art on the modeling end is fairly limited. Surface and bottom loss models in the 10–50 kHz band are available in HFEVA [8]. However, the dynamics of the ocean impulse response (usually characterized by the so-called scattering function) is also important. There are very few measurements of the scattering function in the ocean.

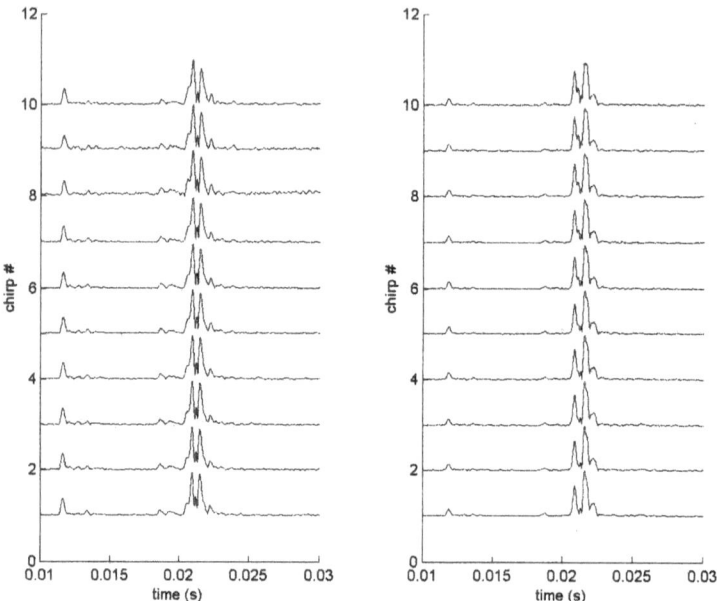

Figure 5. Successive snapshots of the impulse response taken every 250 ms. Right-panel recordings begin 2 minutes after left (from SignalEx/Point Loma, San Diego, California).

Acknowledgements

This work was supported by the Office of Naval Research 321OM (Ocean Modeling and Prediction). Additional support was provided by 321OA (Ocean Acoustics).

References

1. J. Proakis, *Digital Communications* (McGraw Hill, New York, 1995).
2. D.B. Kilfoyle, J.C. Preisig and M. Stojanovic (Eds.), Special issue on underwater acoustic communications, *J. Oceanic Eng.* **25** (2000).
3. V.K. McDonald, J.A. Rice, M.B. Porter and P.A. Baxley, Performance measurements of a diverse collection of undersea, acoustic, communication signals, *Proc. IEEE Oceans'99* (Seattle, Washington, 1999).
4. M.B. Porter, V. McDonald, J. Rice and P. Baxley, SignalEx: Linking environmental acoustics with the signaling schemes, *Proc. IEEE Oceans 2000*.
5. V.K. McDonald, J.A. Rice and C.L. Fletcher, An underwater communication testbed for Telesonar RDT&E. Ocean Community Conference '98, The Marine Technology Society Annual Conference, Baltimore, Nov. 16–19, 1998.
6. J.A. Simmen, S.J. Stanic and R.R. Goodman (Eds.), Special issue on high-frequency acoustics, *J. Oceanic Eng.* **26**, (2001).
7. F.B. Jensen, W.A. Kuperman, M.B. Porter and H. Schmidt, *Computational Ocean Acoustics*, 2nd ed. (Springer-Verlag, New York, 2000).
8. APL-UW High-frequency ocean environmental acoustic models handbook, APL-UW TR9407 (1994).

CHANNEL IMPULSE RESPONSE FLUCTUATIONS AT 6 KHZ IN SHALLOW WATER

W.S. HODGKISS, W.A. KUPERMAN AND D.E. ENSBERG

Marine Physical Laboratory, Scripps Institution of Oceanography,
University of California, San Diego, La Jolla CA 92093-0701, USA
E-mail: [wsh,wak,dave]@mpl.ucsd.edu

A shallow water (~100 m) experiment was carried out to measure the stability of forward transmissions at 6 kHz over a 6 km propagation path. The fixed source, fixed receiving array geometry enabled observing environmentally-induced fluctuations in the channel impulse response. The 64-element receiving array had an aperture of 12 m and thus a corresponding vertical angle of arrival resolution of ~1°. Source transmissions were of duration 20 min and consisted of multiple subcomponents. Of particular interest here are the 2 kHz bandwidth, 1 s duration FM chirps which were transmitted continuously for 5 min and have been matched filtered to yield channel impulse response structure. In addition to CTDs taken in the region between the source and receiving array, a thermistor string at the receiving array site provided continuous measurements of water column temperature fluctuations. Discussed in this paper is the time-evolving structure of the channel impulse response observed from both individual array elements as well as array beams which clearly show significant, environmentally-induced fluctuations over the 5 min duration of these transmissions. Also discussed is the relationship these fluctuations have to the characteristic ray path structure of the channel.

1 Introduction

The shallow water environment can be quite dynamic. In addition to the background internal wave field, rapid water column temperature fluctuations generated by the internal tide as it progresses along the continental shelf have a substantial impact on high frequency acoustic transmissions.

In this paper, we present results from a shallow water (~100 m) fixed source, fixed receiving array experiment carried out to measure the stability of forward transmissions at 6 kHz over a 6 km propagation path. First, we will overview the experiment and the background internal wave-related temperature fluctuation characteristics. Next, we will describe the general multipath structure between the source and receiving array. Lastly, we will focus on a 5 min period and show the time-evolving impulse response at a single element and how the impulse response fluctuations manifest themselves in arrival angle vs. travel time as well as wavefronts observed across the receiving array.

N.G. Pace and F.B. Jensen (eds.), Impact of Littoral Environmental Variability on Acoustic Predictions and Sonar Performance, 295-302.

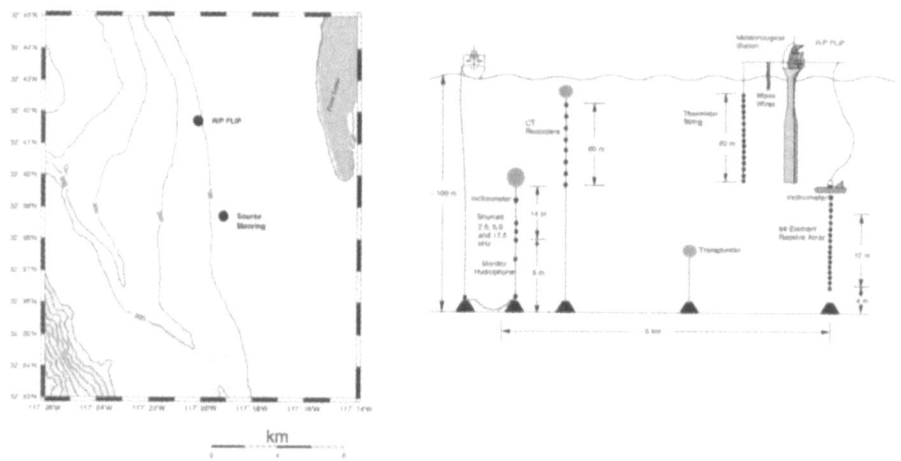

Figure 1. Experiment overview showing the source mooring and receiving array locations (left panel) and the hardware placement in the water column (right panel).

2 Experiment

The experiment was conducted off the coast of San Diego, CA, in October 1997 (Fig. 1a). An overview of the experiment showing the environmental and acoustic instrumentation is shown in Fig. 1b. The fixed source, fixed receiving array geometry enabled observing environmentally-induced fluctuations of the channel impulse response over the 6 km propagation path. The 64-element receiving array had an aperture of 12 m and thus a corresponding vertical angle of arrival resolution of ~1° at 6 kHz. Both the receiving array and source mooring were deployed near the seafloor. Source transmissions were of duration 20 min and consisted of multiple subcomponents. Of particular interest here are the 2 kHz bandwidth, 1-s duration FM chirps which were transmitted continuously for 5 min and have been matched filtered to yield the time-evolving channel impulse response structure. Additional results from an analysis of acoustic communications transmissions centered at 18 kHz are given in [1].

In addition to the acoustic instrumentation, environmental measurements were made throughout the experiment. Along with CTDs being deployed frequently in the area, a 16-element thermistor string was deployed from the R/P FLIP and measured the time-evolving temperature structure from the mixed layer through the main thermocline (10–70 m) where internal wave activity typically is most prevalent. Figure 2 shows this structure throughout the duration of the experiment.

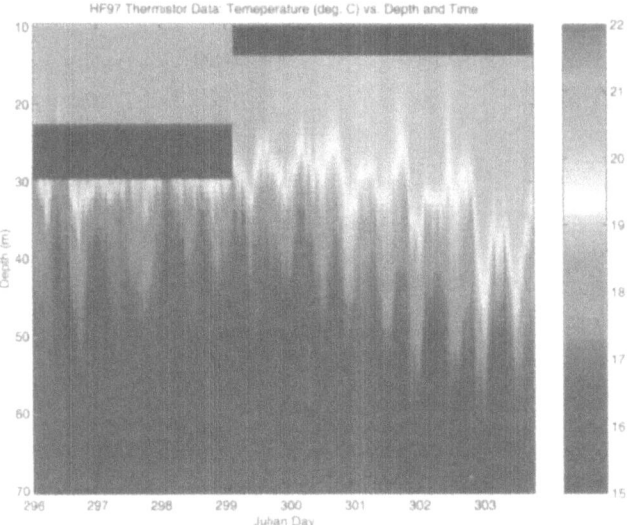

Figure 2. Time-evolving temperature structure measured by the thermistor string from the mixed layer through the main thermocline (10–70 m).

3   Multipath structure

The general nature of the sound speed profile observed during the experiment is shown in Fig. 3a which was derived from one of the CTD casts. The mixed layer is relatively deep followed by a sharp thermocline. Of note is a second thermocline in the 70–90 m depth range. A range-independent ray trace of all significant eigenrays from the source to the receiving array is shown in Fig. 3b. As is evident, one group of rays turns over in the lower thermocline, another group turns over in the upper thermocline, and lastly a group of rays interacts with the sea surface.

Another useful display of the eigenrays is to plot their transmission loss (TL) vs. arrival time at the array and their vertical angle of arrival vs. arrival time. These are shown in Fig. 4

4   Channel impulse response

The matched filtered FM chirp transmissions enable observing the time-evolving channel impulse response with a 1-s update rate. A 5 min period of relatively high dynamics was selected for detailed analysis (JD 301: 2207-2212 UTC). The water column temperature structure over the 16 hour interval bracketing this period is shown in Fig. 5.

4.1   Single-Element Impulse Response

The time-evolving channel impulse response from the source to array element #1 (highest in the water column) is shown in Fig. 6. The general structure is consistent with the eigenray predictions in Fig. 4a. Although the single-element impulse response is

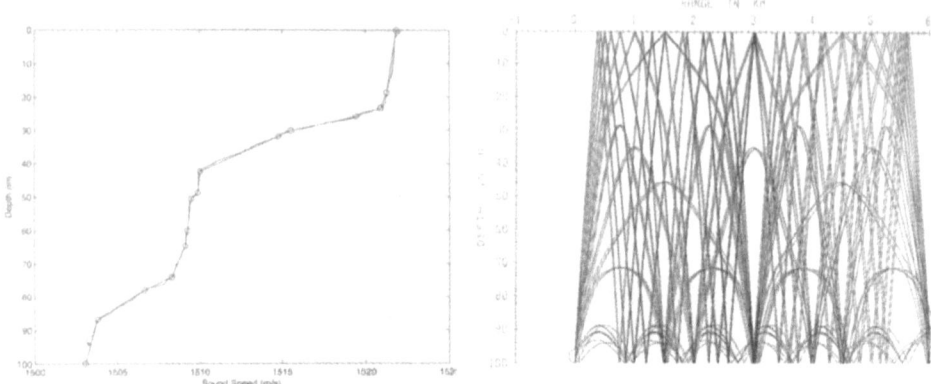

Figure 3. Typical sound speed profile (left panel) and corresponding ray trace of all significant eigenrays from the source to the receiving array (right panel).

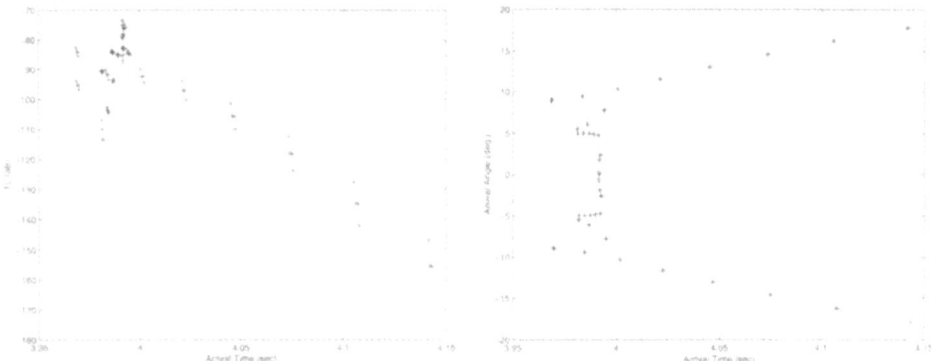

Figure 4. Eigenray transmission loss vs. arrival time (left panel) and vertical angle of arrival vs. arrival time (right panel).

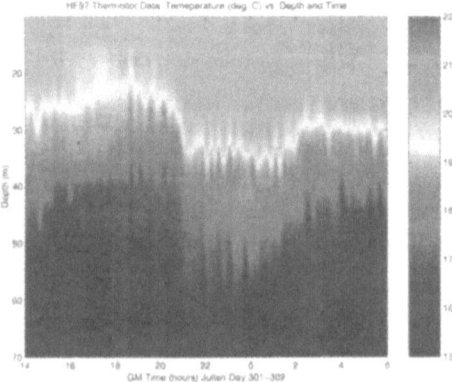

Figure 5. Water column temperature structure over the 16 hour period bracketing the 5 min period selected for detailed analysis (JD 301: 2207-2212 UTC).

relatively stable for a few seconds to a few tens of seconds at a time, there are substantial fluctuations over the total 5 min observation period.

4.2   Arrival Angle vs. Travel Time

The transmissions were observed on a 64-element array. Thus, we can decompose the channel impulse response structure in both space and time. The average arrival angle vs. travel time structure over the entire 5 min observation period is shown in Fig. 7. Although fluctuations are evident in the average, the underlying structure is consistent with the eigenray predictions in Fig. 4b. Individual examples of the spatial structure of the channel impulse response are shown in Fig. 8 where the results represent transmissions 4 min apart.

4.3   Wavefront Arrivals

In addition to the spatial decomposition provided by the beamforming results, displays of the wavefront arrivals also are informative. For the same transmissions displayed in Fig. 8, the wavefront arrivals are shown in Fig. 9.

5   Summary

The fixed source, fixed receiving array transmissions at 6 kHz discussed here provide a detailed look at the time-evolving dynamic structure of the shallow water channel impulse response. In this case, the water column was characterized by having a relatively deep mixed layer followed by a sharp upper thermocline. Of note was a second thermocline in the 70–90 m depth range. Thus, there are ray groups which turn over in the lower thermocline, turn over in the upper thermocline, and interact with the sea surface. Fluctuations of the channel impulse response relate directly to the dynamics of each of these regions of the water column. For the data analyzed, the impulse response structure is somewhat stable for a few seconds to a few tens of seconds at a time but over the total observation period of 5 min there are substantial fluctuations.

Acknowledgements

This research was supported by the U.S. Office of Naval Research.

References

1. Carbone, N.M. and Hodgkiss, W.S., Effects of tidally driven temperature fluctuations on shallow–water acoustic communications at 18 kHz, *IEEE J. Oceanic Eng.* **25**, 84–94 (2000).

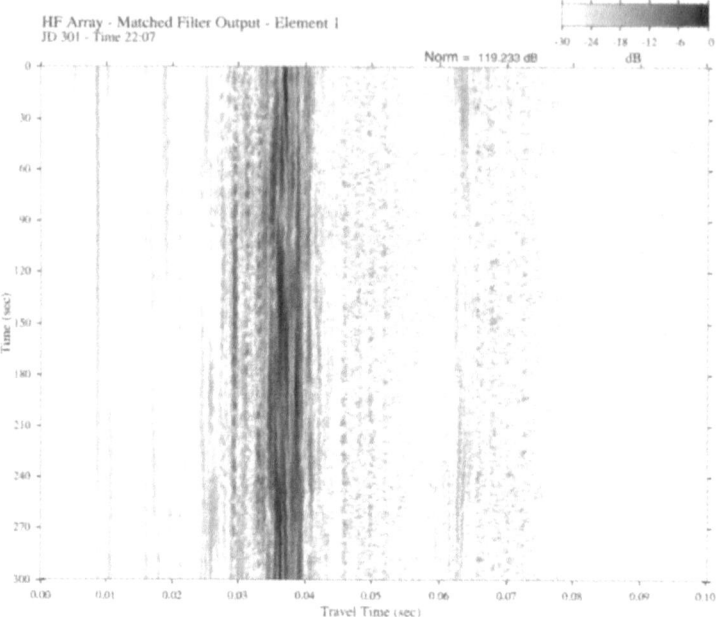

Figure 6. Time-evolving channel impulse response from the source to array element #1.

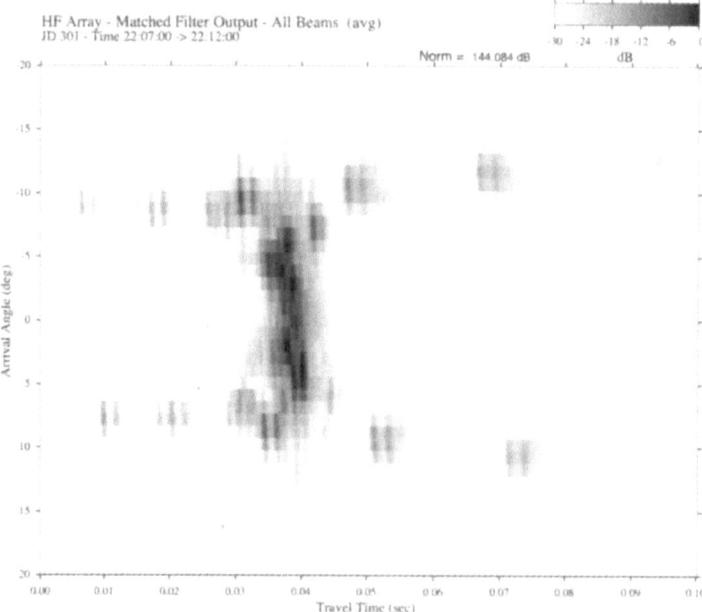

Figure 7. Arrival angle vs. travel time structure averaged over the 5 min period.

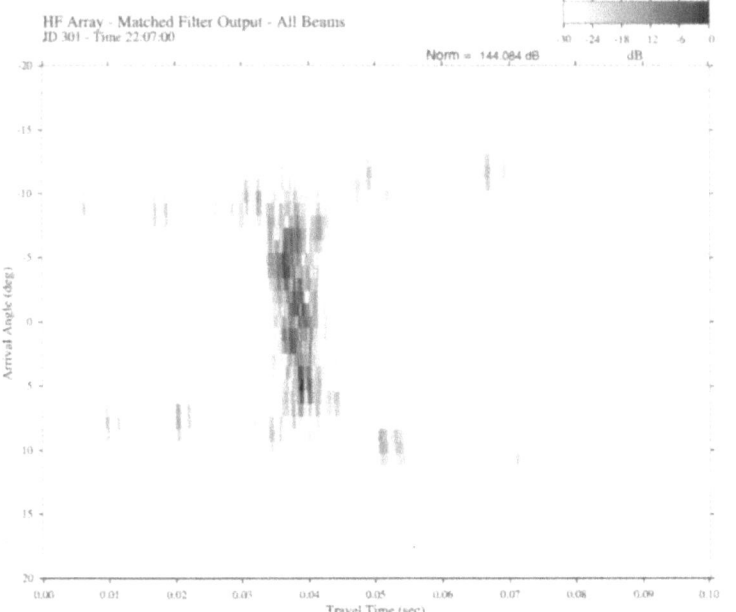

Figure 8a. Arrival angle vs. travel time (JD 301: 2207 UTC).

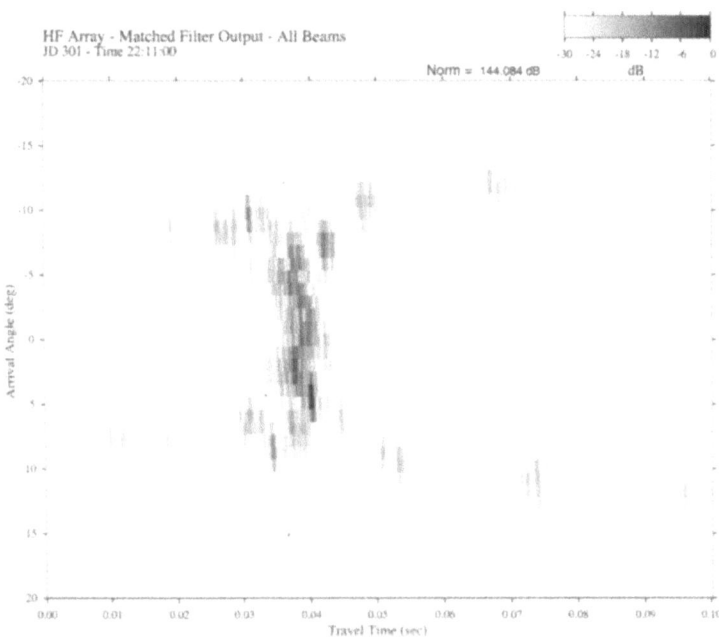

Figure 8b. Arrival angle vs. travel time (JD 301: 2211 UTC).

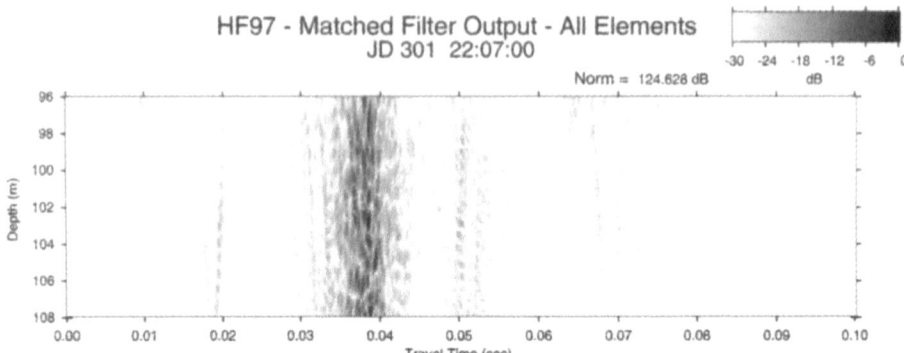

Figure 9a. Wavefront arrival structure across the receiving array (JD 301: 2207 UTC).

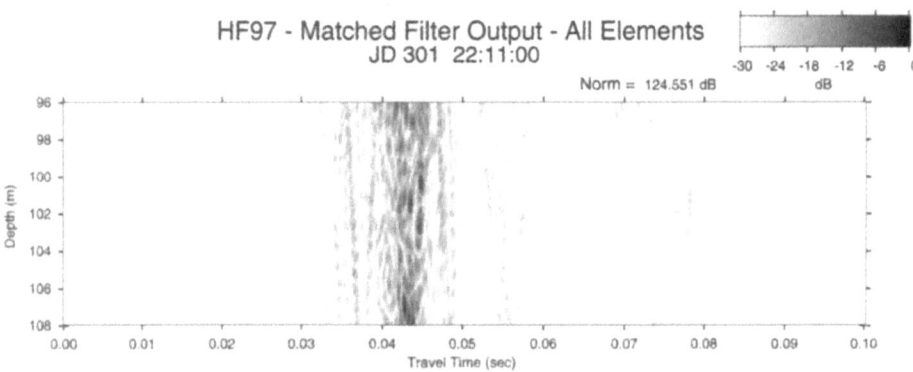

Figure 9b. Wavefront arrival structure across the receiving array (JD 301: 2211 UTC).

HIGH RESOLUTION ANALYSIS OF EIGENRAY GAIN PERTURBATIONS IN ULTRA-SHALLOW WATER

S.M. SIMMONS, O.R. HINTON, A.E. ADAMS, B.S. SHARIF AND J.A. NEASHAM

Department of Electrical and Electronic Engineering Merz Court,
University of Newcastle upon Tyne, Newcastle upon Tyne, NE3 1DQ,, UK
E-mail: s.m.simmons@ncl.ac.uk

This paper presents results obtained from the acoustic analysis of an ultra-shallow channel of around 40 m in the North Sea over ranges of 0.9 km and 3.0 km. It is shown how the channel is characterised by severe multipath propagation and signal fading. A low frequency phenomenon is described that was apparent in all of the arrival paths to the receiver array over these distances. The article then demonstrates how these frequencies could arise as a result of an oscillation of the transmitter, possibly related to the swell of the channel.

1 Introduction

The results and analysis presented in this paper are taken from the data obtained during the first LOTUS (Long range Telemetry in Ultra-Shallow channels) project sea trials held in the summer of 1999 [1]. The purpose of this project was to develop a system of digital communication between deployed underwater units. These units could either be mounted on the sea floor, free floating or attached to an un-tethered underwater vehicle. It was envisaged that the distance between the nodes would be far greater than the depth of the water column. The sea trials in 1999 were carried out in the North Sea off the North-East Coast of England over a period of five days. The average depth of the sea was around 40 metres for the chosen deployments, with transmission distances of 0.8–10 km. The shallow depth of the water combined with the medium to long ranges produced hostile multipath effects as well as variable direct paths due to the seasonal conditions.

On each day of the trials, two transmitter units were deployed at different locations. The transmitter hydrophones were approximately 5 m above the sea floor and transmitted autonomously throughout the day. The receiver structure consisted of an array of 8 hydrophones. The bottom element of the array was 1 m above the sea floor, with an equal vertical spacing of 3 m between the remaining single hydrophones and a central horizontal structure of four of the hydrophones. The data was recorded on one of the support vessels that had a link to the receiver array, which was deployed at the beginning of the experiment. A representation of the transmitter and receiver structures is shown in Fig. 1.

N.G. Pace and F.B. Jensen (eds.), Impact of Littoral Environmental Variability on Acoustic Predictions and Sonar Performance, 303-310.
© 2002 *Kluwer Academic Publishers.*

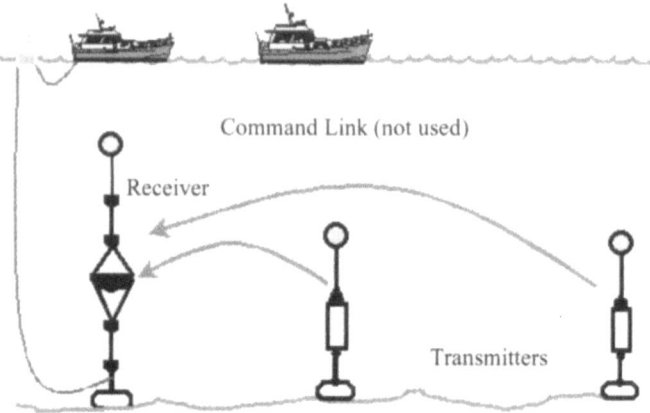

Figure 1. Transmitter and receiver structures.

2   Chirp analysis

During each day of the trials the two transmitters generated a wide range of data formats to assess the multi-user capability of the communications network. At the start of each day a chirp signal was transmitted for around half an hour. This was to provide data from which the statistical properties of the fading, multipath channel could be analysed. The chirp signal was designed to enable a high-resolution analysis of the channel and also to ensure that the current channel response is well separated from the echoes of the previous transmissions. The frequency of the chirp was 8–12 kHz over a period of 12.5 ms throughout the trials. The repetition rate of the chirp was 25 ms on the first two days (0.9 km and 3 km range) and 50 ms on the final three days (3 km, 5 km and 8 km range). The signal was transmitted at a level of 186 dB (re 1 μPa at 1m) and sampled at 48 kHz.

Figure 2(a) shows the received signal at the bottom hydrophone element over a period of 100ms on the first day of the trials transmitted over a range of 0.9 km and Fig. 1(b) shows the signal after cross-correlation with the chirp.

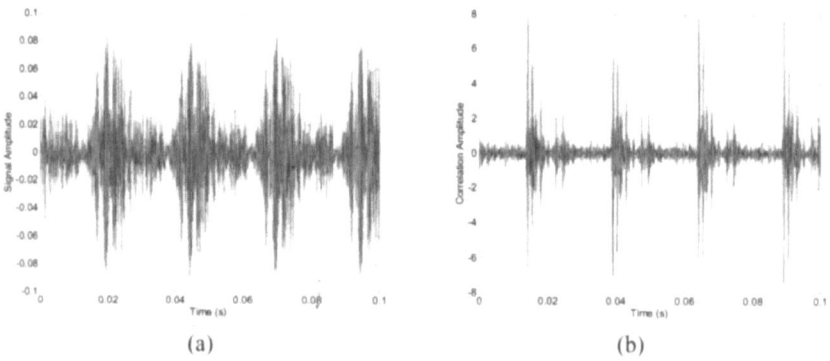

(a)   (b)

Figure 2. (a) Received signal and (b) cross-correlation with the chirp.

It was assumed that the peak of the cross-correlation in the periodicity of the chirp represents the direct path gain at that instant. Hence by taking the values of the maximum peaks it is possible to infer the variation of the direct path gain with time. The sampling period is the inverse of the chirp repetition rate; giving a high resolution of 40 Hz. Figure 3(a) shows the direct path correlation amplitude over a period of 25 s for the bottom receiver element. Low frequency periodic behaviour at around 10 s can be observed in the direct path gain. Also present is higher frequency periodic behaviour, which was analysed using the Wigner-Ville transform with a Choi kernel.

The transformation is shown in Fig. 3(b) corresponding to the direct path gain between 3 s and 15 s from Fig. 3(a) after filtering of the low frequency oscillations. The periodicity varies approximately sinusoidally between 2 Hz and 10 Hz. The period of this change in frequency correlates with the low frequency oscillation of the direct path gain of around 10 s periodicity. During the dry tests prior to the sea trials, oscillations were observed in the peak of the correlation due to the difference in the transmitter and receiver clock crystals. It is this phenomenon that is responsible for the oscillations in the frequency band 2–10 Hz shown in Fig. 3.

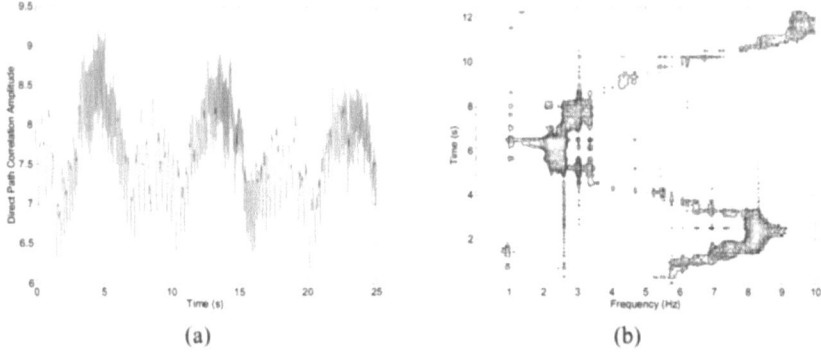

(a)   (b)

Figure 3. (a) Direct path gain of the bottom element of the receiver array and (b) the Wigner-Ville time-frequency transformation of the gain.

The impulse response estimates were filtered above 2 Hz and squared to separate the effects induced the channel properties from the artefact effects of the difference in the two crystal oscillator frequencies. Figure 4(a) shows a 100 ms series of the filtered chirps. The impulse responses were then time aligned, assuming that the transmitter and receiver crystals were operating at exactly 48 kHz. Figure 4(b) shows a contour plot of a section of the aligned responses over a period of 12.5 s for the lower receiver element. It can be seen from the plot that the responses are misaligned due to the differences in the sampling frequencies. Further to the gradient of the peak values, an oscillation can be observed with a period of around 10 s. The oscillation of the position of the peak values was extracted and is shown in Fig. 5. This oscillation appears to be a variation in the time of arrival of the direct path. As can be seen from Fig. 4(b) this oscillation also occurs in the multipath arrivals. Although not shown here, the oscillations were also present in the responses obtained from the other 7 elements of the array.

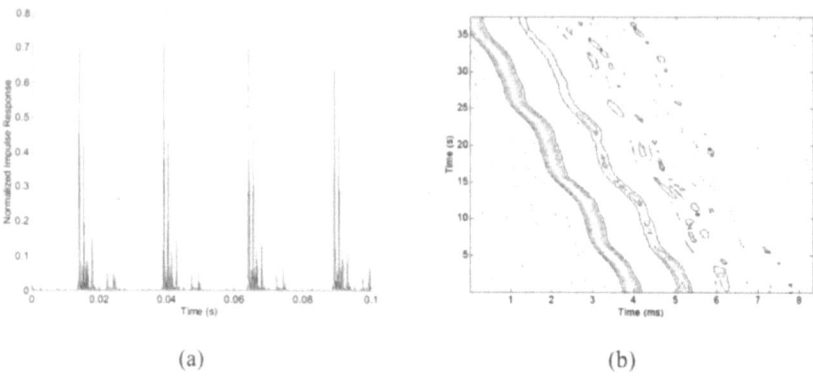

(a)   (b)

Figure 4. (a) Normalized impulse responses after squaring and filtering and (b) a contour plot of the responses after time alignment for the bottom receiver element.

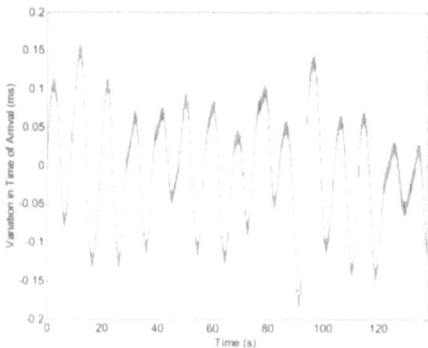

Figure 5. The apparent variation of the time of arrival of the lowest receiver element direct path.

The gradient of the time of arrival shown in Fig. 4(b) corresponds to a difference in the transmitter and receiver crystal oscillators of just under 5 Hz. The combination of this frequency with the variation in the time of arrival shown in Fig. 5 accounts for the variation in the artefact frequency shown in Fig. 3(b).

The plots of Figs. 6(a) and 6(b) show the variation in the direct path for the bottom receiver element for a period of over 2 minutes at a range of 0.9 km and 3.0 km respectively. At the shorter range an oscillation in the amplitude can be observed with a period of around 10 s. At the longer range there still appears to be low frequency oscillations, although not as regular and pronounced as at the shorter range. Figures 7(a) and 7(b) show the power spectra of the direct path gain at a range of 0.9 km and 3.0 km respectively, estimated using the data shown in Fig. 6. The spectrum shows the presence of the 0.1 Hz period as well as what appears to be a second harmonic at 0.2 Hz for the shorter range. The power spectrum at the larger range displays similar behaviour in terms of low frequency components, although at a slightly lower frequency and lower power due to the greater attenuation.

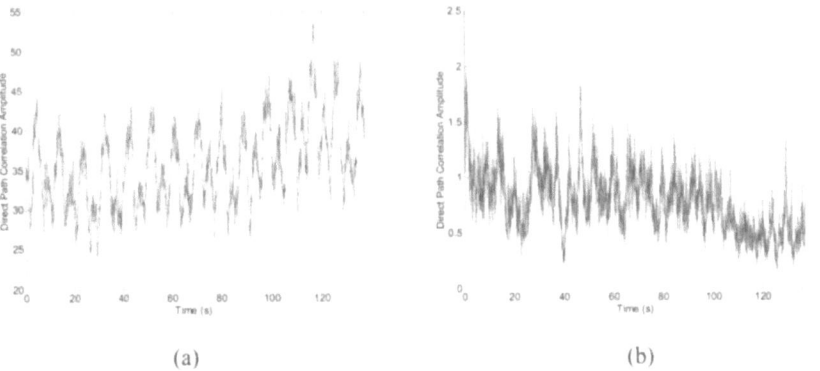

(a)   (b)

Figure 6. Direct path amplitude variation over 130 s for the bottom receiver element at a range of (a) 0.9 km and (b) 3.0 km.

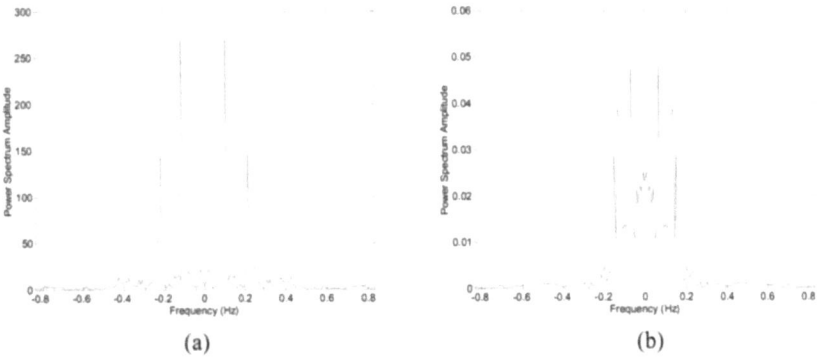

(a)   (b)

Figure 7. The power spectrum of the direct path for ranges of (a) 0.9 km and (b) 3.0 km.

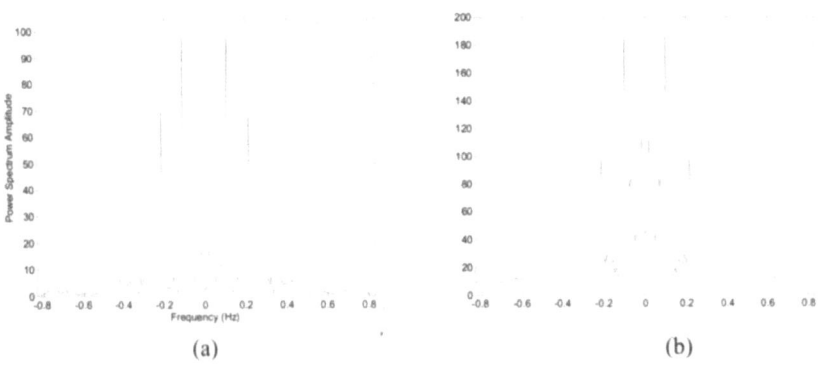

(a)   (b)

Figure 8. (a) The Fourier transform of the cross-covariance of (a) the direct path at the bottom and top receivers and (b) the direct path and second dominant arrival at the bottom element.

Figure 8(a) shows the Fourier transforms of the cross-covariance of the direct path at the bottom and top receivers at a range of 0.9 km. The plot is very similar to the power spectrum of Fig. 7(a). Hence there is a high degree of correlation between the received signals at the different elements due to the large wavelength. The same correlation can be seen in the Fourier transform of the cross-covariance of the direct path at the bottom receiver with the second significant arrival. All of the direct paths and later arrivals were found to be highly correlated for all of the arrays.

Figure 9 shows the evolution of the multipath signals over a period of 500 s for four of the receiver elements. The contour plots were obtained by aligning the direct path at each element. The direct path appears at around a 2 ms (aribitrary) delay from the beginning of the sequence. This direct path is not visible on Fig. 9(b) as the amplitude is too small to show up on the plot. Further to the direct paths there are two distinguishable arrivals visible. The delay of the first of these arrivals decreases as the height above the sea floor increases, indicating that it is a surface reflection. The delay of the second dominant arrival after the direct path behaves in the opposite manner, indicating it has been reflected off the sea floor. Closer inspection of Fig. 9(d) for the top element shows the presence of an arrival between the direct path and the first dominant arrival from the other plots. This arrival has been reflected from the sea floor and has been lost on the other plots as it is too close to the direct path and the resolution is not fine enough. The most noticeable feature of the plots shown in Fig. 9 is the large fading of the later arrivals over long periods.

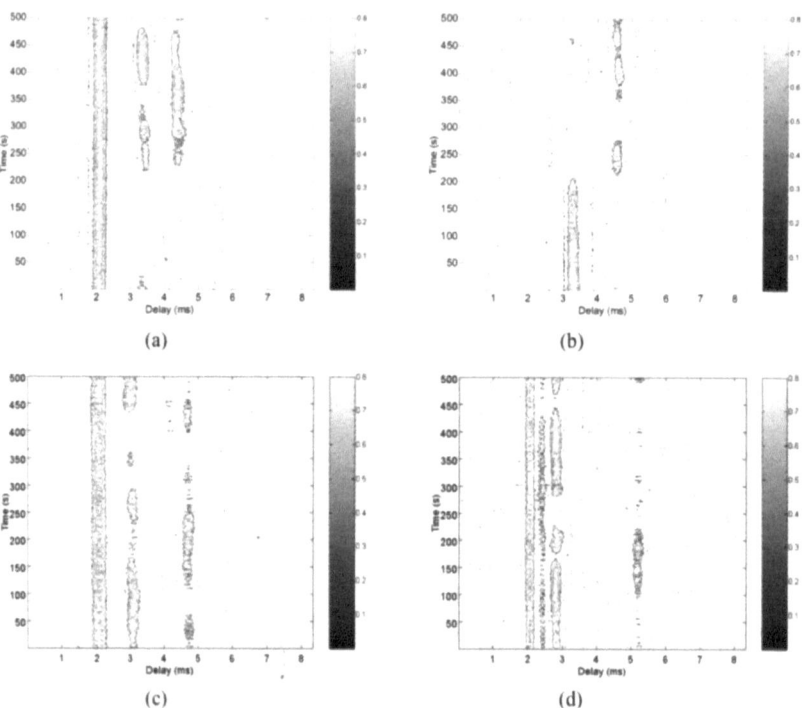

Figure 9. Evolution of the multipath propagation paths over 500 s for receiver elements at (a) 1 m, (b) 4 m, (c) 8 m and (d) 14 m above the sea floor.

Figure 10 shows the direct path over 500 s for the bottom elements at a range of 0.9 km and 3.0 km. The amplitude of the direct path tended to remain fairly constant for each element at the shorter range and varied substantially at the longer range. The amplitudes at the different elements however tended to vary a great deal from one another. The large degree of fading at the 3.0 km range suggests that the arrival is in fact a reflection, as it exhibits similar fading behaviour to the reflected paths at 0.9 km. It is possible for no direct path exists between the transmitter and the receiver at the time of year of the trials due to the downward refraction caused by the sound speed profile. However, this is not verifiable as no sound speed recordings were taken.

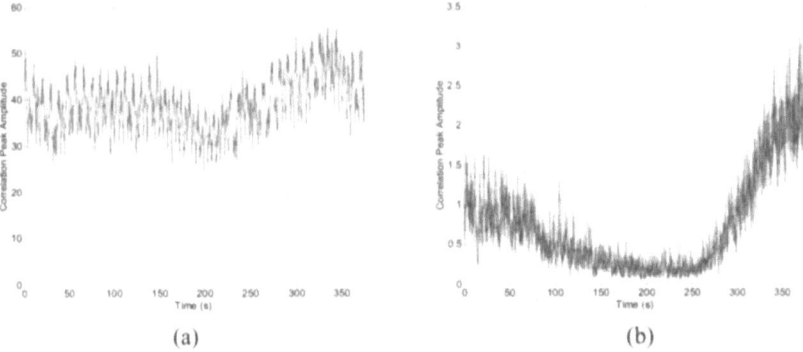

(a)   (b)

Figure 10. Variation of the direct path gain for the lowest receiver element over 375 s for (a) day 1 of the trials at 0.9 km range and (b) day 2 at 3.0 km range.

3   Discussion

Figure 11 shows the normalised power obtained from a simulation of the correlation of a signal consisting of two chirps with different delays. It can be seen that for certain differences in delays between the two chirps a considerable amount of signal fading occurs. The maximum fading occurs when there is a lag of about 0.05 ms corresponding to an underwater path difference of around 70 mm. Such a delay corresponds to the geometry of the transmitter and receiver arrays used in the trials. Hence it is thought that the fading seen in the direct path of Fig. 9(b) is attributable to this phenomenon.

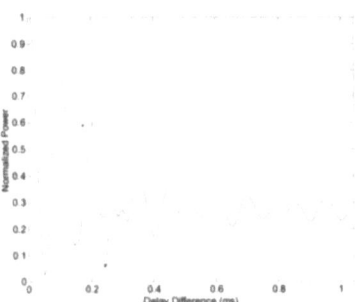

Figure 11. The normalized peak correlation amplitude of the sum of two chirps with different delays.

Low frequency oscillations appear in the amplitude of all of the arrivals for both ranges examined in this paper. The frequency of the oscillations remains virtually constant at the range of 0.9 km and less so at the longer range of 3.0 km. The period of the major component of the oscillations correlates with what appears to be a variation in the travel time delay of all of the paths. Possible mechanisms for the generation of the travel time delay perturbations include the crystal oscillator behaviour, a change in the sound speed and travel path by mechanisms such as the swell and finally an oscillation of either the transmitter and/or the receiver array.

The sea state at the time of the trials was Beaufort state 1 with a 1 to 1.5 m swell having a 10 s to 15 s period. Hence the variation in the travel time could be associated with the swell. One possible mechanism for the generation of the travel-time delays is an oscillation of the transmitter. A peak-peak oscillation of around 0.3 m of the transmitter would produce travel time delays equivalent to those observed. It is unlikely the receiver is oscillating, as the frequency shift of the artefact frequency was the same for all of the elements [2]. The effect of the increase of depth due to the swell is not enough to account for a sufficient change in sound speed on its own. The most overriding piece of evidence is the correlation between the travel-time delay and the amplitude oscillations in the paths. As demonstrated in Fig. 11 a small change in the time delay between the arrivals of two chirps can result in large fading. If all of the travel-time delays oscillated with exactly the same amplitude then no amplitude oscillations would occur. Hence it is necessary for at least one of the paths to have a travel-time oscillation that is different in magnitude to the others. This difference in the time of arrival variation magnitude is possible from the geometry of the system of the trials, especially if it was correlated with the swell (resulting in a slightly greater oscillation amplitude for surface reflections). If the crystal oscillators had frequency variations that create an impression of a variation in the travel-time delay then there would be no change in the amplitude of the paths as all the delays would be equal.

4   Conclusions

It has been shown from experimental data that severe multipath and fading occur in the channel used in the LOTUS sea trials for the chirp signals used to infer the channel properties. It is postulated that the fading of the chirp combined with a low frequency oscillation of the transmitter unit, possibly related to the swell, creates the low frequency phenomena observed in the arrival paths.

References

1. Adams, A.E., Hinton, O.R., Sharif, B.S., Salles, G., Orr, N. and Tsiminedis, C., An experiment in sub-sea networks – The LOTUS sea trials. In *Proc. 5th European Conference on Underwater Acoustics*, Lyon, France, 10–13 July, 2000.
2. Salles, G., Adams, A.E., Hinton, O.R., Sharif, B.S., Orr, N. and Tsiminedis, C., High resolution analysis of ultra-shallow water acoustic travel time and gain perturbations. In *Proc. Oceans'2000*, Rhode Island, USA, 17–20 Sep., 2000.

IMPACTS OF FLOW VARIABILITY ON FIXED SIDE-LOOKING 100 KHZ SONAR PERFORMANCE IN A SHALLOW CHANNEL

M. TREVORROW

Defence Research & Development Canada – Atlantic
PO Box 1012, Dartmouth, Nova Scotia, Canada
E-mail: mark.trevorrow@drea.dnd.ca

A 100 kHz side-looking sonar was operated nearly continuously from September 1996 to the end of May 1997 in Drogden Channel, near Copenhagen, Denmark. This busy shipping channel, 1-km-wide by 12-m-deep, connects the Baltic Sea with the North Sea through the Kattegat. The purpose of this operation was to maintain surveillance for migratory herring during the construction of a bridge and tunnel, however a variety of shipping traffic was also observed. Water temperature, salinity, and current profiles along with surface meteorology were simultaneously monitored at this site. Under normal, weakly stratified flow conditions, ships and fish schools were observed up to 400 m range, detectable against a background of low-grazing-angle seabed backscatter. Occasional saline intrusions were observed to create strong upward-refracting conditions that significantly restricted the available range for target detection. Example echograms and reverberation results from normal and upward-refracting conditions are shown. Ray-tracing analysis is used to account for the observed backscattered reverberation.

1   Introduction

Side-looking, high-frequency sonars offer a promising approach for continuous monitoring of strategic waterways at ranges of 100's to 1000's of meters. The purpose of such monitoring might be to quantify migration by fish or marine mammals, or in a defense scenario for the detection of autonomous underwater vehicles, torpedoes, or SCUBA divers. Side-looking sonar installations potentially offer continuous, full-water-depth surveillance, detecting targets by their motion relative to the nominally time-invariant boundary reverberation. However, probing close to the boundaries in stratified, shallow-water environments necessarily incurs reverberative interference and acoustic propagation effects. The purpose of this present work is to examine the reverberation levels observed during a typical side-looking sonar deployment, and to show that rapid changes in water properties can drastically alter the reverberation background. An earlier paper (Pedersen & Trevorrow 1999) discussed monitoring and detectability issues specific to fish schools.

These sonar installations were deployed and operated during 1996–97 in Drogden Channel as part of an environmental monitoring program during the construction of a bridge and tunnel connecting Copenhagen, Denmark and Malmö, Sweden. This shallow channel (<14 m deep) is a busy shipping lane connecting the Baltic Sea with the Øresund and Kattegat. A key feature of this project was the continuous monitoring of water temperature, salinity, and current profiles alongside the sonar installation. This allows a

N.G. Pace and F.B. Jensen (eds.), Impact of Littoral Environmental Variability on Acoustic Predictions and Sonar Performance, 311-318.

more complete investigation of the environmental influences on acoustic propagation and reverberation.

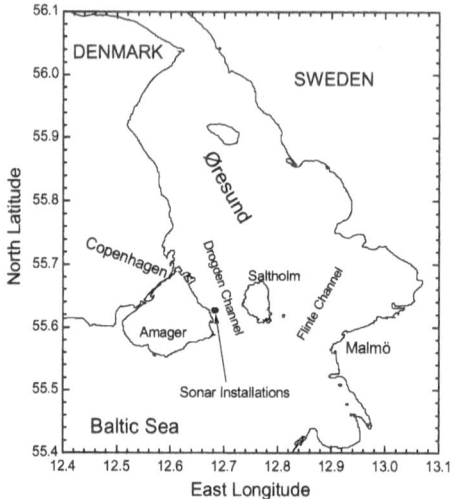

Figure 1. Map of the Øresund showing the location of the sonar deployment in Drogden Channel.

2   Description of site and instrumentation

The sonar monitoring program was conducted across the Drogden navigation channel between the islands of Amager and Saltholm (see Fig. 1) near Copenhagen, DK. The sonar was installed in September 1996 on the western edge of Drogden channel, approximately 150 m East of the Nordre-Røse lighthouse. This location was near the narrowest part of the main channel, which was approximately 1 km wide by 11 to 14 m deep. Two 100 kHz sidescan transducers (separate transmit and receive) were mounted on a tripod 1.2 m above the seabed in total water depth of 10.2 m. Tidal height variations in this area are negligible. The sidescan transducers had fan-shaped beams 3° by 60° (total angle to -3 dB). These sonars were mounted with their wide beam axes horizontal and oriented eastward, i.e. perpendicular to the navigation channel. This allowed a wide surveillance area nominally confined to a region within 2 to 5 m above the seabed, with small (<4°) seabed grazing angles. An underwater cable connected the sonars to the data acquisition system inside the lighthouse. The lighthouse station consisted of a BioSonics model 101 sonar transceiver, two networked personal computers (PC's), and a radio-modem for telemetering data back to the onshore base station. The sonar receiver provided a 20log[r] time-variable gain compensation followed by mix-down to an 8 kHz carrier and amplitude detection. The amplitude-detected signal was then sampled at 1000 samples per second (0.74 m spatial sampling resolution) using a PC-based analog to digital converter with 16-bit resolution. A pulse length of 2 ms was used, yielding an acoustic resolution of 1.45 m, transmitted once every 2 s. The data acquisition PC processed the digital data in real time, generating images of backscattered intensity versus range and time, which were then radio-telemetered to shore and printed. The raw data were normally purged a few hours after processing, however during March and April of 1997 most of the raw digital data was stored to enable later analysis.

Prior to deployment, the complete sidescan system was calibrated utilizing the back-scatter from Tungsten-Carbide target spheres as reference (following Vagle *et al.* 1996). Including the calibration results, a scattering level *Equivalent* to *Target Strength* (*ETS*, in dB re 1m²) within 0.74 m range cells centered at range, *r*, can be given by the sonar equation (e.g. Medwin & Clay, 1998)

$$ETS = K + 20 \cdot \log_{10}[A(r)] - TVG(r) + 40 \cdot \log_{10}[r] + 2 \cdot \alpha \cdot r, \qquad (1)$$

where K is the calibration coefficient (which includes transmit power, transducer sensitivity, waveform detection, and A/D conversion factors), $A(r)$ is the echo amplitude in digital counts, $TVG(r)$ is the receiver time-varying gain in dB, and α is the acoustic absorption in seawater (0.015 dB·m⁻¹ at 100 kHz under these water conditions). This equation would yield the true backscatter target strength if correction could be made for the beam deviation loss, i.e. if the target location with respect to the sonar beam was known. Another important characteristic is the limitations imposed by systemic noise. There are two types of noise voltages: i) transducer, pre-amplifier and cable noise (i.e. pre-*TVG*) with amplitude V_0, and ii) pre-digitization noise with amplitude V_1. An overall noise threshold can then be taken as the incoherent sum of the time-averaged noise root-mean-square voltages, i.e.

$$V_N^2(r) = <V_0^2> \cdot 10^{[TVG(r)/10]} + <V_1^2>, \qquad (2)$$

where <> denotes time-averaging. This can be converted to digital counts, $A(r)$, and substituted back into Eq.(1) to calculate a noise threshold. Measured values of $\sqrt{<V_0^2>}$ and $\sqrt{<V_1^2>}$ were 0.85 and 10 mV, respectively.

For environmental monitoring purposes a number of thermistor and conductivity (TC) sensors and an acoustic Doppler current profiler (ADCP) were operated in the channel. One of the TC moorings was located approximately 25 m to the east of the sonar, and is visible as a discrete target in all sonar records. This TC mooring had sensors at 5 depths: 2.9, 4.4, 5.6, 7.4, and 9.0 m. The ADCP was bottom mounted at a distance of 130 m from the sonar, and had a similar TC sensor (nominally 10.6 m deep). All sensors were sampled at 30 minute intervals. In general the Drogden channel waters were characterized by either a northward flow of relatively fresh Baltic Sea water or a southward flow of more saline water from the northern Øresund and the Kattegat. These flow regimes were not dominantly tidal, but driven by wind forcing and seasonal fluctuations. Figure 2 presents surface and bottom temperature, salinity, and along-channel (north-south) current data for March, 1997. Although the temperature of both water types was similar (between 2.5°C and 4.5°C during March), the typical salinity of the Kattegat water was 25 psu as compared to the out-flowing Baltic Sea water near 10 psu. Under normal flow conditions the waters were relatively homogeneous in both temperature and salinity. However, during transition periods between the flow regimes or during very weak Baltic outflows, a stratified flow regime sometimes occurred. These events are clearly evident in Fig. 2 as large differences between surface and bottom salinity, for example on March 1st, 6th, 8th, 11th, and 15th. During March 1997 this stratified flow regime occurred 26% of the time. The acoustic implications of these changes in flow will be discussed in the next two sections.

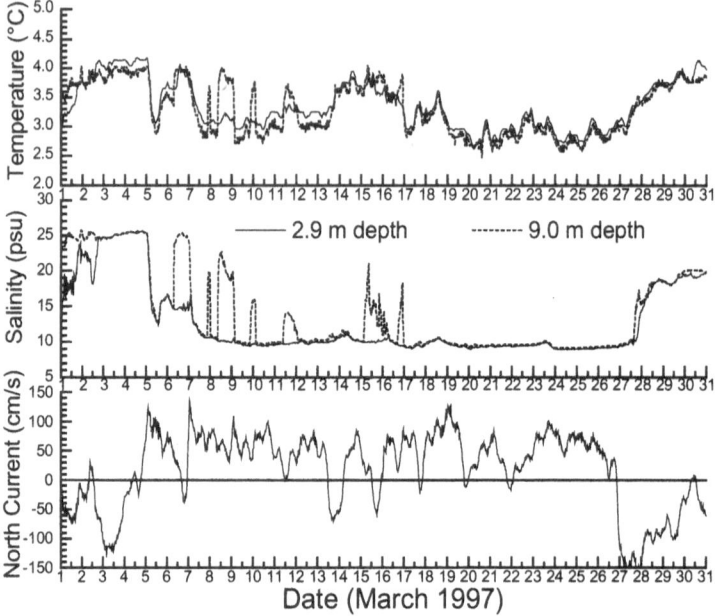

Figure 2. Comparison of near-surface (2.9 m depth) and near-bottom (9 m depth) temperature, salinity, and along-channel (North) current in Drogden Channel during March 1997.

3   Reverberation measurements and propagation modeling

The dominant reverberation source for this bottom-mounted sonar was low-grazing-angle seabed backscatter, which provides a background against which targets must be detected. However, occasional changes in water stratification drastically altered this situation. Figure 3 shows a 100 kHz sonar echogram from a period when the reverberation made a transition from seabed to surface scattering regimes. This transition occurred quite rapidly near 0300UT March 15th, simultaneous with a higher salinity near-bottom intrusion shown in Fig. 2. This salinity stratification created an upward-refracting sound speed gradient of up to 1.8 s[-1]. Prior to the transition, relatively strong seabed backscatter was seen in the range interval 50 to 225 m and beyond 330 m. The nominally time-invariant seabed reverberation exhibited distinct lines due to echoes from discrete seabed targets (presumably boulders). The rapid drop in seabed reverberation near 225 m marked the edge of a roughly 100-m-wide by 3-m-deep gully in the channel bottom. At the beginning of the transition, the seabed reverberation at ranges >330 m disappeared first. After the transition, the reverberation switched from the seabed regime towards a more homogeneous surface scattering regime. This surface reverberation had a broad peak roughly 100 m wide which began at longer range and then moved inwards to 150 m range as the upward-refracting stratification intensified. The most likely source for this near-surface backscatter was air bubbles injected by white-capping processes. Also identifiable in Fig. 3 are the signatures of several ships. These ship signatures exhibited three general characteristics: i) a direct echo from the hull which followed a hyperbolic trajectory in range vs. time as the ship traversed the horizontally wide beam (highly compressed in time in Fig. 3), ii) an intense noise event assumed to be caused by propeller cavitation, and iii) a strong backscatter region up to 40 m wide following the ship due to the injection of air bubbles within the wake. These wakes dissipated over a period of 6 to 10 minutes as the bubbles rose to the surface and/or dissolved (see Trevorrow et al. 1994). The typical *ETS* of these hull echoes, taken at the point of closest approach, was +12 to +15 dB (re 1 m^2). The bubbly wakes had a typical *ETS* near +6 to +15 dB. At closer ranges schools of herring were observed with *ETS* near −15 dB (Pedersen & Trevorrow 1999). Such *ETS* values made ship targets visible under most reverberation conditions.

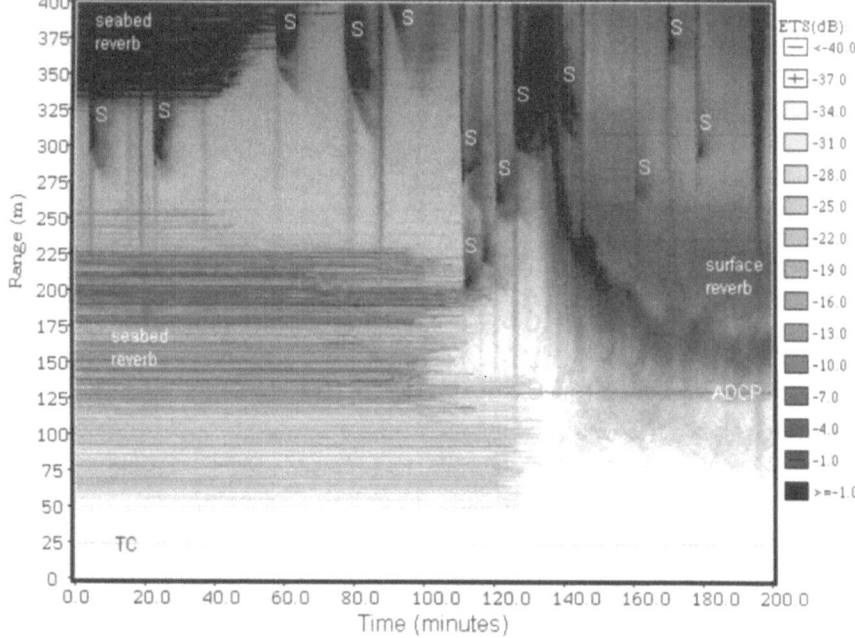

Figure 3. Cross-channel range vs. time echogram starting 0106UT March 15th, 1997. Regions of seabed and surface reverberation are indicated, with labels S indicating ships, TC the echo from temperature-conductivity sensor mooring, and ADCP the echo from a surface float at that location.

Figure 4 compares the seabed and surface reverberation regimes before and after the saline intrusion event shown in Fig. 3. Normally, the flow regime was characterized by weakly downward-refracting conditions, so that the sonar (1.2 m above the bottom) insonified the seabed from 20 to 225 m range (the edge of a central gully) and again beyond 330 m range. In the first 225 m the backscatter levels increased rapidly up to approximately 0 dB (re 1 m^2) due to the relatively large area of seabed sampled by the wide horizontal beam. On the far side of the gully the reverberation level reached roughly +10 dB, again a result of the 60° horizontal sonar aperture. The central gully region (225–330 m range) lay in an acoustic shadow so that the reverberation level approached the systemic noise. In contrast, the upward-refracting case exhibited a reduced reverberation level at closer range (up to 120 m) due to a minimal interaction with either the surface or seabed. Between 120 and 240 m range the upward-refracting reverberation level increased strongly to a level near 0 dB, similar to the seabed reverberation, and then fell away beyond 250 m range. In both cases the reverberation curves show a strong peak at 30 m range due to the T/C mooring. Finally, note a distinct contrast in *texture* between the two curves, with the seabed reverberation seemingly composed of discrete lines while the surface reverberation was relatively homogenous.

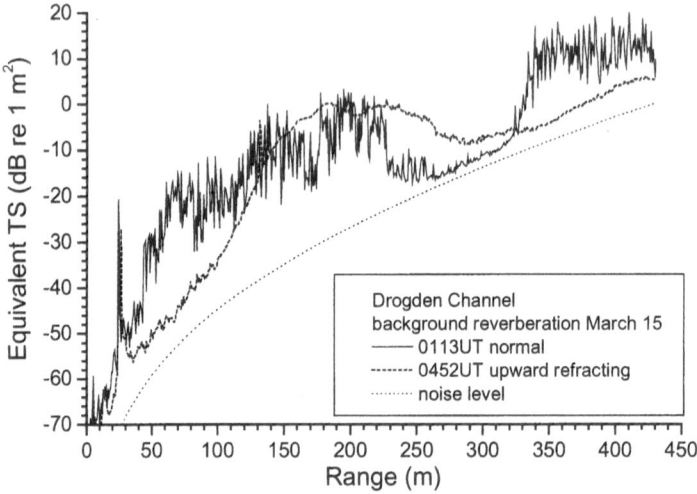

Figure 4. Comparison of seabed and surface dominated reverberation levels vs. range, averaged over 10-minute intervals before and after the transition event at 0300UT March 15th.

For 100 kHz sonars it is appropriate to model acoustic propagation using ray-tracing. This propagation analysis allows determination of the effective insonified volume, the transmission losses to a given region, and the grazing angles at the surface and seabed. A ray-tracing code due to Bowlin *et al.* (1992) was used to calculate sound pressure level as a function of range and depth for a sonar geometry, bathymetry, sound speed profile, and seabed reflection loss. Figure 5 shows a comparison of the propagation results between the normal and saline intrusion regimes. In this analysis, eigenrays connecting the source and a range-depth matrix (10 m range by 0.25 m depth) were calculated, with the intensity from multi-paths (if present) summed incoherently. Eigenrays were calculated within a launch angle interval ±3.5° from horizontal, corresponding to the –20 dB point of the transducer beam pattern. The sound speed profiles for the normal and saline intrusion cases were derived from the moored T/C data, assuming no range-dependence in water properties. Corrections were made for the transducer beam-pattern, seawater absorption loss, a surface reflection loss of 1 dB per bounce, and a seabed forward reflection loss vs. grazing angle calculated using classical two-layer interfacial theory (e.g. Medwin and Clay, 1998), with a 2 dB loss at grazing angles below critical. The sediments in Drogden Channel were comprised of coarse, gravelly sands with assumed density and sound speed of 1900 kg·m^{-3} and 1800 m·s^{-1}. This yields a reflection critical angle of 37° and normal-incidence reflection loss of 8.4 dB. Given the near-horizontal sonar geometry, all seabed reflections were sub-critical.

Figure 5a shows the ray-tracing calculation for the normal, homogenous flow regime. In this case the vertically-narrow beam is largely confined near the seabed, with a monotonically decreasing sound pressure level with range due to acoustic absorption. This analysis confirms the existence of low-grazing-angle (<3.5°) backscatter from the seabed from roughly 20 to 225 m and again after 330 m on the far side of the central gully. These ranges of seabed interaction agree with the measured reverberation vs. range curve shown in Fig. 4. Additionally, the ray-tracing analysis provides insight on the multi-path structure, for example beyond 50 m range the direct path is supplemented by a low-grazing-angle bottom-reflected path. Surface-reflected paths exist beyond 180 m range, and are responsible for insonification of the far slope of the central gully near 320 m range.

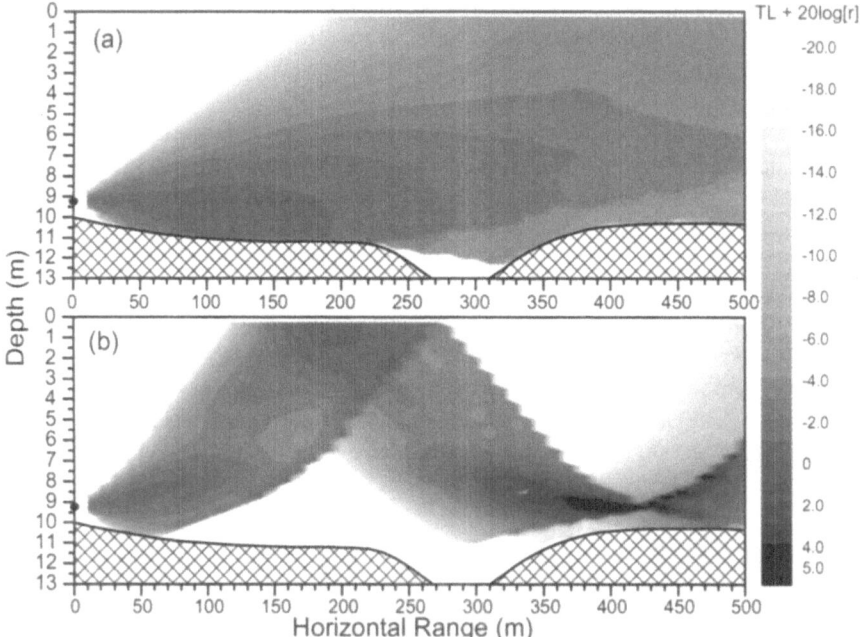

Figure 5. Ray-tracing predictions of normalized sound pressure level (one-way transmission loss + 20log[r] with 0 dB source level) vs. range and depth in Drogden Channel. Rays launched ±3.5° from horizontal. (a) normal flow regime. (b) saline intrusion event.

In contrast, the ray-tracing result for the upward-refracting conditions (Fig. 5b) shows a strong surface reflection and scattering region between 120 m and 280 m, with very little seabed interaction. Overall, there are significant *shadow zones* where targets would be undetectable, particularly near the seabed, making these propagation conditions less suitable for acoustic surveillance. Within this near-surface reflection/scattering region the grazing angle is 4° to 5° and there is a modest acoustic convergence (normalized SPL > 0) near 200 m range. Additionally, near 9 m depth and 350–480 m range there is a very strong (normalized SPL > 10 dB) acoustic convergence region. However this convergence region has only a minimal seabed intersection so that only a low reverberation level is created. Similar to the normal flow case, the propagation analysis is in accordance with the measured upward-refracting reverberation vs. range curve shown in Fig. 4.

4   Discussion

This project demonstrated the feasibility of continuous surveillance of a strategic channel using a fixed sonar installation. This rudimentary system was successful in detection of ships, yachts, and fish schools, suggesting the utility of such a system for surveillance against AUVs, torpedoes, and divers. However, it was found that target detectability was strongly limited by boundary reverberation, particularly from the seabed. This implies that the location for any future sonar installations should consider seabed sediment type, favoring muddy, silty, and fine sandy seabeds which have weaker backscatter properties. The relatively wide horizontal aperture (60°) of these sidescan transducers made them particularly susceptible to boundary reverberation, however this is also something that could be readily improved in future sonar installations.

Under normal flow conditions targets such as ships and fish schools were detectable at ranges up to 400 m using their motion relative to the nominally time-invariant seabed reverberation. This occurred even at relatively low signal-to-reverberation ratios (<10 dB). However, saline intrusion events occurring approximately 26% of the time drastically changed the reverberation conditions, creating shadow zones and making target detection using background referencing more difficult. Clearly, the importance of boundary reverberation and the drastic impact of changing water stratification necessitates the measurement of water properties alongside the sonar installation. Moreover, these measurements should be coupled with acoustic propagation analyses in order to understand and predict the impact on sonar performance, specifically to predict the sonar insonified volume, the existence of shadow zones, and expected reverberation levels.

Acknowledgements

The Danish Environmental Protection Agency, Kontroll- och Styrgruppen för Öresundsforbindelsen (KSÖ Sweden) and Øresundskonsortiet A/S are acknowledged for their support of the field program throughout 1995-98. The author is indebted to Dr. Bjarke Pedersen of LICEngineering A/S for his efforts in installing and maintaining the sonar system during the field trials.

References

1. Bowlin, J., Spiesberger, J., Duda, T. and Freitag, L., Ocean acoustical ray-tracing software RAY. Tech. Rep. WHOI-93-10, Woods Hole Oceanographic Institution, Woods Hole (1992) 49 pp.
2. Medwin, H. and Clay, C., *Fundamentals of Acoustical Oceanography* (Academic Press, San Diego, 1998).
3. Pedersen, B. and Trevorrow, M., Continuous monitoring of fish in a shallow channel using a fixed horizontal sonar, *J. Acoust. Soc. Am.* **105**, 3126–3135 (1999).
4. Trevorrow, M., Vagle, S. and Farmer, D., Acoustic measurements of microbubbles within ship wakes, *J. Acoust. Soc. Am.* **95**, 1922–1930 (1994).
5. Vagle, S., Foote, K., Trevorrow, M. and Farmer, D., Absolute calibrations of monostatic echosounder systems for bubble counting, *IEEE J. Oceanic Eng.* **21**, 298–305 (1996).

CORRELATION BETWEEN SONAR ECHOES AND SEA BOTTOM TOPOGRAPHY

JON WEGGE

Norwegian Defence Research Establishment (FFI), PO Box 115, NO-3191 Horten, Norway
E-mail: jon.wegge@ffi.no

False alarms resulting from sonar signals reflected off the sea floor terrain are of great concern in littoral waters. Series of these alarms often resembles a footprint of the topography. This gives rise to the idea of predicting the positions of such alarms using detailed terrain maps. In addition to the sound speed profile, the success of such a solution depends on the local sea floor composition in addition to the topography and the relative position of the sonar platform. This paper compares high-resolution bathymetric data with the positions of sonar alarms generated from a sequence of pings using a 7 kHz hull mounted sonar. The experiment is based on data recorded in a Norwegian fjord. The results show good correlation between sonar echo clusters and topographic features.

1   Introduction

At an early processing stage, the active sonar signal is passed through a detector to generate detections or alarms, or echoes which is the term used in this paper. As the number of sonar echoes generated during anti-submarine warfare (ASW) operations far outnumber the targets present, most of the echoes are false alarms. The topic of false alarms processing is becoming increasingly important with the increased detection capability of modern long range sonars and the focus on ASW capabilities in littoral waters. More alarms or sonar echoes are generated due to increased noise and reverberation, in particular bottom reverberation in fjords and coastal waters. The reason is the nature of the bottom topography and man-made objects on the sea floor. Although man-made objects like wrecks and pipelines usually generate submarine like echoes, the topographic features will outnumber the man-made objects in most areas.

The intensity of bottom reverberation not only depends on the bottom characteristics, but also on the sea floor depth and the sound speed profile. Some of these influences may be compensated for by adjusting the operation of the sonar, e.g. by tilting the sonar beams. Naval vessels are also maneuvered to reduce the level of reverberation in search areas. This limits the number of possible paths followed by the sonar platform, and enemy submarines may exploit this in order to minimize the probability of detection. A somewhat higher reverberation level could be tolerated if one could discriminate target echoes against echoes from bottom features. Conditions for such a method is a detailed topographic map including sub-bottom information.

N.G. Pace and F.B. Jensen (eds.), Impact of Littoral Environmental Variability on Acoustic Predictions and Sonar Performance, 319-326.

This paper will show correlation between sonar echoes and topography. It will also demonstrate the prediction of sonar echo clusters by applying detailed topographic and oceanographic information to simulations using the LYBIN hydroacoustic model.

2   The experiment

The objective of this work was to conduct a visual investigation of how the sonar echo positions recorded on the Oslo-class frigates' hull mounted sonar (HMS) Spherion, correlate with high resolution sea-floor topography data of. This analysis not only addresses the effect the topography has on echoes, but also the potential effect it has on the generation of tracks. Such tracks are based on series of echoes from topography and may be challenging for an operator to discriminate from real target tracks.

This study operates on a set of topographical data originating from raw data recorded by FFI using multi-beam sonar. Topographic data are compiled from several runs and have a depth resolution of about 1 m, but is also prone to errors in terms of peaks up to 200 m above the sea floor near great rifts and edges. This is clearly shown in the contour graphs in this paper. The range resolution is better than 10 m. No additional information about the bottom or objects on the bottom is taken into account. Comparison between bottom reverberation and topography has earlier been conducted elsewhere by Preston *et al.* at SACLANTCEN [1].

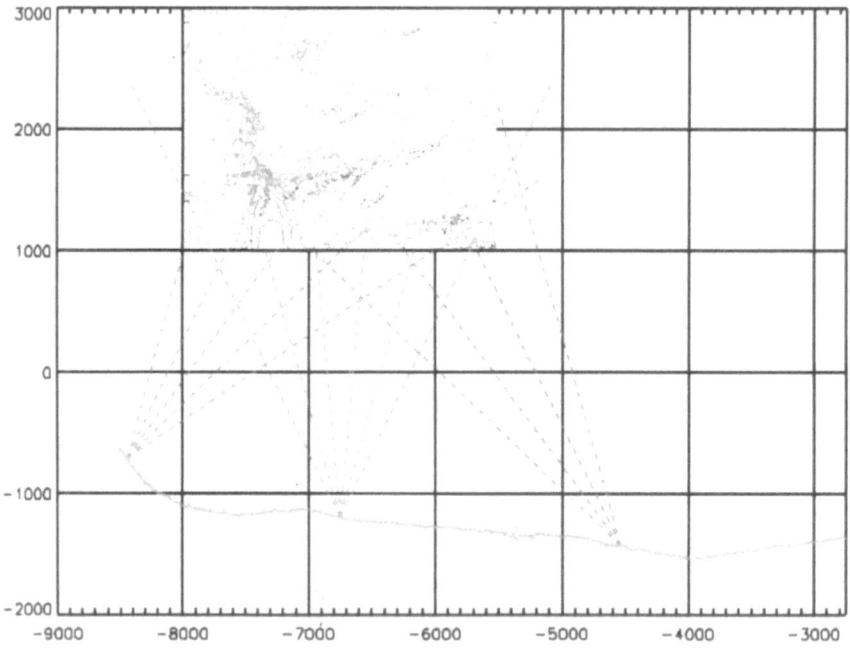

Figure 1. Frigate path is shown at the bottom where beam borders are drawn at pings 300 (right), 550 and 800 (left) as dashed lines in blue, green and red colours respectively. Axis values are in terms of meters from start of sonar run.

The Spherion sonar was delivered by Thales Underwater Systems and has 36 fixed receiver beams, 7 kHz center frequency, 500 Hz bandwidth, approximately 13° horizontal and vertical beamwidth and was doing FM-processing (match filtering), normalization, size filtering, clustering and thresholding to generate the echoes recorded. The data were recorded during a sea trial in April with sea state 1 and a surface duct reaching down to 50 m. Figure 1 shows the run with the sonar platform path at the bottom and the area under observation at the top. The lines extending from the sonar platform, show the direction and distance from the sonar at the different pings.

3   Echoes plotted on top of topographic data

The echo parameters recorded during the sea trial consist of the ping number, type of processing, their displacements from sonar platform in terms of meters in northern and eastern direction in addition to their respective signal-to-noise ratio as estimated by the Spherion sonar processing on board the frigate. As the position of the frigate for each ping was recorded, the absolute position of a reflector, giving rise to an echo, may be estimated. The deviation of the estimated echo position may be up to 100 m in isolated cases. However, the standard deviation is belied to be better than 25 m in both directions.

In the figures displaying the topography in this paper, the contours of the topography are drawn as coloured lines where purple and dark blue contours are depths of about 700 m and deep red contours are 0m topography depth. The contour interval is 20 m. The high-resolution (1 m in depth and 10 m in range) data in this analysis cover an open fjord region measuring 2 by 2.5 km. Outside this area, a range resolution at   50 m is used.

The borders separating the beams of the hull-mounted sonar are drawn for ping 300, 550 and 800 as straight lines in blue, green and red respectively. The same colours are used for plotting the echoes of the corresponding pings and intermediate pings (blue, green, yellow, orange, red).

3.1   Proximity Filtering of Echoes

Echoes estimated from a single sonar ping produce a display filled with almost randomly scattered echoes when plotted according to their relative position to the sonar platform. Many echoes seem to have random positions, they do not normally appear at the same positions in the following pings. This is caused by the combined influence of noise and all classes of reverberation at each ping.

A filtering method was introduced to exclude the echoes not having any close neighbour echoes within the most recent pings. The processing considers each ping and searches for one or more echoes over the ten previous pings within a box measuring 50 by 50 m centered on the echo. If no echoes are found, the echo under analysis will not be drawn. This proximity filtering reduces the number of echoes so that only phenomena repeatedly generating echoes are plotted. Hence echoes caused by more random signal spikes are filtered out. The method of course has a weakness in leaving out any isolated weak signals, which might be a target. This is not considered significant, as the objective of this analysis is to investigate the correlation between topography and sonar echoes from a HMS sonar.

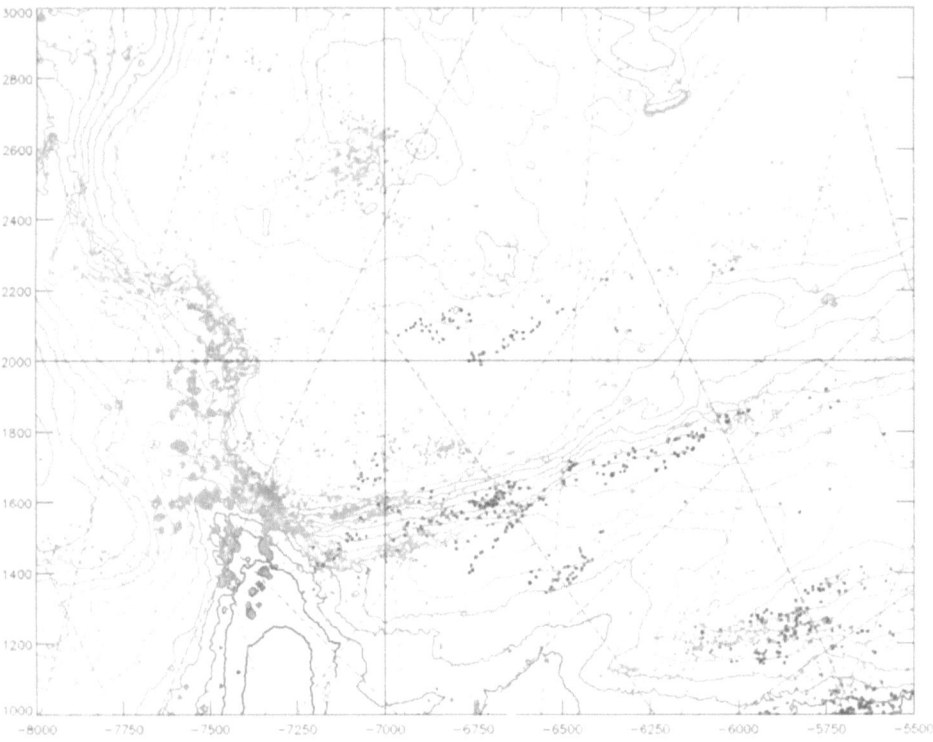

Figure 2. Echoes from ping 300–814 proximity filtered using 50 m and 10 pings and plotted over a 10-m resolution contour map. The echoes are coloured according to their history. Axis values are in terms of meters from start of sonar run.

Among other observations, it is worth mentioning how animations of sequential ping-images were effective for observing trends/dynamics in the scenario as opposed to stills. Including making clear observation of (surface) vessels, it was shown how topography gives rise to false moving targets as the position of sonar platform is changing.

3.2   Detailed Study of Topographic Phenomena

From the area covered by Fig. 2, an area measuring 1000 by 800 m was chosen as an open fjord area in this analysis (Fig. 3). The depth in this region ranges from 120 m (orange) to 500 m (dark blue). The main characteristic of the topography is a plateau in the north which taper off in the south-eastern direction, first more gently, then more sudden. In the southern direction the plateau falls off roughly 200 m in a 30 m range. The whole area is tilted towards the frigate in the run used in this analysis. The distance from the sonar platform to this area ranges from 4 km to about 3 km. All topography charts shown in this report contain data errors that appear as spikes in the topography charts.

Both clear green and red echoes seem to cluster near or on top of sharp edges in the topography. Where the slope is more moderate, the echo density is clearly lower. This can only be shown using charts having high-resolution topography data as used in this paper. This also enables observation of small topographic variations that cause the clustering of echoes. It is also worth noting the low echo density in the flat or close to flat areas where proximity filtering was not applied, which means that mainly reverberation and not noise is the main cause for these echoes.

Analysis has shown that most echoes have a low signal-to-noise ratio (SNR). This is partly caused by their relative position with respect to the sonar platform. Another cause is the channel condition we have which traps a lot of the acoustic energy in a surface channel reaching not deeper than 50 m. As the most shallow area within the window is 200 m, very few strong echoes can be expected. Observations not documented here also revealed that all echoes of medium to high signal to noise ratio appear at the sloped area.

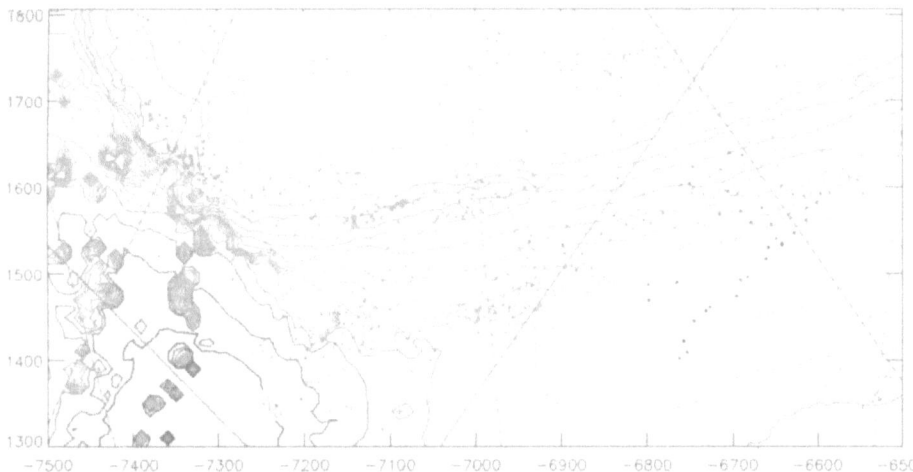

Figure 3. Proximity filtered echoes colour coded according to the time of their ping. The striped lines correspond to the beam borders at ping 300 (blue), 550 and 800 (red). The colour of the beam borders corresponds to the colour of the echoes at time of the ping. Axis values in terms of meters from start of sonar run.

As most echoes here have a low SNR, no alarms would normally be generated when filtering echoes based on their SNR. However, if full sensitivity were desirable, all echoes would be subject of analysis. Then the likelihood of several echoes being present within a small area would have been greater, and likewise the initiation of tracks. This would then have increased the probability of false alarm. Echoes generated by topographic features are the most probable cause for these alarms. By understanding what characteristics of the topography give rise to echoes, one may therefore be able to predict areas where there is a high probability of alarms. This has been observed by animating the pings: the echoes from e.g. a sea-floor slope appear to be moving as the sonar platform moves along. The reason being the point of specular reflection moves as the ship moves past the slope. This might lead to generation of track with realistic speed and heading. When investigating Fig. 3 in detail and taking the uncertainty of echo position into account, it may be suspected that the red echo cluster around x,y-position (-7350, 1650) is caused by the turbulence in the topography in the same area.

324 J. WEGGE

However, hydroacoustic modeling described in the next chapter revealed a minimum for the transmission loss and a peak in reverberation at this depth and range. This causes the energy weighting of the sonar processing to generate echoes.

4 Analysis of hydroacoustic model simulations

The range dependent LYBIN hydroacoustic model was used to simulate sonar performance. Figure 4 shows the simulations for ping no. 800. The bottom profile, which was extending 5 km from the sonar position at 25° bearing, was compiled from both 10 m and 50 m resolution topography data. The figure also contains a plot of the topography and sonar echoes at the left (echoes of ping 800 are red). Estimated noise and reverberation curves are drawn at the top right-hand side. The transmission loss and sound profile are shown on the lower right-hand side. A wind speed of 1 m/s and bottom type 2 (rock-gravel) were used as environmental parameters. Even if much of the energy is trapped in the surface duct (see the transmission loss diagram), enough energy reaches the bottom and makes high bottom reverberation possible.

Apart from exceeding the bottom reverberation at 1700 m, the surface reverberation is 20–25 dB lower than the bottom reverberation from 2 km and outward. Volume reverberation seems to be high over the whole range and dominates up to 2.5 km. Further out in range, the acoustic energy reflected within the surface duct is exceeded by the bottom reflections. Hence, the bottom reverberation dominates.

As the bottom rises sharply, the transmission loss reaches a minimum along the bottom at 2.7 km before it increases again. At 2.7 km the bottom reverberation also peaks which coincide with the clustering of echoes. The peak of the bottom reverberation at 3.7 km is identified as the surface duct sound energy reflecting from the slightly shallower area at this range and bearing. This correlates with the echoes shown in the topography charts.

The bottom profile was smoothed in this investigation to avoid any potential influence caused by the artificial peaks in the topography charts. Despite of the interpolation of the bottom profile within LYBIN, error correction of the topography charts are recommended before a more thorough study of topography correlation of sonar echoes. It is also recommended to increase the range resolution of the model to investigate the details of reverberation data further.

5 Conclusion

Sonar echo data from the Spherion HMS have been analyzed and compared with topographic data. Echoes are observed to cluster where the bottom is sloping steeply towards the sonar and the transmission loss is low. It was not possible to correlate all echoes with topographic features. One reason may be that a contour distance of 20 m was used despite the 1-m resolution of the topography data.

This study was based on only a single set of sonar data from a single hull mounted sonar. The data were recorded in a Norwegian fjord in the month of April. The study has neither considered the accuracy of platform positions or the position accuracy of the sonar echoes recorded. However, any deviation from true position is believed to be insignificant for this study, as there seems to be a close correlation between topography and sonar echo positions.

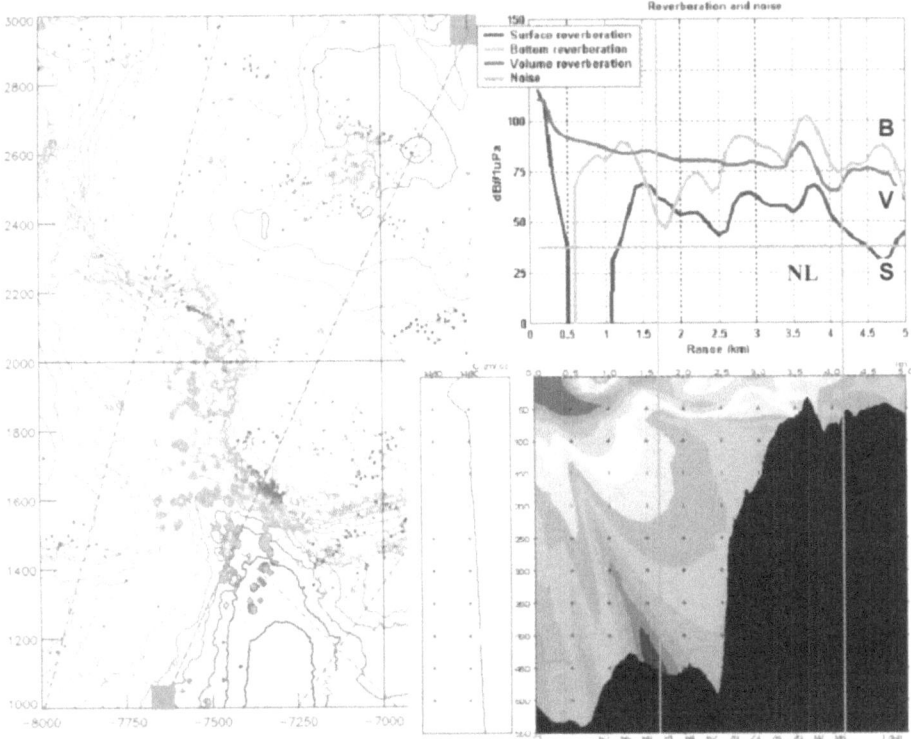

Figure 4. Topography and proximity filtered echoes from Fig. 2 are shown at the left. Curves for noise and volume, surface and bottom reverberation at the top right-hand side (in green, red, blue and orange colours respectively). Below this is a plot of sound speed profile and the transmission loss. The orange coloured vertical lines in the reverberation and transmission loss diagrams signify the start and stop of the red line with orange squares in the topography diagram.

The topography data presented here have a range resolution of 10 m. Even with topography data not as detailed as this, the sub-sea landscape will provide clues to the majority of echoes not caused by the submarine. Large sea-floor landscape types exist which explain most of these false alarms or echo clusters. Examples of such landscape types are local sea-floor peaks, hillsides with favorable reflecting angel or simply sea floor sufficiently shallow to coincide with the surface channel. However, this study has shown the increased capability of explaining potential false alarm and false targets using high-resolution topography information.

This investigation looked into the possibility of correlation sonar echoes with topographic features using only a few numbers of presentation methods. For a more complete manual classification of sonar echoes, a flexible display enabling correlation of sonar echoes with high-resolution topography data and modeled hydro-acoustic data, must be made available. Such a system should be complemented with target tracking and other algorithms for estimating the probability parameters of echoes. It might also prove beneficial to base the processing on a more basic data level than the Spherion sonar echo level used for this study.

Acknowledgements

The author would like to thank Elling Tveit for his guidance and co-operation during the preparation of this paper.

References

1. Preston, J.R., Ellis, D.D., Akal, T. and Desharnais, F., Analysis of low frequency acoustic reverberation in the western Black Sea. SACLANTCEN Report SR-286 (1998).

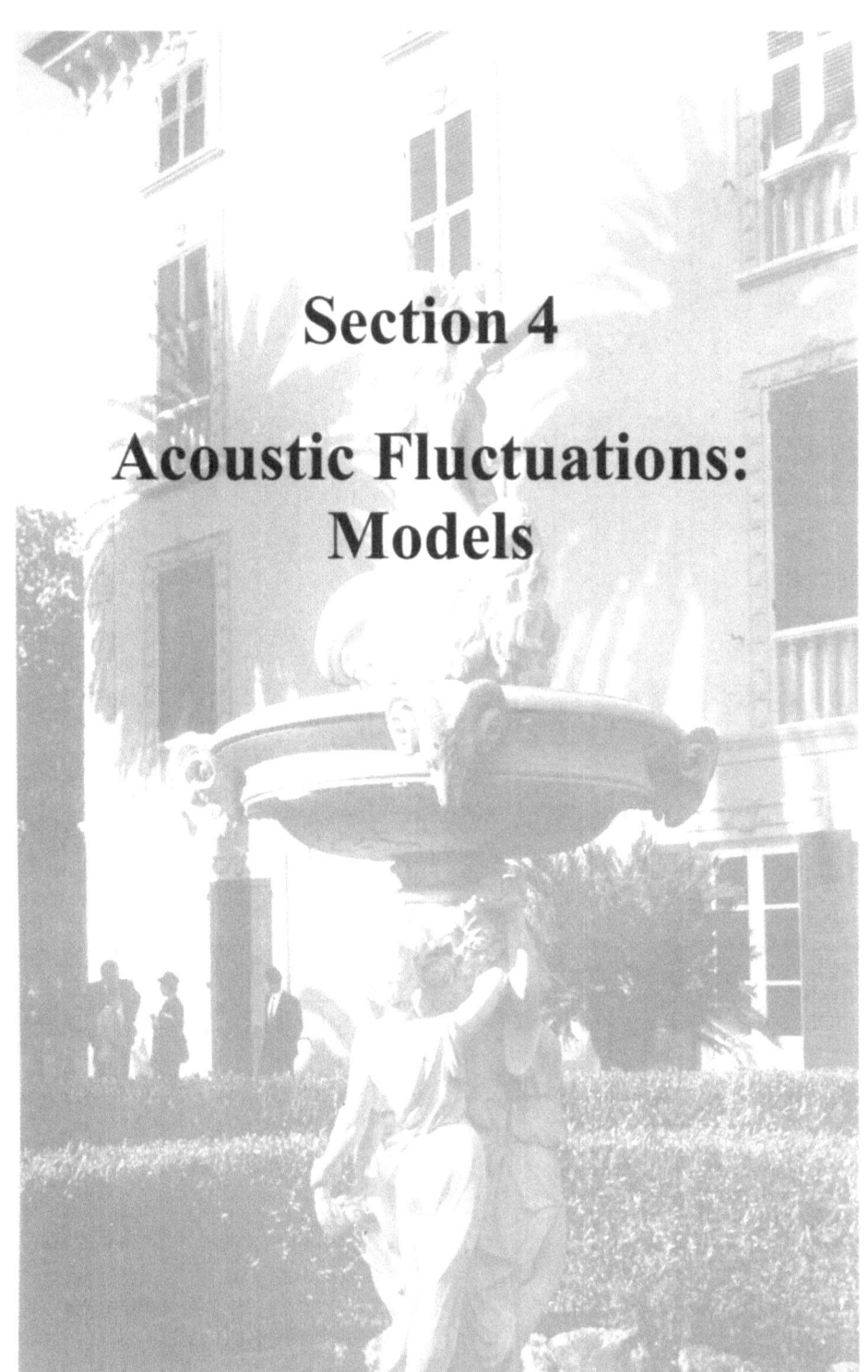

Section 4

Acoustic Fluctuations: Models

ACOUSTIC SCATTERING IN WAVE-COVERED SHALLOW WATER. THE COHERENT FIELD

B.J. USCINSKI

Department of Applied Mathematics and Theoretical Physics,
University of Cambridge, Silver St., Cambridge U.K. CB3 9EW
E-mail:Bju1@damtp.cam.ac.uk

The coupled mode approach to acoustic propagation in shallow water with rough surface and bottom fails when the number of modes becomes too large. A method is presented, based on a set of coupled integral equations, that allows the acoustic field to be calculated for large surface and bottom roughness. Ensemble average forms of these equations are derived for the mean field and a solution sought using Laplace transforms. The method allows us to study how the coherent component of an acoustic field in shallow water is affected by surface waves.

1 Introduction

Sound propagating in shallow water is reflected from both surface and bottom which are then the boundaries of an effective waveguide. Irregular structures in these boundaries, such as surface waves and bottom roughness affect the acoustic field. They impose random modulations on what might otherwise be a regular waveguide diffraction pattern. One method of calculating the acoustic field that can take boundary roughness into account consists of representing the field as a sum of modes [1]. Roughness in the upper and lower boundaries leads to coupling and energy exchange between the modes [2]. This approach can be used successfully provided that the number of modes is not large. It fails in the case of high acoustic frequencies when the wavelength is short and surface waves and bottom roughness produce large scattering effects. The number of modes required in this case is so large that it would be impossible to compute them. This paper describes a non-modal method that allows us to deal with high-frequency acoustic propogation in the shallow water case. The amount of back-scatter is assumed to be small so that the forward propagating parabolic approximation can be used. A set of coupled integral equations describes the propagation [3]. When the ensemble average of the acoustic field is sought these equations can be solved using Laplace transforms. The resulting expressions allow us to study how the mean, or coherent, component of an acoustic field propagating in shallow water is affected by the rough wave-covered sea surface.

2 The shallow water model

The simplified model used to dscribe the shallow water "wave guide" is discussed in dctail elsewheie [3, 4]. The upper boundary is a pressure release surface and has a reflection coefficient $R_s = -1$. The reflection properties of the ocean bottom are, in general,

N.G. Pace and F.B. Jensen (eds.), Impact of Littoral Environmental Variability on Acoustic Predictions and
Sonar Performance, 329-336.
© 2002 *Kluwer Academic Publishers.*

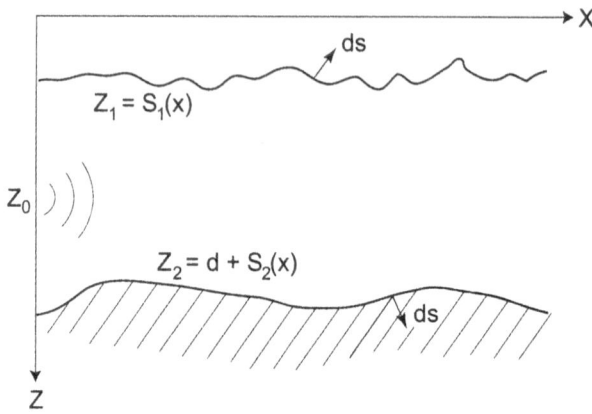

Figure 1. The coordinate system showing the rough surface and bottom together with the source location.

complicated. We thus restrict our treatment to the special cases where the shear-waves propagate at low speed and are radiated away from the liquid-solid interface. The effective acoustic reflection coefficient of such a bottom at low grazing angles has an amplitude somewhat less than unity and a phase that varies between 120° and 180°. The simplest first approximation would thus be to take a bottom reflection coefficient $R_B = -1$ in such cases. This allows the general principles of the method to be demonstrated conveniently and results checked against modal solutions in specific cases. The shallow water channel is shown in Fig. 1. Let (x, z) be a Cartesian system of coordinates. The sea surface and bottom follow the contours

$$Z_1 = S_1(x), \ Z_2 = d + S_2(x) \tag{1}$$

respectively. The source is situated at $(0, z_0)$. Let $p(\mathbf{r})$ be the acoustic pressure at the point $\mathbf{r} = (x, z)$ while $\mathbf{r}' = (x', z(x'))$ lies on the bounding surface. We assume that the acoustic field propagates predominantly in the forward x direction allowing us to write, in the parabolic approximation

$$p(x, z) = E(x, z)e^{ikz} \tag{2}$$

where $E(x, z)$ is the slowly varying envelope of the acoustic pressure field $p(x, z)$ and k is the acoustic wave-number. In the parabolic approximation the integral form of the Helmholtz equation for $E(x,z)$ becomes

$$E(x, z) = \int \{G(\mathbf{r}, \mathbf{r}')E_{z'}(\mathbf{r}') - E(\mathbf{r}')G_{z'}(\mathbf{r}, \mathbf{r}')\} \, ds \tag{3}$$

where G is the parabolic form of the Green's function and the subscript z' denotes the vertical derivative, i.e. with respect to z.

Consider the surface over which Eq. (3) is to be integrated and note that $d\mathbf{s}$ is the outward facing surface element. It consists of four separate sections as shown in Fig. 2. Thus we can write

$$E(x, z) = \int_{S(1)} [\ ] d\mathbf{s} + \int_{S(2)} [\ ] d\mathbf{s} + \int_{S(3)} [\ ] d\mathbf{s} + \int_{S(4)} [\ ] d\mathbf{s} \tag{4}$$

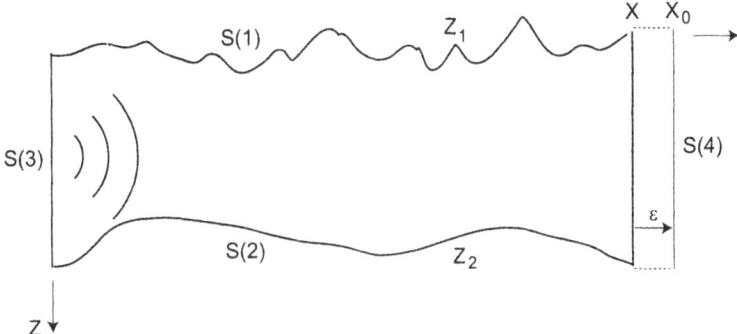

Figure 2. The four surfaces S(1), S(2), S(3) and S(4).

Now consider $S(4)$ over $x_0 = x + \epsilon$. The contribution from this surface is zero, since there is no backward propagation from x_0 to x. Next, let both surfaces S(1) and S(2) move to $\pm\infty$ respectively. The contributions from S(1) and S(2) then vanish leaving

$$E(x,z) = \int_{S(3)} [\ ]d\mathbf{s} = E_{inc}(x,z) \tag{5}$$

Thus the surface S(3) represents the incident field, i.e. that would have been expected from the source in the absence of any surface. We can now rewrite Eq. (4) as

$$E(x,z) - E_{inc}(x,z) = \int_{S(2)} G(\mathbf{r};\mathbf{r}')E_{z'}(\mathbf{r}') - E(\mathbf{r}')G_{z'}(\mathbf{r};\mathbf{r}')dx'$$
$$- \int_{S(1)} G(\mathbf{r};\mathbf{r}')E_{z'}(\mathbf{r}') - E(r')G_{z'}(\mathbf{r};\mathbf{r}')dx' \tag{6}$$

2.1 Case for a Bottom with $R_B = -1$

Now $E(\mathbf{r}')$ is zero on both $S(2)$ and $S(1)$ so Eq. (6) becomes

$$E(x,z) - E_{inc}(x,z) = \int_0^x G(\mathbf{r};x',z_2(x'))E_{z'}(x',z_2(x'))dx'$$
$$- \int_0^x G(\mathbf{r};x',z_1(x'))E_{z'}(x',z_1(x'))dx' \tag{7}$$

On specifying $E(x,z)$ to surfaces z_2 and z_1 respectively, we have

$$E_{inc}(x,z_2(x)) = -\int_0^x (G(x,z_2(x);x',z_2(x'))E_{z'}(x',z_2(x'))$$
$$-G(x,z_2(x);x',z_1(x'))E_{z'}(x',z_1(x'))) \, dx' \tag{8}$$

and

$$E_{inc}(x,z_1(x)) = -\int_0^x (G(x,z_1(x);x',z_2(x'))E_{z'}(x',z_2(x'))$$
$$-G(x,z_1(x);x',z_1(x'))E_{z'}(x',z_1(x'))) \, dx' \tag{9}$$

Equations (8) and (9) are coupled equations for the unknowns $E_{z'}(x, z_2(x))$, $E_{z'}(x, z_1(x))$. These equations can be solved for the two surface derivatives which, when used in Eq. (7), allow us to obtain $E(x, z)$ anywhere in the space between z_1, z_2.

3   The mean field problem

If we wish to obtain the mean or average scattered field by numerical simulations we would have to compute it for many statistically independent realisations of the rough surface $S(x)$ and take the ensemble average. An alternative, and more efficient, approach would be to derive equations for the mean field itself. In order to do this we need to analyse the stochastic properties of both the scattered acoustic field and the rough surface $S(x)$ and see how they are related to each other. First consider the rough surface $S(x)$. The irregular features in $S(x)$ have a certain correlation distance in the x direction of the order of the scale size L, the length of a typical feature in the rough surface. As the acoustic field propogates in the x direction it interacts with the rough surface and its complex amplitude $E(x, z)$ is gradually modulated in an irregular manner. Now consider $E(x', z)$ at some range x' after it has traversed several correlation lengths. The structure in $E(x', z)$ has been caused by the accumulated interactions with the surface up to that range. Now suppose that the surface contour $S(x)$ changed only in the vicinity of x' over a distance of about one correlation length. The change to $E(x', z)$ will be small because its structure has been determined by all the preceding irregularities, so that the effect of the change in $S(x)$ in the final step will be negligible by comparison.

3.1   Statistical Independence of Surface and Field

From the above reasoning we can draw the important conclusion that in this type of low angle propagation the complex acoustic amplitude $E(x', z)$ is statistically independent of the local form of the surface $S(x)$ at any range x', provided that several correlation lengths have been traversed. The following are therefore also valid:

1. Both $E(x', z)$ and $E_{z'}(x', z)$ are statistically independent of $G(x, z; x', S(x'))$ since it is a function of $S(x')$.

2. Likewise $E(x', z)$ and $E_{z'}(x', z)$ are statistically independent of $G(x, S(x); x', S(x'))$.

3. However, in $G(x, S(x); x', S(x'))$ the surface heights $S(x), S(x')$ are not statistically independent of each other but are correlated. This needs to be taken into account when carrying out the statistical averaging.

3.2   The Statistics of the Rough Surface

It is assumed that the surface height $S(x)$ is a stationary random process with Gaussian statistics having zero mean and variance μ^2. The normalised spatial autocorrelation function of $S(x)$ is

$$\rho(x - x') = \frac{< S(x)S(x') >}{\mu^2} \tag{10}$$

These conditions are not unreasonable since many natural surfaces, including that of the wave-covered sea surface, have statistics that are very close to Gaussian. The autocorrelation function $\rho(x - x')$ can be of arbitrary form. In the case of the sea surface it would be that corresponding to a Pierson-Moscowitz wave-height spectrum, for example. The one point probability distribution of surface height S is then

$$P_1(S) = \frac{1}{\sqrt{2\pi}\mu} exp\left\{-\frac{S^2}{2\mu^2}\right\} \tag{11}$$

while the two-point joint probability distribution of

$$S_a = S(x'), \quad S_b = S(x) \tag{12}$$

is

$$P_2(S_a, S_b) = \frac{1}{2\pi\mu^2\sqrt{1-\rho^2}} exp\left\{-\frac{S_a^2 - 2\rho S_a S_b + S_b^2)}{2\mu^2(1-\rho^2)}\right\}. \tag{13}$$

where $\rho = \rho(x - x')$

3.3 Ensemble Averages

We now take the statistical average of Eqs. (7), (8) and (9) using the distributions (11) and (13) for the rough surfaces S_1 and S_2. Remembering that the scattered field

$$E_s(x, z) = E(x, z) - E_{inc}(x, z) \tag{14}$$

and introducing the notation

$$
\begin{aligned}
< E_{inc}(x, z_i(x)) > &= f_i(x) \\
< E_s(x, z) > &= f_s(x, z) \\
< E_{z'}(x, z_i) > &= \phi_i(x) \\
< G(x, z_i(x), x', z_j(x')) > &= G_{ij}(x - x') \\
< G(\mathbf{r}; x', z_i(x')) > &= G_i(x - x', z)
\end{aligned}
\tag{15}
$$

we obtain

$$f_s(x, z) = \int_0^x [\phi_2(x')G_2(x - x'; z) - \phi_1(x')G_1(x - x'; z)]\, dx' \tag{16}$$

$$f_2(x) = -\int_0^x [\phi_2(x')G_{22}(x - x') - \phi_1(x')G_{21}(x - x')]\, dx' \tag{17}$$

$$f_1(x) = -\int_0^x [\phi_2(x')G_{12}(x - x') - \phi(x')G_{11}(x - x')]\, dx' \tag{18}$$

3.4 Scaling

Since it is convenient to work with dimensionless variables the following scaling is now introduced. In the vertical we use the quantity

$$f = \sqrt{\frac{L}{k}} \tag{19}$$

which is seen to be very close to the Fresnel radius for an observer at a range L, the correlation distance of the rough surface. In the horizontal, L itself is the scaling factor. The scaled horizontal and vertical distances are thus

$$X = \frac{x}{L}, \quad Z = \frac{z}{f} \tag{20}$$

and

$$\gamma^2 = \frac{\mu^2 k}{L} \tag{21}$$

which is a variance of the scaled surface height.

4   Solution by Laplace transforms

On taking the Laplace transforms of Eqs. (16)-(19) we obtain

$$F_s(\lambda) = \Phi_2(\lambda)K_2(\lambda) - \Phi_1(\lambda)K_1(\lambda) \tag{22}$$

$$F_2(\lambda) = -\Phi_2(\lambda)K_{22}(\lambda) + \Phi_1(\lambda)K_{21}(\lambda) \tag{23}$$

$$F_1(\lambda) = -\Phi_2(\lambda)K_{12}(\lambda) + \Phi_1(\lambda)K_{11}(\lambda) \tag{24}$$

where

$$F(\lambda) = \int_0^\infty f(X)e^{-\lambda X}dX$$
$$\Phi(\lambda) = \int_0^\infty \phi(X)e^{-\lambda X}dX$$
$$K(\lambda) = \int_0^\infty G(X)e^{-\lambda X}dX \tag{25}$$

the unknown quantities Φ_1, Φ_2 can be obtained by solving equations (23) and (24) and set in (22) to give

$$F_s(\lambda, z) = \frac{K_1(F_2K_{12} - F_1K_{22}) - K_2(F_2K_{11} - F_1K_{21})}{K_{11}K_{22} - K_{21}K_{12}} \tag{26}$$

Finally, the coherent, or mean scattered field $< E_s(X, Z) >$ can be found by taking the inverse Laplace transform of Eq. (26)

$$< E_s(X, Z) >= LT^{-1}[F_s(\lambda, Z)] \tag{27}$$

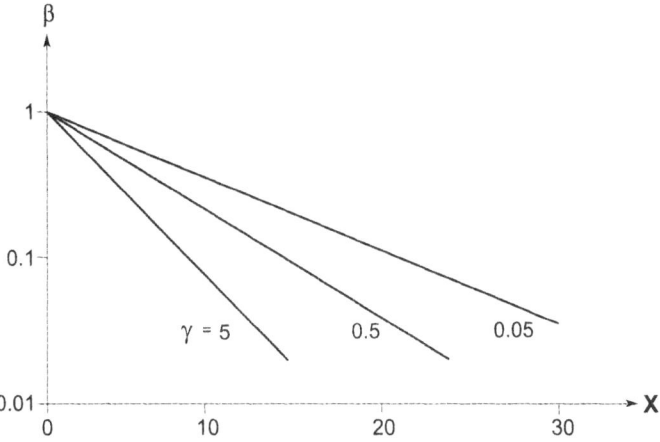

Figure 3. The attenuation factor $\beta(X, Z)$ for a channel with a flat bottom and a surface with an exponential autocorrelation function. Here $\gamma_0 = 0.05, 0.5, 5.0, Z_0 = 16; d = 32$

5   Attenuation factor

We now illustrate the use of the above general results in the special case of a rough surface when the bottom is flat. In order to simplify the calculations we use a surface with an exponented spatial autocorrelation function, $\rho = exp\{-X\}$. This allows us to investigate the overall effect of surface waves without the more complicated algebra that results when a Pierson-Moscowitz autocorrelation function is employed. The effect of different wave heights can be tested by adjusting the scaled variance γ. The inverse Laplace transform was evaluated numerically for a number of different wave heights. The effect of the wave covered surface on $< E_s(x, z) >$ the "coherent" scattered field, is best illustrated by considering the ratio

$$\beta(X, Z) = \frac{< E_s(X, Z) >}{< E_{s0}(X, Z) >} \qquad (28)$$

where $< E_{s0}(X, Z) >$ is the mean scattered field in the absence of surface waves.

The attenuation factor β is shown in Fig. 3 as a function of range X at the centre of the channel for several values of γ^2 corresponding to different surface roughness. We see that the coherent component is progressively attenuated by surface scattering as the range increases, and the higher the waves, the greater this effect.

This paper sets out the general reasoning behind the present approach and the methods used to implement it. The specific forms obtained for the ensemble averages Eq. (15) and their Laplace transforms Eq. (25) are not given here because of space restrictions. They are, however, quite analogous to the expressions obtained in the case of a single rough surface which are set out fully in [5].

References

1. Dolin, L.S. and Nechaev, A.G., Mode description of the interference structure in an acoustic field propagating in a wave guide with statistically rough walls, *izu V.U.Z. Radiofizika* **24**(11), 1337–1344 (1981).

2. Dozier, L.B. and Tappert, F.D., Statistics of normal-mode amplitudes in a random ocean, I. Theory, and II. Computations, *J. Acoust. Soc. Am.* **63**(2), 353–365 (1978) and **64**(2), 533–547 (1978).

3. Uscinksi, B.J., High-frequency propagation in shallow water. The rough waveguide problem, *J. Acoust. Soc. Am.* **98**(5), 2702–2708 (1995).

4. Tindle, C.T. and Zhang, Z.Y., An equivalent fluid approximation for a low shear speed ocean bottom, *J. Acoust. Soc. Am.* **91**, 3248–3256 (1992).

5. Uscinski, B.J. and Stanek, C.J., Acoustic scattering from a rough sea surface: The mean field by the integral equation method, *Waves in Random Media* (to appear 2002).

SIMULATIONS OF TEMPORAL AND SPATIAL VARIABILITY IN SHALLOW WATER PROPAGATION

ERIC I. THORSOS, FRANK S. HENYEY, KEVIN L. WILLIAMS, W. T. ELAM
AND STEPHEN A. REYNOLDS

Applied Physics Laboratory, College of Ocean and Fishery Sciences,
University of Washington, 1013 NE 40th Street, Seattle, WA 98105, USA
E-mail: eit@apl.washington.edu

Propagation in a shallow water waveguide leads to acoustic fields with complex spatial structure, and interaction with time dependent sea surface roughness modifies this structure and introduces temporal variability as well. Results of 2-D simulations will be described for propagation in a shallow water waveguide showing that the effects of temporal variability in field structure due to sea surface roughness are significant at relatively short range, but are much less important at longer ranges. It will also be shown that at longer ranges the effect of boundary roughness has a smoothing effect on the spatial structure of the propagating field.

1 Introduction

It is well known that acoustic propagation in shallow water environments leads to highly variable fields. Spatial variability will arise from complex interference phenomena even in idealized isovelocity waveguides with flat boundaries, and this variability will be modified by effects of the sound speed profile and by acoustic interactions with internal waves and with rough boundaries. Temporal variability will result from interactions with time varying internal waves and surface waves. Acoustic interactions with the bottom are generally understood to determine the important characteristics of shallow water propagation [1–3], especially at low frequencies, but in the mid-frequency region (1–10 kHz) scattering from sea surface roughness may also play an important role in enhancing the bottom interaction.

In this paper, we present two-dimensional simulation results that illustrate the potential effects of forward scattering from sea surface roughness on the spatial and temporal variability of 3 kHz acoustic fields in a shallow water waveguide of 50 m depth. The waveguide is idealized as isovelocity, and the water-sediment interface is taken as flat while the sea surface may be rough.

2 Simulation method

The propagation simulations were done using a wide-angle PE method developed by Rosenberg [4] that we believe accurately accounts for forward scattering from a rough sea surface (in two space dimensions) for the conditions employed in this paper. The Rosenberg propagation model is an extension to a wide-angle PE propagation model developed by Collins [5]. The simulations have been done for a CW source at 3 kHz.

N.G. Pace and F.B. Jensen (eds.), Impact of Littoral Environmental Variability on Acoustic Predictions and Sonar Performance, 337-344.

However, insight on the temporal variability of the field can be obtained from the variability observed among independent realizations of the surface roughness; this will indicate the expected field variability on time scales that extend down to the order of a few seconds.

Other investigators have also developed rough surface PE methods that could be applied to the propagation scenario considered here; these include Dozier [6], Tappert and Nghiem-Phu [7], and Norton *et al.* [8]. In addition, Kuperman and Schmidt [9,10] have developed boundary perturbation methods combined with propagation based on wavenumber integration that could be used to carry out simulations similar to those presented here, though the region of validity of their approximations has not been delimited and to date lower frequencies have been employed (see, for example, Tracy and Schmidt [11]).

For our simulations both the water and the sea floor sediment are taken as homogeneous with sound speeds of 1500 m/s and 1600 m/s, respectively, leading to a critical grazing angle of 20°. The field in the sediment is attenuated at 0.5 dB/λ, and the sediment-to-water density ratio is 2.0. Rough sea surface realizations generated for use in the simulations are consistent with a one-dimensional cut through a two-dimensional isotropic spectrum of a Pierson-Moskowitz form [12]. For examples shown with sea surface roughness, the surface waves are produced by a wind speed of 7.7 m/s (15 knots). A point source is located at the mid depth of 25 m, and a beam pattern has been applied with a full width of 20° and with the beam center aimed up at a 10° grazing angle.

3   Results

Figure 1(a) shows the field intensity in the waveguide out to a range of 30 km for a flat sea surface (at this scale the initial diverging beam cannot be seen). Figure 1(b) gives the corresponding field for one rough surface realization. In the flat sea surface case, most of the energy initially propagating above the critical angle is lost into the bottom in about the first 0.5 km, and from that point on the waveguide field is made up of components with grazing angles less than the critical angle. This field then slowly attenuates due to loss at bottom reflections. (Because a 2-D geometry has been used, the intensity decrease from cylindrical divergence is not included.)

In the case with a rough surface (Fig. 1(b)), scattering from roughness causes some acoustic energy to be continually shifted to grazing angles above the critical angle, where it can then be lost into the sea floor at subsequent bottom interactions. (See [3] for related observations.) After the first few kilometers a point is reached where most of the remaining energy is propagating at very low grazing angles (a few degrees or less) and surface scattering becomes much less effective in causing additional loss to occur at the bottom.

The effect of the rough sea surface is to simplify the field structure at longer ranges by stripping away (at shorter ranges) the higher angle energy more efficiently than bottom attenuation alone does for a waveguide with flat boundaries. This effect can be seen more clearly in Fig. 2, which shows the 5 km range interval from 23 to 28 km from Fig. 1. The effect of the rough sea surface is to smooth the spatial structure of the propagating field at these longer ranges, and therefore reduce the spatial variability.

A normal mode picture is particularly useful for discussing the fields shown in Figs. 1 and 2. When the PE produced fields are projected onto the normal modes to obtain the mode amplitudes as functions of range, it is evident that the net effect of rough surface scattering and subsequent energy loss in the sediment is to strip away the higher modes much more rapidly than occurs for the flat surface case. The remaining field at long range is then made up of a relatively small number of low-order modes, which leads to a relatively smooth spatial field structure. These low-order modes correspond to fields with such low grazing angles at the surface that surface scattering can effectively be neglected. In the particular example shown, with the source at mid depth, odd-order modes dominate the field structure. At the range of Fig. 2 the field would be well described by the lowest 5 or 6 odd-order modes for the rough surface case, but for the flat surface many more modes would be required to accurately represent the field. (For this waveguide at 3 kHz, there are 72 discrete normal modes if attenuation is ignored.)

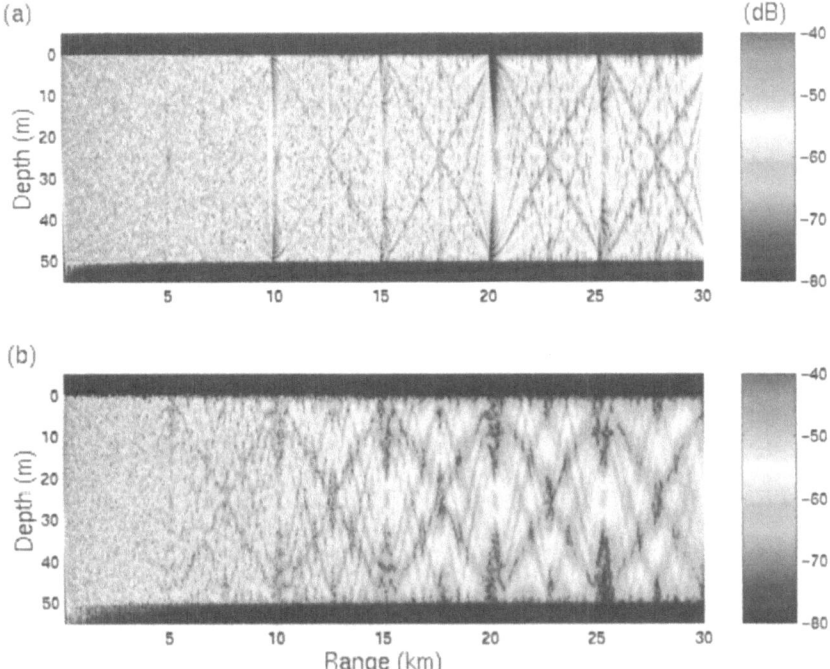

Figure 1. Intensity levels for a 3 kHz source at 25 m depth for a flat sea surface (a) and a rough sea surface (b). The rough sea surface corresponds to a wind speed of 7.7 m/s (15 knots).

The refocusing of the field at about every 5 km is a particular attribute of this isovelocity case with the source at mid depth, since to a very good approximation these are the locations where the odd-order modes all come back into phase. Only small changes in the sound speed profile or source depth will eliminate this regular focusing, and therefore the particular field structure shown is not typical. However, the effects on field variability due to scattering from the sea surface described in this paper are believed to be broadly applicable.

The field structure will have temporal variability as the sea surface roughness evolves in time. We can see the relative level of these temporal changes as a function of range by examining the difference in the field structure for two independent surface realizations. Figure 3 shows such a field difference normalized to remove the average decrease in the field with range. (In particular, the square of the field difference was normalized at each range by the average over the two realizations of the vertically averaged field squared.) The major spatial features present in Figs. 1 and 2 are removed in this difference, since they are part of the coherent field, as will be discussed shortly, and subtract out. What remains is a more uniform distribution that indicates the relative temporal fluctuation level as a function of range. At short range the temporal fluctuations are at about the same level as the field itself, while at longer range they are much less significant. The decrease in relative fluctuation level averaged over depth is about 9 dB over the range shown. Therefore, the effects of sea surface scattering on temporal variability, as well as spatial variability, decreases significantly with range for the conditions being considered here.

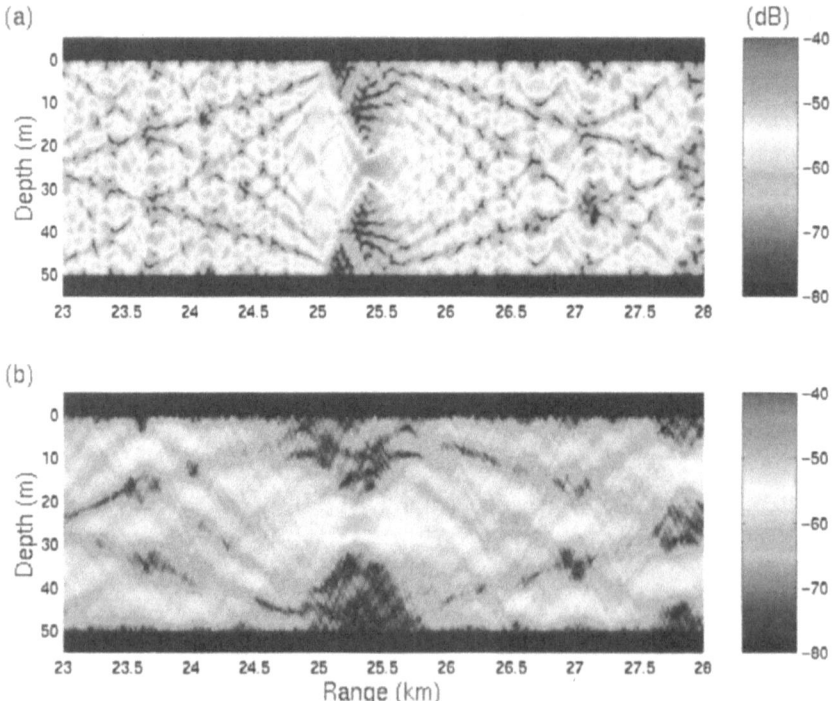

Figure 2. Expanded view of comparison shown in Fig. 1 in 23–28 km range interval.

Note that a horizontal banded structure is evident in Fig. 3. Presumably this is related to modal structure, but it is found to differ when other pairs of realizations are differenced (with varying numbers of bands present), and therefore it is a surface realization dependent structure.

It is also noteworthy that there is no evidence in Fig. 3 for growth with range in the scintillation index, as predicted by Creamer [13] and verified with simulations [14] for the case of intensity fluctuations induced by internal waves. The situation with surface scattering, however, is quite different than for scattering from internal waves. Internal wave couple to all modes about the same, but surface waves couple preferentially to higher modes because of the higher grazing angles at the surface.

The component of the propagating field not subject to temporal fluctuations is given by the coherent field, obtained by averaging results for the complex pressure field over an ensemble of rough surface realizations; the fluctuating component then averages out. The coherent intensity for our 3 kHz example with a wind speed of 7.7 m/s is shown in Fig. 4(a). The total intensity (Fig. 4(b)) is obtained by averaging the field squared over the 50 surface ensemble and therefore includes the fluctuating scattering field. The deep nulls in the coherent intensity are partially filled in by the scattered intensity in Fig. 4(b), but even so at long range the dominant features are well represented by the coherent intensity. This is advantageous, since the coherent field is far easier to model than the total field, using methods such as described in [10]. At short range, however, the scattered intensity makes a significant contribution to the total intensity, as would be expected from Fig. 3 since temporal fluctuations are important there as well. Therefore, accurate acoustic modeling in this region requires a method that can predict the total acoustic field, including the temporally fluctuating component.

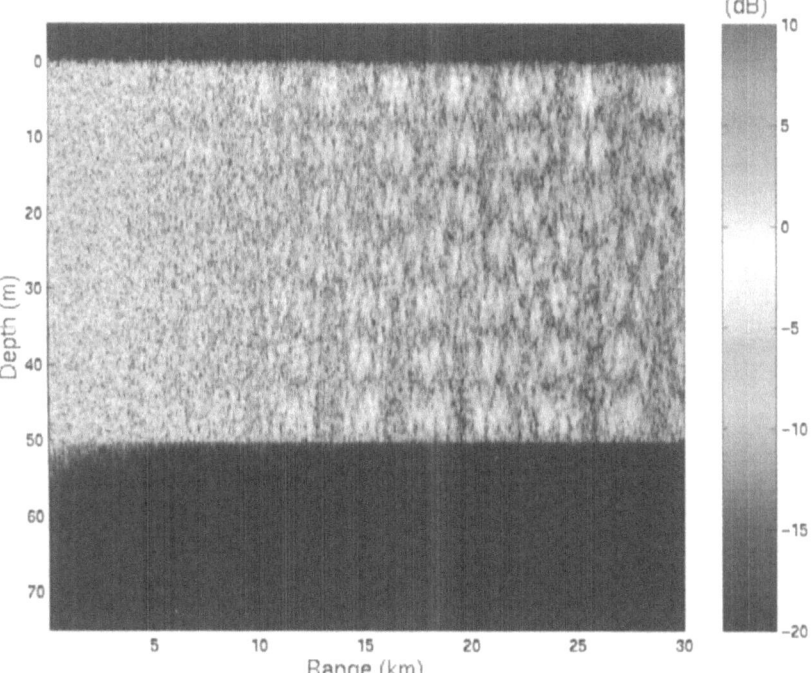

Figure 3. Relative difference between fields for two independent surface realizations.

The relative level of the coherent field as a function of range and depth can be illustrated with a "signal-to-noise" ratio defined in the following way. The "signal" is taken as the coherent intensity given by $|\langle p \rangle|^2$, where $\langle \ \rangle$ denotes averaging over an ensemble of rough surfaces. The "noise" is the average incoherent intensity arising from surface scattering, and is given by $\langle |p - \langle p \rangle|^2 \rangle$. The noise level is therefore a measure of the fluctuating intensity, and the signal-to-noise ratio gives the relative level of the temporally constant to temporally fluctuating components. This signal-to-noise ratio (SNR) for the propagation example being considered is shown in Fig. 5. The SNR (when averaged over depth) increases from about 0 dB to about 12 dB as the range increases, illustrating again the decreasing importance of temporally fluctuating fields due to surface scattering as the range increases.

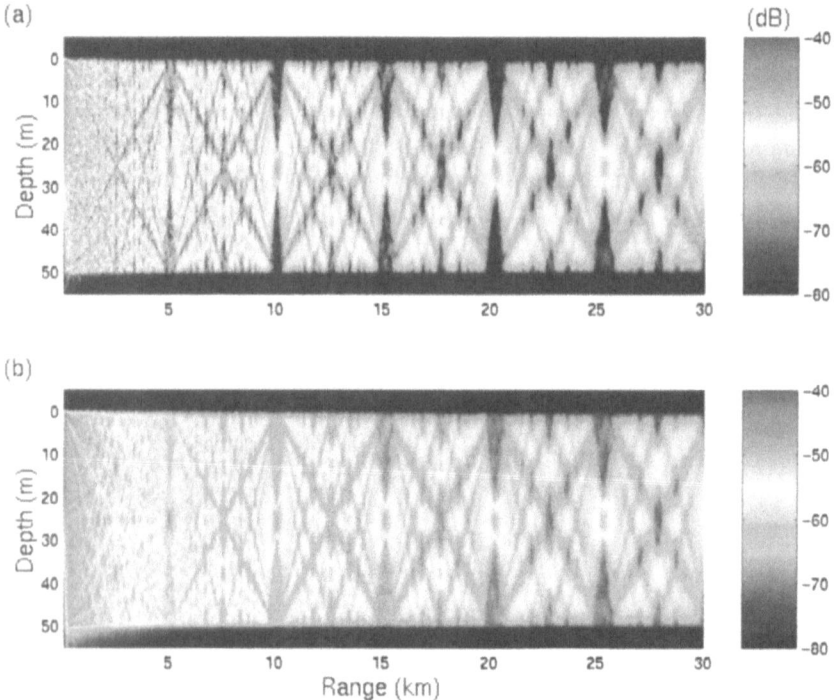

Figure 4. Coherent intensity (a) and total intensity (b) obtained by averaging over the results for 50 rough surface realizations.

It should be noted that in some of the low level waveguide nulls and in essentially all of the sediment the SNRs shown in Fig. 5 are not accurate, and only represent an upper bound on these levels. In these regions the coherent intensity level is quite low, and a 50 surface coherent average is not sufficient to adequately average out the incoherent components to obtain the true coherent level. If the number of realizations used in the averaging were to increase, the SNR in these regions would decrease further.

4 Discussion

A simplified propagation environment has been used in this paper to illustrate the manner in which surface scattering may decrease the complexity of the acoustic field in shallow water propagation. The spatial variability is shown to decrease with range much faster than it would in the absence of surface scattering. The temporal variability, which is caused only by surface scattering in this idealized model, also decreases with range to the point that it may become unimportant. As shown by Fig. 5, the coherent field becomes a relatively good approximation to the total field at long ranges, simplifying the task of modeling the acoustic field.

Though the environment was simplified by assuming the water-sediment interface is flat and the water column sound speed is independent of depth, we suspect the basic features shown will persist in some more realistic environments, such as with the inclusion of small-scale bottom roughness. Because the effect of boundary roughness is to deplete propagating components at all but the lowest grazing angles, it can be anticipated that the effects of internal waves on shallow water propagation may become relatively more important at longer ranges. These topics will be investigated in future work.

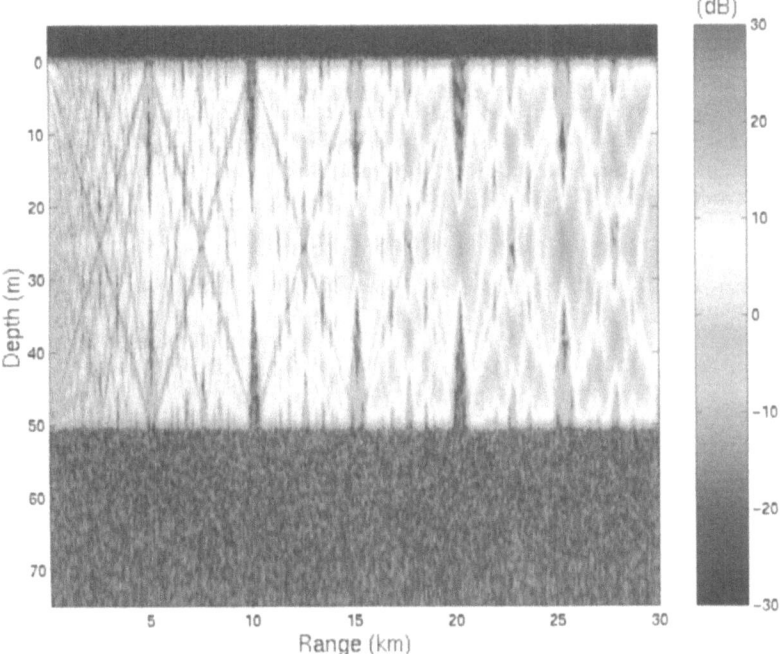

Figure 5. "Signal-to-noise" ratio for propagating field as defined in text.

Acknowledgement

This work was supported by the U.S. Office of Naval Research.

References

1. Akal, T., Sea floor effects on shallow-water acoustic propagation. In *Bottom-Interacting Ocean Acoustics*, edited by W.A. Kuperman and F. B. Jensen (Plenum Press, New York, 1980) pp. 557–575.
2. Jensen, F.B. and Kuperman, W.A., Optimum frequency of propagation in shallow water environments, *J. Acoust. Soc. Am.* **73**, 813–819 (1983).
3. Jensen, F.B., Recent progress in shallow-water acoustic modeling. In *Shallow-Water Acoustics*, edited by R. Zhang and J. Zhou (China Ocean Press, Beijing, 1997) pp. 43–48.
4. Rosenberg, A.D., A new rough surface parabolic equation program for computing low-frequency acoustic forward scattering from the ocean surface, *J. Acoust. Soc. Am.* **105**, 144–153 (1999).
5. Collins, M.D., Generalization of the split-step Padé solution, *J. Acoust. Soc. Am.* **93**, 1736–1742 (1993).
6. Dozier, L.B., PERUSE: A numerical treatment of rough surface scattering for the parabolic wave equation, *J. Acoust. Soc. Am.* **75**, 1415–1432 (1984).
7. Tappert, F. and Nghiem-Phu, L., A new split step Fourier algorithm for solving the parabolic wave equation with rough surface scattering, *J. Acoust. Soc. Am.*, Suppl. 1, **77**, S101 (1985).
8. Norton, G.V., Novarini, J.C. and Keiffer, R.S., Coupling scattering from the sea surface to a one-way marching propagation model via conformal mapping: Validation, *J. Acoust. Soc. Am.* **97**, 2173–2180 (1995).
9. Kuperman, W.A. and Schmidt, H., Self-consistent perturbation approach to rough surface scattering in stratified elastic media, *J. Acoust. Soc. Am.* **86**, 1511–1522 (1989).
10. Schmidt, H. and Kuperman, W.A., Spectral representations of rough interface reverberation in stratified ocean waveguides, *J. Acoust. Soc. Am.* **97**, 2199–2209 (1995).
11. Tracy, B.H. and Schmidt, H., Seismo-acoustic field statistics in shallow water, *IEEE J. Oceanic Eng.* **26**, 317–331 (1997).
12. Thorsos, E.I., Acoustic scattering from a "Pierson-Moskowitz" sea surface, *J. Acoust. Soc. Am.* **88**, 335–349 (1990).
13. Creamer, D.B., Scintillating shallow-water waveguides, *J. Acoust. Soc. Am.* **99**, 2825–2838 (1996).
14. Tielburger, D., Finette, S. and Wolf, S., Acoustic propagation through an internal wave field in a shallow water waveguide, *J. Acoust. Soc. Am.* **101**, 789–808 (1996).

ASSESSING THE VARIABILITY OF NEAR-BOUNDARY SURFACE AND VOLUME REVERBERATION USING PHYSICS-BASED SCATTERING MODELS

ROGER C. GAUSS, JOSEPH M. FIALKOWSKI AND DANIEL WURMSER
Naval Research Laboratory, Code 7144, Washington, DC 20375-5350, USA
E-mail: roger.gauss@nrl.navy.mil

The increased importance of responding to regional conflicts has focused Navy attention on littoral waters, with active sonar expected to be a favored mode of operation. Major performance drivers of such systems are the acoustic interactions with the ocean boundaries and fish. The vicinity of the air-sea interface is in particular a complex mix of scattering by surface roughness and scattering from bubble clouds and fish, coupled with boundary-interference effects. The Naval Research Laboratory has recently developed broadband, physics-based scattering strength models that both unify and advance our understanding of boundary scattering at low frequencies (< 5 kHz) by providing a physical basis for isolating scattering mechanisms. In this paper, these models are used to assess both the sensitivity of scattering strength to environmental variables and their utility as tools for estimating these variables. These efforts are supported by a series of data-model comparisons that demonstrate both the environmental variability of acoustic response with frequency and scattering angle, and the importance of using physics-based tools to predict these responses.

1 Introduction

For a low- (50–1000 Hz) or mid-frequency (1–5 kHz) active sonar, scattering from the ocean boundaries and biologics, coupled with propagation conditions, can severely limit the detectability of returns from features of interest. Furthermore, reverberation levels can vary dramatically, depending on the local geology, oceanography, and biology. Hence, making accurate predictions of active sonar performance will in turn depend on finding suitable models that accurately describe the scattering.

The Naval Research Laboratory (NRL) has been developing physics-based models of scattering strength [1,2]. By having a physics basis, the models allow extrapolation in frequency and to any 3-D scattering geometry. The models have proved essential for isolating scattering mechanisms, and so further the understanding of the complex acoustic interaction processes at the ocean boundaries.

This paper uses several of these physics-based models to explore the sensitivity in the upper ocean of surface and volume (fish) scattering strength to the grazing angle, the acoustic frequency, biological descriptors of the fish, and physical descriptors of the environment. In this case, the total scattering strength (in dB) is

$$SS = 10 \cdot \log_{10} \left(\sigma_{int} + \sigma_{bub} + \sigma_{fish} \right), \tag{1}$$

N.G. Pace and F.B. Jensen (eds.), Impact of Littoral Environmental Variability on Acoustic Predictions and Sonar Performance, 345-352.

where σ is the scattering cross section per unit area, and $\sigma_{int}, \sigma_{bub}, \sigma_{fish}$ represent the contributions due to the rough air-sea interface, bubble clouds and fish, respectively. (MKS units will be used throughout this paper.)

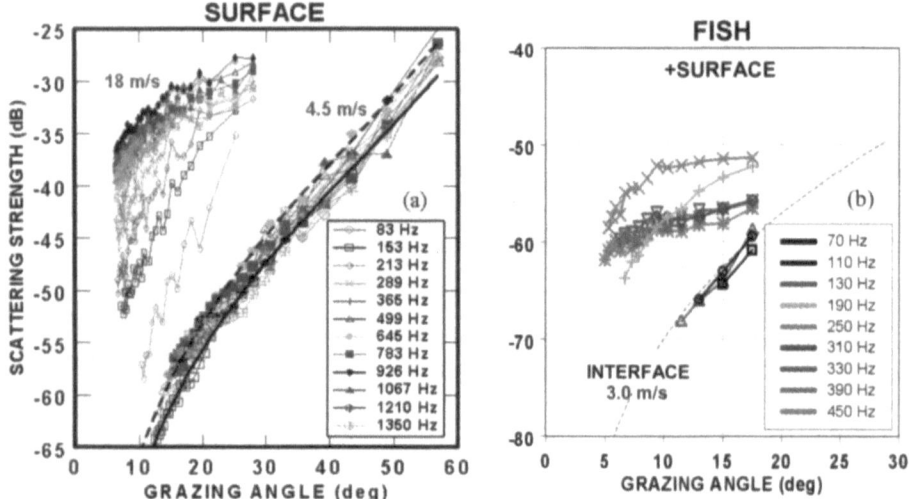

Figure 1. Measured backscattering strength vs. grazing angle and frequency: (a) surface scattering at two wind speeds (February 1992) and (b) surface + salmon scattering (May 1984).

Figure 1 illustrates some of the variability of SS that can be expected due to environmental and biological factors. Shown are at-sea data collected using SUS charges in the Gulf of Alaska [1]. Figure 1a illustrates the dramatic differences in surface scattering strength between low and high sea states. The narrow band of curves corresponds to a wind speed of 4.5 m/s and exhibits almost no frequency dependence, while the set of curves to the left corresponds to a wind speed of 17.9 m/s and exhibits a strong, non-monotonic — peak at ~925 Hz — frequency dependence, with levels elevated up to 30 dB over the lower wind speed data. Figure 1b shows the interesting behavior when fish are added to the mix. In this low-wind speed case, scattering from salmon exhibits a complex frequency behavior, with levels elevated up to 20 dB over air-water-interface scattering at low grazing angles above 130 Hz.

This paper begins with a discussion of the scattering characteristics of the ocean surface (air-sea interface and bubble clouds), followed by a discussion of the scattering characteristics of dispersed bladdered fish near the ocean surface. For simplicity, in this paper we restrict ourselves to monostatic backscattering geometries as they will still illustrate the key relationships. We end with a few comments and recommendations.

2 Surface scattering

Surface scattering is caused by the interaction of acoustic energy with environmental features at or near the ocean surface. The dynamic nature of the air-sea boundary interaction zone complicates this process. As the winds and seas increase, air becomes entrained by breaking waves in the form of subsurface bubbles. Under these conditions, both the rough air-sea interface and bubble clouds may contribute to the acoustic scattering. (For surface scattering strength (SSS), we set $\sigma_{fish} \equiv 0$ in Eq. (1).)

2.1   Scattering Model

Interface scattering strength σ_{int} is well modeled by lowest-order small slope theory [3,4], which requires as input the surface-wave roughness spectrum. The sea surface contains many scales of roughness, from the long gravity waves to the short capillary waves. Scattering from a rough interface is proportional to the spectral density at the Bragg wavelength with modifications due to tilt and modulation by longer waves. While a variety of directional surface-roughness spectral models are available, for this paper we assume an isotropic, pure power-law spectral model:

$$S(K) = A_S U / K^{\gamma_2} \quad , \tag{2}$$

where K is the surface wavenumber and U is the wind speed (in m/s) at an elevation of 10 m. With this spectral model, σ_{int} depends primarily on three environmental parameters: U, A_S, and γ_2. Typical open-ocean values of the latter two parameters are: $\gamma_2 \in (3.4, 4)$ and $A_S \in (5 \times 10^{-5}, 20 \times 10^{-5})$ m^3-s. Best-fit values to low wind-speed Critical Sea Test (CST)-7 data were: $\gamma_2 = 3.8$ and $A_S = 19 \times 10^{-5}$ m^3-s.

A semi-empirical approach is used to model σ_{bub}. In the ocean, breaking waves generate subsurface bubbles whose properties are governed by advective transport, gas dissolution, and buoyancy. At low frequencies (~5 kHz or less), acoustic scattering from bubbles depends primarily on the air-void fraction, and not on the details of the bubble distribution. Our semi-empirical model derives from a stochastic model of Gilbert [5], replacing some of its terms with ones whose parameters were empirically determined from a variety of open-ocean data. The monostatic result is:

$$\sigma_{bub} = \frac{0.006 d^{5.15} k_0^{3.4} \sin^4 \theta}{(1 + d^2 k_0^2 \sin^2 \theta)(1 + 4 d^2 k_0^2 \sin^2 \theta)} \quad , \tag{3}$$

where $k_0 = 2\pi f / c_0$ (with f the acoustic frequency and c_0 the sound speed in bubble-free water), and d is the air-void fraction e-folding depth, quadratically related to the wind speed U [5] via an empirical formula of Farmer and Vagle (derived from CST-7 data) [6]. Thus, σ_{bub} depends primarily on one environmental parameter: U.

2.2   Parameter Study

Figure 2 plots the dependence of SSS on grazing angle, frequency and wind speed. In this study, we set: $\gamma_2 = 3.9$ and $A_S = 2 \times 10^{-4}$ m^3-s. This figure illustrates that bubble clouds become an increasingly important driver of backscattering strength with both increasing frequency and wind speed, and with decreasing grazing angle. Figures 2a-c show the strong dependence of σ_{bub} on wind speed and, below ~1 kHz, on frequency. It has a fairly flat dependence on grazing angle, especially during appreciable winds. In contrast, Figs. 2d-f show that σ_{int} has a very strong dependence on grazing angle, but a relatively weak dependence on frequency and wind speed. As a result, when wave breaking is significant, *SSS* can have a complex dependence on angle, frequency and wind speed (Figs. 2g-i).

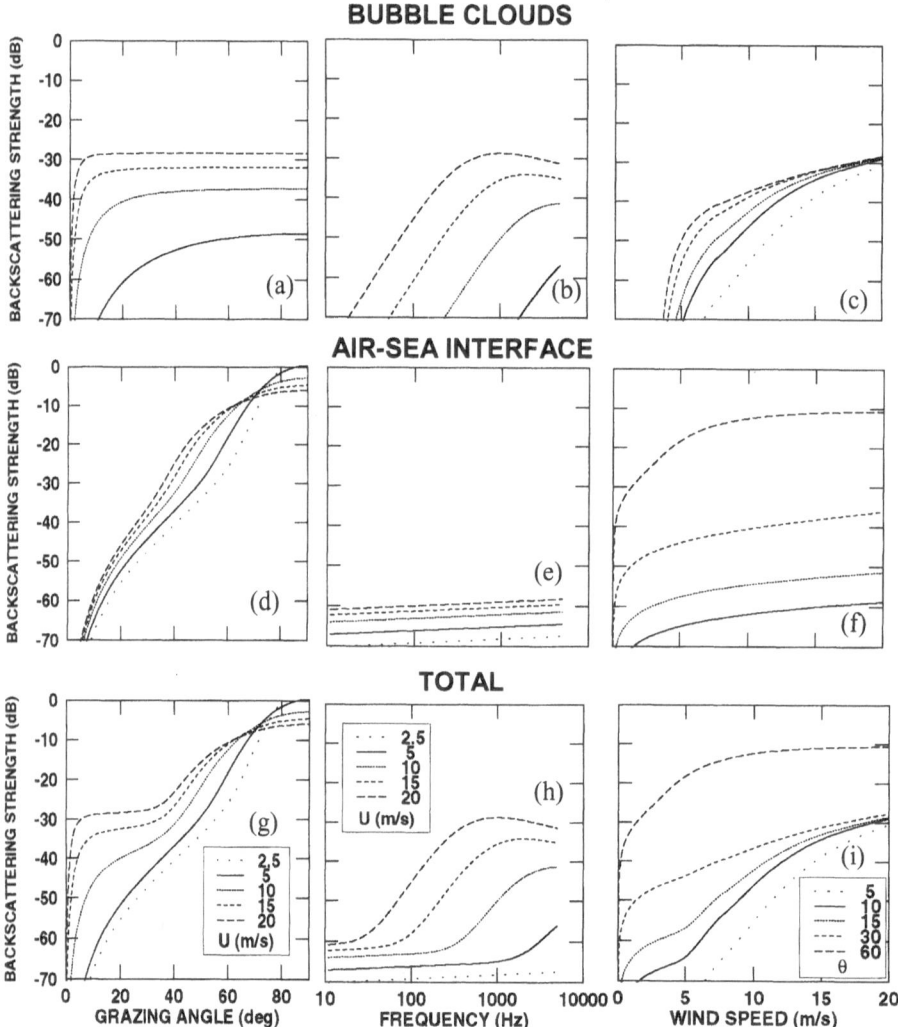

Figure 2. Monostatic predictions of scattering strength versus grazing angle, frequency and wind speed for: (a)-(c) bubble clouds only, (d)-(f) interface only, and (g)-(i) bubble clouds + interface.

The model for σ_{int} is more complex than σ_{bub} in that it also depends on the roughness spectrum through A_S and γ_2 as illustrated in Fig. 3. Figure 3c shows that as γ_2 decreases from 4, σ_{int} exhibits increasing frequency dependence. (With our choice of S in Eq. (2), σ_{int} depends on frequency only through its dependence on γ_2 and is frequency independent if $\gamma_2 = 4$ [4].) The other figures show generally monotonic increases of σ_{int} with decreasing γ_2 and increasing A_S, but that this can depend on the range of angles, wind speeds, and frequencies under consideration. In general, γ_2 drives the frequency dependence, and both γ_2 and A_S the level. The range of possibilities will expand when more sophisticated spectral models are considered in the future.

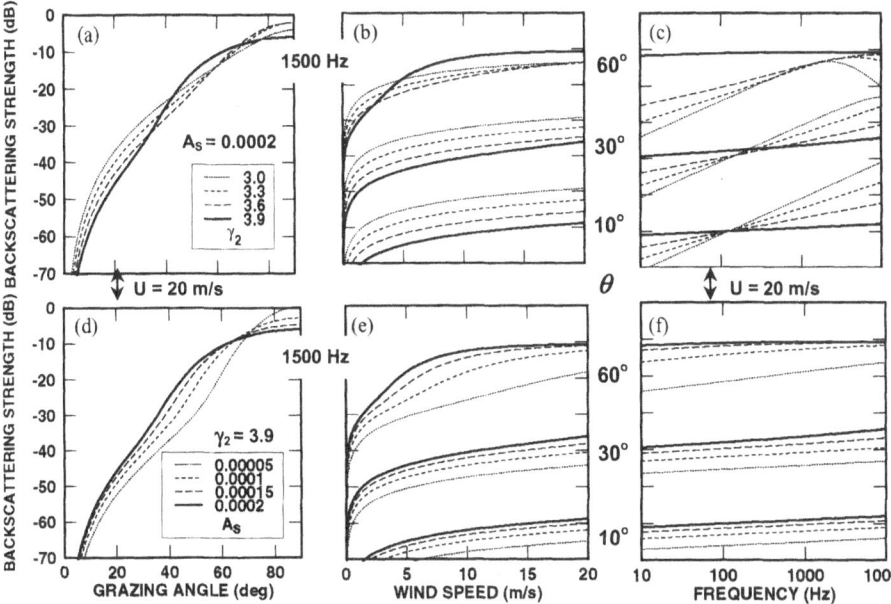

Figure 3. Monostatic predictions of interface scattering strength as a function of grazing angle, wind speed and frequency, for two sets of surface-wave spectral variables: (a)-(c) and (d)-(f).

3 Volume scattering

Due to their variety and dynamic nature, estimating the scattering contributions of fish is particularly challenging. When fish are well separated from the ocean surface or bottom, recognizable broadband acoustic signatures identifying their presence and strength have been observed [1]. However, when fish are in the vicinity of an ocean boundary, as is common in the littoral, these characteristic fish signatures can undergo significant modification due to boundary-interference effects.

3.1   Scattering Model

Below 10 kHz, the primary scattering mechanism of most fish is their air-filled swimbladder [7], typically occupying just ~5% of a fish's volume. The acoustic response depends on the bladder size, which in turn primarily depends upon the fish's size and depth. (The frequency response changes with depth as its bladder compresses due to the increased water pressure at the deeper depths—e.g., the single salmon of Fig. 4a.) For a layer of dispersed fish, we take their total scattering to be the incoherent sum of scattering from the individuals, and so depends on their depths, sizes and total number.

When a fish is near an ocean boundary, the scattering picture increases in complexity. Besides backscattering from the rough air-sea interface, fish backscatter energy to a receiver along multiple paths. The relative time delay of these various paths generates a (Lloyd-mirror) pattern of constructive and destructive interference, the intensity of which depends strongly on the surface grazing angle θ, the distance of the scatterer from the boundary, and the acoustic frequency. This can significantly alter the free-field fish's backscattered intensity (by a factor between 0 and 16—Fig. 4b), leading to a rich variety of frequency and grazing-angle behaviors, especially at low grazing angles:

$$\sigma_{fish} = 16 \int_{z_1}^{z_2} \rho(z) \overline{\tilde{\sigma}}_{bladder}(f, z) \sin^4(k_0 z \sin \theta) dz \qquad (4)$$

for bladdered fish of density ρ and mean target strength (TS) $\overline{\tilde{\sigma}}_{bladder}$ (m^2) in layer(s) covering depths z_1 to z_2 ($z_2 > z_1$). Figure 4c shows an example of this modification at $\theta = 10$ degrees for the single salmon of Fig. 4a (assuming a swimbladder radius of $r_0 = 0.015$ m). (Note the enhanced scale of Fig. 4c.)

For practical applications of this model as a kernel in a reverberation model, a layer of fish is treated as locally spatially (and temporally) uniform, which in turn is treated as an effective modifier of boundary conditions. Hence, it can be used like a surface scattering strength, with no need to introduce a separate layer into the modeled waveguide. (A corresponding formula for fish near the ocean bottom may be found in Ref. 2.)

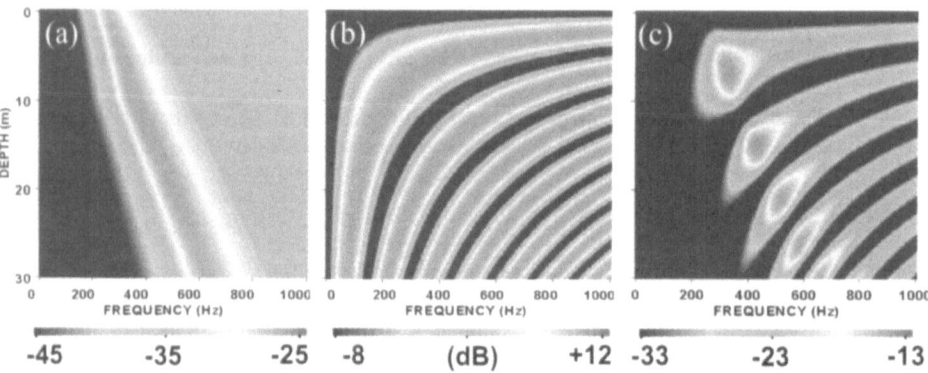

Figure 4. Monostatic predictions of near-surface fish scattering. (a) Free-field salmon scattering; and at $\theta = 10°$: (b) surface acoustic interference pattern and (c) near-surface salmon scattering.

Figure 1b showed a real-world example of backscattering from salmon in the presence of the air-sea interface in the Gulf of Alaska. Figure 5a shows our model prediction for these data. Using our model, salmon depths were inferred to be 2.5 to 7 meters. (The contribution of the rough interface is included in the modeling and is shown as a dashed line in Fig. 1b.) This data-model comparison shows that even at relatively low densities—a few hundred individuals per square kilometer during this measurement [1], fish near resonance can dominate interface scattering at low grazing angles.

3.2   Parameter Study

Figures 5b-c show how this picture can change when the depth of the salmon layer changes. These correspond to typical CST-7 nighttime and daytime depth ranges, respectively. (Here, we now have $U = 5$ m/s for the *SSS* contribution.) This shows when the fish are very shallow, the Lloyd-mirror pattern suppresses the resonance pattern (and the data resemble interface scattering). In contrast, when the salmon are deeper and spread over a greater depth range, their grazing-angle behavior becomes flat with a clear resonance (in this case at ~500 Hz).

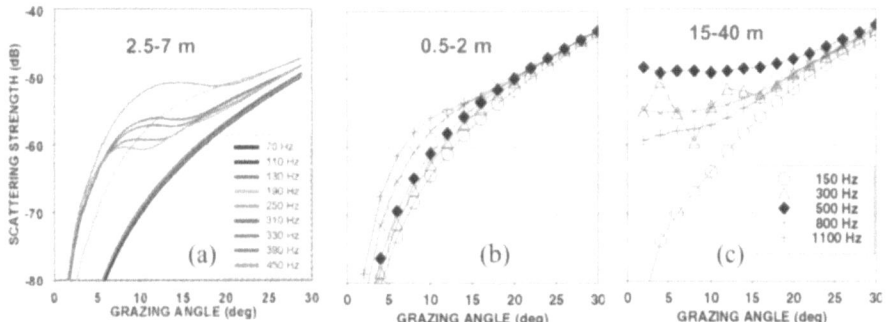

Figure 5. Monostatic predictions of near-surface salmon + interface scattering for 3 depth ranges.

Figure 6a-b illustrates the sensitivity of the TS of a particular fish, the salmon, to its size (as parameterized by r_0) and depth, respectively. Figure 6a shows that for a given depth, there is a strong sensitivity to fish size. That said, often an ensemble of fish of a given species in the ocean are of comparable size so that their depths become more of a driving factor. Figures 5 and 6b explore this for the salmon, and Fig. 6c for another species, the rockfish [2]. While size and depth are key parameters for a given species, the general volume scattering picture is even more complicated, as TS depends on more than just z and r_0 [2]. Adding to the acoustical complexity are their species-dependent diurnal and seasonal behaviors, as well as their temporal and spatial variability.

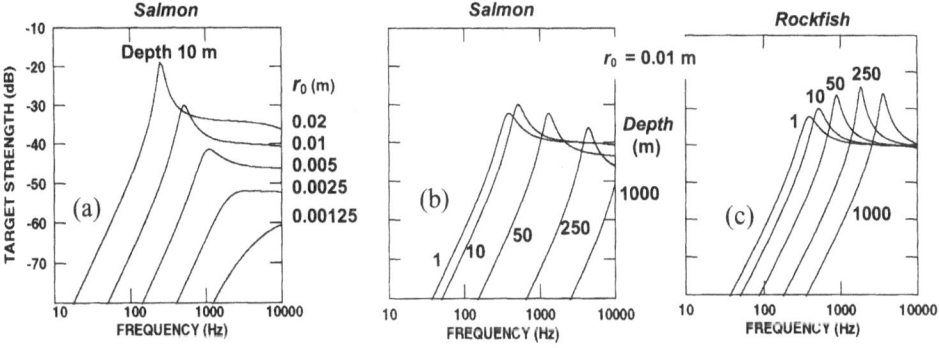

Figure 6. Modeled free-field fish TS vs. frequency for (a) 5 swimbladder radii at a depth of 10 m and at (b)-(c) 5 depths for a swimbladder radius of 0.01 m.

4   Discussion

In general, measures of sonar performance depend nonlinearly on the reverberation, which in turn depends nonlinearly on environmental variables. A key benefit of the models presented is that they allow a systematic determination of the relative influence of these environmental inputs on the strength of the acoustic scattering. In turn, used as scattering submodels in reverberation models, they allow a more accurate estimation of the relative influence of the environment on sonar performance. Furthermore, by independently varying the values of the environmental parameters, the resultant impact on the scattering can be estimated in a statistical sense. Hence, the relative variability or uncertainty in sonar performance can be assessed, and the expected variance modeled.

While these models promise improved predictions of *mean* scattering levels, technical issues remain. A primary need is high-quality acoustic and environmental/biological data to provide the ground truth necessary to rigorously evaluate the models, and to assess the generality and limitations of their physical assumptions. Additional needs include a deeper understanding of fish behavior and the physical properties of bubbles, and robust methods to statistically measure/assess them in situ.

We close with some recommendations for any scattering measurement:

- Maximize the frequency and grazing-angle coverage to sort out scattering mechanisms and help invert for environmental parameters.
- Perform day/night measurements to help sort out the fish contributions.

Acknowledgements

This work was supported by the Office of Naval Research. The authors are grateful for continuing technical discussions on fish acoustics with Dr. Redwood W. Nero (NRL).

References

1. Gauss, R.C., Wurmser, D., Nero, R.W. and Fialkowski, J.M., New bistatic models for predicting bottom, surface, and volume scattering strengths. In *Proc. 28th Meeting of The Technical Cooperation Program, Maritime Systems Group, Technical Panel Nine (TTCP MAR TP-9)*, NRL, Washington, DC (1999).
2. Gauss, R.C., Gragg, R.F., Wurmser, D., Nero, R.W. and Fialkowski J.M., Improved formulas for estimating bistatic bottom, surface, and volume scattering strengths. In *Proc. 30th Meeting of the TTCP MAR TP-9*, DREA, Dartmouth, NS, Canada (2001).
3. Dashen, R., Henyey, F.S. and Wurmser, D., Calculations of acoustic scattering from the ocean surface, *J. Acoust. Soc. Am.* **88**, 310–323 (1990).
4. Gragg, R.F., Wurmser, D. and Gauss, R.C., Small-slope scattering from rough elastic ocean floors: General theory and computational algorithm, *J. Acoust. Soc. Am.* **110**, 2878–2901 (2001).
5. Gauss, R.C. and Fialkowski, J.M., A broadband model for predicting bistatic surface scattering strengths. In *Proc. 5th European Conference on Underwater Acoustics*, edited by M.E. Zakharia *et al.* (European Commission, Luxembourg, 2000) Vol. 2, pp. 1165–1170.
6. Farmer, D.M. and Vagle, S., Inst. of Ocean Sci., Sidney, BC, Canada (private comm., 1998).
7. Love, R.H., Resonant acoustic scattering by swimbladder-bearing fish, *J. Acoust. Soc. Am.* **64**, 571–580 (1978).

MODELING PROPAGATION AND REVERBERATION SENSITIVITY TO OCEANOGRAPHIC AND SEABED VARIABILITY

KEVIN D. LEPAGE

SACLANT Undersea Research Centre, Viale San Bartolomeo 400, 19138 La Spezia, Italy
E-mail: lepage@saclantc.nato.int

The propagation of bottom and oceanographic variability through to the variability of acoustic transmissions and reverberation is evaluated with a simple adiabatic model interacting with Gaussian distributed uncertainty in a narrow band. Results show that there is significant sensitivity of time series and reverberation uncertainty to different types of environmental uncertainty. For propagation over uncertain bottoms, we show that it is that later part of the time series, corresponding to the highest angle energy reflecting most often off the surface and bottom, which is most sensitive to bottom uncertainty. This implies that the higher reverberation from the highest grazing angles is also the most uncertain. Conversely, it is the lowest angle arrivals which are most sensitive to uncertainty in the sound speed profile. These controlling principles are intuitive and are predicted in closed form with the theory.

1 Introduction

The effects of oceanographic and seafloor variability on acoustic propagation and reverberation in shallow water waveguides are of interest in the context of sonar performance uncertainty. In shallow water the downward refracting nature of the waveguide causes significant bottom interaction, so to the extent that the general background properties of the bottom sediments are unknown, significant variability in the forward propagation of energy to and from scatterers is to be expected. The same is of course also expected from oceanographic variability. In order to propagate uncertainty in the background properties of the ocean and the bottom through to uncertainty in acoustic propagation and reverberation an acoustic modeling approach is required which treats the propagation and reverberation stochastically. One common approach is to use one of the available "high fidelity" acoustic models to compute realizations of propagation or reverberation over a sample of oceanographic or bottom variability. This Monte-Carlo approach has the advantage that as long as the acoustic models are accurate and the underlying samples from which the oceanographic or bottom ensemble is drawn is known, then all the statistics of the desired property may be estimated. However, more rapid insight into the controlling parameters of oceanographic and bottom variability can be gained from a simpler, lower fidelity approach which is derived on more restrictive assumptions regarding both the distributions from which the bottom variability is drawn and on the simplicity of the acoustic propagation.

Here the lower fidelity approach is taken to understanding the effects of bottom vari-

N.G. Pace and F.B. Jensen (eds.), Impact of Littoral Environmental Variability on Acoustic Predictions and Sonar Performance, 353.360.
© 2002 *Kluwer Academic Publishers.*

ability on acoustic propagation and reverberation. We begin with the simplest useful parameterization of waveguide variability as a Gaussian distributed process which is adequately described by second order statistics and a correlation length scale. We then evaluate the effects of this variability on adiabatic propagation and reverberation in a narrow band. Using this approach it is possible to derive closed form expressions for the expected value of received intensity [1,2]. The results show that significant understanding of the sensitivity of temporal propagation and reverberation to different types of environmental variability can be gained using the closed form expressions.

2   Examples of variability in propagation

We evaluate the variability of propagation in the 140 m deep shallow water waveguide illustrated in Fig. 1. A downward refracting sound speed profile overlies a slow sediment layer 5 m thick with a background sound speed of 1482 m/s, a density of 1 g/cm^3 and bulk attenuation of 0.06 dB/λ. The sediment lies over a 1562 m/s basement with a density of 1.8 g/cm^3 and an attenuation of 0.1 dB/λ. Time series excited by a 20 m source operating at 500 Hz with 25 Hz of bandwidth are predicted over the full water column at a range of 15 km for ideal propagation and for propagation through oceanographic and over bottom variability.

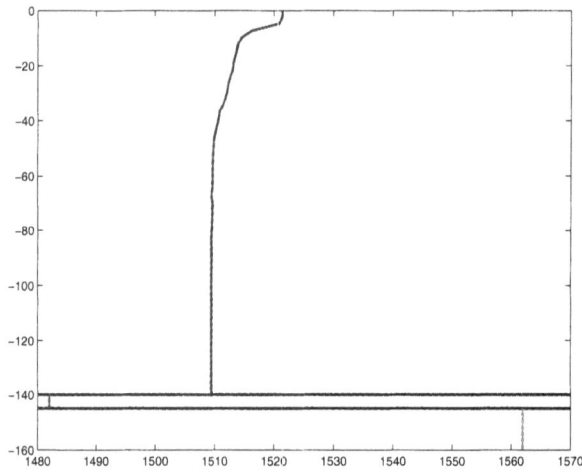

Figure 1. Shallow water environment used in the study.

The first case we study is propagation through internal waves. These are characterized vertically by the EOF mode shapes illustrated in the left panel of Fig. 2 and horizontally by the correlation length scales shown in the right hand panel. The most energetic EOF shapes have the longest correlation length scales and the fewest number of vertical oscillations, while the weakest have the shortest correlation length scales and the most vertical oscillations. Horizontal correlation length scales range from over 3 km for the first EOF to less than 100 for the ninth. The standard deviation of the sound speed defect associated with each EOF (not shown) ranges from 2.5 m/s for the first EOF to 0.13 m/s for the ninth.

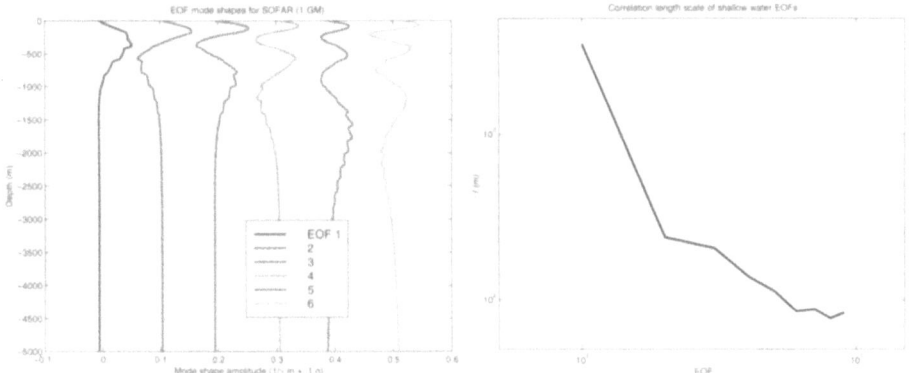

Figure 2. Mode shapes of the first 6 EOFs for the shallow water sound speed profile (left) at 1 GM. The right plot shows the corresponding correlation length scales.

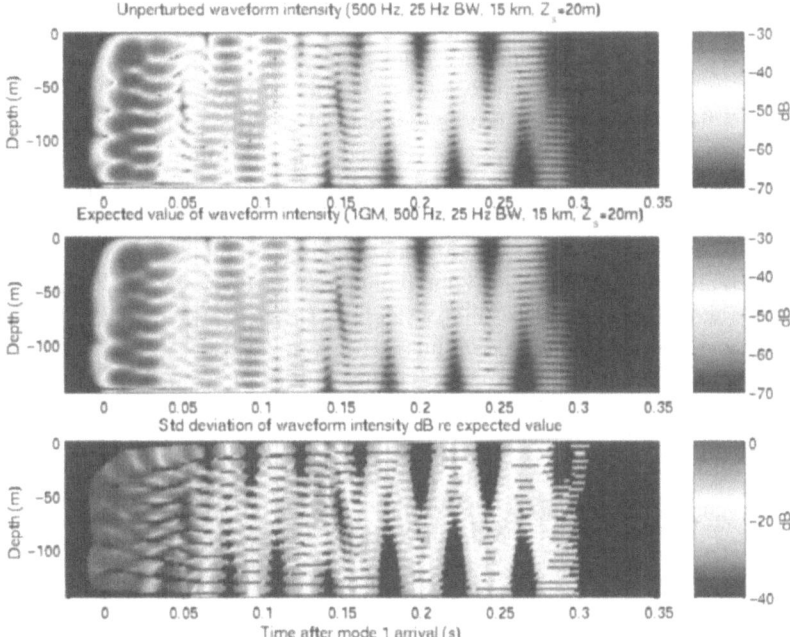

Figure 3. Variability of acoustic propagation in the presence of internal waves. Top: Unperturbed waveforms at 500 Hz. Middle: Expected value averaged over all internal wave realizations. Bottom: Standard deviation of intensity over 50 realizations normalized by the average intensity from the middle plot.

The expected value of the intensity received at a range of 15 km as a function of depth and time is illustrated in Fig. 3. The top panel shows the expected value in the absence of variability, *i.e.* the intensity in an ideal waveguide. The middle panel shows the expected value of the intensity averaged over all internal wave realizations. The bottom panel shown the standard deviation of the intensity estimated over an ensemble of 50 Monte-Carlo realizations of an internal wave field generated using the EOFs, dB

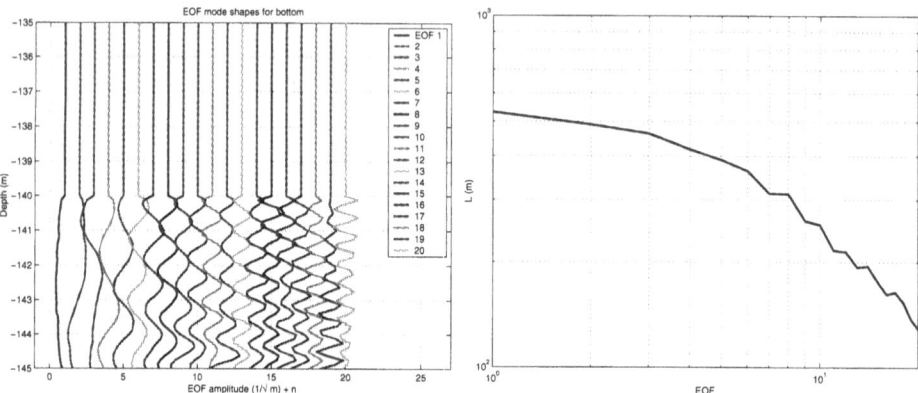

Figure 4. Mode shapes of the first 20 EOFs for sediment sound speed defects conforming to a power law distribution. The right plot shows the corresponding correlation length scales.

re the expected value of the intensity from panel 2. The first panel showns that the time series in a shallow water waveguide is a decresendo with the lowest order modes arriving first followed by the surface-bottom multiples. With a 1562 m/s basement sound speed, the length of the coda is restricted to 300 ms at 15 km range. The middle and bottom panel show that the first arrivals are the least predictable, while the later arrivals are highly predictable. This is understood from the fact that the earliest arrivals travel along the direction of the largest correlation length scale of the internal wave induced variability, while the steeper surface-bottom multiples travel increasingly along the shorter vertical correlation length scale of the variability. This is the predominant characteristic of shallow water propagation through internal wave fields.

We now investigate how bottom variability affects the propagation. In Fig. 4 the EOFs and horizontal length scales of a 1 m/s rms bottom sound speed defect realization are shown. The realization was generated from an anisotropic two dimensional power spectrum conforming to a power law asymptotic to k^{-3} with horizontal and vertical correlation length scales of $\ell_x = 500$ m and $\ell_z = 0.3$ m. As with the internal waves, the most energetic EOFs have the fewest vertical oscillations and the longest correlation length scale. However, the roll-off rate of the correlation length scale from a nominal value of 500 m is much slower, as the underlying process has only the one horizontal correlation length scale. The power in the various EOFs (not shown) also falls off more slowly, with a standard deviation of 0.35 m/s for the first EOF and 0.09 for the twentieth.

In Fig. 5 the variability in the bottom is propagated through to the acoustic uncertainty. As before, the top panel shows the unperturbed time series. The middle panel shows the expected value of the intensity averaged over the ensemble of bottom sound speed perturbations conforming to the EOF decomposition shown in Fig. 4 with a 20 m/s sediment sound speed standard deviation. Here it seen that it is the later arrivals, which interact more strongly with the bottom, which are uncertain. The bottom panel, which shows the ratio of the standard deviation of the intensity to its expected value, also shows that the later arrivals can have an uncertainty which equals their average intensity.

Finally, we evaluate the effect of fluctuations of attenuation in the sediment with a standard deviation of 0.018 dB/λ. The result is shown in Fig. 6, where it is seen that

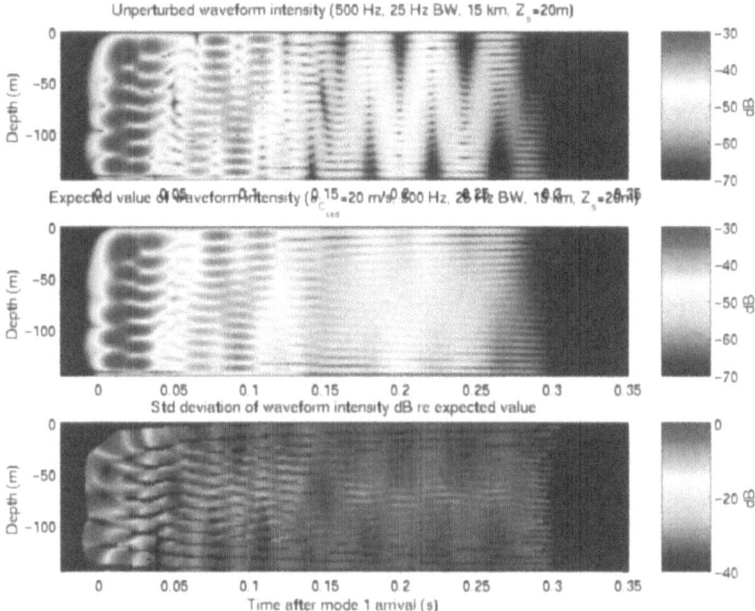

Figure 5. Variability associated with 20 m/s standard deviation in the sediment sound speed.

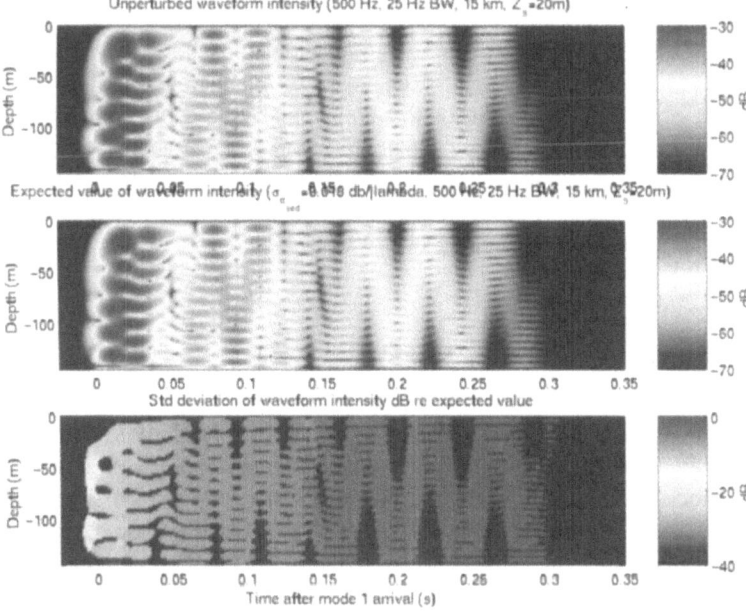

Figure 6. Variability associated with 0.018 dB/λ standard deviation in the background attenuation of the sediment.

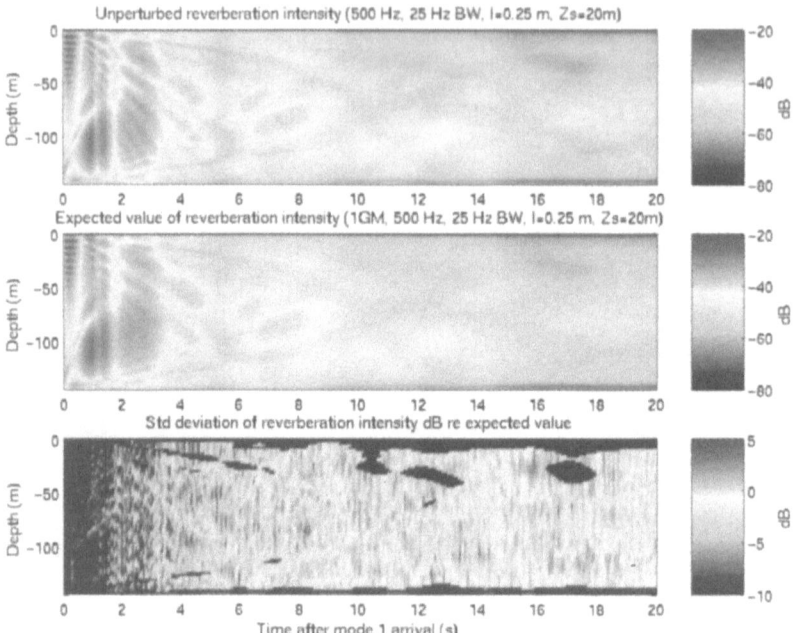

Figure 7. Reverberation variability caused by water column sound speed perturbations associated with unity Garrett-Munk internal wave spectrum.

while the expected value of the intensity is the same as the unperturbed value, the standard deviation can again approach the expected value itself, especially at late times, while at early time the standard deviation can be as much as 10 dB lower. The two results for the sensitivity of received intensity to bottom perturbations clearly show that it is the later arrivals, corresponding to the bottom interacting modes or the steepest ray paths with the most boundary interactions, which are the most sensitive to bottom uncertainty. This intuitively obvious result can be quantified using the closed form expressions as a function of the acoustic normal mode properties of the waveguide and the EOF modes shapes, correlation length scales and power.

3   Examples of variability in reverberation

We here show some preliminary results for variability in coherent reverberation for the same waveguide and source characteristics as were used to study variability in propagation. Reverberation in the case of uncertainty in water column or sediment properties has two sources of variability about the mean intensity: the first is standard deviation of reverberation due to the fact that the intensity is a pseudo-random process typically distributed Rayleigh in amplitude for Gaussian distributed scatterer amplitudes [3,4], and the second is associated with the uncertainty in the underlying watercolumn or sediment properties themselves for fixed realizations of scatterer distributions. Here we evaluate the latter alone, *i.e.* the standard deviation of the reverberation intensity for fixed scatterer realizations due to watercolumn variability alone.

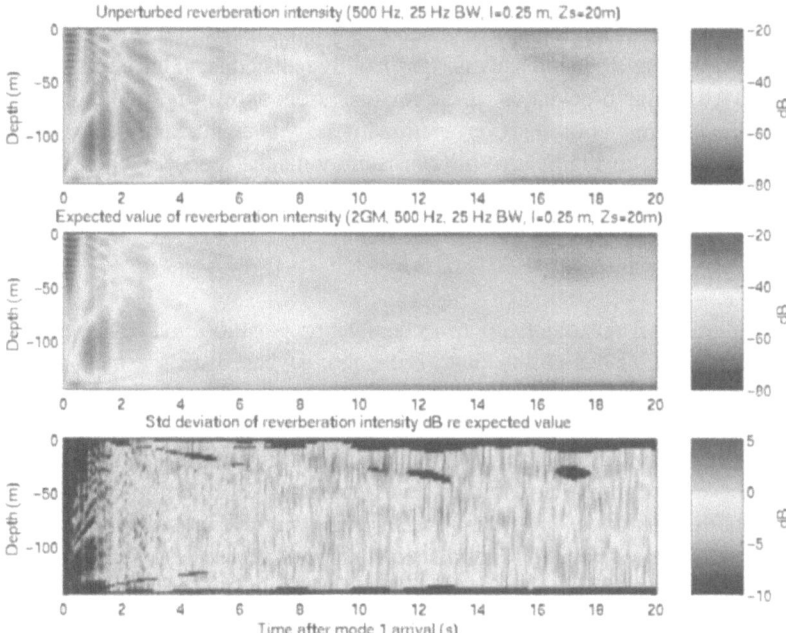

Figure 8. Reverberation variability caused by water column sound speed perturbations associated with twice the intensity of a Garrett-Munk internal wave spectrum.

In Fig. 7 the deviations of the reverberation intensity from the ideal case are shown for water column sound speed perturbations caused by internal waves whos EOF properties are shown in Fig. 2. The top panel shows the reverberation intensity as a function of time and depth as received on a monostatic vertical line array. As with the propagation studies, the source depth is 20 m and the source pulse has a center frequency of 500 Hz, but in this case the bandwidth is 50 Hz, giving a "resolution cell" on the seafloor of 15 m. The reverberation is caused by interaction with scatterers on the sediment-water interface with an rms amplitude of 1 and a correlation length scale of 0.25 m.

The middle panel shows the expected value of the reverberation intensity averaged over the ensemble of oceanographic variability caused by the internal waves. One sees that most of the details of the depth-time arrival structure are preserved, with the exception of some higher frequency modulations at mid-depth between 6 and 8 s, and less deep nulls at late time. The bottom panel shows the standard deviation of the intensity dB *re* the expected value of the intensity from the second panel, estimated by a Monte-Carlo average over 10 oceanographic realizations. With such a small sample size the true spatial structure of the standard deviation is not resolved: however it is a sufficient sample size to see that the reverberation time series at early time is highly stable or robust to oceanographic variability, while at late time is is less so, with the standard deviation often exceeding the mean value itself, especially in regions where the average intensity is lower.

In Fig. 8 the effects on the reverberation are evaluated for a doubling of the intensity of the internal wave spectrum. The middle panel shows the expected value of the now significantly modified coherent structure of the reverberation. The results show that the coherent structure is all but totally smoothed out at late time, and that even at early times from 2 s onward the average intensity is significantly reduced from the unperturbed results. The normalized standard deviation of the reverberation, shown in the bottom panel, also shows that the uncertainty of the reverberation intensity is significantly enhanced over the 1 Garrett-Munk case in Fig. 7, a result consistent with the reduced coherent component shown in the middle panel.

4 Conclusions

Two theories for the uncertainty of time domain propagation and reverberation in the presence of waveguide variability have been derived based on the method of normal modes, the narrow band approximation, perturbation expansions of the deviations of the wavenumbers and modal slownesses to the environmental perturbations, and an EOF decomposition of the oceanographic or bottom variability. The results for time series show that the character of the temporal development of uncertainty is highly dependent on the spatial distribution of the environmental uncertainty. For oceanographic variability associated with internal waves, the results show that it is the earliest, lowest angle arrivals which are the most uncertain. This is consistent with the understanding that the correlation length scales of the environmental variability are several orders of magnitude greater in the horizontal direction then they are in the vertical direction, causing uncertainty to accumulate more rapidly for low angle propagation. Conversely, results for bottom variability show that it is the latest arrivals, corresponding to the highest order surface-bottom multiples, which are most sensitive to the bottom variability, consistent with intuition.

Results obtained with the new theory of reverberation variability show promise for evaluating the relative sensitivity of reverberation to oceanographic and bottom variability. Preliminary results, restricted to evaluating the sensitivity of reverberation to oceanographic variability, show that the coherent, predictable structure of the reverberation time series is reduced as the oceanographic variability is increased, and that the standard deviation of the reverberation is correspondingly increased. Future work will be directed to further exploiting the model to evaluate the relative sensitivity of reverberation to oceanographic and bottom uncertainty.

References

1. Krolik, J.L., Matched field minimum variance beamforming in a random ocean channel, *J. Acoust. Soc. Am.* **92**, 1408–1419 (1992).
2. LePage, K.D., Acoustic time series variability and time reversal mirror defocusing due to cumulative effects of water column variability, *J. Comp. Acoust.* **9**, 1455–1474 (2001).
3. LePage, K.D., Bottom reverberation in shallow water: Coherent properties as a function of bandwidth, waveguide characteristics, and scatterer distributions, *J. Acoust. Soc. Am.* **106**, 3240–3254 (1999).
4. Abraham, D.A., Modeling non-Rayleigh reverberation. Rep. SR-266, SACLANT Undersea Research Centre, La Spezia, Italy, 1997.

UNCERTAINTY IN REVERBERATION MODELLING AND A RELATED EXPERIMENT

C.H. HARRISON, M. PRIOR AND A. BALDACCI

SACLANT Undersea Research Centre, Viale S. Bartolomeo 400, 19138 La Spezia, Italy.
E-mail: harrison@saclantc.nato.int

A serious stumbling block in modelling reverberation is the uncertainty in bottom scattering strength, and in the absence of detailed surveys we are still faced with many possible physical mechanisms. For instance, bottom penetration at low sound speeds allows scattering from buried layers and volume scattering which distorts the angle dependence. Typically in shallow water the propagation fall-off with range is controlled by 'mode-stripping', and the joint scattering and propagation angle dependence has a profound effect on the reverberation. Some predictions of the reverberation and signal-to-noise are made using very simple analytical calculations, and it is shown that there is a regime where signal-to-reverberation is a constant, independent of range, if we assume mode-stripping and Lambert's law! To check these ideas an experiment is proposed to investigate simultaneously the reverberation intensity and one-way path intensity as functions of range and vertical angle. By comparison of the two trends vs range it is possible to estimate the angle dependence of the scattering. The relative importance of position-dependent scattering strength, bottom shape, scattering law, and propagation are considered. These are backed up by calculations from a new multistatic sonar performance model SUPREMO.

1 Introduction

Reverberation is strongly influenced by a number of distinguishable scattering and propagation mechanisms. Amongst the scattering phenomena are: angle-dependence of the scattering law, geographic variation of the scattering strength, and local tilt of bottom facets. Propagation phenomena include range-dependent propagation intensity and range-dependent arrival angle at the scatterer.

So, can we predict diffuse reverberation when so many things are going on? Some of these quantities are more susceptible to surveying than others. Some can already be deduced from existing known data.

From the opposite point of view, given a (non-local) reverberation measurement, is it possible to tell what mechanism is what? Can we uniquely determine scattering properties?

On the face of it there is enormous scope for mathematical complication. But in this paper we try to gain some insight into the relative importance of the mechanisms by first taking an analytical approach. The importance of a mechanism can be judged by its impact on the signal-to-reverberation-ratio (SNR) directly. Finally we return to a numerical model to look at some more detailed effects.

N.G. Pace and F.B. Jensen (eds.), Impact of Littoral Environmental Variability on Acoustic Predictions and Sonar Performance, 361-368.

2   Anlytical (closed-form) reverberation and implications for SNR

Propagation in shallow water tends to follow a mode-stripping law, culminating in a single mode at long ranges [1]. The reason for this is that the bottom loss (in dB) almost always obeys a linear law with angle until the critical angle θ_c (if it exists) $R = \alpha_{dB}\theta$. Then the number of reflections is determined by the ray cycle distance r_c which for isovelocity is given in terms of the water depth H by $r_c = 2H/\tan\theta$. This results in an exponential decay in range but a gaussian angle distribution $\exp(-\alpha\theta^2 r/2H)$. By considering either a sum of eigenrays or a sum of modes [2] the final result for isovelocity can be shown to be a soluble integral over ray angles

$$I = \frac{2}{rH}\int_0^{\theta_c}\exp(-\frac{\alpha\theta^2}{2H}r)\,d\theta = \sqrt{\frac{2\pi}{H\alpha r^3}}\ \mathrm{erf}(\sqrt{\alpha r/2H}\ \theta_c) \tag{1}$$

When the square root term in the error function is large (i.e. the gaussian is narrow compared with the critical angle) the intensity reduces to [1]

$$I = \sqrt{\frac{2\pi}{H\alpha r^3}} = \sqrt{\frac{20\pi \log(e)}{H\alpha_{dB} r^3}} \tag{2}$$

which is the familiar three-halves law or 15 log(r). Closed-form solutions also exist [2] with smooth transition to single mode propagation (an exponential decay in range), and these effects can be seen in some of the later plots here.

For a monostatic sonar the result of two-way propagation with a point target of target strength S_T this becomes

$$I = \frac{2\pi S_T}{\alpha H r^3}\left(\mathrm{erf}(\sqrt{\alpha r/2H}\ \theta_c)\right)^2 \tag{3}$$

The equivalent reverberation with angle-independent scattering strength S_B and elementary scattering area (determined by the spatial pulse length $p=ct_p/2$ and the spatial horizontal beam width $r\Phi$) is

$$I = \frac{2\pi S_B\Phi p}{\alpha H r^2}\left(\mathrm{erf}(\sqrt{\alpha r/2H}\ \theta_c)\right)^2 \tag{4}$$

It inevitably falls off more slowly than the target echo because of the widening of the scattering area with range. Eventually at long range reverberation will win, and at the transition there is a "reverberation limit".

Reverberation with Lambert's Law scatterers behaves differently. Assuming that the scattering strength is separable in the incoming (θ_1) and outgoing (θ_2) ray angles

$$S = \mu \sin(\theta_1)\sin(\theta_2) \tag{5}$$

the intensity integrals also separate

$$I = \frac{4}{r^2 H^2} \int_0^{\theta_c} \int_0^{\theta_c} \sin\theta_1 \sin\theta_2 \exp(-\frac{\alpha\theta_1^2}{2H}r) \exp(-\frac{\alpha\theta_2^2}{2H}r) d\theta_1 d\theta_2 \ \mu \, r\Phi p \qquad (6)$$

to give a closed form solution for the reverberation

$$I = \left(\frac{2}{rH} \int_0^{\theta_c} \theta \exp(-\frac{\alpha\theta^2}{2H}r) \, d\theta\right)^2 \mu \, r\Phi p$$

$$= \frac{4\mu}{\alpha^2 r^3} \Phi p \left(1 - \exp(-\alpha \, r\theta_c^2 / 2H)\right)^2 \qquad (7)$$

In Eq. (3) we saw that the mode-stripping three-halves range law resulted in r^{-3} in two-way propagation when the gaussian was much narrower than the critical angle. Interestingly the Lambert's law reverberation in Eq. (7) shows exactly the same range dependence in this case; the usual proportionality of scattering area to range that alters the range-dependence in Eq. (4) is exactly counteracted by mode-stripping in this new integral. At shorter ranges where the critical angle truncates the gaussian the error function of Eqs. (3) and (4) is replaced by a (1−exponential) term.

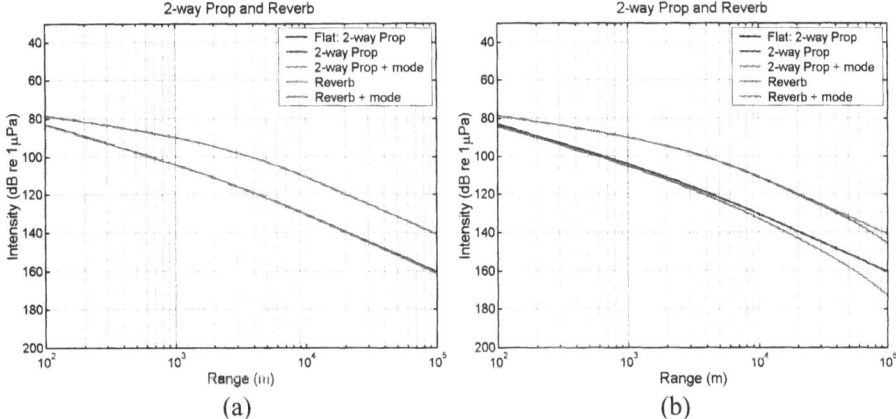

Figure 1. Target (two-way propagation) and reverberation: flat bottom, 100m depth, (a) 1000 Hz, (b) 100 Hz. Both frames show general formula and HF approximation for comparison. The source term and the $\mu\Phi p$ terms have been set to unity in this example.

Another interesting point is that the dependence of reverberation on bottom loss α is stronger than that of the target. In fact low bottom loss benefits the reverberation more than the target so the *SNR is improved by high bottom loss*!

Figure 1 shows the behaviour of these formulae for target and reverberation at 1 kHz and 100 Hz. Note that $\mu\Phi p$ has been set to unity since we are only interested in shapes of the curves. The regime of constant SNR is seen at long ranges where target and reverberation are parallel. Single mode effects are seen in the right hand frame.

Although these formulae assume isovelocity it is easy to see that refraction simply introduces the possibility of ducts at low angles; angles steep enough to hit the boundaries and therefore reverberate are still attenuated in more or less the same way [2].

It is shown in [2] that these formulae can be extended to slowly varying range-dependent environments by using ray invariant $H \sin(\theta) = $ constant [3] to cope with the angle changes. The results are written in terms of water depth at the source H_s and receiver H_r, the shallowest depth H_c, and an intermediate 'effective' depth H_{eff} [2,3] defined in terms of the depth profile $H(r)$.

$$H_{eff} = (H_r^{\ 2} H_s^{\ 2} / r) \int_0^r \frac{dr'}{H^3(r')} \qquad (8)$$

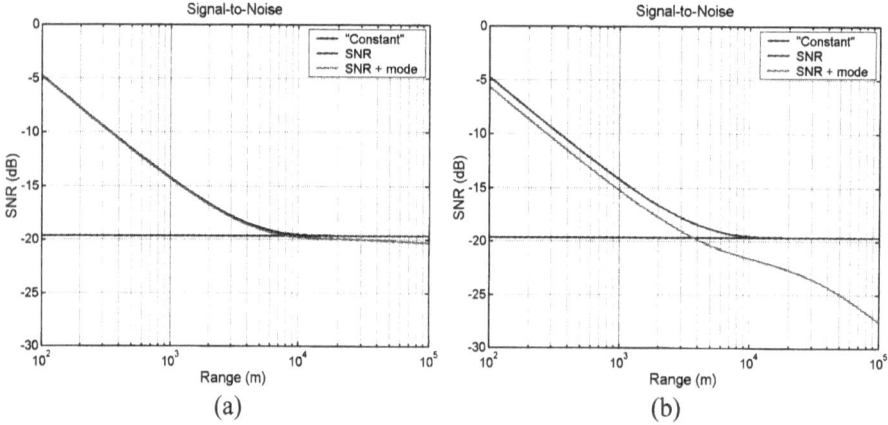

Figure 2 Signal-to-reverberation ratios: flat bottom, 100 m depth, (a) 1000 Hz, (b) 100 Hz. Blue line is large critical angle approximation; green line is arbitrary critical angle; red line is no approximation.

The target echo becomes

$$I = \frac{2\pi}{\alpha r^3 H_{eff}} \left(\mathrm{erf}\{ \sqrt{\frac{\alpha \, r H_{eff}}{2}} \frac{\theta_c H_c}{H_r H_s} \} \right)^2 S_T \qquad (9)$$

Reverberation with angle-independent scatterers becomes

$$I = \frac{2\pi}{\alpha r^2 H_{eff}} \left(\mathrm{erf}\{ \sqrt{\frac{\alpha \, r H_{eff}}{2}} \frac{\theta_c H_c}{H_r H_s} \} \right)^2 S_B \, \Phi p \qquad (10)$$

and reverberation with Lambert's law becomes

$$I = \frac{4H_s^{\ 2}}{\alpha^2 r^3 H_{eff}^{\ 2}} \left(1 - \exp\{ -\theta_c^{\ 2} H_c^{\ 2} \alpha r H_{eff} / (2 H_r^{\ 2} H_s^{\ 2}) \} \right)^2 \mu \, \Phi p \qquad (11)$$

Surprisingly, propagation is only weakly affected because, for instance, the increased bottom loss up-slope tends to be counteracted by the concentration of flux into a shallower water depth. The main effect on Lambert's law reverberation is the change of scattering angle rather than incident intensity, which manifests itself in Eq. (11) as the $(H_s/H_{eff})^2$ term. Up-slope reverberation is therefore generally stronger than down-slope, as one might guess. However the difference is small; for instance on a uniform slope halving the depth results in only a 2.5 dB increase, and doubling it results in a 3.5 dB decrease (H_{eff} is just the average of depths at source and receiver for a uniform slope).

3 A proposed experiment

The fact that we expect different range dependences according to the scattering law angle dependence suggests a possible experimental way of determining the scattering law. Two related methods are proposed, and it is hoped that they will be tried this year. There is one experimental arrangement but two sets of measurements.

3.1 Compare Reverberation and Propagation Range-Dependence

In principle, referring to Fig. 1 a single shot and a single monostatic hydrophone are sufficient to obtain the entire reverberation curve. To obtain the 'target' curve we need to move the source and receiver apart, measuring transmission loss as we go. Doubling the dBs gives a 'point target' echo level. The parallel region on the right of the figure shows that we expect the same range dependence for 'target' and 'reverberation' *if* there is Lambert's law. Otherwise we need to try different analytical forms for the law until the trend is explained. Uncertainty in bottom reflection loss is unimportant for this ratio. Given ideal experimental conditions, no numerical modelling is required. However one can think of many practical problems such as finding an environment that is range-independent for, say, 50 km radius (in all directions) from the source. In addition there is the problem of ducted propagation. The above analysis is strictly for isovelocity, although as noted, there is not much difference for reverberation with or without refraction, but we need to be careful that, if, for instance, there is downward refraction the source and receiver are either both in the duct or both out. Also we need to remember that we are interested in the bottom scattering law rather than the surface one. These latter problems are controlled and possibly solved by using a vertical array that can separate ducted and completely reflected arrivals.

3.2 Reverberation and Propagation Angle-Dependence

If we use a vertical array as receiver in the same experiment we have access to vertical arrival angle for the reverberation as well as for the propagation. This means we have access to the *integrands* of Eq. (1) and (partially) Eq. (6). In more general terms we can write the angle- and range-dependent propagation and reverberation responses as

$$dI_P(\theta, r) = P(\theta, r)\, d\theta \qquad (12)$$

$$dI_R(\theta_1, r) = s(\theta_1) P(\theta_1, r)\, d\theta_1 \int s(\theta_2) P(\theta_2, r)\, d\theta_2 = s(\theta_1) P(\theta_1, r)\, d\theta_1\, F(r) \quad (13)$$

where $s(\theta)$ is the unknown scattering strength and $P(\theta, r)$ and $F(r)$ are initially unknown functions ($F(r)$ is defined by the right hand side of Eq. (13)). The array measurements with moving source (dI_P) determine P completely. The single shot monostatic reverberation measurement (dI_R) determines $s \times P \times F$, but we now know P, and s depends only on θ while F depends only on r. This means that s and F are separable *without a model*! Taking each range in turn, the ratio of the responses always has the angle-dependence of the scattering strength s but with a range-dependent multiplier F.

$$\frac{dI_R(\theta, r)}{dI_P(\theta, r)} = s(\theta)F(r) \tag{14}$$

If we take logs of both sides to obtain, say $A(\theta, r) = B(\theta) + C(r)$, then by taking the average over all ranges we get an estimate of the function $B(\theta)$, i.e. $s^*(\theta)$. Similarly we obtain an estimate of $F^*(r)$ by taking the average over all angles, but there is still an unknown numerical factor which could be attributed to either term. Introducing a factor γ such that $s = s^* \gamma$ and $F = F^*/\gamma$ we can use the definition of F implicit in Eq. (13) to solve for γ.

$$\gamma^2 = \frac{F^*(r)}{\int s^*(\theta)P(\theta, r)\, d\theta} \tag{15}$$

Thus we obtain absolute, separated values for $s(\theta)$ and $F(r)$. Although γ appears to be a function of r it ought to be flat. Knowing $F(r)$ we can also extract bottom reflection parameters such as α and hence geoacoustic parameters by well-known methods.

4   Reverberation examples using the model SUPREMO

Returning to the original question of which phenomena need to be modelled we illustrate some of them here using a new multistatic sonar performance model SUPREMO [4]. The message is that there are many mechanisms that we can model, but given experimental measurements it may still not be possible to identify them.

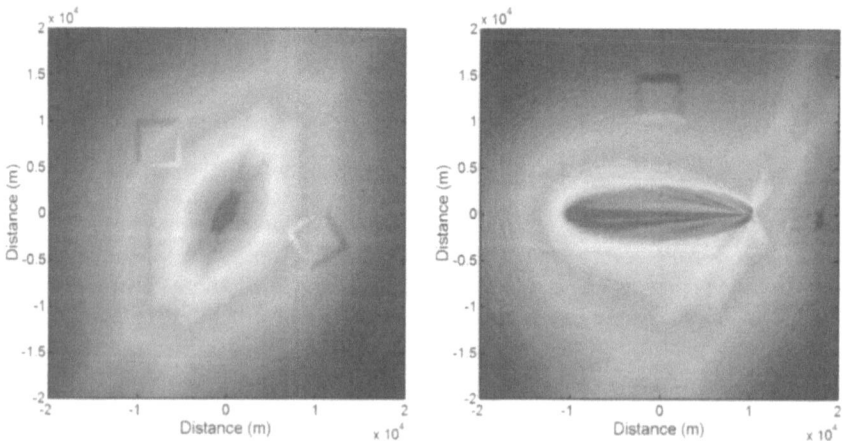

Figure 3. Monostatic and bistatic reverberation from a square seamount modelled with SUPREMO. Ambiguous beam returns can be seen on the right of each frame.

For example, a black and white photograph of craters on the moon could be interpreted as geographic changes in scattering strength rather than sloping facets!

4.1   Scattering: Bottom Topography Constructed from Multiple Sloping Facets

Figure 3 shows the reverberation response of a monostatic and a bistatic towed array sonar to a square seamount. One can easily see a bright front side, dark back side, and additional ambiguous image. SUPREMO has a "plug-in" propagation section, and this demonstration used only a range-independent model since the path on the way out is indeed range-independent.

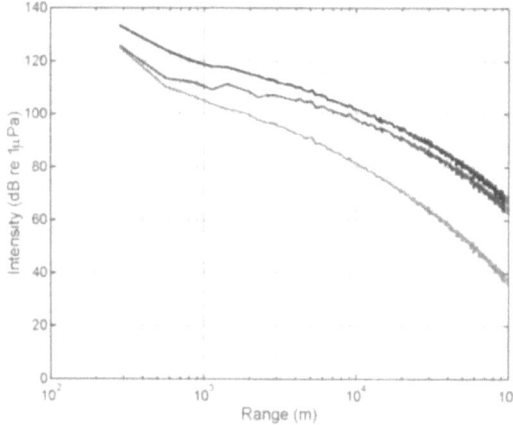

Figure 4. Lambert's law (lower line – red), Lommel-Seeliger (middle line – green), and angle-independent (upper line – blue) reverberation with SUPREMO.

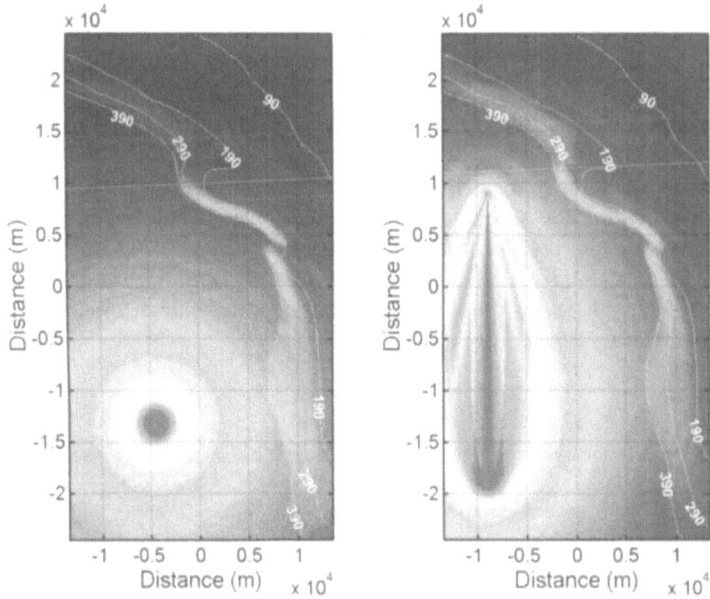

Figure 5. Monostatic and bistatic reverberation from the Ragusa Ridge (depth contours overlaid). Reverberation is underestimated because SUPREMO included a tilted-facet bottom but used a "plug-in" range-independent propagation model.

4.2   Scattering Law Angle Dependence

Figure 4 shows the effect of three different scattering angle laws (Lambert, Lommel-Seeliger, and angle-independent) on an otherwise horizontal seabed.

4.3   Realistic Bathymetry

Figure 5 shows the effect of tilted-facet scatterers on reverberation from the Ragusa Ridge south of Sicily. Although echoes follow the contours their intensities are underestimated because SUPREMO has been run with a range-independent propagation model. The analytical calculations above suggest that we can probably ignore the additional effects of the increased bottom loss up-slope since there is a counteracting concentration of flux in the shallower water. However, they also suggest that the raw multiplier $(H_s/H_r)^2$ would be an overestimate. This could be confirmed with a range-dependent propagation module at a future date.

5   Conclusions

There are a number of scattering and propagation effects that are mathematically distinguishable, for instance, the scattering law angle dependence, local tilt of bottom facets, scattering strength variation, ray arrival angle (as modified by water depth), and propagation intensity. We have used an analytical approach to find first order dependences, then a numerical approach to look at detailed mechanisms.

The implications for coping with environmental uncertainty when predicting reverberation are that scattering is more important than propagation effects. Subjectively ranking the effects we have scattering law, local bottom tilt, and absolute scattering strength. Range-dependent propagation intensity effects are probably less important although the accompanying (predictable) arrival angle change is important. An experiment is proposed that could help to solve the scattering law problem by deducing the angle-dependence and absolute value directly from long range VLA measurements.

The same ranking of effects also has implications for the design of models such as SUPREMO. For instance, one could spend a lot of computational effort on range-dependent propagation modelling when a range-independent code with a simple correction term for depth profile could conceivably do just as well. The pay-off would be that trials planning and operational research studies using such models could then deal with moving platforms and targets in range-dependent environments with acceptable computation times.

References

1. Weston, D.E., Intensity-range relations in oceanographic acoustics, *J. Sound and Vib.* **18**, 271–287 (1971).
2. Harrison, C.H., Reverberation and signal-excess with mode-stripping and Lambert's law. Rep. SR-356, SACLANT Undersea Research Centre (2002).
3. Weston, D.E., Propagation in water with uniform sound velocity but variable-depth lossy bottom, *J. Sound and Vib.* **47**, 473–483 (1976).
4. Harrison, C.H., Prior, M. and Baldacci, A., Multistatic reverberation and system modelling using SUPREMO. In *Proc. 6th European Conference on Underwater Acoustics*, Gdansk, Poland (2002).

STATISTICS OF THE WAVEGUIDE INVARIANT
DISTRIBUTION IN A RANDOM OCEAN

DANIEL ROUSEFF

*Applied Physics Laboratory, College of Ocean and Fishery Sciences,
University of Washington, 1013 NE 40th St., Seattle, WA 98105, USA
E-mail: rouseff@apl.washington.edu*

Brekhovskikh and Lysanov popularized the concept of an ocean "waveguide invariant" in the second edition of their book. The showed how acoustic intensity, mapped in range and frequency, would exhibit streaks of high correlation, and they further described the slope of these striations in terms of an invariant parameter. In shallow water, the description works best at moderate frequencies and when the sound speed profile changes only gradually in range and depth. In more complicated scenarios, the concept of a scalar invariant can be generalized to allow a distribution of values. Previously, the effects of shallow water internal waves on the distribution were studied by numerical simulation. In the present work, equations are derived relating the statistics of the waveguide distribution to the statistics of the random ocean. The formulation makes use of a stochastic coupled normal mode model to describe the acoustic propagation.

1   Introduction

Assume an acoustic source is transmitting in shallow water with the resulting field measured on a horizontal array. At sufficient range, the ocean acts as an acoustic waveguide supporting the propagating acoustic modes. Assume the array is oriented at end-fire from the source and that the measured field is processed incoherently. If the measured intensity is plotted versus frequency and distance along the array, the resulting image will exhibit striations, nearly parallel contours of relatively high intensity. The observed striations are a consequence of interference between the propagating acoustic modes. Chuprov [1] related the slope of the striations, $d\omega/dr$, to the range r and the frequency ω via the parameter β:

$$\beta = \frac{r}{\omega}\frac{d\omega}{dr}.$$ (1)

Chuprov showed how, in some sense, beta is an invariant quantity. Brekhovskikh and Lysanov popularized the concept of an ocean "waveguide invariant" in the second edition of their book [2].

Equation (1) is useful for studying contour plots of measured data. For analytical studies, it is preferable to express beta directly in terms of either the intensity I or

N.G. Pace and F.B. Jensen (eds.), Impact of Littoral Environmental Variability on Acoustic Predictions and Sonar Performance, 369-376.

quantities related to the acoustic modes. One can also write the waveguide invariant beta as

$$\beta = -\frac{r}{\omega}\frac{\partial I/\partial r}{\partial I/\partial \omega} = -\frac{d(1/v)}{d(1/u)}, \qquad (2)$$

where u is the group velocity and v the phase velocity of the pertinent acoustic modes.

The ocean environment will never be perfectly independent of range. Range dependence in the water column might be caused by ocean internal waves. Roughness at the water-sediment interface or in the sea surface would introduce range-dependence, as would variability in the sediment. Random variability in the environment will introduce random variability in the acoustic field with consequent effect on the propagating acoustic modes. From Eq. (2), the waveguide invariant beta will also be a random quantity. Often there is a statistical model for the randomness in the ocean. One would like to relate the statistical models for the environment to a statistical description for observable acoustic quantities. In the present case, the goal is to estimate relevant statistics for the waveguide invariant.

Unfortunately, the two forms for the waveguide invariant given in Eq. (2) are not amenable to a statistical analysis. The simplest quantity of interest would be the mean of beta, $<\beta>$. It follows from Eq. (2) that $<\beta>$ is given by the average of a ratio. The average of a ratio is not something that is commonly calculated; only in special cases, for example, can the average of a ratio be replaced by the ratio of averaged quantities [3].

Recently, the waveguide invariant was reformulated as a distribution. Rather than assign a single scalar, a distribution of values is produced; effectively, one calculates the "beta content" of a set of measurements. Previously, the effects of shallow water internal waves on the distribution were studied by numerical simulation [4]. In the present work, we show how the formulation is also useful in theoretical studies. We outline the derivation of a formula for the first moment of the waveguide distribution. The derivation makes use of a stochastic coupled normal mode model to describe the acoustic propagation.

2   Waveguide invariant distribution

The experimental observable is the intensity measured at depth z over a horizontal array of length $L = r_{max} - r_{min}$ oriented at end-fire from the source. Consider the measurements in the frequency band $\omega_{min} < \omega < \omega_{max}$. Let $I(r, \omega)$ be the intensity as measured over the finite-length receiving aperture. Define $\tilde{I}(\kappa, \tau)$ as the two-dimensional Fourier transform of the intensity:

$$\tilde{I}(\kappa, \tau) = \int_{\omega_{min}}^{\omega_{max}} \int_{r_{min}}^{r_{max}} I(r, \omega) \exp[i(\kappa r + \omega \tau)] dr\, d\omega. \qquad (3)$$

Define the waveguide invariant distribution as

$$E_\phi = (2\pi)^{-2} \int_{-B}^{B} \left| \tilde{I}(K \cos\phi, K \sin\phi) \right|^2 |K| dK \,. \tag{4}$$

Note that $\tilde{I}(K \cos\phi, K \sin\phi)$ is the transform evaluated along a line passing through the origin in Fourier space and oriented at angle ϕ from the κ axis. The integral has been truncated at some maximum spatial frequency of interest B. The ramp filter $|K|$ emphasizes the higher spatial frequencies. Integrating E_ϕ over all angles would yield, in the sense of Parseval's Theorem, the total energy in the original image [4]. The remaining task is to relate the angle ϕ to β. Assume that the range to the midpoint of the array r_{mid} is large compared to the length of the array L, and that the center frequency ω_{mid} is large compared to the bandwidth. Neglecting the effects of the finite data window, we obtain the mapping [4]

$$\beta = -r_{mid} / (\omega_{mid} \tan\phi) \,. \tag{5}$$

In this approach, beta is treated not as a single number but rather as a distribution. The output of the processing is the distribution plotted versus β. This distribution might be sharply peaked around a single value in which case the traditional notion of β as a scalar would be reasonable.

Figure 1 shows a sample calculation of the waveguide invariant distribution. The scenario involves a Perkeris waveguide of depth 70 m. The sound speed in the water column and in the sediment is 1480 and 1580 m/s, respectively. The bottom loss is 0.1 dB/λ. Both the source and receiver are at depth 50 m. The frequency band extends from 400 to 420 Hz and the range is 10 km. The plot is normalized so that the average value of the distribution over the interval shown is one. The distribution is sharply peaked around the canonical shallow water value $\beta = 1$; see the analysis in Brekhovskikh and Lysanov [2]. If there is a significant sound speed profile and the source and receiver are appropriately positioned, the location of the peak can shift or be lost altogether; see [4] for examples of internal wave effects.

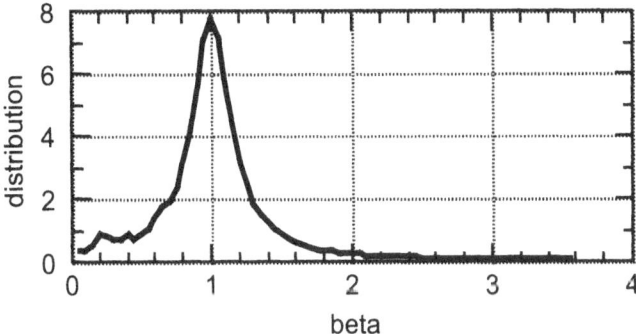

Figure 1. Sample calculation of the waveguide invariant distribution.

3   First moment of the waveguide invariant distribution

In this section, we first express the waveguide invariant distribution in terms of the acoustic normal modes. The resulting expression is a double summation over mode pairs where the weights for the terms depend on the mode amplitudes. Next, the master equations for the stochastic mode amplitudes are outlined. Finally, we take a formal average to get the mean waveguide invariant distribution.

3.1   Modal Formulation for the Waveguide Invariant Distribution

The starting point is a modal representation for the pressure field observed at depth z and at range r from the source

$$p(r,\omega) = \sum_m A_m(r)\Psi_m(z)\big/(\xi_m r)^{1/2} \,, \tag{6}$$

where Ψ_m and ξ_m are the eigenfunction and wavenumber, respectively, of mode m for the unperturbed background sound speed profile $c_0(z)$. For the special case of a range-independent medium, the modal amplitude $A_m(r)$ would be proportional to the eigenfunction evaluated at the source depth z_s:

$$A_m(r) = \Psi_m(z_s)\exp[i(\xi_m + i\alpha_m)r]\,. \quad \text{(Range-independent environment.)} \tag{7}$$

For this special case, there is no coupling of energy between the acoustic modes between the source and the receiving array. The modal attenuation is α_m accounts for energy loss. In the more general case considered here, we allow stochastic mode coupling and defer the specific master equations for $A_m(r)$ to the next section. To simplify the analysis, we assume the modes propagate without coupling or loss over the extent of the aperture. This is reasonable when the extent of the aperture L is small compared to the distance to the source. With the array extending for $r_{\min} < r < r_{\max}$, then over the array

$$A_m(r) = A_m(r_{\min})\exp[i\xi_m(r - r_{\min})]\,. \tag{8}$$

The experimental observable is the intensity $I(r,\omega)$ measured along the end-fire array in the frequency band $\omega_{\min} < \omega < \omega_{\max}$. It follows that

$$I(r,\omega) \equiv pp^* = \sum_{lm} B_{lm}\exp(i\Delta_{lm}r)\,, \tag{9}$$

where $\Delta_{lm} = \xi_l - \xi_m$ is the difference in wavenumber between a pair of modes. The weightings

$$B_{lm} = A_l(r_{\min})A_m^*(r_{\min})\Psi_l(z)\Psi_m(z)\exp[-i(\xi_l - \xi_m)r_{\min})]\big/(\xi_l\xi_m r^2)^{1/2} \tag{10}$$

are derived from Eqs. (6) and (8).

To show how the output of the processing algorithm in Sect. 2 depends on the modal composition of the intensity, substitute Eq. (9) into Eq. (3) and combine with Eq. (4). Rearranging terms yields

$$E_\phi = (2\pi)^{-2} \sum_{lm} \sum_{l'm'} \int_{-B}^{B} |K| dK \int_{\omega_{\min}}^{\omega_{\max}} d\omega \int_{\omega_{\min}}^{\omega_{\max}} d\omega' \int_{r_{\min}}^{r_{\max}} dr \int_{r_{\min}}^{r_{\max}} dr' B_{lm} B_{l'm'}^*$$

$$\times \exp[i(\Delta_{lm} r - \Delta_{l'm'} r')] \exp(iK \sin \phi \omega_d + iK \cos \phi r_d) \tag{11}$$

where the difference coordinates $\omega_d = \omega - \omega'$ and $r_d = r - r'$.

With some manipulation, Eq. (11) can be reduced to a relatively simple formula for the distribution. The derivation closely parallels that for the range-independent case given by Rouseff and Spindel [5]. In the present work, we merely outline the derivation and present the final result; the interested reader is referred to [5] for the details.

The derivation starts by considering the range integrals in Eq. (11). These are rewritten in terms of the difference and average coordinates, r_d and $r_a = (r + r')/2$, respectively. As before, we assume the length of the array L is small compared to the distance to the source which allows us to simplify the limits of integration. Neglecting the weak range-dependence in Eq. (10), the range integrals can then be evaluated. The integration over r_d yields a delta function that can then be used to evaluate the integration over K.

The derivation proceeds with the frequency integrals. It becomes useful to expand the wavenumbers about the center frequency $\omega_{\mathrm{mid}} = (\omega_{\max} + \omega_{\min})/2$. The phase contains terms like

$$\xi_m(\omega) \approx \xi_m(\omega_{\mathrm{mid}}) + \frac{d\xi_m}{d\omega}(\omega - \omega_{\mathrm{mid}}) = \frac{\omega_{\mathrm{mid}}}{v_m} + \frac{(\omega - \omega_{\mathrm{mid}})}{u_m}. \tag{12}$$

where the phase velocity v_m and the group velocity u_m are evaluated at the center frequency. Similar expansions can be developed for the other terms in the phase. It is also useful to define the *local invariant* β_{lm}

$$\beta_{lm} = -\left(1/v_l - 1/v_m\right)\Big/\left(1/u_l - 1/u_m\right), \tag{13}$$

which makes explicit the dependence of the invariant on the specific pair of mode indices. Equation (13) can be viewed as the finite difference approximation to Eq. (2). With these expansions, one can determine the stationary phase points of the remaining integrals. One finds that for most combinations of mode indices, the stationary phase points are outside the regions of integration. The dominant contribution occurs when $l = l'$ and $m = m'$. Retaining only these terms, the frequency integrals can be evaluated in terms of the Fresnel integrals S and C [6]. The final result is

$$E_\beta \equiv E_\phi \left| \frac{d\beta}{d\phi} \right|^{-1} = \tfrac{1}{2} L \omega_{\mathrm{mid}} \sum_{lm} |\beta_{lm}/\beta| |B_{lm}|^2 [F_C^2 + F_S^2], \qquad (14)$$

where the terms involving Fresnel integrals

$$F_C = C(\gamma_+) + C(\gamma_-), \quad F_S = S(\gamma_+) + S(\gamma_-), \qquad (15)$$

have arguments

$$\gamma_\pm = \left[\omega_{\mathrm{mid}} r_{\mathrm{mid}} |g_{lm}| / (\beta\pi) \right]^{1/2} \left[\tfrac{1}{2} Q^{-1} \pm (\beta - \beta_{lm}) \right]. \qquad (16)$$

In Eq. (16), $Q = \omega_{\mathrm{mid}}/(\omega_{\max} - \omega_{\min})$ is the usual "quality factor" that arises when analyzing resonating systems, and $g_{lm} = 1/u_l - 1/u_m$ is the difference in group slowness.

Equation (14) expresses the waveguide invariant distribution as a double summation of terms involving the modal constituents. The behavior of the term in the square brackets is instructive. When the local invariant β_{lm} equals β, this term is maximized and potentially makes a strong contribution to the distribution. As β_{lm} diverges from β, the contribution is reduced. Note that each term in the double summation is weighted by its corresponding $|B_{lm}|^2$. These weightings depend on the mode amplitudes and also contain the depth-dependence in the result; see Eq. (10). For the range-independent problem when Eq. (7) applies, Rouseff and Spindel [5] used Eq. (14) to study the effect of source depth on the distribution. The more general case of a range-dependent environment is considered in this report.

3.2   Stochastic Coupled Mode Equations

It remains to specify the mode amplitudes $A_m(r)$ in Eq. (6). We assume the range dependence is primarily due to random variability in the medium and use the formalism developed by Dozier and Tappert [7]. Dozier later extended the technique to include bottom loss, an important consideration for propagation in shallow water [8]. Creamer used and extended these results to study scintillation in shallow water waveguides [9]. Here, we summarize some of their results pertinent for calculating statistics of the waveguide invariant distribution.

Using the quasi-static and narrow-angle approximations, the mode amplitudes can be shown to satisfy [6-8]

$$\frac{\partial A_m}{\partial r} - i(\xi_m + i\alpha_m) A_m = -i \sum_n \rho_{mn}(r) A_n , \qquad (17)$$

where ρ_{mn} contains the effects of mode coupling. Note that if there is no coupling, the right hand side of Eq. (17) is zero and the solution to the differential equation reduces to

Eq. (7) as it must. The mode coupling terms are related to the perturbations in the index of refraction δc by

$$\rho_{mn} = \frac{k^2}{\sqrt{\xi_m \xi_n}} \int dz \left[\delta c(r,z) / c_0(z) \right] \Psi_m(z) \Psi_n(z) \ . \tag{18}$$

The perturbations δc would typically be modeled as a Gaussian random process. This lets us relate the autocorrelation of the mode coupling terms

$$R_{mn}(r - r') = \left\langle \rho_{mn}(r) \rho_{mn}(r') \right\rangle \tag{19}$$

to a particular model for the fluctuations in the medium. These fluctuations might be due to internal waves or other types of environmental variability. In this short communication, we do not consider specific statistical models for the environment.

3.3   Fourth Moment Equations

The average waveguide invariant distribution follows immediately from Eq. (14). Since all the randomness is contained in the weightings B_{lm},

$$\left\langle E_\beta \right\rangle = \tfrac{1}{2} L \omega_{\text{mid}} \sum_{lm} |\beta_{lm} / \beta| \left\langle |B_{lm}|^2 \right\rangle [F_C^2 + F_S^2] \ . \tag{20}$$

From Eq. (10),

$$\left\langle |B_{lm}|^2 \right\rangle = \left\langle |A_l(r_{\min})|^2 |A_m(r_{\min})|^2 \right\rangle |\Psi_l(z) \Psi_m(z)|^2 \Big/ (\xi_l \xi_m r_{\text{mid}}^2) \ . \tag{21}$$

Note that the average waveguide invariant distribution depends on the fourth moment of the pressure or, more precisely, the second moment of intensity. Making the Markov approximation, Creamer [8] developed the state equations for the second moment of intensity. The state matrix contains projections of the autocorrelation Eq. (19) as well terms related to the modal attenuation α_m. Following Dozier, Creamer diagonalized the state matrix, but this is not necessary for our purposes. One can write the solution to the state equation in terms of the matrix exponential [10]. This lets us calculate the moments in Eq. (21) and hence the average waveguide distribution given by Eq. (20).

4   Discussion

The waveguide invariant distribution is a generalization of the usual scalar invariant. In our previous work, we did numerical simulations to study how the distribution was affected by ocean internal waves [4]. In the present work, we derived an expression for the average distribution when random variability in the medium introduces acoustic mode coupling. The result is quite general and should apply at moderate frequencies whenever the variability in the medium can be modeled as a Gaussian random process. In our future work, we will consider specific models for the ocean variability and do detailed numerical calculations of the mean waveguide invariant distribution.

Acknowledgements

A portion of this research was conducted while the author was a Senior Visiting Fellow in the Department of Applied Mathematics and Theoretical Physics at the University of Cambridge, Cambridge, England. The author thanks Dr. Barry Uscinski and the staff at DAMTP for their hospitality. The author also thanks Dr. Lewis Dozier for providing a copy of reference [8]. This work was supported by the United States Office of Naval Research.

References

1. Chuprov, S.D., Interference structure of a sound field in a layered ocean. In *Ocean Acoustics. Current State*, edited by L.M. Brekhovskikh and I.B. Andreevoi (Nauka, Moscow, 1982) 71–91.
2. Brekhovskikh, L.M. and Lysanov, Y.P., *Fundamentals of Ocean Acoustics* 2nd ed., (Springer, New York, 1991) pp. 140–145.
3. Hart, R.W. and Farrell, R.A., A variational principle for scattering from rough surfaces, *IEEE Trans. Ant. Prop.* **AP-25**, 708–710 (1977).
4. Rouseff, D., Effect of shallow water internal waves on ocean acoustic striation patterns, *Waves in Random Media.* **11**, 377–393 (2001).
5. Rouseff, D. and Spindel, R.C., Modeling the waveguide invariant as a distribution. In *Ocean Acoustic Interference Phenomena and Signal Processing*, edited by W.A. Kuperman and G.L. D'Spain (AIP Press, New York, 2002).
6. Abramowitz, M.A. and Stegun, I.A., *Handbook of Mathematical Functions* (U.S. Govt. Printing Office, Washington, DC, 1964) pp. 300–304.
7. Dozier, L.B. and Tappert, F.D., Statistics of normal-mode amplitudes in a random ocean. I. Theory, *J. Acoust. Soc. Am.* **63**, 353–365 (1978).
8. Dozier, L.B., A coupled-mode model for spatial coherence of bottom-interacting energy. In *Proc. Stochastic Modeling Workshop*, edited by C.W. Spofford and J.M. Hayes (ARL-University of Texas, Austin, 1983).
9. Creamer, D.B., Scintillating shallow-water waveguides, *J. Acoust. Soc. Am.* **99**, 2825–2838 (1996).
10. Kailath, T., *Linear Systems* (Prentice-Hall, Englewood Cliffs, New Jersey, 1980) pp. 160–171.

EFFECTS OF ENVIRONMENTAL VARIABILITY ON FOCUSED ACOUSTIC FIELDS

B. EDWARD MCDONALD, JOE LINGEVITCH AND MICHAEL COLLINS

US Naval Research Laboratory, Washington DC, USA

E-mail: mcdonald@sonar.nrl.navy.mil

A number of experiments have demonstrated the ability to produce tightly focused acoustic fields in the ocean using time reversal mirrors. The resulting focal regions may serve as acoustic probes or communications sites. We will review the effects of various types of environmental variability on the quality of the focus. Among these are bathymetric variation and water column inhomogeneities that may result from internal waves, bubble clouds, fish schools, or water mass intrusions. We will give high resolution simulation results from the RAM code to illustrate and quantify the impact of environmental variability upon the focal region.

1 Introduction

A number of laboratory [1] and ocean experiments [2–5] show that an acoustic field from a point source after propagating through a complex environment can be time reversed and back propagated, resulting in a tightly focused field near the source. The ability to produce a well focused field results from reciprocity, which in turn depends on (1) the propagation medium being nearly static between forward and back propagation; and (2) the medium being relatively free of attenuation. In ocean experiments, the back propagation is carried out with a vertical send-receive array (SRA) some distance from a probe source. If the ocean were azimuthally symmetric about the SRA, the resulting focused field would be a circular annulus centered on the SRA and passing through the probe source. The acoustic field at different azimuths around the annulus might be used to probe the environment and/or targets.

If the environment is not azimuthally symmetric about the SRA, then focal properties will be azimuth dependent. This paper will examine some theoretical and numerical modeling results relating the acoustic variability of the water column and bathymetry to changes in the focal annulus.

2 Theory

Modeling studies for SACLANTCEN's Focused Acoustic Fields 1999 (FAF99) experiment [6] performed off the west coast of Italy in July 1999 implied that the primary effect of azimuthally dependent bathymetry was a range shift of the focal spot. Some simple closed form results for the focal range shift due to environmental variation may be obtained by adiabatic mode theory and the WKB approximation applied to a pressure-release shallow water waveguide [7]. The frequency range used in FAF99 was 3–4 kHz.

N.G. Pace and F.B. Jensen (eds.), Impact of Littoral Environmental Variability on Acoustic Predictions and Sonar Performance, 377-383.
© 2002 *Kluwer Academic Publishers*.

The results presented here are for the center frequency 3500 Hz. We take the acoustic far field of a point source at radian frequency ω to be

$$p \to \sum_n A_n \psi_n(z) e^{i \int_0^r k_n dx} \tag{1}$$

where ψ_n is acoustic eigenmode number n, and mode amplitude A_n is only weakly dependent on r. If we assume for a basic state a range independent isovelocity pressure release waveguide (surface and bottom) with sound speed c and depth D we have

$$k_n^2 = \frac{\omega^2}{c^2} - \left(\frac{n\pi}{D}\right)^2. \tag{2}$$

Focal properties may be determined from the acoustic intensity field,

$$|p|^2 = \sum_n |A_n \psi_n|^2 + 2Re \sum_{n>m} A_n A_m^* \psi_n \psi_m^* e^{i \int_0^r (k_n - k_m) dx} \tag{3}$$

subject to environmental perturbations. From Eq. (2) a depth perturbation

$$D \to D + \delta D \tag{4}$$

implies

$$k_n \to k_n + \delta k_n,$$
$$\delta k_n = \frac{\delta D}{k_n D} \left(\frac{n\pi}{D}\right)^2. \tag{5}$$

The perturbed focal range is that which leaves the interference integral term in Eq. (3) unchanged for mode pairs (n,m):

$$\delta r = -\frac{\int_0^r (\delta k_n - \delta k_m) dx}{k_n - k_m}$$
$$\to -\left(\int_0^r \frac{\partial \delta k_n}{\partial n} dx\right) / \left(\frac{\partial k_n}{\partial n}\right). \tag{6}$$
$$\simeq \frac{2}{D} \int_0^r \delta D dx$$

The last expression is independent of mode number, so it gives the range shift of the intensity of the total acoustic field within the range of validity for linear perturbation theory applied to waveguides [8]. For the environments encountered in FAF99 the final expression in Eq. (6) was found to be within approximately 20% of range shifts determined in modeling studies [6] performed at high resolution with the RAM code [9].

The pressure release waveguide also leads to a simple expression for the change in focal depth as a result of perturbations, as long as mode coupling may be neglected. Since $\psi_n \propto \sin(n\pi z/D)$, then as long as mode amplitudes are preserved in relative proportion, the fractional change in focal depth z_f is equal to the fractional change in D:

$$\delta z_f = z_f \frac{\delta D}{D}. \tag{7}$$

Time reversal of a near-bottom probe source was carried out as part of FAF99 in order to investigate the feasibility of using selectively placed bottom reverberation to probe distant bathymetry [7]. During the modeling studies of [6] it was found that the focal depth behaved in approximate agreement with Eq. (7) for weak changes in bathymetry and for focal spots near the bottom. For mid-water column focii, however, it was noticed that strong bathymetric variation dispersed the mid-water focus to such a degree that its location was not well defined.

3 Simulation

3.1 Bathymetric Variation

Modeling studies are presented here to quantify the level of bathymeric variation at which the approximations made in Eqs. (1–6) begin to fail. Numerical parameters for the modeling are as follows. The vertical grid spacing used in RAM was 5 cm, while the range step was 2 m. We place a point source one meter above bottom of a 112 m deep ocean with a summer sound speed profile appropriate to FAF99 [6] and bottom sound speed 1582 m/s. RAM propagates the signal to an SRA (Fig. 1a) whose aperture is 15–93 m depth as in FAF99. The range to the SRA is taken to be 5 km for the simulations presented here. The complex acoustic field on the SRA is then conjugated and back propagated (Fig. 1b). One sees a strong focal spot near the bottom in Fig. 1b at the probe source range, as verified by the bottom ensonification plot (expressed as transmission loss re 1 m) included in the lower portion of Fig. 1b. Bottom ensonification is determined by resolving the complex pressure field into up and down going waves, then taking the intensity of the downgoing wave at the ocean bottom.

Table 1. Bathymetry for Back Propagation

Range(km)	Depth(m)
0.0	112
3.0	120
4.5	125
6.5	130
7.4	137
10.0	150

Back propagation from the SRA is next considered for an azimuth whose bathymetry is different (Table 1) from that of the foward propagation. Results are shown in Fig. 2. Examination of the bottom ensonification in Fig. 2 reveals that the focus at the bottom has been shifted in range by approximately 1.3 km. The focal shift predicted by Eq. (6) is 926 m, so the simple result of Eq. (6) is in error by approximately 32 percent for this case. In order to determine how the accuracy of Eq. (6) degrades with increasingly strong bathymetric variation, we performed a total of nine simulations similar to those of Figs. 1b and 2, with bathymetry determined by a scaling factor α:

$$D(r) = D_1(0) + \alpha \cdot [D_1(r) - D_1(0)] \qquad (8)$$

where $D_1(r)$ is taken from Table 1. The nine α values used in the simulations are (0, .25, .5, .75, 1, 1.125, 1.25, 1.5, 2). Figure 1b is the range independent result for $\alpha = 0$,

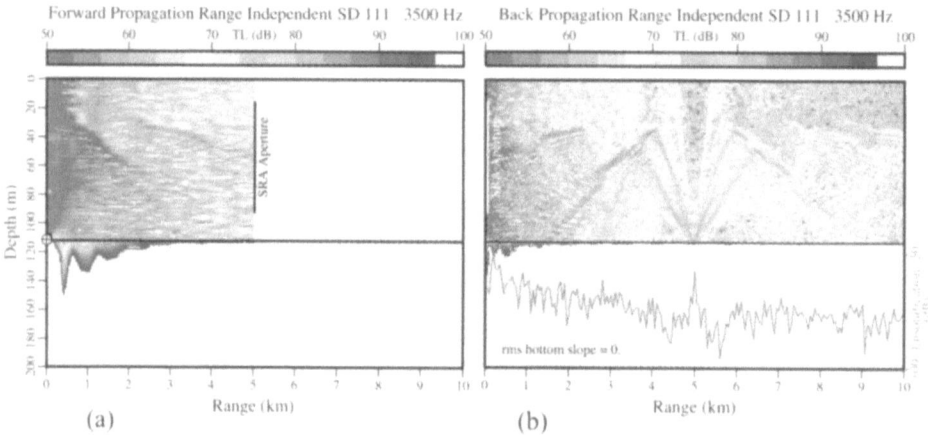

Figure 1. (a): A probe source 1 m from the botton ensonifies the SRA. (b): Back propagation from the SRA to the probe source. The lower curve is the bottom ensonification.

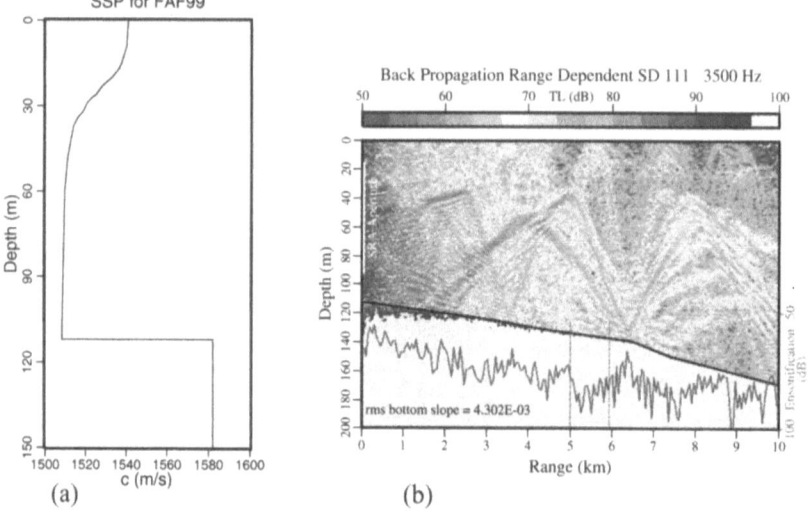

Figure 2. (a): Sound speed profile for Fig. 1. (b): Back propagation from the SRA along a different azimuth from that of the probe source. The lower curve is the bottom ensonification.

and Fig. 2 is equivalent to $\alpha = 1$. We performed seven other back propagations with α values up to 2. Results are shown in Fig. 3 as a function of the root-mean-square slope of the bathymetry between the SRA aperture and the 5 km range of the probe source. The solid curve is the prediction from Eq. (6), the points marked $\times$ are taken from the bottom ensonification curves as in Figs. 1b and 2, and the points marked $\circ$ are taken from the visible focal maximum. For bathymetry slope less than 3×10^{-3}, Eq. (6) is within about 15 percent of the simulation results for focal shift. At greater values of slope, the simulation focal shifts increase at approximately twice the rate predicted by Eq. (6).

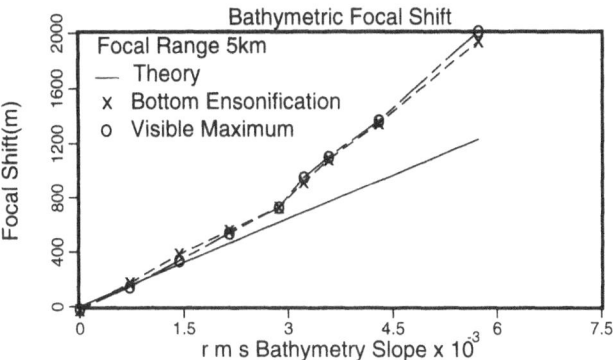

Figure 3. Shift in focal range caused by varying bathymetry.

3.2 Water Column Variation

The dominant variability of focal properties in FAF99 was attributed to internal waves which seemed to appear after wind events. One can get a feel for the internal wave environment by converting the sound speed profile of Fig. 2(a) to a temperature profile assuming salinty of 38 psu. The result is shown in Fig. 4(a) accompanied by the resulting Brunt-Vaisala frequency profile in Fig. 4(b). Although the profiles were taken at times not following wind events, the BV frequency in Fig. 4(b) is of high enough amplitude to indicate a receptive environment for internal waves. Wind events would sharpen the mixed layer and BV profile even further.

Internal wave modes obey the equation [10]

$$q''(z) + k^2 \left(\frac{N^2(z)}{\Omega^2} - 1 \right) q(z) = 0, \tag{9}$$

where the vertical mass flux is $w\rho(z) = q(z)\exp(i(kx - \Omega t))$ and the (radian) BV frequency is given by

$$N^2(z) = \frac{g}{\rho}\frac{d\rho}{dz} \tag{10}$$

with g the acceleration of gravity and z positive downward. In the limit of long horizontal internal waves compared to the width of the thermocline, the modes approach the sharp interface form with dispersion equation [10]

$$\Omega^2 = \frac{1}{2}gk\ln\frac{\rho_2}{\rho_1} \tag{11}$$

where ρ_2 and ρ_1 are the densities below and above the thermocline.

In a range independent waveguide of depth D (Fig. 1) with a sharp interface at depth z_0, internal wave modes have the form

$$q(z) = \begin{cases} \sinh(kz)/\sinh(kz_0), & z < z_0 \\ \sinh(k(D-z))/\sinh(k(D-z_0)), & z > z_0 \end{cases} \tag{12}$$

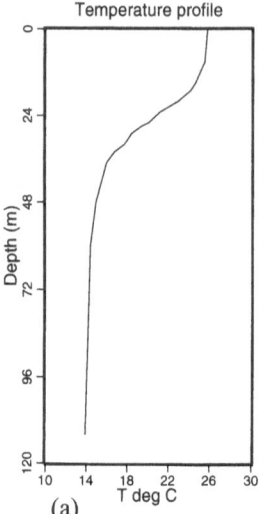
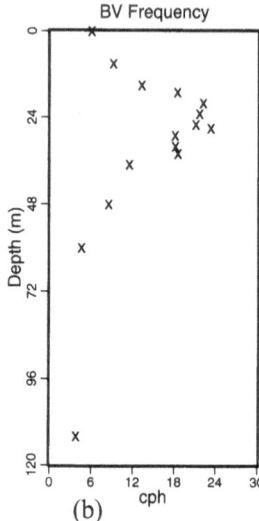

Figure 4. (a): Termperature profile derived from SSP of Fig. 2a. (b): Brunt-Vaisala frequency in cycles per hour.

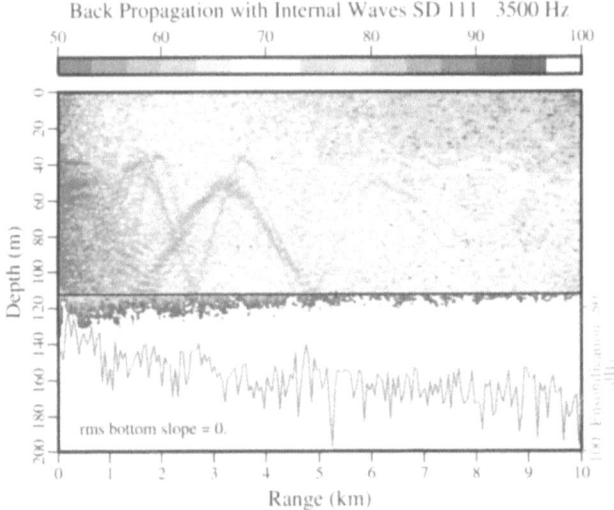

Figure 5. Dispersal of focus by internal waves added to the environment of Fig. 1b. Undisturbed thermocline depth z_0 is 24 m.

A set of three internal waves of the form (12) was introduced into the environment used in Fig. 1b. The horizontal wavelengths were chosen as $100\,\mathrm{m} \times (1, \sqrt{1.1}, \sqrt{1.2})$ with irrational proportions so as to exclude artificial periodicities in the wave train. Amplitudes and phases were randomized and the wave train normalized to root-mean-square thermocline displacement of 5 m. The result of back propagation through these internal waves is shown in Fig. 5. One sees a negative shift in the focal range and a loss of approximately 6db in the peak relative to Fig. 1b. Due to the stochastic nature of internal waves it is much more difficult to extract theoretical estimates for their effect than for the case of bathymetric variation. Not surprisingly though, a vertical variability of order 5 m in the sound speed structure between foward and back propagation seriously disrupts the focal structure of Fig. 1b.

4 Summary

We have shown that simple estimates from waveguide invariant theory applied to bathymetric perturbations of a focused acoustic field are reasonably accurate for mild bathymetric variation. A series of RAM simulations with increasingly steep bathymetry showed (Fig. 3) a clear departure from theory at a root-mean-square bathymetric slope of 3×10^{-3} in a 112 m deep waveguide and focal range of 5 km. The presence of internal waves on a sharp thermocline was found in our simulation to have a strong disruptive effect on focal properties. This is in qualitative agreement with degradations in focal properties following wind events in FAF99. We found in Fig. 5 that a 5 m rms fluctuation in the thermocline had a more disruptive effect on the focus than did a 15 m change in bathymetry (Fig. 2b).

Acknowledgements

Work supported by NRL and the US Office of Naval Research.

References

1. M. Fink, Time-reversed acoustics, *Physics Today* **50**, 34–40 (1997).
2. H.C. Song, W.A. Kuperman and W.S. Hodgkiss, A time-reversal mirror with variable range focusing, *J. Acoust. Soc. Am.* **103**, 3234–3240 (1998).
3. W.S. Hodgkiss, H.C. Song, W.A. Kuperman and T. Akal, A long-range and variable focus phase-conjugation experiment in shallow water, *J. Acoust. Soc. Am.* **105**, 1597 (1999)
4. S. Kim, G.F. Edelmann, W.S. Hodgkiss, W.A. Kuperman, H.C. Song and T. Akal, Spatial resolution of time reversal arrays in shallow water, *J. Acoust. Soc. Am.* **108**, 2606 (2000).
5. C. Holland and B.E. McDonald, Shallow water reverberation from a time reversed mirror. SACLANTCEN Report SR-326 (Dec. 2000).
6. B.E. McDonald and C. Holland, Shallow water reverberation from a time reversed mirror: Data-model comparison, *J. Acoust. Soc. Am.* **109**, 2495 (2001).
7. B.E. McDonald and C. Holland, A Method for rapid bathymetric assessment using reverberation from a time reversed mirror. In *Proc. 17th Intl. Congr. Acoust.* (Univ. of Rome, 2–7 Sep. 2001), Vol. 2, 3_3.pdf, pp. 10–11.
8. G.A. Grachev, Theory of acoustic field invariants in layered waveguides, *Acoust. Phys.* **39**, 33–35 (1993).
9. M.D. Collins, Generalization of the split-step Padé solution, *J. Acoust. Soc. Am.* **96**, 382–385 (1994).
10. J. Lighthill, *Waves in Fluids* (Cambridge Univ. Press, Cambridge, 1978) Sec. 4.3.

EFFECTS OF SOUND SPEED FLUCTUATIONS DUE TO INTERNAL WAVES IN SHALLOW WATER ON HORIZONTAL WAVENUMBER ESTIMATION

KYLE M. BECKER

Acoustics Group, The Pennsylvania State University, Applied Research Laboratory
P.O. Box 30, State College PA 16803, USA
E-mail: kmb166@psu.edu

GEORGE V. FRISK

Dept. of Applied Ocean Physics and Engineering, Woods Hole Oceanographic Institution
MS #11, Woods Hole MA 02543, USA
E-mail: gfrisk@whoi.edu

There is considerable current interest in the influence of water-column variability on acoustic propagation and its effects on geoacoustic inversion. In general, the effect of sound speed fluctuations due to internal waves in the water column is to promote the coupling of energy between propagating acoustic modes. The effects of mode coupling include fluctuations in individual modal amplitudes and arrival times along with time spreading of the original pulses. In contrast to the broadband case, little research has been conducted on the effects of internal waves on cw modal-based inversion methods. These techniques require estimates of the propagating modal eigenvalues for a cw point source field as input data to the inversion algorithm. In much of the literature, wavenumber estimation is performed with the assumption that pressure is given by an adiabatic mode sum. Changes in modal content as a function of range are then attributed to local changes in the waveguide boundaries, specifically, the bottom. For a shallow-water waveguide including internal waves, the adiabatic assumption is violated and estimates of local wavenumber content is affected. This paper addresses the nature of the these affects on the wavenumber estimation problem. In particular, numerical studies of internal wave effects are conducted with respect to identification and bias of individual modes along with the ability to resolve closely spaced eigenvalues. Preliminary results for a weak internal wave field show that mode coupling leads to an enhancement of the wavenumber spectral estimates due to the energizing of weak modes that were previously not excited.

1 Introduction

Sound propagation in range-independent shallow-water waveguides can be concisely represented by the exact Hankel transform relationship between complex pressure, measured as a function of range, and the depth-dependent Green's function, expressed in the horizontal wavenumber domain [1]. Further, from the modal interpretation of this relationship, a linearized inversion method can be developed relating modal eigenvalues, given by discrete horizontal wavenumber values, to sediment geoacoustic properties [2]. A key aspect of this approach is the accurate estimation of the individual modal eigenvalues

385

N.G. Pace and F.B. Jensen (eds.), Impact of Littoral Environmental Variability on Acoustic Predictions and Sonar Performance, 385-392.

contributing to the propagating field. This problem becomes more complicated when spatial dependence enters into the problem, through variation in sediment properties with range [3], or variability in the water column due to sound speed fluctuations or changes in bathymetry.

In the cases of changes in sediment properties or bathymetry that occur slowly over range, horizontal wavenumber estimates can be obtained by assuming regions of local range-independence in the waveguide and applying techniques analogous to short-time Fourier transform methods [4]. However, although the effects of sound speed fluctuations due to internal waves on modal propagation, as well as their corresponding effects on broadband geoacoustic inversion, have been treated in the literature recently (see, for example, [5] and [6]), their effects on horizontal wavenumber estimation have received less attention. Two general assumptions are usually made for wavenumber analysis based on the Hankel transform. The first assumption is the adiabatic mode approximation, which states that no energy is transferred amongst the different modes and that the number of propagating modes remains constant. The second assumption is that the effects of sound speed fluctuations in the water column are minimized due to the averaging effect that occurs when transforming data over sufficiently large apertures.

In the case of internal waves, it is well known that the adiabatic assumption does not apply and that a complex competition between mode coupling and mode stripping occurs [7]. In this paper, a study is presented addressing the wavenumber estimation problem for a range-dependent acoustic field where mode coupling is known to occur. Synthetic complex pressure fields are generated using a range-dependent parabolic equation (PE) model [8] for both a range-independent shallow-water waveguide, and a waveguide with a deterministic internal wave (IW) field that introduces range-dependence into the waveguide through perturbations to a background sound speed profile. Applying short-window methods based on the Hankel transform to the resulting fields, modal content is plotted as a function of range and compared for the different models. In addition, using a fully coupled mode code [9], reference values of modal amplitudes and horizontal wavenumbers are determined as a function of range for both the background and internal wave models. Comparisons are made between estimated wavenumbers and reference values for both acoustic fields. The influence of the internal wave field on wavenumber estimates is examined in terms of the mean and variance of the estimates compared to their reference values. A comparison of modal amplitudes between the different models is also made yielding information about the detectability of particular modes.

2 Methods

For a cw point source, and assuming horizontal stratification, the equations governing sound propagation in a shallow-water waveguide can be reduced to a Hankel transform relationship between the complex-pressure field $p(r; z, z_o)$ and the depth-dependent Green's function $g(k_r; z, z_o)$,

$$p(r; z, z_o) = \int_0^\infty g(k_r; z, z_o) J_0(k_r r) k_r dk_r,$$
$$g(k_r; z, z_o) = \int_0^\infty p(r; z, z_o) J_0(k_r r) r dr. \tag{1}$$

In the above, pressure is measured as a function of range r, and the Green's function is a function of horizontal wavenumber k_r and satisfies the inhomogeneous depth-dependent wave equation, along with impedance boundary conditions at the surface ($z = 0$) and the bottom ($z = h$) [10],

$$\left\{ \frac{d^2}{dz^2} + \rho(z)\frac{d}{dz}\left[\frac{1}{\rho(z)}\frac{d}{dz} \right] + k^2(z) - k_r^2 \right\} g(k_r; z, z_o) = -2\delta(z - z_o),$$

$$g^T(k_r; 0) + \frac{i\mathcal{Z}_T(k_r)}{\omega\rho(0)}\frac{\partial g^T(k_r; 0)}{\partial z} = 0, \qquad (2)$$

$$g^B(k_r; h) + \frac{i\mathcal{Z}_B(k_r)}{\omega\rho(h)}\frac{\partial g^B(k_r; h)}{\partial z} = 0.$$

In Eq. (2), the source and receiver depths, z_o and z, are treated as parameters in the problem, $\rho(z)$ is density, and $k(z) = \omega/c(z)$ is the total wavenumber for the angular frequency ω. The solution to the depth-dependent Eq. (2) is obtained by incorporating into the total solution, the independent solutions $g_T(k_r, z)$ and $g_B(k_r, z)$ which satisfy impedance boundary conditions at the top and bottom, represented by $\mathcal{Z}_T(k_r)$ and $\mathcal{Z}_B(k_r)$, respectively. The general form of the depth-dependent Green's function satisfying both boundary conditions can then be written,

$$g(k_r; z, z_o) = -2\frac{g_T(k_r, z_<)g_B(k_r; z_>)}{W(z_o)}, \qquad 0 \le z \le h, \qquad (3)$$

where $z_< = min(z, z_o)$, $z_> = max(z, z_o)$, and $W(z_o)$ is the Wronskian evaluated at the source location and given by,

$$W(z_o) = g_T(z_o)\frac{\partial g_B}{\partial z}\Big|_{z_o} - \frac{\partial g_T}{\partial z}\Big|_{z_o} g_B(z_o). \qquad (4)$$

For a horizontally stratified ocean, and assuming the top surface satisfies a pressure-release boundary condition, $g(k_r; z, z_o)$ becomes a function of the impedance boundary condition at the water-bottom interface alone. Through the impedance relationship, the Green's function is a complete characterization of the waveguide as needed for the solution of Eq. (1). For shallow-water waveguides, the Green's function is characterized by sharp peaks, or resonances, that occur when the Wronskian in Eq. (3) goes to zero. These peaks occur at distinct values of horizontal wavenumber that correspond to the modal eigenvalues k_n in a modal expansion of the field. The mode sum can be obtained by inserting Eq. (3) into Eq. (1) and applying complex contour integration methods, to obtain

$$p(r, z) = \frac{i\pi}{\rho(z_o)} \sum_{n=1}^{n_{max}} \phi_n(z_o)\phi_n(z)H_0^{(1)}(k_n r) + I(r), \qquad (5)$$

where ϕ_n are mode functions evaluated at the source and receiver depths, n_{max} is the maximum number of propagating modes, and $I(r)$ represents a continuum contribution that can often be ignored for ranges greater than a few water depths from the source.

For analysis purposes, by applying the large argument asymptotic approximation to the Bessel function, $J_0(k_r r)$ in Eq. (1), the Hankel transform operation can be replaced with a Fourier Transform [10],

$$g(k_r; z, z_o) \sim \frac{e^{i\pi/4}}{\sqrt{2\pi k_r}} \int_{-\infty}^{\infty} p(r; z, z_o)\sqrt{r}e^{-ik_r r}dr, k_r r \gg 1, k_r > 0. \qquad (6)$$

The asymptotic form allows the Green's function to be obtained through a Fast Fourier Transform (FFT) of the product $p(r; z, z_o)\sqrt{r}$. This form is particularly useful in that various spectral estimation tools can be applied, including high-resolution methods, to determine wavenumber content. Using these methods, horizontal wavenumbers corresponding to modal eigenvalues are determined from the peak locations in the Green's function estimate.

The relationship between the depth-dependent Green's function and complex pressure field described above is strictly valid for range-independent environments. For the range-dependent case, a convenient representation of the pressure field is given by the adiabatic mode sum expressed as,

$$p(r, z) \sim \frac{\sqrt{2\pi}e^{i\pi/4}}{\rho(0, z_o)} \sum_{n=1}^{n_{max}} \phi_n(0, z_o)\phi_n(r, z) \frac{e^{i\int_0^r k_n(r')dr'}}{\sqrt{\int_0^r k_n(r')dr'}}. \qquad (7)$$

In this representation, mode functions and their corresponding eigenvalues adapt themselves to the local waveguide properties as they change in range. By performing the integral in (6) over a finite range interval given by $-L/2 \leq r \leq L/2$, where L is the aperture length, a local estimate of the Green's function can be obtained, where it is assumed that the waveguide environment is approximately range-independent over L. For a truly adiabatic environment, moving the aperture along in range using short range steps gives a picture of the modal content of the waveguide as it evolves with range [4]. The adiabatic representation provides an intuitive way to understand how modal content evolves with range, but is not a necessary requirement for extracting range-dependent modal information using the Hankel transform. For the synthetic pressure fields generated for this study, the introduction of the internal wave field to the otherwise range-independent waveguide causes energy in the propagating modes to be transferred amongst each other violating the adiabatic assumption. However, estimates of the local depth-dependent Green's function can still be obtained by integrating the field over a finite aperture as described above.

3   Numerical examples and wavenumber estimation

The waveguide environment for the numerical studies was one of several designed as a test bed for benchmarking range-dependent acoustic propagation models [11]. The waveguide for this study was a fluid model with range-independent sediment properties and a constant water/sediment interface depth of 200 m. Density in the water column was 1 g/cm^3 with no attenuation. The bottom had a compressional wave speed of 1700 m/s, a density of 1.5 g/cm^3, and an attenuation of 0.1 dB/λ. The background sound speed profile was generally downward refracting with a slight duct occurring above 26 m depth and was given by

$$\begin{aligned} c(z) &= \quad 1515 + 0.016z, \quad z < 26\ m \\ c(z) &= c_0[1 + a(e^{-b} + b - 1)], \quad z \geq 26\ m, \end{aligned} \qquad (8)$$

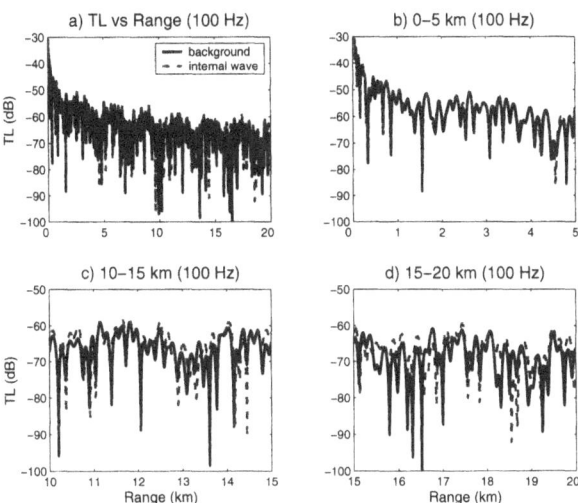

Figure 1. 100 Hz TL plots for internal wave study. 5 km sub-plots show departure of acoustic field for internal waves from range-independent case.

where z is the depth in meters, c_o = 1490 m/s, a = 0.25, and $b = (z - 200)/500$ [12]. To the background sound speed was added a perturbation as a function of depth and range to represent the internal wave field. The sound speed perturbation, δc, was given by,

$$\delta c(z,r) = 4(z/B)e^{-z/B} \sum_{i=1}^{5} cos(K_i r), \qquad (9)$$

where $K_i = 2\pi[2000 - 300(i - 1)]^{-1}$ m^{-1} and B = 25 m [12]. Using this model, the maximum sound speed perturbation is about 7.5 m/s. Using the above models, PECan [8], was used to generate the acoustic field at 100 Hz for both the background and perturbed sound speed environments from 0 to 20 km on a 5 meter range grid. Transmission loss data are shown in Fig. 1 for the full aperture and for several 5 km sub-apertures for a source depth of 45 m and receiver depth of 60 m. The modal interference patterns in the figure result from the constructive and destructive interference of different propagating modes. The differences in interference patterns for the two fields are most evident at long ranges and indicate the influence of the internal waves on the modal interactions. Using the complex pressure fields, wavenumber estimates were obtained using a high resolution sliding widow estimator based on an autoregressive spectral estimator [13]. The estimator is based on a parametric representation of the signal and requires the model order as an input parameter. For the 20 km apertures of the synthetic data fields, wavenumber estimates were made for 3000 m sliding window sub-apertures with a step size of 200 m and a model order of 1/3 the number of data points in each sub-aperture. Additionally, reference values of horizontal wavenumbers for the range-independent waveguide environment with the background sound speed profile were determined using a normal mode code [9]. The resulting wavenumber estimates with the reference results overlayed are shown in Figs. 2 and 3. In the figures, modes identified using the Hankel transform method correspond with the most energetic modes for the acoustic fields. Where there is a match between an estimated wavenumber and a reference value, the reference value

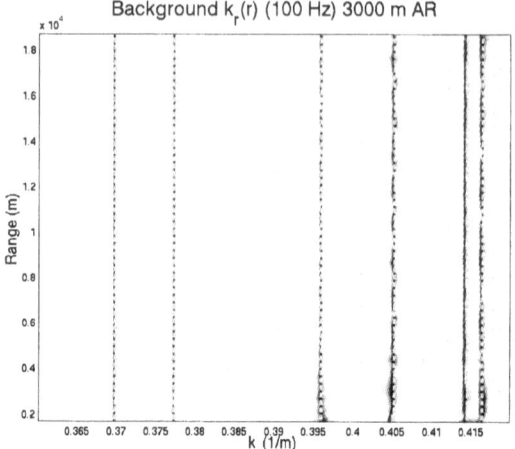

Figure 2. Sliding window wavenumber estimates compared to reference results (dashed) plotted for unperturbed case. The dark lines indicate the locations of the peaks of the Green's function.

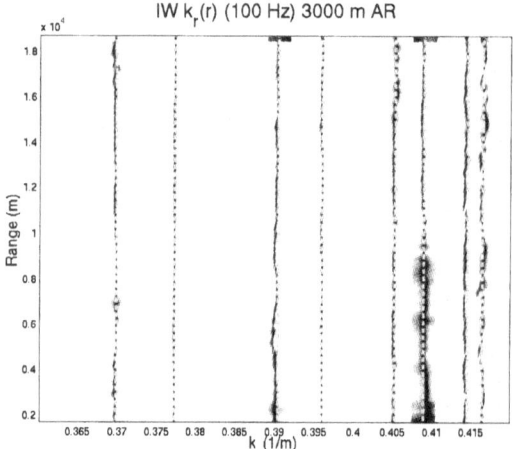

Figure 3. Sliding window wavenumber estimates compared to reference results (dashed) plotted for IW case. The dark lines indicate the locations of the peaks of the Green's function.

is plotted as a dashed line. The plots show that for the unperturbed case, 6 modes are estimated that can be correlated with the reference results. For the IW case, 8 modes are identified that correlate with the reference results. In both cases, the wavenumber estimates are stable with range, which is especially evident for the higher-order modes. This is an indication that the waveguide boundary conditions are range-independent. For the low-order modes, there are small undulations about the background wavenumbers that occur with range. However, there is no apparent wavenumber shift observed for the modes obtained from the two fields. Together, the observations suggest that for the given source/receiver geometry, certain modes are not excited or contain relatively little energy. In order to verify this claim and gain a better understanding of the wavenumber estimates shown, it is necessary to examine both modal amplitudes and horizontal wavenumber content as a function of range.

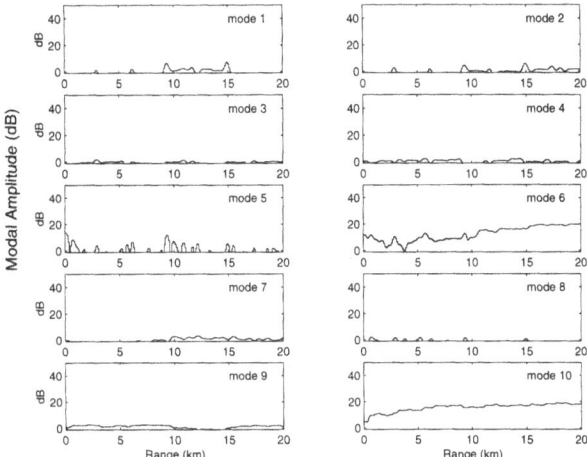

Figure 4. Mode amplitude difference (internal wave - background) vs. range.

To examine mode amplitudes and wavenumbers, the COUPLE [9] range-dependent normal mode code was used. Horizontal wavenumbers and modal amplitudes were output at a receiver depth of 60 m for each range in the problem. The horizontal wavenumbers corresponding to the propagating modes for the background problem are shown as the straight lines in Figs. 2 and 3. For the IW case, discrete values of horizontal wavenumbers were selected from the peak positions of the Green's function at each range. Means and variances over range were calculated for the individual eigenvalue estimates and compared to the background, or reference, eigenvalues. The variances of the individual estimates were small, consistent with the variance expected for the high-resolution estimator, as if it were applied to a signal with additive noise and a high signal to noise ratio (SNR greater than 45 dB) [13]. For the 8 modes identified in the IW case, the biases between the estimates and the background values were less than 1 standard deviation for all modes. This suggests that the mean values for the estimated horizontal wavenumbers could be used as input data for a range-independent geoacoustic inversion algorithm.

Figure 4 shows the difference in individual modal amplitudes between the internal wave field and background field as a function of range for the first 10 propagating modes. The figure shows the influence of the internal wave field on the individual modes. While it is seen that all of the modes are affected, modes 6 and 10 each have a large amount of additional energy in them due to the internal wave field. This accounts for the enhanced number of modes observed in the wavenumber estimates in figure 3, where the coupled mode code verified that modes not observed were only weakly excited in either case. Further, the smaller perturbations seen in the mode amplitudes can account for some of the scintillations observed in the wavenumber estimates.

4 Conclusions

Internal waves greatly influence the propagation of sound in shallow water by redistributing energy amongst different propagating modes. This in turn has an effect on the estimation of local modal content for use in linear inversion algorithms. In the case of the weak internal wave field used as a model for this study, the following conclusions can be drawn. The effect of the internal wave field on local horizontal wavenumber estimates is minimized by the averaging effects controlled by the size of the local aperture used in the processing. A more complete study needs to be done to access an optimal aperture size for a given environment. Finally, the presence of internal waves serves to excite modes that otherwise would not be excited for a particular source receiver geometry and leads to an enhancement of the wavenumber estimates. This enhancement yields additional data which can be used in an inversion algorithm to determine seabed properties.

Acknowledgements

Richard B. Evans was extremely generous with his time in helping to get the coupled mode code up and running. PECan data was generously provided by Gordon Ebbeson of the Defense Research Establishment of the Atlantic (DREA), Canada. This work was partially supported by an ONR Special Postdoctoral Fellowship Award in Ocean Acoustics [Contract No. N00014-02-1-0334].

References

1. G.V. Frisk and J.F. Lynch, Shallow water waveguide characterization using the Hankel transform, *J. Acoust. Soc. Am.* **76**(1), 205–216 (1984).
2. S.D. Rajan, J.F. Lynch and G.V. Frisk, Perturbative inversion methods for obtaining bottom geoacoustic parameters in shallow water, *J. Acoust. Soc. Am.* **82**(3), 998–1017 (1987).
3. G.V. Frisk, J.F. Lynch and S.D. Rajan, Determination of compressional wave speed profiles using modal inverse techniques in a range-dependent environment in Nantucket Sound, *J. Acoust. Soc. Am.* **86**(5), 1928–1938 (1989).
4. K. Ohta and G.V. Frisk, Modal evolution and inversion for seabed geoacoustic properties in weakly range-dependent shallow-water waveguides, *IEEE J. Oceanic Eng.* **22**, 501–521 (1997).
5. J.C. Preisig and T.F. Duda, Coupled acoustic mode propagation through continental-shelf internal solitary waves, *IEEE J. Oceanic Eng.* **22**, (1997).
6. M. Siderius, P.L. Nielsen, J. Sellschopp, M. Snellen and D. Simons, Experimental study of geo-acoustic inversion uncertainty due to ocean sound-speed fluctuations, *J. Acoust. Soc. Am.* **110**(2), 769–781 (2001).
7. D. Tielbürger, S. Finette and S. Wolf, Acoustic propagation through an internal wave field in a shallow water waveguide, *J. Acoust. Soc. Am.* **101**(2), 789–808 (1996).
8. G.H. Brooke, D.J. Thomson and G.R. Ebbeson, PECAN: A Canadian parabolic equation model for underwater sound propagation, *J. Comp. Acoust.* **9**(1), 69–100 (2001).
9. R.B. Evans, COUPLE, 1997 Version (Nov. 20, 1997). ftp://oalib.saic.com/pub/oalib/couple/
10. G.V. Frisk, *Ocean and Seabed Acoustics: A Theory of Wave Propagation* (Prentice Hall, Englewood Cliffs, New Jersey, 1994).
11. Various Authors, Acoustical oceanography and underwater acoustics: Benchmarking range-dependent reference models, chaired by K.B. Smith and A.I. Tolstoy, *J. Acoust. Soc. Am.* **109** (5 pt. 2), 2332–2335 (2001).
12. K.B. Smith, Benchmarking shallow water range-dependent acoustic propagation modeling, Test Case III: Internal waves (Dec. 20, 2000).
    http://web.nps.navy.mil/~kbsmith/Chicago ASA/iws.html
13. K.M. Becker, Geoacoustic inversion in laterally varying shallow-water environments using high-resolution wavenumber estimation, Ph.D. dissertation, MIT/WHOI, WHOI-02-03 (2002).

RELATIVE INFLUENCES OF VARIOUS ENVIRONMENTAL FACTORS ON 50–1000 HZ SOUND PROPAGATION IN SHELF AND SLOPE AREAS

TIMOTHY F. DUDA

Applied Ocean Physics and Engineering Department- MS 11,
Woods Hole Oceanographic Institution, Woods Hole MA 02543,USA
E-mail: tduda@whoi.edu

Within a given continental shelf or slope area acoustic propagation effects from many aspects of the environment sum to give the resultant total environmental effect on sound propagation. Signal parameters influenced by the environment include signal strength (transmission loss), vertical coherence scale, horizontal coherence scale, temporal coherence scale, and further signal details not measured by coherence scales, all as functions of frequency. Environmental (oceanographic) factors include but are not limited to bathymetric slope, episodic high-amplitude internal waves, continuous low-amplitude internal waves, seafloor attenuation, fronts, currents, source depth, and receiver depth. The effects of each of these environmental factors may be investigated individually with simulations or with specialized field experiments, but the results often can't be generalized because of first-order differences between regions. In many situations the various factors do not act independently, but are coupled, and not only do the individual parameters vary but their interactions may also change. An example from our previous work is the altered influence of mode-stripping (bottom interaction) in shallow water when high-amplitude internal waves are either present or absent. This effect is sensitive to source depth, bottom parameters, and wave parameters. This example and other simulated examples of differing dominant parameters at selected locations will be presented and compared with acoustic experiment data.

1 Introduction

Over the past decades many studies have examined acoustic field variability caused by fluctuating oceanographic conditions. Quite often these studies have focused on a single identifiable oceanographic process and have investigated its effects on sound propagation. This approach can effectively decipher and illuminate the physics of the interaction, but is an incomplete treatment of the complex oceanic system, where many competing geophysical processes can play a role simultaneously, and the where the total effect on sound may not be a simple summation of individual effects. Example studies and topics in the shallow water venue are McDaniel and McCammon [1], mode coupling and seabed properties; Shmelerv, Migulin and Petnikov [2], horizontal sound refraction by internal waves; Zhou, Zhang and Rogers [3], scattering and loss induced by groups of sinusoidal internal waves; Lynch *et al.* [4], perturbations from internal waves and tides near a polar front; Creamer [5], coupled mode and sub-bottom loss interaction; Preisig and Duda [6],

N.G. Pace and F.B. Jensen (eds.), Impact of Littoral Environmental Variability on Acoustic Predictions and Sonar Performance, 393-400.

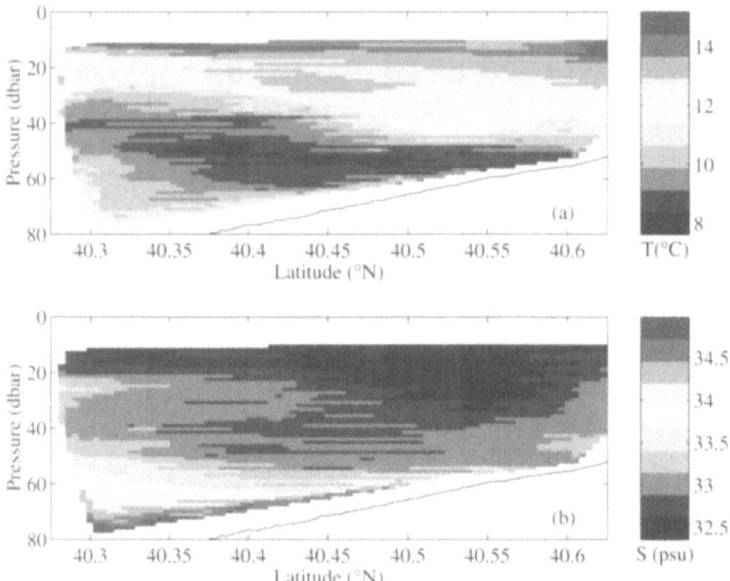

Figure 1. Temperature and salinity in one cross-shore south-to-north transect taken in the ONR Coastal Mixing and Optics Experiment in 1997. The water is warm at the surface, cooler below, and warm near the bottom. The salinity increases with depth in the south. The sound speed near the bottom in the warm salty slope water beneath the front is approximately 12 m/s greater than its minimum value in the water above. The figure is reproduced from published work [12].

individual sech^2-profile high-amplitude internal solitary waves; Tielbuerger, Finette and Wolf [7], propagation through internal waves in shallow water; Duda and Preisig [8], moving groups of sech^2 solitary waves; and Headrick et al. [9, 10], analysis and modeling of scattering by solitary waves in the SWARM experiment.

These and other rigorous studies of isolated processes, or at most a few processes, have advanced our understanding of shallow-water acoustics. On the other hand, making predictions of propagation and fluctuation behavior based on these studies is not simple. Many of the physical processes described by these studies arise only under specific conditions. Furthermore, the occurrence of one or more of these processes may eliminate the possibility of another taking effect. This possibility will be illustrated here with the example of the shelf/slope water front, a ubiquitous feature south of New England, altering the signal-level changing effect of mode coupling by internal solitary waves (ISWs). This example was motivated by observations of the front, appearing as salty and warm near-bottom layers in transects (Fig. 1), and by computational studies of mode coupling by internal solitary waves. The coupling causes variable signal gain (or loss) because of variable energy exchange between modes as a function of wave location, and subsequently altered bottom interaction [6–10].

Some parameters and effects that must be considered in shallow water are listed here: Bathymetric slope, water mass fronts; internal solitary waves and packets; a quasi-stationary internal wave field; internal tides; surface tides; eddy structures; source depth; receiver depth; seabed layering; lateral seabed structure; and seabed material. The "rela-

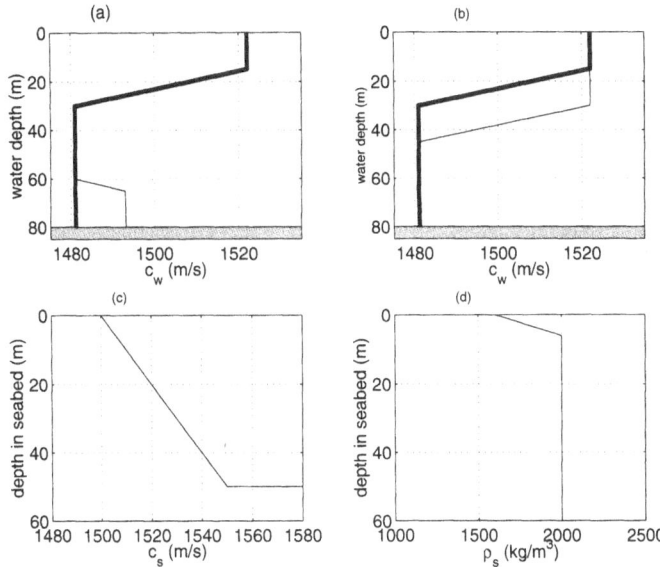

Figure 2. Sound speed and density parameters for the simulation are shown. (a) The background sound speed is shown with the thick line, and the perturbation of a deep warm layer is shown with the thin line. (b) The background sound speed is again shown with the thick line. The thin line shows a perturbed profile from a 15-m amplitude midwater pycnocline displacement, intended to mimic the effect of a 15-m amplitude internal wave of depression. (c) The sound speed in the seabed is shown. (d) The density structure in the seabed is shown.

tive influences" term in the title implies an attempt to order these in importance, but this is beyond the scope of this paper and may vary from place to place, which is entirely the issue.

Previous work [8] has pointed out some interacting parameters. For example, a packet of internal solitary waves was found to provide average signal gain for a shallow source but average signal attenuation for a deep source in the geometry that was considered. In addition, the position of internal waves relative to source and receiver was predicted to be of central importance in their ability to perturb acoustics [8, 10] and offers an explanation of effects observed in the SWARM experiment [10, 11]. Another paper in this volume lists additional interactions encountered in three WHOI experiments [11].

2   Computational example

Interacting effects are illustrated with one computational example. The situation of a deep layer of warm salty water intruding onto the shelf, thereby perturbing signal loss effectiveness of an internal solitary wave packet, is studied using four computational runs. The runs include or exclude each of two water column sound speed structures: An internal solitary wave packet between 13 and 16 km from the sound source, and a warm layer below 60 m depth. Case 1 has no packet and no layer; Case 2 has the packet and no layer; Case 3 has no packet and the layer; Case 4 has both the packet and the layer. Cases 2 and 4 were run multiple times with variable packet location.

The background water sound speed (c_w) structure is shown in Figs. 2a and 2b. An

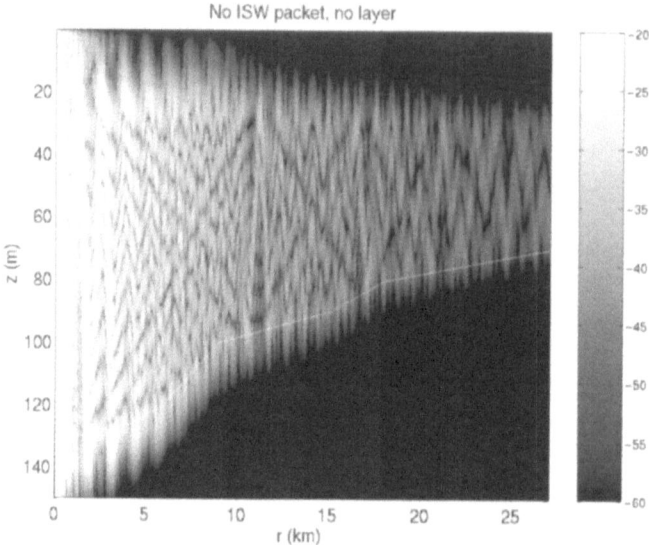

Figure 3. Transmission loss with cylindrical spreading removed (despread level) is plotted on a dB scale versus depth and range for the pycnocline-only situation (Case 1). Despread level is shown because it would not change systematically with range in the absence of sub-bottom loss, so any decreasing trend indicates bottom interaction. The RAM code was used to simulate a 400 Hz CW signal from a source at 20 m depth in 130-m deep water on the edge of the shelf. This is the "control" case of no ISW packet and no deep layer.

upper layer of 1522 m/s transitions between 20 and 30 m depth to a deep layer of 1481 m/s. Below 65 m the sound speed increases linearly at a rate of 1 m/s per 65 m. The warm layer perturbation is shown in Fig. 2a. It is a linear 12 m/s increase of c_w between 60 and 65 m depth, creating a sound channel. The effect of internal waves is displacement of the pycnocline (Fig. 2b). The seabed structures are equal in all situations and are uniform over range in a coordinate frame fixed to the local water/seabed boundary. Figure 2c shows the sound speed in the seabed c_s. It increases from 1500 m/s at the interface for 50 m at the rate of 1 s^{-1}, and increases below that to 2400 m/s at the next computational grid depth. This is intended to crudely imitate a mud and sand layer overlying bedrock. Figure 2d shows the density in the seabed, which increases linearly in the upper 6 m and is constant beneath.

The RAM code [13] was used to solve the parabolic acoustic wave equation. The frequency was 400 Hz. The source was at 20 m depth in 130 m of water. The vertical depth increment was 0.25 meter and the range step was 1 meter. The input and output routines of the original code (RAM version 1.1) were modified: The bottom parameters were fixed with respect to range and were referenced into coordinates starting at the water/seabed boundary, and the complex acoustic field was output. Attenuation in the seafloor was 0.1 dB per wavelength. Further parameters were $c_0 = 1488$ m/s and np = 2. ISW packet dimensions were as in previous work [8], with waves of 15, 12 and 10-m amplitude. Water depth, source depth, packet position and other parameters were chosen arbitrarily.

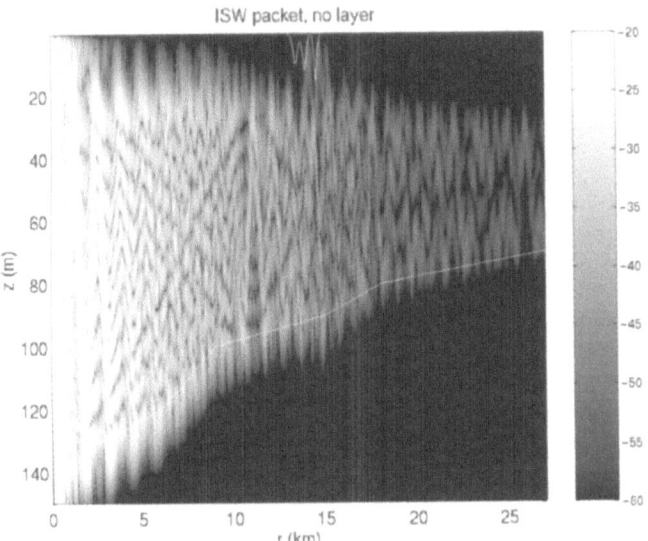

Figure 4. Despread level for a second PE simulation (Case 2) comparable to that depicted in the previous figure. The difference is the addition of an ISW packet of 15, 12 and 10-m amplitude waves (shown schematically at the top) that couple energy into lossy high-order acoustic modes.

3   Discussion

The acoustic pressure magnitudes for each of the four situations are shown in Figs. 3–7. The "despread level" is shown in order to eliminate the effect of cylindrical spreading and to highlight energy loss through sub-bottom interaction. Despread level is computed by multiplying the amplitude by the square root of the range before conversion to decibel.

The control case of no packet and no layer (Case 1, Fig. 3) shows a gradual decline of energy with range and a gradual simplification of vertical acoustic field structure. This occurs because many modes are excited by the 20-m source and mode behavior is adiabatic. The higher order modes are subject to greater loss per km traveled than the lower order modes, creating this field structure that has only the bottom-hugging lower-order modes remaining after 15 km. The addition of an ISW packet at 13.5 km from the source causes modes to couple (Case 2, Fig. 4). Only a few low modes are energized as the sound encounters the packet, and the randomizing effect on mode content causes the modal bandwidth to increase, sending energy upward in the water column. This effect is mentioned elsewhere in this volume [11], and is related to the "near receiver dominance" effect [9–11]. The newly-energized higher modes quickly dissipate downrange, leaving a depleted signal level relative to the control case (also mentioned in [11]). The precise blend of modes that are energized after the coupling is a sensitive function of packet location [8].

The inclusion of the warm layer below 60 m depth, a simple approach to model the effect of the front shown in Fig. 1, yields essentially uniform despread level over range (averaged over mode interference patterns) at ranges greater than 10 km (Case 3, Fig. 5). The few modes persisting to that range have little, if any, interaction with the bottom, in contrast with the attenuating modes of Fig. 3. Including both the 13.5-km ISW packet

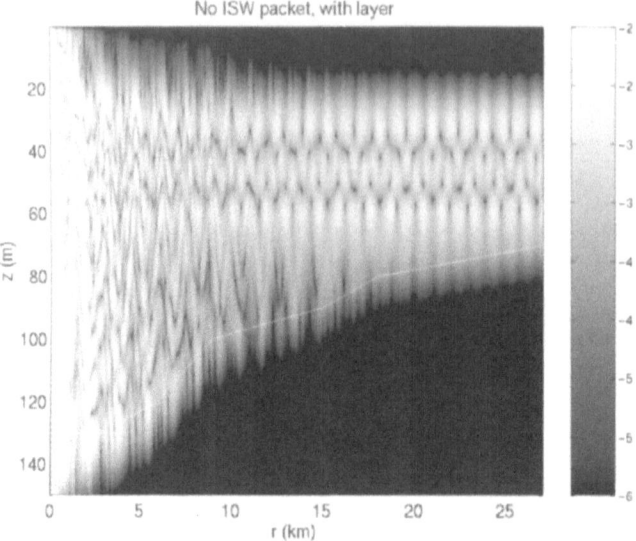

Figure 5. Despread level is shown for the Case 3 PE simulation. The water column has no ISW packet. The warm near-bottom layer is included. Levels are increased over Cases 1 and 2.

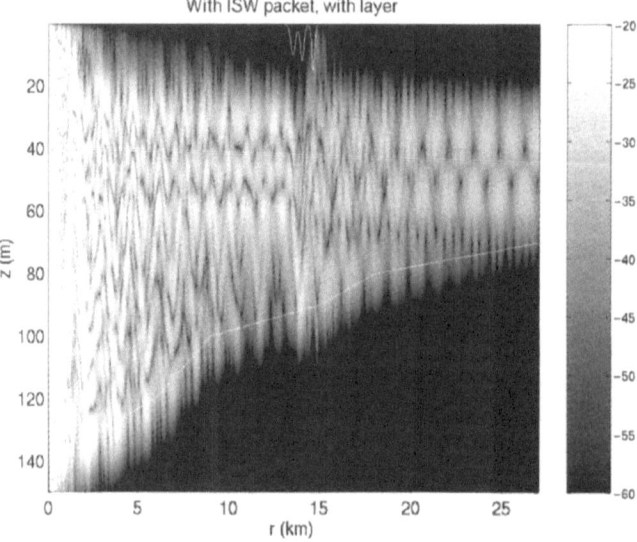

Figure 6. Despread level is shown for a fourth PE simulation (Case 4). The water column includes the ISW packet as does the Fig. 4 case, and the the warm near-bottom layer shown in Fig. 2.

and the layer (Case 4, Fig. 6) reproduces the mode coupling behavior already seen in Fig. 4. The high dB levels in Figs. 3 and 4 also show how the layer serves to reduce sub-bottom loss of acoustic energy.

The ISW mode-coupling effect is seemingly the same with and without the layer, but it is not identical because the signal levels are reduced by different amounts in the

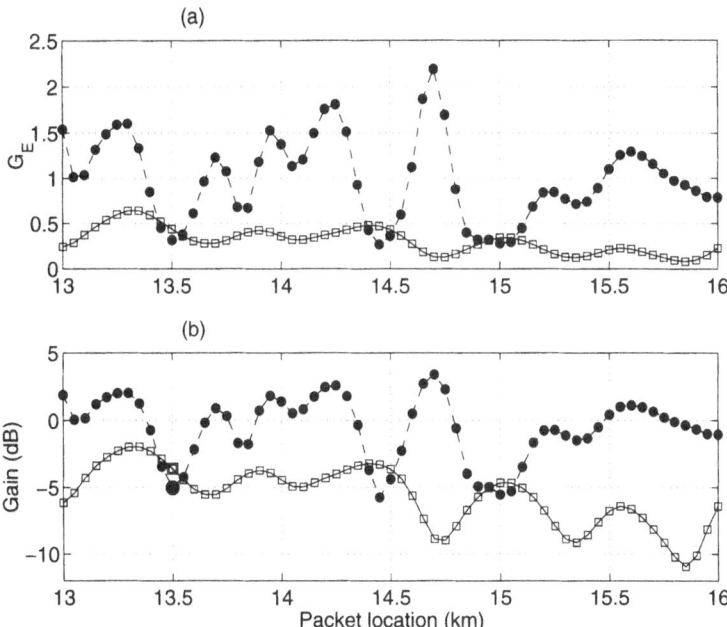

Figure 7. Gain caused by solitons is shown. (a) The filled circles show the ratio of energy in the water at 27 km range for multiple Case-2 runs to the energy of the Case-1 reference run, as a function of packet distance from the source (the only parameter that was varied between runs). The average gain in this case is close to one (0.99) with standard deviation 0.45. The open squares show the ratio of energy for multiple Case-4 runs to the Case 3 reference run. The average gain from the packets is 0.31 with standard deviation 0.14. (b) The data of (a) are shown on a dB scale. The mean Case 4 loss relative to Case 3 converts to −5.1 dB. The values for the computations shown in Figs. 3–7, with packet at 13.5 km from the source, are shown with larger symbols.

two cases. The examples shown are not sufficient to quantify the difference because the signal loss from packets is sensitive to packet position with a scale length of about a kilometer [8]. The position affects the coupling by controlling modal phases. Therefore Cases 2 and 4 were repeated for packet positions of 13 to 16 km from the source, at 50-m increments. Figure 7 shows the gain caused by packets. In the figure, Case 2 results are compared with Case 1 results and Case 4 results are compared with Case 3 results by plotting the ratios of acoustic energy in the water column at 27-km range. The ratio of energy is G_E. The result given by dividing levels shown in Figs. 4 and 3 (−4.8 dB) is given by the large circle in Fig. 7b. The result given by dividing levels shown in Figs. 6 and 5 (−3.8 dB) is given by the large square. The Case 2 runs (no warm layer) give mean gain of about one, with high variability. This means that on average the bottom-loss sensitivity of the mode structure after encountering the packet is the same as before the packet. The Case 4 vs Case 3 ratios G_E are always less that one, averaging 0.31 (−5.1 dB). The explanation for the loss is that water-trapped modes dominate when the sound encounters the packet, so that mode coupling increases the mode bandwidth to include bottom-interacting modes, reducing energy at 27-km range.

Finally, adiabatic mode amplitudes vary with range over a longer scale of about 10 km, so packets near and far from the source will have different effects. Source depth also plays a role by controlling mode excitation. Therefore, this simple example must be examined in greater detail to firm-up the findings. Numerous other examples of interacting features deserve examination.

Acknowledgements

We thank the Office of Naval Research for supporting this research. Mike Collins wrote and distributed the RAM PE code used here in modified form. Jim Preisig and Jim Lynch have been continual collaborators in this work.

References

1. McDaniel, S.T. and D.F. McCammon, Mode coupling and the environmental sensitivity of shallow-water propagation loss predictions, *J. Acoust. Soc. Am.* **82**, 217–223 (1987).
2. Shmelerv, A.Yu., Migulin, A.A. and Petnikov, V.G., Horizontal refraction of low frequency acoustic waves in the arents Sea stationary acoustic track experiment, *J. Acoust. Soc. Am.* **92**, 1003–1007 (1992).
3. Zhou, J., Zhang, X. and Rogers, P.H., Resonant interaction of sound wave with internal solitons in the coastal zone, *J. Acoust. Soc. Am.* **90**, 2042–2054 (1991).
4. Lynch, J.F., Guoliang, J., Pawlowicz, R., Ray, D., Plueddemann, A.J., Chiu, C.-S., Miller, J.H., Bourke, R.H., Parsons, A.R. and Muench, R., Acoustic travel-time perturbations due to shallow-water internal waves and internal tides in the Barents Sea Polar Front: Theory and experiment, *J. Acoust. Soc. Am.* **99** 803–821 (1996).
5. Creamer, D.B., Scintillating shallow-water waveguides, *J. Acoust. Soc. Am.* **99**, 2825–2838 (1996).
6. Preisig, J.C. and Duda, T.F., Coupled acoustic mode propagation through continental-shelf internal solitary waves, *IEEE J. Oceanic Eng.* **22**, 256–269 (1997).
7. Tielburger D., Finette, S. and Wolf, S., Acoustic propagation through the internal wave field in a shallow water waveguide, *J. Acoust. Soc. Am.* **101**, 789–808 (1997).
8. Duda, T.F., and Preisig, J.C., A modeling study of acoustic propagation through moving shallow-water solitary wave packets, *IEEE J. Oceanic Eng.* **24**, 16–32 (1999).
9. Headrick, R.H., Lynch, J.F., Kemp, J.N., Newhall, A.E., von der Heydt, K., Apel, J., Badiey, M., Chiu, C.-S., Finette, S., Orr, M., Pasework, B., Turgut, A., Wolf, S. and Tielbuerger, D., Acoustic normal mode fluctuation statistics in the 1995 SWARM internal wave scattering experiment *J. Acoust. Soc. Am.* **107**, 201–220 (2000).
10. Headrick, R.H., Lynch, J.F., Kemp, J.N., Newhall, A.E., von der Heydt, K., Apel, J., Badiey, M., Chiu, C.-S., Finette, S., Orr, M., Pasework, B., Turgut, A., Wolf, S. and Tielbuerger, D., Modeling mode arrivals in the 1995 SWARM experiment acoustic transmissions, *J. Acoust. Soc. Am.* **107**, 221–236 (2000).
11. Lynch, J., Fredricks, A., Colosi, J., Gawarkiewicz, G., Newhall, A., Chiu, C.-S. and Orr, M., Acoustic effects of environmental variability in the SWARM, PRIMER and ASIAEX experiments. In *Impact of Littoral Environmental Variability on Acoustic Predictions and Sonar Performance,* edited by N.G. Pace and F.B. Jensen (Kluwer Academic, The Netherlands, 2002) pp. 3–10.
12. Rehmann, C.R. and Duda, T.F., Diapycnal diffusivity inferred from scalar microstructure measurements near the New England shelf/slope front, *J. Phys. Oceanogr.* **30**, 1354–1371 (2000).
13. Collins, M.D., Generalization of the split-step Padé solution, *J. Acoust. Soc. Am.* **92**, 382–385 (1993).

SUB-MESOSCALE MODELING OF ENVIRONMENTAL VARIABILITY IN A SHELF-SLOPE REGION AND THE EFFECT ON ACOUSTIC FLUCTUATIONS

STEVEN FINETTE*, THOMAS EVANS** AND COLIN SHEN**

*Acoustics Division, **Remote Sensing Division,
Naval Research Laboratory, Washington DC 20375, USA
E-mail: finette@wave.nrl.navy.mil

A coupled oceanographic/acoustic simulation model is under development for studying the relationship between acoustic field variability and dynamic oceanographic processes in a continental shelf/slope environment. The oceanographic component of the model involves numerical integration of the non-linear hydrodynamic equations of motion describing density, temperature and salinity distributions as a function of space and time. This component includes sub-mesoscale dynamics, allowing for the generation and propagation of non-hydrostatically generated phenomena such as tidally driven internal tides and solitary waves. Results are mapped into the corresponding sound speed distribution, and the resulting set of time evolved sound speed fields is used as input to a wide-angle parabolic equation that computes the acoustic field propagating through the environment. The general approach is discussed, and an illustrative result is presented that links acoustic field variability to specific oceanographic features.

1 Introduction

Modeling acoustic propagation through littoral regions is a difficult problem because of the complex dynamic structure of the temperature and salinity distributions found in such environments. A significant source of this structure is tidal forcing of stratified water at the shelf break which, through buoyancy effects, can induce baroclinic motions in the form of internal tides and solitary waves. In order to properly model acoustic system performance in littoral areas, it is important to develop synthetic sound speed fields that "faithfully" reproduce the proper space-time characteristics of the sound speed variations in the real ocean. The reason is that the sound speed distribution, in conjunction with acoustic source properties, source geometry and waveguide boundary conditions uniquely determines the space-time coherence of the acoustic field and coherence is the fundamental physical property that places restrictions on phase-sensitive sonar processing. For the purpose of studying volumetric contributions to space-time coherence in shallow water, we are developing a hydrodynamic model for the evolution of the sound speed field in a continental shelf-slope environment. An example of acoustic propagation through this shallow water waveguide is presented to illustrate the effect of tidal forcing on transmission loss.

Mesoscale hydrodynamic models are not appropriate in the littoral because they do not allow for small spatial scale (horizontal) ocean dynamics on the order of 50–500 m and below, thus filtering out internal tides, solitary wave production and propagation.

N.G. Pace and F.B. Jensen (eds.), Impact of Littoral Environmental Variability on Acoustic Predictions and Sonar Performance, 401-408.

We discuss below a non-hydrostatic (sub-mesoscale) model of the ocean environment. This model is forced by the M2 tide over a stratified water column with variable bathymetry, generating an internal tide and solitary waves originating at the shelf-break region. As an illustration, a wide-angle parabolic equation is used to propagate the acoustic field emitted by a point source through a set of time-evolved 2-D sound speed snapshots. Internal tides and solitary wave packets are important examples of sub-mesoscale dynamics that play a significant acoustic role in these environments, altering the amplitude and phase of acoustic waves propagating in the waveguide. A number of attempts to model the acoustic wave/internal wave interaction have been made recently [1–7]. The influence of azimuthally anisotropic solitary waves on horizontal array performance has been considered using a data constrained oceanographic model [8,9].

2   Oceanographic and acoustic models

The sub-mesoscale model is based on a vorticity dynamics formulation of the hydrodynamic equations of motion [10]. The model allows for a free ocean surface, stratified density distribution, variable bathymetry, coriolis and tidal forcing. The equations of motion in this formulation are based on the Boussinesq approximation for the density variations and are given by the following expressions.

Sea surface height evolution equation:

$$\frac{\partial \eta}{\partial t} + \vec{U}_\eta \cdot \nabla \eta = \mathrm{w}_\eta, \tag{1}$$

where η is the vertical displacement of the ocean surface from its resting water level, w_η is the surface vertical velocity, $\vec{U}_\eta \equiv (\mathrm{u}_\eta, \mathrm{v}_\eta)$ represents the horizontal surface velocity vector and $\nabla \equiv (\frac{\partial}{\partial x}, \frac{\partial}{\partial y})$.

Surface momentum equation:

$$\left[\frac{D\vec{U}}{Dt} + \vec{f} \times \vec{U} \right]_\eta = -\left(\left[\frac{D\mathrm{w}}{Dt} \right]_\eta + g \right) \frac{\rho_\eta}{\rho_o} \nabla \eta + \left[\mu_H \nabla^2 \vec{U} + \mu_z \frac{\partial^2 \vec{U}}{\partial z^2} \right]_\eta, \tag{2}$$

where $\dfrac{D}{Dt} = \dfrac{\partial}{\partial t} + \vec{U}_* \cdot \nabla_*$, $\vec{U}_* = (\mathrm{u}, \mathrm{v}, \mathrm{w})$ is the total velocity vector, $\vec{U} = (\mathrm{u}, \mathrm{v})$ the horizontal velocity vector, g is the gravitational acceleration, $\nabla_*^2 \equiv \dfrac{\partial^2}{\partial x^2} + \dfrac{\partial^2}{\partial y^2} + \dfrac{\partial^2}{\partial z}$, $\vec{f} = (0,0,f)$ describes the Coriolis rotational frequency vector, ρ_η the sea surface density, ρ_o a constant reference density and $\mu_{H,Z}$ represents the horizontal and vertical eddy viscosities.

The interior evolution equation for the horizontal vorticity vector:

$$\frac{D\vec{\zeta}}{Dt} = (\vec{\zeta}_* + \vec{f}) \cdot \nabla_* \vec{U} + \hat{k} \times \frac{\nabla \rho}{\rho_o} + \mu_H \nabla^2 \vec{\zeta} + \mu_z \frac{\partial^2 \vec{\zeta}}{\partial z^2}, \qquad (3)$$

with $\vec{\zeta} = (\zeta_x, \zeta_y)$ the horizontal vorticity vector, $\vec{\zeta}_* = (\zeta_x, \zeta_y, \zeta_z)$ the total vorticity vector, $(\zeta_x, \zeta_y) = (\frac{\partial w}{\partial y} - \frac{\partial v}{\partial z}, \frac{\partial u}{\partial z} - \frac{\partial w}{\partial x})$, $\zeta_z = \frac{\partial v}{\partial x} - \frac{\partial u}{\partial y}$, $\nabla_* \cdot \vec{\zeta}_* = 0$,

and $\hat{k}$ is a vertical unit vector.

The continuity equation:

$$\nabla_* \cdot \vec{U}_* = 0 \qquad (4)$$

Temperature and salinity equations:

$$\frac{D \begin{bmatrix} T \\ S \end{bmatrix}}{Dt} = V_H \left(\frac{\partial^2}{\partial x^2} + \frac{\partial^2}{\partial y^2} \right) \begin{bmatrix} T \\ S \end{bmatrix} + V_z \frac{\partial^2}{\partial z^2} \begin{bmatrix} T \\ S \end{bmatrix} \qquad (5)$$

where T, S represent temperature and salinity respectively, and V_H, V_z are the horizontal and vertical diffusivities. The latter terms are used to stabilize the temperature and salinity fields against high frequency fluctuations in these quantities.

Equation of state (IES 80):

$$\rho(x, y, z) = \Theta(T, S, P) \qquad (6)$$

where P is the pressure. The functional form of Eq. (6) is given in [11].

A split-time semi-Lagrangian technique is used for the integration of the above equations, to achieve both efficiency and accuracy in the modeling of the nonlinear flow effect and the free surface motion. In essence, this technique computes the flow variables, $\vec{U}_\eta$, $\vec{\zeta}$, T and S at fluid particles' positions from the free surface to the bottom and then interpolates the calculated quantities back to a fixed reference grid. The interior velocity field is then calculated from the kinematic relations,

$$\nabla^2 w + \frac{\partial^2 w}{\partial z^2} = \nabla \cdot \hat{k} \times \vec{\zeta}, \quad \frac{\partial w}{\partial y} - \frac{\partial v}{\partial z} = \zeta_x, \quad \text{and} \quad \frac{\partial u}{\partial z} - \frac{\partial w}{\partial x} = \zeta_y, \qquad (7)$$

and ρ from the equation of state. These new velocities and density are used to start the next Lagrangian time step calculation. In the split-time integration, the faster evolving surface motion governed by the surface equations (1) and (2) is integrated at small time steps set by the CFL condition for the surface gravity waves, while the slower evolving interior flow whose time scale is of the order of the buoyancy period is updated only after many small surface time steps. The spatial derivatives are evaluated using higher order finite difference schemes. The solution of the elliptic equation for w uses the second order finite-difference MUDPACK library from NCAR. The ocean model has been validated by comparing its accuracy in simulating flow instabilities in channels against the highly accurate pseudo-spectral calculations as well as testing it against known analytical solutions for waves and currents. The details are to be reported elsewhere (Shen and Evans, in preparation).

The sound speed field $C(x, y, z, t)$ is considered as environmental input data for a wide-angle, split-step parabolic equation algorithm [12] describing acoustic propagation from a point source. Sound speed is computed through its functional dependence on temperature, salinity and pressure [13], with the latter quantities determined throughout the water column from the solution of Eq. (1–6). We invoke the frozen ocean assumption, by which temporal variations in the ocean are considered negligible during the passage of the acoustic wave from source to receiver. This is well satisfied for acoustic propagation over 30 km for the interior of the ocean, where typical internal waves have speeds less than 1 m/s. The long wave assumption is made for surface waves: $C = \sqrt{gH} \cong 30$ m/s for a maximum water depth H of 100 m, and the assumption is marginally satisfied there. However, the long M2 tide wavelength implies that it will have only a small vertical component of the surface elevation gradient over the 30 km of acoustic propagation and this surface variation is assumed to not significantly violate the condition.

3   Results

An example of the evolution of the sound speed field through a shelf-break region and acoustic propagation in this environment are presented below. The 2-D simulation of a shelf/shelf-break environment for acoustic modeling is carried out in a 100 km wide domain with the shelf-break modeled by h = -25m[3+tanh((x-50km)/1.25km)], where h is the depth of the bottom below the mean sea level, z=0, and x is the horizontal distance measured from the left boundary, x=0, which is a vertical wall assumed to be the location of the coast. The right boundary at x=100 km is assumed to be the open ocean, and the flow there responds to tidal forcing which is applied at the right boundary by varying the sea surface height sinusoidally with an amplitude of 2 m and period of 12.4 h. No wind forcing is applied, and the surface boundary condition is thus stress-free, u/ z=0= v/ z. The same stress-free condition is also used for the bottom and coast line to eliminate the frictional influence from these boundaries, since the focus of this simulation is the generation and propagation of internal waves in the presence of minimal frictional influence. In follow-up studies, viscous effects from the boundaries are to be considered. The domain is resolved horizontally with 24.4 m grid spacing. A surface and bottom following vertical coordinate system is employed, and so the number of vertical grid points is fixed at 33 both on and off the shelf. A 30 km sub-section of the 100 km oceanographic domain and simulated internal wave structure is used for acoustic computations and is shown in Fig. (1), with the starting range coordinate renumbered to zero. The parameters used in the governing equations are $f=2\pi/12.4$ h, $g=9.8$ m/s^2, $\mu_H=v_H=0.5$ m^2/s, $\mu_z=0.05$ m^2/s m, $v_z=10^{-4}$ m^2/s.

A depth dependent temperature field describing a summer thermocline is used as a starting environment, with salinity chosen as a constant 35 ppt. The profile was obtained from the SWARM95 data set [2]; for simplicity, it is assumed to be range independent at the beginning of the simulation, when the model ocean is in a "resting" state. An acoustic source of frequency 400 Hz is placed at a depth of 30 m at the range origin and the acoustic field is propagated over a 30 km range from the shelf region downslope through the shelf-break area.

The figure below gives an example of the evolving ocean sound speed environment covering a 7.5 h period, along with the corresponding acoustic transmission loss for the selected environmental snapshots. Geometric spreading has been removed to emphasize environmental variability. Note that white in the transmission loss figures represents losss greater than 70 dB. Time is measured from the starting point (zero hours) as indicated on the plots. The initial off-shore flow causes a significant variation of the thermocline to appear over the shelf break at t=5 h which, through the corresponding density perturbations, generates a baroclinic tide that propagates outward in both directions (t=6 h) from the shelf-break. The internal tide transfers energy to higher spatial frequencies in the form of two solitary wave packets (t=7.5 h), propagating away from their generation site at the shelf-break.

In this simulation, the solitary wave packets are dominated by the first internal wave mode. Both adiabatic propagation and mode coupling can play a role in acoustic transmission for this environment. There are two potential sources of acoustic mode coupling. One is associated with the range dependent bathymetry and may be causing conversion of higher order modes to lower order modes starting around 20 km from the acoustic source, resulting in a high loss region in the upper 20–25 m of the water column at ranges greater than 22 km. However, the relative importance of the mode coupling and adiabatic terms in this region would have to be assesesed by modal decomposition of the field; that analysis is beyond the scope of this paper.

The second source of acoustic mode coupling involves the range dependent variations of the thermocline induced by tidal forcing. Transmission loss variability beyond a range of 22 km is evident in the modal interference pattern between the 5 h and 6 h snapshots and is caused by the transformation of the depression into an internal tide propagating on and off the shelf.

The most significant variation occurs between 6 and 7.5 h and is induced by the solitary wave packet propagating on the shelf, causing a mean drop in transmission loss of about 8 dB across the shelf break. The seawardc propagating packet does not contibute to this enhancement because it resides at depths less than about 25 m, where only weak acoustic energy levels are available for interacting with the wave packet.

Acknowledgment

This research was supported with funds from the Office of Naval Research.

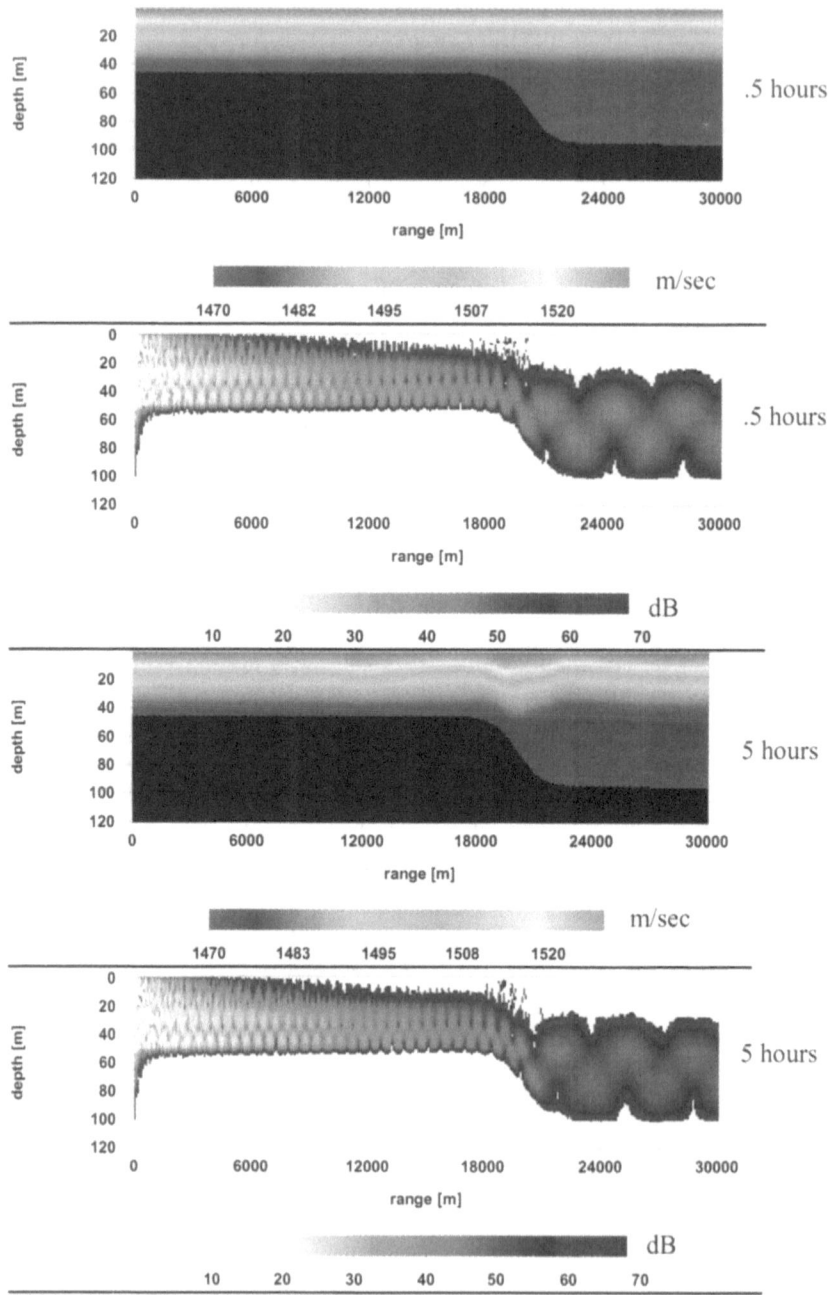

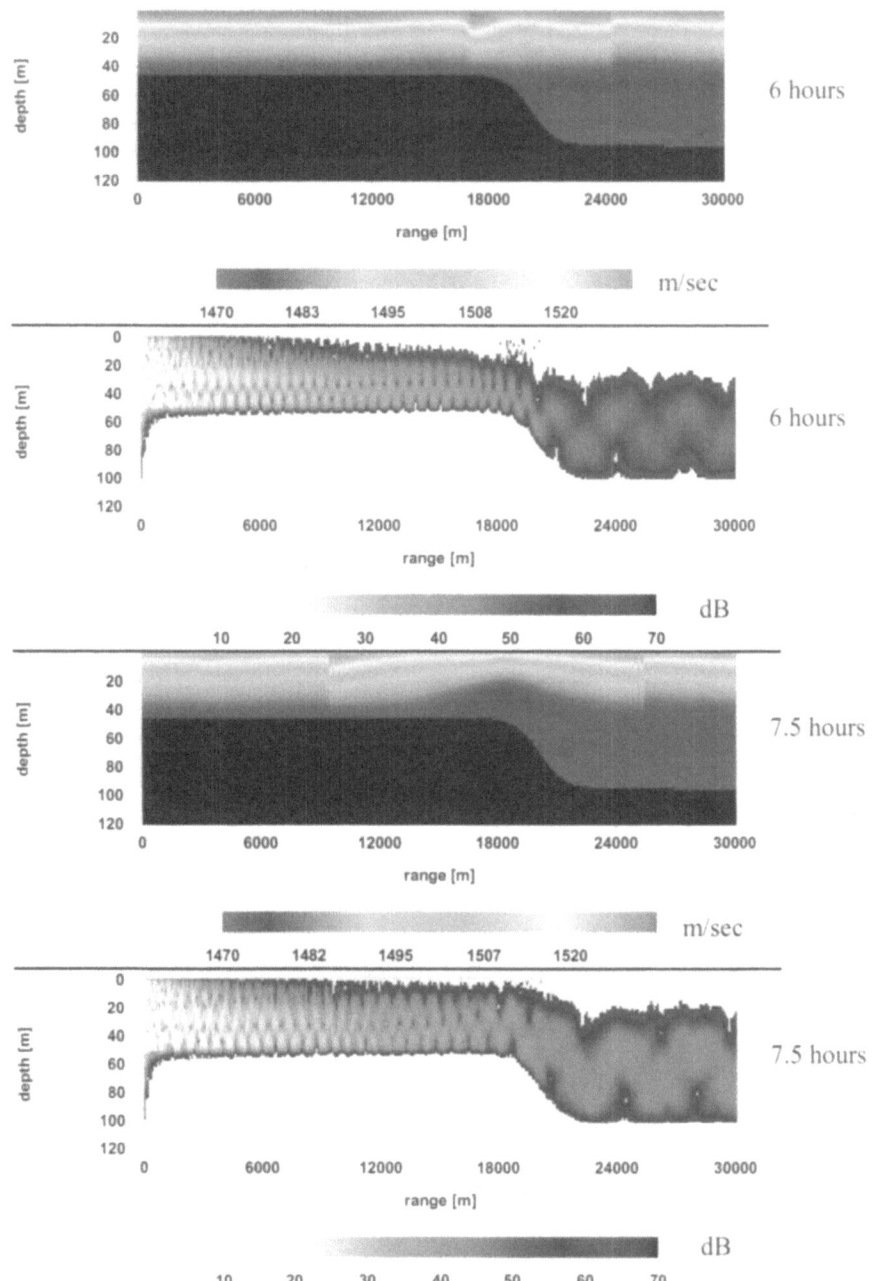

Figure 1. Sound speed fields and corresponding acoustic transmission loss for a 400 Hz point source placed at 30 m depth and zero range, for selected environmental snapshots. A 30 km sub-section of the sound speed field was used in the acoustic computations; the range axis is relabeled.

References

1. Tielbuerger, D., Finette, S. and Wolf, S., Acoustic propagation through an internal wave field in a shallow water waveguide, *J.Acoust. Soc. Am.* **101**, 789–808 (1997).

2. Finette, S., Orr, M.H., Turgut, A., Apel, J., Badiey, M., Chiu, C.-S., Headrick, R.H., Kemp, J.N., Lynch, J.F., Newhall, A.E., von der Heydt, K., Pasewark, B., Wolf, S.N. and Tielbuerger D., Acoustic field variability induced by time-evolving internal wave fields, *J. Acoust. Soc. Am.* **108**, 957–972 (2000).

3. Oba, R. and Finette, S., Acoustic propagation through anisotropic internal wave fields: Transmission loss, cross-range coherence, and horizontal refraction, *J. Acoust. Soc. Am.* **111**, 769–784 (2002).

4. Preisig, J.C. and Duda, T.F., Coupled acoustic mode propagation through continental-shelf internal solitary waves, *IEEE J. Ocean. Eng.* **22**, 256–269 (1997).

5. Duda, T.F. and Preisig, J.C., A modeling study of acoustic propagation through moving shallow water solitary wave packets, *IEEE J. Ocean. Eng.* **24**, 16–32 (1999).

6. Katsnel'son, B.G. and Pereselkov, S.V., Low-frequency horizontal acoustic refraction caused by internal wave solitons in a shallow sea, *Acoust. Phys.* **46**, 684–691 (2000).

7. Rubenstein, D., Observations of cnoidal internal waves and their effect on acoustic propagation in shallow water, *IEEE J. Ocean. Eng.* **24**, 346–357 (1999).

8. Finette, S. and Oba, R., Horizontal coherence estimates for an internal wave dominated shallow water environment. In *Proc. Fifth European Conference on Underwater Acoustics*, edited by M.E. Zakharia, P. Chevret, and P. Dubail (Luxembourg: Office for Offical Publications of the European Communities, 2000) pp. 151–156.

9. Finette, S. and Oba, R., Horizontal array beamforming in an azimuthally anisotropic internal wave field, *J. Acoust. Soc. Am.* (to be submitted 2002).

10. Shen, C., Constituent boussinesq equations for waves and currents, *J.Phys. Oceanogr.* **31**, 850–859 (2001).

11. Millero, F.J. and Poisson, A., International one-atmosphere equation of state of seawater, *Deep-Sea Res.* **28A**, 625–629 (1981).

12. Collins, M.D., Generalization of the split-step Padé solution, *J. Acoust. Soc. Am.* **96**, 382–385 (1993).

13. Del Grosso, V.A., New equation for the speed of sound in natural waters (with comparisons to other equations), *J. Acoust. Soc. Am.* **56**, 1084–91 (1974).

YELLOW SEA INTERNAL SOLITARY WAVE VARIABILITY

A. WARN-VARNAS, S. CHIN-BING AND D. KING
Naval Research Laboratory, Stennis Space Center, MS 39539, USA
E-mail: varnas@nrlssc.navy.mil

J. HAWKINS
Planning Systems Inc., Slidell, LA 70458, USA

K. LAMB
University of Waterloo, Waterloo, Ontario, Canada N2L3G1

M. TEIXEIRA
Polytechnic University of Puerto Rico, San Juan, PR 00919, USA

Our studies are centered in an area south of the Shandong peninsula where the observations of Zhou, Zhang and Rogers [1] showed an anomalous drop in acoustical intensity at 630 Hz. For this region ocean-acoustic modeling studies are performed in conjunction with available SAR observations of internal solitary waves. Acoustic field intensity calculations show that for some frequencies a redistribution of acoustic energy to higher modes occurs.

1 Introduction

The initial interest in the region of the Yellow Sea south of the Shandong Peninsula arose from acoustical measurements of shallow-water sound propagation. Acoustical measurements performed by Zhou *et al.* [1], over a period of several summers, showed an anomalous drop in acoustical intensity of about 20 dB at a range of 28 km for acoustic frequencies around 630 Hz. The transmission loss was found to be time and direction dependent. The authors postulated the existence of solitary waves in the thermocline and, using a gated sine function representation of them, performed transmission loss calculations using an acoustic parabolic equation (PE) model. The simulation results from this hypothetical case showed that an anomalous transmission loss could occur at a frequency of around 630 Hz when acoustical waves and solitary waves interact. Computer simulations subsequently confirmed [2] that the resonant like transmission loss is caused by an acoustical mode coupling due to the presence of solitary waves, together with a corresponding larger bottom attenuation for the coupled acoustic modes. In the acoustical calculations the existence of solitary waves has so far been only postulated for the area south of the Shandong Peninsula. This paper addresses this issue by considering solitary wave generation and propagation in the region together with an acoustical field interaction.

409

N.G. Pace and F.B. Jensen (eds.), Impact of Littoral Environmental Variability on Acoustic Predictions and Sonar Performance, 409-416.

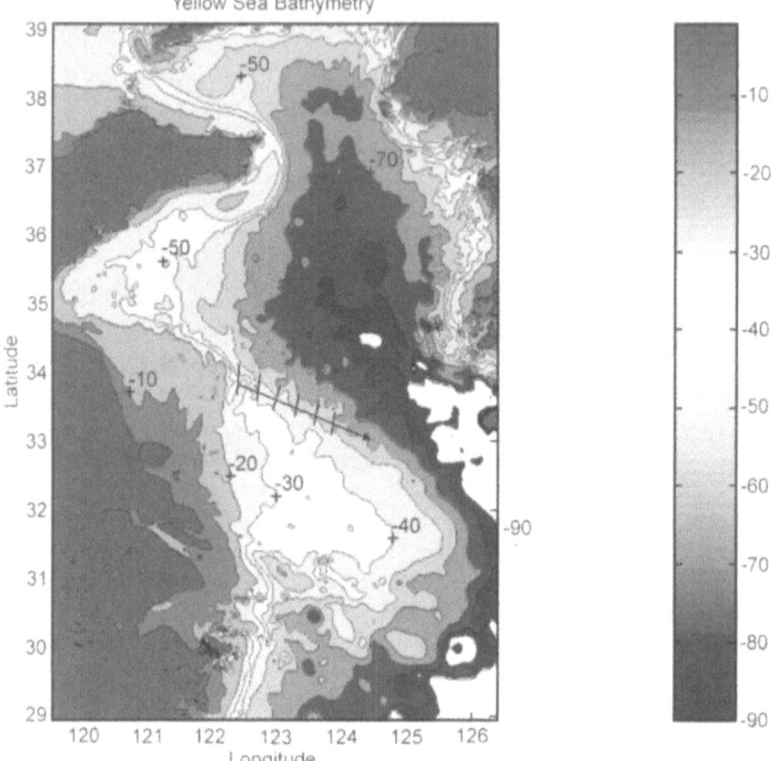

Figure 1. Location of the region south of the Shandong peninsula with the track of RADARSAT1 measurement indicated by the black line. The smaller black lines reflect the orientation of the internal bores and solitary wave trains relative to the direction of propagation.

2   Region

The present study is located south of the Shandong peninsula and will be referred to as the Shandong area, Fig. 1. The arrow shows the direction of an observed solitary wave train with Radarsat1 SAR. At the beginning of the arrow there is a relatively steeper slope at the location where the first internal bore is observed, Digital Atlas of Choi [3]. The lines across the arrow indicate the along crest direction of the wave packets. The variable angle of the lines suggests refraction along the shelf break. We obtained summer SAR observations, for the Shandong area, from Radarsat1 ScanSAR with a 500 by 500 km wide resolution. The observations were acquired on August 8, 1998 and processed at the Alaska SAR Facility. The pixel spacing is 100 m. Figure 1 shows the track location, of the observations, with the topographic features in the background. The results of the Fourier spectral analysis of the SAR images are summarized in Table 1, where the solitary wave trains are labeled from left to right. The listed wavelengths are for the most intense spectral peaks that occur for packets 2, 3, and 4 . For packet 5 there is not enough signal above background for determining a wavelength at which an energy peaking occurs. The

Table 1. Parameters of SAR observations.

Solitary wave train #	Bore 2	P1	P2	P3	P4	P5
Distance (km) 0	36–46	72–93	132	178	226	272
Phase speed C (m/s)	0.8–1.03	0.8–1.05	1.098	1.03	1.075	1.03
Dominant wavelength (m)		630	930	1600	2300	

Fourier analysis reflects the wavelength around which most of the energy is concentrated. The set of wavelengths that we obtained for the solitary wave trains is 630 m, 930 m, 1600 m, and 2700 m, Table 1. The last two wavelengths mark an appreciable increase relative to the first two.

3 Modelling results

The Lamb [4] model is used for simulating the generation and propagation of solitary waves in the Yellow Sea. It consists of the Boussinesq equations with the Coriolis force in a two-dimensional cross-bank plane. In the along-bank direction, the velocity is included but the derivatives are neglected (2.5 dimensional representation). The equations of the model are:

$$\mathbf{V}_t + \mathbf{V} \cdot \nabla \mathbf{V} - f\mathbf{V} \times \mathbf{k} = -\nabla P - \rho g \mathbf{k}$$
$$\rho_t + \mathbf{V} \cdot \nabla \rho = 0$$
$$\nabla \cdot \mathbf{V} = 0 \quad , \tag{1}$$

where $\mathbf{V}$ is the velocity vector, ∇ the gradient operator subscript t denotes the time derivative, ρ the density, P the pressure, g the gravitational constant, f the Coriolis parameter, and $\mathbf{k}$ the unit vector along the z direction that is perpendicular to the surface.

The flow is forced by specifying a semidiurnal tidal velocity at the left boundary of the form $V_t \sin(\omega t)$ where ω is the M_2 tidal frequency assumed to have a 12.4 h period. The strength of the semidiurnal tidal current in the shallow water, V_t, varies between 0.6 and 1.2 m/s, typical of values in the Shandong region.

The parameters for the different model runs are given in Table 2. We consider here case 2. For this case, the pycnocline is at a depth of 15 m, a peak barotropic tidal velocity of 0.7 m/s is used, and the deep water depth is 70 m. The density is specified on the basis of climatology and available data. Each tidal cycle generates a wave propagating on the shelf and a wave packet propagating away from the shelf. This behavior is seen in other areas. At 63 hours or 5.1 semidiurnal tidal cycles into the simulation there are

Table 2. Simulation parameters: h_d is pycnocline depth, H(m) water depth, V_t is the tidal strength, Topo is the topography type a being for cases with a finger ; d_H is the length of the computational domain.

Case	h_d(m)	H(m)	V_t (m/s)	Topo	d_H (km)
1	15	70	1.2	a	150
2	15	70	0.7	a	240
3	15	70	0.35	a	240

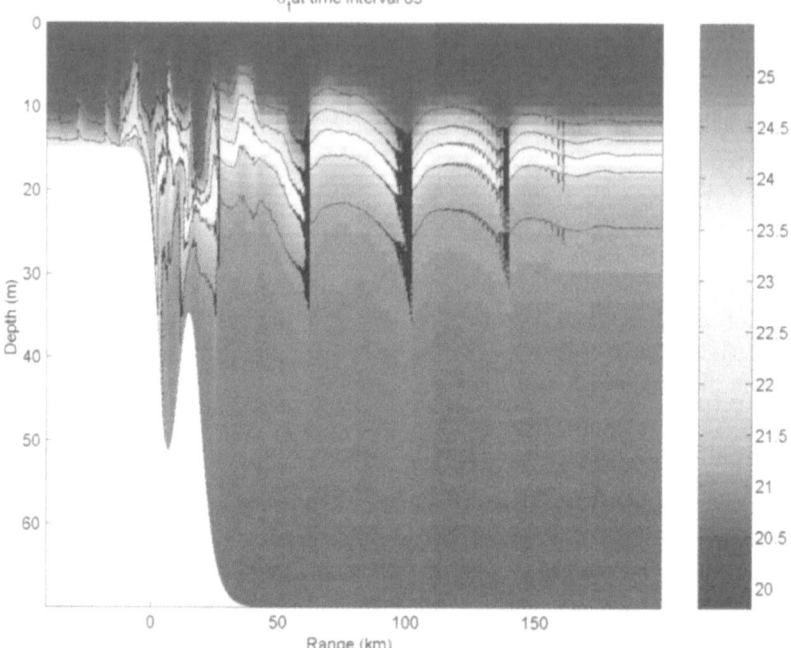

Figure 2. Simulated sigma-t density distributions for case 2 in Table 1 at 5.1 semidiurnal tidal periods.

four well developed wave packets with a fifth starting to form, Fig. 2. Note that the first three wave packets from the shelf show a dramatic increase in amplitude. This is largely due to the response over the shelf edge increase in time, as discussed in Lamb [4]. The third and fourth packets from the shelf are more similar in size (when compared at similar stages of their evolution). The individual waves in each packet grow in size for a while and then start to decay. They also get further apart. This is particularly apparent in the further away packets from the shelf . The decrease in amplitude may be partly due to numerical dissipation.

For comparison with SAR, the tuned simulation with a 15 m pycnocline, case 2 in Table 2, is used. Table 3 shows the calculated wavelengths at the various horizontal locations. At around 100 km the wavelength is 420 m. The measurements, Table 1, show a 630 m wavelength at the location. In the vicinity of 130 km to 140 km the model results yield a wavelength of 810 m, underpredicting the data value of 930 m. At around 170 km the modeled wave train is displaying an increased spacing between waves that is most pronounced towards the back of the wave packet. The resultant wavelength

Table 3. Model results.

Wave train	1	2	3	4	5
Tidal cycle	6.1	6.1	6.1	6.1	6.1
Distance (km)	63	102	143	178	195
Wavelength (m)	335	420	810	2300	3300

due to this increased spacing is around 2300 m, Table 3. The corresponding wavelength in the measurements is around 1600 m, Table 1. This is a situation where the model overpredicts the wavelength instead of underpredicting it. This, also, marks an increase in the measured wavelengths from the previous locations, that indicates wavelengths of 630 m and 930 m with comparable spatial incremental distances. This suggests a change in the behavior of the measured solitary wave trains from type A to type B configuration that results in a sudden increase in wavelength size. The model results at ranges greater than 170 km also indicate such a phenomena.

4 Acoustic model results

The acoustic effects of these solitary waves can be simulated by applying Dr. Michael Collins' acoustic PE propagation model, FEPE, to selected environmental "snapshots" generated by the Lamb model. A selected scenario and the corresponding acoustic simulation are shown in Fig. 3. The upper figure is the ocean environment after 71 hours. This environment was generated by the Lamb model assuming a tidal strength of 0.7 m/s, and validated by comparing with SAR observations. The lower figure shows the acoustic loss that occurs when a 925 Hz acoustic source is placed at the position indicated by the red dot (located on the left hand side of each figure). Clearly, the acoustic transmission is greatly affected by the first two solitary wave packets that are closest to the acoustic source.

Figure 4 (upper figure) shows the transmission loss at a receiver depth of 30 m for the

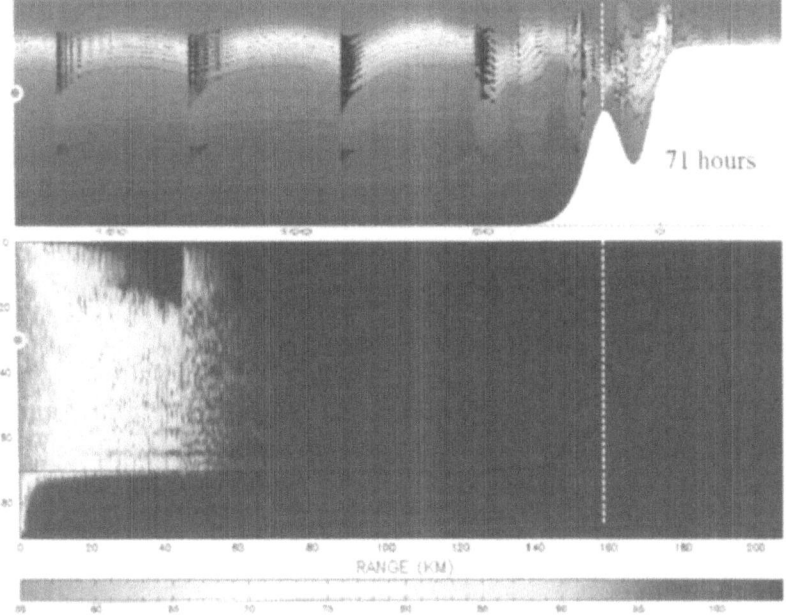

Figure 3. A selected solitary wave packet environment generated by the Lamb mode, and the corresponding acoustic simulation.

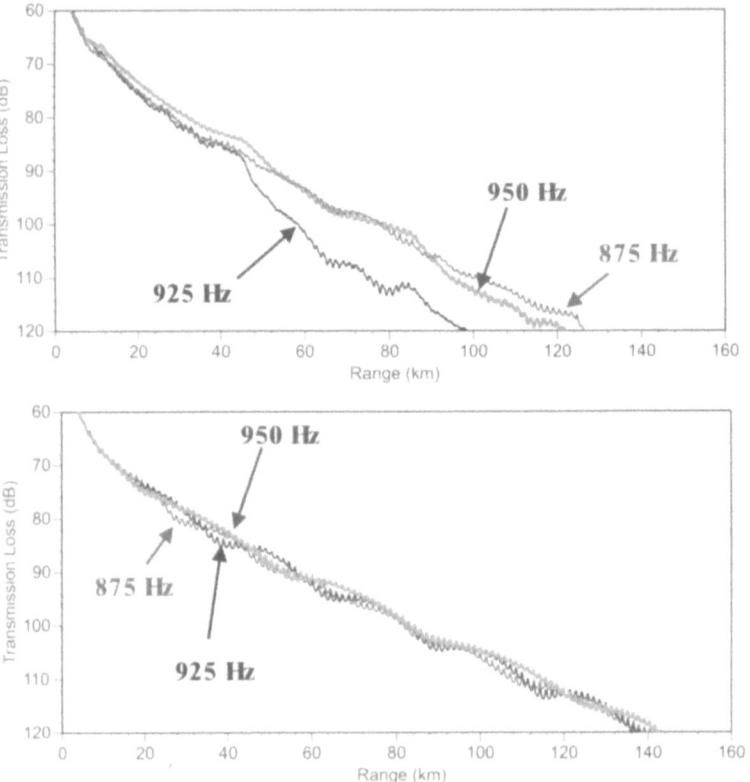

Figure 4. Transmission loss at a receiver depth of 30 m when the solitary wave packet is present (upper figure) and when it is not present (lower figure).

925-Hz case shown in Fig. 3, and for the same scenario, but at two adjacent frequencies, 875 Hz and 950 Hz. There is a loss in transmission at 925 Hz, but not at 875 Hz nor at 950 Hz. The lower figure in Fig. 4 shows the transmission loss for the three source frequencies when the solitary wave packets are removed from the simulation. The loss is virtually identical for the three source frequencies. For the selected environmental scenario and acoustic parameters, there is a significant loss in acoustic signal at 925 Hz that is not seen at surrounding frequencies. This loss is due to the presence of the solitary wave packets. Figure 5 shows the corresponding wave number analysis at 925 Hz (upper figure) and 950 Hz (lower figure) for the simulations with and without the solitary wave packets. The solitary wave packets had only a slight acoustic effect at 950 Hz and this is confirmed in the upper figure of Fig. 5 which shows only slight mode conversion and mode loss. The lower figure of Fig. 5 shows that at 925 Hz the acoustic modes were greatly affected by the presence of the solitary wave packet, with practically every mode experiencing mode conversion and mode loss. Our results tend to confirm the resonance hypothesis of Zhou *et al.* We have performed numerous simulations that indicate that solitary wave packets can cause acoustic mode conversion (from lower-order to higher-order modes) followed by loss due to ocean bottom attenuation (with the higher-order modes having higher bottom attenuation). The results shown in Figs. 3, 4, and 5

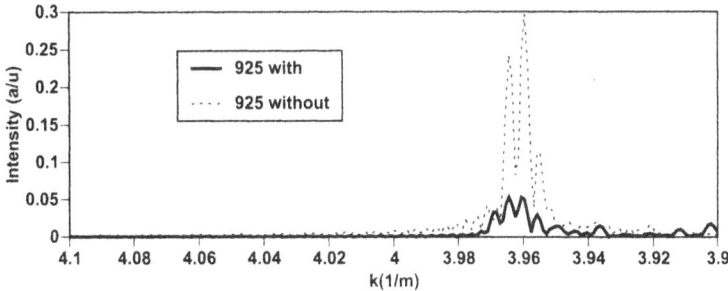

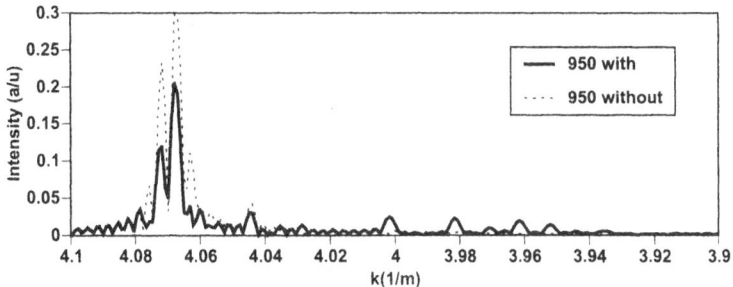

Figure 5. Wave number analysis at 925 Hz (upper figure) and 950 Hz (lower figure) for the simulations with and without the solitary wave packets.

are somewhat different in that higher bottom attenuation is not a required mechanism. Rather, it appears that massive mode conversion occurs, from discrete propagating modes to continuous evanescent modes, resulting in a significant loss in acoustic signal. This new finding is currently under investigation.

5 Conclusion

We have shown that generation and propagation of internal solitary waves can occur along a southeastern track off the Chinese coast located south off the Shandong peninsula. SAR imagery shows the presence of internal solitary waves along the same track and suggest's their generation in the, shallower, shelf break region. Model results indicate generation of internal solitary wave in the same off shelf area. The tuned model simulation and the SAR data both exhibit the presence of two behavior states, A and B that have corresponding solitary wave train characteristics. The two states of behavior A and B could be due to short vs. long time behavior. These states of behavior are evolved by the dynamics of the ocean and the model. The modelled soliton wave amplitudes and wave lengths are within a factor of 2 (or better) of amplitudes and wavelengths derived from SAR data. The simulated phase speeds range from 0.73 m/s to 0.83 m/s. The phase speeds estimated from the measurements range from 0.8 m/s to 1.1 m/s. This suggests that the model formalism does contain dynamics similar to the ocean. Acoustic simulations were performed on several internal solitary wave environments generated by the Lamb model. Large unexpected acoustic losses were observed and were attributable to the solitary wave fields. The results tend to confirm the resonance hypothesis developed by Zhou *et al.*

Acknowledgements

This work was supported by the U. S. Office of Naval Research through the U. S. Naval Research Laboratory base program, PE 62435N. The U. S. Naval Research Laboratory provided technical management.

References

1. Zhou, J.X., Zhang, X.Z. and Rogers, P.H., Resonant interaction of sound wave with internal solitons in coastal zone, *J. Acoust. Soc. Am.* **90**(4), 2042–2054 (1991).
2. Chin-Bing, S.A., King, D.B. and Murphy, J.E., Numerical simulations of lower-frequency acoustic propagation and backscatter from solitary internal waves in a shallow water environment. In *Ocean Reverberation*, edited by D.D. Ellis, J.R. Preston and H.G. Urban (Kluwer Academic Press, Dordrecht, The Netherlands, 1993) pp. 113–118.
3. Choi, B-H., Digital atlas for neighboring seas of Korean Peninsula. Available on compact disk, 1999. E-mail: bchoi@yurim.skku.ac.kr
4. Lamb, K., Numerical experiments of internal wave generation by strong tidal flow across a finite amplitude bank edge, *J. Geophys. Res.* **99**(C1), 848–864 (1994).

FOUR-DIMENSIONAL DATA ASSIMILATION FOR COUPLED PHYSICAL-ACOUSTICAL FIELDS

P.F.J. LERMUSIAUX[1] AND C.-S. CHIU[2]

[1] *Harvard University, DEAS, Pierce Hall G2A, 29 Oxford Street, Cambridge MA 02318, USA*
E-mail: pierrel@pacific.harvard.edu

[2] *Naval Postgraduate School, Monterey, CA 93943, USA*
E-mail: chiu@nps.navy.mil

The estimation of oceanic environmental and acoustical fields is considered as a single coupled data assimilation problem. The four-dimensional data assimilation methodology employed is Error Subspace Statistical Estimation. Environmental fields and their dominant uncertainties are predicted by an ocean dynamical model and transferred to acoustical fields and uncertainties by an acoustic propagation model. The resulting coupled dominant uncertainties define the error subspace. The available physical and acoustical data are then assimilated into the predicted fields in accord with the error subspace and all data uncertainties. The criterion for data assimilation is presently to correct the predicted fields such that the total error variance in the error subspace is minimized. The approach is exemplified for the New England continental shelfbreak region, using data collected during the 1996 Shelfbreak Primer Experiment. The methodology is discussed, computational issues are outlined and the assimilation of model-simulated acoustical data is carried out. Results are encouraging and provide some insights into the dominant variability and uncertainty properties of acoustical fields.

1 Introduction

Ocean acousticians are mainly interested in the distribution and composition of sound pressure fields in the ocean. Physical oceanographers are mainly interested in the oceanic motions and physical properties of the fluid ocean. In both disciplines, the estimation of the variables of interest is challenging because oceanic variability occurs on multiple interactive scales and is difficult to observe. To our knowledge, even though both disciplines employ sophisticated techniques for the estimation of their respective variables, few studies have envisioned a truly coupled four-dimensional estimation, including both the acoustic and oceanic variables in the state vector.

Ocean acoustic wavefields depend on the three-dimensional sound speed field whose evolution is a function of the fluid ocean physics (temperature, salinity, ambient pressure, etc.) and bottom attributes (reflectivity, attenuation, etc.). Due to these dynamical couplings, a joint estimation of acoustical-physical fields is attractive. First, sound waves propagate over long distances in the ocean and acoustic measurements can thus provide valuable integrated oceanic data for physical studies. Similarly, by natural variability, spatial and temporal correlations among environmental properties occur on multiple scales and even sparse measurements of this variability thus provide valuable information for acoustical studies. In fact, accurate physical inputs are necessary for successful acoustic

417

N.G. Pace and F.B. Jensen (eds.), Impact of Littoral Environmental Variability on Acoustic Predictions and Sonar Performance, 417-424.
© *2002 Kluwer Academic Publishers.*

simulations. Finally, even without natural data, acoustical and physical models are sources of coupled data which can be shared to improve estimates.

Considering uncertainties, the mathematical equations used to describe the environment and acoustic properties are approximate, as well as their analytical or numerical solutions. The natural physical and acoustical data are limited in accuracy and coverage. Because of these uncertainties and because of the above dynamical couplings, carrying out a joint estimation is likely to provide substantial advantages. In such an estimation, the sources of information, environmental and acoustical data, and ocean dynamics and sound propagation models, are combined by data assimilation. This combination is optimal in the sense that each information is weighted in accord with its uncertainty. In principle, this process provides better estimates of parameters and properties than can be obtained by using only the observations or models alone. The acoustical data improve physical fields; the physical data improve acoustical fields. Of course, should optimal estimates fail to be accurate, *a priori* assumptions about uncertainties are revised, and models and data sets improved. This manuscript outlines an approach for such four-dimensional (time and space) physical-acoustical estimations via coupled data assimilation and carries out an illustrative example based on data and simulations for the New England continental shelfbreak region.

2 Methodology

Data assimilation [1] combines dynamical models and data sets by quantitative minimization of a criterion or cost function. The links between observational data and dynamical model fields and parameters are provided by measurement models. Since dynamical models, data sets and measurement models are all approximate, they all involve an error component, i.e. the error models. These error models are here stochastic. The dynamical models, data sets and measurement models, and data assimilation scheme are now described.

2.1 Coupled Dynamical Models

Ocean Physics Model. The physical state variables are temperature T, salinity S, velocity $\mathbf{u}$ and pressure p_w. For this study, their mesoscale evolution is computed by the Primitive-Equation model, Eqs. (1)–(7), of the numerical Harvard Ocean Prediction System, e.g. [2]. Atmospheric fluxes based on surface buoy time-series are imposed at the surface. Model parameters and boundary conditions were calibrated based on data and sensitivity studies.

Momentum	$\rho \frac{D\mathbf{u}}{Dt} + 2\rho\,\mathbf{\Omega} \wedge \mathbf{u} = -\nabla p_w + \nabla \cdot \boldsymbol{\tau}^v + \rho\,\mathbf{g}$	(1–3)
Thermal energy	$\rho\,C_p \frac{DT}{Dt} = \nabla \cdot (k\nabla T) + \rho Q$	(4)
Cons. of salt	$\rho \frac{DS}{Dt} = \nabla \cdot (k^s \nabla S) + \rho Q^S$	(5)
Cons. of mass	$\nabla \cdot \mathbf{u} = 0$	(6)
Eqn. of state	$\rho(\mathbf{r}, z, t) = \rho(T, S, p_w)$	(7)
Sound speed eqn.	$c(\mathbf{r}, z, t) = C(T, S, p_w)$	(8)
Wave eqn.	$\nabla^2 p_s(\mathbf{r}, z, t) = \frac{1}{c(\mathbf{r},z,t)} \frac{\partial^2 p_s(\mathbf{r},z,t)}{\partial t^2}$	(9)

Acoustic Model. The acoustic coupled normal mode model [3–5] solves a linearized wave equation, Eq. (9), governing sound pressure p_s whose water-column parameter is the 4D sound-speed field c, Eq. (8). The acoustic pressure is decomposed in the frequency domain into slowly-varying complex envelopes that modulate (mode by mode) analytic, rapidly-varying, adiabatic-mode solutions. Given sound speed, density, attenuation rate and bathymetry vertical cross-sections, the acoustic state is obtained by integrating differential equations governing the complex modal envelopes. Model output contains sound pressure, transmission loss, and travel time, phase and amplitude of the individual modes. With model errors, a stochastic extension of Eqs. (1)–(9) is solved. Presently, only physical model errors are employed: they represent uncertainties due to sub-mesoscales and internal tides not accounted for in the deterministic mesoscale simulations, Eqs. (1)–(7).

2.2   Data Sets and Measurement Models

Presently, *in situ* physical data include profiles of temperature, salinity and velocities. Remotely-sensed data include satellite data (SSH, SST). Before being utilized, the raw measurements from XBTs, CTDs, ADCPs, current meters and satellites are processed via averaging, filtering, de-aliasing and calibration. Acoustic sensor observations are also processed to lead acoustic data such as sound pressure, travel time and transmission loss (TL). These coupled data are linked to the dynamics, Eqs. (1)–(9), by measurement models. Note that in general such models can be sophisticated so as to efficiently link the non-observed state variables in Eqs. (1)–(9) to the observed data and so as to account for all uncertainties, including these that occur in the processing.

2.3   Data Assimilation Approaches: Discrete Equations and Computations

In discrete terms, the physical-acoustical state is represented by a coupled state vector, $\mathbf{x}$, which is evolved from $\mathbf{x}(t_0) = \widehat{\mathbf{x}}_0$ based on, $d\mathbf{x} = \mathcal{M}(\mathbf{x})\, dt + d\boldsymbol{\eta}$, where $\mathcal{M}$ is the coupled model operator and $d\boldsymbol{\eta}$ are stochastic uncertainties. At time t_k, measurement models are of the form, $\mathbf{y}_k = \mathcal{H}(\mathbf{x}_k) + \epsilon_k$, where $\mathbf{y}_k$ is the observed data, $\mathcal{H}$ the measurement model operator and ϵ_k the stochastic uncertainties. The goal of the present four-dimensional data assimilation is to minimize the trace of the *a posteriori* error covariance of the coupled state, $\mathbf{P}_k^p(+)$, i.e. find $\mathbf{x}_k$ such that $J_k = \mathrm{tr}\left[\mathbf{P}_k^p(+)\right]$ is minimized using $[\mathbf{y}_0, ..., \mathbf{y}_k]$.

One, several or all of the acoustic variables can be included in the joint ocean-acoustic state space. Similarly, oceanic and acoustic fields can each be defined on both the physical and acoustical grids. This importantly extends the approach where acoustic computations are restricted to a high-resolution vertical plane while ocean computations are restricted to a lower-resolution volume grid. Solving for the physics and acoustics on both grids by data assimilation then provides internal wave physical resolution along the acoustic paths and range-averaged acoustical resolution on the whole ocean volume. Even though this is an ultimate goal, presently, our new coupled estimation is only illustrated for the ocean and acoustic states on their respective grids.

2.4   Data Assimilation Scheme

The coupled data assimilation methodology for field and uncertainty estimations is Error Subspace Statistical Estimation [6–8] ESSE is based on evolving an error subspace, of variable size, that spans and tracks the scales and processes where dominant errors occur.

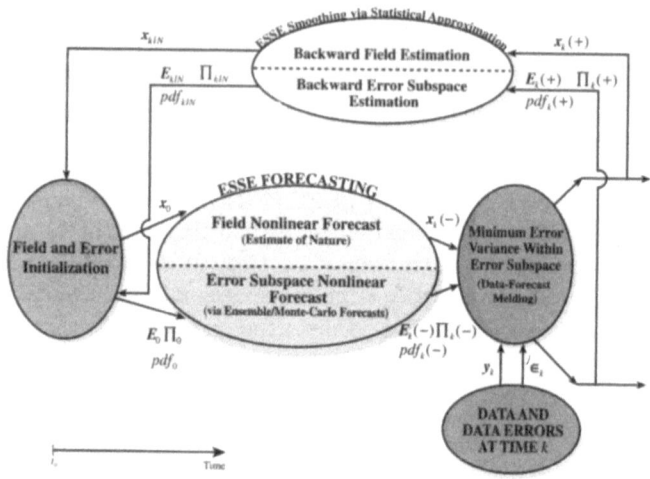

Figure 1. Five main components of the present ESSE system.

With ESSE, the sub-optimal reduction of errors is itself optimal. Presently (Fig. 1), the error subspace is initialized by decomposition on multiple scales [9] and the resulting estimate of the initial error eigendecomposition or error pdf (Fig. 1, blue oval) is used to perturb the initial state x_0. To evolve the physical fields and uncertainties (Fig. 1, light green oval), an ensemble of stochastic ocean model integrations, Eqs. (1)–(7), are carried out in parallel. The ensemble size is controlled by convergence criteria; when satisfied, the ensemble of ocean states leads to the physical forecast of nature $x_k(-)$ and to its error estimate, e.g. the error eigenvectors $E_k(-)$ and eigenvalues $\Pi_k(-)$ obtained by normalized SVD. With these physical fields and uncertainties, one computes an ensemble of 3D sound-speed fields. Each sound-speed realization then enters as a 3D parameter, Eq. (8), in an integration of the acoustic propagation model, Eq. (9). The acoustical ensemble is computed and, as for the physics, its size is controlled by convergence criteria. When satisfied, the acoustical and physical ensembles are concatenated to provide the coupled predicted fields and uncertainties. At this stage, the data and their error estimates (Fig. 1, dark green oval) are employed. Data-forecast misfits are computed and used to correct the predicted fields by minimum error variance estimation in the predicted physical-acoustical error subspace (Fig. 1, red oval). During this melding, acoustical data influence the physical state and vice-versa. The outputs are the *a posteriori* coupled fields $x_k(+)$ and *a posteriori* coupled errors, e.g. $E_k(+), \Pi_k(+)$. *A posteriori* data misfits are then calculated and used for adaptive learning of the dominant errors, e.g. [7]. This learning of errors from misfits can be necessary because error estimates are themselves uncertain. Ultimately, the smoothing via ESSE [8] can be carried out to correct, based on the data at times t_k, the initial coupled fields and uncertainties at t_0: this leads to $x_{0/N}$ and e.g., $E_{0/N}, \Pi_{0/N}$.

3   Illustrative example

The physics considered are the mesoscale dynamics of the Middle Atlantic Bight shelf-break front, including remote influences from the shelf, slope and deep ocean. The

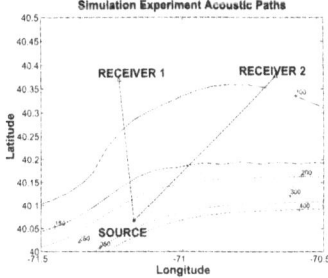

Figure 2. Acoustic paths considered (as in Shelfbreak-PRIMER), overlaid on bathymetry.

acoustics is the transmission of low-frequency sound from the continental slope, through the shelfbreak front, onto the shelf. These dynamics, and also the model parameters, data assimilated in the physical model, and acoustical-physical uncertainties are described in [10].

The coupled assimilation via ESSE is illustrated for the 3D physical fields and 2D transmission loss along an actual Shelfbreak-PRIMER [11] acoustic path (Fig. 2). The 224 Hz source is at 300 m depth. The acoustical data assimilated are simulated towed-array TL data along path 1, i.e. TL1 (the assimilation of simulated VLA's data at receiver 1 was also successful but is not shown). These model data were extracted from a physical-acoustical realization that is independent from the ensemble of 79 simulations carried out during the ESSE computations (Sect. 2.4). This independent realization is called the "true" ocean and such an assimilation exercise is called an "identical twin experiment". Goals in such an experiment are to study the assimilation in an ideal situation and to find out if the *a posteriori* fields become close to the known "true" fields. Presently, TL observations are made at constant 70 m depth, every 50 m from $r = 150$ m to almost receiver 1. These are very sub-sampled data since the (r, z) grid resolution is 5 m by 5 m.

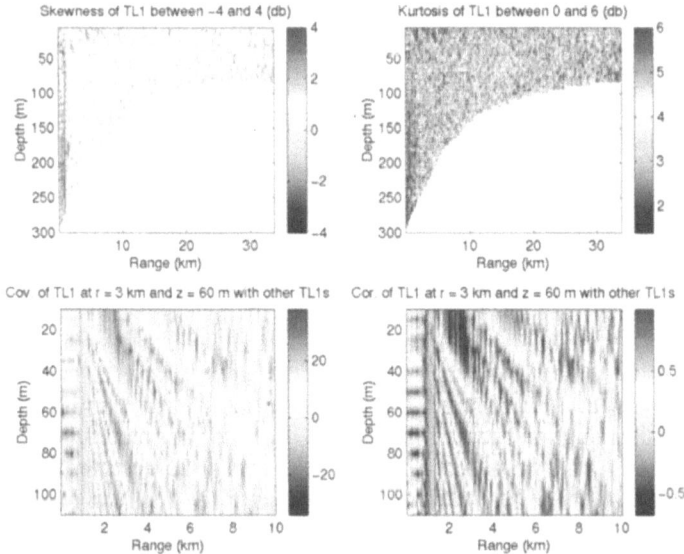

Figure 3. TL error statistics: sample skewness, kurtosis and zoom on a covariance/correlation field.

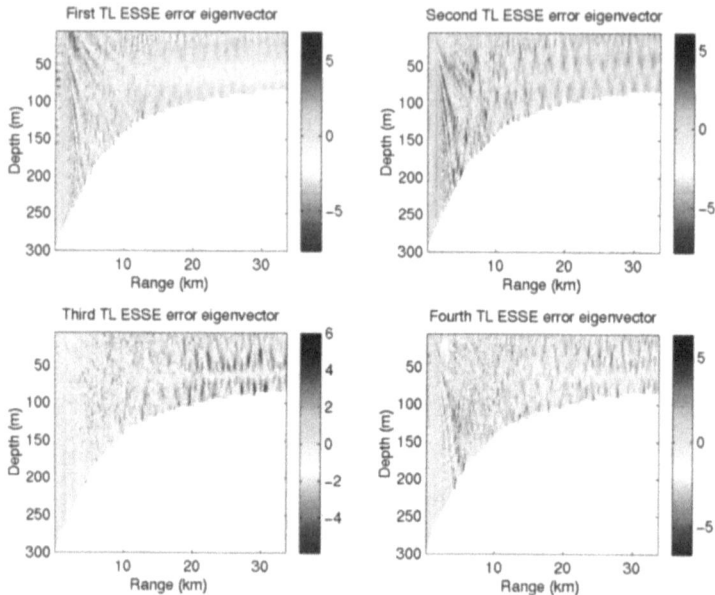

Figure 4. First four ESSE error eigenvectors for TL along section 1.

Figures 3 and 4 illustrate the predicted (*a priori*) TL uncertainties computed by ESSE. Except in the near field (where numerical errors in the acoustical model, Sect. 2.1, are the largest), the skewness and kurtosis (Figs. 3a-b) of the error pdf are patchy on small scales but relatively uniform at larger scales (around 1 and 5, respectively). The sensitivities of this pdf result to the size of the ensemble and to the properties of physical error pdf estimates need to be investigated. A sample estimate of error covariance and correlation functions between TL at ($r =3$ km, $z =6$ m) with other TLs (Figs. 3c-d) clearly shows the influence of acoustic wave patterns as they propagate through the shelfbreak front.

Figure 4 shows the four dominant singular vectors of the 79 TL deviations from the mean (*a priori*) TL. These ESSE estimates of the non-dimensional error eigenvectors account for 12.4, 6.0, 4.2 and 3.6 percent of the 2D acoustic error variance, respectively. They indicate the directions of the acoustic state space with the largest uncertainty. Presently, since we are just before the first assimilation of acoustic data, they also relate to the dominant acoustic variability. The first vector is linked to sub-thermocline propagation of sound onto the shelf, the second to uncertainties at the front due to the locally higher physical variability and the third and fourth (eigenvectors of similar eigenvalues) to successive reflections of sound waves between the thermocline and the bottom/surface.

Figures 5 and 6 illustrate the data assimilation in the predicted error subspace (Figs. 3–4). The simulated true TL, *a priori* (i.e. the mean) TL, *a posteriori* TL and the TL realization closest to the *a posteriori* TL are shown on Fig. 5. Even though the true TL (Fig. 5a) is challenging to retrieve (TL of high-order modal interactions) and the sub-sampled data are limited, the *a posteriori* TL (Fig. 5c) is substantially closer to the true TL than the mean TL (Fig. 5b). From the ensemble of 79 TLs, one can select for best estimate the TL the closest (in some metric sense, here the RMS measure over the r, z grid) to the *a posteriori* TL. This realization (Fig. 5d) is even a bit closer to the true TL than the *a posteriori* TL.

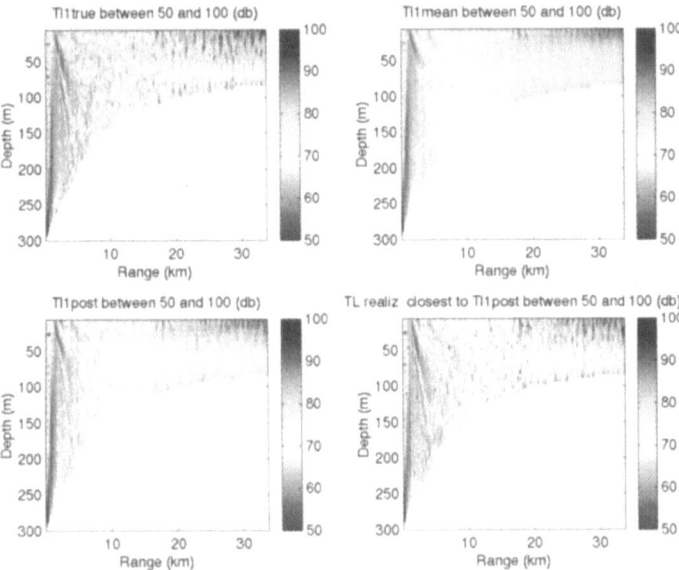

Figure 5. "True" TL, *a priori* TL, *a posteriori* TL and TL realization closest to *a posteriori* TL.

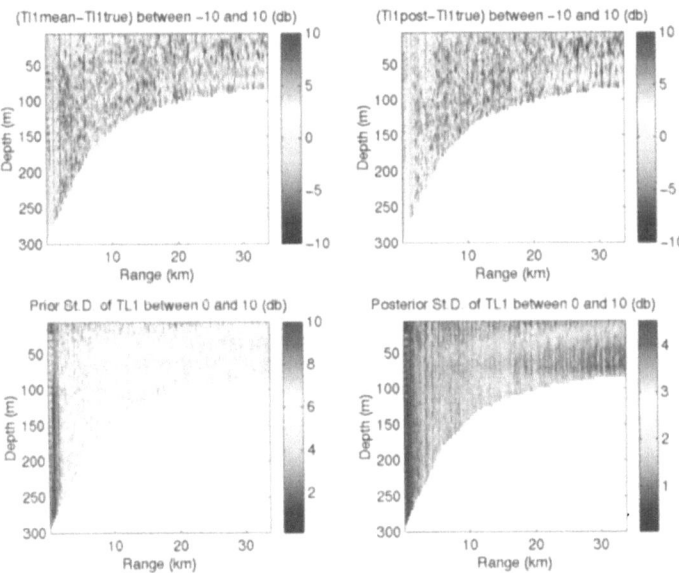

Figure 6. *A posteriori* residuals and *a posteriori* error St.Dv. for TL along section 1.

The differences between the *a priori* and true TLs, and between the *a posteriori* and true TLs, are shown on Figs. 6a–b. The *a posteriori* residuals (Fig. 6b) are much smaller than the *a priori* ones (Fig. 6a) at most locations, except above the thermocline near the surface on the shelf. This is due to the refractive effects of the thermocline (data are below at 70 m) and to the error subspace size (79) which is, based on convergence criteria, too small for accurate correlations everywhere in the large acoustic state ($\sim$4 10^5). With ESSE, error covariances are also estimated: the diagonals of the *a priori* and *a posteriori* error covariances are illustrated on Fig. 6c-d. Overall, these standard deviations agree with the averages of the residuals (note that their accuracy increases with the subspace size). In particular, the expected error along the simulated towed-array at 70 m has been reduced.

4 Conclusions

Coupled four-dimensional data assimilation for physical-acoustical field estimates was carried successfully via Error Subspace Statistical Estimation in the context of an identical twin experiment. Physical uncertainties were transferred to acoustical uncertainties and the dominant acoustical error statistics were decomposed and their properties examined. Results are encouraging and such coupled four-dimensional data assimilations have the potential to provide significant advances in physical and acoustical ocean science.

Acknowledgements

We thank Prof. A.R. Robinson and ONR for support under grants N00014–00–1–0771 and N0001402WR20213. PFJL thanks P.J. Haley and W.G. Leslie for software and data expertise, and P. Elisseeff and H. Schmidt. This is a contribution of the UNITES team.

References

1. Robinson, A.R. and Lermusiaux, P.F.J., Data assimilation in models. In *Encyclopedia of Ocean Sciences* (Academic Press Ltd., London, 2001) pp. 623–634.
2. Robinson, A.R., Physical processes, field estimation and an approach to interdisciplinary ocean modeling, *Earth-Science Rev.* **40**, 3–54 (1996).
3. Chiu, C.-S., Downslope modal energy conversion, *J. Acoust. Soc. Am.* **95**, 1654–1657 (1994).
4. Chiu, C.-S., Miller, J.H., Denner, W.W. and Lynch, J.F., A three-dimensional, broadband, coupled normal-mode sound propagation modeling approach. In *Full Field Inversion Methods in Ocean and Seismic Acoustics,* edited by O. Diachok, A. Caiti, P. Gerstoft and H. Schmidt (Kluwer Academic Publishers, 1995) pp. 57–62.
5. Chiu, C.-S., Miller, J.H. and Lynch, J.F., Forward coupled-mode propagation modeling for coastal acoustic tomography, *J. Acoust. Soc. Am.* **99**(2), 793–802 (1996).
6. Lermusiaux, P.F.J., Data assimilation via Error Subspace Statistical Estimation. Part II: Middle Atlantic Bight shelfbreak front simulations and ESSE validation, *Month. Weather Rev.* **127**(7), 1408–1432 (1999).
7. Lermusiaux, P.F.J., Estimation and study of mesoscale variability in the Strait of Sicily, *Dyn. of Atmos. Oceans* **29**, 255–303 (1999).
8. Lermusiaux, P.F.J. and Robinson, A.R., Data assimilation via error subspace statistical estimation. Part I: Theory and schemes, *Month. Weather Rev.* **127**(7), 1385–1407 (1999).
9. Lermusiaux, P.F.J., Anderson, D.G.M. and Lozano, C.J., On the mapping of multivariate geophysical fields: error and variability subspace estimates, *Q.J.R. Meteorol. Soc.*, April B, 1387–1430 (2000).
10. Lermusiaux, P.F.J., Chiu, C.-S. and Robinson, A.R., Modeling uncertainties in the prediction of the acoustic wavefield in a shelfbreak environment. In *Proc. 5th International Conference on Theoretical and Computational Acoustics*, Beijing, China, May 21–25, 2001.
11. Lynch, J.F., Gawarkiewicz, G.G., Chiu, C.-S., Pickart, R., Miller, J.H., Smith, K.B., Robinson, A., Brink, K., Beardsley, R., Sperry, B. and Potty, G., Shelfbreak PRIMER - An integrated acoustic and oceanographic field study in the mid-Atlantic Bight. In *Shallow-Water Acoustics,* edited by R. Zhang and J. Zhou. (China Ocean Press, 1997) pp. 205–212.

SOURCE LOCALIZATION IN A HIGHLY VARIABLE SHALLOW WATER ENVIRONMENT: RESULTS FROM ASCOT-01

MARTIN SIDERIUS

SAIC, 1299 Prospect St., La Jolla, CA 92037
E-mail: sideriust@saic.com

PETER NIELSEN

SACLANT Undersea Research Centre, Viale S. Bartolomeo 400, 19138 La Spezia, Italy
E-mail: nielsen@saclantc.nato.int

JÜRGEN SELLSCHOPP

Forschungsanstalt der Bundeswehr für Wasserschall und Geophysik,
Klausdorfer Weg 2-24, 24148 Kiel, Germany
E-mail: jsellschopp@bwb.org

Variability in the ocean environment can have a big impact on acoustic propagation. Acoustic receptions often contain multipath contributions with fluctuations that vary significantly from the direct path to the higher order multipath. Matched-field methods take advantage of the multipath to extract information about the source location and seabed properties. Matched-field processing is generally successful in environments that are not highly range dependent and do not vary significantly in time. However, in some cases the environmental conditions are too extreme for good propagation predictions and matched field results suffer. The ASCOT-01 acoustic experiments were conducted in June 2001 specifically to explore the limits of matched field methods in highly variable environments. Measurements were made over several days between a sound source and a moored vertical line array of receivers. Results characterizing the difficulties of matched field processing at this site will be presented. Source localization results using standard estimators will be compared with those using new alternatives intended to be robust against harsh environmental conditions.

1 Introduction

Matched-field processing (MFP) is a beamforming method that can take advantage of multipath propagation environments to extract information about the location of the sound source. It is usually applied in the ocean where multipath can degrade planewave beamforming yet enhance MFP results. The MFP technique uses a numerical propagation model to produce the beamforming replica fields for all possible source locations (instead of using planewave replicas) [1, 2]. With a vertical line array (VLA), MFP beamforming can estimate both source range and depth which also gives it an advantage over planewave beamforming. However, MFP often fails due to a lack of detailed environmental data needed for the propagation models. For instance, if the assumed seabed sound speed

N.G. Pace and F.B. Jensen (eds.), Impact of Littoral Environmental Variability on Acoustic Predictions and Sonar Performance, 425-432.

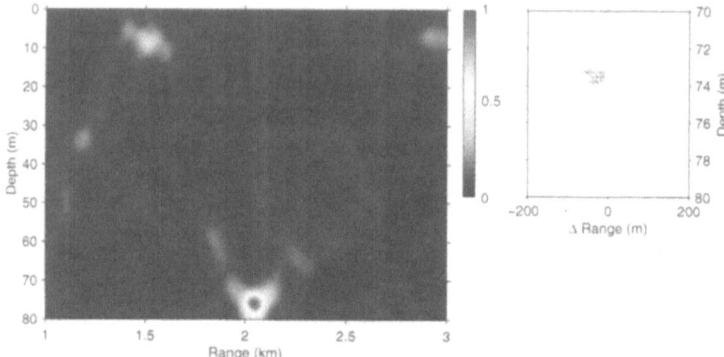

Figure 1. Source localization ambiguity surface for the Advent'99 site. Source position corresponds to the red spot at approximately 2 km range and 74 m depth. Panel on the right shows the position of the highest peak in the ambiguity surface for all data that were processed over the 5 hours of data collected. Because the search contained discrete locations, many source position estimates fall on top of each other. Note: the colors in the left panel are scaled to the maximum value.

is incorrect, the replica fields will contain errors that degrade the beamforming results. This can partially be mitigated by considering a larger set of replicas that includes all possible environments for all possible source positions. This would produce an enormous set of replica fields but, in practice, this can be done with a relatively small number using an optimization, or "focalization" procedure [3]. The focalization procedure searches for environments with replicas yielding a higher beamformer output. The added benefit of focalization is to simultaneously produce an estimate of the unknown environmental properties such as seabed sound speed. When the source location is known, the search is only for environmental properties and this is the basis of MFP geoacoustic inversion.

Matched-field source localization and geoacoustic inversion were tested using data collected during the Advent'99 experiments [4, 5]. These experiments were conducted on the Adventure Bank in the Strait of Sicily in May 1999. Acoustic data were collected on a VLA from sources transmitting in the 200–1600 Hz frequency band. The source was localized using data from transmissions taken over 5 hours containing 6 tones (200–700 Hz in 100 Hz increments) and linear frequency modulated (LFM) sweeps (200–800 Hz). An example of source localization for the Advent'99 site is shown in Fig. 1. The localization results shown used the multi-tone transmissions but the results were nearly identical for the LFM data. The figure indicates excellent localization of the source using data collected over 5 hours.

In addition to localization, the Advent'99, 2-km data was processed for geoacoustic properties using matched-field inversion. As with the localization results, the estimated geoacoustic properties were extremely stable for all the data inverted over the 5 hours. However, this was not true for the data taken at 10 km. Variability (both temporal and spatial) in the ocean sound speed could not be included in the range-independent replica modeling and this caused errors that were sufficient to prevent the search algorithm from finding the correct seabed properties. The 10-km geoacoustic inversion results differed from those at 2 km and erroneously changed with time as a result of changes in the ocean sound speed profile. It was concluded that unless the impact of variability could be compensated for in the modeling, MFP geoacoustic inversion was better suited for short range measurements [5]. Even with the instability in the 10-km geo-acoustic inversion results the source was correctly localized over most of the 18 hours of data. To summarize the Advent'99 results, the localization was extremely stable for data below 700 Hz and at all ranges (with a few outliers at 10 km range). Recent work with the Advent'99 data has shown that all data could be successfully localized if the ocean sound speed profile is included in the focalization procedure [7]. That is, the ocean variability could be compensated for by optimizing a parameterized ocean sound speed profile. These recent results show good localization results at 10 km even for frequencies as high as 1500 Hz. For this reason the Advent'99 site is considered a relatively forgiving environment for MFP which is not the case for the ASCOT-01 site described in the next section.

2   The ASCOT-01 acoustic experiments

The ASCOT-01 acoustic experiments took place June 12–18, 2001 off the coast of New England. The experiment design was similar to that of Advent'99 and ideal for testing MFP localization and geoacoustic inversion. A 64-element VLA was moored at ranges of approximately 1, 2, 5 and 10 km from the sound source. The water depth averaged 102 m and the VLA spanned depths of 28–94 m. The sound source was about 4 m from the bottom as it was mounted in a steel frame tower that was sitting on the seabed. The NATO Research Vessel *ALLIANCE* was used to both power the sound source and receive the VLA data by radio telemetry. To capture the oceanographic conditions, extensive environmental measurements were made: three moored thermistor strings measured the temperature profile and an Acoustic Doppler Current Profiler (ADCP) was used to estimate ocean currents. In addition, a vertical chain containing conductivity, temperature and pressure (CTD) sensors was towed along the acoustic tracks and later processed for sound speed. The CTD chain covered the water column down to about 70 m. Below that depth there was very little change in the sound speed. The ASCOT-01 site had a sound speed profile that changed over depth by about 40 m/s compared to the Advent'99 site that changed by about 5 m/s. A more detailed comparison of the relationship between the acoustic data and environmental condition for these two shallow water sites can be found in Ref. [6].

3   Matched field processing of ASCOT-01 data

Central to MFP is the correlation function that quantifies the agreement between the measured field and the computed replicas. A common correlation function (often called the Bartlett Processor) is given below:

$$B = \frac{1}{N_{\mathrm{F}}} \sum_{j=1}^{N_{\mathrm{F}}} \frac{|\sum_{i=1}^{N_{\mathrm{H}}} p_i(\omega_j) q_i(\omega_j)^*|^2}{\sum_{i=1}^{N_{\mathrm{H}}} |p_i(\omega_j)|^2 \sum_{i=1}^{N_{\mathrm{H}}} |q_i(\omega_j)|^2}, \tag{1}$$

where N_{F} is the number of frequency components, N_{H} is the number of hydrophones and the measured and modeled complex pressure vectors (at frequency ω_j) are p_i and q_i (* denotes the complex conjugate operation). This correlator has a value of 1 for a perfect match between measured data and replica and 0 when uncorrelated. This is the processor used to produce the ambiguity surface in Fig. 1.

A focalization procedure similar to that used to produce Fig. 1 for the Advent'99, 2-km data was applied to the ASCOT-01 data at the same range. That is, replicas were generated

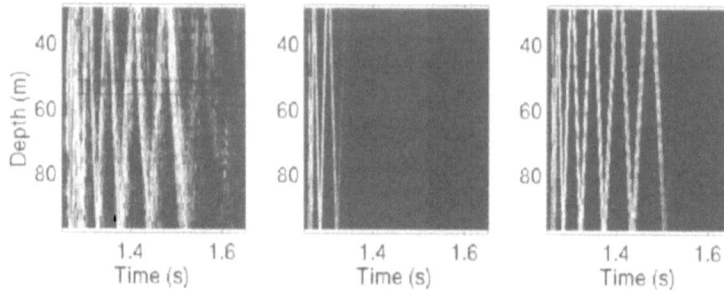

Figure 2. Left panel shows the measured, band limited impulse response data on the VLA as a function of time. Acoustic paths that travel at steeper angles have longer paths and arrive later in time. Middle panel shows the result of simulating the received time series using the environment found in the focalization MFP. Right panel is also a simulation of the received time series except with the seabed fixed to be 1750 m/s. Both simulation use ray-trace propagation model BELLHOP [10]. (Relative amplitudes are shown on a log scale with colors spanning 30 dB).

for all possible source positions as well as for many environmental conditions using the normal mode propagation model SNAP [8]. A genetic algorithm was used to direct the focalization search [9]. As was done for the Advent'99 analysis, the ocean channel was assumed range-independent and focalization included the seabed sound speed, density and attenuation as well as the water depth. Data from the LFM transmissions were used at frequencies of 225–725 Hz in 50 Hz increments (similar results were found using multi-tone data in the same frequency band).

In contrast to the Advent'99 results, the source was *not* correctly localized. The most likely source depth was found to be 27 m instead of 98 m and there were no secondary peaks anywhere near the true source position. Further, the correlation (Eq. (1)) produced a value of only 0.3 compared to 0.8–0.9 for the Advent'99 data. The focalization results for the seabed properties indicated a slow sound speed seabed which was different from that expected (sediment maps of the area showed a highly reflective material typical of a fast sound speed). The seabed sound speed determined from the focalization was 1540 m/s which was near the lower limit of the search interval. A plot of the measured, band limited (200–800 Hz), impulse response on the VLA together with the modeled time series using the focalized environment is shown in the left two panels of Fig. 2. Clearly, the modeled time series does not re-create the arrival pattern seen in the data. The late arrivals correspond to steep angles of propagation. The presence of these implies a highly reflective seabed. Shown in the right panel of Fig. 2 is an improved model for the time series found after just a few attempts at adjusting the sound speed in the seabed (results shown are for 1750 m/s).

Although the data from Advent'99 produced completely stable results for geoacoustic properties and localization, the results for the ASCOT-01 site are very different. It is clear from Fig. 2 that the focalization process did not work and the seabed properties are better modeled with a fast (1750 m/s) sound speed. Would using this *ad hoc* value for the seabed improve MFP localization? The answer to that question is given by the ambiguity surface in right panel in Fig. 3 where a fixed environment was used with seabed sound speed of 1750 m/s. As before, there is no evidence of the source that should appear at a range of about 1.85 km and depth of 98 m. The data collected at 1.85 km was intended

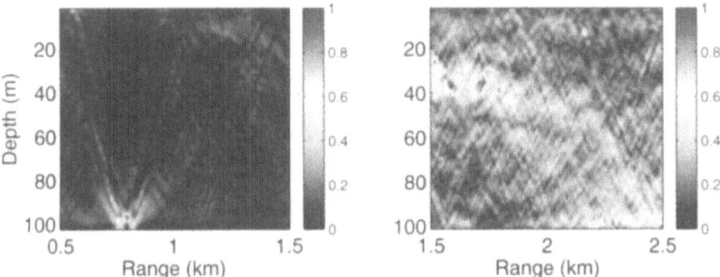

Figure 3. MFP source localization ambiguity surface (using SNAP) for data collected during ASCOT-01. Left panel is the results for the source at about 0.78 km and the right panel for source at about 1.85 km (depth about 98 m). It appears possible to localize the source for the data at 0.78 km but not for 1.85 km.

to provide a sanity check for the MFP localization at close range. However, localization difficulties at that range were already becoming obvious from analysis that took place during the experiments. Therefore, a new sanity check range was chosen at 1 km (which after deployments turned out to be 0.78 km). Localization results for that range provided the sanity check and are shown in the left panel of Fig. 3.

4 An alternative matched field processor

As shown, information can be inferred about the environment by simply making observations about the arrival structure (Fig. 2). To match the extent of multipath arrivals seen with the ASCOT-01 data the seabed must support steep angle propagation and therefore is highly reflective. The measured arrival structure along the array was reasonably well matched using a manual process of changing seabed properties and observing the ray-trace, time series. In this sense the manual process out-performed MFP geoacoustic inversion. The question is: why are the MFP results not producing the correct source location or the correct environment? And, can the information in plots like Fig. 2 be used to better localize and determine geoacoustic properties? The answer to the first question is addressed in Sect. 5 and in Ref. [11]. The second question requires defining a new correlation function that compares quantities similar to those shown in Fig. 2. This requires a cross-correlation of the envelope of the pressure time-series which can be done at each hydrophone depth. To make use of the full array these correlation values can be added together. Or,

$$Corr = \frac{1}{N_{\rm H}} \sum_{i=1}^{N_{\rm H}} \mathcal{R}(|\mathcal{E}(p_i(t_j))|; |\mathcal{E}(q_i(t_j))|)^2, \tag{2}$$

where $N_{\rm H}$ is the number of hydrophones, $p(t_j)$ is the discrete-time, measured pressure, $q(t_j)$ is the modeled time series, $\mathcal{R}$ represents taking the cross-correlation maximum value and $\mathcal{E}$ represents taking the time-series envelope.

The exact same data pings and modeling environment used to create the ambiguity surfaces in Fig. 3 were used with Eq. (2) to make new ambiguity surfaces and these are shown in the top panels of Fig. 4. The source is correctly localized at both .78-km and 2-km ranges. Focalization was *not* used (i.e. the environment was fixed for computing the

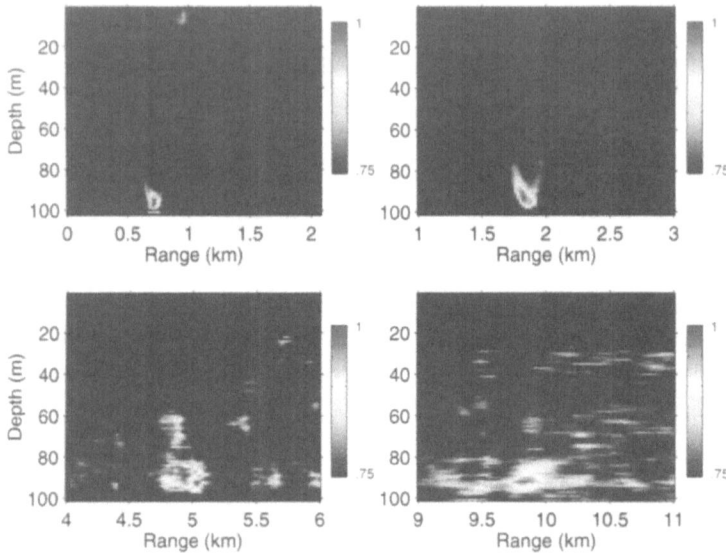

Figure 4. Source localization ambiguity surfaces for ASCOT-01 data taken from 4 source-receiver ranges using Eq. (2). In the top left panel the source is at about 0.78 km, top right about 1.85 km, bottom left about 5.03 km and bottom right about 9.76 km (depth for all is about 98 m).

replicas). Note, that in Fig. 4 a larger search range of possible source locations is used (to see if false peaks were nearby) and the color scale is smaller. The replicas for correlating were generated using the BELLHOP propagation model. The ASCOT-01 environment is probably not an unusual one. Similar observations about robust features of the impulse response have previously been made at different sites and correlation functions of the envelope data were used to estimate source location [12].

The envelope correlation was further tested using longer range data. In the bottom panels of Fig. 4, data is taken with the source about 5 and 10 km from the VLA. In both cases the source is localized correctly.

The localization results shown so far have been taken from relatively low frequency data (200–800 Hz). This band is very relevant to the study of passive sonar applications. However, many active sonar applications extend to frequencies well above this. It is a difficult task to apply standard MFP localization to higher frequency data. That is because accurate modeling is more difficult at higher frequency since the relevant time and length scales of the environment get smaller and require finer sampling. However, observing the arrival patterns on the VLA from data taken centered around 1.2 kHz shows similar arrival features as seen with the lower frequency data. Further, using the BELLHOP ray propagation model, the computation time is about the same as for the low frequency band. Localization results using the envelope of 1.2 kHz (center frequency of 1-s LFM) data collected at 5 km is shown in Fig. 5. As with the low frequency data, the source is localized correctly. Note, that as before no focalization or optimization on the environment or measurement geometry was applied to the processing.

For the localizations previously shown, the source waveform was assumed known and therefore a matched-filter could be applied. This is a reasonable assumption for applications such as geo-acoustic inversion or active sonar. In other applications such

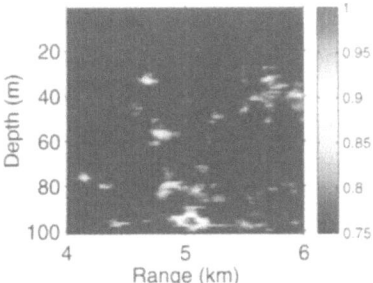

Figure 5. Source localization ambiguity surface for 1.2 kHz (center frequency) ASCOT-01 data using Eq. (2). Results are for the source located about 5.03 km away.

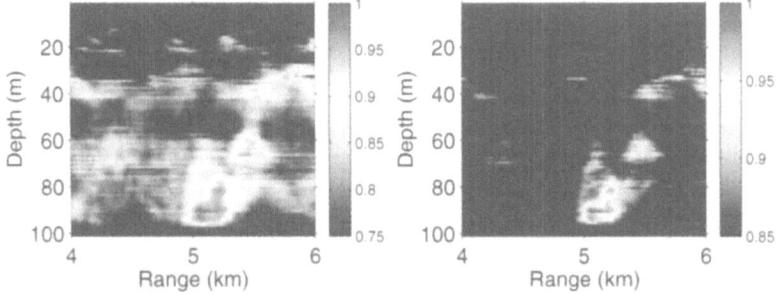

Figure 6. Ambiguity surface for 5.03 km data match-filtered with a single beam time series (200–800 Hz LFM). Color scale for left panel is the same as for Fig. 4 and is decreased slightly on right.

as passive sonar, the source waveform will likely be unknown. Even without prior knowledge of the source waveform, there are several ways to use the envelope correlation for localization (or possibly geoacoustic inversion from sources of opportunity). The envelope correlation uses the relative timing and amplitudes of the multipath arrivals which implies that receptions must be broad-band and pulse compressed. One way to achieve the pulse compression is to auto-correlate or cross-correlate the hydrophone data and compare this with the replica fields processed in a similar way. The auto- or cross-correlations provide pulse compression but introduce additional peaks that appear like multipath. This may introduce some ambiguities but these may not degrade the results since similar features are reproduced with the replica data. This type of processing has been used successfully on hydrophone data taken from a highly sparse array [13].

Another approach to estimating the source waveform is to take advantage of the vertical array and planewave beamform. Taking individual beams will eliminate (or reduce) the multipath interference and the beam time series provides an estimate of the source transmit waveform. This can then be used as the matched-filter and the VLA data can be processed for a pulse-compressed (impulse-like) response. Using the beamforming approach to estimating the source waveform produced an estimated matched-filter arrival pattern that was nearly identical to that using the true matched-filter. In Fig. 6 source localization is shown for this data using exactly the same replica fields as for the bottom left panel of Fig. 4.

5   Discussion and conclusion

While standard MFP processing was successful in the Strait of Sicily, it failed to localize a source just 2 km away for a site off the New England coast. The failure is likely due to the multipath arrivals that are strongly affected by the spatial variability of the ocean. It is not practical (or even possible) to include this variability in the propagation modeling needed for MFP. An alternative process is described that achieves greater stability by correlating the time series envelope. Results presented show correct localization out to 10 km even at frequencies above 1 kHz. Further, this process was shown to be successful at localization even when no prior knowledge of the source transmit waveform was assumed.

Acknowledgments

This work was initiated to support the SACLANTCEN Programme of Work. The analysis was supported by ONR.

References

1. Baggeroer, A.B., Kuperman, W.A. and Mikhalevsky, P.N., An overview of matched field methods in ocean acoustics, *IEEE J. Ocean Eng.* **18**, 401–424 (1993).
2. Tolstoy, A., *Matched-Field Processing for Underwater Acoustics* (World Scientific, Singapore, 1993).
3. Collins, M.D. and Kuperman, W.A., Focalization: Environmental focusing and source localization, *J. Acoust. Soc. Am.* **90**, 1410–1422 (1991).
4. Sellschopp, J., Siderius, M. and Nielsen, P., Advent'99 pre-processed acoustic and environmental cruise data. **CD-35**, SACLANT Undersea Research Centre, La Spezia, Italy (2000).
5. Siderius, M., Nielsen, P., Sellschopp, J., Snellen, M. and Simons, D., Experimental study of geo-acoustic inversion uncertainty due to ocean sound-speed fluctuations, *J. Acoust. Soc. Am.* **110**, 769–781 (2001).
6. Nielsen, P., Siderius, M. and Sellschopp J., Broadband acoustic signal variability in two "typical" shallow-water regions. In *Impact of Littoral Environmental Variability on Acoustic Predictions and Sonar Performance,* edited by N.G. Pace and F.B. Jensen (Kluwer, The Netherlands, 2002) pp. 237–244.
7. Soares, C., Siderius, M. and Jesus, S., Source localization in a time-varying ocean waveguide, *J. Acoust. Soc. Am.* (to appear 2002).
8. Jensen, F.B. and Ferla, M.C., SNAP: The SACLANTCEN normal-mode acoustic propagation model. Report SM-121, SACLANT Undersea Research Centre, La Spezia, Italy (1979).
9. Gerstoft, P., SAGA Users manual 2.0. An inversion software package. Report SM-333, SACLANT Undersea Research Centre, La Spezia, Italy (1997).
10. Porter, M.B., The BELLHOP ray/beam acoustic propagation model. http://oalib.saic.com
11. Sellschopp, J., Nielsen, P. and Siderius, M., Combination of acoustics with high resolution oceanography. In *Impact of Littoral Environmental Variability on Acoustic Predictions and Sonar Performance,* edited by N.G. Pace and F.B. Jensen (Kluwer, The Netherlands, 2002) pp. 19–26.
12. Porter, M.B., Jesus, S., Stephan, Y., Demoulin, X. and Coelho, E., Exploiting reliable features of the ocean channel response. In *Shallow Water Acoustics,* edited by by R. Zhang and J. Zhou (China Ocean Press, Beijing, 1997).
13. Porter, M.B, Hursky, P. and Tiemann, C.O., Model-based tracking for autonomous arrays. In *Proc. MTS/IEEE Oceans 2001,* 786–792 (2001).

EXPERIMENTAL TESTING OF THE BLIND OCEAN ACOUSTIC TOMOGRAPHY CONCEPT

S.M. JESUS AND C. SOARES

SiPLAB-FCT, University of Algarve, Campus de Gambelas, PT-8000 Faro, Portugal
E-mail: {sjesus,csoares}@ualg.pt

J. ONOFRE

Instituto Hidrográfico, Rua das Trinas 49, PT-1000 Lisboa, Portugal
E-mail: mesquita.onofre@hidrografico.pt

E. COELHO

SACLANT Undersea Research Centre, Viale San Bartolomeo 400, 19138 La Spezia, Italy
E-mail: coelho@saclantc.nato.int

P. PICCO

ENEA, Marine Environment Research Centre P.O. Box 224, 19100 La Spezia, Italy.
E-mail: picco@estosf.santateresa.enea.it

Acoustic focalization is a well known concept that aims at estimating source location through the adjustment of multiple environmental parameters. This paper uses the same concept for inverting water column sound speed in a blind fashion, where both source location and source emitted waveform are not known at the receiver - that is Blind Ocean Acoustic Tomography (BOAT). The results obtained with BOAT, using ship noise data received on a vertical line array in a shallow water area off the coast of Portugal, show that it is indeed possible to obtain reliable joint estimates of source location and water column sound speed. During that process, it was shown that source range and depth, and Bartlett power, where good indicators of the degree of focus of the model being used.

1 Introduction

A consistent idea behind ocean acoustic tomography is that source and receiver relative positions, as well as source emitted signal characteristics, should be known to a high degree of precision. Deviations from this assumption generally have a direct impact in the inversion result. From a different perspective, Collins *et al.* [1], suggested that source localization could be greatly facilitated by including additional (known) parameters into the search process in order to allow a better fit of the replica model - that is a technique known as *acoustic focalization*. The same concept has been readily used for generic parameter estimation in [2], for geoacoustic inversion in [3, 4] and for source localization in [5–7].

A rather different concept has been proposed by Jesus *et al.* [8], that attempts to estimate channel propagation physical characteristics together with source properties. By

433

N.G. Pace and F.B. Jensen (eds.), Impact of Littoral Environmental Variability on Acoustic Predictions and Sonar Performance, 433-440.

source properties, it is meant that the source position as well as the source emitted waveform are unknown - this is Blind Ocean Acoustic Tomography (BOAT). Technically, BOAT has little difference from acoustic focalization, apart from the fact that the search space is enlarged to include truelly unkwown geometric, geoacoustic and water column parameters and that source characteristics are also unknown. There are great risks associated with such a global inversion procedure, one of which is that the final result may represent an "equivalent acoustic model" that may be far different from the true environment being sought. This is mainly due to the dramatic increase of the search space dimension and consequently an increase of the number of local maxima of the acoustic based objetive function. It was shown in [8], using active source data, that it is indeed possible to use source location parameters and the Bartlett power as indicators of the degree of "focus" of the environmental model, thus providing reliable water column sound speed estimates, when compared with independent recorded data. This paper pushes even further the concept of unknown source waveform by using acoustic ship noise, instead of deterministic high power source signals, to invert the environmental characteristics of a mildly range-dependent 3.3 km long track over a time interval of 1.5 h. In that regard it extends the preliminary results shown in [9] in the same data set, where a single time slot (2 s duration) at 1 km range and over a range-independent track was successfully inverted.

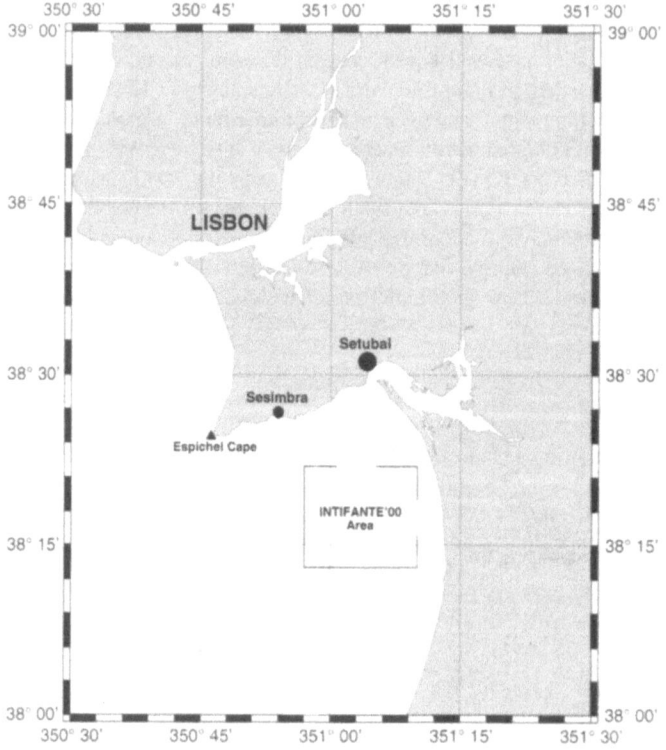

Figure 1. Localization of the INTIFANTE'00 sea trial.

2   The INTIFANTE'00 sea trial and the baseline model

The INTIFANTE'00 sea trial took place during October 2000, near the town of Setúbal, approximately 50 km south from Lisbon, in Portugal (see Fig. 1). An overall description of the sea trial can be found in [10], whereas in this paper the interest will be focused only on Event 6, during which the signals received at the 16-hydrophone vertical line array (VLA) consisted on the noise radiated by the research vessel NRP D. Carlos I, cruising over a mildly range-dependent area up to 3.3 km range from the VLA. The NRP D. Carlos I is a 68 m overall length hydrographic ship with a gross displacement of 2800 tons. Her propulsion is obtained from a double helical diesel-electric engine with a total shaft power of 800 HP, attaining a maximum speed of 11 kn. It should be noted that NRP D. Carlos I was originally built for acoustic surveying so she is supposed to be a rather quiet ship. During Event 6, NRP D. Carlos I performed a triple bow shaped pattern at approximate ranges of 1.2, 2.2 and 3.3 km from the VLA as shown in Fig. 2. A detailed

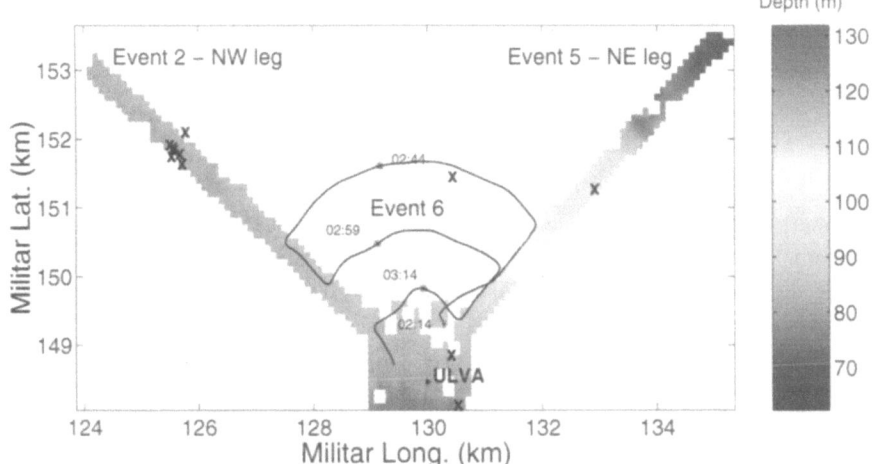

Figure 2. INTIFANTE'00 sea trial Event 6 and site bathymetry. XBT casts locations are marked with **X** and **ULVA** denotes the VLA location.

bathymetry of the area was not available, but approximate bathymetric profiles were made along both the NW and the NE tracks, as shown on Fig. 2. Therefore acoustic propagation between the ship and the VLA is assumed to be slightly downslope range-dependent to the NE, and progressively becoming range-independent, at 120 m water depth, to the NW. The maximum range-dependence is obtained for the 3.5 km range bow, with a maximum water depth difference of 20 m at the NE track. A number of XBT casts were made during the sea trial at various times and locations as marked by the **X** signs on Fig. 2.

Ship's speed and heading, as obtained from GPS, is shown in Fig. 3, plots (a) and (b) respectively. It can be seen that mean ship speed was about 9 kn with several abrupt drops to 7 kn during the ship sharp turns along the triple bow trajectory.

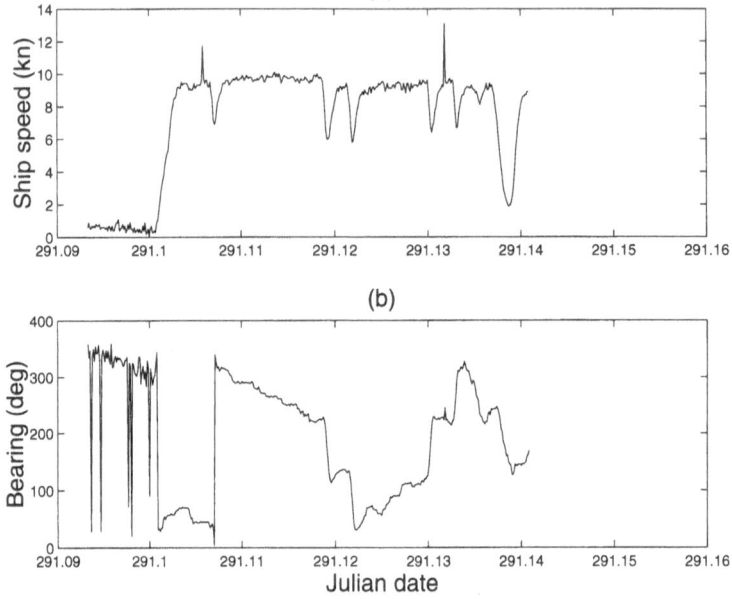

Figure 3. Event 6: GPS measured ship speed (a) and ship heading (b).

2.1   The Baseline Model

An important step towards a successful data inversion relies on the choice of a suitable environmental model. There was no extensive oceanographic or geoacoustic survey concerning the area of Event 6, and therefore, as in previous work [8], the same generic assumptions based on archival data were adopted, giving rise to the baseline model pictured in Fig. 4. Geoacoustic characteristics were empirically drawn from geological tables where the bottom was catalogued as "fine sand layer". A two Empirical Orthogonal Function (EOF) based model was used to represent the water column sound speed evolution through time and space, which coefficients are estimated together with the other

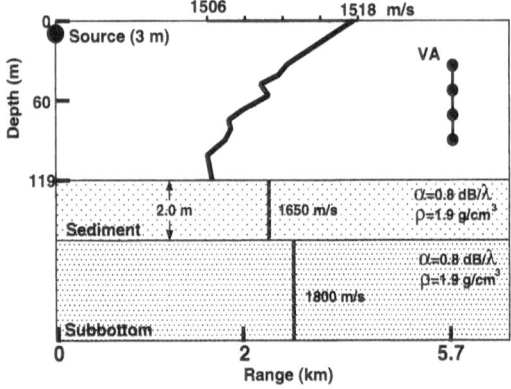

Figure 4. Baseline model for ship noise data inversion during Event 6.

Table 1. Focalization parameters and search intervals: EOF1 (α_1), EOF2 (α_2), source range (sr), source depth (sd), receiver depth (rd), VLA tilt (θ).

Symbol	Unit	Search int./Steps		
α_1	m/s	-20	20	64
α_2	m/s	-20	20	64
sr	km	0.5	3.5	64
sd	m	1	10	32
rd	m	85	95	32
θ	rad	-0.03	0.03	32

parameters. The EOF's are deduced from XBT data taken at locations throughtout the experimental site, thus incorporating space and time variability into the EOF expansion (see **X** signs in Fig. 2). The searched parameters and their respective search intervals are listed in Table 1.

2.2 Ship Radiated Noise

Acoustic signals were received on a moored vertical line array (VLA) with 16 hydrophones, with its shallow and deep hydrophones positioned at nominal depths of 32 and 92 m, respectively. Array depth and tilt were recorded during the experiment. Figure 5 shows the signals received on hydrophone 8 (60 m depth), on panel (a) a time-frequency plot, and in panel (b) a mean power spectrum over the whole event. There are clearly a few characteristic frequencies emerging from the background noise between 250 and 260 and a strong single tone at 359 Hz. There is also a coloured noise spectra in the band 500 to 700 with however, a much lower power. As a preliminary analysis, a power spectrum estimator was run on a 8 s sliding time window throughout all the event duration and the maximum power frequency bins were automatically extracted. Figure 6 shows the selected frequency bins that were then used in the inversion procedure.

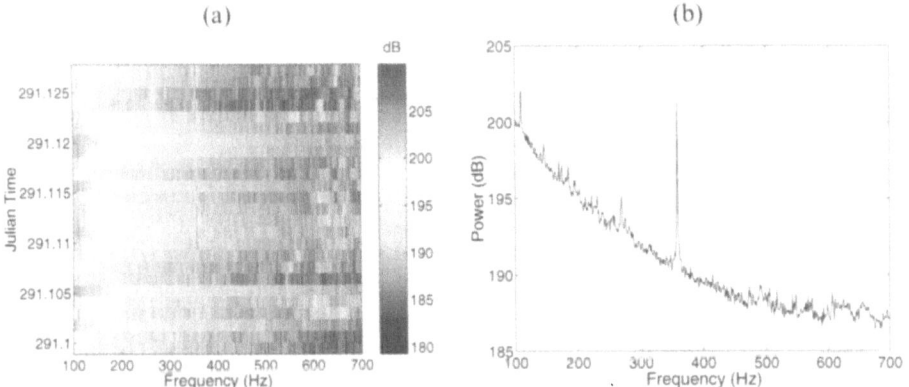

Figure 5. NRP D. Carlos I ship radiated noise received on hydrophone 8: time-frequency plot (a) and mean power spectrum (b).

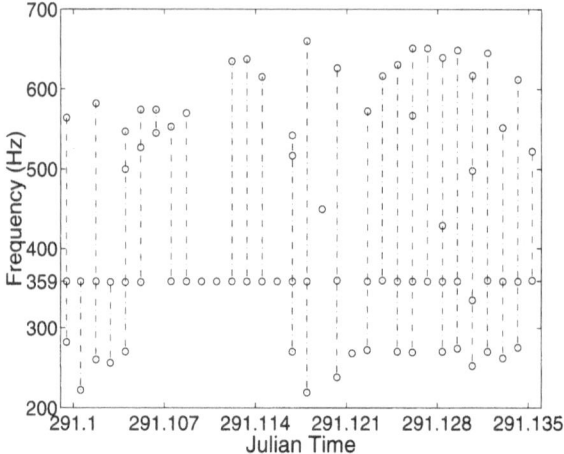

Figure 6. INTIFANTE'00 sea trial, event 6: selected frequency bins for inversion.

3 Inversion results

The inversion methodology was based on a three step procedure: i) preliminary search of the outstanding frequencies in a given time slot, ii) parameter focalization, based on an incoherent broadband Bartlett processor, a C-SNAP forward acoustic model and a GA based optimization and iii) inversion result validation based on model fitness and coherent source range and depth estimates through time. The inversion results are shown in Fig. 7, from (a) to (g) are individual parameter estimates while plot (h) shows the water column sound speed reconstruction based on the EOF linear combination based on parameter estimates (f) and (g).

At first glance the results are poor: Bartlett power is low, always below 0.8; source range and depth, which are leading parameters, show highly incoherent values; and finally the reconstructed sound speed is too variable for such a small time interval (less than 1 h 30 min). Looking more in detail, and comparing plot (a) of Fig. 3 with plot of Fig. 7 the following conclusions can be drawn:

1. For time $\leq$ 291.101 no results can be obtained since the ship is at low speed, Bartlett power is low and parameter estimates are messy.

2. For 291.101 $\leq$ time $\leq$ 291.107, ship speed increases steeply to 9 kn, while heading away from the VLA. Range variation is about 4.6 m/s which, may cause a violation of the stationary assumption during the averaging time. Estimates are also messy during this period.

3. At time = 291.108, the ship makes a sharp turn (slowing down to 7 kn) and initiates the first bow at an approximate range of 3.2 km and at constant speed of 9 kn. At this point the estimation peaks up for a time period between 291.109 to 291.130, i.e. approximately 30 min with a unique exception at 291.119 where there is an estimation loss. That estimation loss curiously corresponds to another sharp turn when the ship heads towards the array between the 3.2 km and the 2.2 km bow. With that exception it can be noticed that range estimates generally coincide to the GPS measurements. There is a slight range error at the begining of the 3.3 km bow

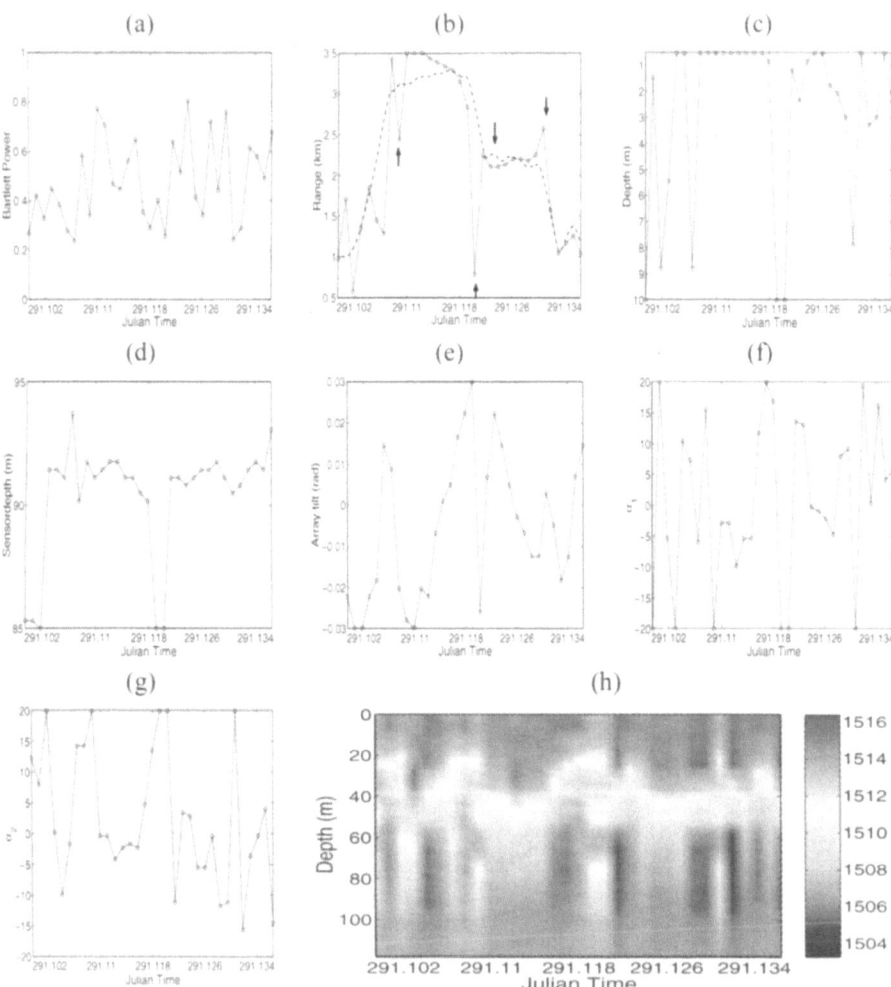

Figure 7. Focalization results for Event 6: Bartlett power (a), source range and GPS measured range (dashline) - arrows indicate sharp ship turns (b), source depth (c), receiver depth (d), VLA tilt (e), EOF coefficient 1 (f), EOF coefficient 2 (g) and reconstructed sound speed (h).

(t=291.11) possibly due to range-dependency mismatch which is the strongest at this point (20 m depth difference between source and array position). Sensor depth and array tilt show credible values with an interesting behaviour of the later that varies from -0.03 to +0.03 almost linearly and possibly due to a change of the source view angle due to the 45 degree turn along the bow shaped track.

4. At time 291.131 there is another sharp turn towards the array with a speed drop and another estimation loss.

5. For the remaining few minutes the parameter estimation resumes during the closest range bow at 1.2 km from the VLA.

As a final comment, it can be added that the reconstructed sound speed suffers from the consecutive losses of estimation that fully coincide with the estimation losses verified in the other curves and are directly related to ship maneuvering and speed drops.

4   Conclusion

Ocean tomography with sources of opportunity has been a longely sought dream for acoustic oceanographers. This paper presents a preliminary result obtained in a shallow water area off the continental coast of Portugal during the INTIFANTE'00 sea trial, that proves the feasibility of the BOAT concept using ship radiated noise as input signal for tomographic inversion. Although the duration of the acoustic data was manifestly too short for a complete oceanographic validation, it was clearly seen that the estimates could be validated by the conjugation of the Bartlett power and obtaining coherent source range and depth values. Using this three parameters it becomes clear when the environment is "in focus" and when it is "out of focus". The computational effort was reasonabe due to an efficient carry on of model information from one time sample to the next in order to allow a simplified search.

Acknowledgements

This work was supported by FCT, Portugal, under projects INTIMATE (2/2.1/MAR/1698/95) and ATOMS (PDCTM/P/MAR/15296/1999). The authors are also in debt to the crew of NRP D. Carlos I of IH, that made the sea trial successful.

References

1. Collins, M.D. and Kuperman, W.A., Focalization: Environmental focusing and source local- ization, *J. Acoust. Soc. Am.* **90**(3), 1410–1422 (1991).
2. Gerstoft, P. and Gingras, D., Parameter estimation using multi-frequency range-dependent acoustic data in shallow water, *J. Acoust. Soc. Am.* **99**(5), 2839–2850 (1996).
3. Gerstoft, P., Inversion of seismoacoustic data using genetic algorithms and *a posteriori* prob- ability distributions, *J. Acoust. Soc. Am.* **95**(2), 770–782 (1994).
4. Hermand, J.-P. and Gerstoft, P., Inversion of broad-band multitone acoustic data from the YELLOW SHARK summer experiments, *IEEE J. Oceanic Eng.* **21**(4), 324–364 (1996).
5. Soares, C., Waldhorst, A. and Jesus S., Matched field processing: Environmental focusing and source tracking with application to the North Elba data set. In *Proc. MTS/IEEE Oceans'99*, Seattle, Washington, USA (1999), pp. 1598–1602.
6. Soares, C.J., Siderius, M. and Jesus, S.M., Matched-field source localization in the Strait of Sicily, *J. Acoust. Soc. Am.* (in press 2002).
7. Soares, C., Siderius, M. and Jesus, S., High frequency source localization in the Strait of Sicily. In *Proc. MTS/IEEE Oceans 2001*, Honolulu, Hawaii, USA (2001).
8. Jesus, S.M., Coares, C., Onofre, J. and Picco P., Blind ocean acoustic tomography: Exper- imental results on the INTIFANTE'00 data set. In *Proc. 6th European Conference on Underwater Acoustics*, Gdansk, Poland (June 2002).
9. De Marinis, E., Gasparini, O., Picco, P., Jesus, S., Crise, A. and Salon, S., Passive ocean acoustic tomography: theory and experiment. In *Proc. 6th European Conference on Un- derwater Acoustics*, Gdansk, Poland (June 2002).
10. Jesus, S.M, Coelho, E., Onofre, J., Picco, P., Soares, C. and Lopes, C., The INTIFANTE'00 sea trial: preliminary source localization and ocean tomography data analysis. In *Proc. MTS/IEEE Oceans 2001*, Honolulu, Hawaii, USA (Nov. 2001).

BENCHMARKING GEOACOUSTIC INVERSION METHODS FOR RANGE DEPENDENT WAVEGUIDES

N. ROSS CHAPMAN

School of Earth and Ocean Sciences, University of Victoria, Victoria, B.C., Canada
E-mail: chapman@uvic.ca

S. CHIN-BING AND D. KING

Naval Research Laboratory, Stennis Space Centre, MS, USA

R.B. EVANS

Science Applications International, Corp.
23 Clara Drive, Suite 206, Mystic, CT 06355, USA

Over the past decade, inversion methods have been developed to provide information about unknown bottom environments from acoustic field data. An effective inversion must provide both an estimate of the bottom parameters, and a measure of the uncertainty of the estimated values. This paper summarizes results from the ONR Geoacoustic Inversion Techniques Workshop that was held to benchmark present day inversion methods for estimating geoacoustic profiles in shallow water. The format of the workshop was a blind test to estimate unknown geoacoustic profiles by inversion of synthetic acoustic field data. The fields were calculated using coupled normal modes for three range-dependent test cases: a monotonic slope; a shelf break; and a fault intrusion in the sediment. Geoacoustic profiles were generated to simulate sand, silt and mud sediments. Several different approaches were presented for inverting the acoustic field data: model-based matched field methods; perturbative methods; methods using transmission loss data; methods using horizontal array information. New methods were also discussed to formalize the measure of uncertainty in the inversion. Comparisons between the different inversions are discussed in terms of a metric based transmission loss calculated using the inverted profiles. The results demonstrate the effectiveness of present day inversion techniques, and indicate the limits of their capabilities for range-dependent waveguides.

1   Introduction

The variability in the geological structure and in the geophysical properties of the ocean bottom are known to have a significant effect on sonar performance in shallow water. Sonar performance prediction systems that make use of geoacoustic information are limited by the lack of high-resolution geophysical databases to describe the spatial variability of the bottom properties in shallow water, where variations over scales of the order of 100 metres are known to exist. To address this deficiency, comprehensive experimental survey programs have been initiated to acquire geoacoustic data, and model based inversion methods have been developed for estimating geoacoustic model

441

N.G. Pace and F.B. Jensen (eds.), Impact of Littoral Environmental Variability on Acoustic Predictions and Sonar Performance, 441-448.

parameters from acoustic field data or from quantities derived from the fields themselves. Examples of different inversion techniques include nonlinear methods based on matched field processing [1,2], and linearized perturbative approaches [3]. There is an extensive literature that describes applications of different inversion methods to experimental data for estimating geoacoustic model parameters.

With few exceptions, the inversions were generally designed to provide only the estimates of a set of model parameters that defined a specific geoacoustic model for the experimental site. However, the complete solution of the inverse problem requires not only the estimates of the model parameter values, but also a measure of the uncertainty of the estimates. For model-based sonar performance predictions, the parameter uncertainties provide essential information to set confidence limits for predictions of transmission loss and other quantities that affect detection or localization performance.

In order to obtain a reasonable measure of the errors, it is necessary to recognize the sources of error in geophysical inverse methods. For most cases, the uncertainties due to errors in the theory and the model are far more significant than errors that arise due to inaccurate data. Examples of the former type of error are: mismatch in the geoacoustic model of the actual environment; mismatch in the experimental geometry; and errors in the acoustic propagation model. Since many of the model parameters are correlated, the presence of mismatch can have a serious impact on the accuracy of the estimates, especially in nonlinear inversion methods that search the model parameter space. A simple example is the acoustic mirage effect that occurs in matched field source localization if there is mismatch in the water depth [4].

This paper presents the results of a Geoacoustic Inversion Techniques Workshop that was sponsored by ONR and SPAWAR to benchmark the accuracy and determine the efficiency of inversion techniques that had been developed for estimating geoacoustic model parameters in range-dependent shallow water environments. The approach in the benchmark process was to compare the inversion performance of the different methods against a standard set of range-dependent test cases. The workshop was the second formal exercise for benchmarking geoacoustic inversion techniques. The first workshop, held in 1997, carried out the initial phase to compare inversion performance of different methods against test cases that were range independent [5]. All the test cases were synthetic, based on a known form for the geoacoustic model. In the second workshop, both synthetic test cases and real data were provided, but we focus on only the synthetic cases in this paper.

2   Workshop format

The benchmarking process described here was carried out using synthetic acoustic pressure fields that were calculated for known values of the geoacoustic model parameters for three range-dependent test case environments. The workshop format was a blind test: participants were provided with only the acoustic field data, and were not given the input parameters for the geoacoustic models. The tasks for the participants were to invert the information in the acoustic fields to obtain estimates of the geoacoustic model parameters of the environment, and to determine a measure of the uncertainty of the estimates. A calibration test case was provided so that the participants could benchmark the propagation model that was used for calculating the acoustic fields in their inversions. For this case, the input geoacoustic profile was also provided to the users.

The acoustic data for the test cases were provided in the frequency domain as spectral components of the acoustic pressure field. This approach eliminates a signal processing step that is normally carried out with time series data from experiments, but was adopted for convenience in generating the large quantity of test case data. The acoustic fields were calculated initially by the coupled normal mode propagation model, COUPLE [6], as a benchmark, and these results were then duplicated by the parabolic equation code, RAM [7]. All subsequent calculations for the workshop test case database were done using RAM, to take advantage for its computational efficiency compared to that for COUPLE. This approach provided accurate solutions for the range-dependent test case environments. However, the environments were restricted to fluid media.

The synthetic pressure fields were generated over a wide frequency band for vertical and horizontal array geometries in the water layer, so that participants could design their own 'experiment' to invert the data, using a subset of the available information that was appropriate for their particular inversion method. The frequency band was from 25–199 Hz in 1-Hz steps, and from 200–500 Hz in 5-Hz steps. The spatial samples were provided as horizontal arrays at depths of 25 m and 85 m in 5–m range steps from 5 m to 5000 m, and as vertical arrays with sensors from 20 m to 80 m in 1-m increments, at ranges from 500 m to 5000 m in 500-m steps. The test case environments were defined in a 5-km shallow water waveguide. For each test case, the source depth was 20 m, and the water layer sound speed profile was downward refracting and range-independent, according to

$$c_w(z) = 1495.0 - 0.04z \tag{1}$$

where z is the depth in metres. This information was provided to the participants. Bathymetry was also provided, but with an uncertainty of ± 1 m.

There was no attempt in this workshop to add either incoherent or coherent noise to the calculated spectral component data. However, since there is certain to be some degree of geoacoustic model mismatch, errors due to this type of noise are inherent in any of the nonlinear inversion methods that are posed as optimization problems. Formal approaches t determine uncertainties due to theory and model noise have been proposed by Gerstoft and Mecklenbrauker [8] and Dosso et al. [9].

3 Geoacoustic model

The geoacoustic model consisted of three components: a water layer with known profile parameters; a series of fluid sediment layers; and a fluid basement halfspace. The geoacoustic profile in the bottom layers was specified by the density, ρ, and the sound speed, c, and attenuation, α, of compressional waves in each layer. The sediment profile was an N-layer model, for which the number of layers and the geoacoustic profile within each layer were unknown. Because the actual form of the multi-parameter geoacoustic model was not known beforehand, the challenges in inverting the synthetic test cases were similar to those for real data inversions. It was expected that the participants would at best invert an approximation to the actual geoacoustic model that was consistent with the information used in the inversion. The estimated profiles could then be tested to determine the range of validity for other scenarios of sound frequency and source/receiver geometry.

The geoacoustic profiles in the sediment were designed for specific sediment types such as sand, silt or mud, using the empirical relationships of Bachman [10]. The sediment material is described by the grain size parameter phi, given by $-\log_2 d$ where d is the average grain size of the sediment particles. For sands, phi < 3.25 and for silts, phi > 5.75. Phi values between these limits were classed as mixtures, and the geoacoustic properties were determined according to weighted proportions of sand and silt. The weights were defined by a silt factor ϕ_s given by:

$$\phi_s = (phi - 3.25)/(5.75 - 3.25) . \tag{2}$$

The sound speed and attenuation profiles were given by:

- Sands: $c(z) = c(0) \times (20z)^{0.015}$

  $$\alpha(z) = (0.23 + 0.268 \times phi) \times (20z)^{-1/6} \tag{3}$$

- Silts: $c(z) = c(0) + 0.712z$

  $$\alpha(z) = 0.297 - 0.035 \times phi$$

where c(0) is the sound speed in the sediment at the sea floor according to:

$$c(0) = SSR \times c_w, \tag{4}$$

and $SSR = 1.180 - 0.034 phi + 0.0013 phi^2$ for both sand and silt [10].

The density at the sea floor is $\rho = (22.85 - phi)/10.25$, and is approximately constant with depth for both sands and silts over the shallow depths in the test cases. The number of layers in the N-layer sediment profile was selected randomly from a uniform distribution between [1–8], and the thickness of each layer was selected from a uniform distribution from 1 to 5 m.

4   Range-dependent test case environments

Three test cases were designed to simulate:

- a monotonic downslope environment with a shallow slope (Fig. 1);

The sediment profile for this case consisted of a thin mud/silt layer (phi = 6.6) over five layers of faster coarse sand (phi = 3.2), overlying a mudstone basement. The total sediment thickness was 26.1 m.

- a shelf break environment (Fig. 2)

The sediment model for this case consisted of five layers of silty sand (phi = 5.25, 80% silt), overlying a sandstone basement. The sediment layer was 17.9 m thick. In these first two cases, the sediment layers were parallel to the bottom slope.

- an intrusion of basement material in the sediment layer to simulate an uplifted fault (Fig. 3).

The geoacoustic profile in the third case was range dependent The sediment consisted of a thin, slow velocity mud layer (phi = 7.5) over a sand-silt mixture of six layers (phi = 4.2, 38% silt), overlying a mudstone basement that was uplifted to replace the sand silt layers for a portion of the range. The sediment thickness was 22.8 m.

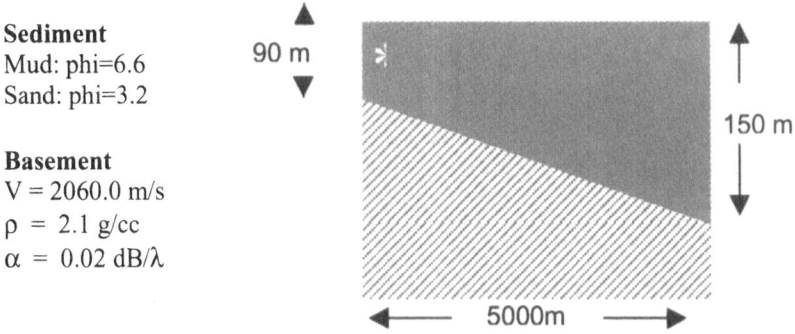

Sediment
Mud: phi=6.6
Sand: phi=3.2

Basement
V = 2060.0 m/s
ρ = 2.1 g/cc
α = 0.02 dB/λ

Figure 1. Geoacoustic model for Test Case 1: Monotonic downslope.

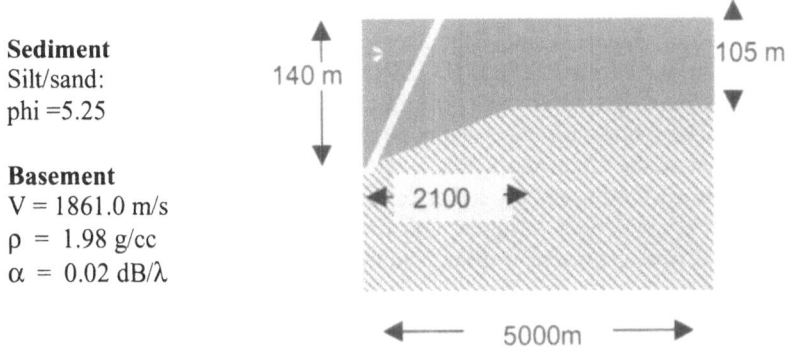

Sediment
Silt/sand:
phi =5.25

Basement
V = 1861.0 m/s
ρ = 1.98 g/cc
α = 0.02 dB/λ

Figure 2. Geoacoustic model for Test Case 2: Shelf break.

5   Inversion results for Test Case 1

Participants presented inversions using full wave signal processing methods (MFP) with either vertical or horizontal array data; methods that used transmission loss data; and specialized techniques such as plane wave beamforming and waveguide invariants. Several different forward propagation models were used that proved to be effective for range-dependent environments: parabolic equation, ray theory and adiabatic normal modes. Although none of the participants recovered the exact form of the profile, the sound speed profile was very well approximated by the estimated profiles. Density and attenuation profiles were generally not as well estimated. For the MF inversions, new hybrid search algorithms and inversions in reparameterized model spaces were introduced that significantly improved the efficiency of the search process.

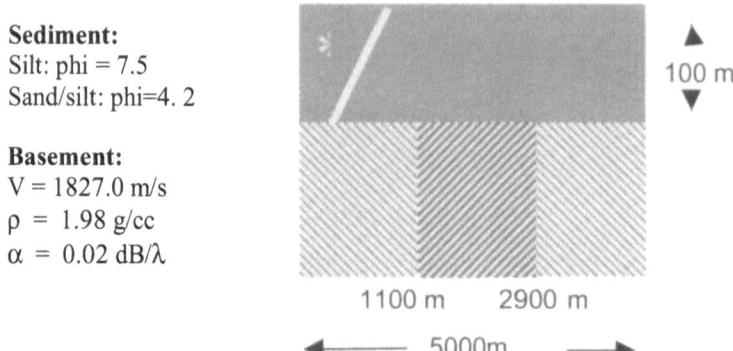

Sediment:
Silt: phi = 7.5
Sand/silt: phi=4. 2

Basement:
V = 1827.0 m/s
ρ = 1.98 g/cc
α = 0.02 dB/λ

Figure 3. Geoacoustic model for Test Case 3: An intrusion of basement material in the sediment to simulate an uplifted fault structure.

The metric used to quantify the performance of the different inversion methods was a simple comparison between transmission loss calculated using the estimated profiles and those for the actual profile, for a different scenario of source/receiver (80m/80m depths) geometry.   Figure 4 shows the average rms transmission loss difference from 0–5 km at frequencies from 25–800 Hz. The most accurate results were obtained using MF inversions based on nonlinear optimization techniques, with either horizontal or vertical array data (shown by the + symbols). For these inversions, the differences are slightly smaller at the frequencies that were used in the inversions, generally from 50–300 Hz. However, the analysis indicates that the estimated profiles provide accurate transmission loss predictions at higher and lower frequencies as well. The differences for profiles that were estimated from transmission loss data inversions (triangles) were generally much greater.

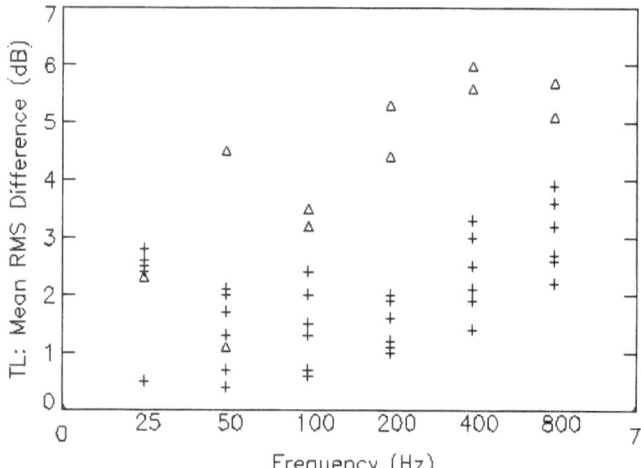

Figure 4.   Mean rms differences between transmission loss (TL) calculated for the estimated profiles and the actual profile, for a source and receiver depths of 80 m.

Although the density and attenuations were not as well estimated, the analysis summarized in Fig. 4 suggests that information about these quantities may not be as critical as the knowledge of the sound speed profile in predicting transmission loss for this environment. A measure of the sensitivity of each parameter was determined in some of the inversion methods that used global search processes. An example is shown in the scatter plots in Fig. 5, where the cost function is plotted versus parameter value for one of the inversions that used a genetic algorithm in the SAGA inversion package. The distributions in each panel indicate the range of values of each model parameter that provide good fits to the acoustic field data. For sensitive parameters such as the sound speeds or the thickness of the first sediment layer, the distributions are peaked, whereas for insensitive parameters such as the attenuations or the sediment density, the distributions tend to be flat. Although these displays are not true representations of the *a posteriori* marginal densities for the parameters, they do provide a means to determine which parameters were well estimated in the inversion.

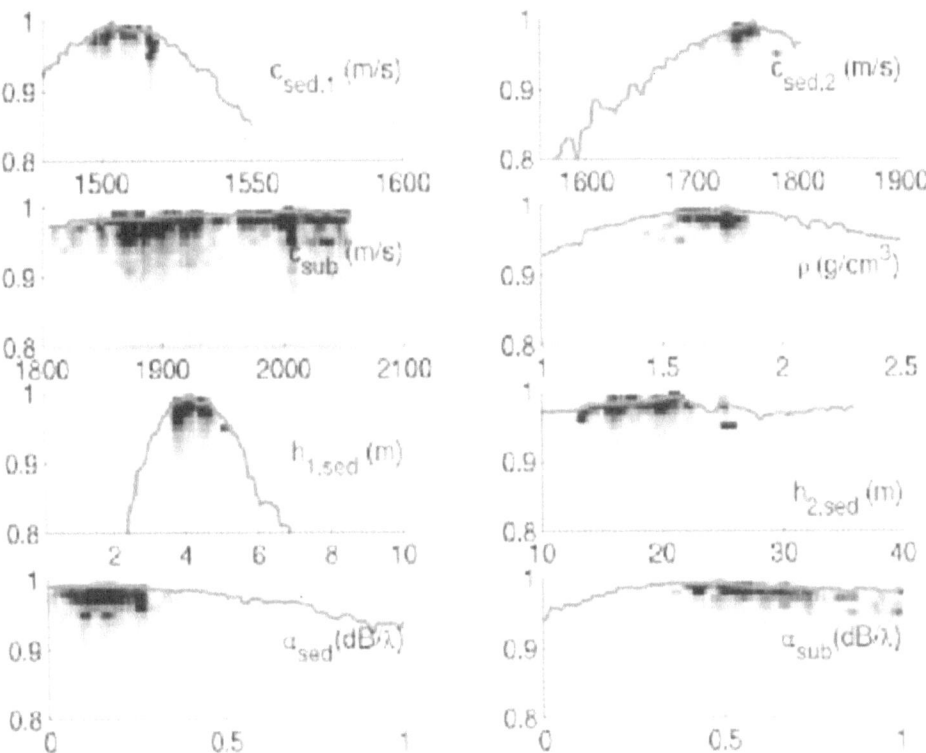

Figure 5. Scatter plots of the cost function versus parameter values for geoacoustic models that were sampled in the global search process. The distributions in each panel are normalized independently to the maximum value. (After Siderius *et al.*)

6 Summary

Performance of present day geoacoustic inversion techniques was assessed in a benchmark workshop, using realistic test cases that were generated for range-dependent shallow water environments. Analysis of the results for the first test case, a monotonic slope environment, indicated:

- Inversion methods were capable of estimating highly accurate approximations to the sound speed profile in the bottom. The most accurate results were obtained by matched field inversions that used global search processes. These techniques were also capable of generating effective measures of the parameter sensitivity and the uncertainty of the estimates.

- The estimated geoacoustic profiles provided the basis for accurate predictions of transmission loss for independent scenarios of source/receiver geometry and sound frequency.

- Estimates of density and attenuation profiles were not as accurate. However, the accuracy of these parameters does not appear to be critical for making accurate predictions of transmission loss for use in sonar performance analysis.

Acknowledgements

This work was supported by ONR (NRC) and SPAWAR.

References

1. Lindsay, C.E. and Chapman, N.R., Matched field inversion for geoacoustic parameters using adaptive simulated annealing, *IEEE J. Ocean. Eng.* **18**, 224–231 (1993).
2. Gerstoft, P., Inversion of seismoacoustic data using genetic algorithms and *a posteriori* probability distributions, *J. Acoust. Soc. Am.* **95**, 770–782 (1994).
3. Frisk, G.V., Lynch, J.F. and Rajan, S.D., Determination of compressional wave speed using modal inverse techniques in a range dependent environment, *J. Acoust. Soc. Am.* **86**, 1928–1939 (1989).
4. D'Spain, G.L., Murray, J.J., Hodgkiss, W.S., Booth, N.O. and Schey, P.W., Mirages in shallow water matched field processing, *J. Acoust. Soc. Am.* **105**, 3245–3265 (1998).
5. Tolstoy, A., Chapman, N.R. and Brooke, G.E., Workshop97: Benchmarking for geoacoustic inversion in shallow water, *J. Comp. Acoust.* **6**, 1–28 (1998).
6. Evans, R.B., A coupled normal mode solution for acoustic propagation in a waveguide with stepwise variations of a penetrable bottom, *J. Acoust. Soc. Am.* **74**, 188–195 (1988).
7. Collins, M.D., A split-step Padé solution for the parabolic equation method, *J. Acoust. Soc. Am.* **93**, 1736–1742 (1998).
8. Gerstoft, P. and Mecklenbrauker, C.F., Ocean acoustic inversion with estimation of *a posteriori* probability distributions, *J. Acoust. Soc. Am.* **104**, 808–819 (1998).
9. Dosso, D.E., Quantifying uncertainties in geoacoustic inversion I: A fast Gibbs sampler approach, *J. Acoust. Soc. Am.* **111**, 129–142 (2001).
10. Bachman, R.T., Estimating velocity ratio in sediments, *J. Acoust. Soc. Am.* **86**, 2029–2032 (1989).

ADJOINT-ASSISTED INVERSION FOR SHALLOW WATER ENVIRONMENT PARAMETERS

PAUL HURSKY AND MICHAEL B. PORTER

Science Applications International Corporation
1299 Prospect Street, Suite 305, La Jolla, CA 92037, USA
E-mail: paul.hursky@saic.com, michael.b.porter@saic.com

BRUCE D. CORNUELLE, W.S. HODGKISS AND W.A. KUPERMAN

Scripps Institution of Oceanography
University of California in San Diego, La Jolla, CA 92093-0701, USA
E-mail: bdc@ucsd.edu, wsh@mpl.ucsd.edu, wak@mpl.ucsd.edu

The adjoint of a forward model can back-propagate mismatch between observations and their predictions and produce the corrections to the forward model inputs that caused the mismatch. As an example of this process, the adjoint of a parabolic equation propagation model is used to invert errors in pressure predictions at a receiver for sound speed perturbations due to internal tides.

1 Introduction

Using the adjoint of a forward model has the potential to sharply reduce the number of modeling runs usually needed to achieve an inversion. Typically, an inversion process varies the parameters of a forward model, running the forward model for many candidate sets of parameter values until the forward model matches the data. Unfortunately, this often requires many runs to adequately search the space of unknown parameters. We present an alternative technique based on the adjoint of the forward model. The adjoint model back-propagates a mismatch between model predictions and measured observations, producing corrections to model input parameters along the trajectory of the forward model. A single run of the adjoint model thus duplicates many forward modeling runs. In this paper, we will use the adjoint of a parabolic equation propagation model to invert for sound speed perturbations due to internal tides.

Adjoint methods have been used in many fields. Reference [1] suggested adjoint methods for tomography. References [2, 3] present how adjoint methods can be used to assimilate data into oceanographic models. Reference [6] derives the adjoint of the Helmholtz equation in terms of continuous variables and discusses the connection between adjoint techniques and time reversal. The observed field is a superposition of a baseline field due to the presumed medium and a perturbed field due to the unknown medium perturbations. The adjoint model back-propagates (time-reverses) the perturbed field to the unknown medium perturbations (viewed as sources of diffraction). Reference [7] shows how adjoints can be used to calculate Fréchet derivatives used to solve inverse problems of the sort we address. References [4, 5] present how adjoints arise in optimal

449

N.G. Pace and F.B. Jensen (eds.), Impact of Littoral Environmental Variability on Acoustic Predictions and Sonar Performance, 449-456.
© 2002 *Kluwer Academic Publishers.*

control theory, where their use is known as the Pontryagin Principle.

In Sect. 2, we derive a tangent linear model for the parabolic equation. In Sect. 3, we show how the adjoint of this model can be used to solve acoustic inverse problems. In Sect. 4, we show in simulation how these models can be used to estimate the sound speed perturbations caused by internal tides.

2 Tangent linear model for the parabolic equation

The standard homogeneous PE equation (with no source in the medium) is

$$2ik_0 \frac{\partial p}{\partial r} + \frac{\partial^2 p}{\partial z^2} + k_0^2 \left(n^2 - 1 \right) p = 0. \tag{1}$$

Expanding this equation in terms of perturbations in pressure p and index of refraction squared n^2 to first order in ε,

$$n^2 = n_0^2 + \varepsilon n_1^2,$$

$$p = p_0 + \varepsilon p_1 + \ldots,$$

yields

$$2ik_0 \frac{\partial p_1}{\partial r} + \frac{\partial^2 p_1}{\partial z^2} + k_0^2 \left[n_0^2(r, z) - 1 \right] p_1 = -k_0^2 n_1^2(r, z) p_0. \tag{2}$$

We will use a discrete formulation of Eqs. (1) and (2) based on the implicit finite differences scheme described in Sect. 6.6 of Ref. [8]. A finite difference approximation to Eq. (1) (the unperturbed problem, with $\varepsilon = 0$) produces a marching solution for the zeroth-order pressures $\mathbf{p}_0$,

$$\mathbf{p}_0(r + \delta r) = \mathbf{F}(r)\mathbf{p}_0(r), \tag{3}$$

where $\mathbf{p}_0$ is a vector sampled in depth. Matrix $\mathbf{F}(r)$ is a symmetric, tri-diagonal matrix with diagonal elements

$$-\frac{2}{h^2} + k_0^2 \left(n_0^2(r, z) - 1 \right),$$

and super-diagonal and sub-diagonal elements $\frac{1}{h^2}$. The diagonal elements contain $n_0^2(r, z)$, the zeroth-order index of refraction squared, which varies with range and depth. Matrix $\mathbf{F}(r)$ is used to propagate pressure vectors (sampled in depth) one range step at a time, given an initial pressure (or starter field). Equation (2) (the first-order perturbation terms, of order ε) generates a marching solution for the first-order pressure vectors $\mathbf{p}_1$,

$$\mathbf{p}_1(r + \delta r) = \mathbf{F}(r)\mathbf{p}_1(r) + \mathbf{G}(r)\mathbf{u}(r). \tag{4}$$

$\mathbf{F}(r)$ is the same as in Eq. (3). $\mathbf{G}(r)$ is a diagonal matrix with values $-k_0^2 \mathbf{p}_0(r)$ (i.e. the zeroth-order pressures sampled in depth at a particular range). Vector $\mathbf{u}(r)$ contains

$n_1^2(r, z)$ sampled in depth. Equation (4) has a forcing function proportional to the zeroth-order pressure $\mathbf{p}_0(r)$ and the first-order perturbation to the index of refraction squared $n_1^2(r, z)$. Matrices $\mathbf{F}(r)$ and $\mathbf{G}(r)$ define our tangent linear model and also form the basis of our adjoint model in the following sections.

3 Using an adjoint model to solve acoustic inverse problems

We will use Eq. (4) to formulate an adjoint method in terms of first-order perturbations to the pressures and the environmental parameters. We will no longer use subscripts 0 and 1 to indicate the order of the perturbation. All $\mathbf{p}_r$ and $\mathbf{u}_r$ will be first-order perturbations, or corrections to zeroth-order quantities calculated using our initial guess at the environmental parameters we are inverting for. These $\mathbf{p}_r$ and $\mathbf{u}_r$ will be vectors, sampled in depth, with subscripts r that indicate range indexes. Equation (4) in terms of this notation is

$$\mathbf{p}_{r+1} = \mathbf{F}_r \mathbf{p}_r + \mathbf{G}_r \mathbf{u}_r.$$

Note both $\mathbf{F}_r$ and $\mathbf{G}_r$ are functions of range. Matrix $\mathbf{F}_r$ propagates the pressure correction vector $\mathbf{p}_r$ at range index r one range step to $\mathbf{p}_{r+1}$. The environmental parameter correction vector $\mathbf{u}_r$ influences the propagation of the pressure via a known matrix $\mathbf{G}_r$. The ith element of vector $\mathbf{p}_r$ contains the pressure correction at range index r at the ith sampled depth. Similarly for vector $\mathbf{u}_r$. Matrices $\mathbf{F}$ and $\mathbf{G}$, derived in Sect. 2, form a tangent linear model of the original, non-linear PE model. $\mathbf{F}$ and $\mathbf{G}$ are functions of the zeroth-order environmental parameters and the zeroth-order pressures (calculated using the original non-linear propagation model with the zeroth-order environmental parameters as inputs). The vectors $\mathbf{p}_r$ (sampled in depth) are the pressure increments calculated by the tangent linear model $\mathbf{F}$ and $\mathbf{G}$ as corrections to the zeroth-order pressures due to the environmental correction vectors $\mathbf{u}_r$ (also sampled in depth).

To solve for $\mathbf{u}_r$, we formulate an objective function $J(\mathbf{p}, \mathbf{u}, \lambda)$ to be minimized:

$$J(\mathbf{p}, \mathbf{u}, \lambda) = \frac{1}{2}(\mathbf{p}_N - \mathbf{m}_N)^2 + \sum_{r=1}^{N} \lambda_r^T (\mathbf{p}_r - \mathbf{F}_{r-1}\mathbf{p}_{r-1} - \mathbf{G}_{r-1}\mathbf{u}_{r-1}) + \frac{1}{2} \sum_{r=0}^{N-1} \mathbf{u}_r^2. \quad (5)$$

The first term in J seeks to minimize the mismatch between the measured pressure increment $\mathbf{m}_N$ and the modeled pressure increment $\mathbf{p}_N$, both at range index N. Since we are dealing with first-order terms, $\mathbf{m}_N$ is the difference between the measured pressure and the zeroth-order pressure prediction. The modeled pressure $\mathbf{p}_N$ is calculated by propagating $\mathbf{p}_r$ from the source to the receiver using our tangent linear model $\{\mathbf{F}_r, \mathbf{G}_r\}$ with the environmental parameter corrections $\mathbf{u}_r$ as driving functions. Given a solution for $\mathbf{u}_r$, the zeroth-order pressure plus the pressure correction $\mathbf{p}_N$ calculated using $\mathbf{u}_r$ should reproduce the measured pressure. The second term uses Lagrange multipliers λ_r (vectors at each range, sampled in depth) to enforce the hard constraint that the $\mathbf{p}_r$ and $\mathbf{u}_r$ must be consistent with the model $\{\mathbf{F}_r, \mathbf{G}_r\}$. The third term is a regularizing term to minimize the amplitude of the environmental perturbations.

Admittedly, this is an unusual way to formulate an inverse problem. We have set up a large number of unknowns in all the intermediate $\mathbf{p}_r$, in addition to the already large number of unknowns $\mathbf{u}_r$. We will show how minimizing the objective function above leads to an iteration that seems to be a much more direct way of inverting for the $\mathbf{u}_r$ than repeatedly running the forward model to explore the surface J as a function of $\mathbf{u}_r$.

Note that J is a function of $\mathbf{u}_r$ at all ranges and depths, so its minimization has the potential to resolve range-dependent features. We will demonstrate inversions using measurements at a single frequency and a single source depth for range-independent features in Sect. 4. To resolve range-dependent features requires a richer set of measurements, using more sources and a wider band.

The partial derivatives of $J(\mathbf{p}, \mathbf{u}, \lambda)$ are

$$\frac{\partial J}{\partial \mathbf{p}_N} = \mathbf{p}_N - \mathbf{m}_N + \lambda_N, \tag{6}$$

$$\frac{\partial J}{\partial \mathbf{p}_r} = \lambda_r - \mathbf{F}_r^T \lambda_{r+1}, \tag{7}$$

$$\frac{\partial J}{\partial \mathbf{u}_r} = \mathbf{u}_r - \mathbf{G}_r^T \lambda_{r+1}. \tag{8}$$

Setting the partial with respect to $\mathbf{p}_N$ to zero in Eq. (6) yields

$$\lambda_N = \mathbf{m}_N - \mathbf{p}_N, \tag{9}$$

initializing λ_N to the mismatch between measured and modeled pressures at range N. Setting the partials with respect to $\mathbf{p}_r$ to zero in Eq. (7) produces a recursion relation,

$$\lambda_r = \mathbf{F}_r^T \lambda_{r+1}, \tag{10}$$

that enables us to propagate the Lagrange multipliers λ_r from the receiver to the source. The λ_r can be viewed as a field propagated by the adjoint model. The starter field of the adjoint model, given by Eq. (9), is the mismatch in our observations at the receiver (i.e. mismatch with our predictions, produced by our zeroth-order model with the zeroth-order environmental parameters as inputs). We are inverting for the corrections $\mathbf{u}_r$ to these zeroth-order environmental parameters that will account for this mismatch. Setting the partials with respect to the $\mathbf{u}_r$ to zero in Eq. (8) yields

$$\mathbf{u}_r = \mathbf{G}_r^T \lambda_{r+1}, \tag{11}$$

producing equations for $\mathbf{u}_r$ at each range index r in terms of the Lagrange multipliers λ_{r+1}.

Equations (9), (10), and (11), in the order presented, can be used to calculate $\mathbf{u}_r$, the first-order corrections to the environmental parameters driving our forward model. However, in the non-linear problem we address, the gradients used to derive these equations may only be accurate in a small neighborhood about the zeroth-order pressure predictions. As a result, we used these equations as the basis for an iterative procedure which is able to follow the curvature of our objective surface J that will inevitably arise in some configurations. At each iteration, we calculate a new set of $\mathbf{u}_r$, given the current $\{\mathbf{F}_r, \mathbf{G}_r\}$.

We use these $\mathbf{u}_r$ to adjust the environmental parameters that are then used to re-calculate a new set of zeroth-order pressures at all ranges (either using the tangent linear model, or the original non-linear model). If these calculated pressures match our measurements, we have found a solution to our problem. If not, the new zeroth-order pressures and the corrected environmental parameters are embedded in $\mathbf{F}$ and $\mathbf{G}$, and the adjoint model is used to calculate another set of corrections to the environmental parameters.

The tangent linear model can be used to calculate the forward sensitivity in a problem (e.g. how sensitive $\mathbf{p}_N$ is to perturbations $\mathbf{p}_0$ and $\mathbf{u}_r$). The adjoint model can be used to calculate the backward sensitivity in a problem (e.g. how sensitive $\mathbf{p}_0$ and $\mathbf{u}_r$ are to perturbations $\mathbf{p}_N$).

4   Inverting for INTIMATE 96 internal tides (simulated results)

We will demonstrate the adjoint method described in Sect. 3 on an ensemble of sound speed profiles measured during the INTIMATE 96 experiment (see [9]) when the passage of internal tides was clearly visible (see Fig. 2). We use our PE model to synthesize pressure measurements at 400 Hertz on a vertical line array at a range of 2 km from a source at depth of 50 meters, using each individual profile to generate a measured pressure vector. Each of these pressure vectors was inverted to estimate the sound speed profile which was used to synthesize it, using the iterative process outlined in Sect. 3 (using tangent linear and adjoint models derived for our PE model). The mean profile of the entire ensemble served as the initial guess for each inversion. We set up our inversions to solve for coefficients of empirical orthogonal functions (EOFs), averaged over the deviations from the mean profile.

In Sect. 4.1, because our inversion process is essentially a steepest descent method, we use our PE tangent linear model to assure ourselves that we are reasonably close to a solution. In Sect. 4.2, we show the results of our inversions. Note that we are using *synthetic* acoustic data to demonstrate the feasibility of inverting for a sequence of internal tides measured during the INTIMATE 96 experiment. The experiment configuration was fixed-fixed, with the line joining the source and receiver perpendicular to the passage of the internal tides, so the inversion was formulated to be *range-independent*. Each profile from the INTIMATE 96 sequence was estimated by a separate inversion.

4.1   Tangent Linear Modeling to Verify we are in Linear Regime

To assess the linearity of our experimental configuration, we compare pressures produced by our fully non-linear PE model and its tangent linear version, given perturbations on the order of those actually observed during INTIMATE 96. We refer to the fully non-linear PE propagation model as $p = F(c)$. We refer to the tangent linear model as $\delta p = \delta F(c_0, \delta c)$, where we have written δF as a function of both c_0, the baseline profile, and δc, its perturbation. Note the tangent linear model depends upon both c_0 and δc. The baseline profile is the mean profile calculated from the entire set of available sound speed profile measurements during the INTIMATE 96 experiment. We chose the profile that deviated the most from the mean profile as a test case. We wanted to see how closely the tangent linear model matched the original PE model in predicting pressure perturbations in the INTIMATE 96 configuration. We ran the original PE model on our

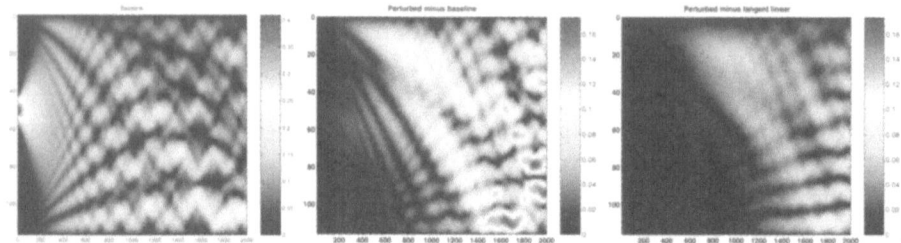

Figure 1. Baseline pressure (left image). Magnitude differences between perturbed and baseline pressures (center image) and between perturbed and tangent linear pressures (right image). Center and right images have the same color scale.

baseline sound speed profile c_0 and on our perturbed sound speed profile $c_1 = c_0 + \delta c$, producing p_0 and p_1. We ran our tangent linear model to produce δp, to see if it would reproduce the perturbation in pressure, $p_1 - p_0$, due to the perturbation in sound speed δc. Equations (12) through (15),

$$p_0 = F(c_0), \tag{12}$$

$$p_1 = F(c_0 + \delta c), \tag{13}$$

$$\delta p = \delta F(c_0, \delta c), \tag{14}$$

$$p_2 = p_0 + \delta p, \tag{15}$$

summarize how the three relevant pressures, p_0, p_1, and p_2, were calculated. Figure 1, containing three images, shows several combinations of these pressures. The left image shows the baseline pressure magnitude, p_0. The following two images have the same color scale, so that their values can be compared. The center and right images show magnitudes of pressure differences. The center image shows the perturbed minus baseline pressures, $|p_1 - p_0|$. The right image shows perturbed minus tangent linear pressures, $|p_1 - p_2|$. The significantly lower amplitudes in the right image indicate that the tangent linear model is able to match the predictions of the fully non-linear PE model reasonably well, at least given the size of the sound speed perturbations and the source-receiver range in our configuration. The quality of the tangent linear model degrades with increasing range, but as we will see in the next section, we are still close enough to a solution that our iterative process resolves the internal tides.

These results indicate that although the problem we are addressing is not strictly linear, it remains reasonable to attempt our iterative process, which we expect can tolerate slightly non-linear problems, because it presumably can follow a non-linear basin of attraction if started out close enough to the final solution point.

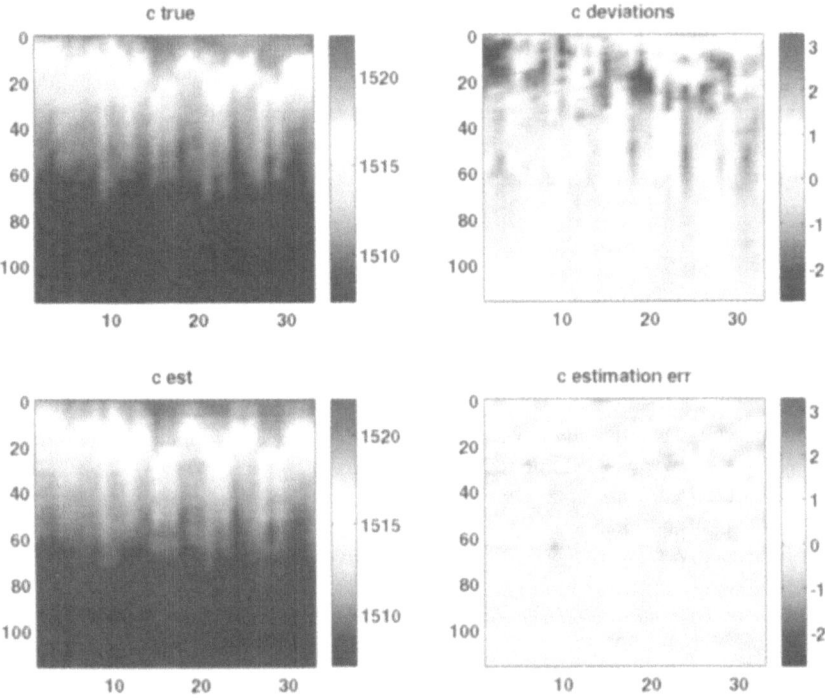

Figure 2. Inversion results for internal tides measured during INTIMATE 96 experiment (with simulated acoustic data). Upper left image shows the true sound speed profiles versus time. The lower left shows the estimated profiles versus time. The upper right image shows the sound speed profile deviations from the mean profile (what we solve for directly using the adjoint method). The lower right image shows the estimation errors (the difference between the upper left and lower left images).

4.2 Adjoint Iterative Process Results

Figure 2 shows the result of applying the process described in Sect. 3 to each of a sequence of profiles measured during the INTIMATE 96 experiment. The entire sequence of measured profiles is shown in the upper left hand plot, with the horizontal and vertical axes corresponding to time and depth. Each profile was processed independently of the others. We ran our adjoint-based inversion for 50 iterations on each profile, using the mean profile as an initial guess. The resulting estimated profiles are shown in the lower left plot. Clearly, the coarse features have been resolved. For a more quantitative assessment, we show the two plots on the right. The upper right plot shows the deviations from the mean profile (the corrections we actually invert for). The lower rght plot shows the estimation errors (i.e. the difference between the measured and estimated profiles). Both right-hand plots have the same color scale. The relatively smaller magnitudes of the estimation errors compared to the deviations from the mean indicate that our adjoint process has done a good job resolving the internal tides during this interval.

5 Conclusions

We have shown how the adjoint of a parabolic equation forward model can be used to invert pressure measurements for sound speed perturbations in the water column. The adjoint technique we have presented uses far fewer propagation model runs than techniques currently being used, in which the forward model is run for each candidate point in a high-dimensional search space.

Acknowledgments

We thank Aaron Thode for pointing out Refs. [4, 6, 7].

References

1. W. Munk, P. Worcester and C. Wunsch, *Ocean Acoustic Tomography* (Cambridge University Press, New York, 1995) pp. 318–319.
2. C. Wunsch, *The Ocean Circulation Inverse Problem* (Cambridge University Press, New York, NY, 1996) pp. 362–391.
3. A.F. Bennett, *Inverse Methods in Physical Oceanography,* (Cambridge University Press, New York, NY, 1992) Chap. 5, pp. 112–135.
4. D.E. Kirk, *Optimal Control Theory* (Prentice-Hall, Englewood Cliffs, NJ, 1970) pp. 184–240.
5. I.M. Gelfand and S.V. Fomin, *Calculus of Variations* (Prentice-Hall, Englewood Cliffs, NJ, 1963,) pp. 218–225.
6. A. Tarantola, Inversion of seismic reflection data in the acoustic approximation, *Geophysics* **49**, 1259–1266 (1984).
7. S.J. Norton, Iterative inverse scattering algorithms: Methods of computing Fréchet derivatives, *J. Acoust. Soc. Am.* **106**, 2653–2660 (1999).
8. F.B. Jensen, W.A. Kuperman, M.B. Porter and H. Schmidt, *Computational Ocean Acoustics* (AIP Press, Woodbury, NY, 1994) pp. 366–375.
9. Y. Stéphan, X. Démoulin, T. Folégot, S.M. Jesus, M.B. Porter and E. Coelho, Acoustical effects of internal tides on shallow water propagation: an overview of the INTIMATE96 experiment. In *Experimental Acoustic Inversion Methods for Exploration of the Shallow Water Environment*, edited by A. Caiti, J.-P. Hermand, S.M. Jesus and M.B. Porter (Kluwer Academic Publishers, Dordrecht, The Netherlands, 2000) pp. 19–38.

TIDAL EFFECTS ON MFP VIA THE INTIMATE96 TEST

A. TOLSTOY

ATolstoy Sciences, 8610 Battailles Ct., Annandale, VA 22003, USA
E-mail: atolstoy@ieee.org

S. JESUS AND O. RODRÍGUEZ

SiPLAB-FCT, University of Algarve, Faro, Portugal
E-mail: {sjesus,orodrig}@ualg.pt

Examining Intimate96 field data we see clearly the effects of tidal changes, i.e., of changing water depths for a bottom-tethered vertical array at mid-frequencies (300 to 800 Hz). This work will examine the sensitivity of such data via Matched Field Processing (MFP) to tidal changes where the depth varies ± 1.0 m from the nominal of 135 m. Is it possible to invert such data to estimate water depth as a function of time (tides)? Are the data dominated by source range estimates where water depths are known to shift in a predictable fashion as a function of source range errors? Results reported here will be for simulated data.

1 Introduction

Matched Field Processing [1] (MFP) is known to be sensitive to numerous environmental parameters as well as to source location, as seen clearly in the combination test cases SSPMIS, GENLMIS, and SLOPE of the MFP benchmarking workshop of 1993 [2]. Of particular importance are: water depth D (to which many inversion methods are quite sensitive [2,3]) and source range r_{sou} (a dominant parameter of intense interest [4,2,3]). Not only are the accuracies of these two parameters D and r_{sou} critical to MFP performance and to all MFP based inversion techniques, but they are known to be linearly related [5,6]. That is, they are strongly interrelated.

2 Intimate96 test

In 1996 a shallow water test (approximately 135 m of water) was conducted off the coast of Portugal which included a data set for a fixed source location (range of approximately 5.5 km, depth 92 m) and a fixed vertical, bottom-tethered array (3 working phones) monitored over a 24 hour period. The test configuration for this set is indicated in Fig. 1 with more details in [7–10]. The time window was sufficient to demonstrate obvious tidal effects on a broadband of mid frequency data (300 to 800 Hz) while effectively holding all parameters fixed, e.g., r_{sou}, *except* for water depth D. These data (for one phone) are seen in Fig. 2 (see [10]) where the 24 hours of data have been aligned to begin simultaneously (at 0.3 s). We can clearly see the tidal effects, particularly on the later arriving multipaths, e.g., after 0.5 s, where oscillations are evident. After some basic

457

N.G. Pace and F.B. Jensen (eds.), Impact of Littoral Environmental Variability on Acoustic Predictions and Sonar Performance, 457-463.

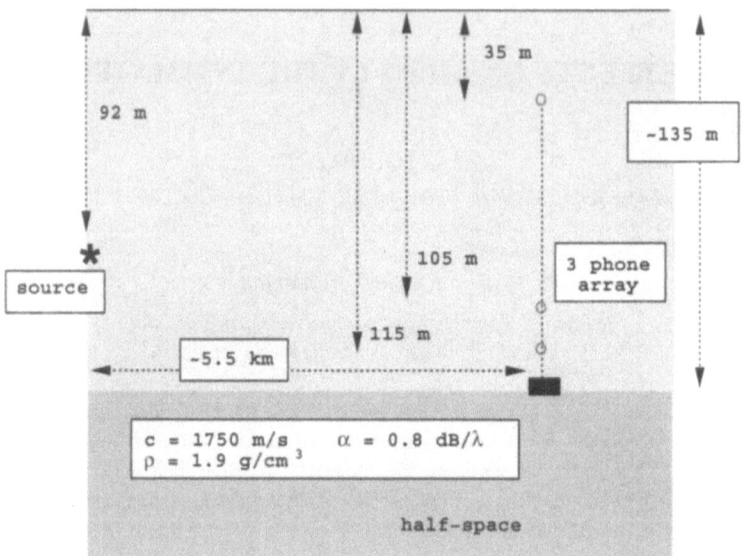

Figure 1. Intimate96 test configuration. Indicated is a source at 92 m depth, in approximately 135 m of water, approximately 5.50 km range from a bottom tethered array of 3 phones at 35, 105, and 115 m depth. The bottom is assumed to be a simple half-space. The sound-speed profile (not shown) is constant with range and refracts energy into the bottom.

MFP processing to produce ambiguity surfaces AMSs) the data (Fig. 3) show similar range fluctuations over 24 hours – if we assume that D is constant while $\hat{r}_{sou}$ varies (Fig. 3, see [10]).

3   Simulated data results

In an effort to examine tidal effects we have simulated the test environment with simple bottom parameters as indicated in Fig. 1, i.e., the bottom is represented as a half-space with parameter values as indicated, and with a downward refracting ocean sound-speed profile. The "data" are generated at a designated frequency by means of RAMGEO, a wide-angle energy-conserving Padé PE propagation model [11, 12]. Increasing attenuation at very deep depths acts as a "false" bottom to prevent the return of bottom penetrating energy, and these "data" have been successfully compared for final confidence to those generated by KRAKEN [12, 13]. The test fields have also been generated by RAMGEO.

3.1   Single Frequency Streaks

An important question becomes: can we resolve, i.e., successfully estimate *simultaneously*, r_{sou} and D? To address this issue, we first examined the source at 320 Hz which assumed data for a true $r_{sou} = 5.50$ km, true $D = 135.0$ m but which allowed the test data to vary in range $5.45 \leq \hat{r}_{sou} \leq 5.85$ km, and in water depth $130.0 \leq \hat{D} \leq 136.0$ m. The high resolution Capon (Minimum Variance: MV) processor [1] results are shown in Fig. 4 where the maximum power 1.10 (for diagonal noise loading 0.1) does appear

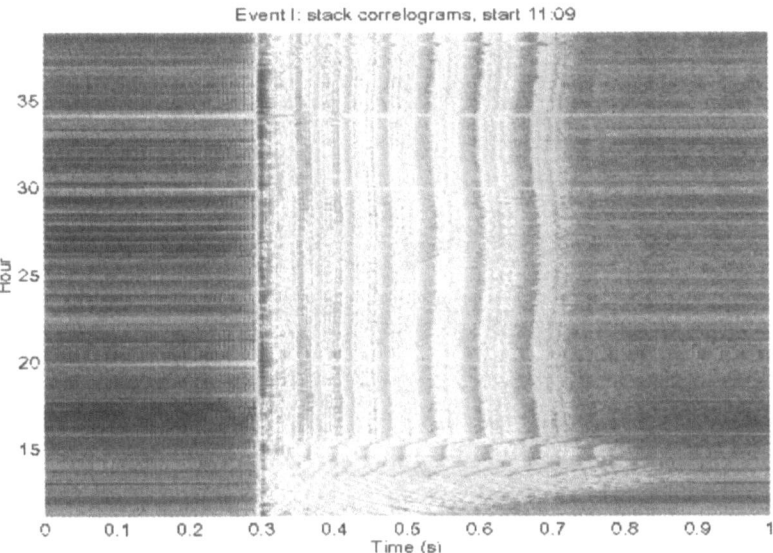

Figure 2. Intimate96 data showing tidal fluctuations for a fixed array and fixed source while time, i.e., water depth, varies over a 24 hour period (courtesy of M.B. Porter).

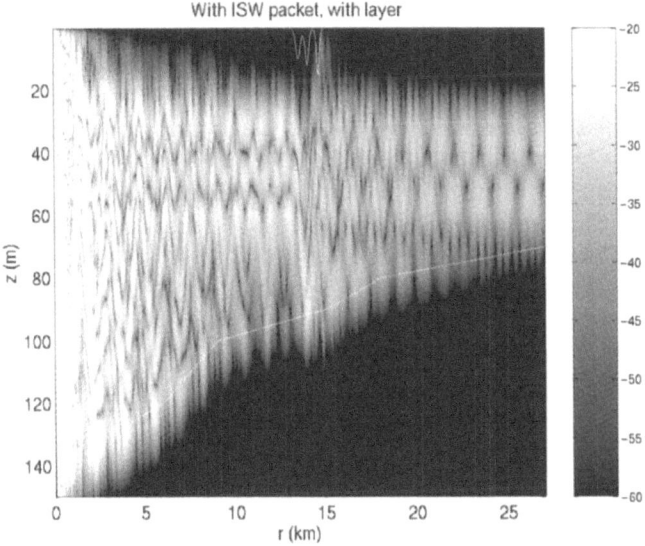

Figure 3. Intimate96 data showing range fluctuations in the AMSs for the data as they vary over a 24 hour period assuming a constant water depth D (courtesy of M.B. Porter).

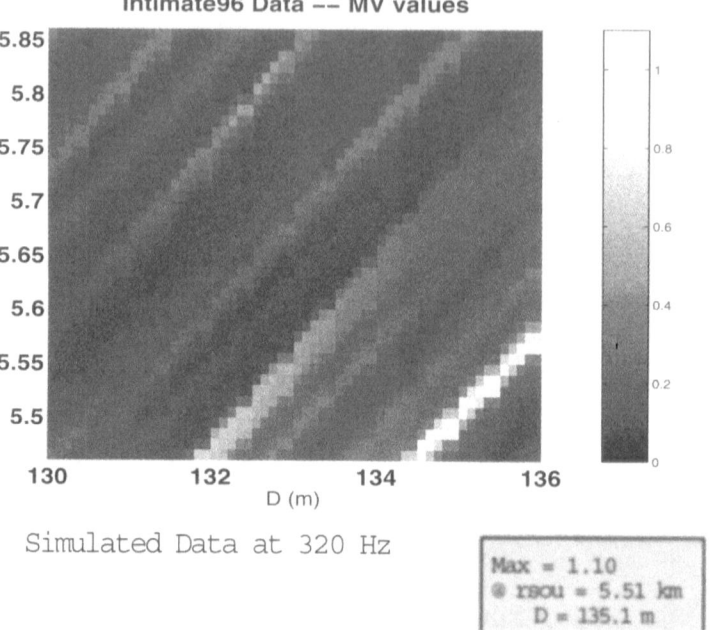

Figure 4. AMS for MV processor for simulated data at 320 Hz. The true source range $r_{sou} = 5.50$ km, true water depth $D = 135$ m. We note the linear relationships between range and water depth.

at the correct range and water depth, as expected. However, we note that there are other estimated ranges and water depths for which the MV power is also quite high, i.e., near 1.10. In particular, these ranges and depths appear along a line passing through the true values. Without knowing either the true source range or the true water depth at a given time, it is impossible to select the true pair of unknown values. We note also that there are other "streaks" in this MV plot, i.e., other $(\hat{r}_{sou}, \hat{D})$ linear combinations with similar slopes, indicating that the usual sidelobes (other phantom solutions) also show similar linear relationships between range and water depth. We see that in general as $\hat{D}$ increases so will $\hat{r}_{sou}$. Moreover, we note that if $\hat{D}$ changes by +0.1 m (a very small depth change), then $\hat{r}_{sou}$ will change by +10.0 m, a factor of 100.

3.2 Multiple Frequency Streaks

Next, we consider another frequency: 360 Hz. This frequency is also fairly low, but we again see (Fig. 5) the same linear relationship that we saw earlier (Fig. 4) between $\hat{r}_{sou}$ and $\hat{D}$ through the true source range and water depth (5.50 km range, 135 m depth). However, the "sidelobes", i.e., the other streaks, are now at *different* locations. Thus, a broadband BB (multi-frequency) sum of these AMSs will help to suppress these other false solutions. Unfortunately, a BB sum will not help us to select the true range-depth pair along the line through the true solution. That is, we are still unable to simultaneously determine r_{sou} and D.

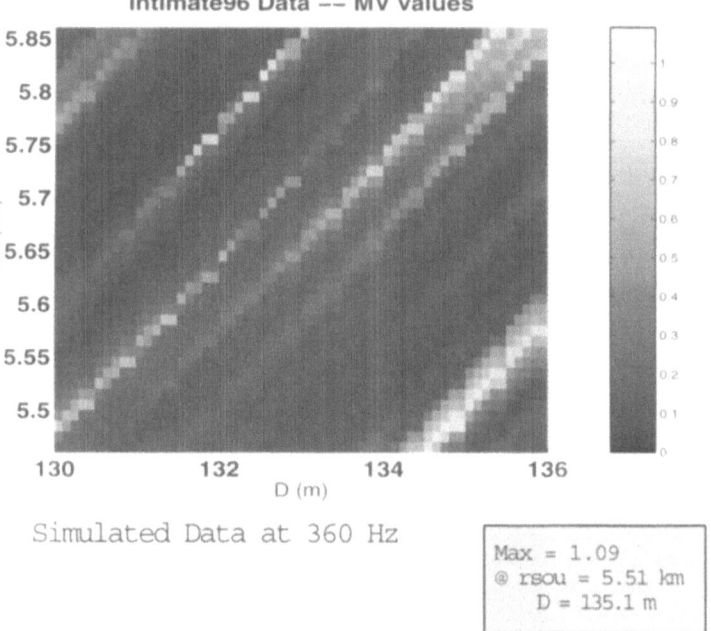

Figure 5. AMS for MV processor for simulated data at 360 Hz. The true source range $r_{sou} = 5.50$ km, true water depth $D = 135$ m. We note the linear relationships between range and water depth. We note also that the sidelobes, i.e., streaks not intersecting the true source range and water depth, are *different* from those at 320 Hz.

3.3 Changing Water Depths

Finally, let us examine the effect of changing water depth on the AMSs. In Fig. 6 we see abbreviated AMSs ($5.46 \leq \hat{r}_{sou} \leq 5.56$km, $130 \leq \hat{D} \leq 136$ m) at 320 Hz for a variety of "true" water depths mimicking tidal effects. In particular, in the top panel of Fig. 6 the true water depth is 135 m (see Fig. 4) while at next panel the true water depth is now 134.9 m. We note that the sidelobes (streaks) have changed similarly to those which occur at another frequency. However, we still have the same difficulty finding the true source and water depth simultaneously. That is, the true source range and water depth still lie on the same line as before but shifted according to depth (or equivalently, range). Thus, the linear relationship between source range and water depth has not changed. We still can *not* resolve the source range versus water depth ambiguity.

4   Conclusions

We conclude from simulations that:

- there is a linear relationship seen in MFP AMSs between water depth $\hat{D}$ and source range $\hat{r}_{sou}$ (consistent with earlier findings by DelBalzo *et al.*);

- an increase in $\hat{D}$ results in a predictable increase in $\hat{r}_{sou}$ as estimated by MFP (factor of 100);

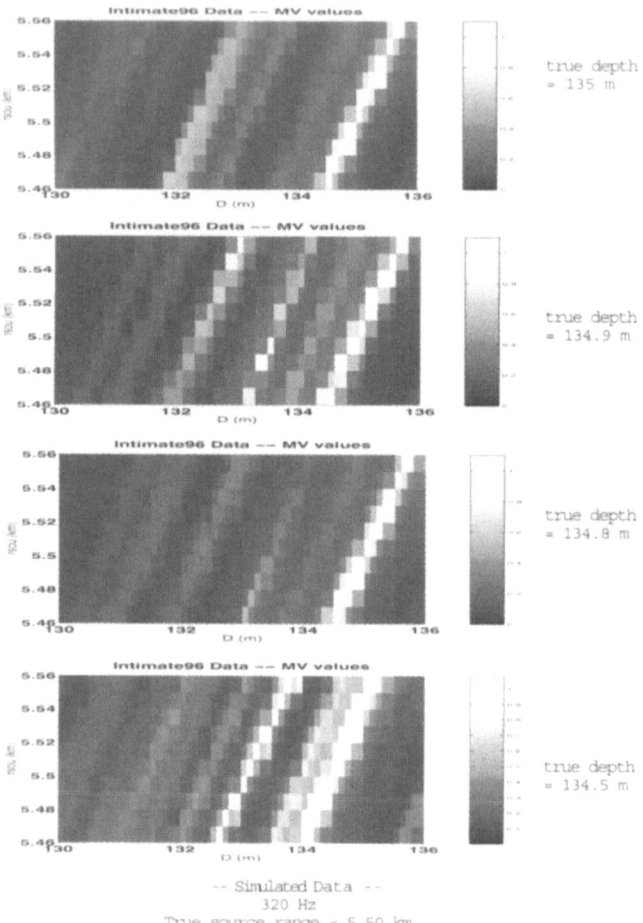

Figure 6. AMSs for MV processor for simulated data at 320 Hz for a source at 5.50 km range and for a variety of true water depths (as indicated).

- multiple frequencies summed incoherently can reduce sidelobe "streaks" in $(\hat{D}, \hat{r}_{sou})$ AMSs but cannot resolve the true (D, r_{sou}), i.e., beyond the basic linear relationship intrinsic to the waveguide;
- multiple time samples allowing for changing D show changing sidelobe "streaks" similar to the behavior seen in multiple frequencies;
- multiple time samples allowing for changing D show translation shifts in the basic linear relationship but still cannot resolve (D, r_{sou}), i.e., the *slope* of the line does not change.

Acknowledgements

Author A.T. thanks ONR for their continued support.

References

1. Tolstoy, A., *Matched Field Processing in Underwater Acoustics* (World Scientific Publishing, Singapore, 1993).

2. Porter, M.B. and Tolstoy, A., The matched field processing benchmark problems, *J. Comp. Acoust.* **2**(3), 161–185 (1994).

3. Tolstoy, A., Chapman, N.R. and Brooke, G., Workshop97: Benchmarking for geoacoustic inversion in shallow water, *J. Comp. Acoust.* **(6)**(1&2), 1–28 (1998).

4. Baggeroer, A.B., Kuperman, W.A. and Schmidt, H., Matched field processing: Source localization in correlated noise as an optimum parameter estimation problem, *J. Acoust. Soc. Am.* **83**, 571–587 (1988).

5. DelBalzo, D.R., Feuillade, C. and Rowe, M.M., Effects of water-depth mismatch on matched-field localization in shallow water, *J. Acoust. Soc. Am.* **83**, 2180–2185 (1988).

6. Weinberg, N.L., Clark, J.G. and Flanagan, R.P., Internal tidal influence on deep-ocean acoustic ray propagation, *J. Acoust. Soc. Am.* **56**(2), 447–458 (1974).

7. Jesus, S.M., Porter, M.B., Stephan, Y., Demoulin, X., Rodrigues, O.C. and Coelho, E., Single hydrophone source localization, *IEEE J. Oceanic Eng.* **25**(3), 337–346 (2000).

8. Porter, M.B., Jesus, S.M., Stephan, Y., Coelho, E., Demoulin, X., Single-hydrophone source tracking in a variable environment. In *Proc. 4th European Conference on Underwater Acoustics*, Rome, Italy (1998) pp. 575–580.

9. Porter, M.B., Jesus, S., Stephan, Y., Demoulin, X. and Coelho, E., Exploiting reliable features of the ocean channel response. In *Shallow Water Acoustics,* edited by by R. Zhang and J. Zhou (China Ocean Press, Beijing, 1997).

10. Porter, M.B, Stephan, Y., Demoulin, X., Jesus, S.M. and Coelho, E., Shallow-water tracking in the Sea of Nazare. In *Proc. Undersea Technology'98,* (IEEE Ocean Engineering Society, Tokyo, Japan, 1998).

11. Collins, M.D., A split-step Padé solution for the parabolic equation method, *J. Acoust. Soc. Am.* **93**, 1736–1742 (1993).

12. Jensen, F.B., Kuperman, W.A., Porter, M.B. and Schmidt, H., *Computational Ocean Acoustics* (Amer. Inst. Physics, New York, 1994).

13. Porter, M.B., The KRAKEN normal mode program. Report SM-245, SACLANT Undersea Research Centre, La Spezia, Italy (1991).

MULTIPATH EFFECT ON DPCA MICRONAVIGATION OF A SYNTHETIC APERTURE SONAR

L. WANG, G. DAVIES, A. BELLETTINI AND M. PINTO
SACLANT Undersea Research Centre, Viale San Bartolomeo 400, 19138 La Spezia, Italy
E-mail: wang@saclantc.nato.int

The Displaced Phase Centre Antenna (DPCA) technique makes use of the correlation of sea bottom back scattering to determine the trajectory of a Synthetic Aperture Sonar (SAS). Multipath has been identified as the main environmental factor degrading the accuracy of DPCA in shallow water environments. An experiment has been carried out with Saclantcen's 100 kHz multi-aspect SAS demonstrator deployed from R/V Alliance. The 256 channel receive array was placed, both horizontally and vertically, at various depths in shallow water channels, in order to quantify the relative levels of the multipath signals as a function of depth, range, sea state and bottom type. The ping to ping signal correlation was measured and compared with the results from a sonar performance model.

1 Introduction

The performance of DPCA micronavigation for SAS applications has been investigated extensively both in theory and experiment in recent years [1–3]. The technique makes use of the correlation of the sea bottom back scattering to estimate the displacement of the sonar between successive pings. The main environmental factor determining the navigation accuracy achievable for a given sonar system is ultimately the signal to noise ratio. The signal is the direct sea bottom back scattering, while the noise consists of background noise of the sea, system noise, surface and volume reverberation and multipath interference such as surface reflected bottom scattering etc.

When SAS is used in shallow and very shallow waters, the multipath interference may be a dominant source of noise. Thus the accuracy of the DPCA micronavigation is degraded, possibly limiting the achievable SAS length. In addition, the multipath will also degrade the quality of the SAS imagery, leading to ghost targets and loss in image contrast (e.g. filling in of shadows). Thus multipath could be the most important environmental factor limiting the range achievable by an SAS system in shallow water.

An experiment has been carried out to investigate the effects of multipath on SAS in shallow water channels with different water depths and bottom characteristics. A 100 kHz sonar with a total receiver aperture of 1.92 m was deployed from R/V Alliance in horizontal and vertical configurations. The data from vertical transmissions provide very revealing information about the shallow water channel because of the high angular resolution of the sonar. The signal level and ping to ping correlation were predicted with a new sonar performance model, ESPRESSO, and compared with the experimental results.

N.G. Pace and F.B. Jensen (eds.), Impact of Littoral Environmental Variability on Acoustic Predictions and Sonar Performance, 465-472.
© 2002 *Kluwer Academic Publishers.*

2 Experiment

The experiment was carried out in shallow water near La Spezia (Italy) in two areas, off Tino island (referred to as area I in the following) and Monesteroli (area II), from the middle of October to early November, 2001. It was one part of a larger experiment designed to further the understanding of SAS micronavigation, combining data-driven techniques such as DPCA and Aided Inertial Navigation Systems. The main piece of equipment used is the MASAI (Multi Aspect Synthetic Aperture sonar Imaging) towbody, which consists of a 100 kHz multi-element sonar array and a high grade inertial navigation system (INS). Figure 1 shows the deployment of the MASAI towbody from R/V Alliance.

2.1 MASAI System

The MASAI array consists of 256 receiver elements at a 7.5 mm spacing to form an aperture of 1.92 m. The signal used for MASAI'01 was a 10 ms chirp with 100 kHz centre frequency and 10 kHz bandwidth. The centre 64 channels were used for transmission and defocused to give a beam width of 40°. The vertical (resp. horizontal) beamwidth of the elements was about 40° (resp. 100°). The source level was 206 dB re 1μPa at one meter. The depression angle of the sonar was 22° in the horizontal configuration.

Figure 1. Deployment of the MASAI system from the Alliance.

The INS used in the MASAI system is a strapdown inertial navigation system with ring laser gyros. The INS was aided by a 1.2 MHz DVL (Workhorse Navigator by RDI), a pressure sensor and a GPS. The aiding was peformed with NAVLAB, a software developed by the Norwegian Defense Research Establishment.

2.2   Environmental Data

The shallow water channels in the location of the trial were flat with a water depth about 26 m in area I and 32 m in area II. Core samples were taken in the areas in order to determine the bottom characteristics. Two methods were used to obtain the sound speed and density in the sediments. One was to derive the parameters from grain size of the sediments and, the other one was to measure directly from the core samples. Table 1 shows the results of grain size analysis for the cores of the top 6 cm in the two areas. It was observed that the sediment in area I was harder than in area II. Sound speed and density in the sediments measured directly from the core samples are given in Table 2. It can be seen that the acoustic impedance of the sediment in area I was much higher than that in area II.

Wind speed was measured by a meteobuoy during the trial. The measured wind speeds at the time of sonar transmissions (most of the time corresponding to sea state in the range 2–3) were used in the sonar model for surface scattering strength and surface reflection loss. A wave rider was also deployed to measure the wave height.

A CTD was used throughout the trial to obtain the sound speed profile in the water column in both area I and area II. An isovelocity profile of about 1526 m/s was found for the whole period.

Table 1. Grain size analysis

Depth (cm)	Tino (area I)		Monasteroli (area II)	
	Sand/silt/clay	ϕ	Sand/silt/clay	ϕ
1.25	45.1/46.7/8.1	5.02	35.9/40.2/23.9	6.02
3.75	60.3/33.3/6.0	4.29	29.5/42.8/27.7	6.49
6.25	45.2/49.5/5.2	4.95	29.6/39.5/26.8	6.17

Table 2. Measured sound speed and density

Depth (cm)	Tino		Monasteroli	
	Sound speed (m/s)	Density (g/cm^3)	Sound speed (m/s)	Density (g/cm^3)
2	1649.66	1.94	1549.81	1.74
3	1643.48	1.92	1550.35	1.72
4	1643.92	1.92	1552.14	1.72
5	1644.47	1.93	1557.18	1.79
6	1639.45	1.91	1556.65	1.79

3   Sonar model

The modelled data presented in this paper were obtained using a tool being developed at SACLANTCEN called ESPRESSO. The tool includes a reverberation model based on the technique of "Geometric beam tracing" [4,5], originally developed to model propagation loss. The adaptation of this technique to reverberation modelling is described in [6]. The reverberation model approximates the seabed, sea surface, and water column as lines of scatterers; the reverberation contribution from each scatterer is allocated to a time bin to create a reverberation time series. The acoustic models used to obtain scattering strengths and reflection losses are described in [7] (the bistatic version of the seabed scattering strength model was used). The absorption coefficient in water was obtained using the algorithm described in [8].

4   Experimental and modelling results

Although a large amount of data was collected with the sonar in a horizontal configuration at fixed positions in both areas during the trial, the vertical configuration was used only in area II. The results presented here are all from the data obtained in area II since the data from the vertical configuration provided more detailed information about the channel.

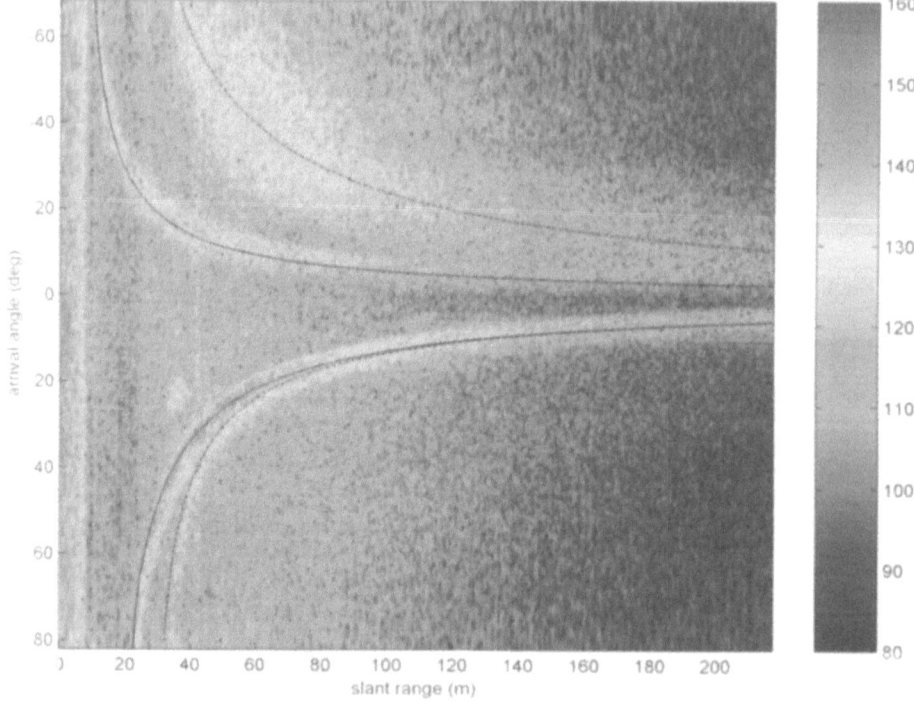

Figure 2. Beamformed sonar data with the array in vertical configuration, showing the arrival angles of the different paths measured with respect to the vertical.

4.1   Sound Field in the Vertical Plane

In order to study the details of the sound field in the channels, the MASAI sonar was deployed vertically in area II. The various arrivals can be examined by beamforming the signals received by the 256 channels, with an angular resolution of 0.45°. One of the measured sound fields as a function of arrival angle and range is shown in Fig. 2. The sonar was lowered with ropes from the ship up to a depth of 9.7 m. The transmission angle was depressed by 6.8 degrees from the horizontal direction, as it was measured by the INS. This unwanted misalignment was compensated at first order[1] by shifting the y-axis origin of Fig. 2.

The lines in the figure indicate where the different arrivals are expected for the given geometry. The continuous lines are for the direct bottom and direct surface scattering paths, while the dashed lines give the paths of the specular surface reflection path of the bottom scattering and its reciprocal path. The signal levels from higher order mutipaths are too low to be observed.

The measured field reveals that, for a fixed return time, there is a diffuse surface bounce that spreads much more than what can be explained by the depression angle from the vertical. This spread is likely to be due to the roughness of the sea surface which produced non-specular surface reflections from bottom scatterers at closer range than that corresponding to the specular path. These non-specular returns came from angles of arrival both larger and smaller than that of the specular path.

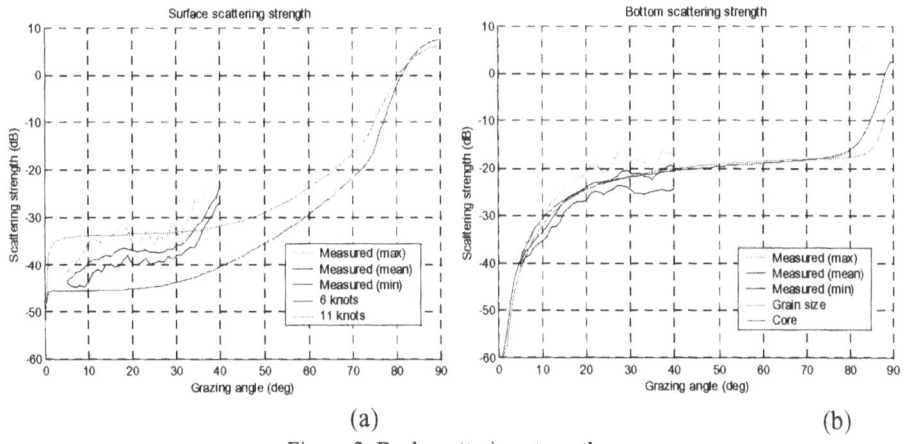

(a)   (b)

Figure 3. Back scattering strength.

4.2   Measurement of Back Scattering Strength from Sea Surface and Bottom

Direct measurement of the back scattering strength from sea surface was made possible by the vertical configuration. Figure 3(a) shows the measured surface back scattering strength as a function of grazing angle using the vertical configuration of the sonar vs the

---

[1] The depression angle causes also a spread in grazing angle (of the same order of magnitude as the depression angle) for fixed range due to the large horizontal beamwidth.

theoretical model [7]. The measured surface back scattering is plotted in three curves with minimum, mean and maximum values to indicate the spread of the data. The measured average wind speed was 11 knots over the period of sonar transmission. Two curves from the model are given in the figure for wind speed 6 and 11 knots for comparison. The surface scattering strength was almost constant for grazing angles less than 30°. The increase of surface scattering strength measured at large grazing angles is due to the artifact of assuming a sinc beam pattern for the transmitter in the calculation of the scattering strength. The near constant surface back scattering suggests the main mechanism of the scattering was due to air bubbles below the surface induced by wind.

The measured bottom back scattering strength is shown in Fig. 3(b) with two different model predictions. In the first (green line) sound speed and density derived from logarithmic grain size ϕ in Table 1 were used (i.e., c = 1506.6 m/s, ρ = 1.183 g/cm^3). In the second (red line) sound speed and density measured from the core sample in table 2 used (i.e., c = 1549.8 m/s, ρ = 1.743 g/cm^3). The sediment volume scattering parameter and loss tangent were adjusted to fit the data. The bottom scattering strength predicted with the measured sound speed and density from the core is higher than that using derived results from the grain size analysis. It is likely that the sediment in the core was compressed during the core sampling process, giving an increase of sound speed and density in the core samples. It is expected that the true value may be some where in between the values obtained by the two methods.

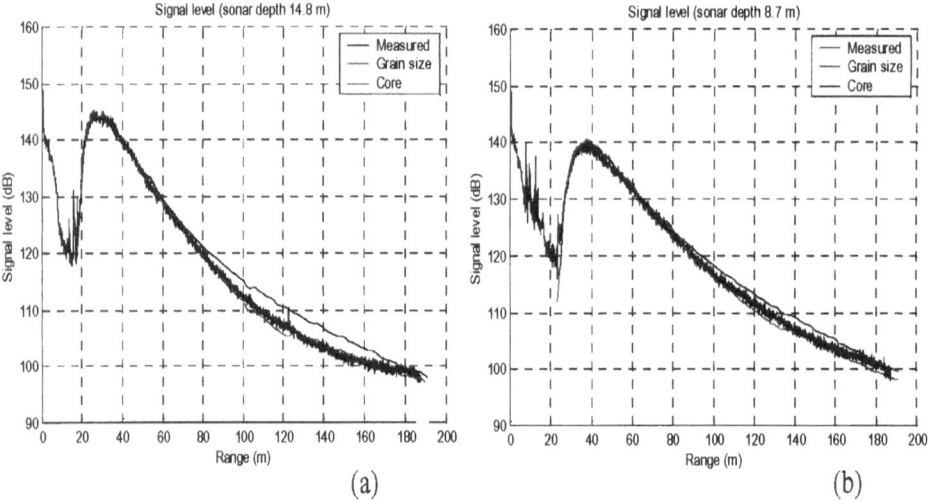

(a)   (b)

4.3   Signal Level at Various Sonar Depths

Figure 4 shows the received signal from one receiver channel as a function of range at two different sonar depths, one at about middle depth, the other one at shallower depth. The signal level predicted by ESPRESSO is in overall agreement with the measured results, although the model takes only specular reflection into account. Applying the bottom parameters obtained from grain size analysis results in a better prediction of the signal level, while using the direct measured parameters from the core sample results in a prediction higher than the measured data at longer ranges.

4.4   Ping to Ping Correlation at Various Sonar Depths

DPCA makes use of the correlation of direct sea bottom back scattering between successive pings to estimate the displacement of the sonar. The received signal by the MASAI sonar at ping k can be expressed as

$$X^k(t)=s^k(t)+n^k(t) \tag{1}$$

where $s^k(t)$ is the direct bottom scattering and $n^k(t)$ is the interference including background noise and multipath signals.

The predicted values of the correlation coefficient μ, derived from the modelled signal to noise ratio, are plotted in Fig. 5 together with the corresponding experimental results. It is seen that there are large variations between the two sets of modelled data and that the agreement between the modelled and measured data is not very good.

For both models, the correlation peaks at close range, before the onset of multipathing and then decreases as the specular multipath enters into the sonar through the sidelobes of the vertical beam pattern. The correlation then increases again, as the specular multipath falls into the null of the vertical beampattern, to again decrease with range as this nulling effect diminishes. In addition both sets of modelled data predict the onset of higher order multipath at long ranges, which is more important for the model with higher impedance and sound speed in the sediment (in particular giving a higher critical angle).

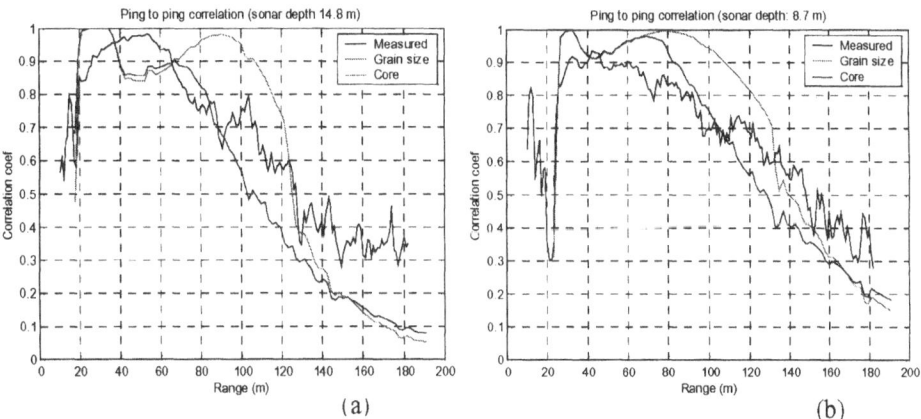

Figure 5. Signal correlation at different sonar depths.

In the experimental data, the correlation peaks at close range. The low values obtained at very short range are probably due to baseline decorrelation, resulting from residual motion of the sonar between pings and the high grazing angles of operation. The peak correlation coefficient increases with the sonar depth, as a result of the reduced sea surface interactions. The correlation then seems to decrease monotonically with range. The increase predicted by the models, due to the nulling of the specular multipath by the vertical beampattern, is not observed. This is indicative of the diffuse nature of the multipathing. The decrease in the correlation with range is faster for the deeper sonar position [Fig. 5(a)] than for the shallower one [Fig. 5(b)], predominantly as a result of the reduction in the grazing angle for given slant range.

5   Summary

The effects of multipath on DPCA micronavigation were studied experimentally. Comparisons between the experimental results and sonar model predictions seem to indicate that the diffuse surface reflection significantly affected the signal correlation. More analysis is required to confirm this assumption. It may be necessary to extend sonar models beyond the specular reflection scenario to take this effect into account.

Acknowledgements

We thank all the people involved in the MASAI'01 experiment. We also thank Eric Pouliquen for his help.

References

1. Bellettini, A. and Pinto, M., Theoretical accuracy of synthetic aperture sonar micronavigation using a displaced phase centre antenna, *IEEE J. Oceanic Eng.*, (submitted 2001).
2. Bellettini, A. and Pinto, M., Experimental results of a 100 kHz multi-aspect synthetic aperture sonar. In *Proc. 5emes Journees d'Etudes Acoustique Sous-Marine*, Brest, France (Dec. 2000).
3. Wang, L., Bellettini, A., Hollett, R., Tesei, A. and Pinto, M., Interferometric SAS and INS aided SAS imaging. In *Proc. Oceans'01*, Hawaii (2001).
4. Porter M.B. and Bucker H.P., Gaussian beam tracing for computing ocean acoustic fields, *J. Acoust. Soc. Am.* **82**(4), 1349–1359 (1987).
5. Porter, M.B. and Liu, Y.C., Finite element ray tracing. In *Theoretical and Computational Acoustics,* - Vol. 2, edited by D. Lee and M. H. Schultz (World Scientific Publishing, 1994.) pp. 947–956.
6. Meyer, M. and Davies, G.L., Beam tracing techniques for high-frequency reverberation modelling. In *Proc. 6th European Conference on Underwater Acoustics*, Gdansk, Poland (2002).
7. APL-UW High-frequency ocean environmental acoustic models handbook. Technical Report 9407, Applied Physics Laboratory, University of Washington, October 1994.
8. Fisher, F.H. and Simmons, V.P., Sound absorption in sea water, *J. Acoust. Soc. Am.* **62**(3), 558–564 (1977).

SEA SURFACE SIMULATOR FOR TESTING A SYNTHETIC APERTURE SONAR

B. DAVIS

University of Arizona, Tucson, AZ

P. GOUGH

University of Canterbury, Christchurch, New Zealand
E-mail: gough@elec.canterbury.ac.nz

B. HUNT

University of Arizona, Tucson, AZ

With the move to use side-looking imaging sonars in very shallow waters as a component part of MCM operations, synthetic aperture sonars (SAS) appear to have some advantages over a conventional real aperture side-looking sonars. One significant advantage of SAS is that it is quite resilant to image degradation caused by surface backscatter and surface multipath. The processing in all SAS imaging algorithm assumes the only thing moving between transmitted pings is the sonar platform. Since the algorithm uses coherent integration to assemble the final image, any movement of the sea surface between pings destroys the ping-to-ping coherence of the surface multipath as well as the ping-to-ping surface backscattered return. To move towards understanding just how effective a SAS is at supressing backscatter and surface multipath, we first need to model the moving sea surface in a believable way and establish just how the sound reflects off the undersurface of the sea. This paper first describes a commonly-used physically justifiable sea-surface autocorrelation function that accounts for wind direction, wave height, wave period and wave velocity. From this autocorrelation function, a statistically appropriate random wave surface is generated which evolves in both time and space. Finally in a first attempt to model the shallow-water sea surface multipath problem, a set of impulse responses are generated from this wave-surface as it evolves in time increments equal to the pulse repetition period. Here we model an isotropic one-way (reflected) acoustic path from the target at a depth of seven metres to the sonar platform at a depth of five metres separated by 25 m with the surface above the path covering an area of 160 m (cross-track) by 60 m (along-track) and we ignore any seafloor multipath. Two sea-surface reflection/scattering mechanisms are used in this model. In the first, each surface facet acts as a diffraction-limited aperture and in the second, each facet acts as a Lambertian reflector. These descibe two limiting situations 1) when the acoustic wavelength is small compared with the roughness of any facet and 2) when the surface roughness is a significant proportion of the acoustic wavelength. Concentrating on the diffraction-limited model, we show the effect of surface multipath on the raw data collected by a SAS and its effect on the processed image. We also make some estimates of the signal to clutter ratio improvements as a function of the number of hits on target.

N.G. Pace and F.B. Jensen (eds.), Impact of Littoral Environmental Variability on Acoustic Predictions and Sonar Performance, 473-480.
© 2002 *Kluwer Academic Publishers.*

1 Introduction

The imaging fidelity of any standard side-looking sonar is degraded by sea surface backscatter and sea-surface multipath reflections as well as seafloor multipath reflections as shown in Fig. 1, however, surface backscatter and multipath are different in that they change with time. By using synthetic aperture sonar (SAS) techniques, we can use the time-variable nature of the sea surface to our advantage.

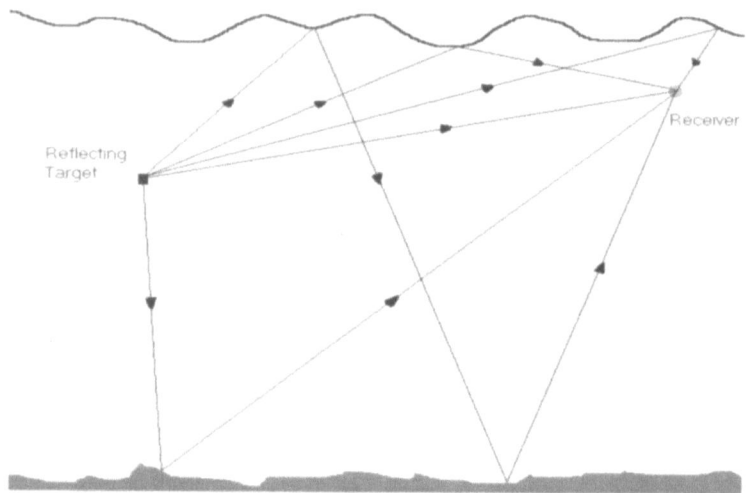

Figure 1. Illustrating sea surface and seafloor multi-path reflections.

All SAS systems record the pulse echo returns from each transmitted ping in both amplitude and phase so they can, by using coherent integration, compute an image from a contiguous collection of ping echoes as if it came from a much larger physical aperture. For SAS to work there are some critical assumptions. The first is that the platform moves in a predicable and usually linear track. A combination of highly accurate navigation units and autofocus techniques (sometimes called micro navigation) can now correct the problems caused by nonlinear track. The second assumption, and the one pertinent to this paper, is that the only thing moving during the collection of the data is the sonar platform. Since the sea-surface is clearly moving, it is important to establish how this movement affects the data collected and more importantly how it effects the final image. To do this, we normally simulate the complete sonar data collection and imaging system both with and without the sea-surface effects but to do this we need to model the sea-surface in a believable way. Unfortunately if all the multipath effects are included in the simulation, the model becomes extremely complicated so we restrict out simulations to consider only the effects of sea-surface multipath on the reflected echoes. That is we assume the vertical beamwidth of the projector is small enough to eliminate any surface backscatter and sea-surface multipath on the outward leg of the acoustic path and that there are no sea-floor multipath effects. Despite these limitations, we believe the simulations show realistic effects of sea-surface multipath and how it effects the SAS imaging process.

2 The autocorrelation of the sea surface

A useful model of the space and time autocorrelation of the sea surface already exists and we use it here without proof [1]

$$R_S(\zeta, \eta, \tau) = A_D^2 \exp\left[\frac{-(\zeta - u_G\tau)^2}{L_w^2}\right] \exp\left[-\left(\frac{\eta}{L_c}\right)^2\right] \cos[K_w\zeta - \Omega\tau] \quad (1)$$

with related variables -

w	windward direction
c	cross wind direction
ζ	windward ordinate in m
η	crosswind ordinate in m
τ	temporal delay ordinate in s
A_D	RMS displacement amplitude in m
L_w	windward correlation length in m
L_c	crosswind correlation length in m
L_t	temporal correlation time in s
K_w	spatial frequency in windward direction in m^{-1}
Ω	temporal frequency in s^{-1}
$u_G = \frac{L_x}{L_t}$	group velocity of waves in ms^{-1}
$u_P = \frac{\Omega}{K_w}$	phase velocity of waves in ms^{-1}.

Basically the wave crests are assumed to be correlated in the windward and crosswind direction and move with an average velocity in the windward direction. Here w, c represent the windward and crosswind coordinates which will eventually be rotated into the imaging coordinates of x and y. In addition, measurements show that the distribution of displacements at a given point is approximately Gaussian [1] and stationarity is also assumed. A single example of a sea surface is shown in Fig. 2 with the following parameters: crosswind correlation length 12 m, windward correlation length 7 m, dominant windward wavenumber 2 m^{-1}, temporal frequency 2 Hz, RMS wave amplitude 0.2 m and wind direction 126° to the x-axis.

3 Tiling the surface

The simulated sea surface needs to be tiled into a contiguous set of reflecting facets. The easiest way to do this is to tile the surface into triangular surface elements; the three vertices of the triangle being defined by three x, y, z coordinates. The three vertices then define a tilted triangular facet. This can be specified to any aspect ratio down to the smallest facet delimited by the rectangular sampling grid with separation Δx and Δy.

Having constructed a single realisation of the sea-surface, $s(w, c, t)$, and having determined the distance (and so delay time) from the transmitter to the target and from the target to the surface facet at x_0, y_0 then from the surface to the receiver as well as the vector-dependent facet gain term, $G_L(x_0, y_0)$ (for Lambertian scattering) or $G_D(x_0, y_0)$ for diffraction-limited scattering), it remains to add the contributions from all the surface facets to calculate the impulse response of the sea surface.

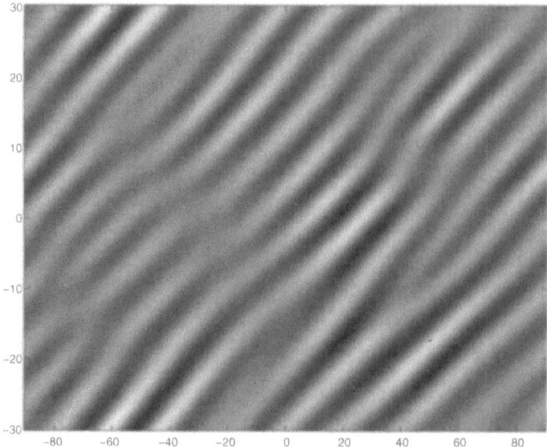

Figure 2. Example of a typical simulated sea surface.

4   Simulation results

Before we simulate the effects of time-varying multipath on SAS imaging, it is useful to
visualise the modulus of the facet gain term $|G_D|$ and $|G_L|$ as a function of x_0, y_0 as well
as its impulse response for a specific surface condition. A useful limiting case to check is
that of a "flat" sea-surface. Clearly for Lambertian scattering to exist at all, there would
be some capillary waves present to drive the scattering mechanism. Since there are far
too many parameters to perturb, we chose to model a specific situation with a target at
a depth of 10 m with a receiver at a depth of 10 m, the two separated by 100 m using
a centre transmitted frequency of 100 kHz. As expected, the Lambertian model showed
little or no variation with facet size and carrier frequency whereas the diffraction-limited
model showed a change in the sinc pattern as would be expected by changing the ratio
of wavelength to aperture size.

When the suface has some non-zero wave height, some interesting behaviour is revealed. Figure 3 shows the gain-modulus image for a sea surface with a Lambertian
scattering model. This image has the same general trend as the "flat" surface case but
with the wave structure imposing lengths of minimal or zero gain. These structures tend
to be more dense away from the area directly between the target and receiver.

The gain-modulus image for a sea surface using the diffraction-limited model is shown
in Fig. 4. It resembles its flat surface counterpart but does have significant effects present
attributable to the wave structure. Like the Lambertian case, the tilts of the facets have
the effect of reducing the gain in certain places.

5   Modelling a SAS

First we make the assumption that there is no movement during the transmission and
reception of a single pulse and that all movements of the surface and the sonar are
condensed into a single instantaneous period between the last echo return of one ping and
the onset of transmission of the next ping. This is known as the "stop and hop" scenario.
This does ignore any temporal Doppler effects that occur due to movement during a pulse.

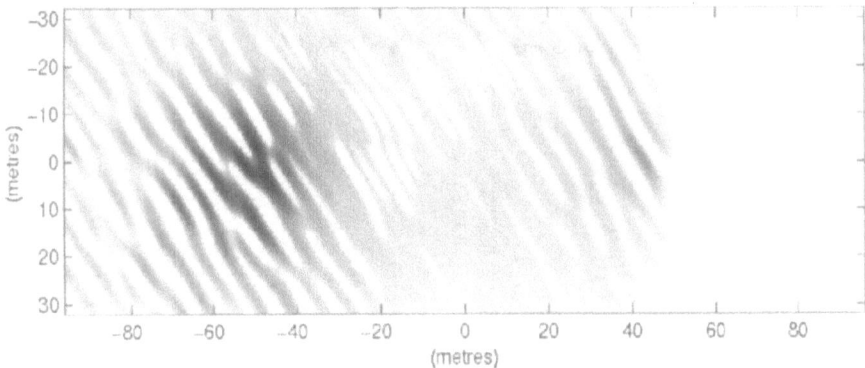

Figure 3. Gain-modulus image for a sea surface with Lambertian scattering.

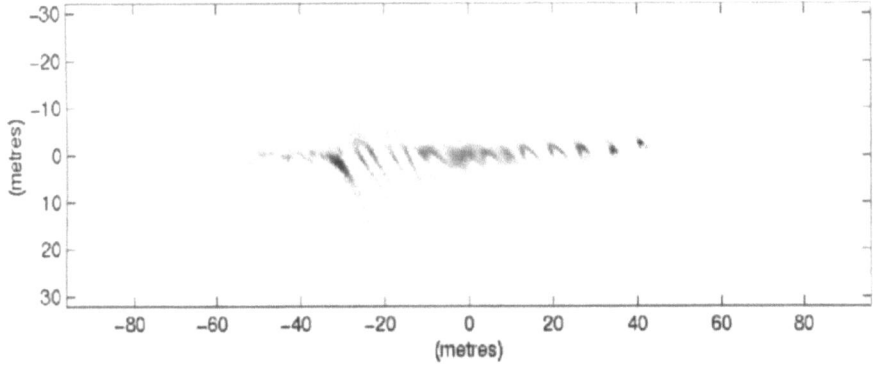

Figure 4. Gain-modulus image for a sea surface with diffraction-limited scattering.

Since existing SAS seldom operate at maximum unambiguous ranges of more than 200 m, they mostly use a pulse repetition period of shorter than 300 ms consequently we consider "stop and hop" scenario accurate enough to model surface multipath effects.

To model the surface effects on the imaging process, we proceed in the following fashion. A point target layout is selected which represents a typical target location and depth along with the SAS parameters and the surface conditions needed to compute the sea surface autocorrelation function. The surface realisation $s(w, c, t)$ is computed and we record the impulse response for the particular position of the sonar relative to the target(s). Then the sonar is moved by Δu and the surface evolved in time by $\Delta T = \Delta u / v_s$ where v_s is the forward velocity of the sonar platform. This process is repeated for every ping as the sonar traverses a single pass of the target field. It is also repeated over a range of differing surface conditions using both Lambertian and diffraction limited scattering.

Using a simulated target field of three point-reflectors gives a basis for comparison and Fig. 5 shows the intensity of raw data echo returns displayed in dB intensity to bring out the features normally concealed in a linear display. Now a surface multipath using the diffraction model is factored into the data and is shown in Fig. 6. Note the waveheight in the sea surface, 1 m, is enough to show the existance of the multipath but not so much

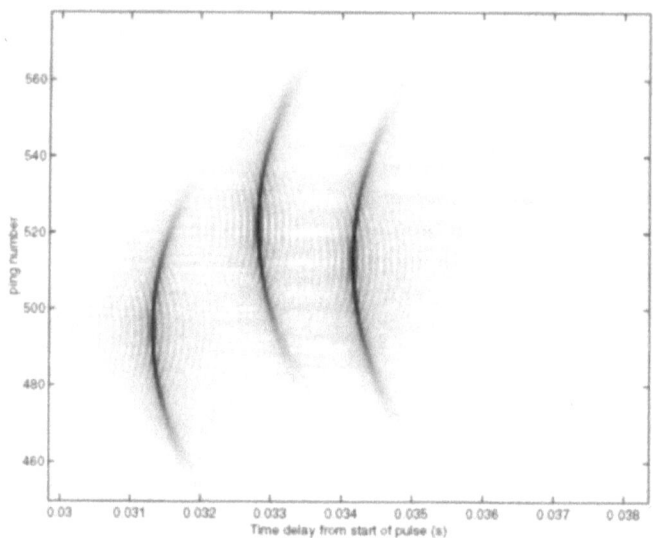

Figure 5. Image intensity of raw data (dB scale) for 3 point refectors with no multipath.

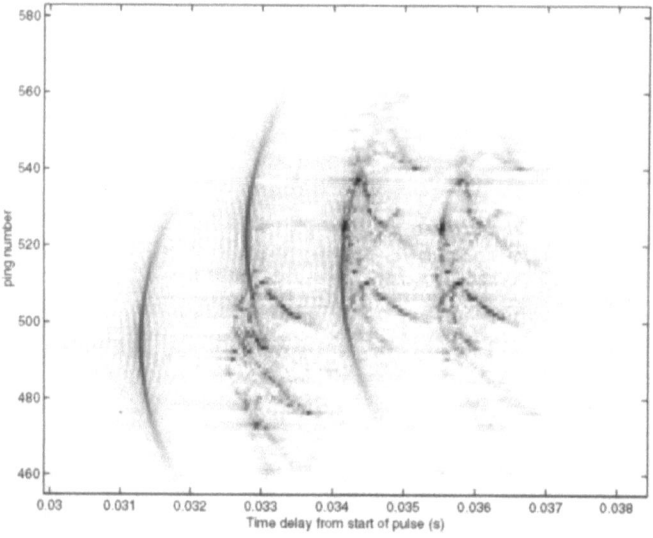

Figure 6. Image intensity of the same raw data corrupted by surface multipath.

as to overwhelm the underlaying raw data. Also note that the multipath appears to have produced some "structured" echoes that could easliy be missinterpreted. This is most easily seen in that there appears to be a fourth target to the right of the central target. In addition the multipath of the upper target overlays the central target and if the multipath were stronger, would conceal it.

To get some estimate of the all-important signal-to-clutter ratio (SCR), we can take an ensemble line scan in range (i.e that is along the fast-time or cross-track axis) through

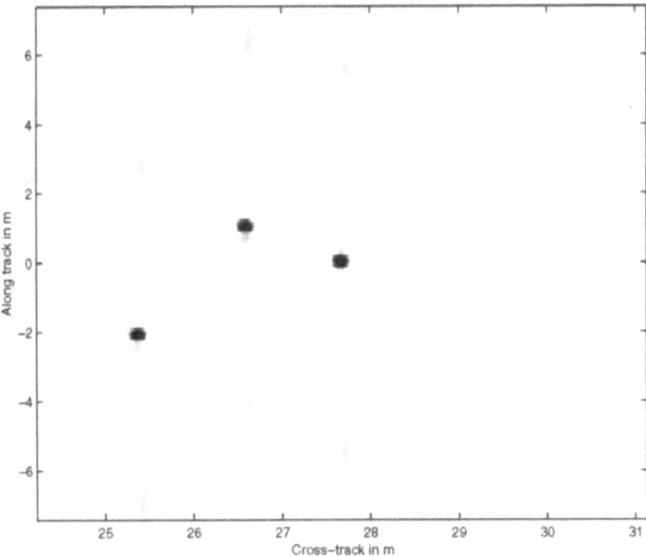

Figure 7. Image intensity of azimuth compressed data (dB scale) for 3 point refectors with no multipath.

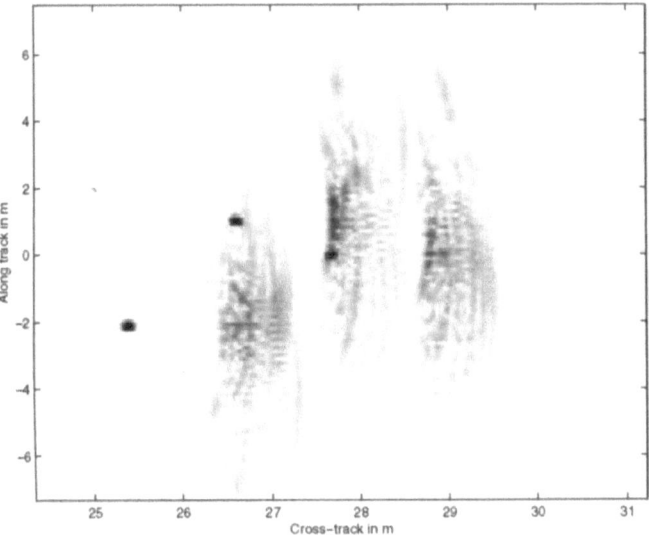

Figure 8. Image intensity of same azimuth compressed data corrupted by surface multipath.

the centre of the target echo shown in Fig. 6. The SCR can be estimated from the RMS value of the target's direct path return followed by the RMS surface multipath return. By looking closely at the clutter surrounding the central unprocessed return, we estimate the SCR caused solely by multipath (since there is no other clutter mechanism) to be about 10 dB.

Now the raw data is processed to a final image intensity and displayed in Fig. 8. Note there are about p = 40 pings on target so if we were to expect all the target returns to add coherently i.e., a 20 log (p) increase in peak intensity, and all the multipath clutter to add non-coherently by 20 log $(p^{0.5})$, we would expect a 16 dB improvement in SCR. So by taking the RMS value of the pixels surrounding the central target in the processed image, we estimate the SCR to be about 20 dB; a 10 db improvement over the 10 dB SCR of the raw data. In this way we can get a quantitative estimate of the improvement over the raw data image as a function of hits on target. Stated briefly

$$\text{SCR improvement} \approx 10 \log (p^{0.65}). \tag{2}$$

which indicates the surface multipath has some correlation from ping-to-ping.

6 Conclusions

In shallow waters, clutter caused by surface multipath limits the image quality of all side-lookings sonars. However, SAS have a significant advantage in the the echo returns from many pings are processed coherently which means that there will be some improvement in the SCR in the processed image. The actual degree of improvement in SCR ratio is dependent on how fast the surface changes with respect to the pulse repetition period but, for our hypothetical suface, target layout and sonar parameters, the improvement appears to be about 10 log $(p^{0.65})$.

References

1. H. Medwin and C.S. Clay, *Fundamentals of Acoustical Oceanography* (Academic Press, San Diego, CA, 1998).
2. J.C. Novarini and J.W. Caruthers, Numerical modeling of acoustic-wave scattering from randomly rough surfaces: An image model, *J. Acoust. Soc. Am.* **53**(3), 876–884 (1973).
3. J.C. Novarini and H. Medwin, Diffraction, reflection, and interference during near-grazing and near-normal ocean surface backscattering, *J. Acoust. Soc. Am.* **64**(1), 260–268 (1978).
4. W.A. Kinney, C.S. Clay and G.A. Sandness, Scattering from a corrugated surface: Comparison between experiment, Helmholtz-Kirchoff theory, and the facet-ensemble method, *J. Acoust. Soc. Am.* **73**(1), 183–194 (1983).
5. D.W. Hawkins and P.T. Gough, Recent sea trials of a synthetic aperture sonar. In *Proc. Institute of Acoustics* **17**(8), 1–10 (Dec. 1995).
6. D.W. Hawkins, Synthetic aperture imaging algorithms: With application to wide bandwidth sonar. Ph.D. Thesis, Electrical and Electronic Engineering, University of Canterbury, Christchurch, New Zealand (Oct. 1996).
7. P.T. Gough and D.W. Hawkins, Unified framework for modern synthetic aperture imaging algorithms, *Int. J. Imaging Systems and Technology* **8**, 343–358 (1997).
8. M. Soumekh, *Fourier Array Imaging* (Prentice Hall, Englewood Cliffs, NJ, 1994).
9. J.R. Schott, *Remote Sensing: The Image Chain Approach* (Oxford University Press Inc., New York, NY, 1997).
10. J.W. Goodman, *Introduction to Fourier Optics*, 2nd ed. (McGraw Hill, New York, NY, 1968).
11. P. Beckmann and A. Spizzichino, *The Scattering of Electromagnetic Waves from Rough Surfaces* (Macmillan, New York, NY, 1963).

USING A FACETED ROUGH SURFACE ENVIRONMENTAL MODEL TO SIMULATE SHALLOW-WATER SAS IMAGERY

A.J. HUNTER, M.P. HAYES AND P.T. GOUGH

Acoustics Research Group, Dept. Electrical and Computer Engineering,
University of Canterbury, Private Bag 4800, Christchurch, New Zealand
E-mail: {a.hunter,m.hayes,p.gough}@elec.canterbury.ac.nz

Synthetic Aperture Sonar (SAS) is an extension of the conventional side-looking sonar technique for higher resolution underwater imaging. In SAS, as with conventional sonar, it is often difficult to obtain ground-truth data to compare with the reconstructed imagery. Thus, realistic simulation is invaluable for development of SAS algorithms. SAS simulation models are complicated by the required coherent processing and the larger beamwidths compared with conventional sonar of the same resolution. In particular, the large beamwidths and high-resolution imagery require the acoustic response of large insonified areas to be simulated down to a very small scale. In this paper, we present a simulation model capable of producing realistic SAS imagery of three-dimensional shallow-water environments. The simulation is conducted in the temporal frequency domain and is based on a faceted representation of the sea-floor and targets, with the sea-floor facets having a roughness component generated using a fractional Brownian motion processes. The acoustic response from the facets is determined using the Kirchhoff method, which is extended for fast simulation using facets with small-scale roughness. Occlusions and multiple scattering are resolved using the ray-tracing technique of geometric optics. Results so far indicate we have constructed an algorithm capable of modelling 3-D targets on a rough sea-floor that appear similar to actual data recorded here and elsewhere.

1 Introduction

Synthetic Aperture Sonar (SAS) differs from conventional side-looking sonar by using both the magnitude and phase of the acoustic response as well as the forward motion of the sonar platform in order to artificially synthesise a larger aperture. This coherent processing allows apertures many times the size of the physical aperture to be generated. Consequently, SAS can achieve improved range-independent azimuth resolution over conventional sonar [1].

Successful reconstruction of SAS imagery relies on the assumption that the sonar platform moves at a known speed along a linear path. Obviously in the case of a free-towed platform, this is not always the case. High-precision navigation hardware and autofocus techniques are used to determine and correct any deviation from the ideal path. It is useful in developing these techniques to have ground-truth data to compare with the reconstructed imagery. Similarly the recent trend in bathymetric SAS research also benefits from ground-truth data. However, in underwater acoustics this data is often difficult to obtain. Thus, a realistic simulation model is required to test and calibrate

481

N.G. Pace and F.B. Jensen (eds.), Impact of Littoral Environmental Variability on Acoustic Predictions and Sonar Performance, 481-488.

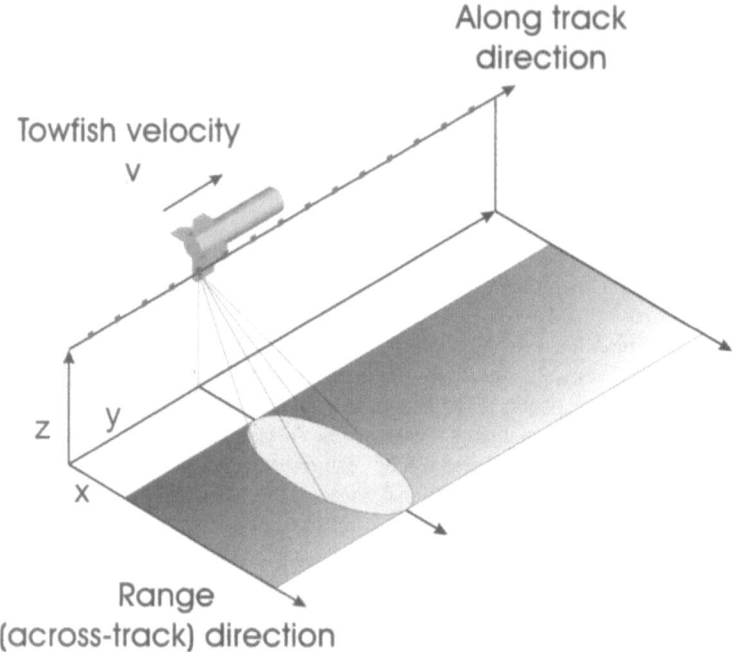

Figure 1. Typical shallow-water SAS imaging geometry.

reconstruction and autofocus algorithms.

SAS simulation models are complicated by the required coherent processing and the larger beamwidths compared with conventional sonar of the same resolution. In particular the large beamwidths and high resolution imagery require the acoustic response of large insonified areas to be simulated down to a very small scale. Rather than to be comprehensive, the goal of the simulation model presented here is to generate reasonable echo data for a broadband widebeam sonar that incorporates aspect dependent scattering, shadowing, sea-floor reverberation, and towfish motion. Aspects such as multiple scattering have been neglected in favour of computational speed.

The approach we use is to model the sea-floor and targets as a collection of facets, small enough so that we can use the Fraunhofer approximation of Fourier optics [2]. However, unlike other SAS simulations [3], we also model coherent sea-floor reverberation by assigning a correlated rough surface to each facet.

2   Imaging geometry

The typical shallow-water SAS imaging geometry is illustrated in Fig. 1. This is a side-looking geometry whereby the free-towed platform travels along an ideally linear path parallel with the strip-map imaged region. When imaging in shallow water environments, the platform sits roughly mid-water to give the best sea-floor coverage. In order to determine the orientation of the transducer beam patterns with respect to the imaged targets, the orientation and location of each transducer must be determined as the sonar moves along-track. Due to the free-towed platform, this is complicated by the additional motions of pitch, roll, yaw, heave, surge, and sway.

Assuming a single projector and a single hydrophone transducer located at ${}^t\mathbf{x}_p = ({}^tx_p, {}^ty_p, {}^tz_p)$ and ${}^t\mathbf{x}_h = ({}^tx_h, {}^ty_h, {}^tz_h)$, respectively, with respect to the towfish coordinate frame, the transducer positions in the world coordinate frame are given by

$$\mathbf{x}_p(t) = \mathbf{x}_t(t) + {}^w\mathbf{R}_t(t)\,{}^t\mathbf{x}_p, \tag{1}$$

$$\mathbf{x}_h(t) = \mathbf{x}_t(t) + {}^w\mathbf{R}_t(t)\,{}^t\mathbf{x}_h, \tag{2}$$

where ${}^w\mathbf{R}_t(t)$ is the rotation matrix describing the orientation of the towfish with respect to the world coordinate frame and $\mathbf{x}_t(t)$ is the displacement of the towfish in world coordinates.

Now consider a scatterer centred at a position $\mathbf{x}_s$ in world coordinates (typically this would be on the sea-floor at $(x_s, y_s, -D)$ where D is the water depth). The ranges from the target to the projector and hydrophone are given by

$$r_{ps}(t) = |\mathbf{r}_{ps}(t)| = |\mathbf{x}_p(t) - \mathbf{x}_s|, \tag{3}$$

$$r_{hs}(t) = |\mathbf{r}_{hs}(t)| = |\mathbf{x}_h(t) - \mathbf{x}_s|, \tag{4}$$

respectively. From these the direction cosines are found using

$$\cos \beta_{ps}(t) = \left({}^w\mathbf{R}_t(t)\,{}^t\mathbf{R}_p\right)^T \cdot \hat{\mathbf{r}}_{ps}(t), \tag{5}$$

$$\cos \beta_{hs}(t) = \left({}^w\mathbf{R}_t(t)\,{}^t\mathbf{R}_h\right)^T \cdot \hat{\mathbf{r}}_{hs}(t), \tag{6}$$

where the matrices ${}^t\mathbf{R}_p$ and ${}^t\mathbf{R}_h$ describe the respective rotations of the projector and hydrophone axes in the coordinate frame of the towfish. The direction cosines describe the orientation of the beam patterns with respect to the imaged targets and are used together with the range expressions to evaluate target responses in the acoustic model.

3 Acoustic scattering model

The acoustic model employs the Kirchhoff scattering method and resolves occlusion by ray-tracing [4]. The Kirchhoff method is selected due to its computational simplicity. However, it is still too demanding for SAS simulations of complicated environments. Thus, the method is extended with an efficient approximation for scattering from rough facets.

3.1 The Kirchhoff Method

The Kirchhoff method (also known as the tangent-plane method or the method of physical optics) gives a good first-order approximation to the field scattered by a rough interface, provided the surface is comparatively smooth compared to the wavelength [5]. At high frequencies this method is easily extended to include the effects of shadowing and multiple scattering using the ray-tracing technique of geometric optics.

The penultimate step in the Kirchhoff method is the assumption that the rough surface can be modelled locally by a planar interface that is independent of the rest of the surface. Thus, over a local region the total field Ψ is given by

$$\Psi = \Psi_i + \Psi_s = (1 + R)\,\Psi_i, \tag{7}$$

where Ψ_i and Ψ_s are the incident and scattered fields, and R is the reflection coefficient that is a function of the incident field angle (we ignore mode conversion and subsea penetration of sound). The acoustic response of a planar surface element Σ can then be determined by substitution of Eq. (7) into the Helmholtz-Kirchhoff equation [2]

$$\Psi_s = \int_\Sigma (1 - R)\, G \nabla \Psi_i \cdot \hat{\mathbf{n}} - (1 + R)\, \Psi_i \nabla G \cdot \hat{\mathbf{n}}\, d\mathbf{S} \qquad (8)$$

where G is the Green's function, Σ is the domain of the planar surface element, and $\hat{\mathbf{n}}$ is the surface normal on Σ. If we assume that the medium above the sea-floor is homogeneous we can use the free space Green's function for G. We usually operate in shallow water harbour conditions where the assumption of a homogeneous medium is sufficient in winter when the temperature gradient is small. Otherwise it is necessary to use high frequency approximations for G given an estimate of the sound speed profile [4].

3.2 Acoustic Response of a Faceted Target

The objects and surfaces in a typical underwater environment (sea-floor and targets) can be conveniently represented as a collection of facets. The acoustic response of these facets can then be determined using the Kirchhoff method. Provided the facets are small enough, the reflection coefficient R can be considered a constant for each facet. Then, assuming a homogeneous medium, the field at a point several wavelengths away from the facet can be approximated by

$$\Psi_s(\mathbf{x}_h|\mathbf{x}_p, f) \approx j2\pi(f/c)\left[(1 - R)\cos\beta_{ps} \cdot \hat{\mathbf{n}} - (1 + R)\cos\beta_{hs} \cdot \hat{\mathbf{n}}\right]$$
$$\times \int_\Sigma G(\mathbf{x} - \mathbf{x}_h, f)G(\mathbf{x}_p - \mathbf{x}, f)\, d\mathbf{S} \qquad (9)$$

where $\mathbf{x}_p$ and $\mathbf{x}_h$ are the positions of the respective projector and hydrophone transducers. Now provided each facet is in the far-field of the sonar transducers, we can employ the Fraunhofer approximation to simplify the integral of Eq. (9) to a 2-D Fourier transform. The resultant echo signal from each visible facet thus has the form

$$E_s(\mathbf{x}_h|\mathbf{x}_p, f) = j2\pi(f/c)S(f)\, G(\mathbf{x}_s - \mathbf{x}_h, f)G(\mathbf{x}_p - \mathbf{x}_s, f)$$
$$\times B_p\left((f/c)\cos\beta_{ps}\right) B_h\left((f/c)\cos\beta_{hs}\right)$$
$$\times B_s\left((f/c)\left(\cos\beta_{ps} + \cos\beta_{hs}\right)\right) K_s(\beta_{ps}, \beta_{hs}), \qquad (10)$$

where $S(f)$ is the temporal spectrum of the transmitted signal, $\mathbf{x}_s$ is the centre of the facet, B_p, B_h, and B_s are the beampatterns of the projector, hydrophone and facet respectively, and $K_s(\beta_{ps}, \beta_{hs})$ describes the scattering amplitude from the facet given by the factor in square brackets in Eq. (9).

The beampatterns are obtained by taking the Fourier transform of the aperture functions. In the cases of typical rectangular transducers the beampattern is the familiar sinc function. However, it is not possible to tesselate an arbitrary object into rectangular facets. Thus, triangular facets are employed instead.

The total scattered field is determined by summing the responses from each of the visible facets at each receiver position, where the visibility of each facet is determined using ray-tracing. This models the effects of occlusion but is only valid for high frequencies.

3.3   Scattering from Rough Facets

High resolution SAS imagery is obtained using large beamwidths and to model seafloor reverberation we must consider large insonified areas down to a very small scale. Therefore, representing surface roughness for a typical shallow-water environment using the simple Kirchhoff method becomes computationally impractical due to the extremely large number of facets required. Instead we use larger facets and assume that the roughness on each facet is small so that no points on the facet are occluded by each other and that multiple scattering is negligible. We can then determine a frequency dependent beampattern for each facet. This can be rapidly computed for each facet using FFTs and interpolated for each field point.

4   Sea-floor model

A realistic model of the sea-floor and its roughness is important for modelling the sea-floor reverberation. We model the sea-floor as a two-dimensional fractional Brownian motion process since this has been shown to be a good model of natural surfaces over a wide range of scales [6]. The correlated surface roughness was generated using the midpoint displacement algorithm [7], adapted for an arbitrary structure function (incremental variance function). This can generate fractional Brownian motion processes more accurately than frequency domain methods often employed [8]. It also is better suited for generating very large rough surfaces.

We assume an isotropic sea-floor roughness although it is straightforward to extend the midpoint displacement algorithm to model anisotropic roughness. It is also straightforward to adapt the algorithm to use the von Kármán correlation function proposed by Goff and Jordan [9] to model large scale sea-floor roughness.

5   Implementation

The described simulation model has been implemented in the Python programming language with C implementations for the computationally intensive routines (Fourier transforms, interpolations, etc.). Python is a freely available modern scripting language suited for interactive development of algorithms [10].

An advantage of the facet model is that it is ideally suited for parallel computing—the response from each facet can be computed independently and then linearly combined. We run our simulations on a cluster of PCs running the *Mosix* kernel, a variation of the *Linux* kernel that can dynamically migrate processes across the cluster to distribute the load.

6   Results

To validate the model, a number of seafloor scenes were modelled with a variety of targets. For example, Fig. 2 shows a rendered image of a rough seafloor with a superimposed cylindrical target and four cubical targets. The cylinder and cubes were modelled using a collection of triangular facets (either smooth or rough). The sea-floor also was modelled using triangular facets but with rough surfaces generated from a fractional Brownian motion process with a Hurst parameter of 0.35 using the midpoint displacement algorithm.

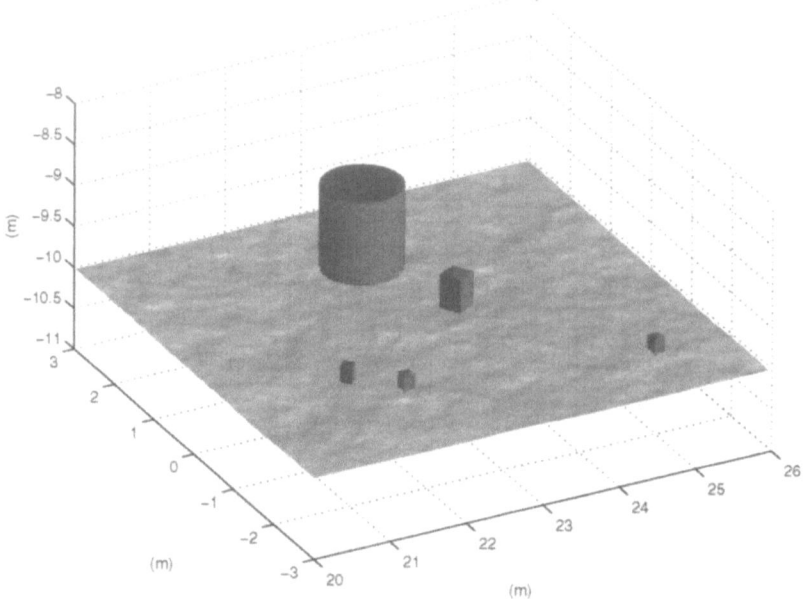

Figure 2. Example of a typical sea-floor with fractional Brownian motion statistics with a Hurst parameter of 0.35 with a smooth cylindrical target and four smooth cubical targets.

The sonar we simulated has the parameters of the KiwiSAS-III sonar [11]. This is a free-towed, short range, synthetic aperture sonar that operates in two frequency bands 20–40 kHz and 90–110 kHz giving a range resolution in each band of nominally 35 mm. The sonar has a single projector with three vertically displaced hydrophones (for interferometric bathymetry [12]), each of the order of 300 m in length giving a theoretical along-track resolution of 150 mm. Fig. 3 shows the pulse compressed echo data simulated for this sonar at 100 kHz using the sea-floor/target model shown in Fig. 2. The rough sea-floor model can be seen to produce speckle with a size commensurate with the resolution of the sonar. The scattering from the targets has generated a few highlights and some shadowing of the sea-floor. After synthetic aperture reconstruction, the shadows become more apparent as shown in Fig. 4.

7 Conclusion

A simulation model has been proposed based on the Kirchhoff approximation. The simple Kirchhoff approximation has been modified such that surfaces with roughness may be implemented rapidly. The model provides aspect dependent scattering, occlusions, and coherent speckle from rough surfaces. However, there are a number of limitations with the model. The small-slope approximation is violated when the surface roughness has correlation lengths shorter than the wavelength [5, 6]. Multiple scattering, low frequency diffraction effects around targets, and sea-surface scattering are neglected. Future enhancements are to correct some of these problems; for example, we envisage modelling multipath from strong scatterers using recursive ray-tracing and to incorporate scattering from the sea-surface [13].

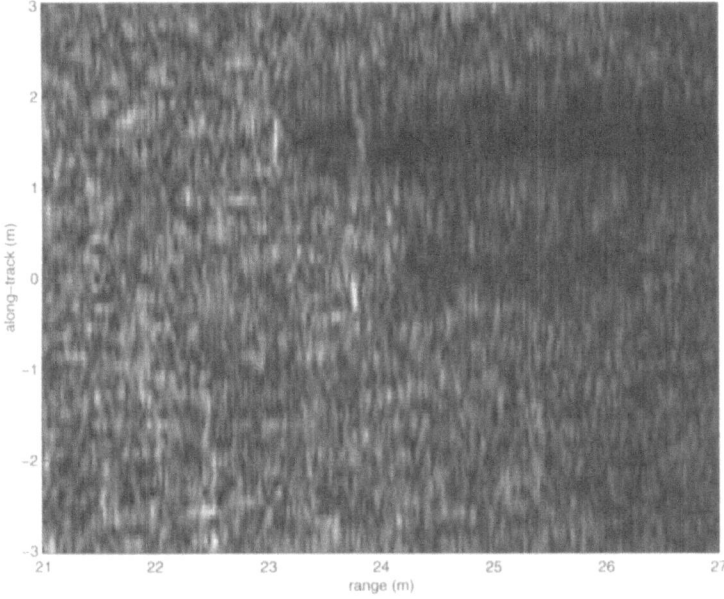

Figure 3. Simulated pulse compressed echo data for scene shown in Fig. 2.

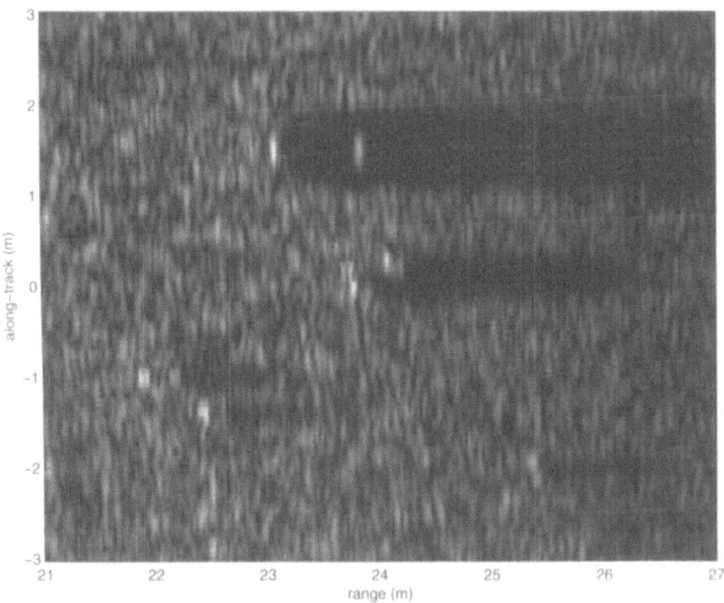

Figure 4. Synthetic aperture reconstructed image for scene shown in Fig. 2.

Acknowledgements

Alan Hunter thanks the University of Canterbury for his Doctoral Scholarship.

References

1. D.W. Hawkins and P.T. Gough, Imaging algorithms for a strip-map synthetic aperture sonar: Minimizing the effects of aperture errors and aperture undersampling, *IEEE J. Oceanic Eng.* **22**(1), 27–39 (1997).
2. J.W. Goodman, *Fourier Optics* (McGraw-Hill, New York, 1968).
3. G.S. Sammelmann. Propagation and scattering in very shallow water. In *Proc. IEEE Oceans2001*, 337–344 (2001).
4. L.J. Ziomek, *Fundamentals of Acoustic Field Theory and Space-Time Signal Processing* (CRC Press, 1995).
5. P. Beckmann and A. Spizzichino, *The Scattering of Electromagnetic Waves from Rough Surfaces* (Artech House, 1987).
6. G. Franceschetti, A. Iodice, M. Migliaccio and D. Riccio, Scattering from natural rough surface modeled by fractional Brownian motion two-dimensional processes, *IEEE Trans. Antennas and Propagation* **47**(9), 1405–1415 (1999).
7. C.M. Harding, R.A. Johnston and R.G. Lane, Fast simulation of a Kolmogorov phase screen, *Applied Optics* **38**(11), 2161–2170 (1999).
8. R.G. Lane, A. Glindemann and J.C. Dainty, Simulation of a Kolmogorov phase screen, *Waves in Random Media* **2**, 209–224 (1992).
9. J.A. Goff and T.H. Jordan, Stochastic modeling of seafloor morphology: Inversion of sea beam data for second-order statistics, *J. Geophys. Res.* **93**(B11), 13589–13608 (1988).
10. M. Lutz and D. Ascher, *Learning Python* (O'Reilly, 1999).
11. D.W. Hawkins and P.T. Gough, Recent sea trials of a synthetic aperture sonar. In *Proc. Institute of Acoustics* **17**(8), 1–10 (1995).
12. M.P. Hayes, P.J. Barclay and P.T. Gough, Test results from a multi-frequency bathymetric synthetic aperture sonar. In *Proc. IEEE Oceans2001*, Honolulu, Hawaii (Nov. 2001) pp. 1682–1687.
13. B. Davis, P. Gough and B. Hunt, Sea surface simulator for testing a synthetic aperture sonar. In *Impact of Littoral Environmental Variability on Acoustic Predictions and Sonar Performance*, edited by N.G. Pace and F.B. Jensen (Kluwer Academic, The Netherlands, 2002) pp. 473–480.

A STUDY OF PING-TO-PING COHERENCE
OF THE SEABED RESPONSE

L. PAUTET, E. POULIQUEN AND G. CANEPA

SACLANT Undersea Research Centre, Viale San Bartolomeo 400, 19038 La Spezia, Italy
E-mail: pautet@saclantc.nato.int

Coherence in acoustic backscatter is used to perform reverberation based micronavigation in SAS imaging. To evaluate limits of ping-to-ping coherence, this paper examines how interface characteristics and acoustic source parameters influence ping-to-ping signal stability. To study these effects independently from other medium variations, a time series model was developed. This model, BORIS-SSA, is an improved version of BORIS which applies the fourth order Small Slope Approximation for treating interface scattering.

1 Introduction

When performing high frequency Synthetic Aperture Sonar (SAS) imaging, platform and medium instabilities of fractions of wavelengths should be avoided or compensated for. If not, signals from successive pings do not sum coherently and produce poor quality images. Platform motion can be compensated for by the use of phase compensation micronavigation using the Displaced Phase Center Antenna (DPCA) technique [1]. This technique requires minimum coherence between successive pings in order to estimate phase error. When successive pings are incoherent, it becomes impossible to compare them and to deduce the movements of the physical antenna. Estimating the effects of platform movements and medium variations on signal fluctuations is essential. An experimental study, MAPLE 2001, was performed in July 2001 off the coast of Halifax to evaluate limits of ping-to-ping coherence [2]. An acoustic transducer mounted on a 5 m high tower was pinging repeatedly at the seafloor while a pan and tilt mechanism changed the heading of the source. This experimental configuration was designed to reproduce the conditions of a platform enduring yaw which is the most difficult motion to measure using DPCA. One of the results of this experiment is that successive signals rapidly loose coherence (after a heading variation larger than $1°$). The angular spread of the scatterers is much smaller than the $16°$ beam aperture used in the experiment. The conclusion is that the persistence of coherence depends on the size and number of scatterers. Strong scatterers provide omnidirectional response but smaller ones combine to form an interference structure which is directional. In most cases, there is no dominant scatterer and the seabed response is highly directional. It was found, that this property depends on source properties (pulse, beam aperture, position), and on interface properties (roughness, heterogeneity, level of clutter). In order to study how these factors influence the angular response of the seafloor and to multiply scattering scenarios, a time-series snapshot model, valid at low grazing angles and high frequency, was developed. BORIS-SSA is based on the Small Slope

N.G. Pace and F.B. Jensen (eds.), Impact of Littoral Environmental Variability on Acoustic Predictions and
Sonar Performance, 489-496.

Approximation. After briefly describing the concept of BORIS-SSA, this paper presents some preliminary results on signal fluctuations from a moving source.

2   BORIS-SSA

The model BORIS was developed primarily to compute time-domain seafloor backscatter at normal incidence [3, 4]. The input parameters of the model are the source parameters and a realization of the seabed having some predefined statistics. Contributions from each element of the interface are calculated using the Kirchhoff approximation (KA), whereas contributions from the volume are treated under the small perturbation theory. The KA is valid at normal incidence but is often inaccurate at low grazing angles [5]. The small slope approximation (SSA) however is a promising method which shows, in the 1-D case, good agreement with exact solutions at all angles [6, 7]. The SSA takes the form of a series expansion in generalized surface slope. The two orders commonly used are the second (SSA-2) and the fourth (SSA-4) orders. The order refers to the development order of the surface slope in the surface scattering strength expression. Figure 1 shows a comparison between the exact solution, the KA, the SSA-2 and the SSA-4 for a power law interface (*i.e.*, an interface having a power spectral density proportional to $\frac{\beta}{(k^2+k_c^2)^{-\gamma/2}}$ for $k < k_{max}$, where k is the wave number and k_c is an arbitrary cutt-off frequency). The coefficient of proportionality is adjusted to obtain the correct RMS-height for the interface. In Fig. 1, the agreement between the exact solution and the SSA-4 is very good at all angles. This justifies the use of the SSA-4 in place of the KA in BORIS as in BORIS-SSA. The mathematical development of this approach will be described in a further paper.

3   Simulations

A comparison between simulations obtained using BORIS-SSA and signals acquired from an acoustic source mounted on a tower [2] is made. A description of the experimental configuration is shown in Fig. 2. In the simulation, a transducer transmits 1 ms bursts with a carrier frequency of 100 kHz and a TX/RX Gaussian beam aperture of $16°$. The source is located 5 m above the seafloor and transmits with a nominal grazing angle of $50°$. The interface is generated with a power spectral density being a modified power law ($\gamma = 3.2$, $k_{max} = 1000$ rad/m, $k_c = 10$ rad/m) with a RMS-height of 1 cm. Figure 3 shows a realization of a simulated signal. A tile of the interface is shown in Fig. 4.

Simulated signals are displayed in a polar plot as the source changes its heading (Fig. 5). Figure 6 presents a similar plot of real signals recorded at sea. The angular spread of scatterers in the simulated data appears larger than in the real data but shows a significantly smaller spread than intuitively expected with a source beam aperture of $16°$. The discrepancy between the real and simulated seafloor responses could be caused by the different statistical properties of the water-sediment interface generated by BORIS-SSA compared to the actual interface around the tower. Interface generation will be improved in the near future. However, BORIS-SSA appears to be an accurate model applicable to scenarios encountered by SAS systems using DPCA.

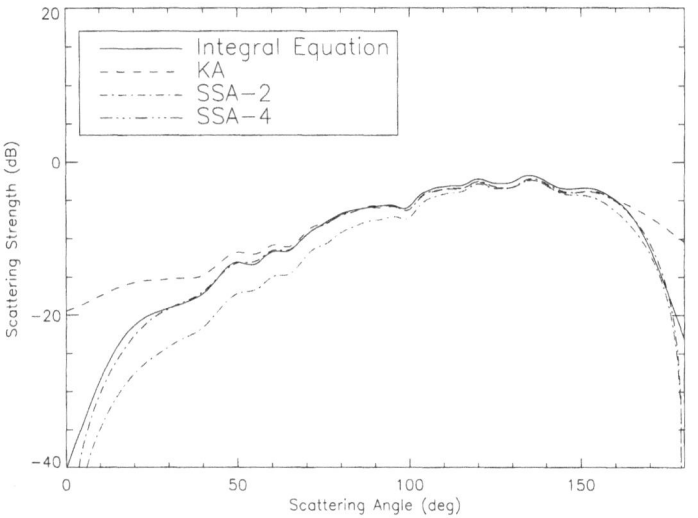

Figure 1. Comparison between scattering strength computed using the integral equation and three approached solutions obtained using the Kirchhoff approximation (KA) and the second and fourth order small slope approximations (SSA-2 and SSA-4 respectively). Scattering strength is the result of an averaging over 50 interface realizations. Each interface is obtained using a modified "power-law" power spectral density - $\gamma = 3.2$, $k_{max} = 1000$ rad/m, $k_c = 10$ rad/m with 2 cm RMS-height. The incident angle is $45°$. The fourth order SSA is accurate even at lower grazing angles.

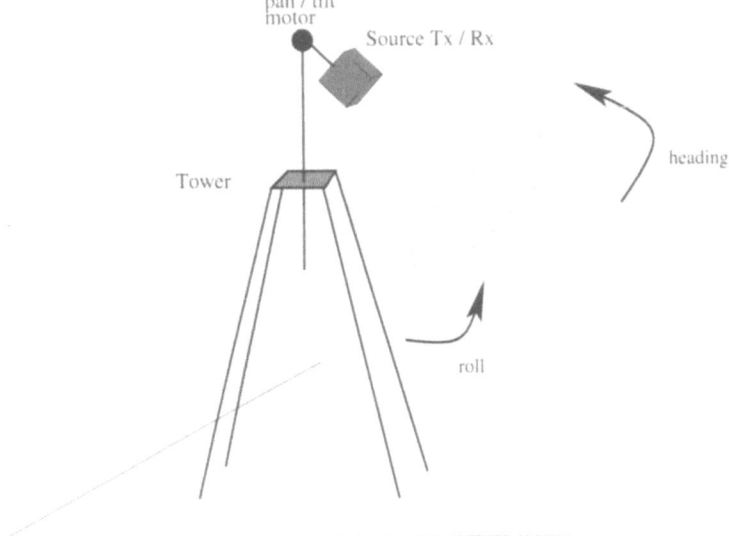

Figure 2. Experimental configuration with an acoustic source located on a 5 m high tower. A pan/tilt motor allows the source to change heading and roll.

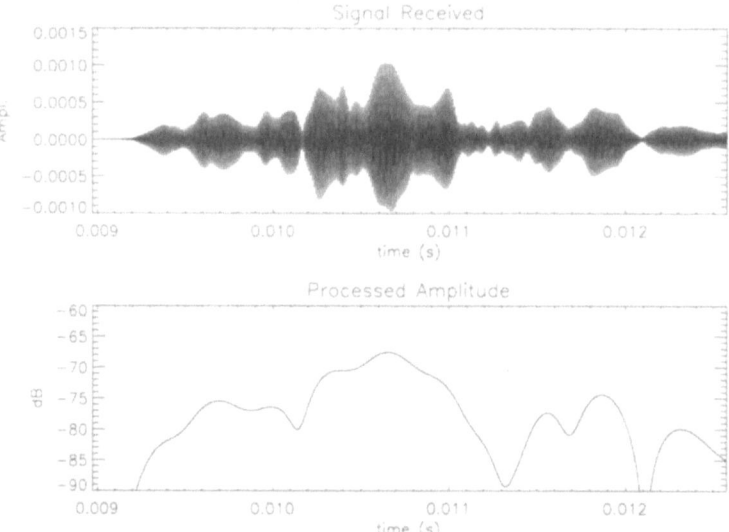

Figure 3. Example of simulated data obtained by a source/receiver located 5 m above the seabed. The nominal grazing angle is 50°. The pulse is 1ms long with a carrier frequency of 100 kHz. The source beam aperture is 16°. Scales are arbitrary.

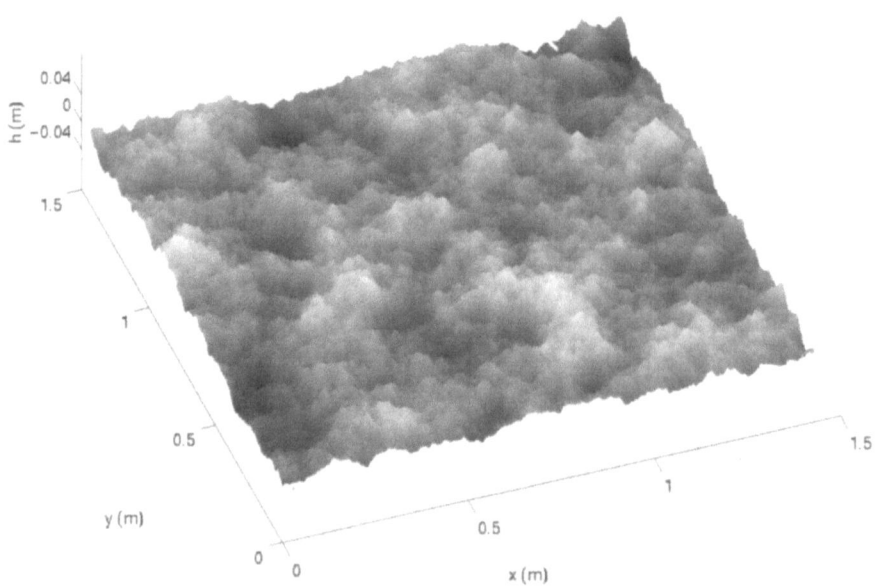

Figure 4. Tile of the generated interface used in the simulation. The power spectral density is a power-law ($\gamma = 3.2$, $k_{max} = 1000$ rad/m, $k_c = 10$ rad/m). The RMS-height is 1 cm.

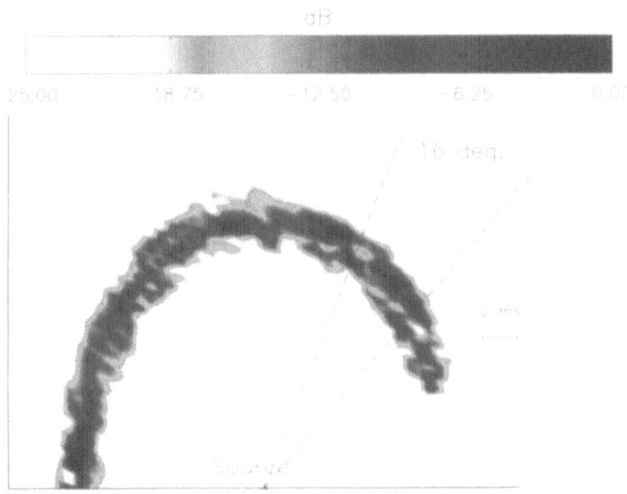

Figure 5. Polar plot of the simulated signals obtained using the interface in Fig. 4. Intensity is coded in color. The absolute level is arbitrary.

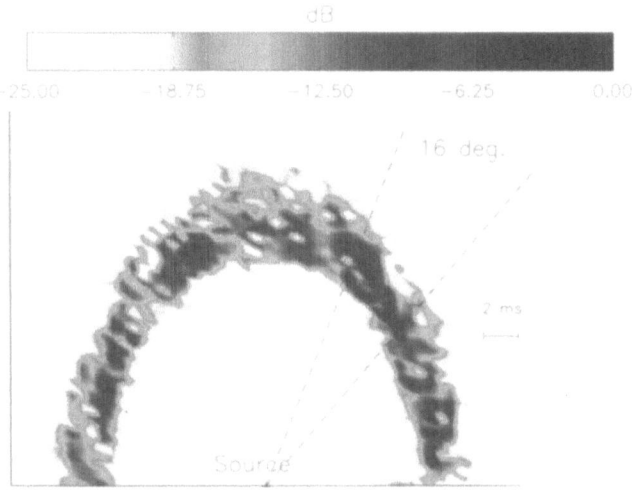

Figure 6. Polar plot of real data received from the acoustic transducer mounted on a tower under similar conditions of roll and height as in Fig. 5. The absolute level is arbitrary.

4   Ping-to-ping coherence

Considering the heading 90° as a reference, the amplitudes obtained at 90±4° and 90±8° are plotted in Fig. 7. The high amplitude observed at $t = 0.0103$ s at the 90° heading, is not present at other headings. At a heading of 86°, this high amplitude is masked by a stronger amplitude (caused by a dominant scatterer) centered around $t = 0.0011$ s.

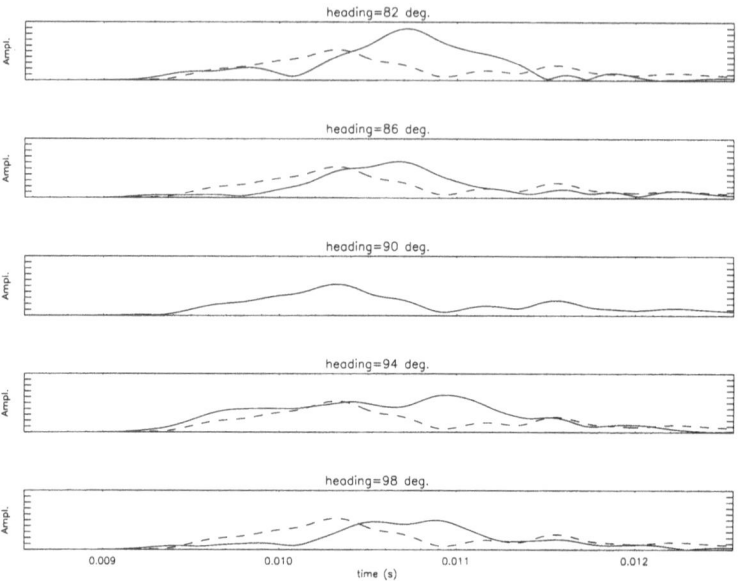

Figure 7. Simulated data for 5 pings panning a total of 16°. The reference ping is at a heading of 90° (represented as a dashed line in the other plots). The pulse length is 1 ms and the interface RMS-height is 1 cm.

This is in agreement with conclusions of the experimental study [2] which state that the coherent summation of seabed contribution is strongly dependent on heading. When a shorter pulse is used, less scatterers are integrated in the seafloor response. Coherence is expected to hold over a greater range of yaw. Figure 8 is similar to Fig. 7 but with a 0.4 ms pulse instead of a 1 ms pulse. The interface is the same. Three independent scatterers can be identified in the signal at a heading of 90°. These three scatterers can also be identified at the heading of 86°. So it appears that a shorter pulse leads to a greater ping-to-ping coherence. In Fig. 9, the interface RMS-height is doubled with respect to Fig. 7. The scatterer at $t = 0.013$ s has now become a very dominant scatterer. It can be identified in all the other pings and in particular in the 94° and 98° pings. As the interface becomes rougher, the number of scatterers increases, but some scatterers stand above the average amplitude. It was noted in the experiment that strong scatterers would remain coherent over a large angular range. In this case, the presence of a dominant scatterer improves ping-to-ping coherence.

5 Conclusion

Conclusions from observation of the signals acquired from a tower, especially on the impact of the number and size of scatterers on the ping-to-ping coherence [2] seem to be confirmed by BORIS-SSA simulations. Dominant interface features (e.g., strong facets) scatter omnidirectionnaly which preserve strong ping-to-ping coherence. When the number of scatterers that are integrated in the response increases (e.g., long pulse, many scatterers per unit surface), the seabed response becomes more directional because an interference structure caused by a unique combination of scatterers is formed.

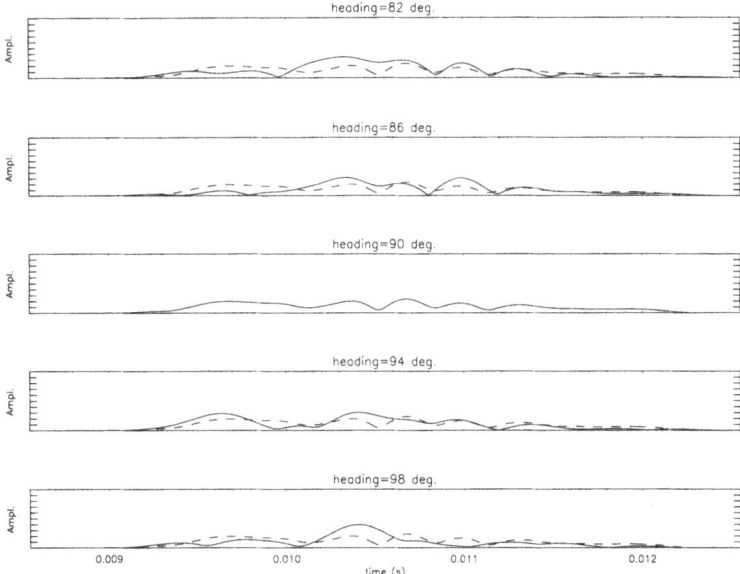

Figure 8. Simulated data for 5 pings panning a total of 16°. The reference ping is at a heading of 90° (represented as a dashed line in the other plots). The pulse length is reduced to 0.4 ms compared to 1 ms in Fig. 7.

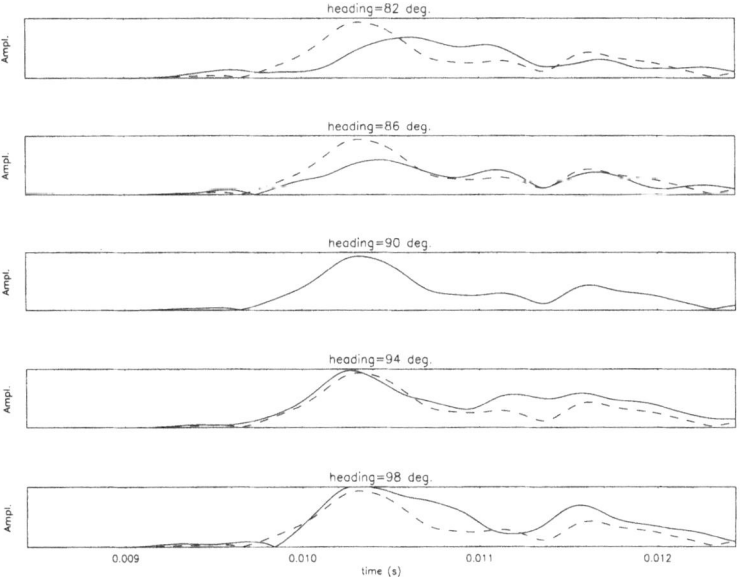

Figure 9. Simulated data for 5 pings panning a total of 16°. The reference ping is at a heading of 90° (represented as a dashed line in the other plots). The interface RMS-height is multiplied by 2 with respect to the simulations in Fig. 7.

In terms of model improvements, generated interfaces will be improved to account for the presence of discrete scatterers, patchiness, adequate interface spectral density, volume scattering, etc. In parallel, comparisons with other acquired signals will be made in the near future.

References

1. Sheriff, R.W., Synthetic aperture beamforming with automatic phase compensation for high frequency sonars, *Symp. on Autonomous Underwater Vehicules*, 236–245 (1992).
2. Pautet, L., Pouliquen, E. and Crawford, A., Experimental study of fluctuations in coherent backscattering. In *Proc. 6th European Conference on Underwater Acoustics*, Gdansk, Poland (2002).
3. Pouliquen, E., Bergem, O. and Pace, N.G., Time-evolution modeling of seafloor scatter. I. Concept, *J. Acoust. Soc. Am.* **105**, 3136–3141 (1999).
4. Bergem, O., Pouliquen, E., Canepa, G. and Pace, N.G., Time-evolution modeling of seafloor scatter. II. Numerical and experimental evaluation, *J. Acoust. Soc. Am.* **105**, 3142–3150 (1999).
5. Thorsos, E.I., The validity of the Kirchhoff approximation for rough surface scattering using a Gaussian roughness spectrum, *J. Acoust. Soc. Am.* **83**, 78–82 (1988).
6. Thorsos, E.I. and Broschat, S.L., An investigation of the small slope approximation for scattering from rough surfaces. Part I. Theory, *J. Acoust. Soc. Am.* **97**, 2082–2093 (1995).
7. Thorsos, E.I. and Broschat, S.L., An investigation of the small slope approximation for scattering from rough surfaces. Part II. Numerical studies, *J. Acoust. Soc. Am.* **101**, 2615–2625 (1997).

VARIABILITY OF THE ACOUSTIC RESPONSE FROM SPHERICAL SHELLS BURIED IN THE SEABED BY MODEL-BASED ANALYSIS OF AT-SEA DATA

A. TESEI, A. MAGUER AND W.L.J. FOX

SACLANT Undersea Research Centre, Viale S. Bartolomeo 400, 19138 La Spezia, Italy
E-mail: tesei@saclantc.nato.int, maguer.alain@tms-pty.com, warren@apl.washington.edu

R. LIM

CSS/Dahlgren Naval Surface Warfare Centre, Panama City, Florida 32407-7001, USA
E-mail: LimRA@ncsc.navy.mil

H. SCHMIDT

Dept. of Ocean Engineering, MIT, Cambridge, MA 02139, USA
E-mail: henrik@keel.mit.edu

The acoustic response from elastic spherical shells buried in the seabed is studied in the bandwidth 1–15 kHz as a function of changing environmental parameters. The targets are either partially or completely buried in the seabed or in the free field. Particular attention has been paid to the elastic surface-guided waves circulating around the target, which are expected to provide significant features for target classification. The variations of their dynamics with the environment were studied by comparing at-sea monostatic data with appropriate models. Three identical spherical shells were measured under different conditions during the GOATS experiment conducted jointly by SACLANTCEN and MIT in 1998. This paper reports the data interpretation results compared with theoretical expectations demonstrating the strong influence of the environment on target response.

1 Introduction

Mines that are completely buried in sandy bottoms are generally considered to be undetectable and unclassifiable by conventional high frequency (> 50 kHz) minehunting sonars due primarily to the low levels of energy that are transmitted into the sediment at these frequencies at the low grazing angles used. As the attenuative effect of the sediment is less at lower frequencies, the possibility of using much lower frequency sonars, with higher fractional bandwidth (1–15 kHz), has been investigated for detection and classification of proud and buried mines. At high frequency, the acoustic scattering by mine-sized targets (~ 1 m) is well described by geometrical theory of diffraction, but at lower frequencies ($f < 30$ kHz), man-made targets in particular may support the excitation of strong structural waves or resonances that are consistent with their typical structural symmetries and can represent potentially significant classification clues.

Past experimental and modelling work has provided evidence for the excitation of structural waves in completely buried spherical shells, including investigations on the

N.G. Pace and F.B. Jensen (eds.), Impact of Littoral Environmental Variability on Acoustic Predictions and Sonar Performance, 497-504.

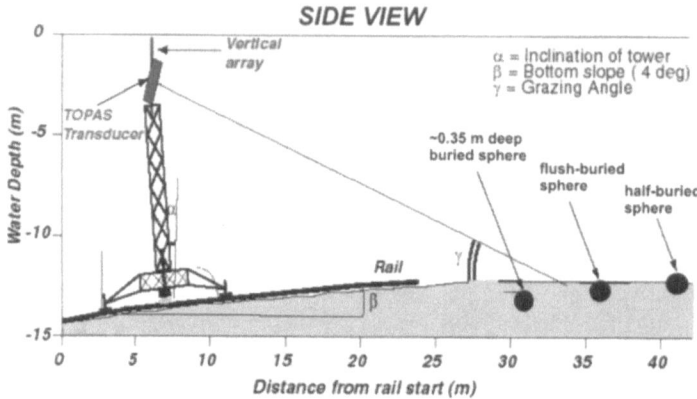

Figure 1. Experimental geometry.

sensitivity of the target response to burial depth, sediment type and grazing angle [1]. The excitation of target resonances for flush and partially buried targets has been investigated using new modelling capabilities [2]. This paper focuses on the dynamics of the predicted structural waves of a thin-walled, steel spherical shell in the bandwidth 2–15 kHz. Appropriate models of elastic waves dynamics [3] are also used. The target typology is ideal for the study as the free-field scattering physics is relatively simple to model and well understood. Theoretical considerations will be validated with at-sea measurements acquired during GOATS'98. This paper will be limited to the analysis of backscattering with insonification above critical grazing for three identical spherical shells: one half buried, one flush buried, and one completely buried.

2 Experimental setup

The GOATS'98 experiment [4] was carried out on a sandy bottom in 12–15 m water off the island of Elba, Italy. The TOPAS source was used to insonify the targets with a highly directional beam in the frequency range 2–16 kHz (secondary frequency) with a vertical beamwidth of 3°–5° and horizontal beamwidth of about 8°. The far field is estimated to start at about 35 m in front of the transducer while the volume of nonlinear interaction is estimated to extend for the first 11 m. The experimental configuration is shown in Fig.1. In order to acquire data from various source-receiver geometries, the transmitter was mounted on a 10 m tower, which in turn was mounted on a 24 m linear rail on the bottom. A linear receiving array of 16 hydrophones (94 mm-spaced) was mounted vertically in a near-monostatic configuration. Three identical spherical shells were deployed in line with the rail at different burial depths (about 35 cm into the sediment, flush and half-buried). One of the shells was also measured suspended in the water column. The shells were air-filled, thin-walled, steel, nominally of 53 cm radius and 3 cm wall thickness, and with a steel lug for deployment. As the spheres were constructed by welding two hemispherical shells together, there was the possibility of thickness nonuniformity.

The average density of the sediment was 1.91 g/cm^3. The sediment propagation loss was estimated to be 0.5 dB/λ, the sound speed 1640 m/s [4] and the sand critical angle 22°.

3   Theoretical concepts and models

Scattering models for a spherical shell either in the free field or buried in the seabed are used for the study of acoustic variability with the environment. Models are developed in the frequency domain and provide the scatterer transfer function. The modelled and measured spherical shells have the same nominal elastic parameters (steel compressional speed $c_p = 5950\,\text{m/s}$, shear speed $c_s = 3240\,\text{m/s}$, density $\rho = 7.7\,\text{g/cm}^3$), the density of the sea water is set to $1\,\text{g/cm}^3$ and its sound speed is set to the measured value $1520\,\text{m/s}$.

Under free field conditions the acoustic pressure scattered by an elastic fluid-filled spherical shell insonified by an incident plane wave is represented by a partial wave series (PWS) model. In the case of fully and partly buried spherical shells the scattering model used is based on transition- (T-) matrix solutions for the scattered field [1, 2].

In the low-to-medium frequency range, for thin air-filled spherical shells, the elastic contribution to scattering is due to the lowest-order flexural and compressional waves of the shell [1]. Their resonance frequencies appear as sharp dips or peaks of the scattering spectral response, and will be predicted through the models developed in [3] based on the application of the shell theories with modified inertia to the scattering of a thin-walled spherical shell. The symmetric (or compressional) S_0 **Lamb-type wave** is supersonic, almost non-dispersive, and travels in the shell with phase and group speeds asymptotically tending to the shell material membrane speed, c^*_{shell}. The antisymmetric (or flexural), A_0 Lamb-type wave of a spherical shell in vacuum bifurcates into two dispersive waves upon fluid loading. Of the two, the wave that more strongly influences the acoustic scattering amplitude, hence that is selected as a potentially robust classification clue, is the subsonic A_{0-} **Lamb-type wave**. At low frequencies (until its phase speed approaches the outer-medium sound speed at the so-called coincidence frequency, f_c), it is flexural in nature. Around the coincidence frequency the A_{0-} wave starts to behave like a fluid-borne wave, becoming difficult to detect with increasing frequency because of increased radiation damping. Its group speed reaches its maximum at the coincidence frequency.

First we will study a sphere completely buried in different *sediment types* (Fig. 2). Assuming that the water-sediment interface cannot significantly influence the target response for completely-buried cases at supercritical grazing, the sphere is simulated as flush-buried in order to minimize signal loss due to burial. Theoretically [1, 2], the dynamics and energetic contribution to backscattering of the S_0 wave should not be unduly influenced by the external fluid, or by burial, except for a slight shift towards lower frequencies of its first modes as the external density increases. The shift should decrease as the modal order (and frequency) increases. Similarly, at low frequency the effect of fluid loading on the dynamics of the A_{0-} wave is essentially inertial: its first free-field modes are predicted to shift to lower frequencies as the exterior density increases. As the A_{0-} wave becomes fluid-borne in nature, it is predicted to be much influenced by the exterior, as its phase speed tends to the exterior sound speed.

The effect of *burial depth* will be considered in the following (Fig. 3). At low frequencies, a shift of the first S_0 and A_{0-} wave modes to lower frequencies is expected with burial, due to the significantly greater inertial loading of the shell in the sediment possessing grater relative density than water. The shift should increase with the percentage of target surface in contact with sediment. In the coincidence frequency region, the phase speed of the A_{0-} wave approaches the exterior sound speed, which is higher when the

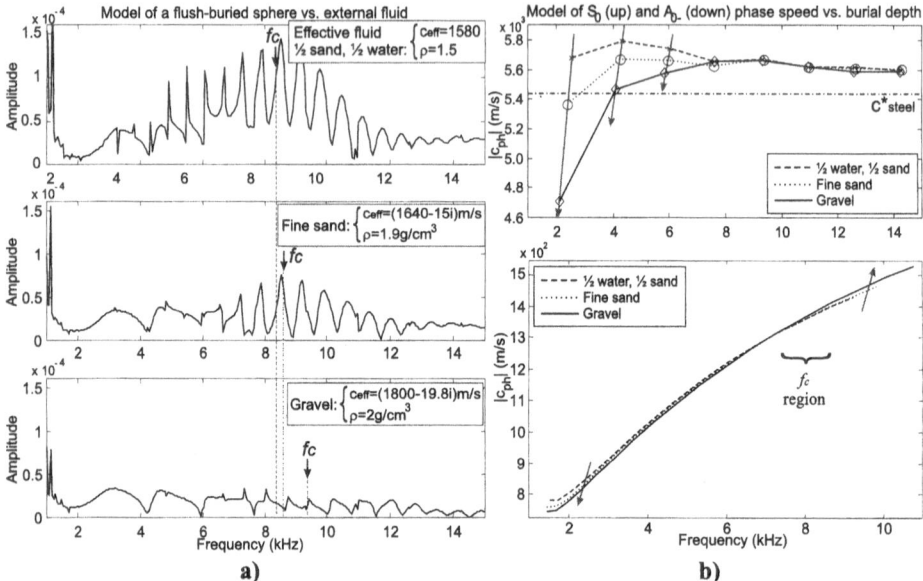

Figure 2. a) Amplitude of the form function of a buried spherical shell as the external fluid changes. b) Model of the related dispersion curves of S_0 and A_{0-} waves.

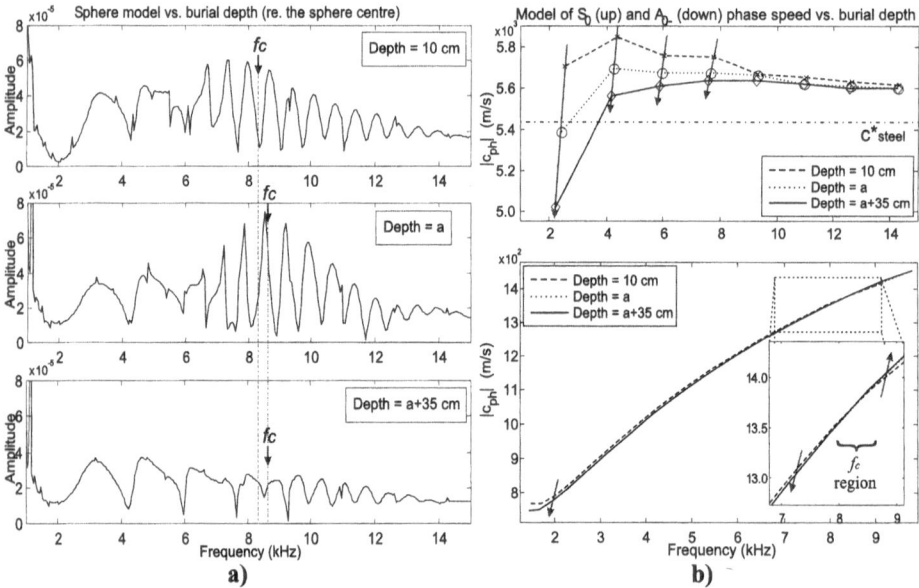

Figure 3. a) Amplitude of the form function of a spherical shell as burial depth changes from partially to deeply buried. b) Model of the related dispersion curves of S_0 and A_{0-} waves.

target is completely buried. This means the modes of the A_{0-} wave that are excited around f_c are expected to experience an upward shift upon burial. For partial burial, we hypothesize that these modes will again shift consistent with the percentage of target surface in contact with sediment. Hence, the simple empirical formula $c_{ext} \approx f_c 2\pi d$

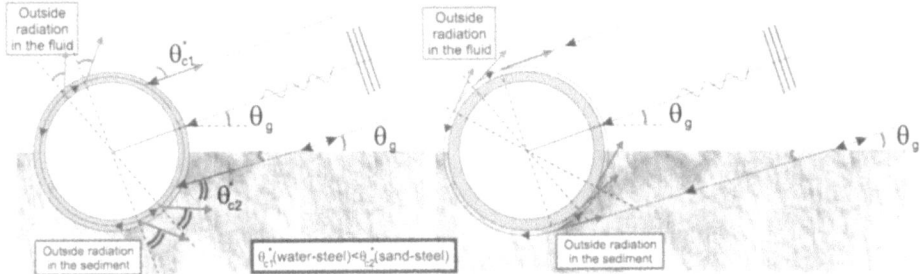

Figure 4. Ray diagram of the travel path of the Lamb-type S_0 (left) and A_{0-} (right) waves in the case of partial burial. The travel path of the A_{0-} wave is drawn outside the shell in order to emphasize its fluid-borne nature around the coincidence frequency. The wall thickness is not in scale.

(being c_{ext} the external medium loading the shell), which was shown to be valid for steel and similar materials is extended to the case of partial burial. If the shell is loaded by more than one fluid the target exterior will be treated as an effective medium characterized by an effective sound speed c_{ext}^{eff} such that:

$$c_{ext}^{eff} \approx f_c 2\pi d. \tag{1}$$

This assumption should be reasonable for determining the dispersion characteristics of the S_0 and A_{0-} waves if the dominant contributions to the backscatter from these waves are from complete circumnavigations of the shell. Under these circumstances, one might also expect c_{ext}^{eff} to be defined by a weighted harmonic average of the sound speeds of the two exterior fluids in order to account for propagation of the exterior diffracted field through both fluids. Similarly the effective external density ρ_{ext}^{eff} will be defined as an average of the water and sediment densities, ρ_w and ρ_s, respectively weighted by the fractions V_w and V_s of the total volume V of the sphere loaded by water and sediment:

$$c_{ext}^{eff} = L/(L_w/C_w + L_s/C_s), \quad \rho_{ext}^{eff} = (V_w \rho_w + V_s \rho_s)/V, \tag{2}$$

where L_w and L_s are the wave pathlengths around the sphere in water and sediment respectively, $L = L_w + L_s$, C_w the water sound speed and C_s the sediment sound speed. On the basis of the above definitions, the models used for the dispersion curves of the S_0 and A_{0-} waves [3], originally developed in the free field, are applied also to the case of a partly-buried sphere. A simplified scheme of the travel paths of the S_0 and A_{0-} elastic waves around the coincidence frequency is shown in Fig. 4 in the case of partial burial.

The dynamics of these waves are not expected to be significantly influenced by the *grazing angle* of sound on the seabed, as shown by the simulations in Fig. 5. Even in the case of subcritical insonification, no significant change in the resonance locations is detected. For this reason we will compare the dispersion curves of waves scattered by spheres buried at various depths, even if measured at different grazing angles, which will be maintained low in order to limit reverberation, but above the sediment critical angle in order to increase penetration.

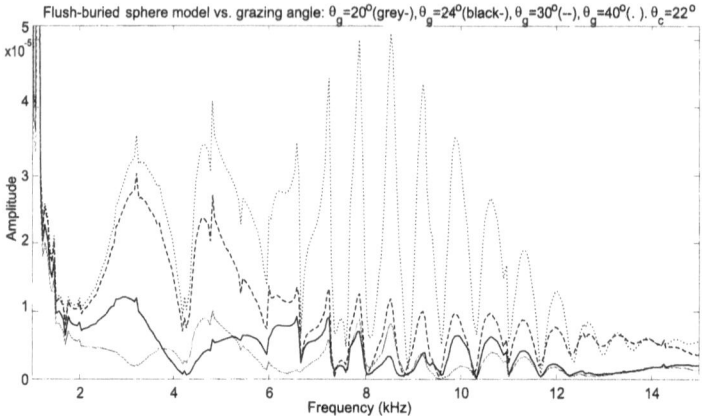

Figure 5. Amplitude of the form function of a buried sphere as the grazing angle θ_g varies.

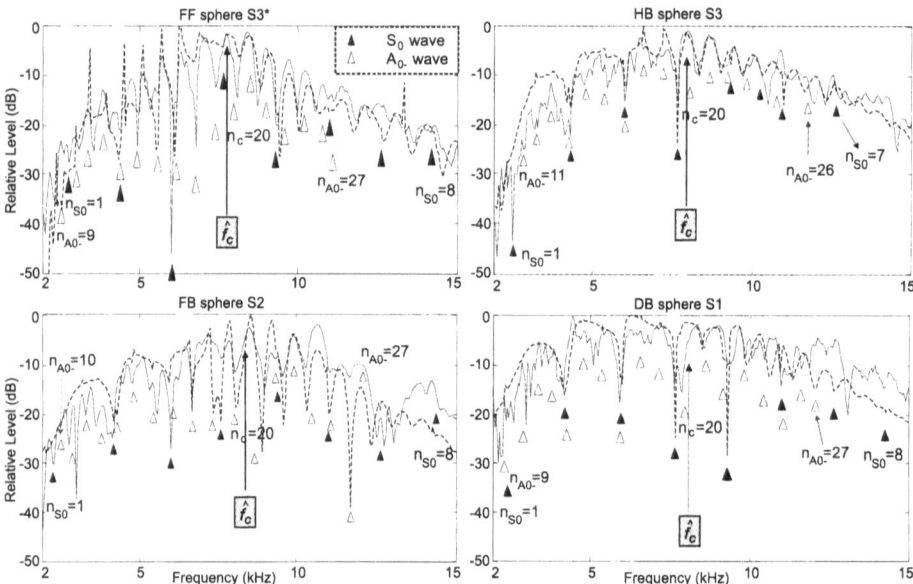

Figure 6. Model-data comparison of the spectral response by free-field, half-, flush- and deeply-buried spheres. The results of resonance mode extraction and identification are superimposed.

4   Experimental results

For model-data comparison the simulated transfer functions are convolved with the model of the incident pulse and inverse transformed in time. A Ricker function centered at 8 kHz is used for the incident pulse. The data selected are the aligned coherent average of 50 pings of the beamformed acquisitions by the vertical array. While the free-field target could be measured in the far field of the source so that the far-field, free-field model presented in Sect. 3 is applicable, the proud and buried targets were measured in the near-field of the TOPAS and the T-matrix model was adapted to simulate directional sound beams in the near-field of the source. The model-data comparison is shown in Fig. 6 for the free-field, partially-, flush- and deeply-buried spheres in the frequency domain. The

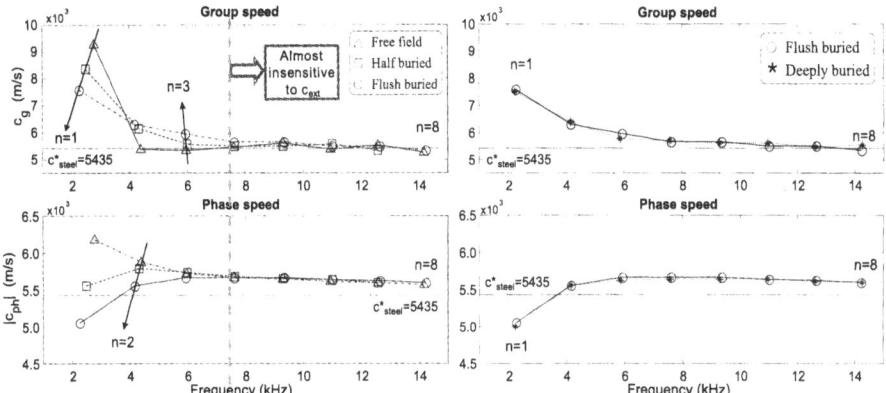

Figure 7. Experimental dispersion curves of S_0 wave as the sphere burial depth varies.

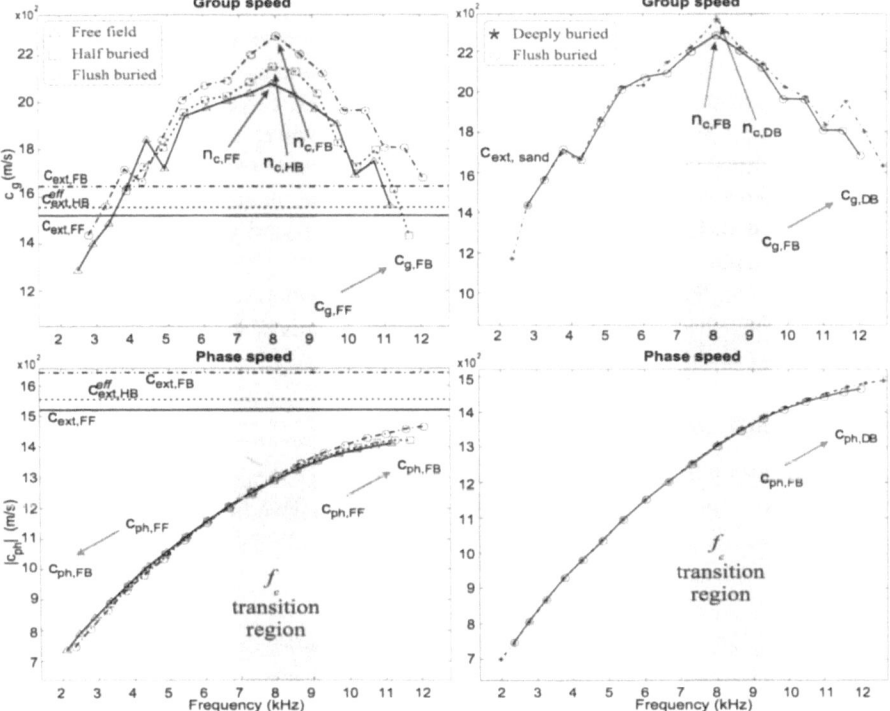

Figure 8. Experimental dispersion curves of A_{0-} wave as the sphere burial depth varies.

S_0 and A_{0-} wave modes are extracted by the approach described in [5], and identified on the basis of model-data comparison.

The free-field (FF) spherical shell was measured at a range of 35 m. Model-data agreement is generally good, except for mismatch in the mid-frequency region, presumably due to the sphere nonuniform wall thickness, and evident also in the other data sets. For the half-buried (HB) shell measurements, the grazing angle was about 26° and the range 22 m. The best fit is obtained for a burial depth of 10.6 cm below the sphere equator and a sand sound speed set to 1647 m/s. The estimate of the effective external sound speed obtained from Eq. (1) is 1555 m/s. This result validates the hypothesis that even when the A_{0-} wave becomes fluid-borne in nature, it continues to revolve around the whole spherical shell. For the flush-buried spherical shell, the measured grazing angle was 35° and the range 18 m. Equation (1) provides the outer medium speed estimate of 1652 m/s. For the measurements of the deeply-buried spherical shell, the grazing angle is about 42° and the range 16 m. The best fit with the model was found by setting the burial depth to 35 cm from the top of the target and the sand sound speed to 1652 m/s. Due to the attenuation caused by propagation through the sand, the A_{0-} wave level decreases significantly. In all buried cases, a significant mismatch can be noticed at low-frequency (for $f < 2.5$ kHz) and beyond 13 kHz presumably due to a significant decrease of the signal-to-reverberation ratio.

The wave analysis is performed in terms of wave speed dispersion curves (Figs. 7 and 8). The free-field, half-buried, and flush-buried cases are compared first in order to analyze the wave characteristics as the percentage of shell surface loaded by the sediment increases from 0 to 100. The trend of the S_0 curves (Fig. 7) are in agreement with theory and the models used in Sect. 3. Between the flush-buried and deeply-buried cases the changes in wave dynamics are slight [1] as the corresponding dispersion curves almost coincide. The nature of the A_{0-} wave (Fig. 8) has been correctly predicted by theory and models. As for the S_0 wave, also for the A_{0-} wave, the changes in dynamics between the flush-buried and deeply-buried cases are slight. Only a slight shift towards higher frequencies of the deeply-buried A_{0-} highest-order modes is detectable, which is in agreement with the slightly greater value of the sand sound speed estimated from the deeply-buried target with respect to the flush-buried case.

In conclusion, the at-sea data analysis, successfully supported by theoretical expectations and models, has allowed the evaluation of the sensitivity of elastic target scattering to burial at low-to-medium frequency.

Acknowledgements

The authors wish to thank E. Bovio, M. Mazzi and M. De Grandi, who contributed greatly to the success of the experiment. Thanks also to the Engineering and Technology Dept. of SACLANTCEN and the *MANNING* crew, for their professional support in the trial.

References

1. Lim, R., Lopes, J.L., Hackman, R.H. and Todoroff, D.G., Scattering by objects buried in underwater sediments: Theory and experiment, *J. Acoust. Soc. Am.* **93**(4),1762–1783 (1993).

2. Lim, R., Scattering by partially buried shells. In *Proc. ICA/ASA Meeting* (AIP, New York, 1998) pp. 501–502.

3. Kaplunov, J.D., Kossovich, L.Yu. and Nolde, E.V., *Dynamics of Thin Walled Elastic Bodies* (Academic Press, San Diego, 1998).

4. Maguer, A., Fox, W.L.J., Schmidt, H., Pouliquen, E. and Bovio, E., Mechanisms for subcritical penetration into a sandy bottom: Experimental and modeling results, *J. Acoust. Soc. Am.* **107**(3), 1215–1226 (2000).

5. Tesei, A., Fox, W.L.J., Maguer, A. and Løvik, A., Target parameter estimation using resonance scattering analysis applied to air-filled, cylindrical shells in water, *J. Acoust. Soc. Am.* **108**(6), 2891–2900 (2000).

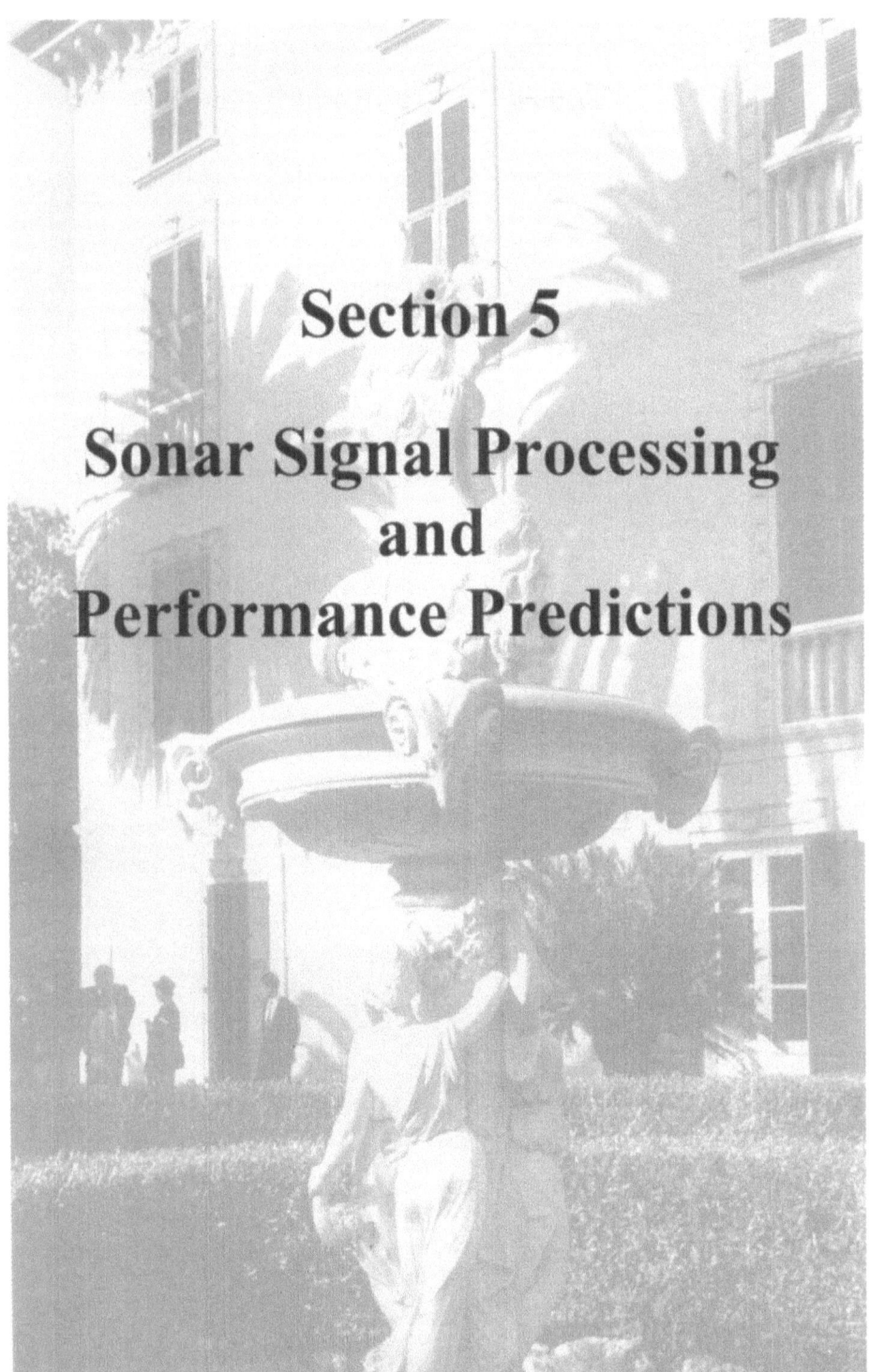

Section 5

Sonar Signal Processing
and
Performance Predictions

PERFORMANCE BOUNDS ON THE DETECTION AND LOCALIZATION IN A STOCHASTIC OCEAN

ARTHUR B. BAGGEROER AND HENRIK SCHMIDT

Massachusetts Institute of Technology, Cambridge MA 02090, USA
E-mail: abb@boreas.mit.edu

Methods for predicting the performance of a sonar for both and localization with a *known* propagation and clutter distribution are well known. Many oceanographic processes such as internal waves and bottom roughness require a stochastic description since they are dynamic and/or have a fine scale beyond which measurements are unrealistic. These and other processes randomize the temporal and spatial structure of signals and a single degree of freedom model for a replica is no longer applicable; hence, virtually all the single replica processing is not optimum. For example, one can have a power flux across and array from a signal with good SNR, yet still not be able to localize because the phase structure has been randomized too much by the stochastic processes of the ocean. Here we introduce a stochastic model for propagation from which the number of degrees of freedom in a signal observed by a sonar array. The objective is a model which reasonably represents the observed signal as observed by a sonar. We then apply these models to developing Chernoff bounds for detection and Cramer-Rao for localization. These models also indicate new receiver structures for a passive sonar which exploits the uncertainty rather than allow it to degrade from the optimal performance. Generally, we can infer that if the eigenspectrum of the observed signal has a "red" structure, which is the case in a partially saturated ocean, one can recover some of the performance compared to a deterministic propagation model. If the eigenspectrum, however, approaches "white" spectrum, then the performance degrades to simple energy detection and localization is not possible.

1 Introduction

There are many models for characterizing the random components of the ocean. For well known examples, there are Pierson-Moscovitz spectra the ocean surface [1], Garrett-Munk spectra for internal waves, [2], and correlation functions for seafloor roughness [3]. Similarly, there are very large number of theories for predicting the coherent and incoherent components of scattered fields. Predicting the performance of a sonar system, however, is difficult because these models do not lead to representations used in detection and estimation analysis. Some theories that do lead to useful representations are the temporal-spatial correlation for propagation fluctuations through internal waves [4] and the propagation of second moments by the parabolic equation [5]. The former uses a ray parameterization and assumes the rays are uncorrelated. which is questionable at low frequencies. The latter does lead to full field results, however, implementations the equation for the second moment has not attained the maturity of those for field itself, for example the sophisticated codes of Collins [6].

N.G. Pace and F.B. Jensen (eds.), Impact of Littoral Environmental Variability on Acoustic Predictions and Sonar Performance, 507-514.

Performance predictions for detection and localization are generally based on like-lihood ratios tests (LRT's) [7]. For Gaussian statistics these lead to a combination of linear and quadratic operations upon the observed data. This contrasts with some of the theoretical and observed statistics for the probability density function (PDF's) at a single receiver; however, it is not clear that the same data could be well represented by a Rician distribution where a phase randomized mean is a component of the data as well. More-over, the authors are not aware of any multivariate PDF's as needed for a sensor array in the literature and these are essential for performance predictions for data from an array of sensors. Consequently, we use a Gaussian vector model for the data which also leads to Wishart PDF's for the sample covariance of the observed data. Essentially, our arguments for this model is that tractable analytic results which incorporate all the propagation and stochastic model effects and the current literature cannot make a firm case for alternative models because of approximations employed and/or cannot be distinguished from vector Gaussian model or generalizations of it such as incorporating a mean or using a Rician PDF.

Currently, the complexity of range dependent propagation with uncertain perturba-tion introduced by an environmental model is sufficiently complicated that the statistics required by a detection or localization performance can only be provided by estimated moments such as the sample covariance. In general, there are no tractable analytic meth-ods for computing these moments except in the single replica, or one degree of freedom, for the signal which corresponds to no uncertainty in the propagation. We consequently use sample covariances obtained by running a range dependent propagation code such as FEPE (RAM) or range dependent OASIS and accumulating the results for different samples of the ocean obtained using uncertainty models of the ocean [6, 8].

Finally, the results below pertain to a passive sonar model such as represented by classical beamforming or Matched Field Processing. Similar results can be derived for active systems, or Matched Field Tomography or Ocean Acoustic Tomography, but they are not included because of space restrictions imposed on the papers. Moreover, we concentrate on presenting the theoretical framework with examples to be subsequently published.

2 Stochastic ocean model

Virtually all sonars are implemented in the frequency domain by transforming a sequence of L data segments, possibly overlapping, to form a "snapshot" vector for an array with N sensors given by

$$\mathbf{R}^l(f) = \int_{T_l - \frac{T}{2}}^{T_l + \frac{T}{2}} w(t - T_l) \mathbf{r}(t) e^{-j2\pi ft} dt$$

where

$\mathbf{R}^l(f)$ is the $N x 1$ vector of the l the "snapshot;"

$\mathbf{r}(t)$ is the vector of observations;

$w(t)$ is a window, or taper, function controlling bandwidth and sidelobes;

T_l is the center of the window for l the "snapshot";

T is the duration of the window.

There are a number of issues in the windowing operation including the spread in group delays and transit time across an array for the signal, doppler effects due to source/receiver motion and environmental processes such as internal waves and the overall stationarity of the field. [9]. These effects lead what is termed "doppler spreading" in detection and estimation theory and are important issues for the performance of sonar systems; however, other than noting that source/receiver motion dominates variability except in fixed-fixed systems, we do not address these issues in uncertainty.

Calculating the performance of detection and localization for a sonar requires the mean vector and covariance matrix of $\mathbf{R}^l(f)$. If the additional assumption of Gaussianity is applied, then analytic expressions for the false alarm and detection probabilities as well as localization accuracy can be determined. For a stochastic ocean we use the following model

$$\mathbf{R}^l(f) = \sum_i^{N_{dof}} \tilde{B}_i^l(f)\mathbf{G}_i(f, \mathbf{a}) + N_i^l(f),$$

where

N_{dof} is the number of degrees of freedom for the representation;

$\mathbf{G}_i(f, \mathbf{a}, i = 1, N_{dof}$ is a sequence of "Green's functions;"

$\tilde{B}_i^l(f)$ are complex Gaussian random variables of the snapshots which are identically distributed with zero mean and covariance $\Sigma_B(f)$;

$N_i^l(f)$ are additive noise components with spectral covariance $\mathbf{S}_n(f)$;

$\mathbf{a}$ is a parameter vector which representing both localization (range, bearing, depth) and environmental parameters (sound speeds, layer thicknesses, random models).

Some comments regarding the model are appropriate here. First, there are numerous results which argue a different PDF for the marginal, *i.e.* first order density, of a signal propagating through a random ocean. The log-normal distribution, for example, is often cited for the envelope which differs from a Raleigh PDF resulting from a Gaussian. Our argument is that these results are all for a single path and observation; when one considers a realistic ocean with many paths and a typical detection or localization process where there is averaging across a large number "snapshots," the central limit theorem rapidly takes over and one cannot distinguish the differences. Moreover, there are no multivariate distributions consistent with these marginal distributions. Next, for the case $N_{dof} = 1$ corresponds to a single, spatially coherent propagation path and reduces to the model used for classical Matched Field Processing (MFP). Finally, the "Green's functions," $\mathbf{G}_i(f, \mathbf{a}$ are not truly Green's functions in the classical usage. Such functions are well defined only a given propagation environment, not an ensemble of them. They are intended to represent a set of functions which span the covariance representation for the signal which is required to incorporates the uncertainty. Examples of this are discussed below.

The ensemble covariances of the signal and noise are the important quantities for performance prediction algorithms. For this we define the $N x N_{dof}$ Green's function matrix [a]

$$\mathcal{G}(f, \mathbf{a}) = [\ \mathbf{G}_1(f, \mathbf{a}) \mid \mathbf{G}_1(f, \mathbf{a}) \mid \cdots \mid \mathbf{G}_{dof}(f, \mathbf{a})\ ].$$

---

[a] A^H indicates the complex transpose of the matrix A.

This leads to the following for the ensemble covariance of the "snapshots"

$$\mathbf{K}_r(f) = \mathcal{G}(f, \mathbf{a}) \Sigma_B(f) \mathcal{G}^H(f, \mathbf{a}) + \mathbf{K}_n(f).$$

We note the $\Sigma_B(f)$ need not, and in most cases, is not diagonal, so the model can include phenomena such as ray/mode correlation. Furthermore, $\mathbf{S}_n(f)$ can also be an arbitrary, for example, as derived from a Kuperman-Ingenito noise model [10].

3   Green's function representations

There are several possible ways to develop the above representation for the covariance of a signal propagating in a stochastic ocean. The tradeoff depends upon implementation of a numerical model for estimating the sample covariance as well as the covariance matrix $\Sigma_B(f)$. Ideally, one wants a representation with this covariance to be very close to a diagonal matrix. (An exact diagonal matrix corresponds to adiabatic propagation.) In the this sections we indicate possible choices for the representation of $\mathbf{G}_1(f, \mathbf{a})$.

For a range independent environment the Green's function can be represented as a sum of normal modes. Extensions to range dependent propagation lead to coupled mode codes. The basic concept is to expand the output in terms of the normal modes at the array receiver. The weightings of the modes form a random vector with a covariance which is a transformation on $\Sigma_B(f)$. If there is no modal coupling, *i.e.* this matrix has rank one whereas if the propagation saturates this matrix has full effective rank of N_{dof}.

A second alternative is to apply a singular decomposition to the array receiver output. In this approach the eigenvectors are orthonormal and $\Sigma_B(f)$ is diagonal. The eigenvalue distibution represents the effective number of degrees of freedom in the signal. For example, if there is no modal coupling, the there is only one non zero eigenvalue and the eigenvector corresponds to the adiabatic solution to wave equation [11]. In general, this leads to the above representation where the eigenvectors are the $\mathbf{G}_i(f, \mathbf{a})$ and a diagonal covariance matrix for $\Sigma_B(f)$.

Finally, the simplest and most practical representation for the array receiver is to form a set of preformed beams which form a complete orthogonal basis at the array receiver. In this case the representation is in terms of the beam steering vector and the covariance matrix represents the intrabeam correlation. For example, phase coupled multipath as occurs in a deterministic ocean leads to full coherence among the respective beams whereas uncorrelated multipath corresponding to saturation leads to zero coherence.

4   Detection performance and the Chernoff bound for a stochastic ocean

The fundamental question here is how to employ the above models to make performance predictions for a sonar system. As posed, this is a spatial extension to the so called detection and estimation of a random signal embedded in additive noise problem. This has been examined in the signal and array processing literature [7], so we extend the results in this literature.

Optimal detection uses a likelihood ratio test to call the presence or absence of a target. The key performance parameters are the detection probability, P_D and false alarm probability, P_F. Except in the case of a diagonal Σ_B, which leads to ξ^2 probability densities, P_D and P_F can only be approximated albeit as closely as one needs with Chernoff bounds. In the interest of space we omit derivations and give just results.

For the binary detection case with a given value of **a** we have two hypotheses:

$$H_1 \; : \; \mathbf{K}_R((f)|H_1) = \mathcal{G}(f,\mathbf{a})\Sigma_B(f)\mathcal{G}^H(f,\mathbf{a}) + \mathbf{K}_n(f) \;\; \text{signal plus noise}$$

$$H_0 \; : \; \mathbf{K}_R((f)|H_0) = \mathbf{K}_n(f) \qquad\qquad\qquad\qquad\quad \text{noise only}$$

In M-ary formulation where one formulates a hypothesis test over all the parameters **a** such as typically done for MFP ambiguity surfaces. For the Chernoff bound and approximation for P_D and P_F we define the semi-invariant of the likelihood ratio conditioned on hypothesis H_0. This is given by

$$\mu(s) = L \int_W \ln(|\mathbf{K}_1|^{s-1}|\mathbf{K}_0|^{-s} + |s\mathbf{K}_0 + (1-s)\mathbf{K}_1|df$$

where

> W is the bandwidth of the signal;
>
> s is a free parameter with $0 \leq s \leq 1$ which can be used to optimize the bounds (In the low SNR case typical of sonar detection, $s = 1/2$.);
>
> $\mathbf{K}_1 \;\; = \;\; \mathbf{K}_R((f)|H_1)$ is a shortened notation for the covariance on the signal present hypothesis;
>
> $\mathbf{K}_0 \;\; = \;\; \mathbf{K}_R((f)|H_0)$ is a shortened notation for the covariance on the null hypothesis.

All the quantities involved can be determined from the problem formulation depending upon the Green's function representation. One can readily derive the following expressions for P_D and P_F [7]

$$P_F = \frac{1}{\sqrt{2\pi s^2 \ddot{\mu}(s)}} e^{\mu(s) - s\dot{\mu}(s)}$$

and

$$P_D = 1 - \frac{1}{\sqrt{2\pi(1-s)^2 \ddot{\mu}(s)}} e^{\mu(s) + (1-s)\dot{\mu}(s)}$$

By sweeping s over its range $[0,1]$, one can sweep out the entire Receiver Operating Characteristic (ROC) for the problem.

We summarize the approach: 1) A model which allows more than one spatial degree of freedom to incorporate the uncertainty in the propagation. 2) We suggested several characterizations -normal modes, SVD's or preformed beams. Generally, these need to estimated by simulation since the theory for propagation in a stochastic cannot now provide the needed covariances. 3) Once these covariances are known, well established bounds from detection theory can be applied. The novel item is the model to represent a stochastic ocean in a format that sonar predictions can be done. All can be addresses directly by numerical methods.

5 Bounds on localization and tomography for a stochastic ocean

Determining the obtainable accuracy of parameter estimates with MFP has been extensively studied. There have been two approaches -extensive simulations [12] and analytic approaches using bounds [13–16]. While simulations are useful, much depends on the algorithm and its implementation; in addition, there is a complicated interplay among the parameters which is hard to extract from simulations.

The bounds performance of a parameter estimation with a single degree of freedom with no parameter model mismatch have three regions of interest as indicated in figure below (from [16]). At $SNR's > SNR1$ the performance is well predicted by a linearized

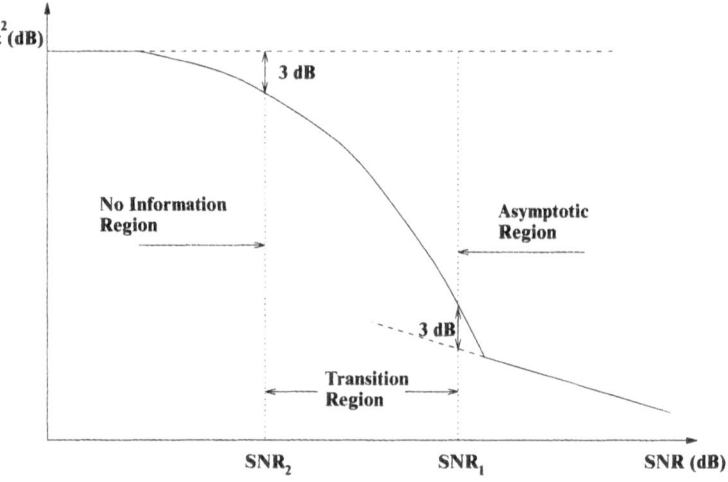

Figure 1. Performance realms in parameter estimation

analysis as given by the Cramer-Rao bound [13, 14]. At medium levels $SNR2 < SNR < SNR1$ the performance enters an "asymptotic region" where sidelobes play the dominant role [15, 16]. In this region nonlinear bounds such as the Weiss-Weinstein, Ziv-Zakai or Barankin bounds have been used successfully. Finally, for $SNR < SNR2$ there is no information in the observed signal and one has only *a priori* knowledge.

In this section we present generalizations of the results of Baggeroer and Schmidt [13] to the model for a stochastic ocean indicated above. For this we define the following quantities [b]:

$$\mathbf{d}^2(f, \mathbf{a}) = \mathcal{G}^\dagger(f, \mathbf{a})\mathbf{S}_n^{-1}(f)\mathcal{G}(f, \mathbf{a}),$$

$$\mathbf{l}_i(f, \mathbf{a}) = \mathcal{G}^\dagger(f, \mathbf{a})\mathbf{S}_n^{-1}(f)\frac{\partial \mathcal{G}(f, \mathbf{a})}{\partial a_i},$$

$$\mathbf{l}_{i,j}(f, \mathbf{a}) = \frac{\partial \mathcal{G}^\dagger(f, \mathbf{a})}{\partial a_i}\mathbf{S}_n^{-1}(f)\partial \mathcal{G}(f, \mathbf{a})\partial a_j,$$

$$\Gamma(f, \mathbf{a}) = 2\left[I + \mathbf{d}^2(f, \mathbf{a})\mathbf{S}_b(f)\right]^{-1}.$$

---

[b]The several spectral quantities scale to the covariance by multiplying by T, the window duration, *i.e.* $\mathbf{K}_x = T\mathbf{S}_x$.

The Cramer-Rao bound is expressed in terms of the Fisher Information Matrix, $\mathbf{J}$. Asymptotically, the performance of *any* parameter estimation algorithm is given the variance being lower bounded as follows:

$$\sigma^2_{a_{i,i}} \geq [\mathbf{J}^{-1}]_{i,i}$$

where

$\sigma^2_{a_{i,i}}$ is the variance, or the inverse of the output SNR for the i the parameter;

$[\mathbf{J}^{-1}]_{i,i}$ is the i the diagonal matrix of the inverse of the Fisher information matrix. Furthermore, one can demonstrate the the cross terms $\mathbf{J}^{-1}_{i,j}$ indicate the coupling among the parameters. This is important because it permits one to search for a minimal parameterization of the models which improves the accuracy of most estimation algorithms.

The only remaining issue is to determine an expression for the elements of the Fisher Information Matrix. One can derive the following result for this [17]

$$\mathbf{J}_{i,j} =$$

$$T \int_{\Delta W} Tr \left[\mathbf{S}_b(f)\Gamma(f,\mathbf{a}) \left[\left[Ev(\mathbf{l}_{i,j}(f,\mathbf{a}))\mathbf{S}_b(f)\mathbf{d}^2(f,\mathbf{a}) - Ev(\mathbf{l}_i^\dagger(f,\mathbf{a})\mathbf{S}_b(f)\mathbf{l}_j(f,\mathbf{a})) \right] \right. \right.$$

$$\left. \left. + (Ev(\mathbf{l}_i(f,\mathbf{a}))\mathbf{S}_b(f)\Gamma(f,\mathbf{a})Ev(\mathbf{l}_j(f,\mathbf{a}))) \right] \right] df.$$

[c] Again, all these quantities can be directly computed numerically. One can readily verify that this reduces to the earlier derived single replica results for $N_{dof} = 1$. These sections of the above can be interpreted fairly easily. The first line indicates the "convexity" of the ambiguity normalized to an SNR. The second line indicates the effect of increasing the observed power as a result of changing a parameter. Both lines are scaled by the a SNR factor which increases with SNR for $SNR < 1$ while at high SNR's it approaches unity. This is a common characteristic of Gaussian detection and estimation. Overall, we have a method for quantifying the obtainable performance for parameter estimation in a stochastic ocean. Bounds for the performance in the asymptotic region can also be derived, however, they are computationally extensive at the current time.

6 Summary

We have two important results in this paper: i) We have a robust model for a stochastic ocean and is coupled to the propagation physics; however, quantities for it now need to be found by simulation using one of the range dependent codes available. ii) We have given methods for analyzing detection, *i.e.* the ROC curve, and for bounding the performance of any estimation algorithm in the high SNR region and where there is no unmodeled mismatch.

---

[c]The "even" part of a matrix is given by:

$$Ev(\mathbf{X}) = \frac{\mathbf{X} + \mathbf{X}^\dagger}{2}.$$

Acknowledgments

This work was supported by the Ocean Acoustic and Underwater Signal Processing Codes at ONR. In addition, the first author was supported by his SECNAV/CNO Chair for Ocean Science funded by ONR.

References

1. Neumann, G. and Pierson, W.J., Jr., *Principles of Physical Oceanography* (Prentice-Hall Inc., Englewood Cliffs, NJ, 1966).
2. Garrett, C. and Munk, W.H, Space-time scales of internal waves, *Geophys. Fluid Dynamics* **3**, 225–264 (1972).
3. Ogilvy, J.A., *Theory of Wave Scattering from Rough, Random Surfaces* (1997).
4. Flatté, S.M., Dashen, R., Munk, W.H., Watson, K.M., and Zachariasen, F., *Sound Transmission Through a Fluctuating Ocean* (Cambridge University Press, Cambridge, U.K., 1979).
5. Ishimaru, A., *Wave Propagation and Scattering in Random Media* (IEEE Press, New York, 1997) (reprint of 1978 original, Oxford University Press).
6. Collins, M. and Westwood, E., A higher order energy-conserving parabolic equation for range-dependent ocean depth, sound speed and density, *J. Acoust. Soc. Am.* **89**, 1068–1075 (1991).
7. Van Trees, H.L., *Detection, Estimation and Modulation Theory*, Parts I and III (John Wiley and Sons, 1970) (reprinted 2002).
8. Schmidt, H., Seong, W. and Goh, J.T., Spectral super-element approach to range-dependent ocean acoustic modeling, *J. Acoust. Soc. Am.* **98**, 465–472 (1995).
9. Baggeroer, A.B. and Cox, H., Passive sonar limits upon nulling multiple moving ships with large aperture arrays. In *Proc. 1999 Asilomar Conference on Signals and Systems* (IEEE Press, New York, 1999).
10. Kuperman, W.A. and Ingenito, F., Spatial correlation of surface generated noise in a stratified ocean, *J. Acoust. Soc. Am.* **67**, 1988–1996 (1980).
11. Jensen, F.B., Kuperman, W.A., Porter, M.B. and Schmidt, H., *Computational Ocean Acoustics* (AIP Press, New York, 1994).
12. Tolstoy, A., *Matched Field Processing in Underwater Acoustics* (World Scientific Pub. Inc, Singapore, 1993).
13. Baggeroer, A.B. and Schmidt, H., Parameter estimation theory bounds and the accuracy of full field inversions. In *Full Field Inversion Methods in Ocean and Seismo-Acoustics*, edited by O. Diachok, A. Caiti, P. Gerstoft and H. Schmidt (Kluwer Academic, The Netherlands, 1995) pp. 79–84.
14. Krolik, J. and Narasimhan, Performance bounds on acoustic thermometry of ocean climate in the presence of mesoscale sound variability, *J. Acoust. Soc. Am.* **99**, 254–265 (1996).
15. Bell, K.L., Ephraim, Y. and Van Trees, H.L., Explicit Ziv-Zakai lower bounds for bearing estimation, *IEEE Trans. Signal Processing* **44**, 2810–2824 (1996).
16. Xu, W., Performance bounds on matched field methods for source localization and estimation of environmental parameters. Ph.D. Thesis, Mass. Inst. of Tech. and Woods Hole Oceanographic Inst. (June 2001).
17. Baggeroer, A.B. and Schmidt, H. Parameter coupling and resolution bounds for source localization and ocean acoustic tomography. MIT Ocean Acoustics Group Internal Memorandum (to be submitted to *J. Acoust. Soc. Am.*).

ROBUST ADAPTIVE PROCESSING IN LITTORAL REGIONS WITH ENVIRONMENTAL UNCERTAINTY

LISA M. ZURK, NIGEL LEE AND BRIAN TRACEY

MIT Lincoln Laboratory, 244 Wood St., Lexington MA 02420, USA
E-mail: zurk@ll.mit.edu

One of the main challenges in shallow water passive sonar is the complex propagation physics for both target and interferer signatures. Matched Field Processing (MFP) addresses this challenge by incorporating a propagation model into the steering vector calculation, but it is extremely sensitive to inaccuracies in the knowledge of the underwater environment. Robust algorithms have been developed to address the losses due to environmental mismatch and to target motion. To address the losses associated with target motion, a motion compensation algorithm based on the invariance principle has been developed. A second algorithm exploits the presence of a strong source to obtain a steering vector which can then be transformed to form beams at other ranges and depths.

1 Introduction

Detection and localization of quiet targets in littoral regions presents a challenging problem both because of the complicated acoustic propagation that occurs and the prevalence of loud surface ship interference. Matched Field Processing (MFP) can help address the first concern by using a propagation model to determine the steering vectors, thus providing optimal array gain and localization accuracy. Adaptive MFP (AMFP) can provide the ability to null surface interference, particularly when an array has vertical aperture that allows discrimination of surface and submerged sources. Under ideal situations, AMFP can provide super-resolution and add 10–20 dB interference suppression.

However, performance gains from AMFP have yet to be realized in practice, for several reasons. Perhaps the most important limitation is that precise information on the underwater channel is generally not available. The mismatch between the computed and actual array steering vectors can result in loss of array gain and - for adaptive processing - significant target self-nulling. A second factor influencing the performance of AMFP is the motion of the targets and interferers which introduces additional signal loss, smearing of source peaks, and consumption of adaptive degrees of freedom.

In this paper, we begin by examining the detection and localization performance for stationary sources as a function of the beamformer. We then quantify the effects of target and interferer motion. Finally, we present two methods that address two of these loss mechanisms. The first method uses the invariance principle to compensate for target motion and decrease signal gain degradation. The second method uses the observed response from a loud source and applies a depth shifting operation to construct a steering vector at the depth of interest.

N.G. Pace and F.B. Jensen (eds.), Impact of Littoral Environmental Variability on Acoustic Predictions and Sonar Performance, 515-522.
© 2002 *Kluwer Academic Publishers.*

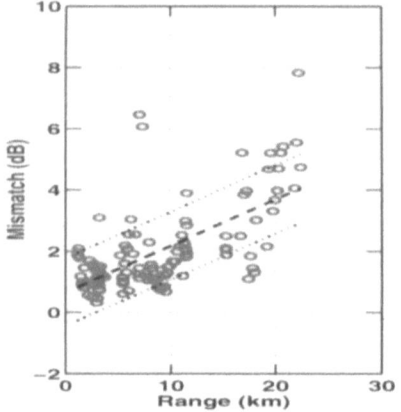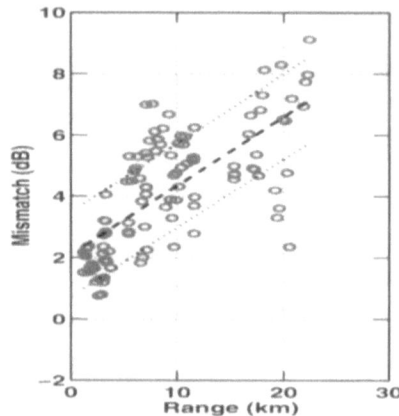

Figure 1. Mismatch from SBCX data as a function of source range for towed source tones at 94 Hz (lefthand plot) and 235 Hz (righthand plot).

2   System performance

MFP suffers signal gain degradation due to imperfect knowledge of the underwater channel. Range-focused single-path beamformers (RFBF) suffer loss due to approximation of a multipath envionment as a single path one. The losses of both of these are a function of the array topology, the position of the target, and the underwater environment. For sonar systems, it is useful to understand the expected losses to determine the appropriate beamformer for the best detection performance (i.e., minimal losses) in a given target region.

Monte Carlo simulations that incorporate the expected error in the shallow water environment were used to determine MFP mismatch loss. These simulations indicate that the loss grows as a function of target range. If one models the error as a perturbation of the modal wavenumbers, it can be easily understood that the total error will accumulate as the range to the source grows. This has been verified with experimental data from the Santa Barbara Channel Experiment, which is shown in Fig. 1, where the measured mismatch to a towed source are plotted for the tone at 94 Hz (left) and 235 Hz (right).

RFBF simulations show that the losses do not grow as a function of target range, implying that there is a range at which the losses from MFP exceed those of RFBF. RFBF losses are also greatest at endfire, where exposure to the unaccounted-for modal structure produces the greatest error. This is particularly true for arrays with vertical aperture, where one sees splitting of target energy into different beams (i.e., "mode splitting").

The implications of the above are that standard MFP might provide acceptable detection performance at close ranges and in endfire regions, but in other regions mismatch losses can be excessive and application of a RFBF may provide superior detection performance .

When localization performance is considered, MFP provides vastly increased localization accuracy. In the littoral region, where interferers are typically on the surface and targets of interest are submerged, the ability to determine depth (in particular) allows one to easily classify sources. This same discrimination capability allows adaptive processors to cancel interference while retaining target. Drawbacks of such fine resolution are the large number of beams required to cover an area and the susceptibility to motion losses.

MFP range and depth resolution does not depend on the array but depend on the wavelengths of the propagating modes. This is because MFP exploits the phasing between the modes to determine source position, which produces superior performance. For RFBF, the resolution is inversely dependent on the horizontal and vertical aperture, respectively. As an example, consider a 50 Hz source in a 200 m iso-speed channel. The MFP range resolution is approximately 46 m. For an array using range focusing, this resolution could only be achieved (assuming a 5 km target range) if the array had 4000 m of horizontal aperture.

In a typical underwater environment, targets, interferers, and often the receive array are moving in time. This is problematic for adaptive beamformers which commonly estimate the covariance matrix by averaging over some observation period during which stationarity is assumed. The total observation time is dictated by the number of elements in the array and the coherent integration time. As just discussed, large arrays with many elements provide fine resolution, so a moving source will transit more beams during a given observation period. Thus, volumetric arrays have a particularly challenging motion problem, since they require a large number of snapshots, leaving them vulnerable to motion losses. Many approaches for addressing this problem have been proposed, such as reduced-dimension processing, time-varying pre-filtering of the data, and sub-aperture processing. For moving targets, the motion spreads the target energy across multiple beams, thus decreasing the signal excess and degrading target localization accuracy. An algorithm for addressing target motion is presented in the following section.

3   Robust algorithms

Many algorithms that provide robustness to either environmental mismatch or source motion have been devised. In this section we discuss two novel techniques. The first, an invariance motion compensation algorithm, exploits the approximate invariance of the modal intensities to construct a covariance matrix in which the target motion is compensated. The algorithm is an extension of model-based compensation algorithms that have been demonstrated in the past [1, 2] with the important difference that this version does not rely on environmental knowledge. The second algorithm considered uses the observed response from what will be termed a "guide source" to determine the steering vector for a given location. With knowledge of the guide source location, this vector can then be shifted in depth to provide a steering vector for beamforming to alternate depths. Thus, an accurate replica vector is obtained without the need for environmental knowledge or the use of a propagation model.

3.1   Invariance Motion Compensation

As previously discussed, long observation times may be necessary for adaptive processing but can result in signal gain degradation for moving targets. The goal of the invariance motion compensation is to provide a means of constructing a covariance matrix for adaptive processing in which the target appears as a stationary source, perhaps at a position that is at the mid-point of its track during the observation time. Previous methods presented by the authors [2] accomplished this by adjusting each data snapshot according to a propagation model. While good motion compensation was demonstrated with this method, it suffered from the potential inaccuracies introduced by environmental uncertainties and it was computationally expensive. To address these shortcomings, we present here a method that compensates for motion in a computationally feasible manner and without the requirement of environmental information. The normal mode representation of the field at frequency ω and time t that is present on a sensor at depth z due to a source at a range r and depth z_s can be written as:

$$p(z_s, z, r; \omega, t) = C \sum_{m=1}^{M} \psi_m(z_s)\psi_m(z) \frac{e^{i(k_m(\omega)+i\alpha_m(\omega))r(t)}}{\sqrt{k_m(\omega)r(t)}} \tag{1}$$

where ψ_m is the mode function, k_m is the horizontal wavenumber, and α_m is the attenuation constant of the mth mode; M is the total number of propagating modes; and C is a constant that includes the source amplitude. In the above, we consider a small frequency range ($\Delta\omega/\omega \leq 10\%$) and thus suppress the weak dependence of the mode functions on frequency. We further consider source and receiver depths that are constant over time. Using Eq. (1), the i, jth component of the time-averaged covariance matrix can be written as

$$\hat{K}_{i,j} = \frac{1}{L} \sum_{l=1}^{L} p(z_s, z_i, r_i; \omega, t) p^H(z_s, z_j, r_j; \omega, t) \tag{2}$$

when averaging over L snapshots. It can be seen from Eqs. (1) and (2) that a source moving in range introduces a time-varying phase term of $k_m(\omega)r_i(t_l) - k_n(\omega)r_j(t_l)$. This phase introduces decorrelation into the estimated covariance and results in loss of signal energy. The objective of this algorithm is to choose data samples in time and frequency so that this phase remains constant over the L snapshots.

To find the stationary phase point, we utilize the invariance principle [3] which can be written in terms of a wavenumber difference χ_{mn} as

$$\beta = \frac{r}{\omega}\frac{dr}{d\omega} = -\frac{\chi_{mn}(\omega)/\omega}{d\chi_{mn}(\omega)/d\omega}, \quad \chi_{mn}(\omega) = k_m(\omega) - k_n(\omega) \approx \chi_{mn_0}\omega^{-1/\beta} \tag{3}$$

For many shallow water regions of interest, the value of beta has been shown to be approximately equal to one, and thus Eq. (3) provides a simple relation between the frequency and range variation of the pressure field. Consider first a vertical line array (VLA) so that $r_j = r_i$ for all i, j and the time-verying phase can be written as

$$(k_m(\omega) - k_n(\omega))r(t) = \chi_{mn}(\omega)r(t). \tag{4}$$

To maintain a constant phase in Eq. (4) we want to find the frequency offset $\Delta\omega$ that satisfies

$$\chi_{mn}(\omega_0)r(t_0) = \chi_{mn}(\omega_0 + \Delta\omega)r(t) \tag{5}$$

at all times t. The solution can be written as

$$\omega(t) = \omega_0 + \Delta\omega(t) = \omega_0 \frac{r(t)}{r(t_0)} \tag{6}$$

which represents the time-varying frequency at which to acquire data snapshots. Note that this solution requires knowledge of the time-dependent source range relative to the initial range. As this is generally not known, the algorithm can be implemented to search

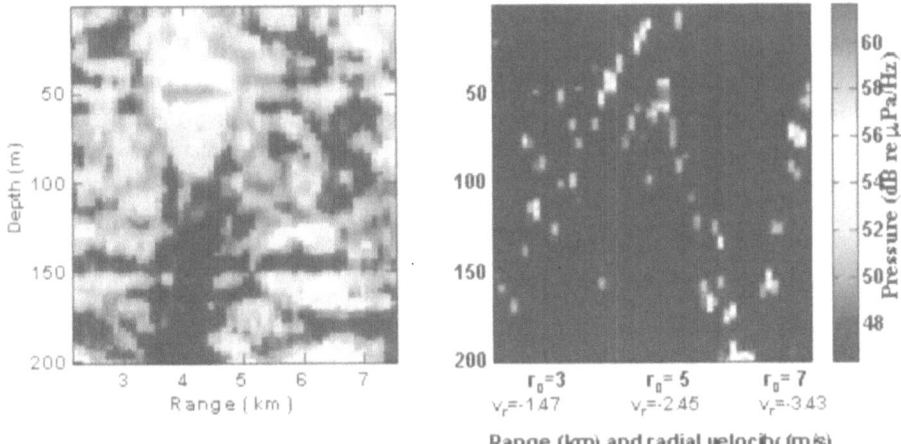

Figure 2. Range-depth ambiguity surfaces for uncompensated motion (left) and results from invariance compensation (right).

over motion hypotheses, where the correct track will provide optimal signal gain. The output of the processor will thus provide concurrent target detection, localization, and tracking. The search can be optimized by assuming that the motion is linear, in which cases each motion hypothesis collapses to a slope in time-frequency space along which to acquire snapshots. The motion-invariant covariance estimation can be written as:

$$\hat{K} = \frac{1}{L} \sum_l \mathbf{x}(\omega(t_l), t_l)\mathbf{x}(\omega(t_l), t_l) \tag{7}$$

where $\omega(t_l)$ is given by Eq. (6). Thus, the invariance motion compensation algorithm consists of the following steps: 1) hypothesize a target motion hypothesis, 2) collect data samples across time and frequency to construct a covariance matrix as given in Eq. (7), and 3) apply the adaptive beamformer in the standard fashion.

The result of applying motion compensation can be seen in the simulation results shown in Fig. 2. The left-hand plot shows the AMFP range-depth ambiguity plot from an incoherent average over 30 frequencies 153–183 Hz with an eight minute observation time and 8 sec coherent processing windows. The simulation was for a 50-m deep source moving 3 m/s in the 200-m Santa Barbara Channel environment with a 30-element VLA and a 60 dB white noise background. The target motion causes smearing of the peak and loss of target energy. The right-hand plot is the AMFP output for the same data when invariance motion compensation is applied. In this result, the target energy is focused at a single spatial location resulting in higher signal energy, lower noise background, and better source localization.

For a horizontal line array (HLA), the range appearing in the phase term is element dependent. The compensation algorithm then includes pre-multiplying each snapshot by a frequency dependent phase adjustment. This introduces one additional computational step and also necessitates knowing the mean wavenumber at each frequency.

The invariance motion compensation algorithm presented here provides a method for correcting for target motion in a computationally efficient manner and without requiring environmental information for the compensation. This correction produces a stationary covariance matrix which can then be used for adaptive processing. One factor worth noting is that the compensation is "tuned" for a given source motion. Additional sources with different motion characteristics will not be compensated for and will be de-focused from the process. For sufficiently long observation intervals (which is the situation this algorithm is intended for) this de-focusing can serve to decrease interferer energy in any one beam.

3.2 Depth Shifting of a Guide Source

In many situations of interest, strong acoustic sources are present in the environment and their locations are known. Some of these could be sources of opportunity such as surface ships, whose position might be obtained from an off-board sensor or from an airborne surveillance platform. We will term these loud sources as *guide sources* since they can be used to determine the acoustic response across a given array. This response is immediately known for the source-receiver path of the guide source, but it still unknown for sources at alternate locations. One approach to address this is to use the response from the guide source and invert for the geophysical parameters that describe the environment (sound speed, sediment densities, etc.). This approach has been investigated with some success in the underwater community, but it suffers from the large number of ambiguities and the computational complexity of the inversion.

The alternate approach presented here instead attempts to determine the response of a source by translating the observed response from the guide source. In an ideal case, this translation would be independent of the environment and hence would not necessitate environmental knowledge. In previous work [4], translation of a response from one source range to another was demonstrated by utilizing multi-frequency data and the invariance principle described above. However, if the guide is a surface ship and the target of interest is a submerged source, a method of translating in depth (as opposed to range) is desired. In this section a method of "depth shifting" the guide source response using a fully-spanning VLA is presented.

For a VLA, output from a conventional matched field processor can be written as

$$I(z_s, r) = |\mathbf{w}^H(z_s, r) \cdot \mathbf{x}(z_s, r)|^2 \tag{8}$$

where $\mathbf{x}(z_s, r)$ is the received data from a source at depth z_s and range r. For a conventional processor, $\mathbf{w}(z_s, r)$ is the normalized replica vector computed at the same range and depth. For depth shifting, the weight vector is instead a translated version of a data observation from the guide source. For a guide at depth z_g and range r_g, it can be written as

$$\mathbf{w}^H(z_s, r) = \bar{A}(z_s, r, z_g, r_g)\mathbf{x}(z_g, r_g) \tag{9}$$
$$A_{i,j} = i - j \text{ for } i = j \pm 1 \text{ or } 0 \text{ otherwise}$$

In the following, we briefly motivate the above choice of the translation matrix $\bar{A}$ whose form computes a centered finite difference in the depth dimension. The depth dependence in Eq. (1) is contained in the mode functions, which are a function of both the horizontal modal wavenumber and source depth. If we make the assumption that the dependence is on the *product* of the two, we can consider $\Psi_m(z)$ as a function of the form $f(k_m z)$ and write

$$\frac{d}{dz}f(k_m z) = g(k_m z)[k_m + \frac{dk_m}{dz}z] \sim g(k_m z)k_m \quad (10)$$

$$\frac{d}{dk_m}f(k_m z) = g(k_m z)z = \frac{d}{dz}f(k_m z)\frac{z}{k_m} \sim \frac{(f(k_{m+1}, z) - f(k_{m-1}, z))}{(k_{m+1} - k_{m-1})}\frac{z}{k_m}$$

where the last equation follows from a centered finite difference approximation. If we then apply the mode orthogonality condition and assume that the horizontal wavenumbers have a linear dependence on mode number at small vertical angles, we arrive at the discrete version of a new (but approximate) orthogonality expression for the mode functions:

$$\sum_0^D \frac{d}{dz}f(k_l z)f(k_m z)z\,dz = \frac{1}{2}[\delta(m - l - 1) - \delta(l - m - 1)]l \quad (11)$$

The above equation suggests that the inner product between the pressure across a VLA and the discrete difference of these pressures will produce a Dirac delta function between neighboring modes. The implications of this can be seen by writing $I(z_s, r)$ with $\bar{w}(z_s, r)$ given by Eq. (10) and using Eq. (11) to arrive at

$$I(z_s, r) = \left| \sum_m^M \Psi_m(z_s)\Psi_m(z_g)e^{i(k_m r - k_{m+1} r_g)} - m\Psi_{m+1}(z_s)\Psi_m(z_g)e^{i(k_{m+1} r - k_m r_g)} \right|^2$$

$$(12)$$

This differs from the conventional MFP expression because the quantity is evaluated at mode indices m and $m + 1$ (the conventional expression contains only the index m for a fully-spanning VLA). Thus, the maximum output does not occur when $z_g = z_s$ and $r_g = r$, but for some other "shifted" location. As an example, if we use the analytic expressions for an isospeed waveguide we can show that (for $r = r_g$) the maximal output occurs when

$$z_s \sim z_g \pm \frac{\pi}{Dk_0}r = z_g \pm \frac{c}{2fD}r. \quad (13)$$

This relationship is illustrated in Fig. 3 where the MFP output $I(z_s, r)$ is shown for a 235 Hz simulation in the Santa Barbara Channel environment with $r_g = r = 2.5$ km. The output is plotted as a function of the guide source depth and the target depth, z_g and z_s, respectively. If conventional processing had been applied, the plot would have high energy only along the diagonal where the guide and target depths are equal. In contrast, for the depth-shifted output, the high energy occurs along the ridges defined by $z_s = z_g \pm 40$. This value is the solution obtained from Eq. (13), even though the environment is not an isospeed channel (it has a downward refracting water column, with two sediment layers). This suggests that the relationship between the guide depth and the shifted depth may be only weakly sensitive to details of the environment.

The right-hand plot in Fig. 3 is the same simulation but with a 20% steeper sound speed profile, a faster bottom layer, and an additional sediment layer. Although some of the power has dissipated, the result shows the same general peak structure, as did simulation results for an isospeed and Pekeris waveguide (not shown).

The above algorithm provides a method of using a known response (e.g. steering vector) from one depth and translating it to another depth by utilizing a VLA. It relies on a new mode orthogonality relationship which shifts the modes that contribute to the

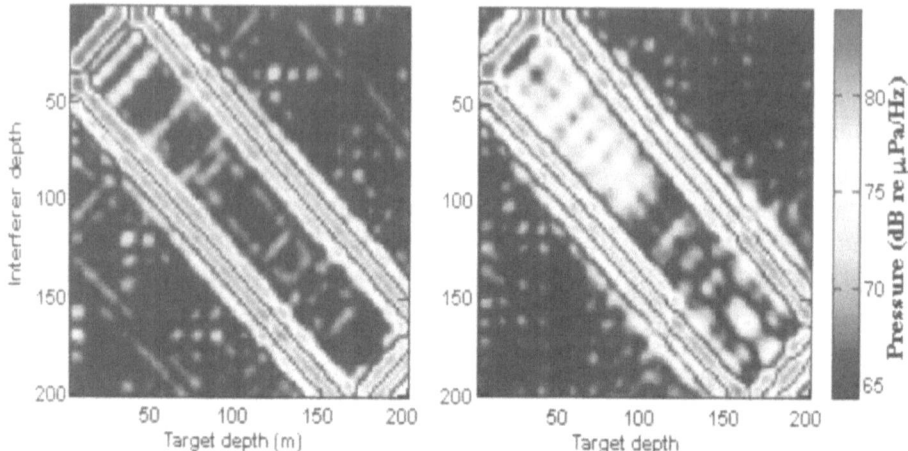

Figure 3. MFP output from depth-shifting guide source algorithm.

MFP output. Initial investigation of this relationship indicate it is insensitive to the fine features of the environment, but further examination is required. For arrays that do not have sufficient vertical aperture, there may be alternate methods of accomplishing the mode shift and the resultant depth shift.

4 Conclusion

Standard Matched Field Processors hold the promise of increased target gain and localization accuracy, but their performance is strongly dependent on fine knowledge of the underwater environment. In this work, two methods of providing robustness are presented. Invariance motion compensation utilizes the invariance principle to compensate for target motion in a robust and computationally feasible method. The depth shifting algorithm utilizes a new, approximate mode orthogonality relationship to translate the response from a known guide source for beamforming to alternate target locations.

Acknowledgements

This work was sponsored by DARPA-ATO under Air Force contract F19628-01-C-0002. Opinions, interpretations, conclusions, and recommendations are those of the authors and are not necessarily endorsed by the Department of Defense.

References

1. Zurk, L.M., Lee, N. and Ward, J., Adaptive matched field processing of moving sources in the Santa Barbara Channel experiment, *J. Acoust. Soc. Am.* (submitted June 2001).
2. Zurk, L.M., Lee, N. and Ward, J., 3D Adaptive matched field processing for a moving source in a shallow water channel. In *Proc. IEEE Oceans '99* **34**, 728–731 (1999).
3. Brekhovskikh, L.M. and Lysanov, Y.P., *Fundamentals of Ocean Acoustics*, 2nd ed. (Springer-Verlag, Berlin, 1991) pp. 139.
4. Song, H.C., Kuperman, W.A. and Hodgkiss, W.S., A time reversal mirror with variable range focusing, *J. Acoust. Soc. Am.* **103**, 3234 (1998).

A ROBUST MODEL-BASED ALGORITHM FOR LOCALIZING MARINE MAMMAL TRANSIENTS

CHRISTOPHER O. TIEMANN AND MICHAEL B. PORTER

Science Applications International Corporation
1299 Prospect St, Suite 303, La Jolla, CA 92037, USA
E-mail: christopher.o.tiemann@saic.com

JOHN A. HILDEBRAND

Scripps Institution of Oceanography
University of California at San Diego, La Jolla, CA 92093, USA

Given the increasing interest in the effects of sound on marine mammals, a new model-based algorithm for localizing transient noises around a sparse, widely-distributed array of receivers has been developed and tested with real acoustic data. The robustness of the algorithm is illustrated in its application to two different scenarios with equal success: localizing humpback whales near Hawaii and blue whales near California. The algorithm is novel in its use of acoustic propagation modeling and in construction of an ambiguity surface to identify the most probable whale location in a horizontal plane around an array. A description of the algorithm and examples of its application to marine mammal localization are provided.

1 Introduction

In 1996, twelve beaked whales stranded themselves in Greek coastal waters while a low-frequency active sonar system was being tested by NATO. In March 2000, a similar event happened in the Bahamas during a U.S. naval exercise [1]. While bi-catch (incidental capture by fishing vessels) is probably the major threat to marine mammals, these stranding incidents have generated an increased sensitivity to the possible effects of sound on the marine environment.

Three directions are being pursued to mitigate these problems: 1) study the behavior of marine mammals, 2) detect the presence of marine mammals in an area where sonar is being operated, and 3) adjust power output or operating areas of sonar systems. Our research here focuses principally on the first two areas and especially on navy ranges. Interestingly, the calls of many of these animals are often easily heard underwater. Indeed some marine mammals call at levels easily exceeding operational sonar (for instance, the bottlenose dolphin is estimated to produce clicks at about 220 dB) [2]. This suggests the possibility of using passive acoustic techniques to both detect the presence of marine mammals and observe their behavior. From a signal processing point of view, the problem is one of detecting and localizing transients on an extremely sparse (i.e. widely distributed) set of hydrophones, but such a technique must be robust against variability typical of the littoral environment in which many marine mammals exist.

N.G. Pace and F.B. Jensen (eds.), Impact of Littoral Environmental Variability on Acoustic Predictions and Sonar Performance, 523-530.

We developed an algorithm for this purpose and initially applied it to tracking humpback whales near the Pacific Missile Range Facility (PMRF) hydrophone array in waters near Kauai, Hawaii [3]. More recently, we have had the opportunity to test the algorithm using data from ocean bottom seismometers (OBS) deployed in shallow waters off the coast of southern California where the key species of interest is blue whales. The calls of blue whales are very loud, low frequency rumbles with a strong 10–20 Hz component and duration of about 20 sec. This is quite different from the brief (1–2 sec), high-frequency (0.2–4 kHz) calls of the humpbacks. Nevertheless, the same algorithm provides reliable tracking of both species despite the markedly different call characteristics and environments under study. In the remaining sections, we present the localization method and describe its application to both sites.

2   Localization algorithm

In past studies of marine mammal behavior, the typical method of localizing animals through passive acoustic means used hyperbolic fixing [4–9]. This technique uses the measured difference in arrival time (or time-lag) of a whale call recorded on multiple hydrophone pairs to produce intersecting hyperbolas indicating the animal's position. However, the assumption of a constant soundspeed inherent to hyperbolic fixing can sometimes prevent an accurate localization, particularly at longer ranges [3].

The algorithm we present here has several novel features and advantages over other localization methods. It is based on an acoustic propagation model to account for variations in soundspeed and bathymetry, thus eliminating the constant soundspeed errors associated with hyperbolic fixing. Its output is a graphical display that easily conveys the source location and confidence in the estimate. It adds robustness against environmental variability and acoustic multipath by performing some processing in the spectral domain [6,8]. Lastly, the algorithm is also suitable for continuous, real-time operation without user interaction.

The localization algorithm consists of two main components: spectral pattern correlation to calculate pair-wise time-lags, plus an ambiguity surface construction to generate a location estimate.

2.1   Spectrogram Correlation

Measuring time-lags between whale call arrivals for all hydrophone pair combinations is a critical step in the algorithm. Typically this is done through cross correlation, but whether the correlation should be performed on the original waveforms or their spectrograms is open to debate [6,8]. In analysis of data from PMRF, best results were obtained by following an example of spectral shape correlation described by Seem and Rowe [10]. Spectrograms from two hydrophones are digitized, i.e. converted to two levels of intensity (on or off) based on a data-adaptive threshold. As two digitized spectrograms are shifted past each other, correlation is performed very quickly by a logical AND of the overlapping regions. Summing the overlapping pixels provides a correlation score whose maximum determines the time-lag between channels.

To demonstrate the strength of the spectral correlation method, an example of cross correlator output using both waveform and spectral correlation is shown in Fig. 1. The data are from two hydrophones of the PMRF data set for minute 20:16 on March 22,

2001. A time window 10 seconds long extracts data subsets to use with each correlation; the window advances in 1-second increments, calculating a time-lag at each step. The waveform correlation's time-lag estimates are quite variable over the minute, perhaps due to interferers such as other distant animals singing simultaneous songs [6]. Nevertheless, the spectral correlation process correctly extracts the pair-wise time-lag of 7 seconds during periods when a whale is clearly singing.

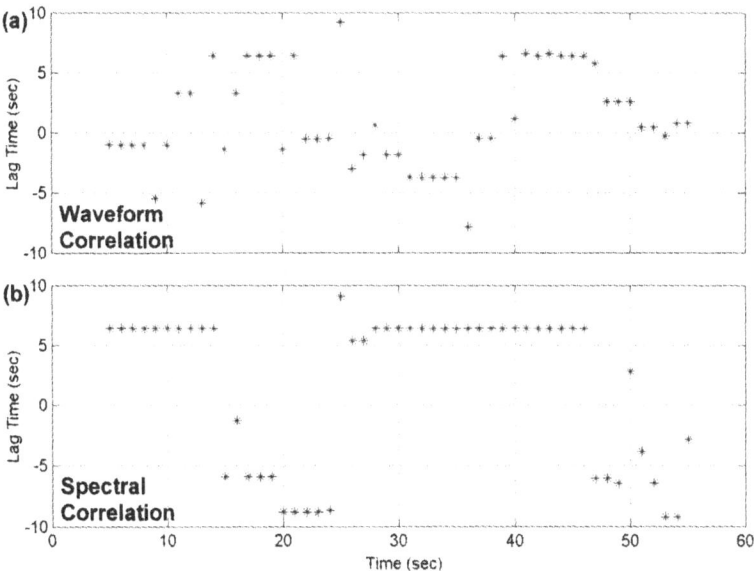

Figure 1. Time-lag measurements from waveform (a) and digitized spectrogram (b) cross correlation. Data is from PMRF hydrophones 2 and 5 for minute 20:16 on 3/22/01. The consistency of the spectral correlator made it the preferred method for time-lag measurements.

2.2   Ambiguity Surface Construction

A singing whale is localized through the construction of an ambiguity surface, which is a probabilistic indicator of source position made through the comparison of measured time-lags ('data') to predicted time-lags ('replicas'). The first step in creation of the replica is to calculate direct path acoustic travel times from a grid of possible source positions within a 30 km square area to every hydrophone. The acoustic propagation model BELLHOP [11] was used to calculate travel times as it can account for depth-dependent soundspeed profiles and varying receiver depths. Historic average soundspeed profiles were used in the acoustic modeling. When localizing humpback whales, a 500-Hz source at 10-m depth was assumed during replica generation; for blue whales a 30-Hz source at 35-m depth was assumed. The replica is computed by calculating the difference in travel times from a hypothesized source position to a pair of receivers.

For each candidate source position, the predicted time-lag that would be seen by a receiver pair is then compared to the measured time-lag for that pair to determine the likelihood that the source is at that position. A likelihood score is then scaled according to the acoustic transmission loss predicted by BELLHOP, thus minimizing the likelihood of a detection at long range from the array. Likelihood scores are then assembled on a two-dimensional horizontal plane around the array, completing one ambiguity surface. Ambiguity surfaces based on time-lag estimates with high correlation scores are summed to make an overall ambiguity surface where source location estimates common to many receiver pairs stack to form a peak. The ambiguity surface peak is declared the best estimate of source position.

3   Localizations at PMRF

The Pacific Missile Range Facility has an underwater array of over 100 bottom mounted hydrophones, and a system has been implemented for transferring acoustic data from 6 hydrophones to the Maui High-Performance Computing Center (MHPCC) for analysis. The hydrophones are spaced 5–20 km apart and are shown on the contour map of Fig. 2. Two days of acoustic data from March 2001 were used for algorithm development. Humpback whale songs were heard on every hydrophone channel at all times of day; sometimes multiple marine mammals could be heard on the same channel simultaneously.

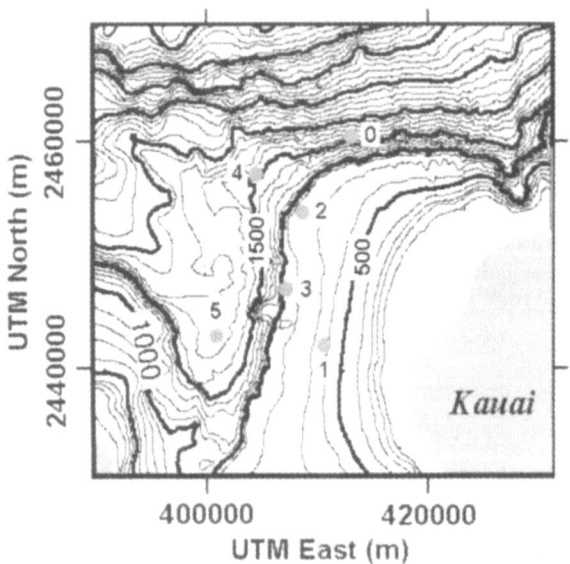

Figure 2. Bathymetry contours (m) and hydrophone locations (0–5) at the Pacific Missile Range Facility. Axes are for UTM Zone 4.

An example of a humpback whale localization through ambiguity surface construction is shown in Fig. 3. The ambiguity surface represents a 30-km square area around the PMRF array with hydrophone positions labeled. Bright areas on the surface represent likely whale positions, with the peak indicated by a crosshair designating the animal's location. The ambiguity surfaces reveal patterns that resemble hyperbolas, but they have been thickened and stacked so that the most probable source location can be automatically extracted. This analysis was applied to many time segments throughout the data set, and in every case, a source was localized by the contribution of four or more receiver pairs.

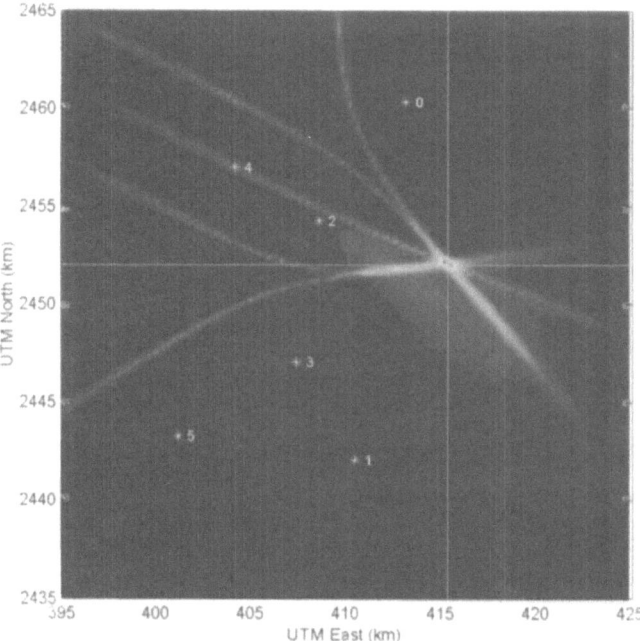

Figure 3. Ambiguity surface showing a plan view of the waters around the PMRF array with hydrophone positions (0–5) indicated. Axes are for UTM Zone 4. Intensity peak and crosshair indicate whale position estimate. Data is from 3/22/01 20:16.

4 Localizations at San Clemente

Twelve hours of data from four seismometers west of San Clemente Island were used for further algorithm testing. The seismometers were deployed on the seafloor at about 200 m depth in a 3 km square, as shown in Fig. 4. The long, low frequency calls of blue whales were heard on every channel during most of the data set. Because of the much longer call duration of blue whales, the correlation window size was increased from 10 seconds to 60 seconds. Doing so allows the spectral correlator to see more distinguishing spectral patterns with each iteration. This was the only change made to the algorithm. After replicas were generated for the new array geometry, the localization process was run again.

An example ambiguity surface showing a single blue whale localization appears in Fig. 5. This surface represents a 20 km square plan view around the seismometer array. Again, the bright peak and crosshair indicate a likely whale location. During times when only one blue whale was heard singing, it was easy to follow its motion through the array with consecutive localizations.

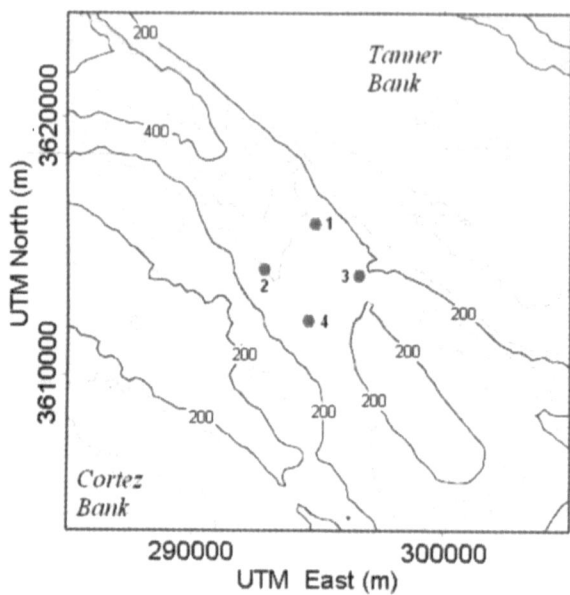

Figure 4.   Bathymetry contours (m) and seismometer locations (1-4) near San Clemente Island off southern California. Axes are for UTM Zone 11.

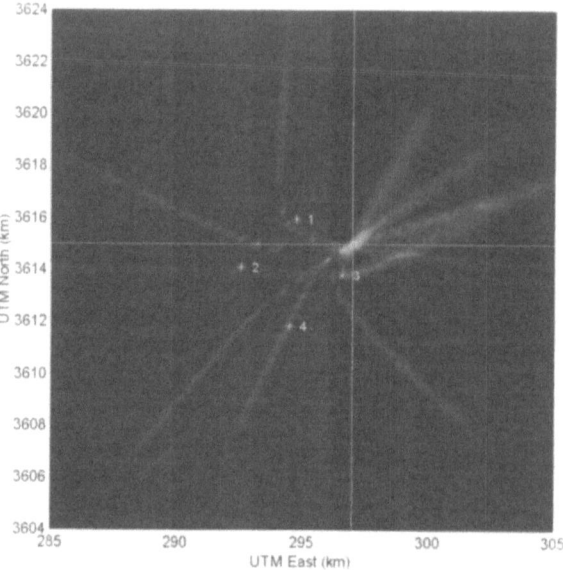

Figure 5.   Ambiguity surface showing a plan view of the waters around the San Clemente array with seismometer positions (1–4) indicated. Axes are for UTM Zone 11.   Intensity peak and crosshair indicate whale position estimate. Data is from 8/28/01 05:36.

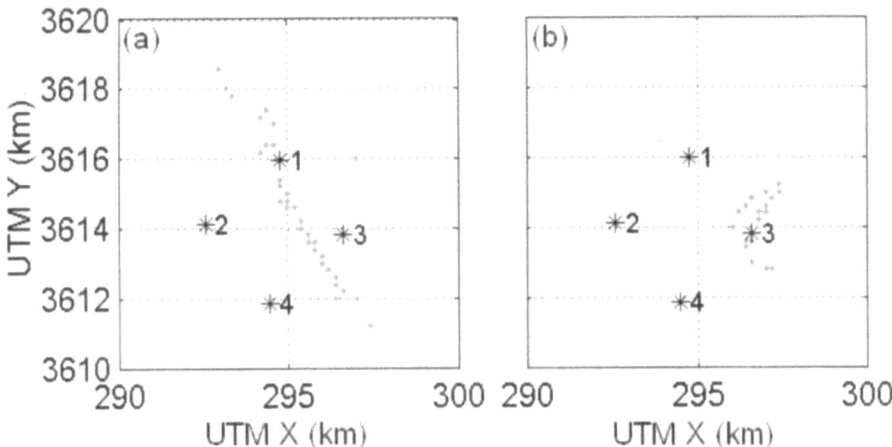

Figure 6. Plan view of the waters around the San Clemente array with seismometer positions (1–4) indicated. Axes are for UTM Zone 11. Points show blue whale location estimates from 8/28/01. (a) From 03:00 to 04:13 a whale moved from northwest to southeast over the array. (b) From 10:10 to 12:00 a whale swam northeast over sensor 3, then reversed direction.

To illustrate this, Fig. 6 shows two track estimates from August 28, 2001, on 10 km square plan views around the seismometer array. In Fig. 6a, a whale is swimming across the array from north to south at approximately 7 km/h. Figure 6b tracks a whale reversing direction.

5 Discussion

The model-based localization technique described here could be a valuable tool for those wishing to study marine mammal behavior or minimize the effects of sound on such animals. When applied to long records, the algorithm has the potential to deliver useful information of whale behavior, such as in visualizing migration paths. However, its real benefit could be for those interested in mammal mitigation issues through continuous, real-time, automated monitoring applications. In tests with the PMRF data set, the computations could be completed within the data update period, and setting high thresholds on scoring would minimize chances of false alarms.

The algorithm's success in two different scenarios hints at its robustness against environment, target type, and array geometry. It is also readily portable to any array shape and may work equally well localizing other transient audible targets, both natural and manmade. Because localization is now limited to a single assumed depth, future enhancements could include taking better advantage of depth-dependent multipath structure to generate a depth estimate. Confirmation of the algorithm through non-acoustic means, such as by concurrent visual observations, is another goal.

Acknowledgements

We gratefully acknowledge the help of the following people from PMRF, MHPCC, and SAIC in providing and preparing the Hawaii data for use: Jim Hagar, Robert Desonia, D.J. Fabozzi, and Paul Hursky. Thanks to Allan Sauter of SIO for preparing the San Clemente OBS data for use. Neil Frazer and Herb Freese contributed in many helpful discussions, and Richard Bachman provided environmental characterizations of both sites. This work was supported by the National Defense Center of Excellence for Research in Ocean Sciences (CEROS contract 47316). Additional support was provided by Office of Naval Research (ONR contract N00014-00-D-0115).

References

1. Joint interim report: Bahamas marine mammal stranding event of 15–16 March 2000 (U.S. Department of Commerce and Department of the Navy, 2001).
2. W.J. Richardson, C.R. Greene, Jr., C.I. Malme and D.H. Thomson, *Marine Mammals and Noise* (Academic Press, 1995) p. 185.
3. C.O. Tiemann, M.B. Porter and L. Neil Frazer, Automated model-based localization of marine mammals near Hawaii. In *Proc. MTS/IEEE Oceans 2001* (Holland Publications, 2001) pp. 1395–1400.
4. S. Mitchell and J. Bower, Localization of animal calls via hyperbolic methods, *J. Acoust. Soc. Am.* **97**, 3352–3353 (1995).
5. A.S. Frankel, C.W. Clark, L.M. Herman and C.M. Gabriele, Spatial distribution, habitat utilization, and social interactions of humpback whales, *Megaptera novaeangliae*, off Hawai'i determined using acoustic and visual techniques, *Can. J. Zool.* **73**, 1134–1146 (1995).
6. V.M. Janik, S.M. Van Parijs and P.M. Thompson, A two-dimensional acoustic localization system for marine mammals, *Mar. Mamm. Sci.* **16**, 437–447 (2000).
7. K.M. Stafford, C.G. Fox and D.S. Clark, Long-range acoustic detection and localization of blue whale calls in the northeast Pacific Ocean, *J. Acoust. Soc. Am.* **104**(6), 3616–3625 (1998).
8. C.W. Clark and W.T. Ellison, Calibration and comparison of acoustic location methods used during the spring migration of the bowhead whale, *Balaena mysticetus*, off Pt. Barrow, Alaska, 1984–1993, *J. Acoust. Soc. Am.* **107**(6), 3509–3517 (2000).
9. C.W. Clark, W.T. Ellison and K. Beeman, Acoustic tracking of migrating bowhead whales. In *Proc. IEEE Oceans 1986*, pp. 341–346 (1986).
10. D.A. Seem and N.C. Rowe, Shape correlation of low-frequency underwater sounds, *J. Acoust. Soc. Am.* **95**(4), 2099–2103 (1994).
11. M.B. Porter and Y.C. Liu, Finite-element ray tracing. In *Proc. International Conference on Theoretical and Computational Acoustics*, edited by D. Lee and M.H. Schultz (World Scientific, 1994) pp. 947–956.

ASSESSMENT OF THE IMPACT OF UNCERTAINTY IN SEABED GEOACOUSTIC PARAMETERS ON PREDICTED SONAR PERFORMANCE

M.K. PRIOR AND C.H. HARRISON

SACLANT Undersea Research Centre, Viale San Bartolomeo 400, 19138 La Spezia, Italy
E-mail: prior@saclantc.nato.int, harrison@saclantc.nato.int

S.G. HEALY

QinetiQ Unit, Southampton Oceanography Centre, Southampton, SO14 3ZH, UK
E-mail: S.Healy@soc.soton.ac.uk

'Uncertainty' can be distinguished from 'variability' as the lack of knowledge of the input variables that a certain calculation must tolerate. In the case studied here, the calculation concerns sonar performance, and uncertainty may stem from many sources including the finite measurement precision of methods used to describe the physical environment. Lack of knowledge of the environmental inputs causes uncertainty in the predicted sonar performance. It is common for this uncertainty to be high but its impact is rarely taken into account. In this paper we take, as an example of an environmental descriptor, plane wave reflection loss already derived from measurements of ambient noise directionality. An inverse method is developed that allows the reflection loss as a function of angle and frequency to be converted to a geophysical description of the seabed in terms of density, sound speed and attenuation. The uncertainty in these geophysical 'inputs', i.e. the "error bar" associated with the inversion, is estimated and the impact of this uncertainty on the prediction of sonar performance is calculated. This is achieved by calculating reverberation and target echo in a series of environments lying within the uncertainty bounds. These calculations are repeated using seabed data derived from a geophysical description of a core. The relative impacts of the uncertainties associated with the two methods of describing the seabed are compared. The calculations performed also take into account uncertainty in sonar-related parameters such as target reflectivity.

1 Introduction

When predicting sonar performance using sonar equation calculations, it is common practise to consider environmental input data as being fixed with no uncertainty. This approach is potentially misleading when one considers the sensitivity of acoustic propagation loss, ambient noise and reverberation to environmental changes and the poor quality of the environmental data that is often available. Sensitivity analyses can be carried out to see how changes in the environmental description affect predicted performance. This is usually not done because of lack of time or insufficient knowledge concerning the spread of environmental parameters.

In this paper, we consider the impact, on one aspect of predicted sonar performance, of uncertainty in the description of the seabed. Two types of seabed

531

N.G. Pace and F.B. Jensen (eds.), Impact of Littoral Environmental Variability on Acoustic Predictions and Sonar Performance, 531-538.
© 2002 *Kluwer Academic Publishers.*

description are used and the impacts of their uncertainties are assessed. These are compared with the impact of uncertainty in a non-environmental parameter, namely the target scattering strength. Calculations are carried out for monostatic and bistatic sonars in a shallow water environment. Sonar performance is quantified via the calculation of regions in which a target, if it were to be present, would result in a positive signal-to-noise ratio (SNR) at the receiver. In the context studied here, "noise" is actually the sum of reverberation and the direct blast from the sonar transmission.

2   Seabed description

Two methods were used to describe the seabed in this study. The first used estimates of plane-wave reflection loss derived from measurements of ambient noise. These reflection losses were input to a genetic algorithm (GA) inversion to produce estimates of the sound speed, density and attenuation of the seabed. The second method used a core sample (core 255 from [1]) taken in the same geographic location as the ambient noise data. This was used to produce a description of the seabed that allowed a range of possible values for the seabed density, sound speed and attenuation to be determined from a database of laboratory measurements on seabed sediment samples. Both descriptions include only compressional wave properties. The methods are now described in more detail.

2.1   Inversion of Reflection Loss Data

Measurements of the ambient noise in the ocean, made on a vertical line array (VLA), can be processed to yield estimates of plane wave reflection loss [2]. While these estimates could be input directly into ray tracing propagation and reverberation models, there is a risk that "glitches" in the data might corrupt such model calculations. Furthermore, it may be that the propagation/reverberation models do not use plane-wave reflection loss directly in their calculations and require instead that the seabed should be described in terms of sound speed, density and attenuation.

To translate from reflection coefficient as a function of angle and frequency to a geoacoustic description, an inverse method was developed. The method used standard techniques [3] based on genetic algorithms but employed a novel method for determining the fitness of each solution. The fitness was determined to be the product of two factors representing the closeness of fit and the geophysical "realness" respectively. The closeness of fit was assessed by a simple root-mean-square (rms) difference between the linear intensity reflection coefficients from the VLA data and Rayleigh reflection coefficients [4] produced using the trial values of density, sound speed and attenuation. The "realness" of each density and sound speed combination was assessed by determining how close the pair lay to a regression curve fitted to a scatter plot of density/sound-speed pairs taken from a database of laboratory measurements made on seabed sediments [5]. The form of the realness function was

$$R = \exp\left[-\left((c - c_\rho)/150\right)^2\right]; \quad c_\rho = 2104.2 - 1.029\rho + 4.55 \times 10^{-4} \rho^2 \qquad (1)$$

where R is the realness, c is the sound speed in m/s, ρ is the density in g/cm^3 and c_ρ is the sound speed determined from the density, using the polynomial fit to the laboratory measurements.

Figure 1 shows a contour plot of the realness function, the data points used to produce the polynomial fit and the fit itself. The figure shows that the realness function is high in regions with many density/sound-speed points and low in regions where the combination of these two closely linked parameters was unphysical. The attenuation of the seabed was left out of the realness function, effectively leaving the search for it unconstrained. This was done because the correlation between attenuation and the other two seabed parameters is usually poor [6].

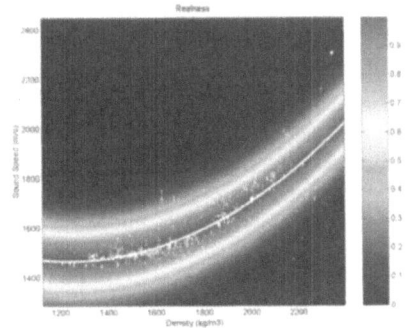

Figure 1. Realness function as contours with density/sound-speed data as points and polynomial fit as line.

The GA inversion searched simultaneously for seven parameters describing a two-layer seabed; sound speed, density and attenuation for each layer and a thickness of the upper layer. Figure 2 shows VLA-measured and inverted values of reflection loss as a function of frequency and angle. The general agreement is shown to be good. The areas of apparently zero loss around the edges of the figures arise due to array limitations (e.g. grating lobes in the top right corner) and were removed from consideration in the inversion. Uncertainty in the seabed description was obtained by running the inversion method ten times to yield ten seabed descriptions.

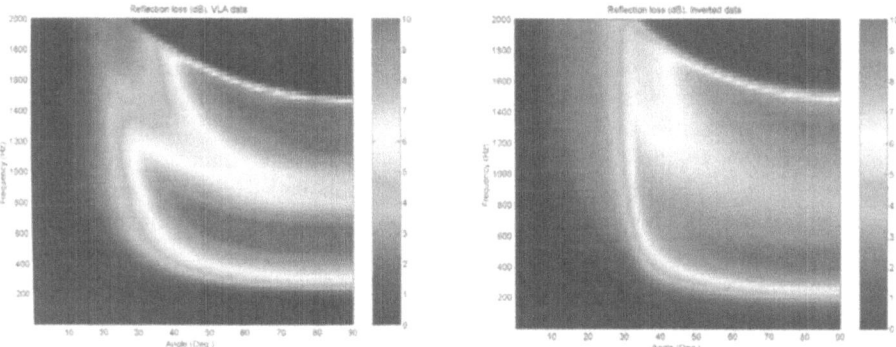

Figure 2. Reflection loss for VLA and inverted data.

2.2   Geophysical Description

In the area of interest, a sediment core had been previously obtained (Core 255 from [1]) and laboratory analysis had indicated that the seabed at the location of the core was silty-sand and sand. A range of sediment properties which may be attributed to sediment of the same general type as Core 255 was obtained to determine the spread of likely values of density, sound speed and attenuation. This information was extracted from the GEOSEIS database [5], a repository of geotechnical, seismo-acoustic and geochemical information on marine sediment and bedrock. 42 sediment records were selected from

GEOSEIS by choosing samples with properties that fitted the description of Core 255. This approach relied on the premise that there were good inter-relationships between sediment physical properties (e.g. porosity, grainsize) and the resulting geoacoustic properties such as velocity and attenuation [6]. Geological materials are generally complex, so samples with identical porosities or grainsizes may have significant differences in structure and composition and so are likely to exhibit a range of possible values of density, velocity and attenuation.

3   Sonar scenario

The impact of seabed uncertainty was translated into a spread of sonar performance using the SUPREMO [7] multistatic model to predict target echo and reverberation in one monostatic and one bistatic scenario. SUPREMO used the Gamaray [8] eigenray model to calculate propagation paths for the prediction of target echo, reverberation and direct blast (in the bistatic case). Gamaray allows the seabed to be described in terms of its sound speed, density and attenuation. The seabed data from both sources was therefore directly inserted into the model.

In both cases, an omnidirectional source was used, transmitting an 80 Hz bandwidth LFM signal centred on 500 Hz. The signal was received by a 64-element horizontal line array with 1.5 m spacing between elements. The target was represented as a cylinder with hemispherical end-caps and had length 100 m and radius 5 m. Receiver and target headings were arranged so that the source signal was specularly reflected and detected by the receiver in its broadside beam. The environment had a 1500 m/s isovelocity water column above a seabed described using data from the two methods. The area of study was a 20 km (x,y) square grid centred on (0,0) with the monostatic sonar placed at this central point. The bistatic scenario had the source at (-5km,0) and the receiver at (5km,0).

Sonar performance was quantified by the SNR, calculated as the highest value of the ratio of the intensities of the target echo and reverberation plus direct blast. It was recognised that a positive SNR alone does not qualify as indicating a detection possibility but the inclusion of sonar-specific terms such as detection threshold [9] was rejected so as to avoid difficulties in widely disseminating the results of the study. In addition to uncertainty arising from the environmental description, it is important to remember that the non-environmental terms under consideration are also not known perfectly. To reflect this, an uncertainty of +/-3 dB was arbitrarily placed on the target strength calculated by SUPREMO.

Quantification of the impact of seabed uncertainty was achieved by calculation of the area in which SNR was positive. Further description of the impact of uncertainty was achieved by determining for each target location whether the SNR was always negative, always positive or changed sign due to the changes in seabed description and target strength. This approach resulted in five different descriptions of the SNR due to a target at each possible location; 1) always positive, 2) always negative, 3) uncertain due to environmental factors, 4) uncertain due to non-environmental factors 5) uncertain due to *both* environmental and non-environmental factors.

4 Results

4.1 Seabed Data

Figure 3 shows the density and attenuation of the seabed plotted as a function of the sound speed for the laboratory measurements and for the results of the inversion.

The results of the inversion can be seen to lie within the range of the laboratory measurements but the attenuation coefficient returned from the inversion is large. The reason for this is not known but it is possible that some process not included in the stratified fluid model of the seabed was interpreted as extra attenuation when the inversion was carried out.

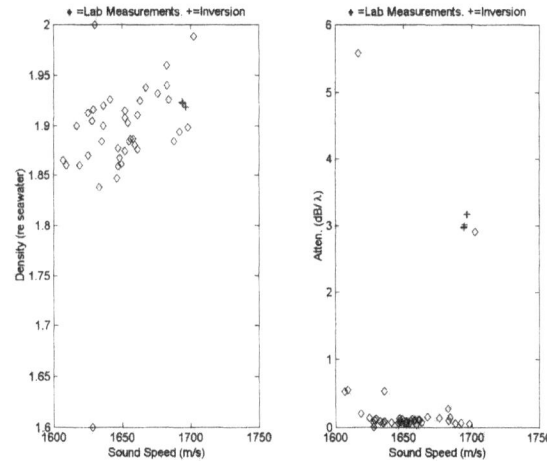

Figure 3. Seabed data from lab measurements and inversion.

4.2 Sonar Performance

Figure 4 shows the impact on sonar performance of uncertainty in the seabed description and target strength with the seabed description derived from the laboratory measurements. Figure 4a shows the SNR averaged over all target positions with mean values plotted as crosses and the error bars indicating the standard deviation from the mean. The x-axis of the plot is the value of the gradient of reflection loss versus angle calculated using Weston's equation [10]

$$\alpha_W = \frac{\rho \alpha_\lambda}{10\pi \log_{10}(e)} (c_w / c)^2 \left(1 - (c_w / c)^2\right)^{-3/2} \tag{2}$$

where α_λ is the attenuation in dB per wavelength and α_W is the gradient (Weston's Alpha). α_W is a useful single value that expresses the impact of the seabed on reflection and Harrison showed [11] that for targets beyond a critical range, R_c,

$$R_c = (8H) / (\alpha_W \theta_c) \tag{3}$$

(where H is the water depth and θ_c is the critical angle) the SNR is a constant value for shallow water conditions. Figure 4a shows that the mean SNR is linked to α_W. SNR remains constant for small values of α_W then increases with it for values beyond −5dB. This value corresponds to R_c being 10 km and the transition occurred when the area of interest lay within the region where SNR was proportional to α_W. The connection between α_W and SNR is further shown in Fig. 4b where the Area Of Positive SNR (AOPS) is shown to increase with α_W.

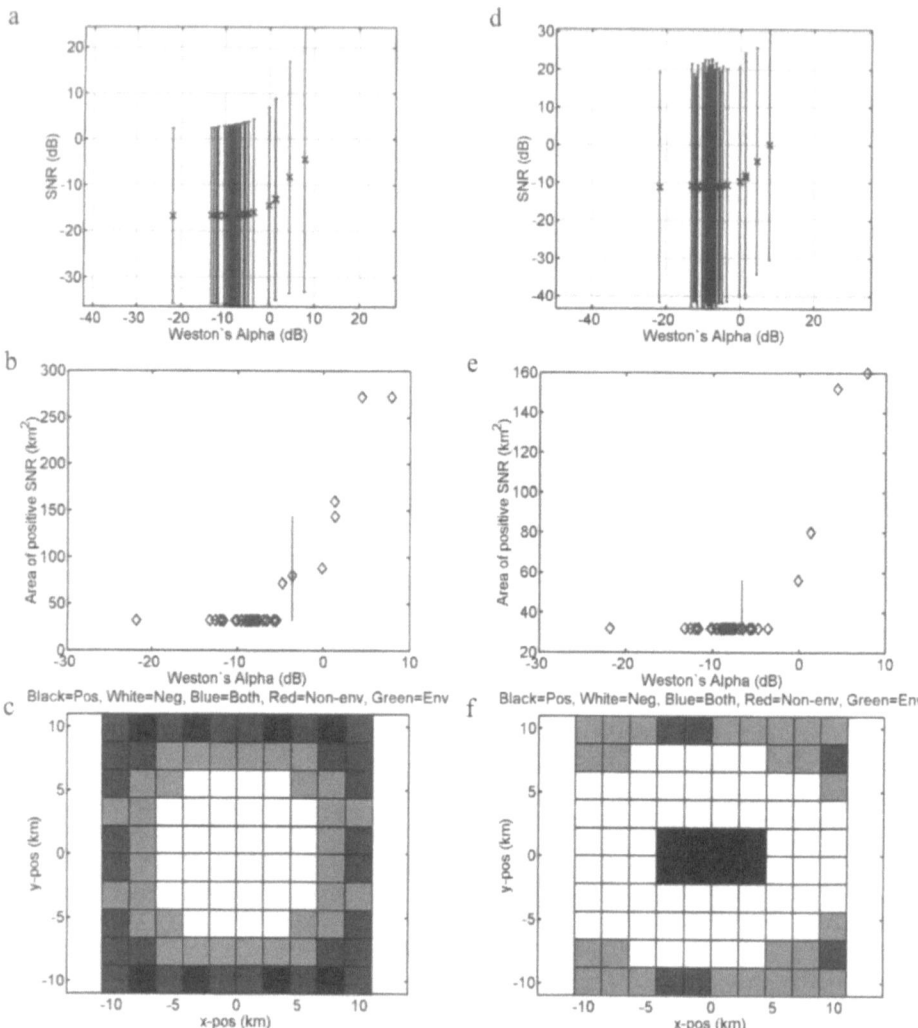

Figure 4. Sonar performance as quantified by SNR (top), area of positive SNR (middle) and SNR description (bottom). Results for laboratory measurements. Monostatic sonar left, bistatic right.

The impact of the seabed uncertainty is high with changes in area of over 200 km². The error bars in Fig. 4b indicate the change in AOPS associated with the +/-3dB uncertainty in target strength, plotted on the median value of AOPS from the environmental uncertainty set. The impact of environmental uncertainty is greater than the non-environmental uncertainty. Figure 4c shows the area of interest with colour coding indicating the nature of the uncertainty in SNR. Black, white, green, red and blue refer to the numbers given in Sect. 3. The dominance of environmental uncertainty is illustrated. Figure 4d-f shows the results for the bistatic configuration and the same trends are shown to be present as for the monostatic case. The main differences are the reduction in AOPS and the bistatic pattern in Fig. 4f.

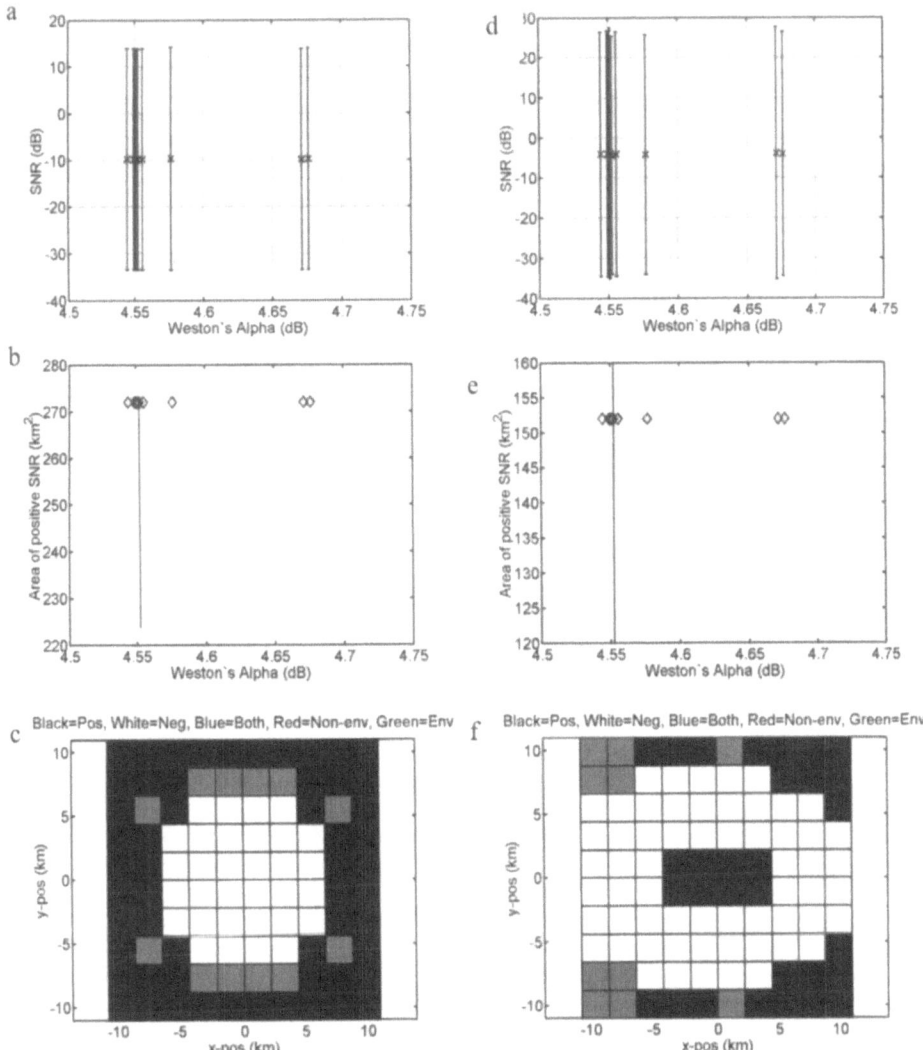

Figure 5. Sonar performance as quantified by SNR (top), area of positive SNR (middle) and SNR description (bottom). Results for inversion. Monostatic sonar left, bistatic right.

Figure 5 shows the sonar performance results for seabed uncertainty arising from the ambient noise inversion method. The smaller uncertainty arising from this method of determining the seabed type is reflected in the very small spread of values of α_W present on the x-axes of Fig. 5a. Non-environmental effects dominate the uncertainty in area. Trends are the same for both monostatic and bistatic sonars with the main difference between the two being the lower AOPS in the latter case.

5   Summary and discussion

The impact on sonar performance of uncertainty in seabed data was shown to differ for the two types of seabed data. The inversion of acoustic measurements was shown to yield a smaller spread in geoacoustic parameters and this spread was shown to predict a negligible spread in sonar performance. This stemmed from the repeatability of the inversion method and from the self-compensating nature of the process. That is, the ambient noise measurements gave reflection loss estimates and the SNR was linked to α_W, a measure of reflection loss. The uncertainty arising from laboratory measurements of sediments resulted in larger uncertainty in SNR. It is important that this uncertainty should be compared with uncertainty arising from target strength and any assessment of the impact of environmental uncertainty should also include estimates of uncertainty in non-environmental factors. The work reported here illustrated the usefulness of accompanying numerical calculations of sonar performance with mathematical analysis. The importance of α_W indicates a possible way in which the impact of uncertainty could be estimated directly without the need for numerical models.

References

1. Tonarelli, B., Turgutcan, F., Max, M.D. and Akal, T., Shallow sediment composition at four localities on the Sicilian-Tunisian platform. In *UNESCO Reports in Marine Science: Geological Development of the Sicilia-Tunisian Platform* **58**, 123–128 (1992).
2. Harrison, C.H. and Simons, D.G., Geoacoustic inversion of ambient noise: A simple method. In *Proc. Institute of Acoustics: Acoustical Oceanography* **23**(2), 91–98 (2001).
3. Houck, C., Joines, J. and Kay, M., A genetic algorithm for function optimization: A Matlab implementation. NCSU-IE TR 95-09 (1995).
4. Brekhovskikh, L. and Lysanov, Yu., *Fundamentals of Ocean Acoustics* (Berlin, Springer-Verlag, 1982).
5. McCann, C., McDermott, I., Grimbleby, L., Marks, S.G., McCann, D.M. and Hughes, B.C., The GEOSEIS database. A study of the acoustic properties of sediments and sedimentary rocks. In *Proc. Oceanology 1994* (1994).
6. Hamilton, E.L., Acoustic properties of sediments. In *Acoustics and Ocean Bottom*, edited by A. Lara, C. Ranz and R. Carbo, pp. 3–58 (1987).
7. Harrison, C.H., Prior, M.K. and Baldacci, A., Multistatic reverberation and system modelling using SUPREMO. In *Proc. 6th European Conference on Underwater Acoustics*, Gdansk, Poland (2002).
8. Westwood, E.K., Tindle, C.T. and Chapman, N.R., A normal mode model for acousto-elastic ocean environments, *J. Acoust. Soc. Am.* **100**, 3631–3645 (1996).
9. Urick, R.J., *Principles of Underwater Sound for Engineers* (McGraw-Hill, 1975).
10. Weston, D.E., Intensity-range relations in oceanographic acoustics, *J. Sound and Vib.* **18**, 271–287 (1971).
11. Harrison, C.H., Reverberation and signal excess with mode stripping and Lambert's law. SACLANTCEN Report (2002).

REVERBERATION ENVELOPE STATISTICS AND THEIR DEPENDENCE ON SONAR BEAMWIDTH AND BANDWIDTH

D.A. ABRAHAM AND A.P. LYONS

The Pennsylvania State University, Applied Research Laboratory
P.O. Box 30, State College, PA 16870, USA
E-mail: d.a.abraham@ieee.org, apl2@psu.edu

In order to combat high reverberation power levels in shallow water operational areas, active sonar systems have employed increased bandwidth transmissions and larger arrays. Both of these techniques have the effect of limiting the contribution of reverberation in each range-bearing resolution cell of the sonar by decreasing the cell size, which can also have an adverse effect on the probability density function (PDF) of the reverberation induced matched filter envelope. This effect is examined using real data in conjunction with a recently developed model [1, 2] predicting that the shape parameter of K-distributed reverberation is proportional to the range-bearing resolution cell size. Estimation of the shape parameter of the K-distribution from real data as a function of the beamwidth of the towed-array receiver confirms this relationship. Although a similar effect may be expected for changes in the bandwidth of the transmit waveform, real data analysis indicates that, as bandwidth increases, the shape parameter estimate first decreases as expected but then increases, implying the data become more Rayleigh-like at higher bandwidths. An explanation for this counterintuitive effect is proffered wherein it is hypothesized that increasing bandwidth over-resolves scatterers in range but not in angle. After accounting for the size of the scatterers with respect to the size of the range resolution cell in the model of [1, 2], the shape parameter of an equivalent K-distribution for circular scatterers is seen to closely resemble the observed data.

1 Introduction

Heavy-tailed, non-Rayleigh reverberation has been observed in active sonar systems with varying array sizes and transmit waveform bandwidths. A descriptive analysis of the effects of changing bandwidth and center frequency (and thus array beamwidth) was presented in [3]. In this paper, real data from SACLANT Undersea Research Centre's SCARAB 1997 sea-trial (C. Holland, Scientist-in-Charge) are analyzed in conjunction with the model developed in [1, 2]. This model assumes that reverberation in a given range-bearing resolution cell arises from a finite number of scatterers that have an exponentially distributed size. The resulting reverberation envelope is then K-distributed with the shape parameter equal to half the number of scatterers in the range-bearing resolution cell and the scale parameter proportional to the average power of a single scatterer. Assuming that the scatterers are uniformly distributed on or in the seafloor, this implies that the shape parameter of K-distributed reverberation is proportional to the size of the range-bearing resolution cell. The size of the range-bearing resolution cell is approximately the range times the beamwidth divided by the bandwidth, the latter two of which

N.G. Pace and F.B. Jensen (eds.), Impact of Littoral Environmental Variability on Acoustic Predictions and Sonar Performance, 539-546.

are considered in this paper. In Section 3, it is seen that estimates of the K-distribution shape parameter change linearly with beamwidth. However, as seen in Section 4, the shape parameter estimates first decrease as bandwidth increases, but then begin to increase, indicating a trend toward more Rayleigh reverberation. It is hypothesized that this results from over-resolving scatterers. Appropriate modification of the model of $[1, 2]$ reveals a trend in the K-distribution shape parameter similar to that observed in the real data as bandwidth increases.

2 Sea-trial description

The data to be analyzed were taken during the SCARAB '97 sea-trial sponsored by the SACLANT Undersea Research Centre in La Spezia, Italy. The trial occurred off the coast of Italy in June 1997 in the Capraia Basin which is north of the Island of Elba. The data analyzed in this paper were taken June 2–3 using the Centre's low-frequency towed hydrophone-array and the Towed Vertically Directive Source (TVDS), which was configured to send a linear frequency modulated (LFM) pulse with a two second duration spanning 450–700 Hz. The subsequent reverberation was recorded from the towed-array for 40 seconds following each of 49 transmissions throughout the basin. Owing to the shallow water environment and downward refracting sound speed profiles that were measured during the sonar data acquisition, the reverberation data are dominated by bottom scattering.

The towed-array receiver was comprised of 128 hydrophones with a 0.5 meter spacing, resulting in a design frequency of 1500 Hz. The array data were beamformed with a hanning window such that the beampatterns of adjacent beams overlapped at their 3 dB down points at 900 Hz, resulting in 54 beams equally spaced in wavenumber, spanning from forward to aft along the array. The signal processing applied to the beamformed data prior to analysis of the reverberation statistics included basebanding, match filtering and normalization.

2.1 Fit of the Rayleigh and K-distributions

Before evaluating the shape parameter of the K-distribution and how it changes with array beamwidth and transmit waveform bandwidth, it is necessary to determine how well fit the data are by the K-distribution. The Kolmogorov-Smirnov (KS) test $[4]$ is applied to the normalized matched filter data to test the ability of the Rayleigh and K-distribution models to represent the observed data. The KS test evaluates the maximum difference between the sample cumulative distribution function (CDF) generated by the data and a test CDF which is, in this case, either the Rayleigh or K-distribution with their parameters estimated from the data being tested. The Rayleigh distribution only depends on its power, which is estimated by the sample intensity (i.e., the average of the matched filter intensity over the window being tested). As the data have already been normalized to have unit power, this should be near one. Estimation of the K-distribution parameters is more involved; the method of moments estimator (MME), as described in $[5]$, has been employed.

For the data under consideration, windows 1000 samples long (4 seconds of data at a 250 Hz sampling rate) with fifty percent overlap are used to estimate the model parameters and then form the KS test statistic. Using the asymptotic p-value of the KS test statistic [4], the data are either accepted as being well fit by the Rayleigh or K-distribution or rejected. At the $p = 0.05$ level, 68% of the data are well fit by the Rayleigh distribution and 97.4% are well fit by the K-distribution. Thus, the K-distribution is accepted as a good model for these data and it is noted that a significant portion of the data are Rayleigh-like. Note that only data between 2 and 40 seconds following the end of signal transmission on each ping are utilized in this and subsequent data analysis. Over these times the data are reverberation limited except on the few beams where nearby shipping dominated the reverberation or where geological features precluded acoustic propagation (e.g., the Island of Capraia causing shadow zones).

3 Beamwidth effect

Based on the model developed in [1, 2], it may be hypothesized that the effective number of scatterers parameter (α) of the K-distribution is proportional to the area of the range-bearing resolution cell. As such, if the beamwidth of the sonar is doubled, then α should also double, regardless of the density of scatterers as long as they are uniformly placed within the range-bearing resolution cell.

The beamwidth of a sonar approximately doubles when the array size is halved. The same effect may be accomplished by coherently summing adjacent beams. The towed-array data have been spatially filtered into 54 beams that overlap at the 3 dB down points of their beampatterns at 900 Hz. The transmit waveform being analyzed ranged from 450 to 700 Hz, at which frequencies the beampatterns of adjacent beams overlap, respectively, at the 0.75 and 1.8 dB down points. A beampattern having width equal to the combined width of the individual beams may be formed without significant destruction of the mainlobe by coherently summing every other beam at this spacing. This does result in a ripple in the mainlobe with less than a one decibel height at 450 Hz and less than a two decibel height at 700 Hz.

The shape parameter of the K-distribution is estimated from either the original beam data or the summed beam data after matched filtering and normalization. For each beam, the data are separated into windows that are 500 samples long (at a sampling rate of 250 Hz this is 2 seconds of data) with an 80% overlap. For the ith window on the jth beam of the pth ping, call this estimate $\hat{\alpha}_k(p, i, j)$ where the index k indicates the span of how many original beams the beam-sum is formed over (beam-spans of $k = 1, 3, 5, 7$, and 9 are analyzed). The estimates of the shape parameter from a single ping of data are displayed in Fig. 1 in the form of histograms for each of the beam-spans. As expected, the data exhibit a trend toward higher values as the beamwidth increases. On the figure this is evident from the median, mean (which was trimmed by removing the largest and smallest 0.5 percent of the values), and the quantity of estimates that exceed 50. It is also clear that there is a wide range of values observed on this single ping, indicating that the density of scatterers (i.e., frequency of occurrence) varies within the geographic region represented by this ping. To remove this variability, the ratio of the estimated shape parameters on the summed beams to that estimated on the individual beams is formed for each individual data window,

$$\Delta_k(p, i, j, j') = \frac{\hat{\alpha}_k(p, i, j)}{\hat{\alpha}_1(p, i, j')} \tag{1}$$

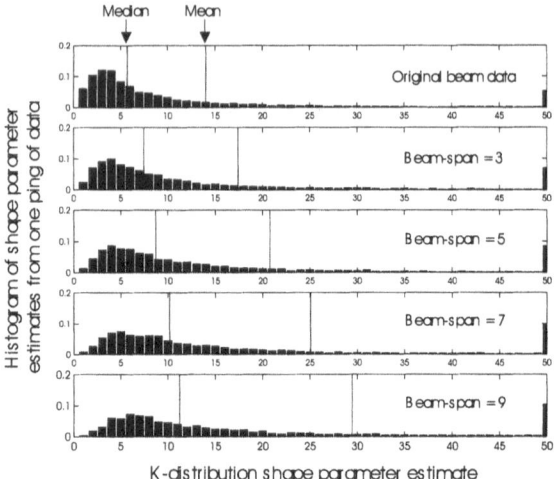

Figure 1. Histograms of the estimates of the K-distribution shape parameter from a single ping for various beamwidths. The expected trend toward higher values of α as beamwidth increases is evident in the mean, median and number of estimates exceeding 50.

where $j' = j, \ldots, j + k - 1$ represent the original beams that span the beamwidth of summed beam j. An average value is formed for each ping and beam-sum,

$$\bar{\Delta}_k(p) = \frac{1}{m} \sum_{i,j,j'} \Delta_k(p, i, j, j') \tag{2}$$

where the summation is over the m cases of the indices (i, j, j') such that the method of moments estimators provide shape parameter estimates for both the individual and summed beams and the estimates satisfy $\hat{\alpha}_k(p, i, j) < 10^4$ and $\hat{\alpha}_1(p, i, j') < 10^4$. Extremely large estimates of α are discarded because of their high variability.

The change in the shape parameter predicted by the model of [1, 2] is exactly the change in beamwidth. This is formed in a similar manner to how the change in the shape parameter is estimated from the data, by taking the average of the change in beamwidth between the summed beams and each of the individual beams for all of the summed beams. Despite the fact that beamwidth varies with arrival angle, the ratios of the beamwidths of the summed to individual beams do not vary significantly. These average ratios are computed for the edges of the frequency band of the transmit waveform (450 and 700 Hz) and also for the peak power frequency of the source (600 Hz) and shown with the average change in the estimated shape parameter of the K-distribution in Fig. 2. In the figure, it is seen that the estimated changes $\bar{\Delta}_k(p)$ fall very close to the predictions and that the average value over all pings has a slope similar to that predicted for the 450 Hz case, but is biased high. This may be explained by first noting that the expected change (1) is formed from the ratio of two random variables, say $\Delta = \frac{X}{Y}$. Assuming that both X and Y are positive random variables, it can be shown (through the use of Jensen's inequality) that $E[\Delta] > \frac{E[X]}{E[Y]}$. This result also requires that the two random variables in the ratio be independent, which is not the case with the real data analyzed. However, for the minimal dependence that is expected, the result should be approximate and does proffer an explanation for the bias seen in the data.

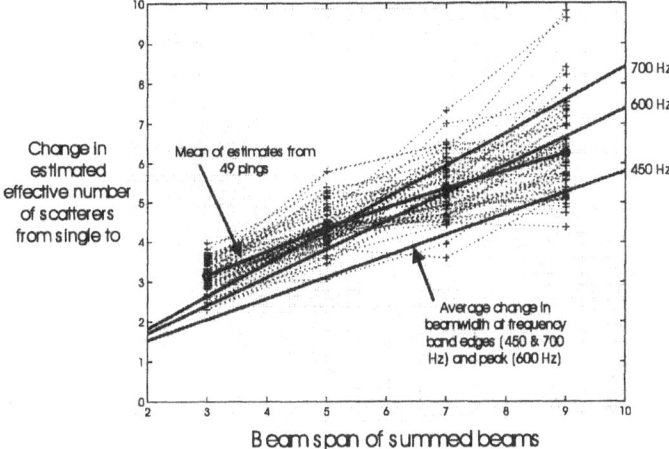

Figure 2. Change in the shape parameter of the K-distribution between summed and individual beams as a function of the span of the beams summed (i.e., beamwidth). The + marks are the average values from each ping. The change as estimated from the data is very similar to that predicted by the change in beamwidth with the exception of an explainable upward bias.

The data shown in Fig. 2 are also seen to increase in variability as the beam-span increases. This may be the result of variability induced by non-stationarity in the frequency of occurrence of scatterers on the bottom over wider angles or may arise from the increased variability of shape parameter estimates when α is large. The latter of which is expected to occur more often for larger beamwidths (i.e., larger range-bearing resolution cell sizes) and when the data are nearly Rayleigh distributed (which is known to occur frequently in this data set).

The close similarity between the estimated and predicted change in the K-distribution shape parameter indicates that, within our ability to estimate it, the shape parameter is proportional to the beamwidth of the towed-array receiver. This also implies that a K-distribution model assuming a finite number of scatterers with the shape parameter tied to the number of scatterers in a sonar range-bearing resolution cell can provide more realistic simulations or predictions of sonar system performance as a function of system parameters such as beamwidth and possibly bandwidth, which will be examined in the next section.

4 Bandwidth effect

The model developed in [1, 2] assumed that the scatterers were fully within the sonar resolution cell. Owing to the asymmetry between the down-range and cross-range extents of the sonar range-bearing resolution cell, this may not always be the case for higher bandwidth transmit waveforms where the scatterers may be over-resolved in range and not in bearing. For example, a sonar system with an array receiver with a 2 degree beamwidth (similar to the array used in the SCARAB sea-trial at broadside) and a 250 Hz bandwidth transmit waveform will have a 3 m down-range extent and a 350 m cross range at 10 km. In this section, the model of [1, 2] is extended to consider the case of over-resolved scatterers. Counterintuitively, the model, and subsequent data analysis, illustrate that when the bandwidth increases to the point of over-resolving scatterers, the reverberation becomes more Rayleigh-like rather than more non-Rayleigh.

The finite-number-of-scatterers model of [1, 2] assumed that the complex envelope of the matched filter output was the sum of n scatterers with random phases,

$$Z = \sum_{i=1}^{n} A_i e^{j\theta_i},\tag{3}$$

where A_i is the contribution from the ith scatterer and θ_i is its phase. If the scatterer amplitudes are not exponentially distributed, the matched filter envelope is not necessarily K-distributed. However, the shape parameter of the K-distribution, as obtained by moment matching, may still be used to represent how similar the PDF is to the Rayleigh distribution (large values yield Rayleigh-like data and small values yield data heavier tailed than Rayleigh). Without significant effort, it can be shown that the shape and scale parameters of the K-distribution, in terms of the moments of the individual scatterer contribution, are

$$\tilde{\alpha} = \frac{n}{2} \left(\frac{4E\left[A_i^2\right]^2}{E\left[A_i^4\right] - 2E\left[A_i^2\right]^2} \right) \quad \text{and} \quad \tilde{\lambda} = E\left[A_i^2\right] \left(\frac{E\left[A_i^4\right]}{2E\left[A_i^2\right]^2} - 1 \right).$$

Evaluation of the moments of A_i for various scatterer shapes and orientations may be computed either analytically or numerically and used to examine how $\tilde{\alpha}$ changes as a function of the down-range extent of the sonar resolution cell (Δ) and the average size of the scatterer (μ). For various scatterer shapes having exponentially distributed area, it was found that $\tilde{\alpha}/\frac{n}{2}$ was unity for large values of $\gamma = \frac{\Delta}{\sqrt{\mu}}$ and increased as γ decreased. This implies that as Δ is decreased (i.e., bandwidth is increased), the effective number of scatterers can increase and the reverberation can become more Rayleigh-like, an unexpected result.

The number of scatterers within a sonar resolution cell (n) depends on the size of the resolution cell and the density and size of the scatterers. Assuming randomly placed scatterers, it can be shown that the average number of scatterers is proportional to γ when γ is large and tapers down to a constant greater than zero when γ is small. Combining this effect with the aforementioned one wherein $\tilde{\alpha}$ increases as γ decreases, we expect that as bandwidth is increased, one might observe $\tilde{\alpha}$ to first decrease when the scatterers are all within the resolution cell and then increase as the scatterers become over-resolved. This trend is observed in the histograms of the K-distribution shape parameter from one ping of data as shown in Fig. 3 where the shape parameter decreases when bandwidth is raised from 0.5 Hz to about 8 Hz, at which point it begins to increase again and the data become more Rayleigh-like.

For each analysis window, the K-distribution shape parameter estimates (if they exist) are normalized by the minimum estimated value over all bandwidths processed and then averaged over all beams for ranges less than 10 km. These are displayed for each of the 49 pings of data in Fig. 4 along with the average value over all pings and that predicted by the over-resolved scatterer model for a circular scatterer. This latter is normalized so that its lowest value is one, which is what was done to the real data, and scaled on its abscissa so that the minimum aligns with the minimum of a least squared error fitting of a quadratic function to the average of the estimated changes in the shape parameter. From this figure it is clear that the data are initially in a stage where increasing bandwidth

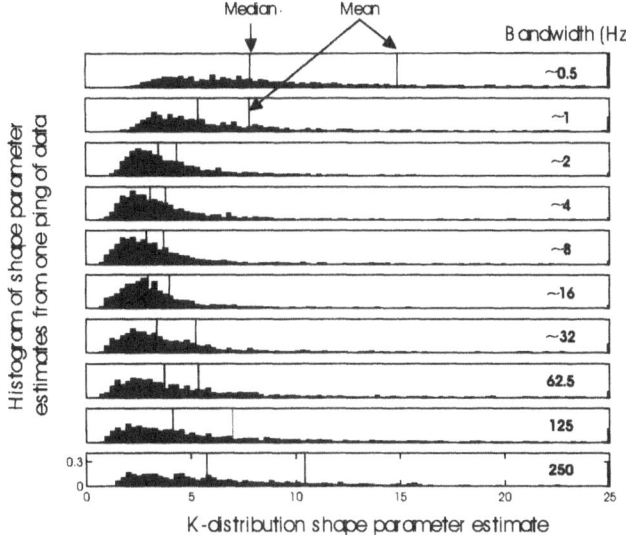

Figure 3. Histograms of the estimates of the K-distribution shape parameter from a single ping for various bandwidths. The expected trend toward lower values of α as bandwidth increases is evident at the lower bandwidths, but does not continue at the higher bandwidths.

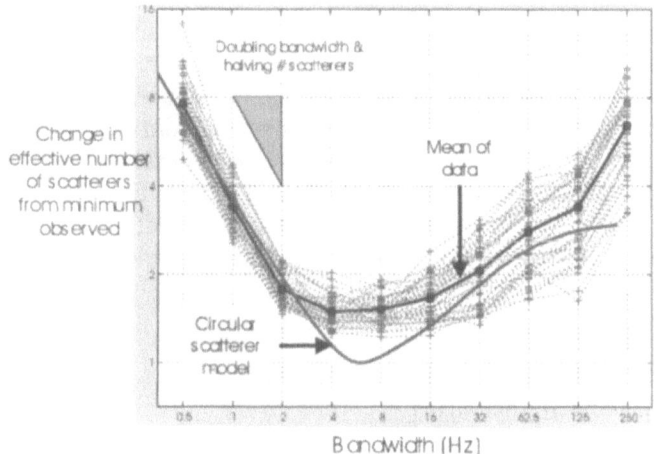

Figure 4. Change in the shape parameter of the K-distribution from the minimum observed over various bandwidths. The + marks are the average values from each ping, the 'o' marks are the averages over all pings for the given bandwidth.

results in a proportionate decrease in the K-distribution shape parameter. However, the shape parameter estimates begin to increase when the bandwidths reach about 8 IIz. The effective number of scatterers ($\tilde{\alpha}$) predicted from a model assuming that circular scatterers are over-resolved well approximates the observed data, despite the obvious oversimplifications of its development. It should be noted that the results presented here are circumstantial and not evidentiary proof that the phenomenon observed in the data arises from over-resolution of scatterers. It is possible that this bandwidth effect is caused by other conditions such as propagation through a shallow water environment, although preliminary investigation does not lend much support to this alternative hypothesis.

Increased bandwidth is expected to result in heavier-tailed reverberation than Rayleigh, not more Rayleigh-like reverberation. Thus, the phenomenon of a return to Rayleigh-like reverberation at high bandwidth that is observed in both the data analysis and model is in some sense counterintuitive. Mathematically, this may arise from an increased rate of convergence of the central limit theorem that results from a reduction in the variability of the individual components. This result might initially lead one to suspect that the Rayleigh-like reverberation at higher bandwidth will result in improved detection performance. However, if over-resolution of the scatterers is the underlying mechanism producing the phenomenon, then it may be expected to induce dependence of the data samples (not necessarily correlation) which will reduce the effectiveness of a detector that incoherently combines (e.g., integrates) the matched filter intensity over multiple ranges to account for spreading of the target echo. Thus, a net gain in detection performance is not guaranteed.

5 Conclusion

In this paper, the statistics of the reverberation induced matched filter envelope were evaluated as a function of the sonar's receive array beamwidth and transmit waveform bandwidth. The statistics were quantified by the shape parameter of the K-distribution, which, when the scatterers are small compared with the sonar's resolution, is proportional to array beamwidth and inversely proportional to the transmit waveform bandwidth. Real data analysis supported both of these results except that at higher bandwidths, the shape parameter was seen to increase indicating a return toward Rayleigh-like reverberation. Accounting for the size of the scatterers in the model produced a similar trend which was seen to well represent the observed data. These results may be used to improve the accuracy of performance prediction and simulation of sonar systems in cluttered environments and possibly to help set optimal sonar bandwidths.

Acknowledgements

This work was sponsored by the Office of Naval Research under grant numbers N00014-02-1-0115 and N00014-01-1-0352.

References

1. Abraham, D.A. and Lyons, A.P., Exponential scattering and K-distributed reverberation. In *Proc. MTS/IEEE Oceans 2001*, 1622–1628 (2001).
2. Abraham, D.A. and Lyons, A.P., Novel physical interpretations of K-distributed reverberation, *IEEE J. Oceanic Eng.* (in review 2002).
3. Abraham, D.A. and Holland, C.W., Statistical analysis of low-frequency active sonar reverberation in shallow water. In *Proc. 4th European Conference on Underwater Acoustics* (1998).
4. Fisz, M., *Probability Theory and Mathematical Statistics*, 3rd ed. (John Wiley & Sons, 1963).
5. Abraham, D.A., Modeling non-Rayleigh reverberation. Rep. SR-266, SACLANT Undersea Research Centre, La Spezia, Italy (1997).

THE ROLE OF NOWCAST AND FORECAST INPUT PARAMETERS FOR RANGE DEPENDENT TRANSMISSION MODELS

JANICE S. SENDT

Thales Underwater Systems Pty, 274 Victoria Road, Rydalmere NSW 2116, Australia
E-mail: Janice.Sendt@au.thalesgroup.com

ADRIAN D. JONES AND JARRAD R. EXELBY

Defence Science and Technology Organisation, P.O.Box 1500, Edinburgh SA 5111, Australia
E-mail: adrian.jones@dsto.defence.gov.au

The Maritime Operations Division of DSTO is assisting the Royal Australian Navy in its assessment of a sonar performance prediction tool for range dependent ocean environments: TESS 2, prepared by Thales Underwater Systems. This assessment has included comparisons between acoustic transmission loss data measured by MOD at shallow ocean sites with range-dependent transmission predictions obtained by TUS. Part of this task has been the inference of the appropriate input parameters from the measured data and comparison with historic data sets. This applies to bathymetry and to seafloor acoustic properties. Examples detailed in this paper indicate the variability in properties which must be described at differing regional locations and the difficulty that can occur at arriving at suitable input parameters from historical data alone. These examples include reference to high resolution bathymetry (ocean depth) data, and the inference of seafloor reflectivity based on received signal data.

1   Introduction

This paper addresses some aspects of a joint DSTO/TUS Pty task for benchmarking the range dependent acoustic transmission loss models used in the TESS 2 software. The TESS 2 software provides performance prediction for the Royal Australian Navy (RAN) platform sonar sensor systems. The part of this assessment which is addressed in this paper has been the comparison of results from TESS 2 with measured results from a considerable number of shallow water sites around Australia which had been collected and analysed by DSTO. Also discussed briefly is an assessment of the potential for an MOD *in-situ* technique to infer seafloor reflectivity at shallow grazing angles and provide input to TESS 2 for regions for which existing holdings of seafloor properties are sparse. The TESS 2 software, in particular, the underwater component called SAGE [1], was developed by TUS Pty. SAGE allows the user to compute sonar performances in detection range for realistic ocean environments located at a user defined latitude and longitude. It achieves its purpose by accessing appropriate internal global databases and supplying the necessary parameters to run range dependent sonar performance models.

N.G. Pace and F.B. Jensen (eds.), Impact of Littoral Environmental Variability on Acoustic Predictions and Sonar Performance, 547-554.

The databases include bathymetry, wave height, wind speed, sound speed, sediment thickness and a global sediment province database (TUS proprietary).

This paper addresses the approach taken in analyzing the input data from three different sites. The data available from the measured results included start-of-track and end-of-track core samples, a number of sound speed profiles along the track and echo soundings at approximately every 1 km for most tracks.

The data available from the SAGE databases included gridded 30" bathymetry and globally gridded 2' sediment province information. The latter provides sediment information along the whole length of the track whilst the core data only provides information at two points on the track. Sound speed versus depth profiles determined from site bathythermograph data were used in preference to the available historic data.

Additionally, a review of the geological work for the area was undertaken to assist in understanding the biogenic and fine detail of the area.

At this point in time, this input data has been used only in the two SAGE transmission loss models, namely RAM [2] which is a parabolic equation (PE) model used for frequencies below about 400 Hz and RAVE which is a ray model (TUS proprietary) used for higher frequencies. The examples shown in this paper, which refer to the use of SAGE data within a TESS 2 framework, are from the RAM model.

2   Large bathymetry variations

At one site, a comparison of the bathymetry data showed an anomalous point where the DSTO echo sounder data recorded a 30 m depth discrepancy (at 8 km along the track) to the gridded data. The DSTO data around this point was more sparse than along the rest of the track, with the previous point 6 km distant and next point 2 km distant from the point in question (see Fig. 1). Agreement along the rest of the track was good. Thus the echo sounder dataset was showing a gully possibly 8 km wide. The question arose as to whether this point actually existed or was a misreading.

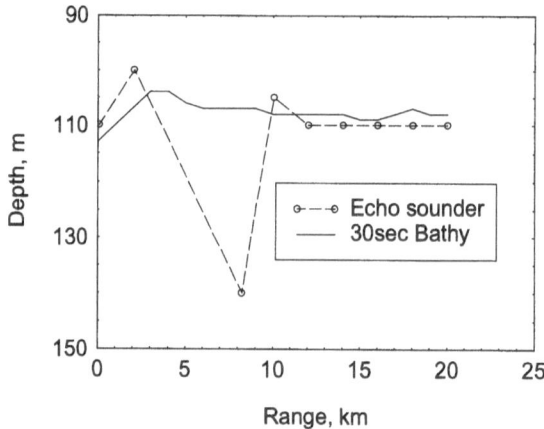

Figure 1. Bathymetry data at first site – echo sounder versus 30 second database.

In turn, a doubt was raised as to whether the gridded data was in error. It was also noted that the impact of this point on transmission loss was large (see Fig. 2). A review of other data for this site showed that the track was crossing a relict river bed and that the echo sounder point was indeed correct. However the agreement between the measured and calculated results showed that the gridded bathymetry data gave the best match. The transmission loss values calculated with the echo sounder data gave a discrepancy of up to 10 dB at the ranges of the bathymetry discrepancy. The transmission loss discrepancy continued at further ranges by producing an "out of phase" transmission loss pattern.

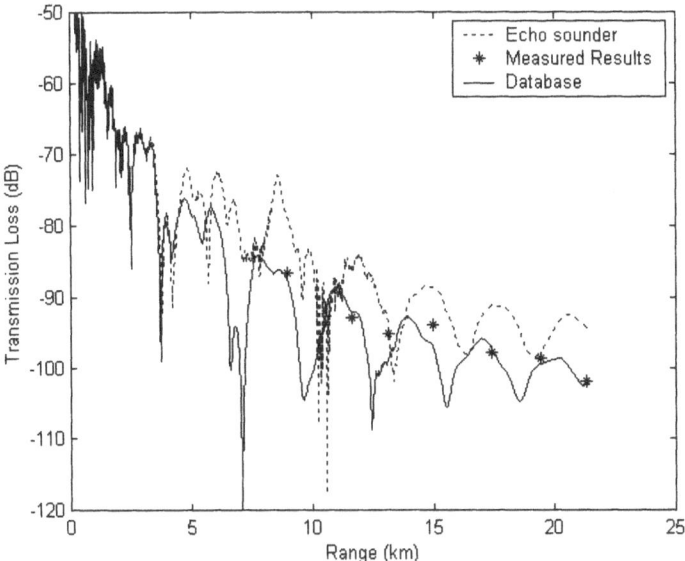

Figure 2. Comparison of measured transmission loss and values calculated using either gridded bathymetry or echo sounder bathymetry with 8 km wide, 30 m deep gully centred 8 km from source. Transmission data at 100 Hz for source and receiver at 18 m depth.

As the river channel does not appear in the gridded data set, its width may be presumed less than 1 km. Accordingly, the width of the gully was reduced to 1 km and the transmission loss re-calculated. Figure 3 shows that the transmission loss results for the revised echo sounder data now show better agreement with the measured results. Reducing the width further would allow the results to converge on the values obtained with the gridded data set. If the measured dataset was larger, it would then be possible to use the transmission loss model to infer the correct bathymetry profile, particularly as there is only a one point discrepancy.

The existence of relict river beds is not an uncommon occurrence on the continental shelf region. The results of this study emphasise the need for point bathymetry data to be used in acoustic transmission loss modelling as well as gridded data. This data should be provided as a "bedform" database and include the location, width and depth of relict river beds.

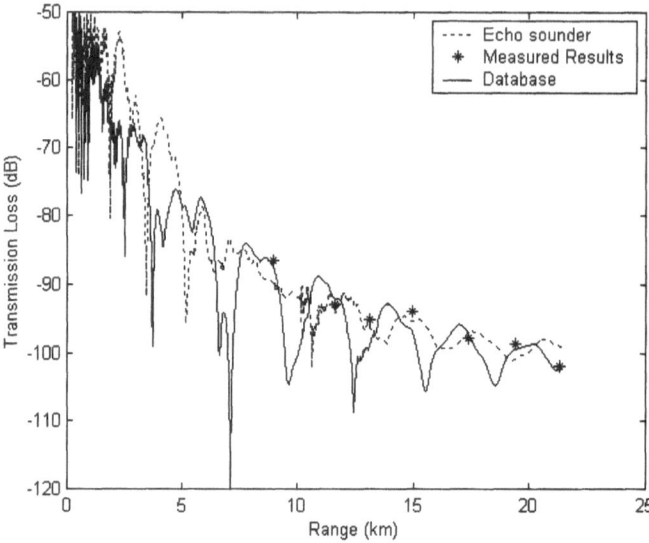

Figure 3. Comparison of measured transmission loss and values calculated using either gridded bathymetry or modified echo sounder bathymetry with 1 km wide, 30 m deep gully centred 8 km from source. Transmission data at 100 Hz for source and receiver at 18 m depth.

3 Biogenic effects

One of the difficulties in predicting transmission loss with reliability in Australian northern waters is the biogenic impact. A grab sample may be analyzed into gravel/sand/mud, based on mean grain size, but when this analysis also includes a calcium content of 90% then the description is insufficient to discriminate between oolitic type grain and the casts of marine mammals. If a porosity measurement has also been taken then this will allow discrimination between the two to be quantified, and if this is not available, then a marine geological description of the area can allow an estimate to be made. Data obtained for a second site illustrates the difficulties which may be encountered.

At the second site, the following data had been recorded across a 20 km track:
1. A surficial sediment grab sample at the start and end of the track.
2. 11 sound velocity profiles along the track.
3. Echo sounder depth data at approximately 300 m intervals (Fig. 4).
4. Transmission loss measurements at known ranges (Fig. 5).

Additional data which was available from the SAGE database:
5. The sediment province data for the track.
6. The calcium carbonate % for the sediment.
7. 30 arc second gridded bathymetry.
8. Sediment thickness data.

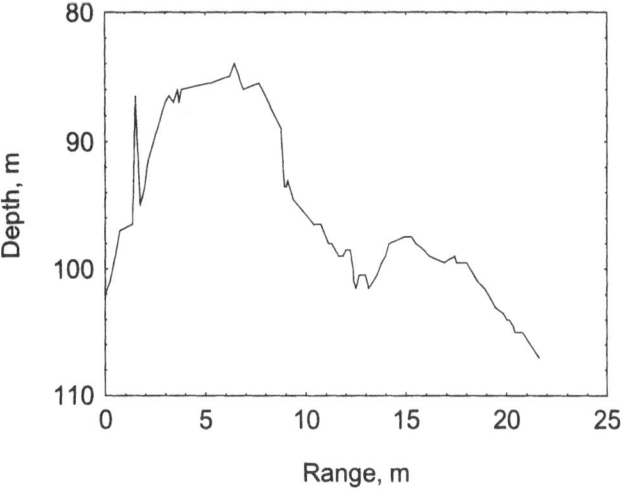

Figure 4. Echo sounder bathymetry along the 20 km track, second site.

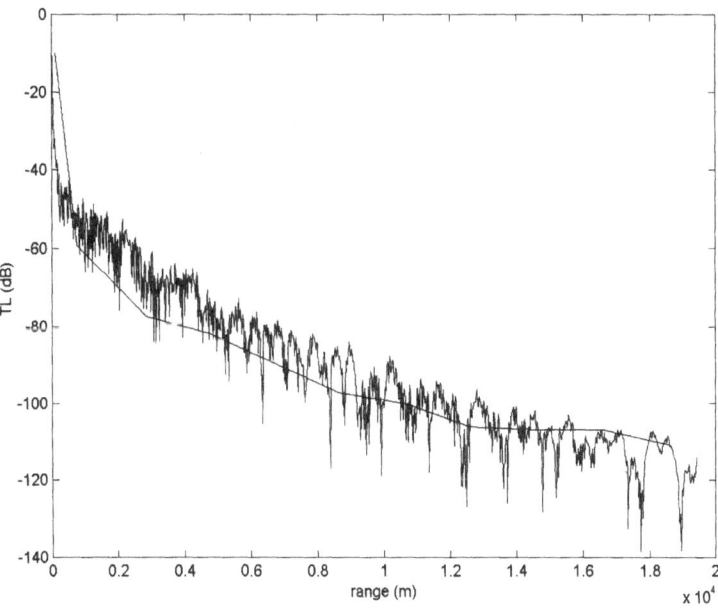

Figure 5. Comparison of predicted and measured transmission loss at 100 Hz for source and receiver at 18 m depth.

The marine geological description for the area stated that there were irregularly spaced sand waves. The high carbonate content of the sediment is due to the pellets of eroded limestones and the relic skeletal debris.

The core data describes the sediment as 95% sand and 95% calcite across the entire track. If the core descriptors alone are used then the seabed would be reflective and the transmission loss would decrease slowly with range. As this was not the case, in fact, at the beginning of the track the transmission as measured is quite lossy (one-third octave data is presented in Fig. 5), then the porosity of the sediment must be higher than would be expected assuming a "standard sand". It may be then be speculated that the calcium carbonate content is due to the skeletal debris. It is also noted that for the deeper bathymetry, further along the track, the seabed becomes more reflective indicative of a reduction in the skeletal debris content of the sediment and an increase in the pellets.

As the transmission loss values at the start of the track indicate that the seabed is acting as an absorbing layer, it is essential to get an estimate on the thickness of the sediment to the acoustic basement as reflections from this layer may impact on the transmission loss results. Shallow water seismic data is available for this area and estimates from the two way travel time based on the speed of sound for the surficial sediment give a sediment depth of 70 m [4]. This value is used as the sediment thickness in the transmission calculation for the data presented in Fig. 5.

4   *In-situ* determination of seafloor reflectivity

In support of optimum sonar performance prediction capability, MOD has on-going programmes of research on rapid environmental sensing techniques, with an emphasis on the determination of effective seafloor parameters. In particular, MOD has developed a unique method for *in-situ* determination of seafloor specular reflectivity [5]. The potential value of this technique as an adjunct to the TESS 2 system is under present investigation. Progress in this work is illustrated below.

To assess the potential for the MOD *in-situ* determination of seafloor reflectivity, three-way comparisons have been carried out between: (i) measured transmission loss data determined by MOD at specified ranges; (ii) predictions of transmission loss obtained by the TESS 2 system; (iii) predictions of transmission loss obtained using inverted seafloor reflectivity as input to the KRAKENC model [6]. The measured data shown in this paper are obtained at a site near the shelf break, for which bathymetry along two tracks is indicated by Fig. 6. Here, Run 5 was along a track which proceeded down the continental slope, whereas Run 6 was an intersecting track which followed the bathymetry contours, thereby retaining near uniform depth. The acoustic data capture was afforded by deployment of sonobuoys and Mk 61 SUS from a P-3C maritime patrol aircraft of the Royal Australian Air Force. For each track, the signal received from a single SUS was input to the MOD algorithm and seafloor reflectivity determined for shallow grazing angles.

The comparisons of transmission loss are shown in Figs. 7 and 8 respectively for Run 6 (KRAKENC run range-independent) and Run 5 (KRAKENC run range-dependent). Here, the measured data was processed in one-third octave bands, the KRAKENC data was obtained by coherent processing at frequencies within each one-third octave and then incoherently averaged, and the TESS 2 (RAM) prediction was obtained coherently at a single frequency (250 Hz).

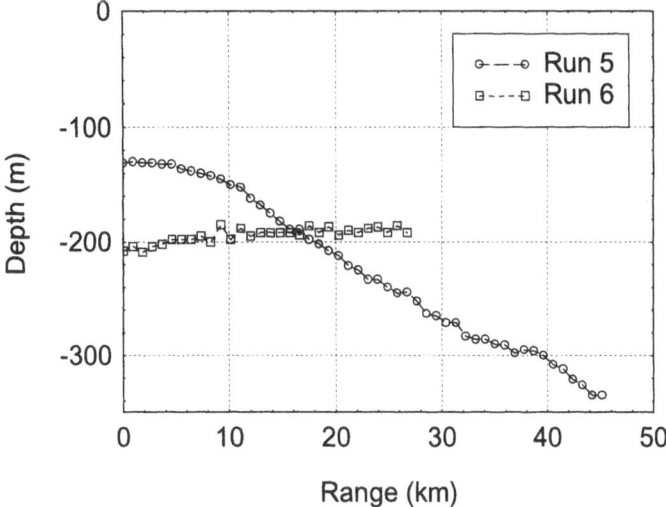

Figure 6. Bathymetry along Run 5 and Run 6 at 3rd site.

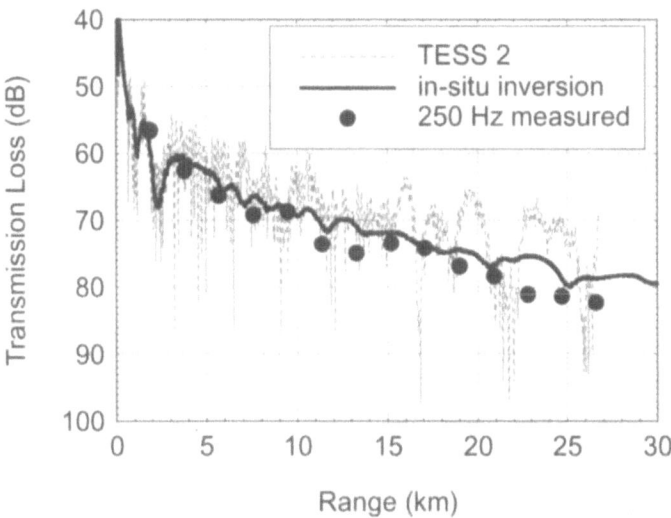

Figure 7. TL measured & predicted, Site 3 Run 6 – range-independent, 250 Hz.

As shown by the data in each of Figs. 7 and 8, the agreement between the TESS 2-predicted and measured transmission loss values is very good for this site. This is presumed due to the fact that considerable seafloor data exists in the historical record from which the SAGE data was derived. Also at each site, the transmission data based on the *in-situ* seabed reflectivity is close to the measured data.

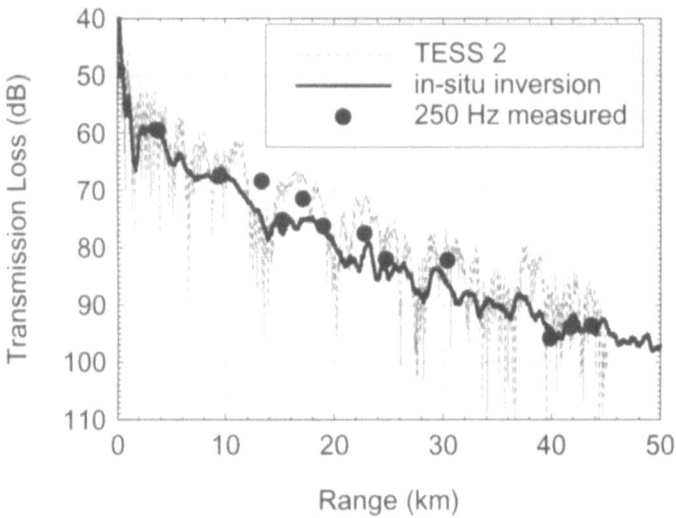

Figure 8. TL measured & predicted, Site 3 Run 5 – range-dependent, 250 Hz.

5 Conclusions

Based on the data presented in this paper, it does appear that operational models for the prediction of sonar system performance may be best employed in a way in which collated or gridded databases of historical information are supplemented, judiciously, by the input of additional quality data for the local region. In shallow oceans, the spatial variability in ocean depth and in seafloor properties requires range-dependent modelling which is supplemented by local sampling when feasible.

Acknowledgements

The authors thank Mr. Paul Clarke for his assistance in the preparation of this paper.

References

1. TESS 2 operator's manual, TUS (2000).
2. Collins, M.D., User's guide for RAM versions 1.0 and 1.0p, available at:
   ftp@ram.nrl.navy.mil
3. Jones, H.A., Marine geology of the Northwest Australian Continental Shelf, Bulletin 136, Bureau of Mineral Resources (1973).
4. Gravity, magnetic and seismic profiles of the Timor Sea, Australian Geological Survey Organisation.
5. Jones, A.D., Bartel, D.W., Clarke, P.A. and Day, G.J., Acoustic inversion for seafloor reflectivity in shallow water environment. In *Proc. UDT Pacific 2000*, Australia, 7– 9 February 2000.
6. Porter, M.B., The KRAKEN normal mode program. Rep. SM-245, SACLANT Undersea Research Centre (1995).

ARE CURRENT ENVIRONMENTAL DATABASES ADEQUATE FOR SONAR PREDICTIONS IN SHALLOW WATER?

CARLO M. FERLA AND FINN B. JENSEN

SACLANT Undersea Research Centre, Viale San Bartolomeo 400, 19138 La Spezia, Italy
E-mail: ferla@saclantc.nato.int, jensen@saclantc.nato.int

The usefulness of environmental databases (bathymetry, sound-speed profile and bottom reflectivity) for sonar performance predictions in high-variability littoral waters has often been questioned. Thus it is conceivable that spatial and temporal averaging of sparsely sampled data could result in data holdings which do not capture some of the acoustically important environmental features required for accurate sonar predictions. To address this issue on a larger geographical scale, SACLANTCEN has undertaken a study using the Allied Environmental Support System (AESS) as the prediction tool and the NATO Standard Oceanographic Data Base (NSODB) as the environmental representation. To quantify prediction errors in selected shallow-water areas as a function of bottom type, water depth, season, frequency, sonar/target depth, etc., SACLANT-CEN's vast broadband transmission-loss database established over the past 30 years will be used as ground truth. Initial results from the Mediterranean and the Norwegian Sea indicate that databank-based performance predictions in shallow water are indeed unreliable and that the weakest link is the bottom-loss information.

1 Introduction

Oceanographic and geophysical data collection programs have been operating for many years with the scope of establishing reliable data sources for sonar predictions on a global scale. Of course, the quality and the temporal and spatial coverage of the data vary significantly from area to area, with the Mediterranean maybe being one of the best mapped seas of the world. In using a gridded database, one has generally no information about the number of original data points and their spatial distribution nor about the measurement accuracy (some data values are derived values based on measuments with different types of instruments). However, despite the uneven data coverage and quality, when no *in situ* measurements are available, environmental database (bathymetry, sound-speed profiles, bottom reflectivity) are routinely used to forecast the sonar range of the day for operating navies. Of course, sonar operators know perfectly well that these databank-based range predictions are not always accurate.

Historically, much of the validation work associated with performance predictions has been carried out in deep water, which was the main operating theater during the Cold War. In deep water the sound speed structure is very stable below a few hundred meters, both temporally and spatially. Also the upper ocean shows less variability than in coastal areas. Moreover, important acoustic paths do not interact with the bottom in deep oceans,

N.G. Pace and F.B. Jensen (eds.), Impact of Littoral Environmental Variability on Acoustic Predictions and Sonar Performance, 555-562.
© 2002 *Kluwer Academic Publishers.*

and, hence, neither the bathymetry nor the bottom reflectivity are critical parameters for sonar performance predictions. The result is that reliable sonar range predictions can indeed be performed in deep water with current databases.

In the past decade the navy operational interest has shifted heavily towards littoral waters, and here both spatial and temporal variability is a limiting factor for database usage. Moreover, important acoustic paths in shallow water all interact with the bottom and, hence, both bathymetry and bottom reflectivity become critical parameters for sonar predictions. SACLANTCEN has pioneered the study of shallow-water acoustics, both experimentally and theoretically, and some of the key issues related to accurate transmission-loss (TL) predictions were reported at conferences dating back to the early 1980s [1, 2].

That current "deep water" databases are inadequate for performance predictions in shallow water is generally accepted, but there has been no attempt to quantify prediction errors on an area-by-area basis and as a function of bottom type, water depth, season, frequency, sonar/target depth, etc. SACLANTCEN has developed a strategy for doing exactly this, which involves using the Centre's vast broadband transmission-loss database established over the past 30 years as ground truth, to which AESS/NSODB predictions will be compared. Detailed analysis will be performed with high-fidelity models from the Centre's model library [3]. Initial results are presented from areas of the Mediterranean and the Norwegian Sea.

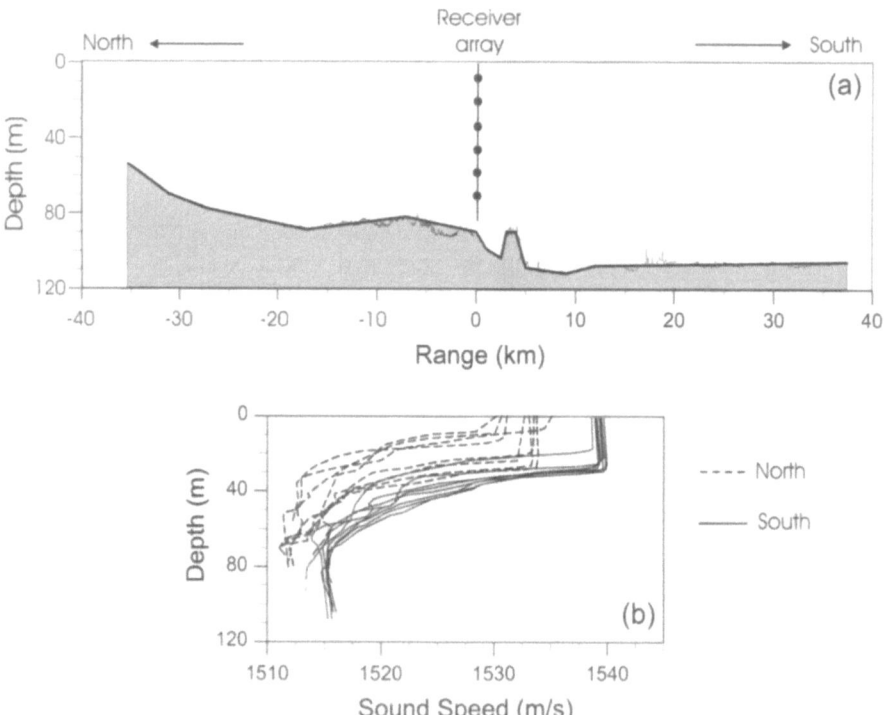

Figure 1. Environment for Strait of Sicily experiment.

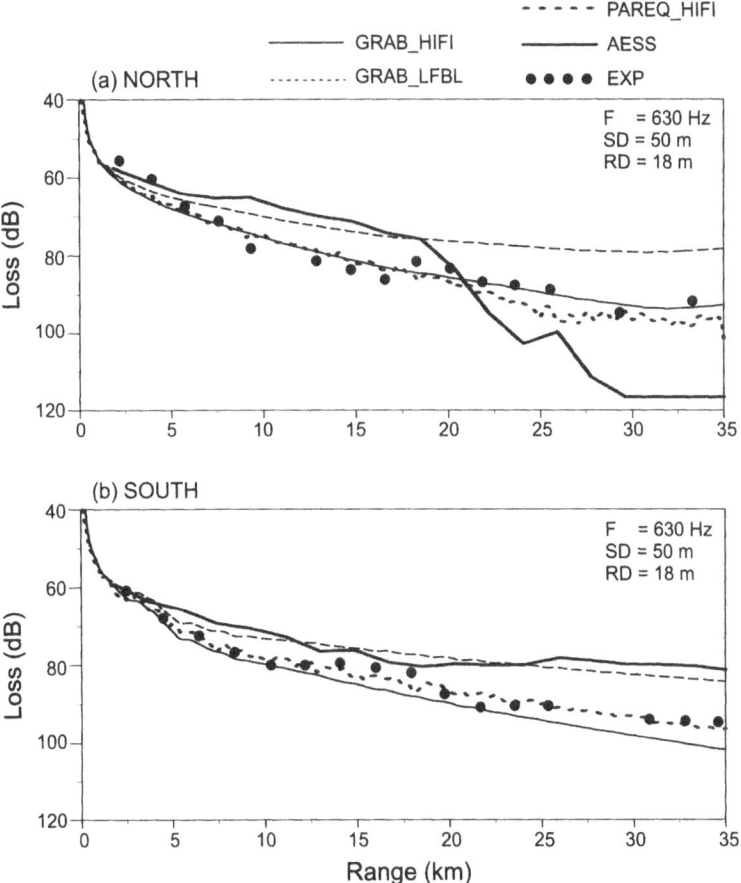

Figure 2. Model-data comparisons at 630 Hz for (a) the north-going track and (b) the south-going track.

2 Strait of Sicily

This data set was collected in September 1996 as part of a major oceanographic/acoustic survey of the Malta Plateau. The measured environmental conditions are given in Fig. 1, indicating that the experiment was carried out with a receiver array suspended in the water column, and with a source ship dropping SUS charges along two 35-km tracks, one to the north into shallower water, and one to the south in almost constant water depth of 100 m. Both the measured bathymetry and the range-smoothed version used as input to the acoustic models are shown in Fig. 1(a). The recorded sound-speed profiles along the tracks during the experiments are shown in Fig. 1(b). Note that the water is warmer and more stable to the south. The highest variability is observed when moving into shallower water on the north-going track. In the acoustic inversion to determine average geoacoustic properties for each track, a single representative profile was selected from each group of profiles shown in Fig. 1(b).

Figure 2 shows model-data comparisons for a frequency of 630 Hz and for a source at 50 m and a receiver at 18 m. The upper graph is for the northern track into shallower water,

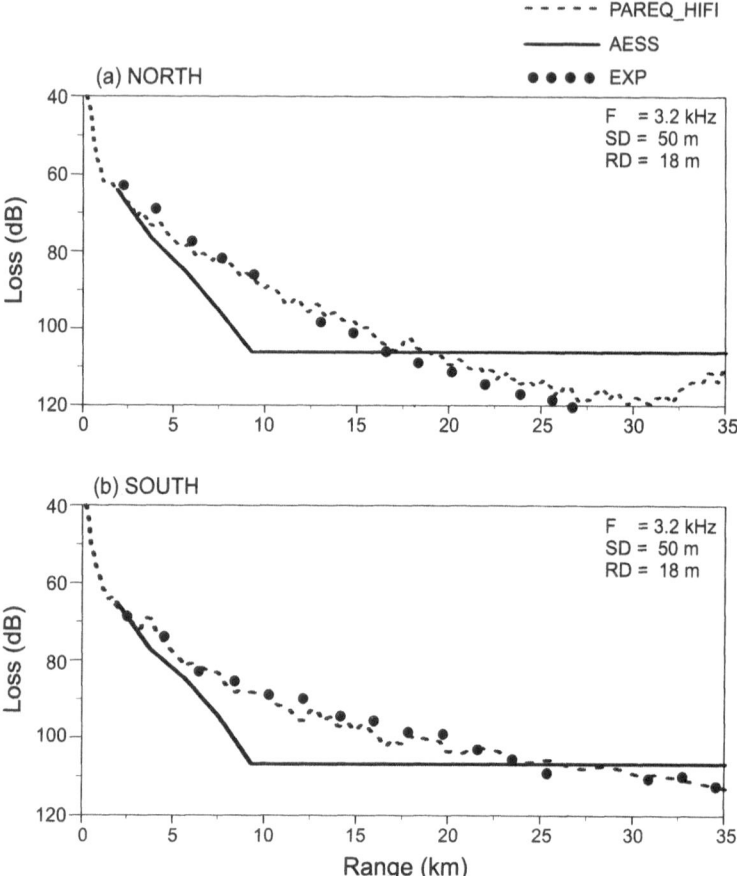

Figure 3. Model-data comparisons at 3.2 kHz for (a) the north-going track and (b) the south-going track.

whereas the lower graph is for the track to the south. The first thing to note in both graphs is the excellent agreement between the high-fidelity model results (PAREQ_HIFI) [4] and the data, which, in turn, provides a measure of confidence in the quality of the data. The kind of agreement seen here was obtained for all hydrophone depths and for frequencies between 100 and 3200 Hz (see Fig. 3 for results at 3.2 kHz). The geoacoustic models derived from the inversions were similar on the two tracks: a 3-m soft top layer with a 1.5% lower sound speed than in the water column near the bottom ($c \approx 1482$ m/s), a density of 1.5 g/cm^3 and an attenuation of 0.1 dB/λ to the north and 0.15 dB/λ to the south. The subbottom was found to have the following properties: $c = 1650$ m/s, $\rho = 1.9$ g/cm^3, $\alpha = 0.5$ dB/λ.

The next step was to obtain AESS predictions based solely on database information. Hence the exact track coordinates were provided and a TL prediction obtained from the ASTRAL model, which has been determined to be the most reliable among the various acoustic models available to the AESS user. The AESS predictions (red curves) in Fig. 2 generally provide too little loss and hence too long sonar ranges. The rapid fall-off beyond 20 km in the upper graph is due to wrong bathymetry values in the database for the shallow end of the north-going track.

To determine which database input from NSODB is primarily causing the prediction error seen here, we designed a control case with the GRAB model [5], which can run inputs directly from the NSODB. First GRAB was run with the same high-fidelity data as PAREQ, and Fig. 2 shows that we obtain consistent answers in good agreement with the data. Next we take just the seasonal mean sound-speed profile from the NSODB, which only changes the GRAB prediction slightly. Similarly, if we use the bathymetry information form the NSODB, we get only slight changes, except beyond 20 km on the northern track, where the database values are much too shallow (< 20 m). Finally, if GRAB is run with the bottom-loss information retrieved from NSODB we obtain the results given by the dashed curves in Figs. 2(a) and (b) (GRAB_LFBL). Here LFBL refers to the low-frequency bottom-loss tables to be applied below 1 kHz. It is clear that the bottom-loss model is responsible for the optimistic prediction ranges obtained with the AESS.

Moving now to the 3.2-kHz results in Fig. 3, we note the excellent agreement between data and the high-fidelity model predictions (PAREQ_HIFI), which are based on the exact same geoacoustic model used for the low-frequency case. The AESS prediction using the high-frequency bottom-loss tables in NSODB provides too high losses and hence too short sonar ranges. The red curve is clearly truncated at a maximum loss of around 105 dB. By running GRAB in a control mode, it is easily shown that the higher losses predicted by the AESS is caused by the bottom-loss model used. Again there is little effect of using database information for sound-speed profile and bathymetry.

In terms of performance predictions in this particular area of the Mediterranean, it is clear that the bottom-loss information is the weakest point of the database. Thus, there is too little bottom loss at low frequencies and too much bottom loss at high frequencies. This, in turn, means that the sonar range predictions are discontinuous around 1 kHz. In practice, the transition is smoothed over a 500 Hz band, but we could still see level differences of tens of decibels by changing the frequency from 1.0 to 1.5 kHz. A single geoacoustic model as used in the HiFi modeling avoids such artifacts.

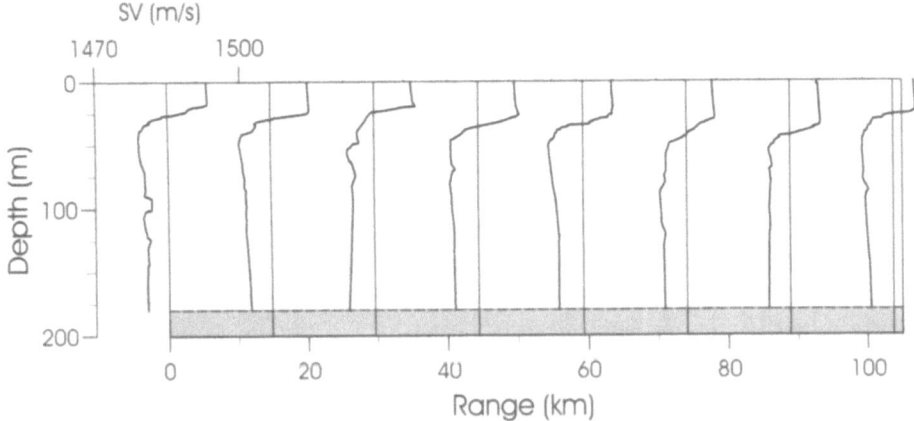

Figure 4. Environment for Norwegian Shelf experiment.

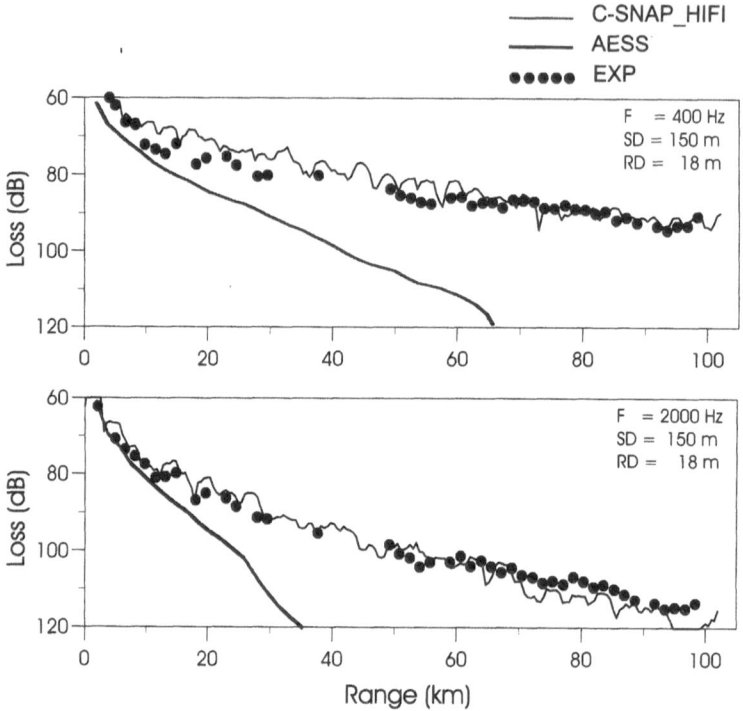

Figure 5. Model-data comparison at 400 and 2000 Hz for a deep source and shallow receiver.

3 Norwegian Shelf

The second data set was collected along the Norwegian west coast in September 1993, again using SUS charges and a vertical hydrophone array for reception. Figure 4 shows a cross-section of the acoustic track which runs parallel to the shelf break in 180 meters of water. There are 8 measured sound-speed profiles with an average spacing of 15 km. There is some variability in the oceanographic conditions along the track, but there is always a well-defined mixed layer (20–30 m deep) followed by a sharp thermocline. All eight profiles were used in the geoacoustic inversion, whereas the water depth was considered constant along the entire track.

Figure 5 shows model-data comparisons at 400 and 2000 Hz for a source at 150 m and a receiver at 18 m. Note the excellent agreement between the high-fidelity model results (C-SNAP_HIFI) [6] and the experimental data. This kind of agreement was observed for all receiver depths and for frequencies between 50 and 2000 Hz, which lends credence to both the quality of the data and the geoacoustic model. A simple homogeneous bottom with $c = 1670$ m/s, $\rho = 2.0$ g/cm^3 and $\alpha = 0.5$ dB/λ was found to adequately represent bottom reflectivity for the entire frequency band.

To assess the quality of AESS-based sonar predictions for this area, we provided track coordinates, source/receiver geometry, frequency and season (September) to the AESS system and requested a TL prediction with the ASTRAL model. The result is shown in Fig. 5 (red curves) and there is clearly too much bottom loss at both frequencies. We have yet to run the control cases with the GRAB model to determine whether the profile

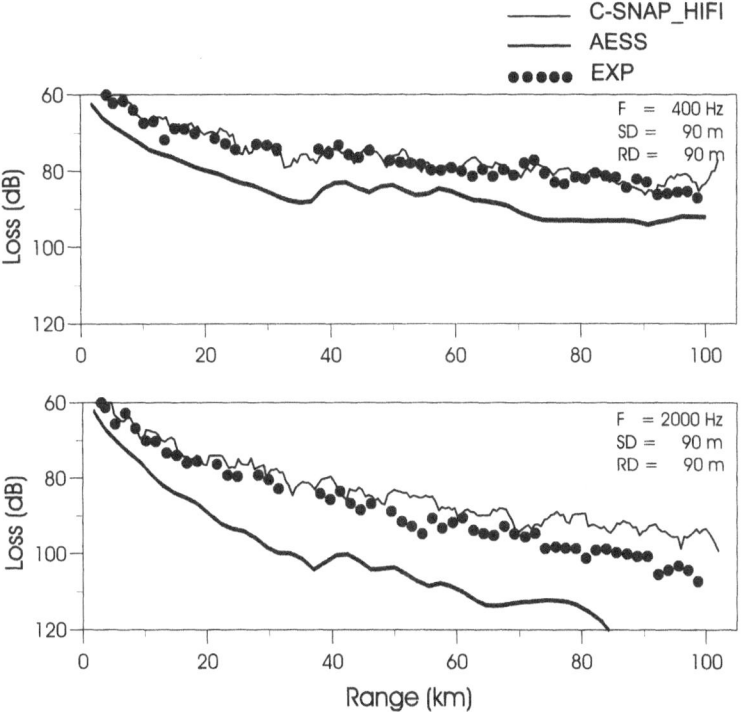

Figure 6. Model-data comparison at 400 and 2000 Hz for a mid-water source and receiver.

and bathymetry information in the NSODB is adequate.

Turning to the case of a mid-water source and receiver (Fig. 6) there is an indication that it is not just the bottom-loss information that is inaccurate in the NSODB. Thus for source and receiver both at 90 m, we should expect excellent propagation conditions with sound being channeled to long distances with little bottom interaction. The AESS result at 400 Hz has the correct shape, but the red curve is displaced down by 10 dB compared to the data. This type of problem is most likely associated with the use of an incorrect sound-speed profile. These issues will all be investigated by running control cases with GRAB for each of the NSODB inputs, i.e. sound-speed profile, bathymetry and bottom reflectivity.

In summarizing the results in Figs. 5 and 6 for the Norwegian Shelf, we note that the AESS always predicts too much loss and hence too short sonar ranges. This behavior is quite different from the Strait of Sicily predictions in Sect. 2, where the low-frequency results showed too little loss, and this despite the fact that the two geoacoustic environments were found to be very similar. Clearly, the effects of inaccurate data in the NSODB on sonar performance predictions can have many different manifestations, and many data sets must be analysed in order to create a statistically significant decision basis for determining the geographical areas and operational situations for which current databases are inadequate for sonar performance predictions.

4   Conclusions

In trying to address the question whether current environmental databases (NSODB) are adequate for sonar performance predictions in shallow water, we have looked at two different geographical areas (the Strait of Sicily and the Norwegian Shelf) and compared three different views of the same acoustic picture:

1. High-quality broadband transmission-loss data representing *ground truth*.

2. High-fidelity TL predictions based on *in situ* oceanographic inputs and inverted geoacoustic information.

3. AESS generated TL predictions based on environmental information from the NSODB.

The picture that emerges from this comparison is rather complex, but the AESS prediction is generally deemed unsatisfactory, due mainly to inaccurate environmental inputs from the NSODB. The bottom-loss information is found to be the weakest link, but also bathymetry and profile information can have adverse effects on the prediction accuracy. More areas and more acoustic track need to be analysed before a general statement can be made about the usefulness of the NSODB for performance predictions in littoral waters.

References

1. Kuperman, W.A. and Jensen, F.B. (eds.), *Bottom-Interacting Ocean Acoustics* (Plenum Press, New York, 1980).
2. Akal, T. and Berkson, J.M. (eds.), *Ocean Seismo-Acoustics* (Plenum Press, New York, 1986).
3. Jensen, F.B., Ferla, C.M., LePage, K.D. and Nielsen, P.L., Acoustic models at SACLANT-CEN: An update. Report SR-354, SACLANT Undersea Research Centre, La Spezia, Italy (2001).
4. Jensen, F.B. and Martinelli, M.G., The SACLANTCEN parabolic equation model (PAREQ). SACLANT Undersea Research Centre, La Spezia, Italy (1985).
5. Weinberg, H. and Keenan, R.E., Gaussian ray bundles for modeling high-frequency propagation loss under shallow-water conditions, *J. Acoust. Soc. Am.* **100**, 1421–1431 (1996).
6. Ferla, C.M., Porter, M.B. and Jensen, F.B., C-SNAP: Coupled SACLANTCEN normal mode acoustic propagation loss model. Report SM-274, SACLANT Undersea Research Centre, La Spezia, Italy (1994).

YELLOW SEA ACOUSTIC UNCERTAINTY
CAUSED BY HYDROGRAPHIC DATA ERROR

PETER C. CHU AND CARLOS J. CINTRON

Naval Postgraduate School, 833 Dyer Road, Monterey CA 93943, USA
E-mail: chu@nps.navy.mil

STEVEN D. HAEGER AND DAVID SCHNEIDER

Naval Oceanographic Office, Stennis Space Center, MS 39529, USA
E-mail: haegers@navo.navy.mil

RUTH E. KEENAN

Scientific Application International Corporation, Mashpee, MA02649, USA
E-mail: rkeenan@capcod.net

DANIEL N. FOX

Naval Research Laboratory, Stennis Space Center, MS 39529, USA
Email: fox@nrlssc.navy.mil

This paper investigates the acoustic uncertainty due to hydrographic data error and in turn to determine the necessity of a near real time ocean analysis capability such as the Naval Oceanographic Office's (NAVOCEANO) Modular Ocean Data Assimilation System (MODAS) model in shallow water (such as the Yellow Sea) mine hunting applications using the Navy's Comprehensive Acoustic Simulation System / Gaussian Ray Bundle (CASS/GRAB) model. To simulate hydrographic data uncertainty, Gausian-type errors (produced using the random number generator in MATLAB) with zero mean and three standard deviations (1 m/s, 5 m/s, and 10 m/s) are added to the sound profile. It is found that the acoustic uncertainty depends on the location of the error and sound sources. It is more sensitive to errors in the isothermal structure in the winter than in the layered structure in the summer.

1 Introduction

The major threats in the littoral are diesel submarines and sea mines. The combination of improvements in noise reducing technology and the development of Air Independent Propulsion (AIP) technology have made diesel submarines very difficult to detect in both the littoral and blue waters. After a weapon platform has detected its targets, the sensors on torpedoes designed for blue water operations are not designed to acquire a target in a reverberation-crippling environment. Recently, the U.S. Navy has focused much of its research and development efforts in designing high frequency sensors and corresponding acoustic models to overcome the threat in the littoral. The Comprehensive Acoustic Simulation System (CASS) using the Gaussian Ray Bundle (GRAB) model is a valuable tool for the AN/SQQ-32 mine hunting detection and classification sonar. The

563

N.G. Pace and F.B. Jensen (eds.), Impact of Littoral Environmental Variability on Acoustic Predictions and Sonar Performance, 563-570.
© 2002 *Kluwer Academic Publishers.*

performance of this model, as in all models, is determined by the accuracy of its inputs such as sea surface conditions, bathymetry, bottom type, and sound speed profiles. Here, the effect of sound speed errors (i.e., hydrographic errors) on the acoustic uncertainty in the Yellow Sea is investigated using CASS/GRAB.

2   Environment of the Yellow Sea

2.1   Geology and Structure

The Yellow Sea is a semi-enclosed basin situated between China and the Korean peninsula with the Bohai Sea to the northwest and the East China Sea to the south. The Yellow Sea is a large shallow water basin covering an area of approximately 295,000 km^2. The water depth over most of the area is less than 50 m (Fig. 1). The bottom sediment of the central and western regions of the Yellow Sea consists primarily of mud and the eastern region is primarily sand. The mud sedimentation in the central and northwestern regions of the Yellow Sea is due to the runoff from the great rivers of China. Four regions with different bottom types were selected for the acoustic model runs in this study (Fig. 2): (a) rock bottom type which is located in the north-central Yellow Sea at 37°-37.5° N, 123°-123.8° E, (b) gravel bottom type which is located in the northern Yellow Sea at 38.4°-39° N, 122°-123° E, (c) sand bottom type which is located in the southeastern Yellow Sea at 35.5°-36.5° N, 124.5°-126.2° E, and (d) mud bottom type which is located in the south-central Yellow Sea at 35°-36.5° N, 123°- 124.5° E.

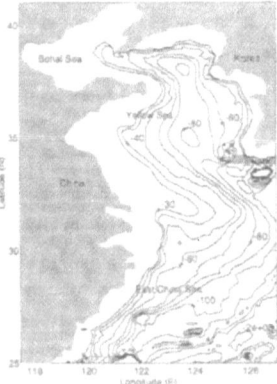

Figure 1. Bottom topography of the Yellow   Figure 2. Yellow Sea bottom sediment chart.
Sea and the surrounding regions.

2.2   Oceanography

The four seasons in the Yellow Sea are defined as follows: the winter months run from January through March; the spring months run from April through June; the summer months run from July through September; and the fall months run from October through December. The two main characteristic temperature profiles of the Yellow Sea are during the winter and the summer months. In the winter months, the temperature profiles throughout the region are characterized as isothermal (Fig. 3a). In the summer months, the temperature profiles throughout the region are characterized by a multi-layer profile consisting of a mixed layer, a thermocline, and a deep layer (Fig. 3b).

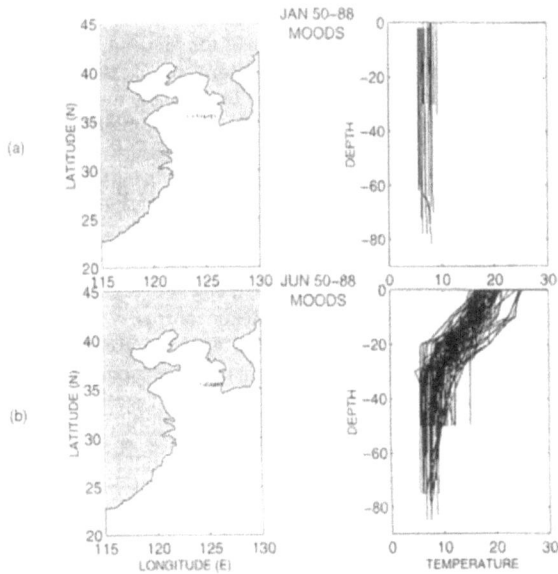

Figure 3. Eastern Yellow Sea (around 36 N) temperature profiles during 1950–1988; (a) January and (b) June. Solid dots show the location of the observation stations (From Chu *et al.* [1]).

3   CASS/GRAB model

CASS/GRAB is an active and passive range dependent propagation, reverberation, and signal excess acoustic model and is accepted as the Navy's standard model. The GRAB model's main function is to calculate eigenrays in range-dependent environments in the frequency band 600 Hz to 100 kHz and to use the eigenrays to calculate propagation loss. The CASS model is the range dependent improvement of the Generic Sonar model (GSM). CASS performs range independent monostatic and bistatic active signal excess calculations. The major difference between the GRAB model and a classic ray path is that the amplitude of the Gaussian ray bundles is global, affecting all depths to some degree, whereas classic ray path amplitudes are local. GRAB calculates amplitude globally by distributing the amplitudes according to the Gaussian equation

$$\Psi_v = \frac{\beta_{v,0}\Gamma_v^{\,2}}{\sqrt{2\pi}\,\sigma_v\,p_{r,v}\,r}\exp\left\{-0.5\left[(z-z_v)/\sigma_v\right]^2\right\}, \qquad \sigma_v = (0.5)(\max(\Delta z, 4\pi\lambda)),$$

where the Γ_v represents losses due to volume attenuation and boundary interaction; σ_v is the effective standard deviation of the Gaussian width; and $\beta_{v,0}$ is a factor that depends only on the source and is chosen so that the energy within a geometric-acoustic ray tube equals the energy within a Gaussian ray bundle. The variable z_v is the depth along the vth test ray at range r, z is the target depth, p_r is the horizontal slowness, Δz is the change in ray depth at constant range due to a change in source angle, and λ is the wavelength. The selection of the effective standard deviation σ_v is the weakest component in providing a firm theoretical basis for the GRAB model. GRAB computes the random or coherent propagation loss from the eigenrays stored in the eigenray file and stores in them in separate pressure files (Aidala *et al.* [2]).

4   Modular Ocean Data Assimilation System

Modular Ocean Data Assimilation System (MODAS), recently developed at the Naval Research Laboratory (NRL), uses a modular approach to generate three-dimensional gridded fields of temperature and salinity. Its data assimilation capabilities may be applied to a wide range of input data, including randomly located in-situ, satellite, and climatological data. Available measurements from any or all of these sources are incorporated into a three-dimensional, smoothly gridded output field of temperature and salinity. MODAS' primary outputs are temperature and salinity fields that may be used to calculate three-dimensional sound speed fields. The sound speed field, in turn, may be used to drive acoustic performance prediction scenarios, including simulations, tactical decision aids, and other capabilities. These are employed in a wide variety of naval applications and tactical decision aids (TDAs) (Fox *et al.* [3]).

5   Acoustic characteristics

On February 15, 2000, the sound speed profile at 36.4°N, 124.4°E (mud bottom) from MODAS is quite uniform. The sound speed decreases 0.2 m/s from the surface to 8.2 ft depth, increases 0.1 m/s from 8.2 ft to 41.0 ft depth, and 0.2 m/s from 41.0 ft to 57.4 ft depth. The weak sound speed minimum at near surface (8.2 ft depth) generates a weak sound channel with sound source at 25 ft (Fig. 4) and 125 ft (Fig. 5).

Table 1. Sound speed from MODAS on February 15, 2000, at 36.4° N, 124.4° E, mud bottom	
Depth (Feet)	M/S
0.00	1479.90
8.20	1479.70
24.60	1479.80
41.00	1479.80
57.40	1480.00
82.00	1480.00
106.60	1480.10
131.20	1480.30
164.00	1480.50
205.00	1480.40
246.00	1480.40

6   Acoustic uncertainties

To simulate hydrographic data uncertainty, a Gausian-type error (produced using the random number generator in MATLAB) with zero mean and three standard deviations (1 m/s, 5 m/s, and 10 m/s) is added to the MODAS sound profiles. All the sound speed profiles with the mud bottom are selected. Two sets of hydrographic data (MODAS and MODAS with error) are inputted into the CASS/GRAB model. The model was integrated with two sound source depths (25 and 125 ft) and two seasons (February 15, 2000 and August 15, 2000) to capture the effect of the hydrographic error on the acoustic uncertainty in the Yellow Sea.

Histograms of the detection range difference (acoustic uncertainty) between control (no error) and sensitivity (with error) runs demonstrate that the acoustic uncertainty has non-Gaussian-type distribution in winter (Fig. 6) and Gaussian-type distribution in summer (Fig. 7) and is much larger in winter than in summer. This indicates that the isothermal structure of the winter profiles is much more susceptible to errors in sound speed. The acoustic uncertainty depends on where the random error is situated in the water column in relation to the position of the source. For a specific profile, if an error of 1 m/s is positioned within approximately 5 feet of the source depth and an error of 10 m/s is positioned greater than the 5 feet of the source depth, the 1 m/s error will have a much greater effect on the acoustic transmission. If the error near the source is positive, the gradient that is formed in the sound speed profile will decrease detection ranges. If the error is negative, the gradient that is formed in the sound speed profile will increase detection ranges.

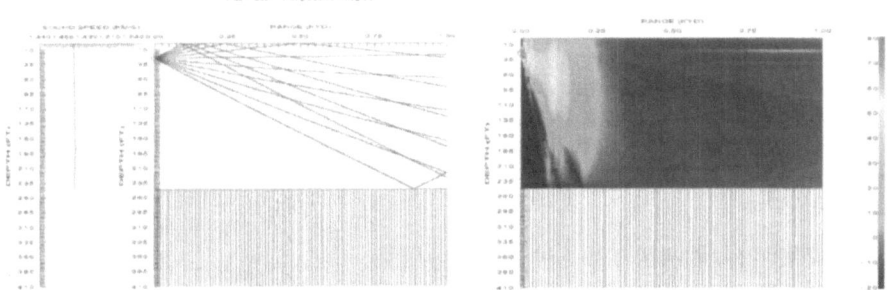

Figure 4. Acoustic transmission with source depth of 25 ft on February 15, 2000 at 36.4 N 124.4 E (mud bottom) using the MODAS data: (a) ray trace, and (b) signal excess. Notice that a maximum detection range is 260 yd near the sound source.

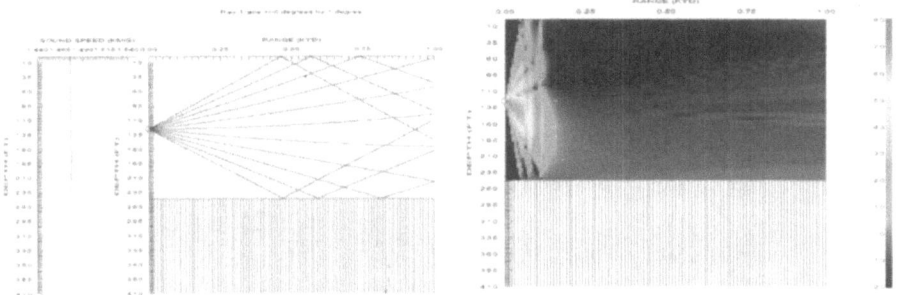

Figure 5. Acoustic transmission with source depth of 125 ft on February 15, 2000 at 36 4 N 124.4 E (mud bottom) using the MODAS data: (a) ray trace, and (b) signal excess. Notice that a maximum detection range is 145 yd near the sound source.

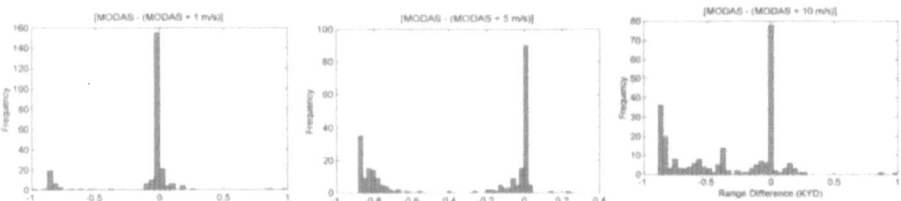

Figure 6. Histograms of the detection range difference caused by the Gaussion-type errors in the sound speed profiles on February 15, 2000 with the mud bottom and 125 ft source depth: (a) MODAS minus MODAS with 1 m/s error, (b) MODAS minus MODAS with 5 m/s error, and (c) MODAS minus MODAS with 10 m/s error.

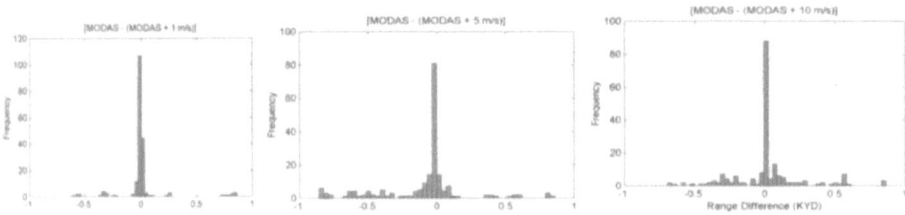

Figure 7. Histograms of the detection range difference caused by the Gaussion-type errors in the sound speed profiles on August 15, 2000 with the mud bottom and 125 ft source depth: (a) MODAS minus MODAS with 1 m/s error, (b) MODAS minus MODAS with 5 m/s error, and (c) MODAS minus MODAS with 10 m/s error.

7 Hydrographic errors at the source depth

When an error (+1 m/s) was added into the MODAS sound speed profile at both source depths (25 and 125 ft), a shadow zone was formed in front of the source that significantly decreased the detection ranges at that depth (Figs. 8 and 9). When an error (–1m/s) was added into the MODAS sound speed profile at both source depths, a strong sound channel formed that dramatically increased detection ranges at that depth (Figs. 10 and 11).

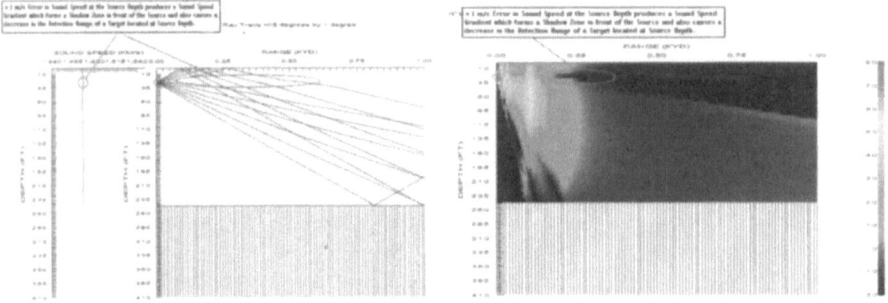

Figure 8. MODAS with +1 m/s sound speed error at the source depth (25 ft) on February 15, 2000, 36.4 N 124.4 E, mud bottom: (a) ray trace, (b) signal excess (maximum detection range at source depth = 175 yd, Amax detection range at the source depth = – 85 yd).

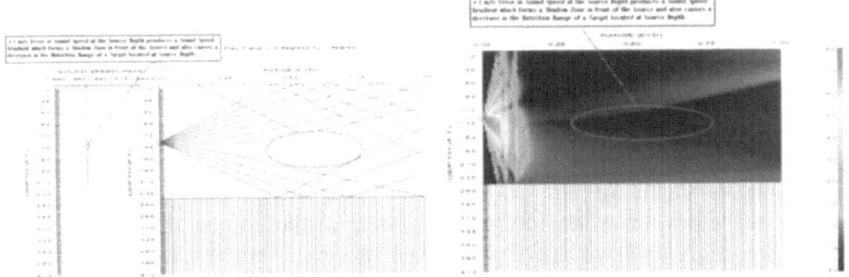

Figure 9. MODAS with +1 m/s sound speed error at source depth (125 ft) on February 15, 2000, 36.4 N 124.4 E, mud bottom: (a) ray trace, and (b) signal excess (maximum detection range near the source depth = 150 yd, Amax detection range near the source depth = −5 yd).

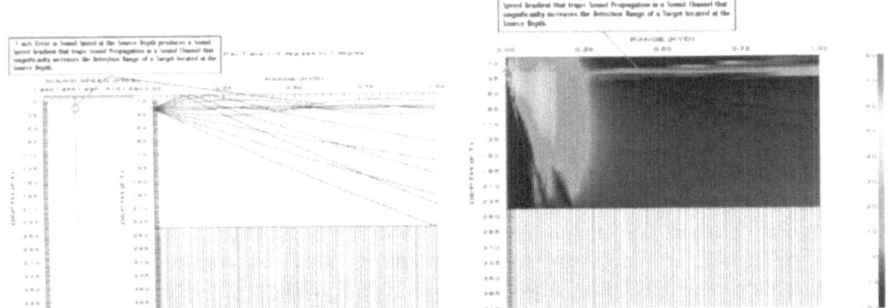

Figure 10. MODAS with −1 m/s sound speed error at the source depth (25 ft) on February 15, 2000, 36.4 N 124.4 E, mud bottom: (a) ray trace, and (b) signal excess (maximum detection range near the source depth >1000 yd, Amax detection range near the source depth >740 yd).

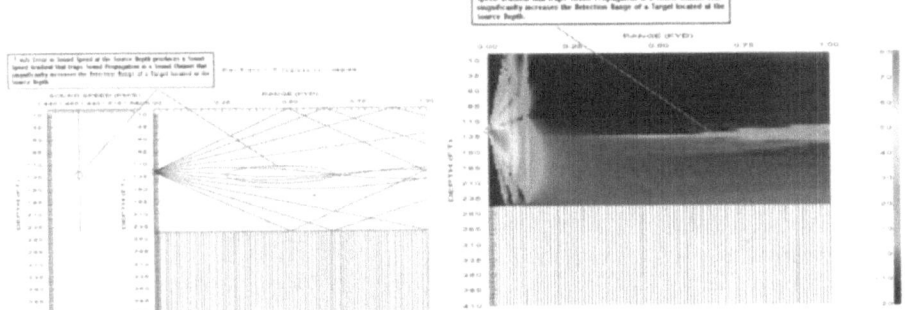

Figure 11. MODAS with -1 m/s sound speed error at the source depth (125 ft) on February 15, 2000, 36.4 N 124.4 E, mud bottom: (a) ray trace, and (b) signal excess (maximum detection range near the source depth >1000 yd, Amax detection range near the source depth >855 yd).

8   Conclusions

(1) The seasonal variation in acoustic transmission in the Yellow Sea for all regions was mainly due to the isothermal structure in the winter and a multi-layer thermal structure in the summer. The acoustic transmission in the winter is shorter due to the

effect of the isothermal structure of the sound speed profile. The acoustic transmission in the summer is significantly longer due to the down bending effects of the multi-layer structure of the sound speed profiles, which produce convergence zone and caustics.

(2) The acoustic uncertainty due to error in sound speed profiles depends on location of that error relative to the sound source. It has non-Gaussian-type distribution in winter and Gaussian-type distribution in summer and is much larger in winter than in summer.

(3) In winter, when an error (+1 m/s) was added into the sound speed profile at the source depth, a shadow zone was formed in front of the source that significantly decreased the detection ranges at that depth. When an error (−1m/s) was added into the sound speed profile at the source depth, a strong sound channel formed that dramatically increased detection ranges at that depth.

Acknowledgements

This work was supported by the Naval Oceanographic Office and the Naval Postgraduate School.

References

1. Chu, P.C., Fralick, C.R., Haeger, S.D. and Carron, M.J, A parametric model for the Yellow Sea thermal variability, *J. Geophys. Res.* **102**, 10499–10507 (1997).
2. Aidala, F.E., Keenan, R.E. and Weinberg, H., Modeling high frequency system performing in shallow-water range-dependent environments with the comprehensive acoustic simulation system (CASS). NUWC Newport Technical Digest, 54–61 (1998).
3. Fox, D.N., Teague, W.J., Barron, C.N., Carnes, M.R. and Lee, C.M., The Modular Ocean Data Assimilation System (MODAS), *J. Atmos. Oceanic Tech.* **19**, 240–252 (2002).
4. Chu, P.C., Wells, S.K., Haeger, S.D., Szczechowski, C. and Carron, M.J., Temporal and spatial scales of the Yellow Sea thermal variability, *J. Geophys. Res.* **102**, 5657–5658 (1997).
5. Keenan, R.E., Weinberg, H. and Aidala, F.E., Software requirements specifications for the Gaussian Ray Bundle (GRAB) eigenray propagation model. Rep. OAML-SRS-74, Systems Integration Division, Stennis Space Center, MS (1999).
6. Keenan, R.E., An introduction to GRAB eigenrays and CASS reverberation and signal excess. Science Applications International Corporation, MA (2000).
7. Software design document for the Gaussian Ray Bundle (GRAB) eigenray propagation model. Rep. OAML-SDD-74, Naval Oceanographic Office, Systems Integration Division, Stennis Space Center, MS (1999).
8. Data base description for the Generalized Digital Environmental Model (GDEM-V) Ver. 2.5. Rep. OAML-DBD-72C, Naval Oceanographic Office, Systems Integration Division, Stennis Space Center, MS (2000).
9. User's manual for the Modular Ocean Data Assimilation System (MODAS) Ver. 2.1. PSI Tech. Report S-285, Stennis Space Center, MS (1999).
10. Confidence level assessment of MODAS, Appendix 1: Upgraded altimetry processing. Naval Research Laboratory, Stennis Space Center, MS (2000).

DETECTION OF SONAR INDUCED MEASUREMENT UNCERTAINTIES IN ENVIRONMENTAL SENSING: A CASE STUDY WITH THE TOROIDAL VOLUME SEARCH SONAR

CHRISTIAN de MOUSTIER[1] AND TIMOTHY C. GALLAUDET[2]

Marine Physical Laboratory, Scripps Institution of Oceanography,
8602 La Jolla Shores Dr., La Jolla CA 92037-0205, USA

Shallow water field measurements of ocean volume and boundary acoustic backscatter, made with the US Navy's 68 kHz Toroidal Volume Search Sonar (TVSS), are used to demonstrate the benefits and side effects of a synoptic 360° vertical viewing field with 120 beams at 3° increments when investigating the spatial and temporal variability of the environment. Boundary backscatter measurements can help identify and quantify uncertainties introduced by the sonar system, thus setting realistic bounds on the spatial and temporal scales of environmental features that can be detected with this multibeam sonar. However, acoustic energy reflected at the boundaries and received in the vertical sidelobes of beams steered in the ocean volume often masks finer volume acoustic reverberation features from scattering layers of zooplankton or resonant micro-bubbles. In such cases, a 3 dimensional image built from a sequence of pings along the sonar's track has proven effective in discriminating sonar induced apparent acoustic variability from environmental variability. A multisector and multi-frequency transmission scheme is proposed to minimize boundary sidelobe interferences.

1   Introduction

Studies of shallow water acoustic variability strive to describe the underlying physical and environmental factors causing the observed variability against which a probability of target detection is sought. Passive sonar measurements rely on discrete spatial and temporal changes in the environment to establish a detection threshold, whereas active sonar measurements rely on disruptions of presumed propagation paths to infer environmental processes at work and the likelihood of target detection. However, assuming that a sound source transmits repeatable and stable acoustic signals somewhere in the water column, the resulting time series of acoustic energy received at hydrophone arrays, co-located with the source or some distance away, will be a combination of reflection and scattering at the sea surface and seafloor boundaries and in the ocean volume, along multiply interfering paths that are often difficult to resolve even when either or both the transmitter and the receiver are directional arrays.

Present addresses: (1) Center for Coastal and Ocean Mapping, Chase Ocean Engineering Lab, University of New Hampshire, Durham, NH 03824-3525, USA, E-mail: cpm@ieee.org; (2) LCDR Tim Gallaudet, Operations Department, USS Kitty Hawk CV-63, FPO AP 96634-2770.

571

N.G. Pace and F.B. Jensen (eds.), Impact of Littoral Environmental Variability on Acoustic Predictions and Sonar Performance, 571-577.
© 2002 *Kluwer Academic Publishers.*

The high spatial resolution of the multi narrow-beam sonar geometry might seem to be a reasonable choice to reduce such multipath interferences. However, we shall show that this geometry introduces its own set of ambiguities that can be mistaken for environmental acoustic variability. We illustrate this point with quasi monostatic acoustic backscatter measurements made by the US Naval Surface Warfare Center: Coastal Systems Station (CSS), Panama City, Florida, during engineering tests of the Toroidal Volume Search Sonar (TVSS). This sonar operates at 68 kHz with two adjacent and co-axial horizontal cylindrical arrays, each 0.53 m in diameter [1]. One is a 32 element projector array meant to produce a toroidal beam 3.7° wide fore-aft and omni-directional in the sonar's roll plane. The other is a 120 element hydrophone array used to form 120 beams at 3° increments in the roll plane [2].

The tests took place in the northeastern Gulf of Mexico, in about 200 m of water depth, over a nearly featureless sandy silt bottom [3,4], and in sea state 1.5 [4]. For the data considered here, the sonar was towed at 78 m depth, 735 m astern a vessel moving at about 4 m/s. Acoustic propagation conditions were controlled by the sound speed vs. depth profile shown in Fig. 1, and the position of the sonar in the water column yielding the monostatic multipath geometry also shown in the figure.

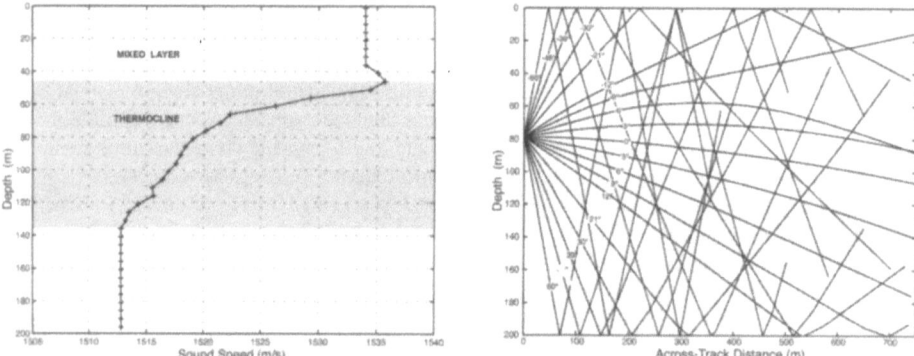

Figure 1. Acoustic propagation conditions at the test site about 120 km southwest of Panama City, Florida: Sound speed profile and rays traced from the 78 m tow depth of the sonar.

In the following we show how the TVSS's 360° multibeam imaging capability for environmental sensing and target detection applications can be strongly biased by sonar induced interferences masquerading as variability in the ocean volume. To minimize these limitations, we propose a multisector transmission geometry with distinct acoustic frequencies for each sector, similar to that used in some modern multibeam swath bathymetry sonars [5], but covering the full 360°.

2   Environmental sensing with the TVSS

The benign environmental conditions at the test site made it possible to highlight the spatial and temporal characteristics of acoustic imagery obtained with the TVSS. For example, an image of average volume acoustic backscattering strength in polar coordinates of elevation angle θ and slant range R from the sonar ($S_v(\theta,R)$), was obtained from 97 successive pings recorded during a straight run of the sonar at constant depth, by stacking and averaging echoes received in each of the 120 beamformed and roll compensated sectors covering 360° in the roll plane (Fig. 2) [6]. With a 1 Hz ping rate and 4 m/s tow speed, this corresponds to averaging over roughly 400 m along track, thus emphasizing volume acoustic backscatter "features" that remain coherent from ping to ping over that distance.

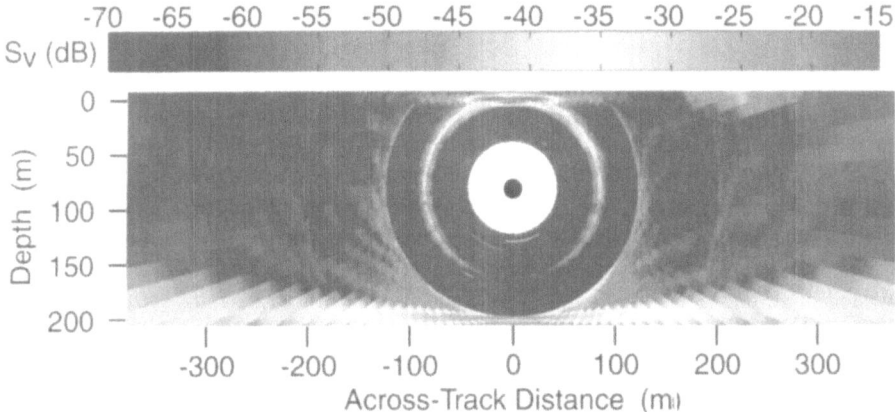

Figure 2. Vertical slice of volume acoustic backscattering strength ($S_v(\theta,R)$) measured in the roll plane of the TVSS (black dot in center of picture) and averaged along track over 97 pings. Echo digitization and recording started at the outer boundary of the central white zone.

In this image, all the beams have the same –3 dB widths of 4.9° (θ_r) in the roll plane and 3.7° (θ_p) in the pitch plane, and they are equally spaced at 3° increments in the roll plane. The transmitted signal is a CW pulse with a bandwidth W of 4.4 kHz. Therefore, with a nominal sound speed C of 1500 m/s, the theoretical volume of a resolution cell for a single ping is given by:

$$V(R) = 2/3 \; \theta_p \; \sin(\theta_r /2) \; ((R+C/2W)^3 - R^3) \quad m^3,$$
$$\approx 1/3 \; \theta_p \theta_r \; ((R+C/2W)^3 - R^3) \quad m^3, \; \text{for } \theta_r < \pi/9 \qquad (1)$$

Hence, for receive beam widths of 20° or less, the resolution cell's volume is directly proportional to the fore-aft transmit beam width and the athwarships receive beam width of the sonar. However, the effective volume of the resolution cell will most likely be somewhat larger than predicted by Eq. (1) because it will depend also on the density, size and aspect ratio of scatterers in each cell [7].

Notable environmental features in Fig. 2 include boundary returns with high volume scattering strength, a small near-surface target visible in two beams at about 210 m horizontal range, a bubble layer from the ship's wake extending 6-8 m below the surface, and biological scattering layers near 45 m depth (more detailed views and analyses of these features are available in refs. [4,6]). Returns from the sea surface are strongest in a ±30° sector, centered on zenith, that corresponds to scattering by resonant micro-bubbles generated in the towing vessel's wake. Observable coherent acoustic backscatter from the sea surface extends to ±45° about zenith and is also due to the ship's wake which exhibits remarkable persistence over time and space. In fact, the small near-surface target, with a volume scattering strength of 25 dB above ambient noise, is due to micro-bubbles from the decaying wake of the ship on a previous parallel track 200 m away [4]. By contrast, bottom returns exhibit a characteristic angular dependence with high backscatter near nadir dropping rapidly by about 20 dB at 15° incidence and leveling off to the edges of the swath shown. Actually, the bottom returns extend beyond the horizontal limits of the plot and cover a sector in excess of ±75° about nadir.

3   Sonar induced variability

Other notable features in Fig. 2 include concentric rings centered at the sonar, and several diagonal lines appearing in the volume with increasing slopes. The two most prominent rings, that are respectively tangent to the sea surface and the bottom, are due to near zenith, respectively nadir, echoes received in the vertical sidelobes of beams pointed in all the other directions. The fainter rings with larger radii are due to similar vertical sidelobe reception of zenith and nadir multiples. The diagonals are also due to sidelobe reception of: (1) the off-nadir bottom echoes for diagonals that appear to originate at nadir on the bottom and remain tangent to that first bottom sidelobe ring, and (2) off-nadir surface-bottom multiples for the steeper diagonals that appear tangent to the next larger ring with radius of about 200 m.

The importance of these artifacts becomes clear when one considers looking for targets or monitoring environmental variability in the water column contained between 120 m and 200 m horizontal range in Fig. 2. Because of the strong angular dependence of near-specular seafloor acoustic backscatter for most sediments [8], changes in bottom relief might eliminate the sidelobe interference or shift its location in the water column over a number of pings, introducing 8–12 dB of variability in the volume backscatter strength measurements that are unrelated to the spatial characteristics of the volume environment.

The beams displayed in Fig. 2 were formed with a resampled Dolph-Chebyshev amplitude shading window [9] for a uniform -30 dB sidelobe reduction. This technique produces the narrowest mainlobe for a given sidelobe reduction level, and the uniform level of its sidelobes facilitates artifacts detection thanks to their common arrival time in all the beams [10]. However more aggressive sidelobe reduction is obviously needed, but the required amplitude shading windows would increase substantially the width of the mainlobe (e.g. > 60% for raised cosine windows with –60dB sidelobe reduction) [11–13], and increase the volume of the resolution cells in the same proportion (Eq. (1)). To reduce sidelobe interference without sacrificing spatial resolution requires a different transmission paradigm as will be discussed in Sect. 4.

With the toroidal transmission pattern in the roll plane of the TVSS, and the subsequent 360° synoptic view of volume acoustic backscatter afforded by its receiver's multi-narrow beam geometry, one can form a 3 dimensional image of the acoustic field over successive pings. Similar 3 dimensional acoustic visualizations, though not with multibeam sonars, have been used effectively in zooplankton patchiness studies (e.g. [14]). In the case of the TVSS, the 3D image has proven useful to identify region of the water column that are free of sidelobe interference [4,6]. One such region is the horizontal slice, shown in Fig. 3, which was taken at 4.7 m depth between 40 m and 140 m horizontal range between the first surface and second bottom induced sidelobe rings (Fig. 2).

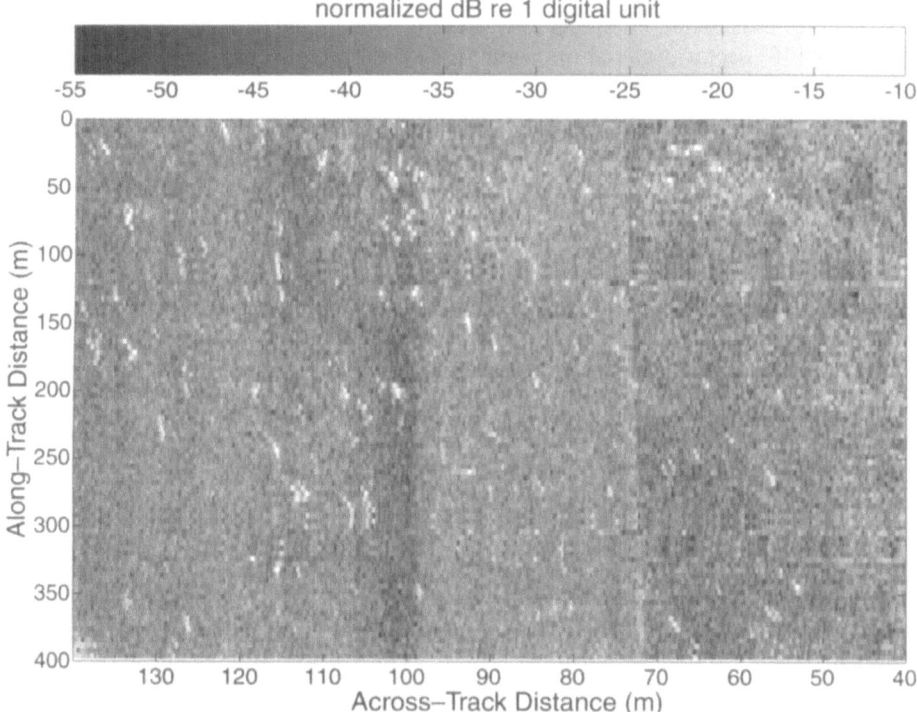

Figure 3. Raster image of a horizontal slice of volume acoustic backscatter 4.7 m below the sea surface. The sonar moves from top to bottom on the right side of this image.

The light features distributed diagonally across the image are most likely due to near surface resonant bubble clouds generated by the wake of the towing vessel. They are 10–15 dB above the background reverberation level, however this background is obscured by along-track bands of pixels with lower volume backscattering strength than their neighbors, resulting from incomplete normalization for the transmit beam pattern.

This illustrates another aspect of multibeam imaging with a sonar whose toroidal transmit beam is not truly uniform. In fact, the TVSS's transmit beam had scallops as deep as 10 dB in the roll plane. Because this transmit beam is not compensated for the instantaneous roll of the sonar, an unnecessary expense for a true toroidal pattern, the highs and lows of the actual transmit beam pattern roll with the sonar and introduce a wavy uncertainty in the volume backscatter data that could affect its statistical analysis (e.g. [15–18]). This uncertainty can only be detected by plotting the data as shown in Fig. 3. Then correction techniques similar to those that have been effective in seafloor acoustic backscatter imagery (c.g. [20,21]) must be applied before any sort of statistical analysis.

4   Multisector transmit geometry

The sidelobe interference patterns shown in Fig. 2 could be reduced substantially, without compromising the volume cell resolution, by modifying the transmit geometry from a toroidal pattern obtained in a single ping to a similar pattern obtained with at least 4 pings transmitted milliseconds apart at different center frequencies in 4 discrete sectors: a 120° sector centered on zenith, a 160° sector centered on nadir and a 120° sector centered on the horizontal on each side of the sonar. The order in which these sectors are transmitted will depend on the position of the sonar in the water column. The volume sectors on either side of the sonar should be transmitted first, using the two lowest center frequencies and the largest source level possible to maximize their range capability. On the other hand, the boundary sectors might require relatively less source level to avoid saturating the receivers with specular surface or bottom backscattering strengths. In addition, the boundary sector with the shortest range should be transmitted last to optimize the ping repetition rate. The four sectors overlap to provide maximum volume coverage individually or combined. However the bandwidths of their transmitted signal should not overlap to maintain good spatial discrimination between the sectors and to avoid inter-sector sidelobe interferences. Real-time roll compensation will be needed during transmit and receive beamforming operations, thus requiring that the sonar's attitude be sampled at about 100 Hz.

5   Conclusions

Applications of the multi-narrow beam sonar geometry are expanding from the now common swath bathymetry usage begun in the mid 1970s, to 3 dimensional imaging of the whole water column as illustrated above. However problem encountered in swath bathymetry (e.g. [19–21]) can be found also, and to a larger extent, in 3 dimensional multibeam imaging. We have used volume acoustic backscattering strength derived from data recorded with the TVSS to show that boundary reflection and scattering can be easily picked up by the vertical sidelobes of beams pointing in the ocean's volume, as well as by non-vertical sidelobes in a multipath environment. Such arrivals could easily be confused for acoustic variability in the medium if a 3 dimensional acoustic image of the water column were not available for inspection prior to statistical analyses of the recorded acoustic reverberation (e.g. [4, 22,23]).

The TVSS was built in the early 1990s and has remained a prototype, but the concept can be generalized to future toroidal sonars provided a different transmission scheme is implemented to avoid boundary sidelobe interference. We proposed transmitting into 4 overlapping sectors, centered respectively at zenith, nadir and on the horizontal, on either side of the sonar. Each sector is assigned a distinct center acoustic frequency and there should be sufficient bandwidth separation between their respective signals to avoid interferences.

Acknowledgments

This work was funded by the US Office of Naval Research under ONR-NRL contract No. N00014-96-1-G913. We wish to thank Maria Kalcic and Sam Tooma (Naval Research Laboratory), CAPT Tim Schnoor USN (ret) (ONR) for their support.

References

1. Volume Search Sonar array program: Preliminary hydrophone test data. Unpublished Tech. Report, Raytheon Co. Submarine Signal Division – Portsmouth RI (1993).
2. McDonald, R.J., Wilbur, J. and Manning, R., Motion-compensated beamforming algorithm for a circular transducer array, *U.S. Navy J. Underwater Acoust.* **47**(2), 905–920 (1997).
3. Stanic, S., Briggs, K.B., Fleischer, P., Ray, R.I and Sawyer, W.B., Shallow-water high-frequency bottom scattering off Panama City, Florida, *J. Acoust. Soc. Am.* **83**(6), 2134–2144 (1988).
4. Gallaudet, T.C., Shallow water acoustic backscatter and reverberation measurements using a 68 kHz cylindrical array, Ph.D. Dissertation, Univ. California San Diego (2001).
5. Kongsberg-Simrad, EM300 & EM120 product descriptions (2001).
6. Gallaudet, T.C. and de Moustier, C., Multibeam volume acoustic backscatter imagery and reverberation measurements in the northeastern Gulf of Mexico, *J. Acoust. Soc. Am.* (in press 2002).
7. Foote, K.G., Acoustic sampling volume, *J. Acoust. Soc. Am.* **90**, 959–964 (1991).
8. APL-UW high frequency ocean environmental acoustics models handbook. Tech. Rep. APL-UW TR 9407, Applied Physics Laboratory – Univ. Washington (1994).
9. Gallaudet, T.C. and de Moustier, C., On optimal amplitude shading for arrays of irregularly spaced or non-coplanar elements, *IEEE J. Oceanic Eng.* **25** , 553–567 (2000).
10. de Moustier, C., Signal processing for swath bathymetry and concurrent seafloor acoustic imaging. In *Acoustic Signal Processing for Ocean Exploration*, edited by J.M.F. Moura and I.M.G. Lourtie (NATO ASI Series – Kluwer, The Netherlands, 1993) pp. 329–354.
11. Sureau, J. and Keeping, K., Sidelobe control in cylindrical arrays, *IEEE Trans. Antennas Propagat.* **30**(5), 1027–1031 (1982).
12. Vu, T.B., Sidelobe control in circular ring array, *IEEE Trans. Antennas Propagat.* **43**(12), 1143–1145 (1993).
13. Mailloux, R.J., *Phased Array Antenna Handbook* (Artech House, Boston, 1994).
14. Greene, C.H., Wiebe, P.H., Pelkie, C., Benfield, M.C. and Popp, J.M., Three-dimensional acoustic visualization of zooplankton patchiness, *Deep-Sea Res. II* **45**, 1201–1217 (1998).
15. McDaniel, S.T., Sea surface reverberation: A review, *J. Acoust. Soc. Am.* **94**, 1905–1922 (1993).
16. Trevorrow, M.V., Vagle, S. and Farmer, D.M., Acoustical measurements of microbubbles within ship wakes, *J. Acoust. Soc. Am.* **95**, 1922–1930 (1994).
17. Medwin H., Acoustic fluctuations due to microbubbles in the near-surface ocean, *J. Acoust. Soc. Am.* **79**, 952–957 (1974).
18. Dahl, P.H. and Plant, W.J., The variability of high-frequency acoustic backscatter from the region near the sea surface, *J. Acoust. Soc. Am.* **101**, 2596–2602 (1997).
19. de Moustier, C. and Kleinrock, M.C., Bathymetric artifacts in Sea Beam data: How to recognize them, what causes them, *J. Geophys. Res.* **91**(B3), 3407–3424 (1986).
20. Hughes-Clarke, J.E., Mayer, L.A. and Wells, D.E., Shallow-water imaging multibeam sonars: A new tool for investigating seafloor processes in the coastal zone and on the continental shelf, *Mar. Geophys. Res.* **18**, 607–629 (1996).
21. Hellequin, L., Lurton, X. and Augustin, J.M., Postprocessing and signal corrections for multibeamechosounder images, *Proc. IEEE Oceans '97* , Vol. 1, 1–4 (1997).
22. Abraham, D.A., Modeling non-Rayleigh reverberation. Rep. SR-266, SACLANT Undersea Research Center, La Spezia, Italy (1997).
23. Middleton, D., New physical-statistical methods and models for clutter and reverberation: The KA-distribution and related probability structures, *IEEE J. Oceanic Eng.* **24**, 261–284 (1999).

ENVIRONMENTAL VARIABILITY
OF THE LBVDS SEA TESTS

S. SUTHERLAND-PIETRZAK AND E. MCCARTHY

Naval Undersea Warfare Center Division, 1170 Howell St., Newport, RI 02842, USA
E-mail: sutherlandsa@npt.nuwc.navy.mil

The Lightweight Broadband Variable Depth Sonar (LBVDS) Demonstration Model (LDM) Sea Test was conducted in September-October 2001 off the eastern coast of the United States. The primary objective of the LBVDS program is to develop and demonstrate a prototype sonar system that will improve the Navy's capability to conduct surface undersea warfare (USW) operations, particularly against low-Doppler threat submarines in highly reverberant, littoral areas. One of the sea test objectives of the LBVDS program was to evaluate the benefits of environmental adaptation. Studies of shallow-water active sonar system performance have indicated that a large loss in potential performance is often incurred because the sonar operating parameters are not matched to the current environment. The LDM system can recommend waveforms and processing parameters that are based on environmental measurements and performance predictions. Acoustic performance predictions and measured results from this LBVDS sea test will be presented.

1 Introduction

Improvement in tactical active sonar detection and classification in shallow water environments is a critical USW need for the Navy's future missions in littoral areas where quiet, slow-moving, diesel-electric submarines are a threat. In order to provide the improved performance for tactical platforms, the LBVDS Program, funded by the Office of Naval Research (ONR), was established to develop both high-energy-density, lead magnesium niobate transduction technology and environmentally adaptive, broadband processing technology.

The Naval Undersea Warfare Center (NUWC) Division, Newport, Rhode Island, and Lockheed Martin Corporation, Syracuse, New York, conducted the LBVDS FY01 demonstration system sea test (Sea Test C) to demonstrate the full capability of the newly developed source, the real-time broadband processing, and the new receiver. The Multi-Function Towed Array (MFTA) was used as the receiver. Both the source and the receiver had variable depth capability. In addition, the system had automated environmental adaptation capabilities that characterized the environment and could be used to continually optimize the system setup and processing to match the changing acoustic conditions. These new capabilities are expected to improve detection, classification, and tracking performance against a deep target (below the surface layer) in shallow water. The sea test objectives were (1) demonstration of variable depth sonar capability, (2) demonstration of broad bandwidth waveforms and processing, and (3) demonstration of the value added of environmental adaptation.

N.G. Pace and F.B. Jensen (eds.), Impact of Littoral Environmental Variability on Acoustic Predictions and Sonar Performance, 579-586.
© 2002 *Kluwer Academic Publishers.*

Sea Test C was the third in the series of tests. The two previous sea tests, Sea Tests A and B (completed in 1997 and 1998, respectively) were conducted with the Broadband Active Sonar Testbed (BAST) which had a transmitter with a lower source level than the LDM. Sea Test A was conducted off the Gulf Coast of Florida. Sea Tests B and C were conducted in the Long Bay area, off the coast of South Carolina (Fig. 1).

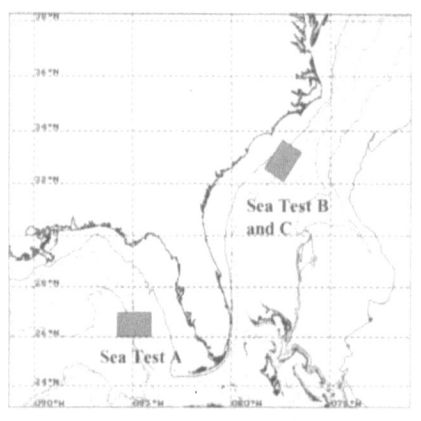

Figure 1. LBVDS Sea Test sites.

The littoral environment of the Long Bay area is a good site for testing sonar performance under the duress of changing environmental conditions because of its oceanographic temporal fluctuations and spatial variations. Sea Test C occurred in the same general location as Sea Test B, but the operational area was expanded to include a moderately reverberant bottom type (sandy) that is more common in the littorals than the highly reverberant bottom type (rocky) encountered in Sea Test B. Most of the testing for Sea Test C occurred in this moderately reverberant environment. Data were acquired with an impressive array of waveform types—varied pulse types, bandwidths, center frequencies, and Doppler sensitivities—at tactical ranges in multiple geometries, i.e., varied source and receiver depths and a target at different depths and aspects.

Detailed results and recommendations from Sea Tests A and B are available in the LBVDS Sea Test A and B final test reports [1–3]. This paper briefly discusses the data processing results from Sea Tests A and B as background and then presents the environmental observations and measurements and acoustic modeling from Sea Test C.

2 Background

The first test, Sea Test A, was conducted in the Gulf of Mexico with a point source target (echo repeater). The test focused primarily on evaluating the benefits of additional bandwidth (BW) in a moderately reverberant environment. Gains were measured at 8 log(BW), and analysis confirmed that the measured gain could be attributed to the increased bandwidth.

The second sea test, Sea Test B, revisited the issue in the highly reverberant acoustic environment of Long Bay, South Carolina. This test focused primarily on bandwidth effects on hyperbolic frequency modulation (HFM) transmissions. A complete set of HFM measurements was made for a specified set of center frequencies and bandwidth combinations. All of these measurements were made with the transmitter at a single depth below the layer.

Analysis of Sea Test B data confirmed that broadband waveforms offer increased detection and tracking performance. Figure 2a shows the gains obtained in Sea Test A against reverberation and ambient noise. Notice that, as expected, increasing the bandwidth does not provide additional gain against an ambient background and may actually degrade system performance. However, in reverberant environments, in both Sea Tests A and B, increasing bandwidth produced significant gains. In Fig. 2b, the signal excess gains against reverberation are also plotted for three different center frequencies (fc): low, medium, and high. In all three cases, the pulse length (T) remained constant, and only the bandwidth for each center frequency changed.

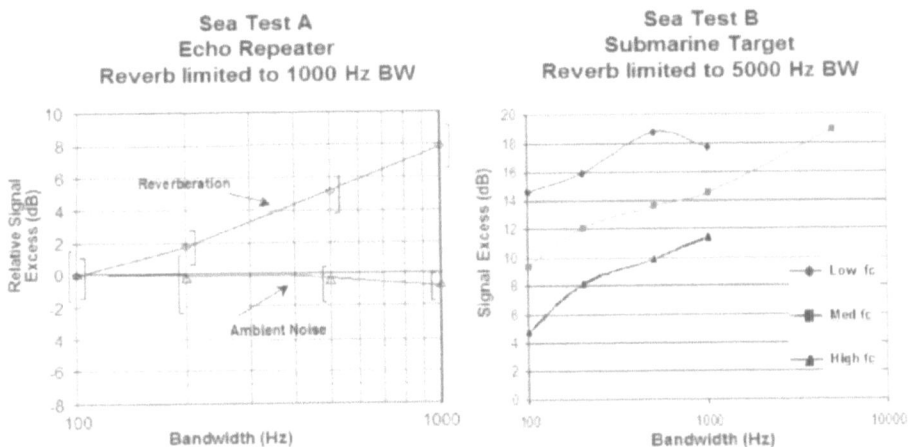

Figure 2a. Sea Test A signal excess.   Figure 2b. Sea Test B signal excess.

The ocean bottom types for Sea Tests A and B were very different and consequently produced very different reverberation backgrounds. Reverberation may also change because of the center frequency of the waveform. In the Sea Test B environment, as can be seen in Fig. 2b, the lower end of the band produced less reverberation. The background levels were closer to ambient. The higher end of the band produced greater reverberation levels, and the performance for this portion of the band was not as good. However, this was not the case for Sea Test A, which was conducted in a more moderate reverberation environment. The reverberation levels for Sea Test A were relatively consistent across the band.

Determining the best center frequency and bandwidth for a specific environment is important for optimizing system performance. The ability to measure the characteristics of the ocean environment as it changes, both spatially and temporally, is therefore essential to sonar system optimization. The Instrumented Tow Cable (ITC) was developed under the LBVDS program to provide a more accurate measure of a key environmental characteristic—the temperature profile of the water column.

The ITC design is based on an optical fiber embedded in the tow cable for the source. Light transmitted in the fiber is used to measure the temperature of the ocean water in situ. An example of the variability of ocean temperature as measured by the ITC is shown in Fig. 3. The measurements were made every 3 minutes in the Hudson Canyon [4]. Given the variation in temperature as a function of range and depth, the measurements provided a more complete environmental picture to support more accurate acoustic modeling and performance prediction. Without the ITC, Fig. 3 would have required over 60 expendable bathythermographs (XBTs).

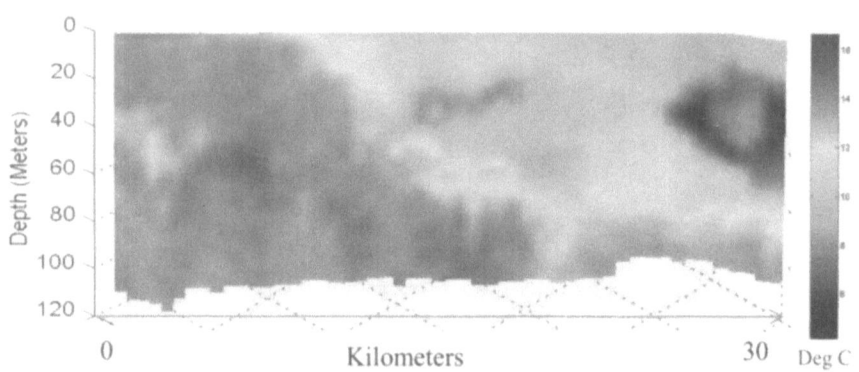

Figure 3. Instrumented Tow Cable (ITC) ocean temperature.

3   Environment

LBVDS Sea Test C required an acoustic environment that was shallow, reverberation limited, and downward refracting, with a flat or gently sloping bottom. The test also required a site with a low occurrence of endangered marine mammals. The site which best met these criteria was Long Bay, South Carolina. This operational area (OPAREA) had been studied in the previous LBVDS test (Sea Test B), and it was the site for two Towed Active Receiver System (TARS) sea tests (June and September 1998) [5] and a Littoral Warfare Advanced Development experiment (LWAD SCV 97) [6].

The bottom is characterized by a sand wedge that slopes gently down from northwest to southeast, with the depth changing from 100 to 1500 m. The sand wedge gives way to a rough, deeply scoured, limestone plateau at 200 to 250 m in depth before the continental shelf plunges into the ocean depths beyond. The plateau has been observed to be predominantly sand covered in the northeast and rocky in the southwest. All test geometries were executed between the 100- and 400-m isobaths. This area is strongly influenced by the Gulf Stream current and the "Charleston bump," which diverts the current and can form a stationary eddy that is almost perfectly centered in the OPAREA. This eddy, consisting of warm Gulf Stream water surrounding a cool, coastal parcel, was observed during the first phase of both LBVDS sea tests. The eddy appeared to travel north (Fig. 4) during the second half of the test, so that fairly homogeneous Gulf Stream water dominated the area during this phase.

The coastal waters of the Long Bay OPAREA were found to be highly variable in both time and space. Figure 5 shows four XBT temperature profiles that represent the basic profile types. Most of the coastward profiles exhibited the strong midcolumn temperature gradient as characterized by profiles #14, #17, and #49. The upper isothermal layer very often formed a deep surface duct, and the deeper isothermal layer had a tendency to form another acoustic duct offset somewhat from the bottom. The fluctuations in depth of the steep temperature gradients between the layers in sea test B were found to follow a Garrett-Munk spectrum for internal waves; continuous temperature data were not available for a similar analysis on Sea Test C. The uniform, downward refracting temperature profile shown by XBT #103 represents the more homogeneous waters of the Gulf Stream. In this case, reflection from the bottom allowed both the downward refracting and layered profiles to form a bottom channel.

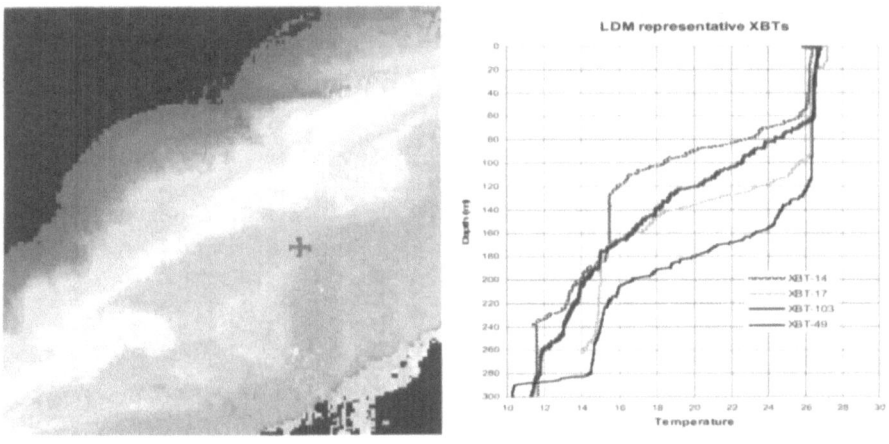

Figure 4. Sea surface temperature, 17 October.   Figure 5. Selected temperature profiles.

4   Acoustic modeling of environmental variations

At sea, the acoustic modeling was performed by the Comprehensive Acoustic System Simulation/Gaussian Ray Bundle (CASS/GRAB) and was used to determine sonar setups in real time. The modeling determined the best waveform, processing parameters, and source and receiver depths for a specified scenario. Post-test results presented in this paper are full bistatic SNRs modeled by CASS/GRAB.

Figures 6, 7, and 8 show the results of modeling performed to analyze the potential system performance as a function of source and receiver depths. The SNR plots are based on a medium center frequency, 2-s pulse length, 400-Hz bandwidth, and omnidirectional sensors. Each pair of plots shows a shallow source and receiver (top) and a deep source and receiver (bottom). For consistency, the profiles were run for the same location (XBT #49 location) but with different XBTs.

Figure 6 shows the SNR for XBT #14. The shallow source and receiver depths are 200 and 300 ft, respectively; the deep source and receiver are at 500 and 600 ft, respectively. This figure shows clearly the presence of a deep duct, whereas XBT #49, in Fig. 7, is dominated by a surface duct. The source and receiver in Fig. 7 are at the same depths as in Fig. 6.

Figure 8 is an example of the predicted SNR for XBT #103. This environment clearly shows a strong surface layer for a shallow source and bottom bounce returns for the deeper source. The source and receiver are at 25 ft in the top plot and at 400 ft in the bottom plot. The source and receiver depth configuration is one of the important environmental adaptation parameters. Clearly a sensor placed deep provides a detection advantage for profiles such as #14 and #103 but does not add significant coverage for the deep layered profiles such as #49. A more detailed analysis of the acoustic modeling will be performed on the Sea Test C data and presented in follow-on analysis efforts.

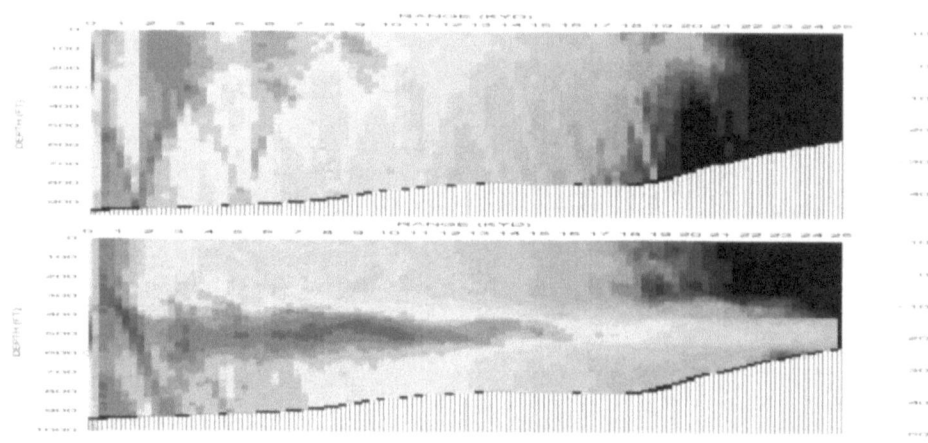

Figure 6. Signal to noise ratio for XBT #14 – shallow (top) and deep (bottom) sensors.

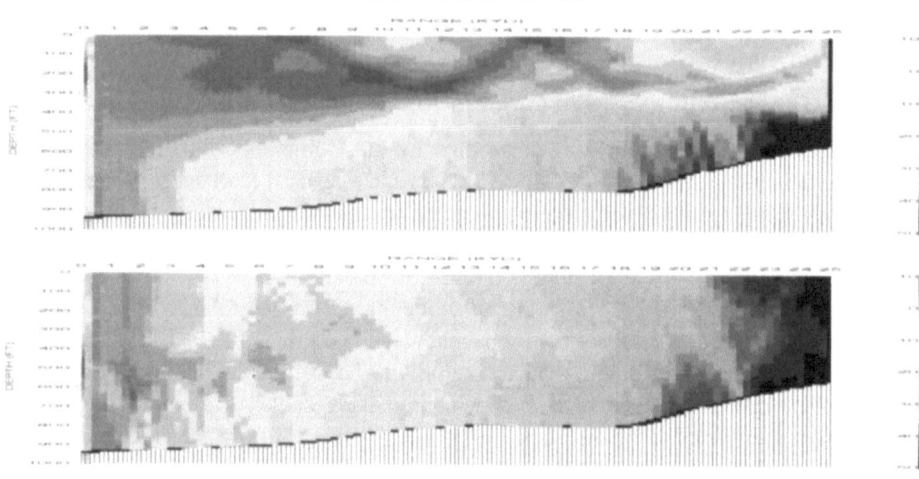

Figure 7. Signal-to-noise ratio for XBT #49 – shallow (top) and deep (bottom) sensors.

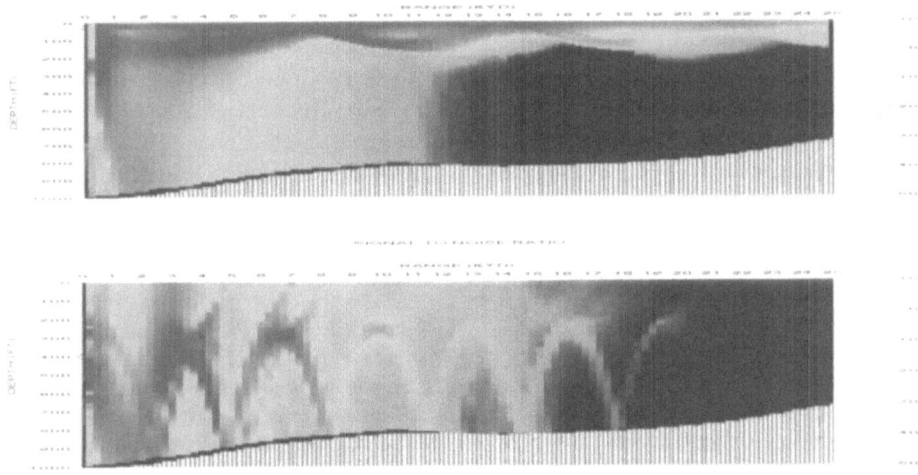

Figure 8. Signal-to-noise ratio XBT#103; S/R at 25 ft (top) and at 400 ft (bottom).

For this environment source and receiver depth significantly affect the system performance. Optimum sonar performance depends on choosing the appropriate depths, as well as the appropriate bandwidth, center frequency and pulse length.

5 Discussion

Both accurate measurement and analysis of the acoustic environment are critical to real-time adaptive setup and operation of sonar systems to achieve optimal sonar performance in littoral waters because these environments change so rapidly and dramatically over both time and space. The sonar performance of waveforms of a given center frequency and bandwidth may vary considerably because of the environment. The results from Sea Tests A and B showed the variability of system performance in reverberation-limited and ambient-noise-limited environments due to bandwidth and center frequency. Against reverberation, the performance improves as a function of increasing bandwidth. In ambient-noise-limited conditions, the coherent gains from increasing the bandwidth will not provide any performance improvement. However, using the bandwidth in a noncoherent method may provide improved performance.

The source and receiver depths are also critical for optimal sonar performance. The acoustic modeling results using the measured XBTs from Sea Test C show that, in certain environments, placing the source and/or receiver below the layer, or closer to the bottom to reduce the grazing angle, should provide improvement over a shallow source and receiver configuration.

The LBVDS data sets will contribute significantly to the effort to determine how to automatically adapt and optimize the search performance of active sonar systems on a USW-capable platform, using in-situ knowledge of the acoustic environment.

Acknowledgments

This work was sponsored by the Office of Naval Research (ONR), program officer Kenneth Dial, and managed by the Naval Sea Systems Command, Program Executive Office for Mine and Undersea Warfare (PEO(MUW)). The Surface Ship USW Combat System Program Office (PMS411) program manager is Greta Conde (PMS411U). The project manager for NUWC Division, Newport, is Maurice Simard (Code 3112).

References

1. Lightweight broadband variable depth sonar, Sea Test A final test report, Vols. I & II. Lockheed Martin Ocean, Radar & Sensor Systems, Syracuse, NY, and Naval Undersea Warfare Center Division, Newport, RI, 9 March 1998 (UNCLASSIFIED).
2. Sutherland, S.A. and Smigel, J.R., Lightweight broadband variable depth sonar, Sea Test B quick-look report. NUWC-NPT Technical Memorandum 98-0075, Naval Undersea Warfare Center Division, Newport, RI, and Lockheed Martin Ocean, Radar & Sensor Systems, Syracuse, NY, 3 June 1998 (UNCLASSIFIED).
3. Sutherland, S.A. and Smigel, J.R., Lightweight broadband variable depth sonar, Sea Test B final test report executive summary. NUWC-NPT Technical Memorandum 99-0094, Naval Undersea Warfare Center Division, Newport, RI, and Lockheed Martin Ocean, Radar & Sensor Systems, Syracuse, NY, 23 April 1999 (UNCLASSIFIED).
4. Regnier, R. and Sundvik, M., Geo clutter cruise, Hudson Canyon. Naval Undersea Warfare Center Division, Newport, RI, 5 May 2001 (UNCLASSIFIED).
5. Charette, R., TARS data analysis overview, 31 March 1999, presented to CAPT W.D. Morris, PMS411. NUWC-NPT presentation, Naval Undersea Warfare Center Division, Newport, RI, 31 March 1999 (UNCLASSIFIED).
6. Wolf, S.N., Pasewark, B.H., Erskine, F.T., McEachern, J.F. and Love, R.H., Overview of the littoral warfare advanced development system concept validation experiment (SCV-97). NRL/MR/7140-99-8375, Naval Research Laboratory, Washington, DC, 30 June 1999 (UNCLASSIFIED).

AREA: ADAPTIVE RAPID ENVIRONMENTAL ASSESSMENT

HENRIK SCHMIDT

Massachusetts Institute of Technology, Cambridge, MA 02139, USA
E-mail: henrik@keel.mit.edu

AREA: Adaptive Rapid Environmental Assessment is a new operational paradigm for minimizing the sonar performance uncertainty in shallow water. The coastal environment is characterized by variability on small spatial scales and short temporal scales, which obstruct the formation of a robust and reliable tactical picture. Thus, a Rapid Environmental Assessment (REA) capability has long been recognized as a tactical need, but its implementation is being constrained by limited in-situ measurement resources. Ocean modeling and data assimilation can produce 4-D field estimates together with their associated uncertainty. However, the resolution is inadequate for direct use in acoustic environment prediction. On the other hand the forecasts can be used to identify features such as fronts and eddies which are critical to the acoustic sonar performance uncertainty and which should therefore be targeted by the REA resources. AREA uses environmental acoustic and sonar models to translate the oceanographic and geophysical parameter uncertainty estimates into PDFs for the appropriate sonar performance metric. These are then used to objectively design survey patterns that target regions and parameters that produce the best possible sonar performance prediction within the actual operational constraints.

1 Introduction

The uncertainty of the acoustic predictability is critical to the dB-budget of classical sonar systems by directly affecting the detection and false alarm probabilities. The uncertainty of the acoustic environment prediction is also one of the major obstacles to adapting new model-based sonar processing frameworks, such as matched field processing (MFP) [1], to the coastal environment.

The acoustic uncertainty associated with spatially and temporally varying sound speed and the random characteristics of the bottom are also of critical influence to acoustic communication systems, which with the integration of new Autonomous Ocean Sampling Network (AOSN) [2] concept in the operational Navy is becoming of increasingly tactical significance. The prediction of the fidelity of the communications link is important for modern adaptive platform behaviors, where a manned or unmanned submersible may seek an optimal depth for using its acoustic communication systems.

The performance prediction of such acoustic systems is dependent on estimates of the statistics of the environmental acoustic parameters. In general, the best we can hope for is knowledge of the second order statistics such as the auto-correlation functions of the sound speed distribution, the bottom density and the irregular bottom profile; as well as the cross correlation functions between these parameters.

N.G. Pace and F.B. Jensen (eds.), Impact of Littoral Environmental Variability on Acoustic Predictions and Sonar Performance, 587-594.

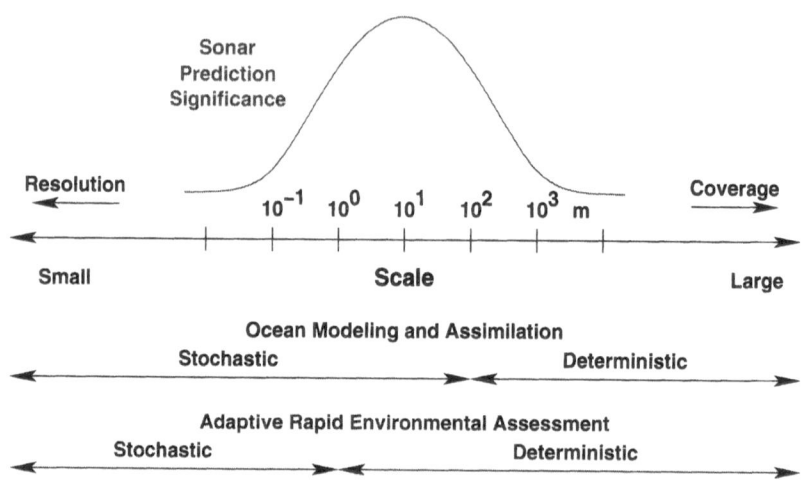

Figure 1. Sonar system performance is dependent on acoustic environment variability over a wide range of scales. Optimal environmental assessment will therefore be a compromise between conflicting requirements of coverage and resolution. By targeting areas of high sensitivity to the sonar system, AREA will shift the deterministic assessment towards smaller scales.

2 Sonar environment variability

The coastal environment has variability over a wide range of spatial and temporal scales. Therefore, the environmental assessment is facing the classical conflict between *resolution*, needed to capture the fine scale variability and *coverage*, needed for the large scale environmental phenomena. Thus, the resources available must be focusing on the scales critical to the specific sonar system, but these may also span a wide range of scales, requiring that either resolution or coverage be sacrificed.

Of particular importance to the sonar detection statistics is the environmental scales centered at the acoustic wavelengths of the sonar systems. Thus, as illustrated in Fig. 1, variability on spatial scales of a few meters may be critical to the acoustic predictability. Smaller scale variability is averaged out by the acoustic wavelength, while larger scale variability can be accurately assessed using satellite remote sensing and traditional oceanographic surveys.

To determine the critical intermediate scales of the environmental variability an in-situ measurement capability has long been recognized as a tactical need. However, its implementation is being constrained by limited resources. Thus, as illustrated in Fig. 1, the compromise between the measurement resolution and the coverage necessarily sacrifices both small and large scale variability of significance. Consequently the limited resources will always have the effect that the ocean environment will be under-sampled, both in terms of coverage and resolution, eliminating the possibility of a true deterministic predictability.

The lower limit of the spatial scales that can be described deterministically depends on the coupling to the acoustic environment and the particular sonar configuration. Thus, for example, it may be a waste to use valuable REA resources on assessing small scale spatial variability of the seabed during the morning where a well developed surface duct exists.

On the other hand such measurements may significantly reduce the prediction uncertainty later in the day, where the surface duct disappears and strong bottom interaction becomes the dominant environmental condition ('*afternoon effect*').

Oceanographic forecasting by modeling and data assimilation can produce 4-D oceanographic field estimates and their associated uncertainties [3, 4]. However, even though highly applicable to a wide range of applications such as coastal environmental management, the uncertainty of the oceanographic forecast is in general inadequate for direct use in acoustic environment prediction frameworks.

Most importantly the spatial and temporal grids are limited by the available computational resources. Even using nested computational grids, spatial scale smaller than several hundred meters in the horizontal, and tens of meters in the vertical cannot be modeled deterministically, as indicated in Fig. 1. Modern modeling and assimilation frameworks have a capability of representing the smaller, sub-grid-scale variability statistically. However, in general these scales are at least an order of magnitude larger than the scales important to the acoustic predictability. Another reason for the inadequacy of the ocean forecasts for sonar performance prediction is the highly sensitive and non-linear relation between the ocean variability and the acoustic environment statistics [5]. This issue is obviously enhanced by the limits in scale imposed by computational constraints. The limited availability of local data for assimilation into the modeling framework severely limits the usefulness of the forecasts to the acoustic environment prediction. New adaptive sampling concepts based on previous forecasts are currently being developed in connection with the emergence of the new AOSN technology [6]. In principle these could be used to deploy the limited tactical resources in a manner which is optimal to the acoustic forecasting [3]. However, the time required to generate the forecasts, design the adaptive sampling patterns, and subsequently assimilate the new data to produce accurate now-casts is orders of magnitude larger than the temporal scales of the littoral ocean. Thus, the computational resources will remain insufficient for the ocean forecasting frameworks to be directly applicable to operational prediction of the acoustic environment without supporting in-situ measurements.

In spite of its limited resolution, ocean forecasting can be extremely useful for providing large-scale coverage, and for identifying region and features with strong variability, such as coastal fronts, which could be targeted by the REA resources. As such ocean forecasting frameworks such as HOPS [4] are cornerstones of the AREA concept, allowing the environmental assessment resources to be deployed in regions of high variability where resolution is crucial, without sacrificing coverage.

Using the ocean forecasts to define optimal deployment strategies for the REA resources, the limit of deterministic characterization may be shifted significantly towards smaller scales. This can actually produce much finer resolution than earlier achieved by the same REA resources because the forecasting framework is providing the coverage, as illustrated in Fig. 1.

Another environmental factor particular to the littoral environment is the significance of the seabed because of the typical downward refracting sound speed profile. Thus, for many sonar scenarios, the spatial variability of the seabed, i.e. roughness and volume inhomogeneities, is far more severe to the acoustic variability than the oceanographic uncertainty. Clearly, this will affect the optimal REA resource allocation. Thus, for example, one available AUV may be more optimally deployed for side-scan/sub-bottom

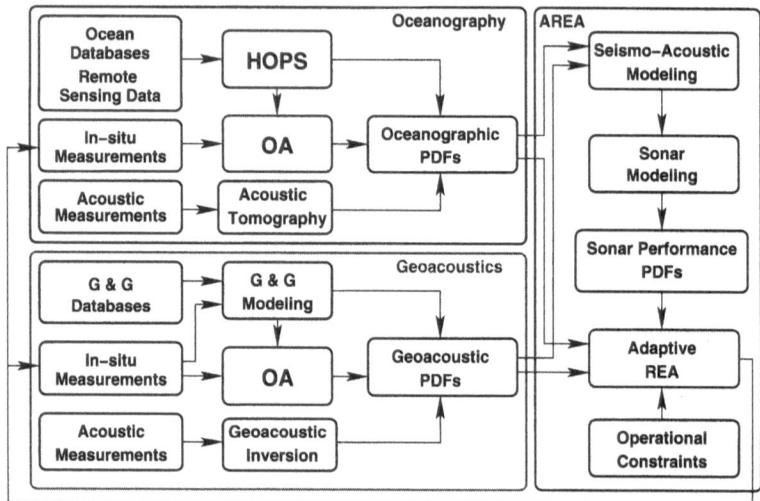

Figure 2. AREA functionality. Fore- and now-casts of the local oceanography and geology are producing spatial and temporal environmental statistics in the form of coupled PDFs of the associated environmental acoustics parameters, subsequently translated toPDFs for the sonar performance. These PDFs then form the basis for optimal deployment of the in-situ REA resources.

profiling than for water column sampling.

Adaptive Rapid Environmental Assessment (AREA) is a probabilistic approach to the adaptive sampling problem of littoral REA. By combining the coverage of coastal ocean forecasting frameworks with adaptive deployment of in-situ measurement resources for high-resolution measurements, AREA is envisioned as a real time tactical tool for not only capturing, but also minimizing the acoustic uncertainty of significance to specific sonar systems.

The AREA framework can also be used to objectively evaluate the performance of new REA concepts, such as e.g Acoustically Focused Ocean Sampling (AFOS) [6] and Acoustic Data Assimilation (ADA) [7] recently developed at MIT. In contrast to Ocean Acoustic Tomography [8], ADA does not require a substantial acoustic network, but allows for even simple point-to-point acoustic transmissions to be assimilated consistently with other oceanographic data, and may therefore become a valuable REA resource, directly reflecting the acoustic uncertainty [7].

3   AREA: Adaptive Rapid Environmental Assessment

The AREA concept is envisioned as an optimal combination of classical environmental assessment based on databases and local measurements, and full-blown forecasting frameworks based on modeling and assimilation with adaptive in-situ sampling. As described above, both of these approaches are resource limited and do not capture the acoustic environment optimally in terms of acoustic uncertainty. AREA instead uses the modeling and assimilation framework to continuously provide an initial estimate of the oceanographic fields and their coupled uncertainties, and their significance to the sonar system dB-budget. These forecasts are then used to identify optimal deployment patterns for the in-situ sampling resources such as XBT, CTD casts and AUV surveys. The functionality

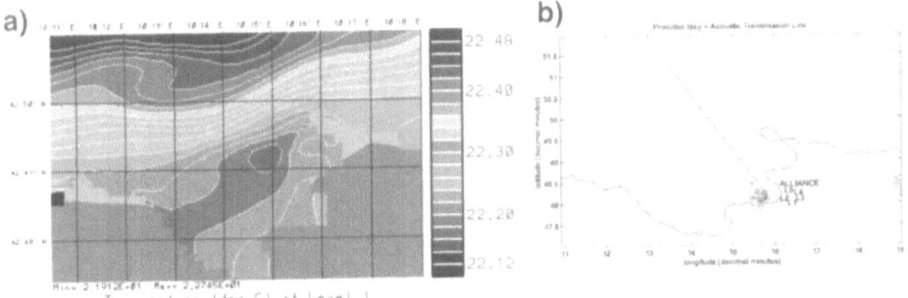

Figure 3. Coastal forecasting during GOATS/MEANS'2000 experiment. A nested implementation of HOPS was used to forecast the temperature field in Procchio Bay, Elba, Italy in October 2000. This field estimate and the associated estimated temperature uncertainty is used to test the AREA concept for optimizing the performance prediction of the transmission loss associated with the insonification a very shallow beach area from an off-shore sonar platform.

of AREA is illustrated in Fig. 2, and is envisioned to proceed as follows

An environmental forecasting framework based on assimilating data from databases, satellite remote sensing and the most recent local REA data to provide estimates of the oceanographic and geophysical fields, e.g. parameterized as Empirical Orthogonal Functions (EOF), and their associated uncertainties. The EOF parameterization also directly characterizes the spatial and temporal scales of the environmental features, needed later for the adaptive objective analysis. Using combined acoustic propagation and sonar models the system sensitivity is then determined, using quantitative sensitivity measures such as the Fisher information matrix, directly quantifying the sonar uncertainty in terms of the uncertainties of the oceanographic parameters [5]. This step produces a sonar performance metric in the form of a sonar PDFs, based on, and consistent with the environmental forecasts.

The sonar PDFs are still in the form of 'uncertainty maps' over the ocean volume, but representing the ultimate effect on the sonar detection dB-budget. The sonar uncertainty PDFs will then be combined with an objective measure of cost associated with the REA resource deployment, to provide a conditional probability distribution (CPD) which directly identifies which environmental parameters must be targeted to minimize the uncertainty of the acoustic prediction within the operational constraints.

This initial CPD is inherently assuming the underlying environmental uncertainty to have Gaussian statistics, an assumption many years of sonar operation experience shows to be unrealistic. To optimize the deployment pattern accordingly, Monte Carlo simulation of the sonar performance is performed. First, the new adaptive sampling patterns are applied to 'virtual oceans' produced by generating random ocean realizations based on the forecast field and error estimates. These are then translated into sonar PDFs applying the environmental acoustic modeling framework. In case the associated sonar performance uncertainty is unsatisfactory, the deployment strategy may be changed, and the procedure repeated. Finally, after the REA measurements have been collected, they are objectively analyzed similarly to the 'virtual' data, using the forecast scales, and passed through the sonar modeling framework, producing sonar performance predictions with significantly sharper uncertainty.

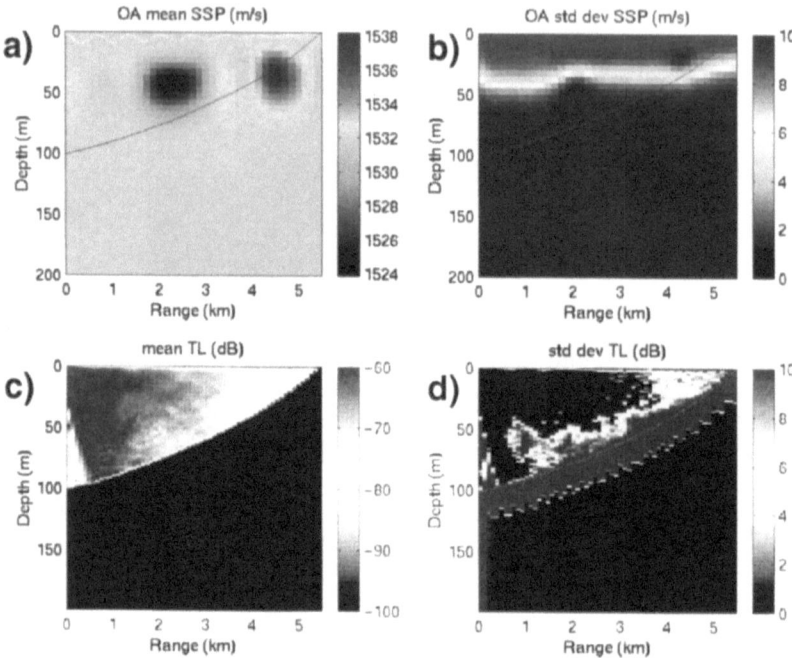

Figure 4. Transmission loss estimate for Procchio Bay track obtained by objective analysis of simulated surface temperature measurements using spatial statistics predicted by HOPS model, followed
by Monte Carlo simulation using PE acoustic propagation model. (a) Temperature estimate along
communication track. (b) Estimated uncertainty of temperature field. (c) Contours of estimated
mean transmission loss in dB. (d) Estimated uncertainty of transmission loss.

4   AREA example

As an example of the functionality of AREA for reducing the uncertainty of acoustic performance prediction, a low-frequency mine-countermeasure sonar scenario is used. The
SACLANTCEN GOATS JRP has as one of its objective to investigate the performance
of new low-frequency, bi- and multi-static sonar concepts, using a powerful sonar for
insonifying the very shallow water near the beach from a position off-shore, with multiple receivers being caried on one or more AUVs operating in the target area as bi-static
platforms [9]. During the SACLANTCEN GOATS'2000 experiment [10] carried out in
collaboration with MIT and Harvard University, among others, the Harvard Ocean Prediction System (HOPS) was used to forecast the temperature and current field in the
bay area which was the focus of the bi-static MCM experiments. Figure 3(a) shows an
example of the temperature estimate at 10 m depth. Figure 3(b) shows a hypothetical
insonification path from an off-shore 3.5 kHz source pointed towards shore in Procchio
Bay.

First the HOPS forecast is used to generate a 'true ocean' by generating a realization
of the forecast statistics. This 'virtual ocean' is then assumed to be 'sampled' by an AUV
using different survey patterns.

In the first example, the AUV is assumed to perform a constant-depth survey close

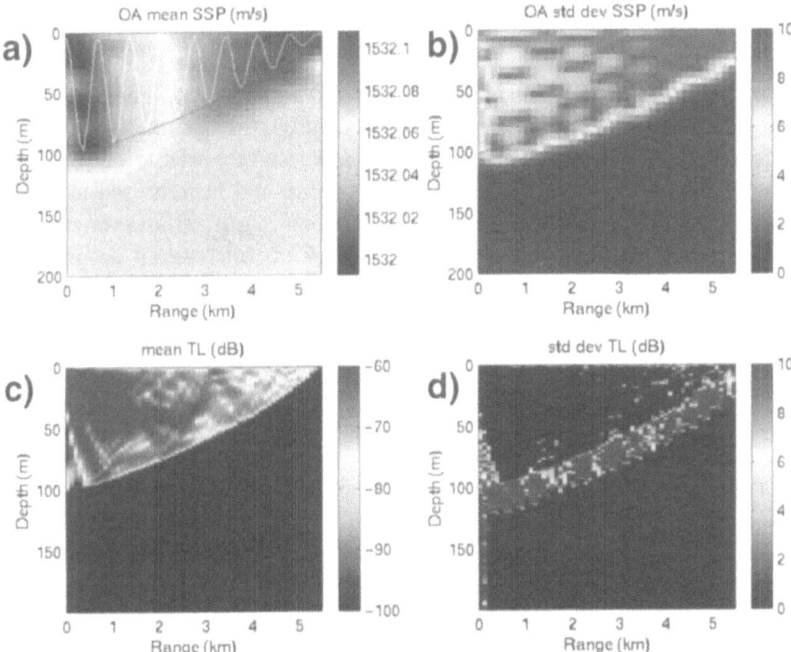

Figure 5. Transmission loss estimate for Procchio Bay track obtained by objective analysis of simulated volume temperature measurements by an AUV performing a 'yoyo' survey pattern, using spatial statistics predicted by HOPS model, followed by Monte Carlo simulation using PE acoustic propagation model. (a) Temperature estimate along communication track. (b) Estimated uncertainty of temperature field. (c) Contours of estimated mean transmission loss in dB. (d) Estimated uncertainty of transmission loss.

to the sea surface, measuring the temperature. These measurements are then objectively analysed using the spatial statistics produced by HOPS to produce the estimate of the sound speed shown in Fig. 4(a), and the associated uncertainty estimate in Fig. 4(b). As expected the sound speed has small uncertainty close to the sea surface where the temperature was measured directly by the AUV, while a large uncertainty is present at depths beyond 10 m, where no direct measurements are made by the AUV.

Using this new sound speed estimate and uncertainty, the expected transmission loss vs range and depth in Fig. 4(c) is obtained using Monte-Carlo simulation, based on 40 realizations, and the associated uncertainty in dB shown in Fig. 4(d). As is clear from Fig. 4(d) the transmission loss predicted for a communication source off-shore to a receiver close to the beach is of order 10 dB. Consequently, the bistatic sonar would be associated with 10–20 dB uncertainty, severely affecting the detection statistics.

Figure 5 shows the corresponding sound speed and transmission loss estimates in the case where one of the AUVs is performing a 'yoyo' survey on its path to the target area, reducing the transmission loss uncertainty everywhere in the water column to less than 1 dB. Even though this hypothetical scenario ignores the temporal variability, it clearly illustrates how the sonar performance prediction may be significantly improved by optimally deploying the resources aavailable for environmental assessment.

5 Conclusion

AREA is a new operational paradigm currently being developed by MIT in a partnership with other research teams involved in the ONR 'Capturing Uncertainty' DRI. Oceanographic forecasts obtained by modeling and assimilation networks such as HOPS are used to provide initial estimates of the environment and its spatial and temporal statistics. Acoustic propagation and sonar performance models are then used to derive the associated PDFs for the performance of the acoustic systems, and in turn used to identify the optimal deployment patterns for the available rapid environmental assessment resources. This paper has demonstrated the concept applied to establishing an accurate estimate of the transmission loss associated with a littoral mine countermeasures scenario, but is equally applicable to modern high-resolution passive sonar processing techniques such as matched field processing, which are even more sensitive to environmental mismatch.

Acknowledgments

The author appreciates the effort of the HOPS group at Harvard for providing the forecasts of the oceanographic field estimates used for the development of AREA. This research is sponsored by the Office of Naval Research under the *Capturing Uncertainty* DRI.

References

1. A.B. Baggeroer, W.A. Kuperman and H. Schmidt, Matched field processing: Source localization in correlated noise as an optimum parameter estimation problem, *J. Acoust. Soc. Am.* **83**, 571–587 (1988).
2. T. Curtin, J.G. Bellingham, J. Catipovic and D. Webb, Autonomous oceanographic sampling networks, *Oceanography* **6**(3), 86–94 (1993).
3. A.R. Robinson, P.F.J. Lermusiaux and N.Q. Sloan, Data assimilation. In *The Sea: The Global Coastal Ocean*, edited by K.H. Brink and A.R. Robinson, Vol. 10, 541–594 (1998).
4. A.R. Robinson, Forecasting and simulating coastal ocean processes and variabilities with the Harvard ocean prediction system. In *Coastal Ocean Prediction*, AGU Coastal and Estuarine Studies Series (American Geophysical Union, 1999) pp. 77–100.
5. H. Schmidt and A.B. Baggeroer, Physics-imposed resolution and robustness issues in seismo-acoustic parameter inversion. In *Full Field Inversion Methods in Ocean and Seismic Acoustics*, edited by O. Diachok, A. Caiti, P. Gerstoft and H. Schmidt (Kluwer Academic Publishers, Dordrecht, The Netherlands, 1994).
6. H. Schmidt, J.G. Bellingham and P. Elisseef, Acoustically focused oceanographic sampling in coastal environments. In *Rapid Environmental Assessment*, edited by E. Pouliquen, A.D. Kirwan and R.T. Pearson. SACLANTCEN Conference Proceedings Series CP-44 (1997).
7. P. Elisseeff, H. Schmidt and W. Xu, Ocean acoustic tomography as a data assimilation problem, *IEEE J. Oceanic Eng.* (in press 2002).
8. W. Munk, P. Worcester and C. Wunsch, *Ocean Acoustic Tomography* (Cambridge University Press, 1995).
9. J.R. Edwards, H. Schmidt and K. LePage, Bistatic synthetic aperture target detection and imaging with an AUV, *IEEE J. Oceanic Eng.* **26**(4), 690–699 (2001).
10. H. Schmidt and E. Bovio, GOATS: Autonomous vehicle networks. In *Proc. Intl. Conf. Maneuvering and Control of Marine Craft*, International Federation for Automation and Control, Aalborg, Denmark, Aug. 23–25, 2000.

ENVIRONMENTALLY ADAPTIVE SONAR CONTROL
IN A TACTICAL SETTING

WARREN L.J. FOX, MEGAN U. HAZEN AND CHRIS J. EGGEN

University of Washington, Applied Physics Laboratory, 1013 NE 40th St., Seattle WA 98105, USA
E-mail: warren@apl.washington.edu

ROBERT J. MARKS II AND MOHAMED A. EL-SHARKAWI

University of Washington, Dept. of Electrical Engineering, Box 352500, Seattle WA 98105, USA

Automatic environmentally adaptive sonar control in littoral regions characterized by high spatial/temporal acoustic variability is an important operational need. An acoustic model-based sonar conroller requires an accurate model of how the sonar would perform in the current environment while in any of its possible configurations. Since high-fidelity acoustic models are computationally intensive, and finding the optimal sonar mode may require a large number of these model runs, such a controller may not be able to provide optimal line-up solutions in tactically useful time frames. We have explored a method of statistically characterizing a given operations area, generating a large ensemble of acoustic model runs, and training specialized artificial neural networks to emulate acoustic model input/output relationships. The neural networks reproduce the acoustic model outputs to a good degree of accuracy in a small fraction of the compute time needed for one of the original model runs. In this paper, the neural network training method is described, examples of neural network performance are given, and an example of controller solutions in a variable environment are presented.

1 Introduction

Naval sonar systems continue to evolve and become more capable, while at the same time becoming more complex to operate. With the emergence of littoral areas as the prime regions of interest, characterized by underwater acoustic environments that change quickly in both the temporal and spatial domains, automatically optimizing sonar line-ups has become a key operational need. The desire is for sonar operators to concentrate on the key tasks of target detection and classification, while the sonar system automatically determines an optimal line-up based on the current goals of the operator and an estimate of the current environment. Also, as autonomous systems are developed for operational use, the need for automation of environmentally adaptive sonar control becomes paramount.

Sonar control schemes generally fall into two categories: rule-based, and acoustic model-based. Rule-based systems are developed by acoustic and sonar system experts, who first define generic sets of environmental conditions, and then apply acoustic modeling techniques and the sonar equation to determine the best line-up for the sonar in those

N.G. Pace and F.B. Jensen (eds.), Impact of Littoral Environmental Variability on Acoustic Predictions and Sonar Performance, 595-602.
© *2002 Kluwer Academic Publishers.*

conditions. In practice, the environment in which the sonar is deployed must be assessed as to which of the generic design environments is closest to the real environment, so that the proper sonar line-up can be set. Although relatively simple and with a low real-time computational burden, rule-based controllers may not be able to take into account all of the environmental variability that may confront the sonar system and its operators.

Model-based controllers embed an acoustic model in the real-time controller. Typically, the best available estimate of the current environment is fed into the controller which makes acoustic performance predictions for the various possible line-ups of the sonar system. The line-up that best satisfies some performance metric (which may change based on the current employment of the sonar) is chosen. Although a model-based controller is more readily able to adapt to finer scale environmental conditions than a rule-based controller, high-fidelity acoustic models are computationally intensive. Assessing the performance of the various modes of the sonar may take too much time to be useful in an environment with high temporal/spatial variability, or may require excessive computing resources.

In this paper, we present a method of training artificial neural networks to emulate the input/output relations of a computationally intensive acoustic model for use in a sonar controller, either shipboard or aboard an autonomous vehicle. The advantage of using the neural networks is that they generate these acoustic model emulations orders of magnitude faster than it would take the original high-fidelity acoustic model to run, using modest computing resources. We describe the basic neural network training methodology, along with some special techniques developed specifically for this application. We also show some examples of the training performance. Finally, we give an example of control solutions obtained using a neural network.

2   Neural network training

2.1   Basic Idea

Neural networks are mathematical constructs loosely modeled on biological neural interconnections [1]. Figure 1 shows a schematic of how the neural network training is performed in this application. In the figure, a multilayer perceptron neural network is established with an input layer, one hidden layer, and an output layer. The number of hidden layers and number of nodes in each layer are design parameters, and must be considered carefully for each application. In Fig. 1, only some of the connections between layers are shown for illustration purposes. In reality, each node in any given layer is connected to every other node in the preceding and following layers.

The input layer contains parameters describing the sonar (e.g., center frequency, bandwidth, vertical steering angle, etc.) and parameters describing the environment (e.g., wind speed, bottom type, sound speed, etc.). These values on the input layer of the neural network are also used as inputs to an acoustic model. In the case of Fig. 1, the model computed signal-to-interference ratio for hypothetical locations of a target in a vertical slice of the ocean. The outputs of the model are assigned to nodes of the output layer of the neural network. The network is then trained using error back-propagation [1], so that when these inputs are presented to the neural network, a forward computation through the neural network reproduces, to some level of fidelity, the outputs of the acoustic model. Neural networks used in this fashion are sometimes referred to as "regression machines,"

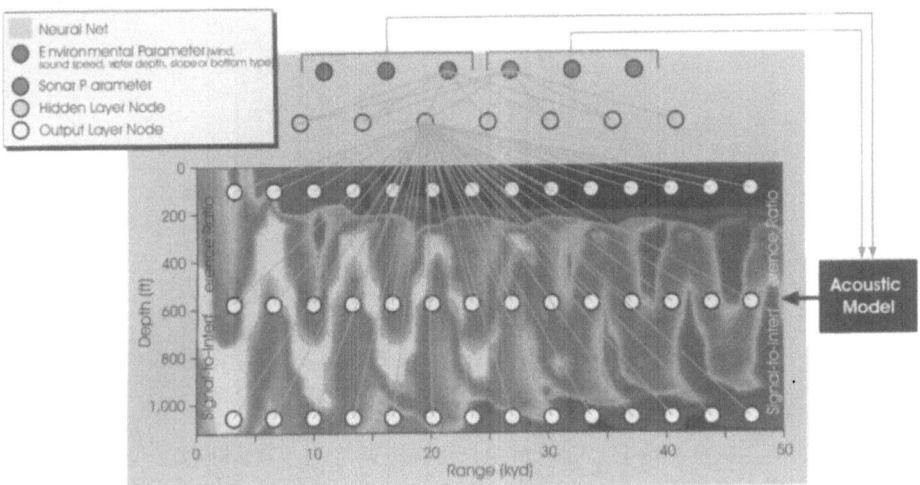

Figure 1. Neural network training on acoustic model output (only a subset of inter-node connections are illustrated).

or as "associative memories."

For neural networks with three hidden layers and roughly 50 nodes per layer, one of these forward computations takes about 5 milliseconds on a standard (circa 2002) desktop workstation. High-fidelity acoustic models may take on the order of 60 seconds for the original computation. In practice, the training data generation and neural network training would take place in laboratories with high-performance computing facilities, and the neural network subroutines would be installed on the sonar platform for use in the real-time sonar controller.

2.2 Training Data

The neural networks are actually trained on an ensemble of acoustic model runs. In order to generate this ensemble, we define a geographic region in which the sonar system will be operating, and statistically characterize both the environmental parameters in this region and the possible settings of the sonar parameters. For example, we may allow wind speed to be a uniform random variable between 0 and 12 m/s, while volume scattering strength is Gaussian with mean -75 dB/m^3 and variance of 5 dB/m^3.

For training the neural network, a standard set of depths at which sound speed in the water column will be specified must be defined. Care must be taken with variables where correlation is important, such as sound speed profile. Simply allowing the various points to vary independently as a function of depth might produce unrealistic sound speed profiles. A technique has been developed [2] for generating realistic sound speed profiles by collecting sets of historical sound speed profiles for an area, computing the covariance matrix, and multiplying the Cholesky factors of the covariance matrix by a set of independent random numbers. The first and second order statistics of the data are thereby maintained in the randomly generated sound speed profile data.

In our development of this technology, we have limited the geographic region being characterized to boxes $O \sim 1°$ square, and generated sets of acoustic models with 20,000-

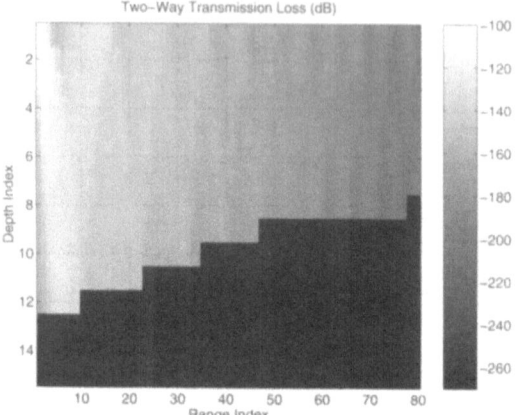

Figure 2. Example of two-way transmission loss used for neural network training.

40,000 members. As the geographic area grows and encompasses more environmental variability, the training set would need to grow and the structure of the neural network may also need to be modified.

2.3 Training for Variable Bathymetry

Figure 2 shows a typical transmission loss plot (two-way) used in neural network training. The dark portion at the bottom of the figure corresponds to the sea bottom, in which we assume the acoustic field levels are very low and are not part of our modelling process. We have made a bilinear approximation to the bottom bathymetry for our current development, meaning that the input layer to the neural network has four nodes corresponding to a description of bathymetry (depth at source, range of breakpoint, depth of breakpoint, and depth at final range). This approximation could be expanded at the cost of more nodes on the input layer, although the bilinear approximation has shown to be fairly robust in our testing [2].

An important issue with the neural network training has to do with the bathymetry. As with the other parameters, the bathymetry is generated randomly for the different members of the ensemble input sets, meaning that certain range-depth pixels would be in the water column for some cases and in the sea bottom for others. Early training efforts simply considered all output pixels in a uniform manner, but the transition from water column to below the seabed was difficult for the networks to learn, resulting in unacceptably high errors. A novel technique for avoiding this problem was developed [3,4], dubbed "don't care training." Here, during error backpropagation training, weights connected to output nodes associated with pixels in the sea bottom are not updated. See references [3,4] for a more complete description of the technique.

3   Neural network results

3.1 Training Data

Figure 3 shows some examples from a neural network trained on modelled two-way tranmission loss for an active sonar. The first column contains four samples from a training

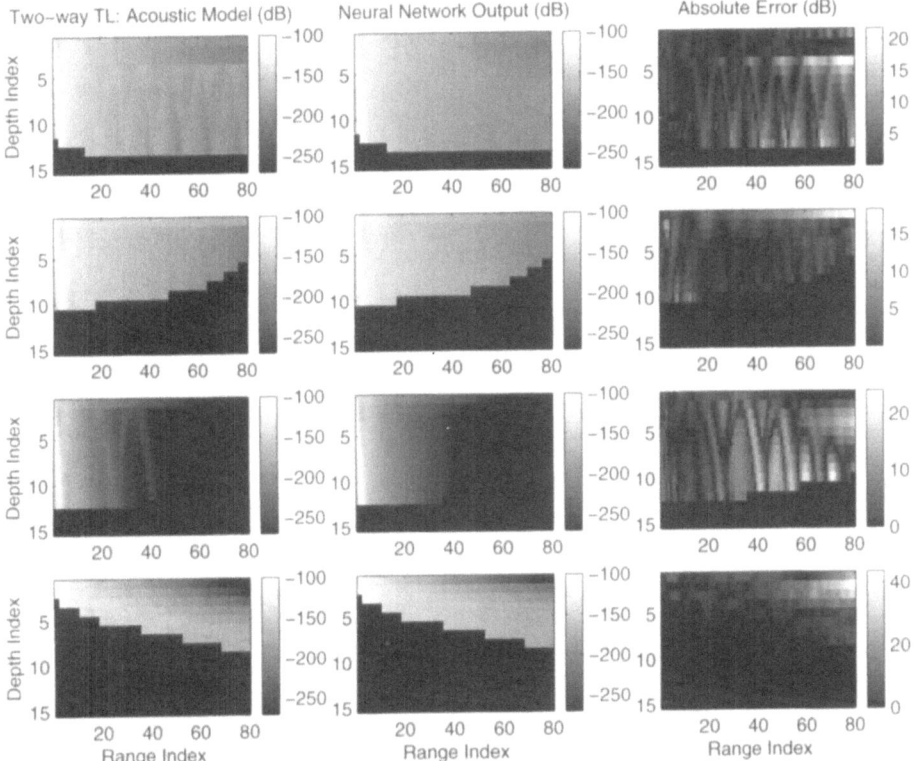

Figure 3. Examples of neural network performance on two-way transmission loss: training data.

set of 40,000 model runs, generated as described in the previous section. The second column contains the corresponding neural network outputs when the model parameters used to generate the results in the first column are placed on the input nodes of the neural network. The third column contains the absolute values of the difference between the first and second column.

The example in the first row is for a case of a downward refracting sound speed profile and a low-loss bottom, resulting in the arching pattern seen in the first plot. The neural network reproduces the pattern early in range, but tends to smear the arch energy in range and depth at longer ranges. The error plot shows the residual of the arching pattern at the longer ranges. One reason for this behavior is the sensivity of the acoustic model to small changes in input values. For example, a small change in sound speed profile can change the exact location in the range/depth plane of the arching patterns, and this sensitivity is difficult to train for. For use in a controller, it may be enough to know that there are significant amounts of energy propagating out to a particular range withough knowing the detailed structure of the acoustic field. Even if the actual acoustic model were embedded in the controller, imprecise knowledge of model input parameters would lead to imprecise location of these types of structures.

The second row is a case of a shallow surface duct and a low-loss bottom resulting in significant sub-duct propagation. The surface duct is fairly well reproduced by the

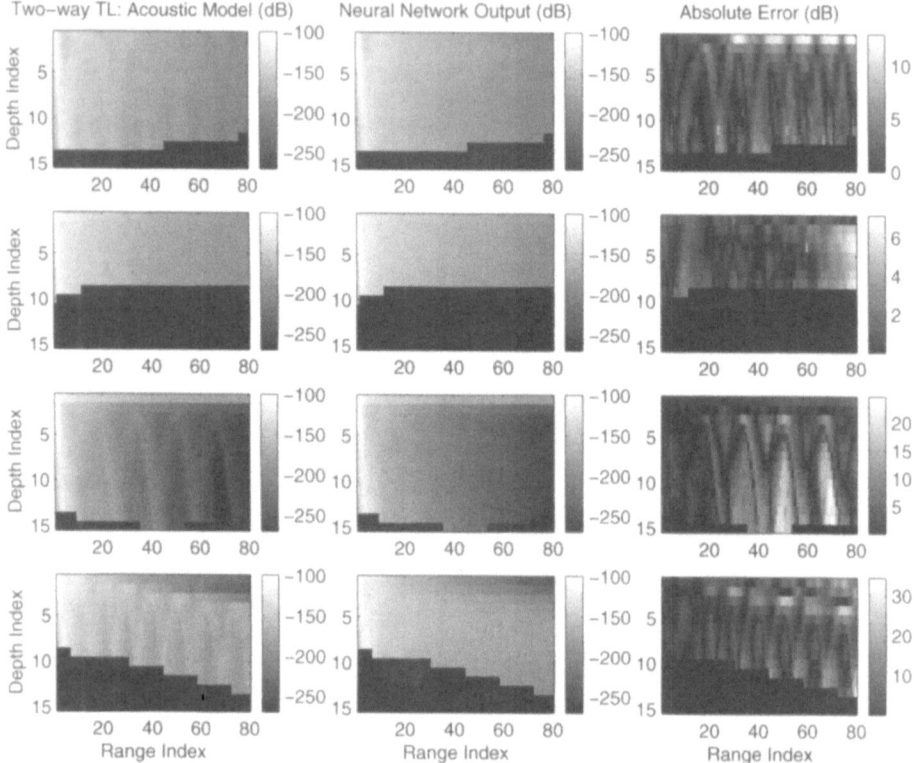

Figure 4. Examples of neural network performance on two-way transmission loss: testing data.

neural network, with increasing errors at long range. Again, for use in a controller, the most important piece of information may be the existence of the duct, not the exact level of the acoustic field in the duct.

The third row is a case of a downward refracting sound speed profile and a high loss bottom. Note that the neural network correctly reproduced the differences in the general trends between the low loss bottom in the first example and the high loss bottom here. Note that an arching pattern also exists here, although it is only evidenced by the residual seen in the error plot. Since it is occurring in an area of very high transmission loss, it would probably not be an issue for a controller.

The fourth row shows an example of down-slope propagation. The neural network reproduces fairly well the general trends of the propagation, with the exception of some high errors at the furthest ranges.

3.2   Testing Data

As with any neural network application, assessment of the performance must be made using testing data, i.e., data that was not used during the training process. Figure 4 is formatted similar to Fig. 3, but has examples of testing data. The input parameters were generated similarly to the training data.

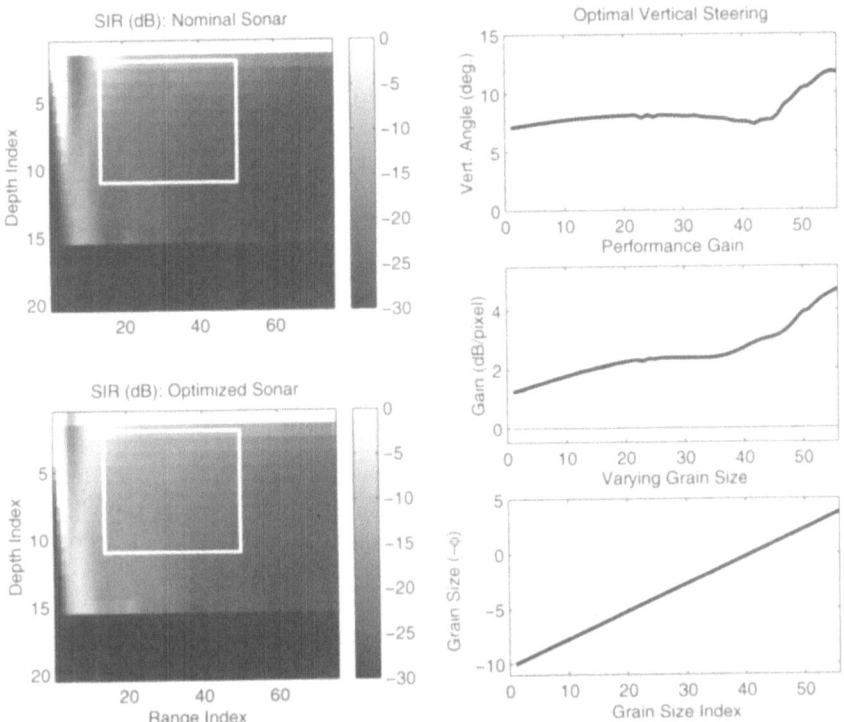

Figure 5. Examples of controller results.

Note that similar characteristics are evident in the training examples: general acoustic trends are represented fairly well, and some details (i.e., arching patterns) are smeared (or averaged) in range and depth. This gives us confidence that the neural network has not over-trained on the data (referred to as "memorizing").

4   Controller results

Figure 5 shows the results of using a neural network trained using the above-described method in a sonar controller [5]. In this case, the data used in the training was signal-to-interference ratio (SIR) for an active sonar, assuming a hypothetical target with a fixed low target strength at all possible locations in the water column.

The environment was held constant except for the bottom type, which was varied from very soft (high loss) to very hard (low loss). The bottom is characterized by the parameter ϕ, where $\phi = -log_2(d)$, where d is the grain size in mm. The parameter $-\phi$ (note the sign change) was used as an input to the acoustic model used for neural network training, and subsequently to the neural network itself on one of the nodes in the input layer. The variation of grain size is shown in the bottom right plot of Fig. 5.

The left two plots of Fig. 5 show range-depth maps of SIR, with a target search area outlined in a white box. This is the region over which the controller was required to obtain the maximum average SIR. The controller was allowed to control the vertical steering angle of the sonar. The top plot shows the SIR obtained for the final $-\phi = 4$

with a nominal sonar steering angle of 3 degrees, and the bottom plot shows the SIR obtained using the controller-optimized steering angle of approximately 12 degrees. The optimal steering angles for all bottom types is plotted in the upper right, and the average gain per pixel obtained by using the optimal steering angle vice the nominal is plotted in the middle axis on the right.

In this example, it is shown that even for softer bottoms, steeper vertical steering angles around 7 degrees are more advantageous, with a sharp increase for harder bottoms, accompanied by higher gains in performance. The key point here, however, is not the specific results for this example, but that the results can be generated in a very short period of time compared to what would be necessary if the actual acoustic model were used. This allows rapid investigation of how sonar systems should be employed in particular environments.

5 Discussion

It is important to keep in mind that the technique described in this paper for emulating acoustic model input/output relations is intended for use in a sonar controller. In other words, the emulation must be "good enough" to put the sonar in the correct configuration. The emulation is not intended to be used for detailed acoustic analysis. As mentioned before, the main benefit of this technique is the reduced computational complexity it brings for real-time applications. The primary thrust of our continuing work is assessing trained neural networks in actual controller scenarios, and comparing the controller solution performance against controllers with embedded acoustic models.

Also of note is that the training and testing examples presented in Sect. 3 are the result of fairly straightforward data set generation and neural network training with the "don't care" technique for range-depth pixels below the sea bottom. We have also developed several techniques whose details are beyond the scope of this paper for improving the performance of the neural networks. One of these involves detailed examination of the input/output sensitivities of the underlying acoustic model (from the generated data set), and insertion of additional data (model runs) where the sensitivities are high. These techniques are also important aspects of our continuing work.

Acknowledgements

This work was performed under sponsorship from the Office of Naval Research, contract number N00014-98-G-0001.

References

1. Reed, R.D. and Marks II, R.J., *Neural Smithing: Supervised Learning in Feedforward Artificial Neural Networks* (MIT Press, Cambridge, MA, 1999).
2. Eggen, C.J., Unpublished APL/UW Memorandum (Oct. 2001).
3. Jung, J.-B., *et al.*, Neural network training for varying output node dimension. In *Proc. Intl. Joint Conf. on Neural Networks* **3**, 1733–1738 (2001).
4. Jung, J.-B., Neural network ensonification emulation: Training and application. Ph.D. Dissertation, University of Washington, Seattle, Washington (2001).
5. Jensen, C.A., *et al.*, Inversion of feedforward neural networks: Algorithms and applications, *Proc. IEEE* **87**(9), 1536–1549 (1999).

TRANSFER OF UNCERTAINTIES THROUGH PHYSICAL-ACOUSTICAL-SONAR END-TO-END SYSTEMS: A CONCEPTUAL BASIS

A.R. ROBINSON[1], P. ABBOT[2], P.F.J. LERMUSIAUX[1] AND L. DILLMAN[2]

[1]*Harvard University, Cambridge, MA 02138, USA*
E-mail: robinson@pacific.deas.harvard.edu, pierrel@pacific.deas.harvard.edu

[2]*Ocean Acoustical Services and Instrumentation Systems (OASIS), Inc.,*
5 Militia Dr., Lexington, MA 02421, USA
E-mail: abbot@oasislex.com, dillman@oasislex.com

An interdisciplinary team of scientists is collaborating to enhance the understanding of the uncertainty in the ocean environment, including the sea bottom, and characterize its impact on tactical system performance. To accomplish these goals quantitatively an end-to-end system approach is necessary. The conceptual basis of this approach and the framework of the end-to-end system, including its components, is the subject of this presentation. Specifically, we present a generic approach to characterize variabilities and uncertainties arising from regional scales and processes, construct uncertainty models for a generic sonar system, and transfer uncertainties from the acoustic environment to the sonar and its signal processing. Illustrative examples are presented to highlight recent progress toward the development of the methodology and components of the system.

1 Introduction

The littoral environment can be highly variable on multiple scales in space and time, and sonar performance can be affected by these inherent variabilities. Uncertainties arise in estimates of oceanic and acoustic fields from imperfect measurements (data errors), imperfect models (model errors), and environmental variabilities not explicitly known. The focus of this paper is to present a conceptual basis to achieve the following: i) develop generic methods to efficiently characterize, parameterize, and prioritize system variabilities and uncertainties arising from regional scales and processes; ii) construct, calibrate, and evaluate uncertainty and variability models for the end-to-end system and it's components to address forward and backward transfer of uncertainties; and, iii) transfer uncertainties from the acoustic environment to the sonar and its signal processing in order to effectively characterize and understand sonar performance and predictions. In order to accomplish these objectives, an end-to-end system approach is necessary.

2 End-to-end system approach

An overview of the forward portion of our approach is shown by the bold arrows in Fig. 1, which connect the environment, acoustics and sonar to applications. The characterization and transfer of uncertainty begins with the environment, in particular

N.G. Pace and F.B. Jensen (eds.), Impact of Littoral Environmental Variability on Acoustic Predictions and Sonar Performance, 603-610.

from the spatial and temporal variability in the physical oceanography and spatial variability in the bottom. These effects impact the acoustics and result in uncertainties in acoustic predictions of propagation loss, reverberation and ambient noise. The uncertainties impact sonar performance, specifically, the ability of the system to detect a target. Ultimately, these methodologies can lead to improvements in scientific and fleet naval applications.

END-TO-END APPROACH

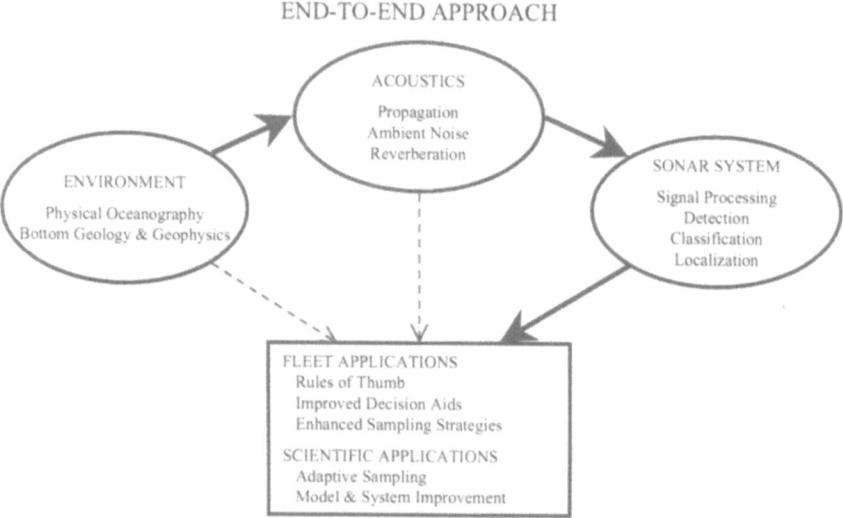

Figure 1. Overview of the end-to-end forward approach for capturing and transferring uncertainty.

The interactions among the different components of the system being more complex than Fig. 1 implies, it is only after multiple discussions on our respective disciplines and the corresponding information exchanges that a comprehensive and schematic picture of the end-to-end system crystallized. Figure 2 schematizes the end-to-end system from the model point of view, where models are used to represent each of the coupled dynamics (boxes) and also the linkages to observation systems (circles). An effort was made to make the diagram exact but as simple as possible. The diagram illustrates the forward transfer of information, including uncertainties, in terms of observed, processed and model data (dots on arrows) and products and applications (diamond). The system concept encompasses the interactions and transfers of information with feedback from: i) observing systems, the information being physical-acoustical-bottom-noise-meteorological-sonar data, ii) coupled dynamical models, the information being physical-acoustical-bottom-noise-sonar state variables and parameters, and, iii) sonar equation models, the information being parameters in sonar equations. Note that the sonar system requirements configures the acoustic component and other system components for any particular problem and thereby sets the scales of uncertainty for the system.

Each of the coupled parameters and fields, i.e., the ocean environment, bottom characterization, noise field, acoustic state and sonar parameters, are classically described by mathematical equations; for example and respectively, by a primitive-equation model, Hamilton model, Wenz model, Helmholtz wave-equations and sonar equations (Sect. 3.2). These models are usually either static or dynamical, and deterministic or statistical. In addition to environmental noise, the sampling, processing and assimilation of data output from observational systems are also sources of uncertainty. First, sensor data are approximate. Second, platforms are of limited coverage in space and time and in variables measured. The linkages between data outputs from observational systems and the coupled model fields also involve substantial processing, thus uncertainties (Sect. 3.1). These linkages are represented by measurement models, which include error estimates. Data assimilation [1], combining data and model estimates to minimize uncertainties, is an integral part of our end-to-end system.

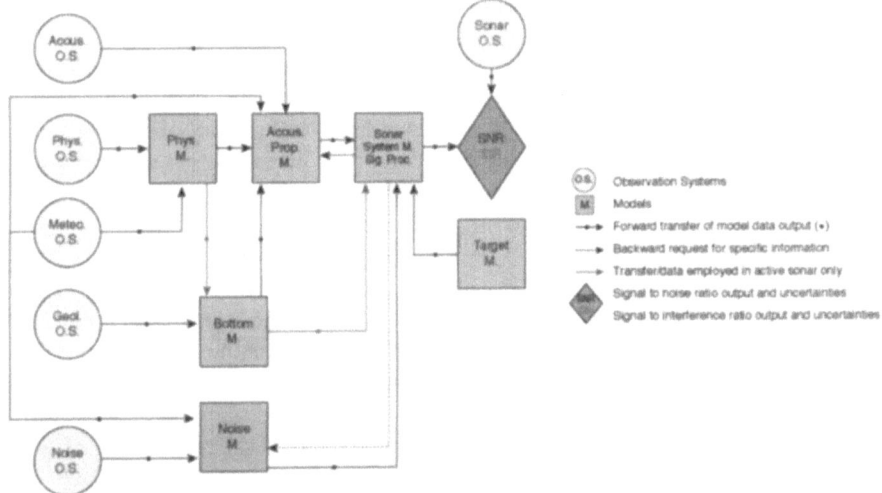

Figure 2. Schematic diagram of the end-to-end system (model point of view).

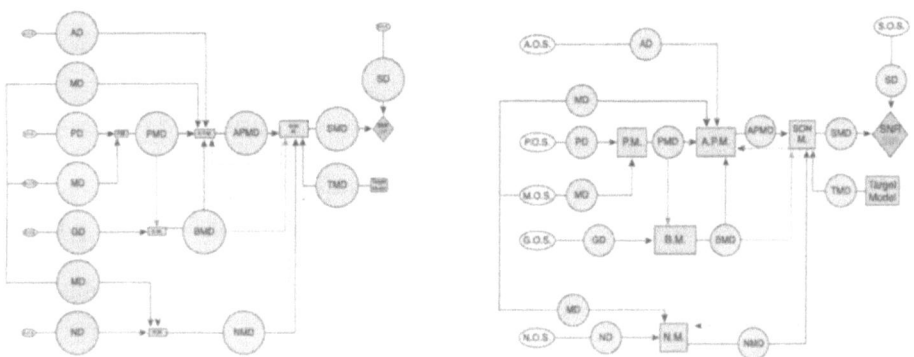

Figure 3. a) As Fig. 2 but from the data point of view; b) as Fig. 2, but with all (data + model) details.

In Fig. 2, each dot on an arrow corresponds to sets of estimated fields and of uncertainties. These information dots are enlarged on Fig. 3a; the detailed model and data diagram is on Fig. 3b. In these diagrams, uncertainties are defined by a representation of the likely errors in the estimated fields, usually in some probabilistic sense. Although uncertainties can be reduced, e.g. via data assimilation, there always remain some irreducible errors which need to be represented. There are several methodologies to represent errors and, for such complex interdisciplinary systems, efficient representations are an issue. An approach that we utilize (Sect. 4) is based on evolving an error subspace, of variable size, that spans and tracks the scales and processes where dominant errors occur. With this technique, Error Subspace Statistical Estimation (ESSE, [2–4]), the sub-optimal reduction of errors is itself optimal. Presently, the error subspace is initialized by decomposition on multiple scales and evolved in time by an ensemble of stochastic model iterations, where the stochastic terms represent model errors. The ensemble size is controlled by convergence criteria and *a posteriori* data residuals are employed for adaptive learning of the dominant errors [5]. This learning of errors from data misfits can be necessary because error estimates are themselves uncertain. In fact, a complete system likely represents each dot on an arrow (Fig. 2) as the sum of the: i) field, ii) its uncertainty and iii) the uncertainty on its uncertainty. Fuzzy information theories based on new uncertainty measures [6] and imprecise probability theories [7] are thus considered.

3   End-to-end components

In this section, we discuss in more detail the end-to–end system components (Figs. 2, 3). The descriptions are limited to systems that may be employed in the current research and are not meant to be exhaustive. The present effort focuses on target detection performance.

3.1   *Observation Systems and Data*

Each observation system (OS on Figs. 2, 3) corresponds to measurement models and uncertainties that describe the direct measurements, the sensors and procedures used to collect the data. In what follows, some examples are listed for illustrative purposes.

Acoustic Observation Systems. Acoustic data are observed and processed to produce estimates such as transmission loss (TL), reverberation, scattering, attenuation etc. In our end-to-end system, we differentiate the direct observations from calculated acoustic parameters. Pertinent information, especially the uncertainty, regarding the collection of these data is contained by the AOS circle. The processed data and the method used to process the data, is included in the acoustic data products (data dot in Fig. 2).

Physical Observation Systems. Physical data (temperature, salinity and velocity profiles) involve processing of raw measurements from XBTs, CTDs, ADCPs and current meters. Averaging, filtering, de-aliasing and calibration is necessary. Radar and satellite data (SSH, SST, etc.) also require processing, especially for use at high resolution in coastal regions.

Geological Observation Systems. Geological data may be processed to produce data products used to characterize the ocean bottom properties such as bathymetry, bottom loss, bottom backscattering, sub bottom profiles, sediment thickness, sediment type, roughness, grain size, density and sound speed in the sediment, etc. Sensors and platforms include core grabbers, seismic surveys and source-receiver arrays.

Noise Observation Systems. Although the noise measurements are usually collected jointly with acoustic measurements, it is differentiated here to allow individual assessments of this important parameter. Noise data may be processed to produce data products to characterize or calculate ambient noise as a function of depth and azimuth, shipping noise, shipping density, sea surface agitation and characteristics, etc.

Sonar Observation Systems. Sonar observations are the observed SIR or SNR output of the sonar system and include the hardware as well as the signal processing characteristics of a given system. In our end-to-end system, the observed sonar SIR and SNR data are compared to modeled SIR and SIR that include the aggregate uncertainty.

3.2 Coupled (Dynamical) Models

The dynamical models and acoustical models are derived from the basic equations of the conservation of momentum, mass, heat, etc. for the fluid continuum (i.e. Navier-Stokes eqs.) appropriate approximations lead to the primitive equations for dynamics [8] and the Helmholtz equations for acoustics [1]. As discussed in [8,9] the ocean environment is a complex system that is inhomogeneous in three spatial dimensions and time. To completely describe this requires a conceptual model capable of handling the intricacies of the real ocean. For most applications further approximations are necessary. Numerical models discretize the approximate physical models for numerical results. Computations are subject to errors associated with the algorithms used and the limitations of the computer. A short overview of relevant coupled dynamical models (Figs. 2, 3) is now provided.

Physical Models. Physical oceanography encompasses a wide range of scales and considers processes from microstructure interactions to climate change. The focus here is on meso- and sub-mesoscale fronts and eddies, tides, internal tides, waves and solitons. Relevant dynamical models are then partial differential equations which include, in order of decreasing complexity: i) primitive-equation models, with or without a free ocean surface, ii) shallow-water models, iii) non-hydrostatic models, and iv) simplified time-independent balance equations such as the geostrophic equilibrium and thermal-wind balance. Most numerical ocean models are finite-difference, but some finite element models are also used. Only a few stochastic error models are employed, e.g. for representing uncertainties due to sub-mesoscale and internal tidal effects in mesoscale resolution models [4], but such stochastic research is growing.

Acoustic Propagation Models. As reviewed in the literature (e.g. [9]), several propagation models may be used in this research. Most existing models are based on similar assumptions of boundary conditions and acoustic frequency, being all solutions or approximations to the Helmholz equation (linearized wave-equation). These models include Ray Models, Parabolic Equation Models, Normal Mode Models and Coupled Mode Models.

Bottom Models and Noise Models. Bottom and sea-state surface models provide boundary conditions to acoustic models. Few studies have quantified the nature of the variability in bottom attribute (attenuation, reflectivity, velocity, density and proxies). Most geological bottom models, e.g. Hamilton model [10], are either a deterministic or a statistical representation of these attributes. For naval applications, several noise models are based on empirical and rule-of-thumb models, e.g. the Wenz [11] ambient noise spectra. However, this is an area of active research and complex models for the 3-D in space, frequency-dependent, ambient noise (shipping, biologics and sea-surface noise) are being developed.

Sonar Models. We focus on target detection and the sonar model is represented by the sonar equations [12,13] that describe signal excess (SIR or SNR) for an active or passive system. The sonar equations describe the interdependence of the environment, target, and sonar system and provide an estimate of the performance of the sonar system.

4   Progress and prospectus

In this section, we illustrate the methodologies that we developed to: i) characterize and transfer environmental uncertainties to acoustical uncertainties [3]; and ii) characterize uncertainty as applied to transmission loss and sonar performance prediction.

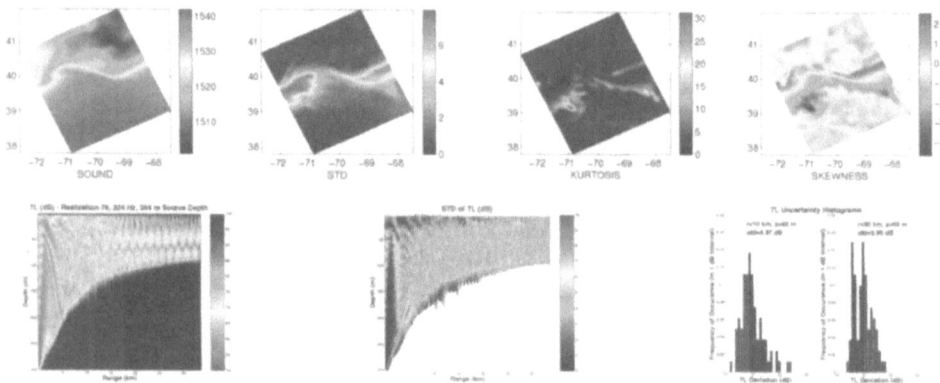

Figure 4. ESSE example of uncertainty transfer from the ocean physics to the acoustics.

Figure 4 illustrates an ESSE simulation of the transfer of environmental uncertainty to acoustic prediction uncertainty in a shelfbreak environment.   Based on observed oceanographic data during the 1996 Shelfbreak Primer Experiment, HOPS was initialized with perturbed fields that are in statistical accord with a realistic error subspace and then integrated to produce 80 realizations of a regional forecast of the sound speed field.   The different realizations of the sound speed were then fed into a sound propagation model to produce realizations of the predicted TL for a low-frequency transmission from the slope, across the shelfbreak, onto the shelf.   The ensemble mean, standard deviation, kurtosis and skewness of the sound speed at the surface level are shown (top).   A realization of TL along the transmission path, its standard deviations and the histograms (PDF estimates) of TL uncertainties at two different locations (shelfbreak and shelf) are shown (bottom), revealing the complexity and inhomogeneity of the uncertainty statistics in this locale.

Next we discuss the use of uncertainty in the development of probabilistic performance predictions, using the metric predictive probability of detection (PPD). Figure 5a illustrates conceptually the PPD, details of which are given in [14]. The system-based environmental PDF is derived in our studies by a comparison of model predictions with system data. A histogram of the differences between the data and the acoustic model is fit with an appropriate distribution to yield the PDF. This PDF represents the uncertainty in the computational modeling process, typically small, and the inherent variability of the environment not contained in the model inputs, which typically is larger. The PPD is a prediction of the system performance versus range. Rather than use a single range value (e.g. "range-of-the-day" or "range-of-the-moment"), the PPD provides the system operator with a probabilistic representation of the system performance. The operator can thus use this information to operate the system more effectively, and can make more informed decisions on search, risk, and expenditure of assets.

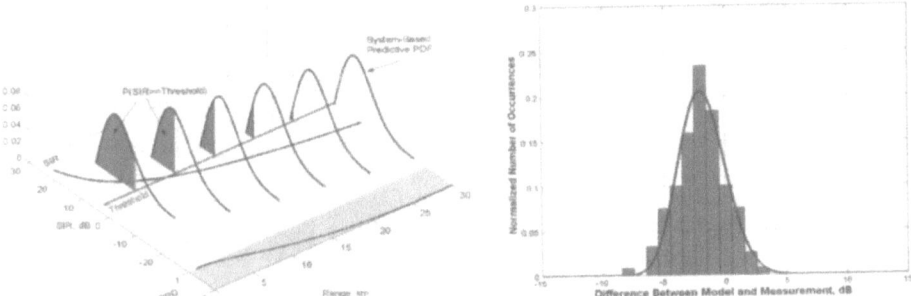

Figure 5. (a) Probabilistic system performance prediction and Predictive Probability of Detection (PPD) vs. range using system-based environmental PDF that incorporates environmental uncertainty. (b) 1-way TL environmental PDF fit to histogram.

Figure 5b illustrates an example of the 1-way TL environmental PDF based on operational experiments in shallow water [14]. This PDF quantifies the TL differences between predictions of the model and the truth of the data, caused by stochastic variability of the environment. Additional PDFs are presently being developed. These will include environmental (ambient noise, reverberation) and non-environmental (source level, target strength, system self-noise).

5 Summary

The end-to-end system encompasses the interactions and transfers of information with feedback from observing systems, coupled dynamical models that result in sonar performance predictions. The linkages and feedback among these different components are now being developed. The end-to-end framework is designed to support the individual components, environmental as well as non-environmental uncertainties (system related) so that an assessment of the dominant mechanism of uncertainty as it affects the SNR/SIR (Fig. 2) can be identified. The ability to assess the importance of the individual uncertainty components in the sonar performance prediction along within its aggregate uncertainty can be an invaluable tool in the development of tactical guidance.

Acknowledgments

We thank the Office of Naval Research for support under grants N00014-00-1-0771 and contract N00014-00-D-0119, and W.G. Leslie for manuscript preparation. PFJL thanks Prof. C.-S. Chiu for his collaboration and M. Armstrong for figure preparation. This is a contribution of the UNITES team.

References

1. Robinson, A.R., Lermusiaux, P.F.J. and Sloan, N.Q., Data assimilation. In *THE SEA: The Global Coastal Ocean*, Vol. 10: Processes and Methods, edited by K.H. Brink and A.R. Robinson (John Wiley and Sons, New York, 1998) pp. 541–594 .

2. Lermusiaux, P.F.J. and Robinson, A.R., Data assimilation via error subspace statistical estimation; Part I: Theory and schemes, *Monthly Weather Review* 127(8), 1385–1407 (1999).

3. Lermusiaux, P.F.J., Anderson, D.G. and Lozano, C.J., On the mapping of multivariate geophysical fields: Error and variability subspace estimates, *Q.J.R. Meteorol. Soc.*, 1387–1430 (2000).

4. Lermusiaux, P.F.J., Chiu, C.-S. and Robinson, A.R., Modeling uncertainties in the prediction of the acoustic wavefield in a shelfbreak environment. In *Proc. 5th International Conference on Theoretical and Computational Acoustics* (2001).

5. Lermusiaux, P.F.J., Estimation and study of mesoscale variability in the Strait of Sicily, *Dynamics of Atmospheres and Oceans*, Special Issue 29, 255–303 (1999).

6. Klir, G.J. and Wierman, M.J., *Uncertainty-Based Information. Elements of Generalized Information Theory*, 2nd ed., Studies in Fuzziness and Soft Computing (Physica-Verlag, 1999).

7. Walley, P., *Statistical Reasoning with Imprecise Probabilities* (Monographs on Statistics and Applied Probability Series, No. 42, 1991).

8. Robinson, A.R. and Lee, D., Ocean variability, acoustic propagation and coupled models. In *Oceanography and Acoustics: Prediction and Propagation Models*, edited by A.R. Robinson and D. Lee (AIP Press, New York, 1994) pp. 1–6.

9. Chin-Bing, S.A., King, D.B. and Boyd, J.D., The effects of ocean environmental variability on underwater acoustic propagation forecasting. In *Oceanography and Acoustics: Prediction and Propagation Models,* edited by A.R. Robinson and D. Lee (AIP Press, New York, 1994).

10. Hamilton, E., Geoacoustic modeling of the sea floor, *J. Acoust. Soc. Am.* 68(5), (1986).

11. Wenz, G., Acoustic ambient noise in the ocean: Spectra and sources, *J. Acoust. Soc. Am.* 34 (12), (1962).

12. Urick, R. J., *Principles of Underwater Sound*, 3rd ed. (McGraw-Hill, New York, 1983).

13. Kinsler, L., Frey, A., Coppens, A. and Sanders J., *Fundamentals of Acoustics*, 3rd ed. (John Wiley & Sons Inc., New York, 1982).

14. Abbot, P. and Dyer, I., Sonar performance predictions based on environmental variability. In *Impact of Littoral Environmental Variability on Acoustic Predictions and Sonar Performance,* edited by N.G. Pace and F.B. Jensen (Kluwer, The Netherlands, 2002) pp. 611–618.

15. Abbot, P., Celuzza, S., Dyer, I. Gomes, B., Fulford, J., Lynch, J., Gawarkiewicz, G., Volak, D., Effects of Korean littoral environment on acoustic propagation, *IEEE J. Oceanic Eng.* 26(2), 266–284 (2001).

16. Abbot, P., Celuzza, S., Dyer, I. Gomes, B., Fulford, J. and Lynch, J., Effects of East China Sea shallow water environment on acoustic propagation, *IEEE J. Oceanic Eng.* (submitted Dec. 2001).

17. Robinson, A.R., Physical processes, field estimation and interdisciplinary ocean modeling, *Earth-Science Reviews* 40, 3–54 (1996).

SONAR PERFORMANCE PREDICTIONS INCORPORATING ENVIRONMENTAL VARIABILITY

PHILIP ABBOT AND IRA DYER[1]
Ocean Acoustical Services and Instrumentation Systems (OASIS), Inc.,
5 Militia Drive, Lexington, MA 02421, USA
E-mail: abbot@oasislex.com

Perfect spatial/temporal knowledge of the ocean environment is rarely available for evaluation of sonar performance. Instead, performance prediction is often done with acoustic models assuming idealized inputs. The current capability of many such models is superb; realistic inputs, however, often are not available, and this is the central focus of the present paper. The prediction of sonar performance using a probability density function (PDF) based on environmental variability is presented. The PDF describes the distribution of the predictive capability of an acoustic model with respect to measurements of actual performance and, therefore, represents the uncertainty in one's ability to model the actual performance of the system. The PDF accounts for the inherent variability of the environment not contained in the model inputs, and is a useful probabilistic description of the environment's intrinsic variability. As examples, two littoral transmission loss data sets are invoked, and other passive sonar inputs are assumed, from which curves of predictive probability of detection (PPD) versus range are presented.

1   Introduction

Sonar performance predictions typically are based on a set of assumptions that attempt to describe the oceanography, bottom and sea surface conditions of the environment under consideration. Perfect spatial and temporal knowledge of the ocean environment at all its relevant acoustic scales is rarely available, however.   Thus, performance is often predicted with idealized environmental inputs (e.g. direction-independent sound speed profiles, or horizontally isotropic bottom properties, etc.).   While the current state of many acoustic models, such as those for transmission loss (TL), is excellent, realistic environmental inputs at all the important acoustic scales often are not available.

The central focus of this paper is the prediction of sonar performance using a probability density function (PDF) based on environmental variability. These environmental PDFs are discussed generally in the next section and are suggested as a useful way to predict performance, albeit probabilistically. This approach is not new [e.g., 1–3], but the growing availability of relevant ocean acoustics data makes its adoption practical.

In this paper we use PDFs for the 1- and 2-way transmission losses (TL) measured in the East China Sea (ECS) and Sea of Japan (SOJ) during the summertime, with downward refracting sound speed profiles.   These provide predictive probability of

---

[1] Also, MIT Department of Ocean Engineering, Cambridge, MA 02139

N.G. Pace and F.B. Jensen (eds.), Impact of Littoral Environmental Variability on Acoustic Predictions and Sonar Performance, 611-618.
© 2002 *Kluwer Academic Publishers.*

detection examples for a simulated system operating in the ECS and SOJ during summer-like conditions.

2   Description of predictive probability of detection based on environmental variability

Figure 1 illustrates conceptually sonar performance prediction, and how environmental uncertainty is incorporated into what we call "predictive probability of detection", or PPD. The environmental PDF shown in the figure is, for a given environment, derived in our studies comparing model predictions with acoustic data. The PDF is a best fit to histogramatic differences between the data and the acoustic model, with inputs idealized as the predictor elects, and is typically represented as an n-th degree-of-freedom (dof) Chi-squared probability density function. This PDF represents the uncertainty in the computational modeling process, which typically can be made small, and the inherent variability of the environment not contained in the model inputs, which typically is larger.

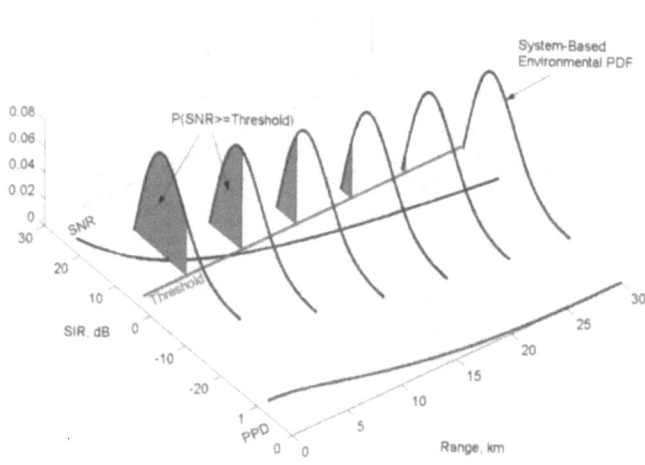

Figure 1. Illustration of probabilistic system performance prediction and Predictive Probability of Detection (PPD) vs. range using system-based environmental PDF which incorporates environmental uncertainty.

Figure 1 is applied by calculating, with a model of choice, the signal-to-noise ratio (SNR) for an active or passive system, as a function of range. The system-based environmental PDF is anchored by setting its mean to the SNR, at each range. Of course, the PDF must be based on the same calculation model. Then, at each range, the area under the PDF above the detection threshold level is computed, and the resulting integrand is the PPD (i.e., the probability that the SNR is greater than or equal to the threshold). In the example, the environmental PDFs are shown at range increments of 5 km with the range of the integrals (as set by the threshold level) shown in red. At the bottom of the plot is the resulting PPD, which is close to unity at close-in ranges and slowly decreases as the range increases.

The PPD shown in Fig. 1 is a prediction of the system performance versus range, and the uncertainties in the model estimate due to environmental variability are accounted for in this function. Rather than use a single range value (e.g. "range-of-the-day" or "range-of-the-moment"), the PPD provides the system operator with a probabilistic representation of the system performance as a function of range. This distribution of ranges is, at least to first order, independent of the acoustic model used, coupled with a data-based statistical description of the uncertainties. The operator can thus use this information to operate the system more effectively, and can make more informed decisions on search, risk, expenditure of assets and assumptions of covertness.

With use of a sonar system example in Sect. 5, the foregoing method is generalized to include non-environmental origins of variability. The resulting method considers both environmental and non-environmental origins, so that the effects of each on the system can be directly compared.

3   Environmental PDFs for 1-way and 2-way TL in East China Sea

Operational acoustics experiments were recently conducted over the frequency range of 25 to 800 Hz in the ECS, in water depth of about 100 m, during downward refracting conditions [4,5]. The TL data were obtained from broadband explosive Signal Underwater Sound (SUS) sources, and omnidirectional hydrophone sonobuoys. The tests were conducted in directions approximately normal and parallel to the bathymetric contours. In all directions, the TL was observed to be generally low. During the tests, the sound speed profiles were downward refracting, with thermocline temporal variations caused by internal tides.

A state-of-the-art TL model was adopted, based on environmental idealizations typical of operational forecasting. The bottom bathymetry and geoacoustic inputs were based on data atlases, the latter assuming a horizontally isotropic sedimentary layer. The comparison between the model and measurements results in the histogram shown in Fig. 2. This is a histogram of the differences between the propagation model and the measured TL data, the latter aggregated over all ranges (≤ 40 km) and encompassing four transmission runs with up to three independent measurement paths per run. The data shown are in an octave band, centered at 400 Hz (282 Hz bandwidth) and integrated over 640 ms. The histogram is described by the mean difference $\mu = -1.0$ dB and the standard deviation of the differences $\sigma = 2.0$ dB. A chi-square density (parameterized by n = 14 dof) is selected to represent the histogram. This is the 1-way TL environmental PDF and it quantifies the differences between predictions of the model and the truth of the data, caused by the stochastic variability of the environment. It illustrates our statistical approach for portrayal of environmental uncertainties that are not captured by a predictive tool restricted to idealized inputs. It was argued in [4] that the primary cause of the TL uncertainties are bottom complexities associated with the variability of the bottom, at spatial scales presumably not resolved in the historical geological data bases.

Recently, the 2-way TL environmental PDF was measured from a towed array system operating in the ECS, also during an operational experiment. During these tests, the acoustic conditions consisted of a broadband sound source, and a known target, with a total 2-way travel distance of 70 km. The location was about 100 km southwest of the 1-way measurement site, and deployed similarly near the ECS shelf-break, under similar downward refracting sound speed profiles, but twenty-four months later. The transmission paths were primarily in the direction parallel to the bathymetry contours. With use of the same processing as in the 1-way experiment, the resulting histogram and Chi-square fit (8 dof) are shown in Fig. 3. Here the data have statistics with $\mu = 0$ and $\sigma = 2.5$ dB.

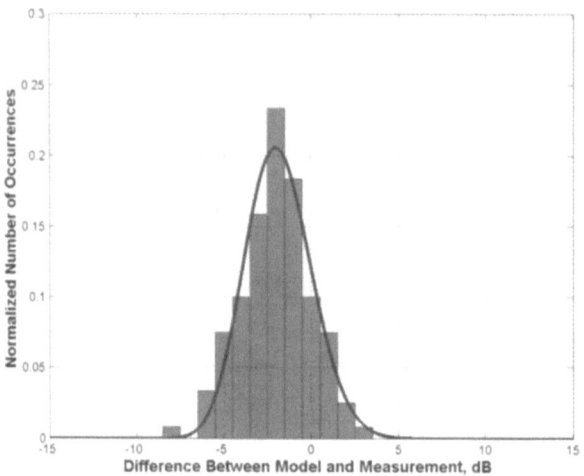

Figure 2. ECS 1-way TL environmental PDF fit to histogram measured at omniphone, 400 Hz, BW = 282 Hz, T = 640 ms, R ≤ 40 km, $\mu = -1.0$ dB, $\sigma = 2.0$ dB [1].

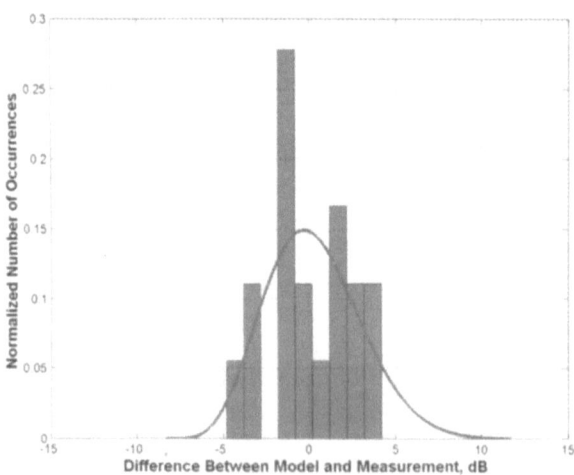

Figure 3. ECS 2-way TL environmental PDF fit to histogram measured at towed array, 400 Hz, BW = 282 Hz, T = 640 ms, R = 70 km, $\mu = 0$, $\sigma = 2.5$ dB.

We assume that the PDF from the out-going (source-to-target) and in-coming (target-to-receiver) TL are random variables and statistically independent. Then, the 2-way PDF will be the convolution of the out-going and in-coming PDFs [6]. In Fig. 4, the convolution of the 1-way PDF (from Fig. 2) with itself is compared with the 2-way TL PDF (from Fig. 3), and these agree remarkably well. This implies that the differences in the PDF from one nearby site, or from one summertime period, to another in the ECS are not large.

TL measurements with a towed array could result in wider PDFs if out-of-plane scattering is prevalent in the propagation paths. The results shown in Fig. 4 suggest that out-of-plane scattering are not important in this environment.

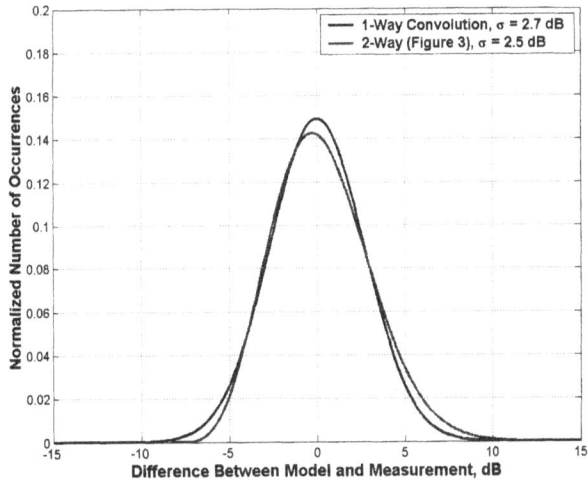

Figure 4. ECS 2-way TL environmental PDF (Fig. 3) comparison with 1-way TL (Fig. 2) convolution, 400 Hz, T = 640 ms, BW = 282 Hz.

4 Environmental PDF for 1-way TL in the Sea of Japan

Similar operational acoustic experiments were conducted off the Coast of Korea, in the Sea of Japan (SOJ) in shallow to intermediate water depths, along the shelf and slope [7], in downward refracting conditions. The sources were SUS and the receivers were standard sonobuoys. These tests were conducted over varying bottom depths and slopes, in directions approximately normal and parallel to the bathymetric contours, with a processing method identical to that described for the ECS tests. Two different source depths were included. Measured TLs were quite high (especially relative to ECS), and were dependent on the direction of propagation. The TL was largest for the up-slope direction (source-to-receiver), for both shallow (18 m) and deep (200 m) sources. The TL was smallest for the cross-slope direction (parallel to the bathymetric contours), and intermediate for the down-slope direction; both these latter tests used shallow sources. We concluded that the bottom conditions, including bathymetry and geoacoustic properties, varied widely within the test area, causing large differences between the TL model predictions and the measured data. We also showed that sound speed variability in the water column had a comparatively weak effect on the TL.

The 1-way TL environmental PDFs were not derived in [7], but are here. Five TL runs in an octave at 400 Hz are used, each encompassing up to four independent paths per run. These were compared with the same state-of-the-art TL model as used with the ECS data, with inputs based on measured range-dependent sound speed profiles, and on atlases covering range-dependent bottom depths and geoacoustic bottom data. Differences between the TL predictions and data are shown as a histogram in Fig. 5, in which the five TL measurement runs have been demeaned and aggregated. The resulting 1-way TL environmental PDF, with a Chi-square fit with 8 dof, has $\sigma = 5.9$ dB, significantly larger than the 2.0 dB found for the ECS.

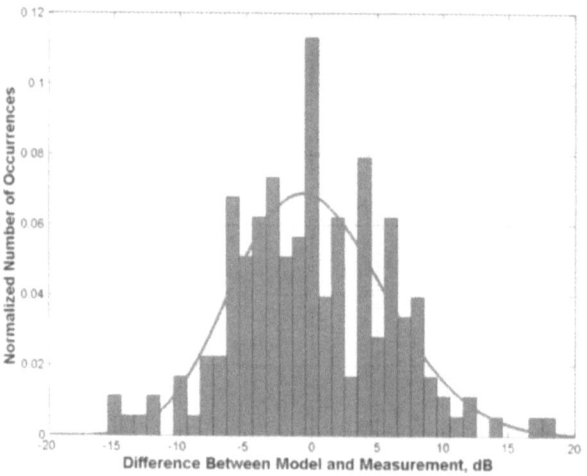

Figure 5. SOJ 1-way TL environmental PDF fit to histogram measured at omniphone, 400 Hz, BW =282 Hz , T = 640 ms, $\mu = 0$, $\sigma = 5.9$ dB (model and measurements given in Ref. [7]).

5 Probability of detection; ECS and SOJ predicted performance

In Fig. 6, we show predictive probability of detection (PPD) curves for a simulated broadband passive sonar operating in the ECS and SOJ environments described in the foregoing. The simulated system operates in the 400 Hz octave band and integrates for 640 ms. (Passive systems analyze signals and noise over much narrower frequency bands and integrate for much longer time periods, but we simplify here in order to match the conditions of the existing data sets). We assume the system's Figure-of-Merit (FOM) = 65 dB, as defined in [6] for a passive system. For each environment, we use two PDFs in the prediction, one consisting of the 1-way TL environmental PDF, the other being a System-Based PDF as discussed in Sect. 2. The latter includes the ambient noise and the source level, in addition to the 1-way TL environmental PDFs. We further: i) take the ambient noise PDF as approximately normal, with $\sigma = 0.4$ dB [8], and ii) assume the source level PDF to be a log-normal density with $\sigma = 3$ dB^2. The 1-way TL

---

^{2} While reasonable, we have no physical basis for this particular assumption on source level, but we use it for illustrative purposes.

environmental PDF is convolved with the ambient noise PDF and the source level PDF to result in the system-based PDF used in the simulation.

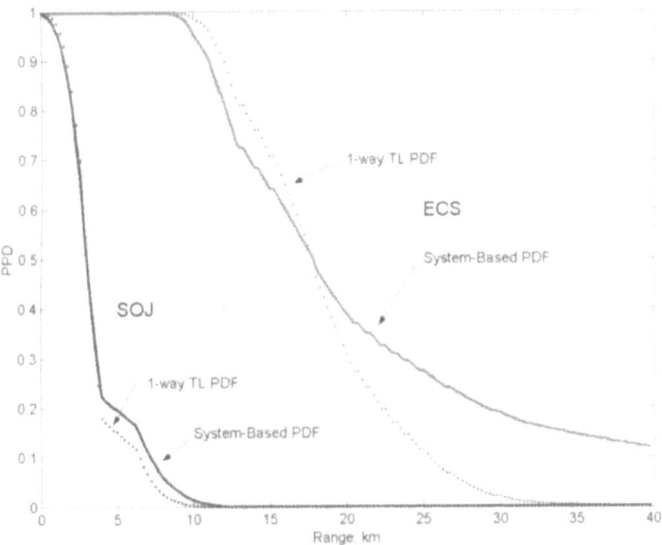

Figure 6. Predictive probability of detection (PPD) vs. range, for simulated passive system operating in the ECS and SOJ (downward refracting sound speed conditions) for FOM = 65 dB, 400 Hz, BW=282 Hz, T=640 ms. Discontinuities in PPD are caused by discontinuities in the underlying predicted TL curves versus range.

Figure 6 illustrates that all classes of variability affect the PPD. Their origins could be environmental (as in TL, ambient noise and, by extension, reverberation) or, non-environmental (as in source level and, by extension, target strength, sonar self-noise, array uncertainties, recognition-differential, and the like). Variability controls the *slope* of PPD versus range in Fig. 6; the larger its total σ, the larger the slope. To be sure, the slope is not a straight line over the entire PPD interval, its shape being controlled near its 0/1 limits by the tails of the component PDFs. Nonetheless, the slope provides an operator with a basis for trading the gradual range-dependence of detection probability with mission desiderata.

Where PPD is small (but not zero) in Fig. 6, the range at a fixed PPD is shown to increase for larger σ. The relative increase for the ECS and SOJ examples is about the same; such increases can also be important in meeting mission desiderata.

The range at which PPD = 0.5 for the two examples, is, in contrast, governed by their respective range-dependent parameters (as in the TL mean and, by extension, in the reverberation mean). Poor mean transmission in the SOJ leads to smaller predicted detection ranges, while good transmission in the ECS leads to larger ranges which, of course, makes the mean TL a dominant effect. That is, variability is important in performance prediction, but the means are more so.

6 Summary and conclusions

Predictive Probability of Detection curves have been shown for a simulated passive system operating under a large time-bandwidth product in ECS and SOJ summertime environments. These curves show significant differences of sonar system operation for the two locations, with simulated detection ranges much greater in ECS. The means of the relevant sonar equation terms retain their dominant role in performance, and variability around the means causes the detection probability to spread in range. The predicted spreads appear to be realistic. Named simply as *slopes* in Sect. 5, these spreads potentially could provide additional detection insights for sonar operators.

The PPD method appears to be useful for incorporating environmental uncertainty into predictions of sonar system performance. Of course, the non-environmental origins of variability must also be considered. Because reasonable judgements can not be made *a priori* as to which of the two classes is more important, both environmental and non-environmental origins should be considered simultaneously in the PPD method. For example, a passive sonar with a significantly smaller time-bandwidth product than the one illustrated here, will have significantly wider TL and ambient noise PDFs, which could make non-environmental uncertainties moot. Further, narrower beamwidths can also result in wider PDFs for the environmental origins of variability.

Acknowledgements

We thank OASIS personnel Chris Emerson and Stephen Celuzza for helping with the data reduction and analyses. The ECS and SOJ omniphone data were acquired from Dave Volak at NAWC. The work is sponsored by ONR in the Uncertainty Program.

References

1. Urick, R., Models for the amplitude fluctuations of narrow-band signals and noise in the sea, *J. Acoust. Soc. Am.* **62**, 878–887 (1977).
2. Worcester, P., Reciprocal acoustic transmission in a midocean environment: Fluctuations, *J. Acoust. Soc. Am.* **66**, 1173–1181 (1979).
3. Gough, E., University of Washington/Applied Physics Laboratory (personal conversation, April 12, 2002).
4. Abbot, P., Celuzza, S., Dyer, I., Gomes, B., Fulford, J. and Lynch, J., Effects of East China Sea shallow water environment on acoustic propagation, *IEEE J. Oceanic Eng.* (submitted Dec. 2001).
5. Volak D., Gindhard, R. and Zeidler, E., OPEX 97 – Monostatic reverberation, transmission loss, initial summary. NAWAIRWARCENADSIVPAX TN-OPEX-98 (April 1998).
6. Burdic, W., *Underwater Acoustic System Analysis* (Prentice-Hall Inc., Englewood Cliffs, New Jersey, 1984).
7. Abbot, P., Celuzza, S., Dyer, I., Gomes, B., Fulford, J., Lynch, J., Gawarkiewicz, G. and Volak, D., Effects of Korean littoral environment on acoustic propagation, *IEEE J. Oceanic Eng.* **26**, 266–284 (2001).
8. Dyer, I., Statistics of sound propagation in the ocean, *J. Acoust. Soc. Am.* **48**, 337–345 (1970).

AUTHOR INDEX